Computer-Aided Analysis of Rigid and Flexible Mechanical Systems

NATO ASI Series

Advanced Science Institutes Series

A Series presenting the results of activities sponsored by the NATO Science Committee, which aims at the dissemination of advanced scientific and technological knowledge, with a view to strengthening links between scientific communities.

The Series is published by an international board of publishers in conjunction with the NATO Scientific Affairs Division

A Life Sciences
B Physics

Plenum Publishing Corporation
London and New York

C Mathematical and Physical Sciences
D Behavioural and Social Sciences
E Applied Sciences

Kluwer Academic Publishers
Dordrecht, Boston and London

F Computer and Systems Sciences
G Ecological Sciences
H Cell Biology
I Global Environmental Change

Springer-Verlag
Berlin, Heidelberg, New York, London, Paris and Tokyo

NATO-PCO-DATA BASE

The electronic index to the NATO ASI Series provides full bibliographical references (with keywords and/or abstracts) to more than 30000 contributions from international scientists published in all sections of the NATO ASI Series.
Access to the NATO-PCO-DATA BASE is possible in two ways:

– via online FILE 128 (NATO-PCO-DATA BASE) hosted by ESRIN, Via Galileo Galilei, I-00044 Frascati, Italy.

– via CD-ROM "NATO-PCO-DATA BASE" with user-friendly retrieval software in English, French and German (© WTV GmbH and DATAWARE Technologies Inc. 1989).

The CD-ROM can be ordered through any member of the Board of Publishers or through NATO-PCO, Overijse, Belgium.

Series E: Applied Sciences - Vol. 268

Computer-Aided Analysis of Rigid and Flexible Mechanical Systems

edited by

Manuel F.O. Seabra Pereira

and

Jorge A.C. Ambrósio

Instituto de Engenharia Mecânica,
Instituto Superior Técnico,
Lisboa, Portugal

Springer Science+Business Media, B.V.

Proceedings of the NATO Advanced Study Institute on
Computer-Aided Analysis of Rigid and Flexible Mechanical Systems
Tróia, Portugal
June 27–July 9, 1993

A C.I.P. Catalogue record for this book is available from the Library of Congress.

ISBN 978-94-010-4508-7 ISBN 978-94-011-1166-9 (eBook)
DOI 10.1007/978-94-011-1166-9

Printed on acid-free paper

NATO-ADVANCED STUDY INSTITUTE

COMPUTER AIDED ANALYSIS OF RIGID AND FLEXIBLE MECHANICAL SYSTEMS

TRÓIA, PORTUGAL
27 JUNE - 9 JULY, 1993

MAIN SPONSORS:

NATO:	North Atlantic Treaty Organization
U.S. Army TARDEC:	United States Army, TARDEC
NSF:	National Science Foundation

OTHER SPONSORS

JNICT:	Junta Nacional de Investigação Científica e Tecnológica
FLAD:	Fundação Luso Americana para o Desenvolvimento
IDMEC:	Instituto de Engenharia Mecânica
CEMUL:	Centro de Mecânica e Materiais da Universidade Técnica de Lisboa

DIRECTOR

Manuel F.O. Seabra Pereira, IDMEC-Instituto Superior Técnico, Portugal

SCIENTIFIC COMMITTEE:

Michel Geradin, Université de Liége, Belgium
Edward J. Haug, University of Iowa, USA
Manfred Hiller, University of Duisburg,Germany
Parviz Nikravesh, University of Arizona, USA

CONTENTS

LIST OF PARTICIPANTS

Lecturers:

Jorge Ambrósio
IDMEC - Pólo do I.S.T., Av. Rovisco Pais, 1096 Lisboa, Portugal

Jorge Angeles
McGill University, 817 Sherbrooke Street West, Montreal, Quebec H3A 2K6, Canada

Alberto Cardona
INTEC, Güemes 3450, 3000 Santa Fé, Argentina

Jeffrey S. Freeman
Center for Computer Aided Design, Coll. of Engineering, University of Iowa, Iowa City, Iowa 525242, USA

Harold P. Frisch
Robotics Branch, Code 714.1, Goddard Space Center, Greenbelt, Maryland 20771, USA

Michel Geradin
LTAS - Université de Líège, Rue Ernest Solvay 21, 4000 Liège, Belgium

Manfred Hiller
Fachgebiet Mechanik, Universitat Duisburg, Lotharstr. 1, 4100 Duisburg, Germany

Ronald L. Huston
Dept. of Mech. and Industrial Eng., Univ. of Cincinatti, Cincinatti, Ohio 45221-0072, USA

J. García de Jalón
CEIT, Pº Manuel de Lardizábal 15, 20009 San Sebastián, Guipúzcoa, Spain

Thomas R. Kane
Division of Applied Mechanics, Durand Bldg, Stanford Univ., Stanford, CA 94305, USA

Parviz Nikravesh
Dept. of Aerospace and Mech. Engng., University of Arizona, Tucson, AZ 85721, USA

Manuel S. Pereira
IDMEC - Pólo do I.S.T., Av. Rovisco Pais, 1096 Lisboa, Portugal

Linda Petzold
Comp.Scie.Dept.(4-192 EE/CS Blg), 200 Union Street SE, Univ.of Minnesota, Minneapolis MN55455, USA

W. Schiehlen
Institut B fur Mechanik, Pfaffenwaldring 9, Univ. Stuttgart, D-7000 Stuttgart 80, Germany

J. De Schutter
K.U.L., Afdel. Mech. Konst. en Prod., Celestijnenlaan 300 B, B-3001 Heverlee, Belgium

Richard Schwertassek
Inst. fur Dynamik der Flugsystem, DLR Oberpfaffenhoffen, D-8031 Wessling, Germany

Ahmed Shabana
University of Illinois at Chicago, Dept. of Mech. Engng.(M/C 251), 2039 Engineering Research Facility, 842 West Taylor Street, Chicago, Illinois 60607-7022, USA

Roger Wehage
System Simulation & Tech. Division, AMSTA-RY, U.S. Army TARDEC, Warren MI 48397-5000, USA

J. Wittenburg
Institut fur Technische Mechanik, Universitat Karlsruhe, D-76139 Karlsruhe, Germany

Participants:

João Abrantes
Laboratório Biomecânica, Faculdade de Motricidade Humana, 1499 Lisboa, Portugal

Venkatesh Agaram
L.O.M., Dept. Mecanique, Swiss Federal Inst. of Techn., CH-1015 Lausanne, Switzerland

Sunil Agrawal
Ohio University, Dept. Mech. Engng., Athens, OH 45701, USA

Cláudia Oitáven Alves
Instituto Gulbenkian de Ciência, Centro de Biologia, R. Quinta Grande, Apartado 14, 2781 Oeiras , Portugal

Martin Anantharaman
Universitat Duisburg, FB 7 FG 16, Lotharstraße 1, 47057 Duisburg, Germany

Kurt S. Anderson
Institut fur Mechanik, TH Darmstadt, Hochschulstrasse 1, D-6100 Darmstadt, Germany

Daniel Bach
Institut fur Mechanik, HG F38.1, ETH Zentrum, CH-8092 Zurich, Switzerland

Behnam Bahr
Wichita State Univ., Mech. Eng. Dept., 1845 N. Fairmount, Wichita, KS67260-0035, USA

J. Infante Barbosa
ENIDH - Escola Náutica, Av. Bonneville Franco, Paço d'Arcos 2780 Oeiras, Portugal

Eduardo Bayo
University of California, Dept. of Mech. Engng., Santa Barbara, CA 93106, USA

M. Belczynski
System Simulation & Tech. Division, U.S. Army TACOM, (AMSTA-RY), Warren, MI 48397-5000, USA

F. Bennis
Laboratoire d'Automatique de Nantes, École Centrale Nantes, URA CNRS 823, 1 Rue de la Noe, 44072 Nantes Cedex 03, France

Roman Bogacz
Inst. of Fundamental Tech. Res., Polish Academy of Sciences, ul. Swietokrzyska 21, 00-049 Warsaw, Poland

Marco Borri
Dipart. di Ingegn. Aerospaziale, Politecnico di Milano, via Golgi 40, 20133 Milano, Italy

Hans Brauchli
Institut fur Mechanik, HG F38.1, ETH Zentrum, CH-8092 Zurich, Switzerland

Pasquale Campanile
C. Ricerche FIAT, Ente Veicoli/Calcoli, Strada Torino 50, 10043 Orbassano (TO), Italy

Marco Ceccarelli
Dept. Industrial Engng., Univ. of Cassino, Via Zamosch 43, 03043 Cassino (FR), Italy

J. Choi
Dept. Mech. Eng (M/C 251)-2039 ERF, Univ. of Illinois at Chicago, Box 4348, Chicago, IL 60680, USA

Kenneth Clark
Mathematical Sc. Div., U.S. Army Res. Office, Research Triangle Park, NC 27709, USA

José Sá da Costa
I.S.T.- DEM, Av. Rovisco Pais, 1096 Lisboa, Portugal

M. A. Crisfield
Imperial Colllege, Dept. Aero., Prince Consort Road, London SW7 2BY, United Kingdom

J. Cuadrado
University of Navarra, Manuel de Lardizabal 15, 20009 San Sebastian, Spain

João P. Dias
Instituto Superior Técnico, Dept. Eng. Mecânica, Av. Rovisco Pais, 1096 Lisboa, Portugal

D. B.Doan
L.T.A.S., Université de Liège, Rue Ernest Solvay 21, 4000 Liège, Belgium

Robert Dombroski
U.S.Army Ardec, SMCAR-FSA-F Bldg 61N, Picatinny Arsenal - NJ 07806-5000, USA

J. Duysens
L.T.A.S., University of Liège, 21, Rue Ernest Solvay 21, B-4000 Liège, Belgium

Haluk Erol
Istanbul Technical University, Faculty of Mech. Engng., Gumussuyu, Istanbul, Turkey

Yanis Erotokritos
Inst. of Sound & Vibration Research, Univ. of Southampton, SO9 5NH, United Kingdom

Christian Goualou
École Centrale de Paris, L.R.S.A. - Res. Lab. for Auto. Safety, Grande Voie des Vignes,
92295 Chatenay-Malabry, France

Simon Guest
Univ Cambridge, Dept.of Eng.,Trumpington Street, Cambridge CB2 1PZ,United Kingdom

John Hansen
Technical University of Denmark, Dept. of Solid Mechanics, Building 421, 2800 Lyngby, Denmark

Michael Hansen
Technical University of Denmark, Dept. of Solid Mechanics, Building 404, 2800 Lyngby, Denmark

Christian Hardell
Lulea Univ. of Technology, Division of Computer Aided Design, S-95187 Lulea, Sweden

S. G. Hutton
Univ. of British Columbia, Dept. of Mech. Engng, 2324 Main Mall, Vancouver, BC ,V6T 1Z4, Canada

How-Young Hwang
1100 Arthur St., Apt. I-1, Iowa City, IA 52240, USA

J. M. Jimenez
University of Navarra, Manuel de Lardizabal 15, 20009 San Sebastian, Spain

Andrés Kecskemethy
Fachgebiet Mechatronik, Universitat Duisburg, Lotharstr. 1, 47057 Duisburg, Germany

Albrecht Keil
Inst. fur Mechatronik, Reichenhainer Str. 88, P.O. Box 408, 0-9022 Chemnitz, Germany

Willy Koppens
TNO Road-Vehicles Research Institute, P.O. Box 6033, 2600 JA Delft, Netherlands

Hamid Lankarani
Wichita State University, Mech. Engng. Dept., Wichita, KS 67260-0035, USA

Dirk Lefeber
Vrije Universiteit Brussel, Pleinlaan 2, 1050 Brussels, Belgium

Tomasz Lekszycki
Inst. Fund. Tech. Research, Swietokrzyska 21, 00-049 Warszawa, Poland

Bill Lipsett
Dept. of Mechanical Engng., University of Alberta, Edmonton Alberta T6G 2G8, Canada

Jean-Pierre Mariot
Lab. de Mécanique Prod. et Mat., Faculté des Sciences - B.P. 535, 72017 Le Mans, France

J. P. Meijaard
Lab. Eng. Mechanics, Delft Univ. of Tech., Mekelweg 2, NL-2628 CD Delft, Netherlands

Frank Melzer
Inst. B of Mechanics, Univ. Stuttgart, Pfaffenwaldring 9, D-70550 Stuttgart, Germany

Arkadiusz Mezyk
Silesian Tech. Univ. Gliwice, Eng. Mech. Dept., Konarskiego 18a, 44-100 Gliwice, Poland

François Minne
Laboratoire de Génie Mécanique, Le MontHouy - B.P. 311, 59304 Valenciennes, France

Carlos A. Mota Soares
IDMEC - Pólo do I.S.T., Av. Rovisco Pais, 1096 Lisboa, Portugal

John McPhee
University of Waterloo, Systems Des.Eng., Ontario, N2L 3G1, Canada

Álvaro Costa Neto
Universidade de São Paulo, Esc. de Eng. de São Carlos, Av. Dr. Carlos Botelho 1465,
C.P. 359, CEP 13560-250, São Carlos - SP, Brasil

Juana Mayo Nuñez
Escuela Sup. de Ing. Indust., Dept. de Ing. Mec. y Materiales, Av. Reina Mercedes s/n, 41012 Sevilla, Spain

Jose Luis H. Oliver
Univ. Politecnica Valencia, Dept. Ingenieria Mecanica, 46022 Valencia, Spain

Verlinden Olivier
Faculté Polytechnique de Mons, 31 Boulevard Dolez, 7000 Mons, Belgium

Martin Otter
Insti. for Robotics and Syst. Dynam., DLR Oberpfaffenhofen, D-8031 Wessling, Germany

Tugrul Ozel
Dokuz Eylul University, Dept. Mech. Engng., Bornova 35100, Izmir, Turkey

Ettore Pennestri
Univ. di Roma "Tor Vergata", Dip. Ingegneria Meccanica, Via Della Ricerca Scientifica, 00133 Roma, Italy

Vincent Pesch
Univ. of Illinois at Urbana-Champaign, 1206 W. Green Street, Urbana, Il 61801, USA

Jean-Marc Pini
Dassault Aviation - Centre Spatial, 17 Av. Didier Daurat, B.P. 23, 31701 Blagnac, France

Enrico Pisino
C. Ricerche FIAT, Ente Veicoli/Calcoli, Strada Torino 50, 10043 Orbassano (TO), Italy

Marco Quadrelli
Comput. Modelling Center, Georgia Institute of Tech. Atlanta, GA 30332-0356, USA

Udo Rein
Inst. A fur Mechanik, Univ. Stuttgart, Pfaffenwaldring 9, D-70550 Stuttgart, Germany

David Russell
Clarkson University, Dept. of Mech. and Aeron. Engng., Old Main, Potsdam, New York 13699-5725, USA

Ernst Dieter Sach
Deutsche Aerospace AG., Dept. RTT22, Postfach 801169, 8000 Munich 80, Germany

Delf Sachau
DLR - Inst. fur Robotik und Syst Dyn., Oberpfaffenhofen, D-8031 - Wessling, Germany

Jean-Claude Samin
Univ. Catholique de Louvain, Place du Levant 2, B-1348 Louvain-la-Neuve, Belgium

Saide Sarigul
Dokuz Eylul University, Dept. Mech. Engng., Bornova 35100, Izmir, Turkey

Ole Ivar Sivertsen
Dept. of Machine Design, Norwegian Institute of Tech., N-7034 Trondheim, Norway

Miguel Sofer
Institut fur Mechanik, HG F38.1, ETH Zentrum, CH-8092 Zurich, Switzerland

Annika Stensson
Lulea Univ. of Technology, Division of Computer Aided Design, S-95187 Lulea, Sweden

Tahsin Sumer
Arçelik A.S., Research Dept., Cayirova 41460, Istanbul, Turkey

G. Tzvetkova
Bulgarian Academy of Sciences, G. Bonchev str., Blok 4, 1113 Sofia, Bulgaria

Ken Willmert
Clarkson University, Dept. of Mech. and Aeron. Engng., Old Main, Potsdam,
New York 13699-5725, USA

Harry Yae
University of Iowa, Dept. of Mech. Engng., Iowa City, Iowa 52242, USA

Yung M. Yoo
Korea Automotive Tech. Inst., 1638-3 Seocho Dong, Seocho Ku - Seoul, Korea

E. V. Zakhariev
Bulgarian Acad of Sciences, Inst.of Mech. & Biomechanics, G.Bontchev Street Bl 4, 1113 Sofia, Bulgaria

PREFACE

This book contains the edited version of the lectures presented at the NATO ADVANCED STUDY INSTITUTE on "COMPUTER AIDED ANALYSIS OF RIGID AND FLEXIBLE MECHANICAL SYSTEMS", held in Tróia, Portugal, from the 27 June to 9 July, 1993, and organized by the Instituto de Engenharia Mecânica, Instituto Superior Técnico. This ASI addressed the state-of-art in the field of multibody dynamics, which is now a well developed subject with a great variety of formalisms, methods and principles. Ninety five participants, from twenty countries, representing academia, industry, government and research institutions attended this Institute. This contributed greatly to the success of the Institute since it encouraged the interchange of experiences between leading scientists and young scholars and promoted discussions that helped to generate new ideas and to define directions of research and future developments.

The full program of the Institute included also contributed presentations made by participants where different topics have been explored. Such topics include: formulations and numerical aspects in rigid and flexible mechanical systems; object-oriented paradigms; optimal design and synthesis; robotics; kinematics; path planning; control; impact dynamics; and several application oriented developments in weapon systems, vehicles and crashworthiness. These papers have been revised and will be published by Kluwer in a special issue of the Journal of Nonlinear Dynamics and in a forthcoming companion book.

This book brings together, in a tutorial and review manner, a comprehensive summary of current work and is therefore suitable for a wide range of interests, ranging from the advanced student to the researchers and implementators concerned with advanced theoretical issues in multibody dynamics. The applicational aspects will help the readers to apprise the different approaches available today and their use and suitability as efficient design tools. This book is organized into five parts, each one addressing the state-of-art techniques and methods in the principal areas of study of the Institute:

Part I	Formulations and Methods in Rigid Multibody Systems
Part II	Dynamics of Flexible Multibody Systems
Part III	Kinematics Aspects of Multibody Systems
Part IV	Computational Techniques and Numerical Methods
Part V	Man/Machine Interaction and Virtual Prototyping

Different techniques that may be used to obtain efficient and general computer based dynamics modeling and simulation algorithms of rigid multibody systems are presented in Part I. The adequate choice of coordinates to represent the dynamics of multibody systems it is still a central issue in multibody dynamics. It is shown here that both recursive and absolute methods have their merits in the response to the growing complexity of mechanical systems. The use of augmented Lagrangian and penalty methods are still popular as a result of their conceptual simplicity and of permitting, with a minimum effort, the calculation of forces associated with the constraints. Projection methods, singular value decomposition and Gaussian triangularization techniques are also used. Here a set of independent coordinates are selected as a sub-set of the dependent ones. Formulations based on independent coordinates present, as main advantage, the reduction of the number of equations to be integrated, improving in the process the numerical stability of the problem. The integration of constraints, generally associated with cut-joints, is still possible. It is also shown that these methodologies are generally very efficient, leading to faster running programs. General graph theoretic approaches are also referred in this part of the book. Paths may be identified to define spanning trees of kinematic and dynamic computational sequences that are used to create efficient algorithms. The degree of parallelism of the equations is identified allowing an efficient implementation in multiprocessor computer platforms.

Finally special emphasis has been given to the automatic generation of the equations of motion of multibody systems, in close form, by means of symbolic manipulators. The computer time needed for the analysis of specific problems is greatly reduced with this approach. Specialized software such as NEWEUL and ROBOTRAN or general programs like MAPLE and MATHEMATICA can be used for this purpose. Extension of this philosophy of development are computer codes included within the framework of Object Oriented Programming. Yet involving a high level of abstraction, they are easy to maintain, comparing to more conventional practices of code development. Regarding implementation and use of tools for modeling, generation, simulation and visualization of multibody systems, the problem of integration of CAD interfaces with current formalisms is also addressed by using of parametric multibody system databases.

Part II includes different formulations and methodologies to incorporate the elastodynamic effects which are treated either using a finite element modeling concept or a moving frame approach. Flexible members are treated in a fully nonlinear manner including geometric stiffening and other second order effects, often encountered in fast speed operations. The more conventional linearized forms of substructuring and modal approaches, which are used to reduce the size of the problem, are reviewed. Nonlinear material behavior issues are also addressed as applied to crashworthiness of vehicles. The scope of application is extended to satellite antenna deployment problems, lightweight robots and vehicle dynamics.

Of special importance to the dynamics of multibody systems are kinematic considerations. Part III is devoted to kinematic issues of particular interest to the general dynamics problem of multibody systems. The concepts of twist and wrench generators and annihilators are developed, which can find extensive applications in the mechanics of grasping and provide a better understanding of problems associated with force and motion control of manipulators. In this part, new approaches to the modeling and motion specification of robotics manipulation tasks involving complex time varying motion constraints are presented, enhancing the state-of-art in complex applications in this area.

Part IV is devoted to the computational challenges in mechanical systems analysis concerned with the reduction of multibody dynamics simulation time. Related developments and applications in solution methods in the fields of numerical methods, new algorithms and emerging new powerful parallel computer platforms are presented and analyzed. Recent contributions to time integration methods applicable to differential algebraic equations and problems related to time integration of flexible systems with high frequency content are thoroughly discussed here. The efficiency and robustness of numerical integration schemes in the modeling of multibody systems are also extensively discussed. New stable formulations for second order integration schemes are discussed as applied to constrained multibody systems. Computational challenges for efficient and reliable numerical methods are underlined. Some contributions on the aspects of stability, conditioning, accuracy and time step control are also included. Finally, ideas to combine the methods available in computer sciences and numerical mathematics are explored with the objective of taking advantage of interdisciplinary approaches in order to achieve, with acceptable accuracy levels, computer execution times beyond the present limits.

Virtual prototyping and human/machine interaction are fundamentally new capabilities that are now becoming available to address human factor based design issues. Part V includes contributions that address the key requirements and recent developments in operator-in-the-loop virtual prototyping capabilities. The objective is to access, in real-time, the behavior of mechanical systems with high level of fidelity in order to ensure valid human-machine interaction prediction. These emerging methodologies provide the framework for vehicle and other kinds of man operated machinery simulators. Biomechanical aspects, including muscleskeleton, and traditional multibody dynamics are pointed out to establish interdisciplinary lines of communication on human behavior analyses and mechanical systems design capabilities with the focus on human performance, ergonomics and cumulative injury potential.

Without the sponsorship and financial support of the Scientific Affairs Division of NATO, the U.S. Army - TARDEC and the National Science Foundation, this ASI and book would not have been possible. The financial support of the Junta Nacional de Investigação Cientifica e Tecnológica, the Fundação Luso-Americana para o Desenvolvimento and the Centro de Mecânica e Materiais da Universidade Técnica de Lisboa is also gratefully acknowledged.

The Institute Director wishes to express his gratitude to Prof. Jorge A.C. Ambrósio for his help and cooperation in organizing the ASI. The editors are indebted to the members of the Organizing Committee, Prof. M. Geradin, Prof. E. Haug, Prof. M. Hiller and Prof. P. Nikravesh and also to Prof. C. Mota Soares for their help, advise and support in organizing the ASI. We would like to thank all lecturers and participants in the Institute for their active participation in the discussions and contributed presentations. The smooth running of the ASI is also a result of the competent work of Ms. Glória Ramos, Miss Alexandra Andrade, and Mr. Amândio Rebelo from Centro de Mecânica e Materiais da Universidade Técnica de Lisboa.

Lisbon, February 1994

Manuel F.O. Seabra Pereira

Jorge A.C. Ambrósio

PART I
Methods in Rigid Multibody Dynamics

CONSTRAINED MULTIBODY DYNAMICS

R. A. WEHAGE and M. J. BELCZYNSKI
US Army TARDEC
System Simulation and Technology Division (AMSTA-RY)
Warren, Michigan 48397-5000
USA

ABSTRACT. This paper presents some techniques that may be used to obtain more efficient and general computer-based dynamics modeling and simulation algorithms with potential real-time applications. Constrained equations of motion are first formulated in an augmented differential-algebraic form using spatial Cartesian and joint coordinates. Spatial algebra and graph theoretic methods allow separation of system topology, kinematic, and inertia properties to obtain generic equation representations. Numerical stability is improved by employing coordinate partitioning or singular value decomposition to define suitable sets of independent variables. Substantial matrix operations during run-time are avoided by employing equation preprocessing to generate explicit expressions for all dependent variables, and coefficients of their first and second time derivatives. The velocity and acceleration coefficients allow explicit elimination of all spatial and dependent joint coordinates yielding a minimal system of highly coupled differential equations. A symbolic recursive algorithm that simultaneously decouples the reduced equations of motion as they are generated, was developed to maximize algorithm parallelism.

1. Introduction

Developing real-time computer algorithms for large-scale, highly constrained mechanical systems is a challenge. Extensive understanding of underlying equation structures, symbolic and numerical procedures, and supporting computer architectures is essential. Equation manipulation and solution procedures may be hand optimized, but most are not easily automated. Problems stem from equation complexity, the diversity of parametric descriptions, topology and interdisciplinary effects, and changing computer hardware and software architectures.

The purpose of this paper is to abstract kinematics and dynamics equations into block matrix structures, and investigate procedures for achieving better performing computer programs. Large scale constrained systems with 70% or more dependent state variables are the grand challenge for real-time simulations because the run-time computation of the dependent variables and terms depending on them represents a significant overhead in evaluating and solving the state equations. A special class of bounded systems composed strictly of lower pair joints is considered because the kinematic equations allow many quantities normally computed at run-time to be precomputed and evaluated using interpolating functions. This can help reduce recursive computational bottlenecks and improve parallel processor performance. One problem with precomputing terms is that quantities generated at successive stages of recursive decoupling will depend on increasing numbers of variables, which increases dimensionality of the interpolating functions. Function evaluation overhead increases exponentially with dimensionality, so tradeoffs between precomputing quantities off-line or evaluating them at run-time must be made.

M. F. O. S. Pereira and J. A. C. Ambrósio (eds.),
Computer-Aided Analysis of Rigid and Flexible Mechanical Systems, 3–29.

2. Approach

Spatial algebra [1, 2] is used to formulate the equations because it allows rotational and translational quantities to be combined into homogeneous forms making symbolic manipulation easier. In addition, a compact notation simplifies the representation of arbitrary kinematic and dynamic quantities. The algebra is developed in sufficient detail to provide background for the derivations. Spatial vectors and transformations in configuration and function spaces are introduced and illustrated [2, 3]. They are used to develop compact equations of motion for an unconstrained rigid body and primitive building blocks for defining arbitrary joints. Block matrix representations for constrained systems composed of any number of joints and rigid bodies are introduced.

A general graph theoretic approach [4-6] facilitates matrix representation of mechanical systems containing arbitrary parametric and topological properties. The topology of a mechanical system model with n_a rigid bodies and $n_a + n_c$ joints is adequately described by a connected graph with n_a nodes corresponding to the bodies, and n_a arcs and n_c chords corresponding to the joints. A system's defining graph contains a connected spanning tree with exactly n_a arcs joining the n_a nodes. The minimal spanning tree corresponds to an equivalent open kinematic-loop mechanical system with only the minimum number of joints necessary to hold all the bodies together. This includes one or more artificial six degree-of-freedom (dof) joints for referencing floating base bodies to an inertial frame. All variables in open kinematic-loop systems are independent.

Incorporating the remaining n_c chords into the spanning tree completes the graph and generates n_c independent circuits. Likewise, adding the remaining n_c joints to a model completes it and generates n_c independent kinematic loops. While this process adds more joint variables to the model, it subsequently reduces overall system dof because kinematic loop constraints cause more joint variables to become dependent than are added. In summary, if $0 \le k_i \le 6$, $i = 1, \ldots, n_a + n_c$ represents the dof allowed by the respective joints, then

$$\mathrm{dof}_c = \sum_{i=1}^{n_a} k_i \tag{1}$$

represents the effective system dof with all chord joints removed. With chord joints included, the system dof becomes [7]

$$\mathrm{dof}_s = \sum_{i=1}^{n_a + n_c} k_i - \mathrm{rank}(\text{constraint Jacobian matrix}) \tag{2}$$

where the constraint Jacobian matrix represents the composite coefficients of the combined system of linearized kinematic constraint equations.

Block matrix representation of the uncoupled equations of motion for the n_a bodies and arc joints, and the n_c chord joints given in a combined Cartesian and joint coordinate space are easily formed. Block Boolean arc and chord connectivity matrices that may contain embedded spatial transformations, are used to couple the joint and body equations of motion through constraint reaction forces into a system of constrained equations of motion. This yields a large system of loosely coupled differential-algebraic equations involving all absolute and joint accelerations, and joint reaction forces that could be solved by sparse matrix and differential-algebraic integration procedures [8, 9]. However, the overhead of solving these equations is too high for real-time simulation applications.

The kinematic loop constraint equations and an iterative Newton-Raphson procedure are used to solve for all dependent joint variables (and subsequently coefficients of their first and second time derivatives) in terms of selected independent variables [2, 10]. This allows all dependent joint variables and their time derivatives to be explicitly eliminated yielding a smaller system of differential-algebraic equations. These equations may also be solved by sparse matrix and differential-algebraic integration procedures, but again it would be impractical for real-time applications. However, it is possible to decouple a further reduced version of these equations using a variant of $O(n^3)$ LU factorization. Such an algorithm is developed in this paper.

Without additional considerations, these efforts would still be insufficient to achieve real-time simulation capability for most large scale system models. Ideally, the equations of motion would be expressed in explicit form as $\ddot{q} = \mathbf{f}(q, \dot{q}, t)$ where $\mathbf{f}(q, \dot{q}, t)$ is easy to evaluate. The elements of $\mathbf{f}$, or at least parts of it, would be precomputed off line as functions of $\mathbf{q}$, $\dot{\mathbf{q}}$ and t. However, this is usually impractical because each element of $\mathbf{f}$ may depend on too many elements of $\mathbf{q}$ and $\dot{\mathbf{q}}$.

The problem could also be formulated directly in factored form $\mathbf{L}(q)\,\mathbf{U}(q)\,\ddot{q} = \mathbf{g}(q, \dot{q}, t)$ where $\mathbf{L}(q)$ and $\mathbf{U}(q)$ are nonsingular lower and upper triangular matrices, respectively. Now the individual terms in $\mathbf{L}(q)$, $\mathbf{U}(q)$ and $\mathbf{g}(q, \dot{q}, t)$ will have fewer variable dependencies and it may be feasible to precompute functional relationships at the expense of additional computational overhead during run-time. In most cases, it is also impractical to find functional relationships for all elements of $\mathbf{L}(q)$, $\mathbf{U}(q)$ and $\mathbf{g}(q, \dot{q}, t)$ because they, too, may depend on a relatively large number of variables. Therefore, some expressions may have to be broken down even further until an optimal compromise is reached between the overhead of evaluating multivariable functions versus computing the quantities at run-time. The primary advantage of using precomputed functions is their potential for eliminating recursive operations that bottleneck parallel processors.

3. Spatial Algebra

3.1. NOTATIONAL CONVENTIONS

The quantities dealt with in this paper are first and second order tensors that are generally represented by column and rectangular matrices, respectively. Column matrices, usually identified by lower-case letters, are used to represent the coordinates of first order tensors. Rectangular and square matrices represent coordinates of second order tensors and are identified by upper-case letters. Stacked or block collections of column matrices are represented by bold lower-case letters, which usually match the corresponding symbols of their constituent column matrices. In a similar manner, bold upper-case letters denote block arrays of matrices that may be block diagonal, triangular, symmetric or irregular.

A first order vector tensor is denoted by a lower-case symbol with an overhead arrow $\vec{a}$ and its three by one column coordinate matrix by an underscore

$$\underline{a} = [a_1, a_2, a_3]^T \tag{3}$$

The matrix representation of vector dot product operation $\vec{a}\cdot$ is represented by $\underline{a}^T$ where superscript "T" denotes matrix transpose and gives a one by three row matrix. A vector cross product operation $\vec{a}\times$ is represented by a three by three skew-symmetric tensor matrix

$$\tilde{\underline{a}} = \begin{bmatrix} 0 & -a_3 & a_2 \\ a_3 & 0 & -a_1 \\ -a_2 & a_1 & 0 \end{bmatrix} \tag{4}$$

Second order vector tensor matrices are identified by underlined symbols that may be lower or upper case.

A first order spatial vector tensor is composed of a rotational vector tensor $\underline{a}_r$ and a translational vector tensor $\underline{a}_t$ where either one or both may be zero. It is represented as a stacked column matrix of the form

$$a = \left[\underline{a}_r^T, \underline{a}_t^T\right]^T \tag{5}$$

Since spatial vectors and tensors are used almost exclusively throughout this paper, they are denoted by plain lower and upper case letters with no underscore. The matrix representation of spatial vector inner product is given by a^T. A spatial vector cross product is represented by a six by six matrix

$$\tilde{a} = \begin{bmatrix} \tilde{\underline{a}}_r & 0 \\ \tilde{\underline{a}}_t & \tilde{\underline{a}}_r \end{bmatrix} \tag{6}$$

that is not skew-symmetric. Spatial vector tensor matrices are identified by symbols that are usually upper-case. Spatial transformations are homogeneous, and can be represented by six by six [1] or four by four [10] matrices. The latter form is desirable when carrying out products associated with successive transformations because they involve fewer multiplications. A circumflex "^" is used to identify the four by four matrices.

Identity and zero matrices are denoted by regular or bold "I" and "0" symbols, respectively. Their dimensions are inferred by adjacency to other matrices. The 0 symbols in sparse matrices are dropped when matrix dimensions can be inferred from other submatrix entries. All matrices conform to the operations implied by the equations, and the superscript and subscript conventions will be consistent. A single superscript will associate the elements in a column matrix with a given coordinate system and a single subscript will indicate block column matrix partitioning. Double superscripts are generally associated with coordinates of transformation matrices or second order tensors. Superscripts and subscripts of quantities being combined must conform to avoid numerical errors. In addition, adjacent superscript and subscript symbols in matrix products must match. However, transpose and inverse effectively interchange matrix rows and columns so the corresponding identifying superscript and subscript symbols on these matrices will be reversed.

Since general spatial displacement transformations are not orthogonal or orthonormal, it is convenient to identify dual spaces called configuration and function space, respectively [3]. Configuration space contains all dimensional quantities such as displacement, velocity and acceleration, and function space contains derived quantities such as force and momentum. A function space quantity is transformed by the inverse transpose "–T" of any matrix that transforms a corresponding configuration space quantity. Reversing the double superscript or double subscript on any transformation matrix is equivalent to inverting that matrix. The inverse and transpose of orthonormal transformations are also equal.

3.2. SPATIAL VECTORS AND TRANSFORMATIONS

Spatial vectors are composed of bound and free vector components. Bound vectors describe quantities such as rotation axes and forces that can be located in space relative to reference points. Free vectors describe quantities such as translational directions and moments that cannot be located relative to reference points.

A bound vector is located relative to a point by specifying its moment about that point. Let $\vec{u}$ denote a unit vector lying on a line, and $\vec{t}$ a vector from some reference point p to any point on the line. Then $\vec{u}$ and its moment $\vec{t}\times\vec{u}$ about p completely define the bound vector. The moment $\vec{t}\times\vec{u}$ is a free vector because it represents a quantity that cannot be associated with any point. If the line lies on p then $\vec{t}\times\vec{u}=0$ and it has no moment about p.

If $\vec{u}$ lies on some point q which has zero velocity and the body rotates around the line defined by $\vec{u}$ with a speed of ω, then $\vec{\omega}=\omega\,\vec{u}$ gives its rotational velocity and $\vec{t}\times\vec{\omega}=\omega\,\vec{t}\times\vec{u}$ gives the translational velocity of any point p fixed in the body. Note that $\vec{\omega}$ is a bound vector and $\vec{t}\times\vec{\omega}$ is a free vector. Likewise, if f denotes the magnitude of a force acting at a point, then $\vec{f}=f\,\vec{u}$ gives its force vector and $\vec{t}\times\vec{f}=f\,\vec{t}\times\vec{u}$ gives its torque vector. In this case, $\vec{f}$ is bound and $\vec{t}\times\vec{f}$ is free. In configuration space, rotational vector quantities are bound and translational vector quantities are free. Conversely, in function space, translational vector quantities are bound and rotational vector quantities are free. In either case, coordinates of the rotational component of a spatial vector are always stacked above coordinates of the translational component.

These ideas are enforced with an example. Figure 1a shows two Cartesian frames labeled α and β where β is moving relative to α. The spatial velocity of β relative to α in β coordinates is

$$v^{\beta}_{\alpha\beta}=\left[\underline{\omega}^{\beta T}_{\alpha\beta},\ \dot{\underline{t}}^{\beta T}_{\alpha\beta}\right]^{T} \tag{7}$$

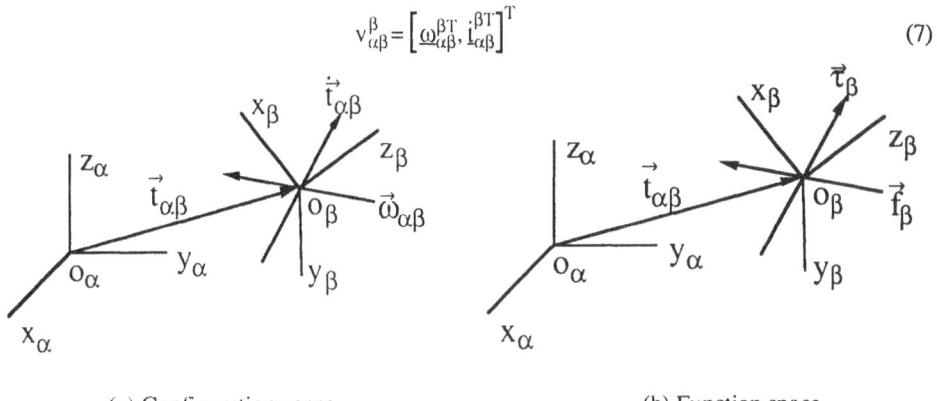

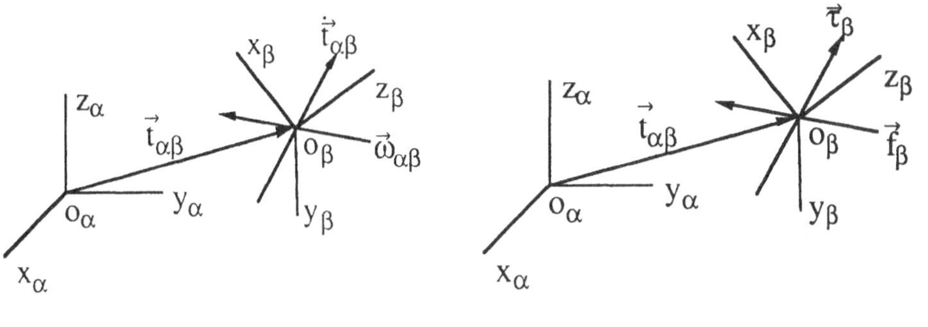

(a) Configuration space (b) Function space

Figure 1. Spatial velocities and forces referenced to frame β.

Superscript β is a reminder that the coordinates are projected onto β axes, and that the reference point is taken at origin O_β where the translational velocity is specified. Double subscript $\alpha\beta$ indicates that this quantity defines the velocity of β relative to α.

Figure 1b shows an equivalent system of forces acting on β that have been reduced to O_β

$$\underline{g}_\beta^\beta = \left[\underline{\tau}_\beta^{\beta T}, \underline{f}_\beta^{\beta T}\right]^T \tag{8}$$

Again superscript β is a reminder that the coordinates have been projected onto β axes and the net force has been specified at O_β where the rotational torque was arbitrarily placed. In this case, the single subscript β indicates that these quantities act on β.

Figures 2a and 2b show the same systems where the spatial velocity and force are specified on β, but at the unique point fixed in it that instantaneously coincides with O_α. Inspection of these

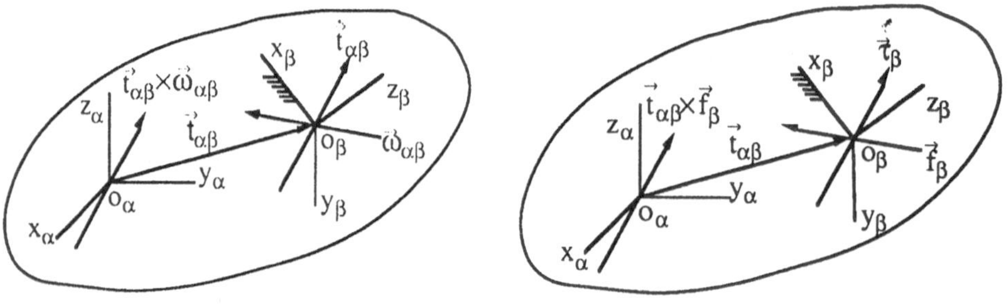

(a) Configuration space (b) Function space

Figure 2. Spatial velocities and forces referenced to frame α.

figures reveals that the relative spatial velocity, which is now expressed in α coordinates is

$$v_{\alpha\beta}^\alpha = \left[\underline{\omega}_{\alpha\beta}^{\alpha T}, \left(\dot{\underline{t}}_{\alpha\beta}^\alpha + \tilde{\underline{t}}_{\alpha\beta}^\alpha \underline{\omega}_{\alpha\beta}^\alpha\right)^T\right]^T \tag{9}$$

and the spatial force is

$$\underline{g}_\beta^\alpha = \left[\left(\underline{\tau}_\beta^\alpha + \tilde{\underline{t}}_{\alpha\beta}^\alpha \underline{f}_\beta^\alpha\right)^T, \underline{f}_\beta^{\alpha T}\right]^T \tag{10}$$

Now, superscript α is a reminder that the reference point coincides with O_α and all vector quantities have been transformed to α. Equations 7 and 9 also imply that $v_{\alpha\beta}^*$ may be interpreted as representing the vector sum of velocity component $\dot{\vec{t}}_{\alpha\beta}$ in a nonrotating frame and the additional velocity component $\vec{t}_{*\beta}\times\vec{\omega}_{\alpha\beta}$ at an arbitrary point * fixed in a frame rotating at $\vec{\omega}_{\alpha\beta}$ around fixed point O_β. Likewise Eqs. 8 and 10 imply that g_β^* represents the vector sum of torque component $\vec{\tau}_\beta$ with the additional moment component $\vec{t}_{*\beta}\times\vec{f}_\beta$ about arbitrary point * from force

$\vec{f}_\beta$ acting at O_β. Before finding functional relationships between the quantities in Eqs. 7 to 10, it will be helpful to develop expressions for homogeneous spatial transformation matrices.

An orthonormal rotational displacement transformation matrix may be expressed as [11]

$$\underline{R}^{\alpha\beta} = I + \sin\theta_{\alpha\beta}\,\tilde{\underline{u}}_o + (1-\cos\theta_{\alpha\beta})\,\tilde{\underline{u}}_o^2 \tag{11}$$

where $\underline{u}_o$ represents a bound unit vector defining the orientation axis between α and β, and $\theta_{\alpha\beta}$ is the rotational displacement of β relative to α around this axis.

Let $\underline{a}^\alpha$ and $\underline{a}^\beta$ denote coordinates of an arbitrary vector $\vec{a}$ in frames α and β, respectively. If

$$\underline{a}^\alpha = \underline{R}^{\alpha\beta}\,\underline{a}^\beta \tag{12}$$

the following similarity transformation holds [12]

$$\tilde{\underline{a}}^\alpha\,\underline{R}^{\alpha\beta} = \underline{R}^{\alpha\beta}\,\tilde{\underline{a}}^\beta \tag{13}$$

Let

$$R^{\alpha\beta} = \begin{bmatrix} \underline{R}^{\alpha\beta} & 0 \\ 0 & \underline{R}^{\alpha\beta} \end{bmatrix} \tag{14}$$

denote an orthonormal spatial rotational displacement transformation matrix. Then Eqs. 12 to 14 may be used to verify that Eqs. 7 and 9 are related by

$$v^\alpha_{\alpha\beta} = D^{\alpha\beta}\,v^\beta_{\alpha\beta} \tag{15}$$

where

$$D^{\alpha\beta} = T^\alpha_{\alpha\beta}R^{\alpha\beta} = R^{\alpha\beta}T^\beta_{\alpha\beta} \tag{16}$$

is a general nonorthonormal spatial displacement transformation matrix and

$$T^*_{\alpha\beta} = \begin{bmatrix} I & 0 \\ \tilde{\underline{l}}^*_{\alpha\beta} & I \end{bmatrix}, \quad * = \alpha, \beta \tag{17}$$

is a nonorthonormal spatial translational displacement transformation matrix.

In a similar manner, Eqs. 8 and 10 are related by

$$g^\alpha_\beta = D^{\alpha\beta-T}\,g^\beta_\beta \tag{18}$$

where

$$D^{\alpha\beta-T} = T_{\alpha\beta}^{\alpha-T} R^{\alpha\beta} = R^{\alpha\beta} T_{\alpha\beta}^{\beta-T} \tag{19}$$

and

$$T_{\alpha\beta}^{*-1} = \begin{bmatrix} I & 0 \\ -\tilde{\iota}_{\alpha\beta}^{*} & I \end{bmatrix}, \; * = \alpha, \beta \tag{20}$$

These transformations preserve the inner product between function and configuration spaces.

Let a_c^α and b_c^α represent the coordinates of arbitrary spatial vectors in configuration space, each referenced to α. Then the matrix representation of spatial cross product (see Eqs. 3 to 6)

$$\tilde{a}_c^\alpha b_c^\alpha = -\tilde{b}_c^\alpha a_c^\alpha \tag{21}$$

holds for the coordinates of any two spatial vectors in configuration space referenced to the same frame. The cross product of a spatial vector with itself is zero. Let b_f^α represent the coordinates of an arbitrary spatial vector in function space, referenced to α. In this case, the following matrix representation of spatial cross product holds

$$-\tilde{a}_c^{\alpha T} b_f^\alpha = \tilde{b}_f^{\alpha T} a_c^\alpha \tag{22}$$

Note the sign reversal and matrix transpose when forming cross products of spatial vectors between the two spaces. Cross products between spatial vectors in function space are not defined.

The spatial transformation matrices presented in Eqs. 14 and 17 may be defined directly in terms of spatial vectors as follows. Let

$$u_o = [\underline{u}_o^T, \underline{0}^T]^T \tag{23}$$

represent the spatial coordinates of a bound unit vector that defines the orientation axis between α and β, and let $\theta_{\alpha\beta}$ be the rotational displacement of β relative to α around this axis. Then (see Eqs. 5, 6, 11 and 14)

$$R^{\alpha\beta} = I + \sin\theta_{\alpha\beta}\, \tilde{u}_o + (1 - \cos\theta_{\alpha\beta})\, \tilde{u}_o^2 \tag{24}$$

In a similar manner, let

$$\iota_{\alpha\beta}^{*} = [\underline{0}^T, \underline{\iota}_{\alpha\beta}^{*T}]^T, \; * = \alpha, \beta \tag{25}$$

represent the spatial coordinates of a free translational vector that defines the translational displacement of β relative to α. Then (see Eqs. 5, 6 and 17)

$$T_{\alpha\beta}^{*} = I + \tilde{\iota}_{\alpha\beta}^{*}, \; * = \alpha, \beta \tag{26}$$

Now identities similar to Eqs. 12 and 13 may be verified. Let a_c^α and a_c^β denote coordinates of an arbitrary configuration space spatial vector in α and β, respectively. Then if

$$a_c^\alpha = D^{\alpha\beta} a_c^\beta \tag{27}$$

the following congruent transformation holds

$$\tilde{a}_c^\alpha D^{\alpha\beta} = D^{\alpha\beta} \tilde{a}_c^\beta \tag{28}$$

In a similar manner, let a_f^α and a_f^β denote coordinates of an arbitrary function space spatial vector in α and β, respectively. If a_f has the form of Eq. 5 and $a_f^* = [\underline{a}_t^T, \underline{a}_r^T]^T$, then

$$a_f^{\alpha *} = D^{\alpha\beta} a_f^{\beta *} \tag{29}$$

and the following congruent transformation also holds

$$\tilde{a}_f^{\alpha *} D^{\alpha\beta} = D^{\alpha\beta} \tilde{a}_f^{\beta *} \tag{30}$$

3.3. SPATIAL VELOCITIES AND ACCELERATIONS

A matrix identity for rotational velocities may be obtained by differentiating the coordinates of an arbitrary vector fixed in a moving frame and equating its matrix coefficients giving [12]

$$\dot{\underline{R}}^{\alpha\beta} = \tilde{\underline{\omega}}_{\alpha\beta}^{\alpha} \underline{R}^{\alpha\beta} = \underline{R}^{\alpha\beta} \tilde{\underline{\omega}}_{\alpha\beta}^{\beta} \tag{31}$$

where

$$\underline{\omega}_{\alpha\beta}^{\alpha} = \underline{R}^{\alpha\beta} \underline{\omega}_{\alpha\beta}^{\beta} \tag{32}$$

defines the rotational velocity of β relative to α (compare with Eqs. 13 and 12.) Using Eq. 15, a similar identity may be derived by differentiating an arbitrary spatial vector fixed in a moving frame and equating its matrix coefficients giving

$$\dot{D}^{\alpha\beta} = \tilde{v}_{\alpha\beta}^{\alpha} D^{\alpha\beta} = D^{\alpha\beta} \tilde{v}_{\alpha\beta}^{\beta} \tag{33}$$

(compare with Eqs. 27 and 28.) Noting that $D^{\alpha\beta-1} = D^{\beta\alpha}$ and $v_{\beta\alpha}^{*} = -v_{\alpha\beta}^{*}$, $* = \alpha, \beta$, Eq. 33 may be used to find the time derivative of the inverse of any spatial transformation matrix as

$$d\left(D^{\alpha\beta-1}\right)/dt = -\tilde{v}_{\alpha\beta}^{\beta} D^{\alpha\beta-1} = -D^{\alpha\beta-1} \tilde{v}_{\alpha\beta}^{\alpha} \tag{34}$$

These identities will be used extensively to simplify many of the following developments.

It is convenient to represent the spatial displacement of an arbitrary frame or body by a sequence of spatial displacements given by homogeneous spatial products. For example, let

$$D^{\alpha\gamma} = D^{\alpha\beta} D^{\beta\gamma} \tag{35}$$

represent the results of two successive spatial displacements. Then Eq. 35 may be differentiated with respected to time giving

$$\dot{D}^{\alpha\gamma} = \dot{D}^{\alpha\beta} D^{\beta\gamma} + D^{\alpha\beta} \dot{D}^{\beta\gamma} \tag{36}$$

which, according to Eq. 33, represents the spatial velocity. Using Eqs. 15, 27, 28 and 33, it follows that

$$\dot{D}^{\alpha\gamma} = \tilde{v}^{\alpha}_{\alpha\gamma} D^{\alpha\gamma} = D^{\alpha\gamma} \tilde{v}^{\gamma}_{\alpha\gamma} \tag{37}$$

where

$$v^{*}_{\alpha\gamma} = v^{*}_{\alpha\beta} + v^{*}_{\beta\gamma}, \; * = \alpha, \beta, \gamma \tag{38}$$

Equations 35 to 38 may be generalized and applied any number of times to write out spatial displacements and velocity equations between any two frames.

Spatial velocities are differentiated to obtain spatial accelerations. Let α be fixed and differentiate $v^{\alpha}_{\alpha\beta}$ in that coordinate system. Differentiating Eq. 15 gives

$$a^{\alpha}_{\alpha\beta} = dv^{\alpha}_{\alpha\beta}/dt = \left[\dot{\underline{\omega}}^{\alpha T}_{\alpha\beta}, \left(\ddot{\underline{i}}^{\alpha}_{\alpha\beta} + \tilde{\underline{i}}^{\alpha}_{\alpha\beta} \dot{\underline{\omega}}^{\alpha}_{\alpha\beta} + \tilde{\dot{\underline{i}}}^{\alpha}_{\alpha\beta} \underline{\omega}^{\alpha}_{\alpha\beta}\right)^T\right]^T \tag{39}$$

which is the spatial acceleration of β relative to fixed α. Observe that $a^{\alpha}_{\alpha\beta}$ contains an extra term that is quadratic in first derivatives.

Demanding a homogeneous transformation of spatial accelerations gives the relation

$$a^{\beta}_{\alpha\beta} \equiv D^{\alpha\beta-1} a^{\alpha}_{\alpha\beta} = \left[\dot{\underline{\omega}}^{\beta T}_{\alpha\beta}, \left(\ddot{\underline{i}}^{\beta}_{\alpha\beta} + \tilde{\dot{\underline{i}}}^{\beta}_{\alpha\beta} \underline{\omega}^{\beta}_{\alpha\beta}\right)^T\right]^T \tag{40}$$

Comparing Eq. 40 with the time derivative of Eq. 7 shows that the defined spatial acceleration, $a^{\beta}_{\alpha\beta}$ does not represent coordinates of the true acceleration of β relative to α because it contains an additional term that is quadratic in first time derivatives. However, the homogeneous transformation in Eq. 40 simplifies the equations of motion and the extra term will pose no problems as long as it is subtracted out when referring to coordinates of the actual acceleration.

3.4. SPATIAL JOINT BUILDING BLOCKS

In idealized mechanical system models, rigid bodies are connected by joints with nondeforming surfaces. Many types of joints may be assembled from a set of primitive joint building blocks, where each joint allows only a single relative displacement between adjacent frames in one of six possible directions. The six directions are defined locally by the orthogonal unit spatial vectors u_1 through u_6 as shown below [5]

$$\overset{\text{Rotation}}{\left[\overbrace{u_1\; u_2\; u_3}\right.}\ \vdots\ \overset{\text{Translation}}{\left.\overbrace{u_4\; u_5\; u_6}\right]} = \left[\begin{array}{ccc|ccc} \underline{u}_1 & \underline{u}_2 & \underline{u}_3 & \underline{0} & \underline{0} & \underline{0} \\ \underline{0} & \underline{0} & \underline{0} & \underline{u}_1 & \underline{u}_2 & \underline{u}_3 \end{array}\right] = \left[\begin{array}{ccc|ccc} 1&0&0&0&0&0\\ 0&1&0&0&0&0\\ 0&0&1&0&0&0\\ 0&0&0&1&0&0\\ 0&0&0&0&1&0\\ 0&0&0&0&0&1 \end{array}\right] \tag{41}$$

Matrices u_1 through u_3 define bound rotational directions around respective common x, y or z-axis pairs. Likewise, u_4 through u_6 define free translational directions along respective common x, y or z-axis pairs. If a primitive joint allows relative motion around or along its axis, then the corresponding spatial unit vector will define that joint's influence coefficient matrix [5, 6]. Whether a primitive joint allows a fixed or a variable displacement, that displacement is given by

$$R_i^{\alpha\beta} = I + \sin\theta_{\alpha\beta}\,\tilde{\underline{u}}_i + \left(1 - \cos\theta_{\alpha\beta}\right)\tilde{\underline{u}}_i^2,\ i = 1, 2, 3 \tag{42}$$

if it is a rotational joint (see Eqs. 5, 6, 23, 24 and 60) or

$$T_{\alpha\beta,i}^{\alpha} = T_{\alpha\beta,i}^{\beta} = I + t_{\alpha\beta}\,\tilde{\underline{u}}_i,\ i = 4, 5, 6 \tag{43}$$

if it is a translational joint (see Eqs. 5, 6, 25, 26 and 61.) The magnitude of rotation is $\theta_{\alpha\beta}$ and the magnitude of translation is $t_{\alpha\beta}$. Any number of constant and variable primitive joint building blocks may be combined sequentially to form numerous joint configurations by multiplying the respective displacement transformation matrices together as illustrated in Eq. 35. This approach allows constant displacement transformations to appear in variable displacement transformations.

Let joint i have $0 \le k_i \le 6$ dof and associated with it, if $k_i > 0$, is a 6 by k_i influence coefficient matrix H_i^{α}, where the individual columns correspond to the primitive joint influence coefficient matrices, each transformed to the common frame α. Also there is a k_i by 1 column matrix of primitive joint rotational and translational variables p_i that correspond to the k_i joint dof. It is assumed that each joint is defined so H_i^{α} will have full column rank. If this composite joint connects frames α and β, there is a spatial displacement matrix $D_i^{\alpha\beta}$ formed from a sequence of primitive spatial displacement products that depend on the p_i variables, as well as zero or more constant primitive displacements. The spatial velocity of β relative to α in α coordinates is

$$v_{\alpha\beta}^{\alpha} = H_i^{\alpha}\,\dot{p}_i \tag{44}$$

Each primitive joint and corresponding spatial displacement or velocity will have an associated direction or orientation. The displacement transformation of any primitive joint oriented opposite to the assumed composite joint orientation must be inverted when forming the product $D_i^{\alpha\beta}$. The inverse may also be affected by reversing the sign on the corresponding variable in p_i. In Eq. 44, this inversion may be accounted for by reversing the sign on the corresponding column of H_i^{α}.

The time derivative of H_i^{α} is required when computing the relative spatial acceleration. To find this derivative, first note that the jth column of H_i^{α}, denoted by H_{ij}^{α} defines one displacement transformation appearing in the product forming $D_i^{\alpha\beta}$ (see Eqs. 41 to 43.) Let

$$H_{ij}^{\alpha} = D_i^{\alpha\gamma} u_{*j} = D_i^{\alpha\delta} u_{*j},\ 1 \le j \le k_i \tag{45}$$

represent the jth primitive joint connecting frames γ and δ that allows one dof, where u_{*j} is one of the six constant spatial unit vectors in Eq. 41. Either transformation is valid because the influence coefficient matrix u_{*j} is invariant under transformation $D_i^{\gamma\delta}$. Now

$$D_i^{\alpha\beta} = D_i^{\alpha\gamma} D_i^{\gamma\beta} = D_i^{\alpha\delta} D_i^{\delta\beta} \tag{46}$$

and if α is fixed

$$\dot{H}_{ij}^{\alpha} = \tilde{v}_{\alpha\gamma}^{\alpha} H_{ij}^{\alpha} = \tilde{v}_{\alpha\delta}^{\alpha} H_{ij}^{\alpha} \tag{47}$$

Derivatives of successive columns of H_i^{α} include additional terms from the relative velocity vector $v_{\alpha\beta}^{\alpha}$. From Eq. 44, note that $v_{\alpha\beta}^{\alpha}$ is the sum of the k_i relative spatial velocities across the k_i primitive joints and $v_{\alpha\gamma}^{\alpha}$ is the sum of the first j of these. For programming purposes, explicit expressions for time derivatives of the influence coefficient matrices may be written out directly and it suffices to leave them in the form $\dot{H}_{ij}^{\alpha}$. With this convention, differentiating Eq. 44 gives

$$a_{\alpha\beta}^{\alpha} = H_i^{\alpha} \ddot{p}_i + \dot{H}_i^{\alpha} \dot{p}_i \tag{48}$$

If α is not fixed, then there will be some frame such as 0 that is, and the above influence coefficient matrices may be transformed to that frame. In this case, Eq. 44 is transformed before differentiating, and as indicated in Eqs. 35 to 38, the additional spatial velocity cross product associated with differentiating this transformation will be included in the influence coefficient matrix derivative.

3.5. SPATIAL EQUATIONS OF MOTION FOR SINGLE BODY

The spatial equations of motion for an unconstrained rigid body may be obtained by differentiating the coordinates of its spatial momentum that have been expressed in a fixed or inertial frame. Figure 3 shows a rigid body with embedded frames α and c where frame c defines the principal centroidal axes.

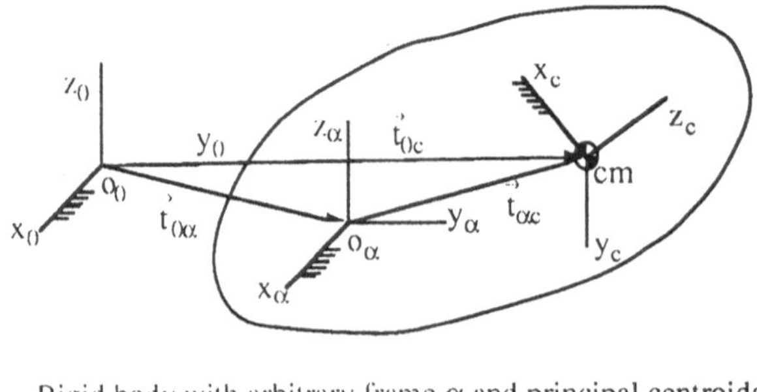

Figure 3. Rigid body with arbitrary frame α and principal centroidal axes c shown relative to inertial frame 0.

Let the constant spatial inertia matrix

$$M_{\alpha}^{cc} = \begin{bmatrix} \underline{M}_{\alpha 2}^{cc} & 0 \\ 0 & \underline{M}_{\alpha 0} \end{bmatrix} \tag{49}$$

represent the centroidal inertia for this body where the respective three by three matrices $\underline{M}_{\alpha 0}$ and $\underline{M}_{\alpha 2}^{cc}$ are the zeroth order (mass) and second order inertia tensors. The body momentum relative to the point fixed in 0 which instantaneously coincides with its center of mass is given by

$$p_{\alpha}^{c} = M_{\alpha}^{cc} v_{0\alpha}^{c} \tag{50}$$

where $v_{0\alpha}^{c} = D^{\alpha c-1} v_{0\alpha}^{\alpha}$ specifies the spatial velocity in principal coordinates. Transforming Eq. 50 to α gives

$$p_{\alpha}^{\alpha} = M_{\alpha}^{\alpha\alpha} v_{0\alpha}^{\alpha} \tag{51}$$

where the constant spatial inertia matrix defined by a congruent transformation

$$M_{\alpha}^{\alpha\alpha} = D^{\alpha c-T} M_{\alpha}^{cc} D^{\alpha c-1} \tag{52}$$

is now expressed in α. In a similar manner, the spatial momentum in frame 0 coordinates is

$$p_{\alpha}^{0} = M_{\alpha}^{00} v_{0\alpha}^{0} \tag{53}$$

where another congruent spatial displacement transformation gives

$$M_{\alpha}^{00} = D^{0\alpha-T} M_{\alpha}^{\alpha\alpha} D^{0\alpha-1} \tag{54}$$

Now Eq. 53 may be differentiated with respect to time to obtain the spatial equations of motion

$$\dot{p}_{\alpha}^{0} = M_{\alpha}^{00} a_{0\alpha}^{0} + \dot{M}_{\alpha}^{00} v_{0\alpha}^{0} \tag{55}$$

Differentiating Eq. 54 with the help of Eq. 34 gives

$$\dot{M}_{\alpha}^{00} = -\tilde{v}_{0\alpha}^{0T} M_{\alpha}^{00} - M_{\alpha}^{00} \tilde{v}_{0\alpha}^{0} \tag{56}$$

Substituting Eq. 56 into Eq. 55 gives the spatial equations of motion for a single rigid body as

$$M_{\alpha}^{00} a_{0\alpha}^{0} = \tilde{v}_{0\alpha}^{0T} p_{\alpha}^{0} + g_{\alpha}^{0} \tag{57}$$

where g_{α}^{0} represents the combined spatial forces acting on the body or an extension thereof, taken at the point which instantaneously coincides with the origin of frame 0.

4. Constrained Equations of Motion

4.1. REPRESENTATION OF SYSTEM TOPOLOGY

It is assumed that motion is measured relative to an inertial frame, denoted by the symbol 0. Thus the absolute displacement, velocity, etc., of every body in the system will be measured relative to 0. However, if the absolute displacement, velocity, etc. of one body in a collection of bodies connected by joints is known, and if the corresponding relative joint displacement quantities are also known, then all absolute quantities may be calculated from this information. Graph theoretic methods are used to provide an organized means to accomplish these calculations.

It is also assumed that constrained mechanical systems are composed of rigid bodies connected by idealized joints with nondeforming surfaces. Furthermore, one body in each kinematically disconnected system (designated as the base body) must be physically connected to 0 or referenced to it by an artificial six dof joint that gives its absolute displacement relative to 0. The remaining bodies in each disconnected system are directly or indirectly connected to the corresponding base body by additional joints. The minimum number of joints (including the artificial six dof ones) necessary to tie all bodies together into a contiguous system (no closed kinematic loops) is exactly equal to the number of bodies in the system. It is convenient to describe the interconnectivity of these n_a bodies through the corresponding n_a joints with a minimum spanning tree. The joints comprising this tree are called arcs and their number, and the corresponding number of bodies are denoted by n_a. The subscript a is used extensively to denote various quantities associated with the spanning tree. Figure 4 shows an example spanning tree where rectangles, solid lines and dashed lines identify respective nodes (bodies), and arcs and chords (joints).

An n_a by n_a Boolean arc connectivity matrix C_a may be devised to associate each arc joint with

the corresponding two bodies (parent and child) joined by it. The base body is at the base of the spanning tree and the other arc joints and bodies radiate outward from there. Assume that each joint has been oriented so its corresponding child body (farther out in the tree) is referenced positively to its parent body across the joint. Now it is possible to associate each of the n_a joints with exactly one and only one unique child body. Furthermore, let the n_a joints and bodies be ordered so that no child appears before its parent in the list. Associate the columns of C_a with the bodies, and the rows with the joints. Within each row, place a "1" in the column corresponding to the child body and "−1" in the column corresponding to the parent body. Frame 0 does not appear in this matrix, so there will be no parent entry for any base body.

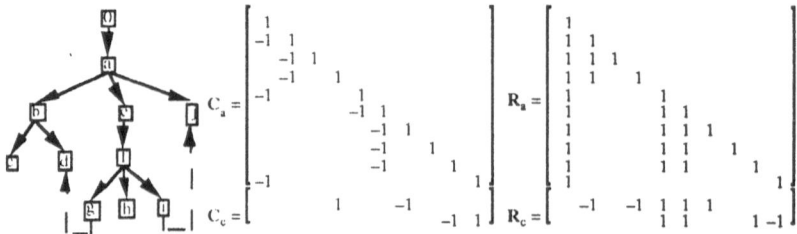

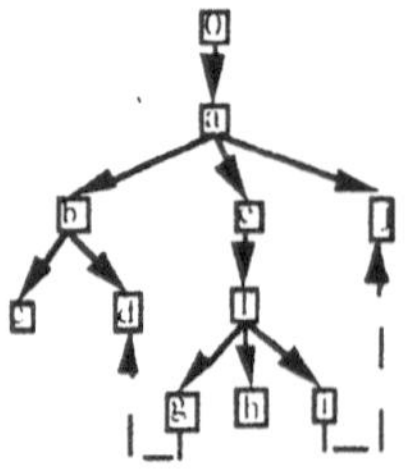

$$C_a = \begin{bmatrix} 1 &&&&&&&&& \\ -1 & 1 &&&&&&&& \\ & -1 & 1 &&&&&&& \\ & -1 && 1 &&&&&& \\ -1 &&&& 1 &&&&& \\ &&&& -1 & 1 &&&& \\ &&&&& -1 & 1 &&& \\ &&&&& -1 && 1 && \\ &&&&& -1 &&& 1 & \\ -1 &&&&&&&&& 1 \end{bmatrix} \qquad R_a = \begin{bmatrix} 1 &&&&&&&&& \\ 1 & 1 &&&&&&&& \\ 1 & 1 & 1 &&&&&&& \\ 1 & 1 && 1 &&&&&& \\ 1 &&&& 1 &&&&& \\ 1 &&&& 1 & 1 &&&& \\ 1 &&&& 1 & 1 & 1 &&& \\ 1 &&&& 1 & 1 && 1 && \\ 1 &&&& 1 & 1 &&& 1 & \\ 1 &&&&&&&&& 1 \end{bmatrix}$$

$$C_c = \begin{bmatrix} &&& 1 &&& -1 &&& \\ &&&&&&&& -1 & 1 \end{bmatrix} \qquad R_c = \begin{bmatrix} & -1 && -1 & 1 & 1 & 1 &&& \\ &&&& 1 & 1 &&& 1 & -1 \end{bmatrix}$$

Figure 4. Spanning tree showing nodes arcs and chords with respective connectivity matrices.

With this convention, C_a is now lower triangular with 1's on the diagonal and has a lower triangular inverse denoted by $\mathbf{R}_a$. In this form $\mathbf{R}_a$ may be written down just as easily as $\mathbf{C}_a$ from the spanning tree representation of the system. Starting from the jth body in the system, place a "1" in each row i, column j position corresponding to each body i that is a descendant of body j. Alternately the nonzero entries in the ith row of $\mathbf{R}_a$ identify the arc joints (and their orientations) which lie in the shortest path from the root of the tree to the ith joint.

The remaining joints in the system (indicated by dashed lines) connect bodies to form closed kinematic loops. These joints are called chords and their number is denoted by the symbol n_c. They generate exactly n_c independent closed kinematic loops. Similar to the above discussion, an n_c by n_a Boolean chord connectivity matrix $\mathbf{C}_c$ may be devised to associate each chord joint with the corresponding two bodies it joins. Assigning an orientation to each joint with associated parent and child bodies, each row of $\mathbf{C}_c$ will contain exactly one "1" and one "–1". Clearly $\mathbf{C}_c$ is rectangular and does not have an inverse. However, it does have a right inverse defined by

$$\mathbf{R}_c = -\,\mathbf{C}_c\,\mathbf{R}_a \tag{58}$$

Similar to $\mathbf{R}_a$ above, the nonzero entries in the kth row of $\mathbf{R}_c$ identify the arc joints and their orientations relative to the kth kinematic loop and corresponding kth chord joint. This convention assures that each chord joint and corresponding kinematic loop have the same orientation.

4.2. ABSOLUTE SPATIAL DISPLACEMENTS IN CONSTRAINED SYSTEMS

The kth chord joint represented by spatial displacement matrix D_k^{ij} connects two tree branches to form a closed kinematic loop. These branches share a common ancestor body, say α. Matrices $D_k^{\alpha i}$ and $D_k^{\alpha j}$ are defined by identifying those joints in each branch corresponding to the nonzero entries in the portions of rows i and j of $\mathbf{R}_a$ located from its diagonal left to, but not including, column α. The constraint loop will be oriented with $D_k^{\alpha i}$ and D_k^{ij}, and opposite to $D_k^{\alpha j}$. The net displacement around any closed loop must be zero, which is described by the matrix product

$$D_k^{\alpha i}\,D_k^{ij}\,D_k^{\alpha j-1} = I \tag{59}$$

Equation 59 defines the loop constraint equations, but it is evaluated using transformation matrices of the form

$$\hat{R}^{\alpha\beta} = \begin{bmatrix} 1 & \underline{0}^T \\ \underline{0} & \underline{R}^{\alpha\beta} \end{bmatrix} \tag{60}$$

$$\hat{T}^{*}_{\alpha\beta} = \begin{bmatrix} 1 & \underline{0}^T \\ \overset{*}{\underline{l}}_{\alpha\beta} & I \end{bmatrix},\; * = \alpha, \beta \tag{61}$$

and

$$\hat{D}^{\alpha\beta} = \hat{T}^{\alpha}_{\alpha\beta}\,\hat{R}^{\alpha\beta} = \hat{R}^{\alpha\beta}\,\hat{T}^{\beta}_{\alpha\beta} \tag{62}$$

Now Eq. 59 may be revised to read

$$\hat{\phi}_k^{\alpha} = \hat{D}_k^{\alpha i} \hat{D}_k^{ij} \hat{D}_k^{\alpha j-1} - I \cong \begin{bmatrix} 0 & \underline{0}^T \\ \underset{\sim}{\phi}_{tk}^{\alpha} & \underset{\sim}{\phi}_{rk}^{\alpha} \end{bmatrix} \tag{63}$$

where $\underset{\sim}{\phi}_{rk}^{\alpha}$ and $\underset{\sim}{\phi}_{tk}^{\alpha}$ correspond to respective rotational and translational constraint violations. Numerical values for all loop constraint equations are obtained from expressions such as Eq. 63.

The constraint equations are nonlinear, and to solve for the dependent joint variables requires a procedure such as Newton-Raphson iteration and a Jacobian matrix. The constraint loop Jacobian matrix may be expressed directly in terms of joint influence coefficient matrices. First let

$$\Delta \mathbf{d}_a^0 = \mathbf{H}_a^0 \Delta \mathbf{p}_a \tag{64}$$

represent a stacked column matrix of small changes in relative displacements between adjacent bodies connected by arc joints where the elements are stacked in the same order as the bodies and joints. Column matrix $\Delta \mathbf{p}_a$ contains small changes in arc joint displacements stacked in the same order. Matrix $\mathbf{H}_a^0$ is a stacked block diagonal matrix of the individual arc joint influence coefficient matrices, all transformed to 0.

Arranging all chord joints in a similar manner gives a second displacement equation

$$\Delta \mathbf{d}_c^0 = \mathbf{H}_c^0 \Delta \mathbf{p}_c \tag{65}$$

where now $\Delta \mathbf{d}_c^0$ is a stacked column matrix of small changes in relative displacements between bodies connected by chord joints, $\Delta \mathbf{p}_c$ contains small changes in all chord joint displacements and matrix $\mathbf{H}_c^0$ is a stacked block diagonal matrix of the chord joint influence coefficients.

Earlier it was noted that the kth row of $\mathbf{R}_c$, as defined in Eq. 58, specifies which arc joints appear in the kinematic loop defined by that row, and it also defines how they are oriented relative to the loop. Let $\mathbf{R}_c$ (and likewise the other topological matrices) be composed of six by six identity matrices replacing the 1's and six by six zero matrices replacing the 0's. The product $\mathbf{R}_c^0 \Delta \mathbf{d}_a^0$ adds up all small arc joint displacements in all constraint loops. Combining these displacements with the remaining small chord joint displacements defined in Eq. 65 gives the total constrained system displacement as

$$\Delta \phi^0 = \mathbf{R}_c^0 \Delta \mathbf{d}_a^0 + \Delta \mathbf{d}_c^0 \tag{66}$$

where $\Delta \phi^0$ represents a stacked column matrix of the first order variations in the constraints. Substituting Eqs. 64 and 65 into Eq. 66 and factoring out the coefficients of $\Delta \mathbf{p}_a$ and $\Delta \mathbf{p}_c$ identifies the loop constraint Jacobian matrix

$$\left[\mathbf{R}_c \mathbf{H}_a^0, \mathbf{H}_c^0\right] \begin{bmatrix} \Delta \mathbf{p}_a \\ \Delta \mathbf{p}_c \end{bmatrix} = \Delta \phi^0 \tag{67}$$

For Newton-Raphson iteration, these equations are revised as

$$\left[R_c H_a^0, H_c^0\right]\begin{bmatrix}\Delta p_a \\ \Delta p_c\end{bmatrix} = -\phi^0 \tag{68}$$

where numeric values for ϕ^0 are obtained from Eq. 63.

It may be necessary to place additional constraints of the general form

$$\psi(p_a, p_c) = 0 \tag{69}$$

on some of the joint variables. The linearized equations for Newton-Raphson iteration are

$$[\psi_a, \psi_c]\begin{bmatrix}\Delta p_a \\ \Delta p_c\end{bmatrix} = -\psi \tag{70}$$

where

$$\psi_a = \partial\psi/\partial p_a \text{ and } \psi_c = \partial\psi/\partial p_c \tag{71}$$

A set of independent variables must be defined before Eqs. 68 and 70 can be applied. Coordinate partitioning may be used to select independent variables from $\mathbf{p_a}$ and $\mathbf{p_c}$ at various points in configuration space. This selection process may be done manually or automatically using LU factorization of the Jacobian matrix of Eqs. 68 and 70 with full row and column pivoting [10, 13, 9]. The resulting independent coordinate definition is implemented by

$$q = \bar{I}_a p_a + \bar{I}_c p_c \tag{72}$$

where $\bar{I}_a$ and $\bar{I}_c$ are Boolean matrices that pick out elements from the joint variable arrays.

Alternately, singular value decomposition (SVD) may be applied to the Jacobian matrix at various points in configuration space to define independent variables. SVD gives the equations of motion better numerical properties than those obtained from coordinate partitioning because the rows of constant matrix $[V_a, V_c]$ are more nearly tangent to the constraint manifold than are the rows of $[\bar{I}_a, \bar{I}_c]$ [14-17]. In this case

$$q = V_a p_a + V_c p_c \tag{73}$$

Observing that the independent variables are held fixed during iteration, the Newton-Raphson algorithm corresponding to Eqs. 68, 70 and 73 may be combined as

$$\begin{bmatrix} R_c H_a^0 & H_c^0 \\ \psi_a & \psi_c \\ V_a & V_c \end{bmatrix}\begin{bmatrix}\Delta p_a^{(k)} \\ \Delta p_c^{(k)}\end{bmatrix} = \begin{bmatrix} -\phi^0 \\ -\psi \\ q - V_a p_a^{(k)} - V_c p_c^{(k)} \end{bmatrix} \tag{74}$$

and

$$\begin{bmatrix} p_a^{(k+1)} \\ p_c^{(k+1)} \end{bmatrix} = \begin{bmatrix} p_a^{(k)} \\ p_c^{(k)} \end{bmatrix} + \begin{bmatrix} \Delta p_a^{(k)} \\ \Delta p_c^{(k)} \end{bmatrix} \tag{75}$$

where k is an iteration counter. Similar equations may be defined using Eq. 72 instead.

Equations 74 and 75 provide an opportunity to numerically precompute all dependent joint variables in terms of the selected independent ones. This is done by sweeping the individual independent variables through their limited domains and generating multidimensional surfaces of the dependent quantities. The surfaces may be interpolated by polynomials or other suitable functions and stored in memory for future run-time evaluation. If such functions have been defined, then $\mathbf{p}_a(\mathbf{q})$ and $\mathbf{p}_c(\mathbf{q})$ are given explicitly in terms of $\mathbf{q}$ and the system configuration may be evaluated during run time without iteration.

4.3. SPATIAL VELOCITIES IN CONSTRAINED SYSTEMS

The relative spatial velocities between the bodies connected by respective arc and chord joints may be written in inertial frame coordinates as (compare with Eqs. 64 and 65)

$$v_a^0 = H_a^0 \dot{p}_a \tag{76}$$

and

$$v_c^0 = H_c^0 \dot{p}_c \tag{77}$$

The absolute spatial velocities of all bodies relative to the inertial frame are collected into a stacked block column matrix v^0. Absolute velocities and the relative velocities in Eqs. 76 and 77 are related by arc and chord connectivity matrices as

$$v^0 = R_a H_a^0 \dot{p}_a \tag{78}$$

and

$$C_c v^0 = H_c^0 \dot{p}_c \tag{79}$$

Substituting Eq. 78 into Eq. 79 and using the identity in Eq. 58 gives (compare with Eq. 67)

$$\left[R_c H_a^0, H_c^0 \right] \begin{bmatrix} \dot{p}_a \\ \dot{p}_c \end{bmatrix} = 0 \tag{80}$$

which is the time derivative of the loop constraint equations. Combining Eq. 80 with the time derivative of Eqs. 69 and 73 gives (compare with Eq. 74)

$$\begin{bmatrix} R_c H_a^0 & H_c^0 \\ \psi_a & \psi_c \\ V_a & V_c \end{bmatrix} \begin{bmatrix} \dot{p}_a \\ \dot{p}_c \end{bmatrix} = \begin{bmatrix} 0 \\ 0 \\ \dot{q} \end{bmatrix} \tag{81}$$

Equation 81 shows that if all independent velocities are specified, the complete system velocity can be computed.

Since $\mathbf{p}_a(\mathbf{q})$ and $\mathbf{p}_c(\mathbf{q})$ are explicit functions of $\mathbf{q}$, let

$$\dot{p}_a = \partial p_a(q)/\partial q\, \dot{q} = \mathbf{B}_a(q)\, \dot{q} \tag{82}$$

and

$$\dot{p}_c = \partial p_c(q)/\partial q\, \dot{q} = \mathbf{B}_c(q)\, \dot{q} \tag{83}$$

Substituting Eqs. 82 and 83 into Eq. 81 and equating coefficients of the independent $\dot{\mathbf{q}}$ gives

$$\begin{bmatrix} \mathbf{R}_c \mathbf{H}_a^0 & \mathbf{H}_c^0 \\ \psi_a & \psi_c \\ \mathbf{V}_a & \mathbf{V}_c \end{bmatrix} \begin{bmatrix} \mathbf{B}_a \\ \mathbf{B}_c \end{bmatrix} = \begin{bmatrix} \mathbf{0} \\ \mathbf{0} \\ \mathbf{I} \end{bmatrix} \tag{84}$$

which may be used to numerically evaluate $\mathbf{B}_a$ and $\mathbf{B}_c$ as explicit functions of $\mathbf{q}$.

4.4. SPATIAL ACCELERATIONS IN CONSTRAINED SYSTEMS

Explicit expressions for spatial accelerations are also required to formulate the equations of motion. Differentiating Eq. 78 gives

$$\mathbf{a}^0 = \mathbf{R}_a \mathbf{H}_a^0 \ddot{p}_a + \mathbf{R}_a \dot{\mathbf{H}}_a^0 \dot{p}_a \tag{85}$$

which shows that $\ddot{\mathbf{p}}_a$ is also required when evaluating $\mathbf{a}^0$. Differentiating Eqs. 82 and 83 gives

$$\ddot{p}_a = \mathbf{B}_a \ddot{q} + \left(\sum_{i=1}^{dof} \partial \mathbf{B}_a/\partial q_i\, \dot{q}_i \right) \dot{q} \tag{86}$$

and

$$\ddot{p}_c = \mathbf{B}_c \ddot{q} + \left(\sum_{i=1}^{dof} \partial \mathbf{B}_c/\partial q_i\, \dot{q}_i \right) \dot{q} \tag{87}$$

Equations 86 and 87 indicate that second partial derivatives of $\mathbf{p}_a(\mathbf{q})$ and $\mathbf{p}_c(\mathbf{q})$ are required when accelerations must be computed. These quantities may be evaluated by taking the partial derivative of Eq. 84 with respect to each of the independent variables giving

$$\begin{bmatrix} \mathbf{R}_c \mathbf{H}_a^0 & \mathbf{H}_c^0 \\ \psi_a & \psi_c \\ \mathbf{V}_a & \mathbf{V}_c \end{bmatrix} \begin{bmatrix} \partial \mathbf{B}_a/\partial q_i \\ \partial \mathbf{B}_c/\partial q_i \end{bmatrix} = \begin{bmatrix} -\mathbf{R}_c\, \partial \mathbf{H}_a^0/\partial q_i\, \mathbf{B}_a - \partial \mathbf{H}_c^0/\partial q_i\, \mathbf{B}_c \\ -\partial \psi_a/\partial q_i\, \mathbf{B}_a - \partial \psi_c/\partial q_i\, \mathbf{B}_c \\ \mathbf{0} \end{bmatrix}, \; i = 1, 2, \ldots, dof \tag{88}$$

Explicit symbolic expressions for the partial derivatives of $\mathbf{H}_a^0$ (also required to evaluate $\dot{\mathbf{H}}_a^0$ in Eq. 85) and $\mathbf{H}_c^0$ may be obtained with the help of Eqs. 33 to 38 (see the discussion in Section 3.4 as well.) The basic identities $\partial \mathbf{H}_a^0/\partial q_i = \partial \dot{\mathbf{H}}_a^0/\partial \dot{q}_i$ and $\partial \mathbf{H}_c^0/\partial q_i = \partial \dot{\mathbf{H}}_c^0/\partial \dot{q}_i$ are helpful in this process.

4.5. AUGMENTED AND REDUCED EQUATIONS OF MOTION

The equations of motion for unconstrained rigid bodies were given in Eq. 57. Now let

$$\mathbf{M}^{00}\mathbf{a}^{0} = \tilde{\mathbf{v}}^{0T}\mathbf{p}^{0} + \mathbf{g}^{0} + \mathbf{C}_a^T\mathbf{f}_a^0 + \mathbf{C}_c^T\mathbf{f}_c^0 \tag{89}$$

represent the composite system of constrained equations of motion. Matrix $\mathbf{M}^{00}$ is a symmetric, stacked block diagonal composition of the body inertia submatrices $\mathbf{M}_\alpha^{00}$ arranged in the same order as the arc joints. Likewise, column matrix $\mathbf{g}^0$ contains the array of spatial forces, $\mathbf{g}_\alpha^0$. Matrix $\tilde{\mathbf{v}}^0$ is a stacked block diagonal composition of the individual six by six spatial velocity cross product submatrices of the form given in Eq. 6. Each joint in the system has a corresponding internal spatial reaction force. The stacked matrix of arc joint reaction forces is denoted as $\mathbf{f}_a^0$ and the matrix of chord reaction forces is represented by $\mathbf{f}_c^0$. The connectivity matrices $\mathbf{C}_a^T$ and $\mathbf{C}_c^T$ place the appropriate joint reaction forces into the correct equations of motion in Eq. 89.

The joint reaction forces in $\mathbf{f}_a^0$ and $\mathbf{f}_c^0$ contain components of forces that are tangent to the joint manifolds and other components that are perpendicular to them. If every joint is workless, then the projection of these reaction forces onto the tangent directions will all be zero. If joints contain active internal forces such as actuators or friction, then the projections will be equal to these quantities. These nonzero internal forces act in the same directions as the corresponding joint displacements $\mathbf{p}_a$ and $\mathbf{p}_c$, and are given as

$$\mathbf{Q}_a = \mathbf{H}_a^{0T}\mathbf{f}_a^0 \tag{90}$$

and

$$\mathbf{Q}_c = \mathbf{H}_c^{0T}\mathbf{f}_c^0 \tag{91}$$

These reaction forces may also be projected onto the independent variable subspace using the velocity coefficient matrices $\mathbf{B}_a$ and $\mathbf{B}_c$ defined earlier as

$$\mathbf{Q}_{qa} = \mathbf{B}_a^T\mathbf{Q}_a = \mathbf{H}_{aq}^{0T}\mathbf{f}_a^0 \tag{92}$$

and

$$\mathbf{Q}_{qc} = \mathbf{B}_c^T\mathbf{Q}_c = \mathbf{H}_{cq}^{0T}\mathbf{f}_c^0 \tag{93}$$

where

$$\mathbf{H}_{aq}^{0} = \mathbf{H}_a^{0}\mathbf{B}_a \tag{94}$$

and

$$\mathbf{H}_{cq}^{0} = \mathbf{H}_c^{0}\mathbf{B}_c \tag{95}$$

Equations 85, 89, 92 and 93 represent the augmented constrained equations of motion. To obtain the reduced equations of motion, first substitute Eq. 85 into Eq. 89, then isolate $\mathbf{f}_a^0$ by inverting $\mathbf{C}_a^T$ and substitute this result into Eq. 92 giving

$$\mathbf{J}\,\ddot{\mathbf{q}} = \mathbf{Q}_{qa} + \left(\mathbf{R}_a\,\mathbf{H}_{aq}^0\right)^T\left(\mathbf{C}_c^T\,\mathbf{f}_c^0 + \mathbf{g}^0 + \tilde{\mathbf{v}}^{0T}\,\mathbf{p}^0 - \mathbf{M}^{00}\,\mathbf{R}_a\,\dot{\mathbf{H}}_{aq}^0\,\dot{\mathbf{q}}\right) \tag{96}$$

where

$$\mathbf{J} = \left(\mathbf{R}_a\,\mathbf{H}_{aq}^0\right)^T\mathbf{M}^{00}\left(\mathbf{R}_a\,\mathbf{H}_{aq}^0\right) \tag{97}$$

The unknown chord joint reaction forces $\mathbf{f}_c^0$ may be eliminated from Eq. 96 using Eqs. 58 and 93 to show that

$$\left(\mathbf{R}_a\,\mathbf{H}_{aq}^0\right)^T\mathbf{C}_c^T\,\mathbf{f}_c^0 = \mathbf{Q}_{qc} \tag{98}$$

Thus Eq. 96 reduces to the final form

$$\mathbf{J}\,\ddot{\mathbf{q}} = \mathbf{Q}_q \tag{99}$$

where $\mathbf{J}$ is given in Eq. 97 and

$$\mathbf{Q}_q = \mathbf{Q}_{qa} + \mathbf{Q}_{qc} + \left(\mathbf{R}_a\,\mathbf{H}_{aq}^0\right)^T\left(\mathbf{g}^0 + \tilde{\mathbf{v}}^{0T}\,\mathbf{p}^0 - \mathbf{M}^{00}\,\mathbf{R}_a\,\dot{\mathbf{H}}_{aq}^0\,\dot{\mathbf{q}}\right) \tag{100}$$

5.0. Recursive Reduction and Uncoupling of Equations of Motion

The methods and techniques used to implement algorithms on the computer for evaluating and uncoupling the equations of motion for highly constrained mechanical systems are crucial to achieving real-time simulation capability. Factorization algorithms that follow minimum path trajectories through a system's topology will have the smallest computational overhead. These algorithms invoke recursive application of matrix projections.

No matter how broad a system's spanning tree, if the nodes are grouped by level, connectivity matrix $\mathbf{C}_a$ may be block partitioned as if representing a single chain. In addition to simplifying algorithm development, this arrangement also maximizes parallelism across the projection front, ensuring the best processor performance. A four level example is used to illustrate the algorithm.

5.1. FOUR LEVEL EXAMPLE

The minimum spanning tree graph for a constrained mechanical system may be arranged many different ways, and it is convenient to order the arc joints so they are grouped by levels according to the arrangement of independent variables within each level. When all arc joints within each level are also adjacent in the arc connectivity matrix, it will partition into contiguous block submatrices making it easier to derive and illustrate the recursive uncoupling algorithm.

Consider a constrained mechanical system containing a number of closed kinematic loops that has been partitioned into four discrete levels from the base to the outer leaves of the tree. The actual spanning tree may have more than four levels so some levels within this partitioning may contain arc joints and bodies from several levels of the spanning tree. The four level partitioning of the arc connectivity matrix is written as

$$\mathbf{C}_a = \begin{bmatrix} \mathbf{C}_{a11} & & & \\ -\mathbf{C}_{a21} & \mathbf{C}_{a22} & & \\ & -\mathbf{C}_{a32} & \mathbf{C}_{a33} & \\ & & -\mathbf{C}_{a43} & \mathbf{C}_{a44} \end{bmatrix} \tag{101}$$

where each matrix on the diagonal is also diagonal or lower triangular and nonsingular. Numerical subscripts on the submatrices indicate their relative positions within the composite connectivity matrix. Negative signs were factored out of the off-diagonal submatrices because every nonzero off-diagonal submatrix has a negative sign in front of it (see the matrix in Fig. 4.)

The various stacked matrices developed earlier may also be block partitioned according to the partitioning in Eq. 101. With suitable arrangement of the arc joints within the spanning tree and partitioning of the independent variables, matrix $\mathbf{H}^0_{aq}$ can always be arranged in lower block triangular form and most of the partitioned block submatrices below the diagonal will be zero.

Equation 101 is modified to simplify the recursive algorithm development by premutiplying by the inverse of its diagonal matrices

$$\overline{\mathbf{R}}_a = \overline{\mathbf{C}}_a^{-1} = \begin{bmatrix} \mathbf{C}_{a11}^{-1} & & & \\ & \mathbf{C}_{a22}^{-1} & & \\ & & \mathbf{C}_{a33}^{-1} & \\ & & & \mathbf{C}_{a44}^{-1} \end{bmatrix} = \begin{bmatrix} \mathbf{R}_{a11} & & & \\ & \mathbf{R}_{a22} & & \\ & & \mathbf{R}_{a33} & \\ & & & \mathbf{R}_{a44} \end{bmatrix} \tag{102}$$

to give a modified connectivity matrix

$$\mathbf{C} = \overline{\mathbf{R}}_a \mathbf{C}_a = \begin{bmatrix} \mathbf{I} & & & \\ -\mathbf{C}_{21} & \mathbf{I} & & \\ & -\mathbf{C}_{32} & \mathbf{I} & \\ & & -\mathbf{C}_{43} & \mathbf{I} \end{bmatrix} \tag{103}$$

5.2. RECURSIVE REDUCTION AND UNCOUPLING ALGORITHM

For the four level example, Eq. 99 is written in block partitioned form as

$$\begin{bmatrix} \mathbf{J}_{11} & \mathbf{J}_{12} & \mathbf{J}_{13} & \mathbf{J}_{14} \\ \mathbf{J}_{21} & \mathbf{J}_{22} & \mathbf{J}_{23} & \mathbf{J}_{24} \\ \mathbf{J}_{31} & \mathbf{J}_{32} & \mathbf{J}_{33} & \mathbf{J}_{34} \\ \mathbf{J}_{41} & \mathbf{J}_{42} & \mathbf{J}_{43} & \mathbf{J}_{44} \end{bmatrix} \begin{bmatrix} \ddot{\mathbf{q}}_1 \\ \ddot{\mathbf{q}}_2 \\ \ddot{\mathbf{q}}_3 \\ \ddot{\mathbf{q}}_4 \end{bmatrix} = \begin{bmatrix} \mathbf{Q}_1 \\ \mathbf{Q}_2 \\ \mathbf{Q}_3 \\ \mathbf{Q}_4 \end{bmatrix} \tag{104}$$

A modified version of block matrix factorization will be used to isolate the variables in Eq. 104. The primary advantage of this algorithm is that the partitioned coefficient matrix does not have to be fully evaluated before the decoupling procedure can begin. The matrix representation of **J** in Eq. 97 is first revised by solving Eq. 103 for

$$\mathbf{R}_a = \mathbf{C}^{-1}\,\bar{\mathbf{R}}_a = \mathbf{R}\,\bar{\mathbf{R}}_a \tag{105}$$

and substituting to give

$$\mathbf{J} = \mathbf{H}^T\,\mathbf{R}^T\,\mathbf{M}\,\mathbf{R}\,\mathbf{H} \tag{106}$$

where

$$\mathbf{M} = \mathbf{M}^{(0)} = \begin{bmatrix} \mathbf{M}_{11} & & & \\ & \mathbf{M}_{22} & & \\ & & \mathbf{M}_{33} & \\ & & & \mathbf{M}_{44} \end{bmatrix} \tag{107}$$

and

$$\mathbf{H} = \bar{\mathbf{R}}_a\,\mathbf{H}^0_{aq} = \begin{bmatrix} \mathbf{H}_{11} & & & \\ \mathbf{H}_{21} & \mathbf{H}_{22} & & \\ \mathbf{H}_{31} & \mathbf{H}_{32} & \mathbf{H}_{33} & \\ \mathbf{H}_{41} & \mathbf{H}_{42} & \mathbf{H}_{43} & \mathbf{H}_{44} \end{bmatrix} \tag{108}$$

As noted earlier, matrix **H** is shown worst case and many of the block matrices below the diagonal may be zero. Now **C**, as defined in Eq. 103, may be factored into the product of elementary matrices that are easily inverted to give

$$\mathbf{R} = \begin{bmatrix} \mathbf{I} & & & \\ & \mathbf{I} & & \\ & & \mathbf{I} & \\ & & \mathbf{C}_{43} & \mathbf{I} \end{bmatrix} \begin{bmatrix} \mathbf{I} & & & \\ & \mathbf{I} & & \\ & \mathbf{C}_{32} & \mathbf{I} & \\ & & & \mathbf{I} \end{bmatrix} \begin{bmatrix} \mathbf{I} & & & \\ \mathbf{C}_{21} & \mathbf{I} & & \\ & & \mathbf{I} & \\ & & & \mathbf{I} \end{bmatrix} \tag{109}$$

The factored form of **R** in Eq. 109 makes the recursive decoupling algorithm easier to derive and understand. The algorithm represents the operations necessary to evaluate **J** in Eq. 106, perform LU factorization and solve for the unknowns. Most significantly, the LU factors of **J** are generated as it is evaluated. As implied by Eq. 109, an n-level system requires n–1 projection operations to evaluate **J** and an additional n–1 projection operations to generate its LU factors. This algorithm works in a pipeline fashion. After completing the first level evaluation of **J**, it then simultaneously works on the second level evaluation and the first level LU factorization. This process continues until the final factorization has been accomplished at the nth step. Rather than 2n–2 steps, this algorithm requires n steps which is significant when parallel processors are used. To keep the algorithm compact, the contents of various matrices are overwritten. The overwriting operation is indicated by the assignment arrow ←.

Assuming an n level system, the algorithm follows:

First perform the initialization

(1) $J_{ij} = H_{ni}^T M_{nn} H_{nj}$ $((i = n, n-1, ..., 1), j = i, i-1, ..., 1)$

(2) $\ddot{q}_n = J_{nn}^{-1} Q_n$

Next, setting k = n−1, the evaluation procedure continues recursively as follows:

(3) $M_{jk} = M_{j,k+1} C_{k+1,k}$ $(j = n, n-1, ..., k+1)$

(4) $M_{kk} \leftarrow M_{kk} + M_{k+1,k}^T C_{k+1,k}$

(5) $J_{ij} \leftarrow J_{ij} + H_{ki}^T M_{kk} H_{kj}((i = k, k-1, ..., 1), j = i, i-1, ..., 1)$

(6) $\bar{J}_{ij} = 0$ $(i, j = k, k-1, ..., 1)$

(7) $\bar{J}_{ij} \leftarrow \bar{J}_{ij} + H_{mi}^T M_{mk} H_{kj}((m = n, n-1, ..., k+1), i, j = k, k-1, ..., 1)$

(8) $J_{ij} \leftarrow J_{ij} + \bar{J}_{ij} + \bar{J}_{ji}^T$ $((i = k, k-1, ..., 1), j = i, i-1, ..., 1)$

(9) $J_{ij} \leftarrow J_{ij} + H_{mi}^T M_{mk} H_{kj}(((m = n, n-1, ..., k+1), i = m, m-1, ..., k+1), j = k, k-1, ..., 1)$

(10)
$$\left.\begin{aligned} &L_{jk} = J_{jj}^{-1} J_{jk} \\ &J_{ik} \leftarrow J_{ik} - J_{ji}^T L_{jk} (i = j-1, j-2, ..., k) \\ &Q_k \leftarrow Q_k - L_{jk}^T Q_j \end{aligned}\right\} \quad (j = n, n-1, ..., k+1)$$

(11) $\ddot{q}_k = J_{kk}^{-1} Q_k$

(12) reduce $k \leftarrow k - 1$

(13) if k = 0, decoupling procedure is completed; recourse to step (15) below.

(14) recourse to step (3), above until completed.

(15) increase $k \leftarrow k + 1$

if k = n, algorithm is completed; stop.

(16) $\ddot{q}_{k+1} \leftarrow \ddot{q}_{k+1} - L_{k+1,j} \ddot{q}_j$ $(j = k, k-1, ..., 1)$

(17) recourse to step (15), above until completed.

This algorithm gives a rough picture of the procedures required to uncouple the equations of motion using block submatrices. The algorithm represents worst case because many of the submatrices of **H** appearing in steps 1, 5, 7 and 9 may be zero. Algorithm changes may be accounted for by modifying the index counters to skip over zero matrices. For effective parallel implementation, the algorithm should not be considered as strictly sequential. For large scale problems, most of the submatrices will be sparse and substantial overhead will be saved by breaking these operations down even further.

In the current formulation, connectivity matrices $C_{k+1,k}$ contain only 1's and 0's, so steps 3 and 4 require simple additions. These connectivity matrices effectively transform child body inertia matrices to conform with parent body inertias so they can be projected or added onto the appropriate parent inertias. In the present form, the inertia and influence coefficient matrices are all expressed in a common frame. The inertia and influence coefficient matrices may also be expressed in local body frames so the terms will involve fewer variables. Now the connectivity matrices $C_{k+1,k}$ will contain local spatial transformation matrices to transform inertia matrices, similar to Eqs. 52 and 54. However, this will make it much easier to analyze the equations and determine variable dependencies for precomputing coefficients. In many applications, this form will result in a higher percentage of the operations involving three or less variables, making interpolation of these quantities practical. This discussion clearly indicates the evolution of hybrid techniques and shows that automated optimization of general system models will be difficult.

Obviously the operations necessary to evaluate Q_q in Eq. 100 must also be taken into account. These equations contain many of the same quantities appearing in **J**, and it will be possible to use interpolating functions here as well. Another factor important to load balancing on parallel processors, is synchronizing the evaluation of these equations with those in the above algorithm.

6. Large Scale Vehicle Example

A large scale vehicle model is presented to illustrate the types of simulations possible using the procedures briefly described in this paper. Figures 5 to 7 show computer generated graphical images of the vehicle system with cutaway views illustrating major steering and suspension components. The system is composed of an eight by eight tractor towing a multiaxle trailer carrying a tracked vehicle. The wheels on tractor axles one and two are steered and all steering and suspension kinematic linkages were accurately modeled. Nonlinear suspension compliance, damping, hysteresis, and jounce and rebound limiters on all three vehicles were accurately modeled with nonlinear functions using measured data. Wheels on all three vehicles are allowed to rotate and leave the surface, and support and tractive tire forces are modeled in all three directions on every wheel. The tractor drive train was not modeled, but the vehicle is propelled by applying equal driving torques to all eight wheels.

The tracked vehicle model, with 40 dof is a fully functional stand-alone system composed of 35 rigid bodies (chassis, 14 road arm/road wheel pairs, 2 drive sprockets, 2 idlers, turret and trunnion,) 35 joints and no closed kinematic loops. The model has massless deformable tracks that support the road wheels and allow it to be propelled and steered through the drive sprockets. It is interfaced with the trailer chassis model through deformable road wheel models that can rotate and slide, develop forces in all three directions, and leave the trailer bed surface. The chassis model is fastened to the trailer bed by deformable chain models. The high resolution vehicle model was used in lieu of a dummy load because its suspension compliance significantly affects the transient loads generated in the trailer suspension and overall system roll stability.

The tractor model, with 23 dof is a fully functional stand-alone system composed of 58 rigid bodies (chassis, 30 suspension elements, 4 steering hubs, 8 wheels, 13 steering linkages and 2 fifth wheel bodies,) 78 joints and 20 closed kinematic loops. The tractor chassis outline was not shown on the graphical image in Fig. 6 to provide a better view of the suspension and steering models as it negotiates a 0.3 meter ramp (the tractor is heading to the right.) Each suspension is composed of a control arm pair connected to a control hub. The front suspensions are supported by fore-aft torsion bars connected between the control arms and chassis as shown in the figure. The rear suspensions are supported by pivoting walking beams. A pitman arm shown in the upper right hand corner provides steering input to the system through the steering kinematic linkages.

The trailer model, with 29 dof is a fully functional stand-alone system composed of 49 rigid bodies (chassis, 36 suspension elements and 12 wheels,) 59 joints and 10 closed kinematic loops. Part of the trailer chassis outline was not shown on the graphical image in Fig. 7 to provide a better view of the suspension as it negotiates a 0.25 meter hole (the trailer is heading to the left.) The suspension is composed of two main beams that pivot on the chassis above axles 2 and support axles 3. Two secondary beams pivot on the main beams and support axles 1 and 2. The main suspension cylinders move vertically relative to the chassis and are connected to transverse axles that can rotate around fore-aft axes to equalize loads on the tires. The beams are connected to the suspensions through connecting link/cushion cylinder arrangements that help absorb road shock. A series of yaw links connected between the chassis and suspension cylinders 1, and between suspension cylinders 2 and 3 prevent the suspensions from steering.

Figure 5. Computer generated image of tractor-trailer system transporting a tracked vehicle.

Figure 6. Cutaway view of the tractor suspension and steering as it negotiates a 0.3 meter ramp.

Figure 7. Cutaway view of the trailer suspension as it negotiates a 0.25 meter hole.

7. Acknowledgments

The authors would like to thank Stacy Budzik, William Veenhuis and David Gunter for developing the vehicle graphics outlines and providing the images in Figures 5, 6 and 7.

8. References

1. Featherstone, W.R. (1984) *Robot Dynamics Algorithms*, Ph.D. dissertation, University of Edinburgh.
2. Wehage, R.A. and Belczynski, M.J. (1992) "High Resolution Vehicle Simulations Using Precomputed Coefficients", in G. Rizzoni, M. El-Gindy, J.Y. Wong and A. Zeid (eds.) *Transportation Systems-ASME*, DCS-Vol 44, 311-325.
3. Altmann, S.L. (1986) *Rotations, Quaternions, and Double Groups*, Clarendon Press, Oxford, 1986.
4. Wittenburg, J. (1977) *Dynamics of Systems of Rigid Bodies*, B.G. Teubner, Stuttgart.
5. Roberson, R.E. and Schwertassek, R. (1988) *Dynamics of Multibody Systems*, Springer-Verlag, New York, NY.
6. Paul, B. (1979) *Kinematics and Dynamics of Planar Machinery*, Prentice-Hall, Inc. Englewood Cliffs, N.J.
7. Freudenstein, F. (1962) "On the Variety of Motions Generated by Mechanisms", *Journal of Engineering for Industry Transactions*, ASME Ser. B, 84, 156-160.
8. Duff, I.S., Erisman, A.M. and Reid, J.K. (1986) *Direct Methods for Sparse Matrices*, Clarendon Press, Oxford.
9. Wehage, R.A. and Haug, E.J. (1982) "Generalized Coordinate Partitioning for Dimension Reduction in Analysis of Constrained Dynamic Systems", *Journal of Mechanical Design*, **104**(4), 785-791.
10. Sheth, P.N. (1972) *A Digital Computer Based Simulation Procedure for Multiple Degree of Freedom Mechanical Systems with Geometric Constraints*, Ph.D. dissertation, The University of Wisconsin, Madison, WI.
11. Shabana, A.A. (1989) *Dynamics of Multibody Systems*, John Wiley & Sons, New York, NY.
12. Nikravesh, P.E., (1988) *Computer-Aided Analysis of Mechanical Systems*, Springer-Verlag, New York, NY.
13. Wehage, R.A. (1980) *Generalized Coordinate Partitioning in Dynamic Analysis of Mechanical Systems*, Ph.D. dissertation, The University of Iowa, Iowa City, IA.
14. Singh, R.P. and Likens, P.W. (1985) "Singular Value Decomposition for Constrained Dynamical Systems", *Journal of Applied Mechanics*, **52**, 943-948.
15. Mani, N.K. (1984) *Use of Singular Value Decomposition for Analysis and Optimization of Mechanical System Dynamics*, Ph.D. dissertation, The University of Iowa, Iowa City, IA.
16. Mani, N.K., Haug, E.J. and Atkinson, K.E. (1985) "Application of Singular Value Decomposition for Analysis of Mechanical System Dynamics", *Journal of Mechanisms, Transmissions, and Automation in Design*, **107**, 82-87.
17. Wehage, R.A. and Loh, W.Y. (1993) "Application of Singular Value Decomposition to Independent Variable Definition in Constrained Mechanical Systems", to be presented at the 1993 ASME Winter Annual Meeting in New Orleans, LA.

CONSTRUCTION OF THE EQUATIONS OF MOTION FOR MULTIBODY DYNAMICS USING POINT AND JOINT COORDINATES

PARVIZ E. NIKRAVESH and HAZEM A. AFFIFI
Department of Aerospace and Mechanical Engineering
University of Arizona
Tucson, AZ 85721 USA

ABSTRACT. A systematic process for constructing the equations of motion for multibody systems containing open or closed kinematic loops is presented. We first illustrate a nonconventional method for describing the configuration of a body in space using a set of dependent point coordinates, instead of the more classical set of translational and rotational body coordinates. Based on this point-coordinate description, body mass and applied loads are distributed to the points. For multibody systems, the equations of motion are constructed as a large set of mixed differential-algebraic equations. For open-loop systems, based on a velocity transformation process, the equations of motion are converted to a minimal set of equations in terms of the joint accelerations. For multibody systems with closed kinematic loops, the equations of motion are first written as a small set of differential-algebraic equations. Then, following a second velocity transformation, these equations are converted to a minimal set of differential equations. The combination of point- and joint-coordinate formulations provides some interesting features.

1. Introduction

The derivation of equations of motion for computational multibody dynamics has been the topic of many research activities. The scope of these activities has been quite broad. Some techniques allow us to generate the equations of motion in terms of a large set of dependent coordinates in the form of a large set of differential-algebraic equations. Other techniques yield the equations of motion as a minimal set of ordinary differential equations. Many other "in between" approaches provide us with various alternatives. Each formulation has its own advantages and disadvantages depending on the application and the needs.

The equations of motion can easily be written in the form of a large set of differential-algebraic equations due to their simplicity and ease of manipulation. The configuration of a rigid body is normally described by a set of translational and rotational coordinates. Then algebraic constraints are introduced to represent the kinematic joints that connect the bodies, and the Lagrange multiplier technique is employed to describe the joint reaction forces. Although these formulations are easy to construct, one of their main drawbacks is their computational inefficiency. A detailed discussion on this type of formulation, which is referred to as the absolute coordinate formulation, can be found in [1].

A method which has the simplicity of the absolute coordinate formulation and it also provides computational efficiency is the so-called joint coordinate formulation [2-4]. In this method a set of relative joint coordinates is defined, and the equations of motion are

M. F. O. S. Pereira and J. A. C. Ambrósio (eds.),
Computer-Aided Analysis of Rigid and Flexible Mechanical Systems, 31–60.

converted from absolute accelerations to joint accelerations. For open-loop systems this process is done in one step, and the resultant equations are equal in number to the number of degrees of freedom of the system. The conversion process can be performed in two steps for systems containing closed loops [4]. It has been demonstrated that the joint coordinate formulation is by far more efficient than the absolute coordinate formulation.

One elegant method for generating the equations of motion for multibody systems has been presented in several papers by Garcia de Jalon et al. [5, 6]. This method takes advantage of a rudimentary idea of describing a body as a collection of points and vectors. The idea may initially appear as a step backward in the evolutionary process of generating the equations of motion. However, the method exhibits many interesting and extremely useful features. The coordinates and components of points and vectors that are defined to describe a body are dependent on each other through kinematic constraints. For example, we may define twelve coordinates and six constraints to describe a free body in space. Furthermore, additional constraints are introduced to represent the kinematic joints interconnecting the rigid bodies. This process yields a large set of loosely coupled differential-algebraic equations of motion. However, these equations can be converted to a minimal or a small set, as in the body-joint coordinate formulation.

In this paper we first present some of the ideas that appear in [4, 5] and a few other papers by the same authors, with a different slant. Although the general ideas are adopted from those references, the methodology of deriving of the equations, and many of the techniques presented in this paper are new [7]. Here, the bodies are described only by points. The mass and the external force associated with each point are determined as a function of the inertial characteristics of the body and the applied forces acting on the body. The equations of motion are derived using the equations of motion for a system of particles and the Lagrange multiplier technique. Then we present a technique based upon a velocity transformation between the point velocities and a set of joint velocities, in order to transform the equations of motion to a smaller set. The resulting equations of motion in terms of the joint accelerations can be expressed in different forms for systems containing open and closed kinematic loops. These equations have all the useful features of the original body-joint coordinate formulation. Furthermore, additional improvement in computational efficiency is made mainly due to the absence of rotational coordinates.

The presentation in this paper is organized in three parts. We first state the equations of motion for a rigid body using the traditional translational and rotational coordinates, referred to as *body* coordinates. Then, we describe the *point* coordinate method, followed by the *joint* coordinate method which is divided into open- and closed-loop sections.

2. Notation

One-dimensional vectors are denoted by lower-case bold-face characters ($\mathbf{q}$, $\dot{\mathbf{r}}$, $\boldsymbol{\omega}$). Matrices are denoted by upper-case bold-face characters ($\mathbf{C}$, $\mathbf{D}$). Scalars are denoted by light-face characters. A right-subscript denotes a body or a joint index. A right-superscript denotes a point or a point index. A left-superscript denotes the index of a reference coordinate system; if the reference system is a nonmoving system, then the left-superscript is omitted. An over-score "tilde" indicates the conversion of a 3-vector to a 3 x 3 skew-symmetric matrix, i.e., if $\mathbf{a} = \begin{bmatrix} a_1 \\ a_2 \\ a_3 \end{bmatrix}$, then $\tilde{\mathbf{a}} = \begin{bmatrix} 0 & -a_3 & a_2 \\ a_3 & 0 & -a_1 \\ -a_2 & a_1 & 0 \end{bmatrix}$.

3. Body Coordinates

Traditionally, for specifying the position of a rigid body in a global nonmoving xyz coordinate system, it has been sufficient to specify the spatial location of the origin and the angular orientation of a body-fixed $\xi\eta\zeta_i$ coordinate system, as shown in Fig. 1. For a typical body *i*, vector $\mathbf{c}_i$ denotes a vector of coordinates that contains a vector of Cartesian translational coordinates $\mathbf{r}_i = [x_i \;\; y_i \;\; z_i]^T$, and a set of rotational coordinates such as Euler angles, Euler parameters, etc. A 3 x 3 rotational transformation matrix $\mathbf{A}_i$ denotes the angular orientation of the $\xi\eta\zeta_i$ relative to the xyz system, which can be expressed in terms of the defined rotational coordinates. With this transformation matrix the components of a vector described in the body-fixed coordinate system can be transformed to the xyz coordinate system as $\mathbf{s}_i = \mathbf{A}_i\, {}^i\mathbf{s}_i$. A vector of velocities for body *i* is defined as $\mathbf{v}_i$, which contains a vector of translational velocities $\dot{\mathbf{r}}_i$ and a vector of angular velocities ω_i. A vector of acceleration for this body is denoted as $\dot{\mathbf{v}}_i$, which contains $\ddot{\mathbf{r}}_i$ and $\dot{\omega}_i$.

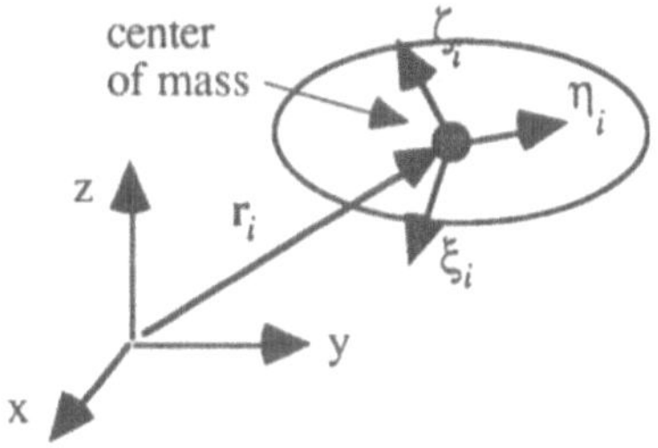

Figure 1. Locating a body in a nonmoving coordinate system.

The Newton-Euler equations of motion for body *i* are written as

$$\begin{bmatrix} m_i \mathbf{I} & \mathbf{0} \\ \mathbf{0} & \mathbf{J}_i \end{bmatrix} \begin{bmatrix} \ddot{\mathbf{r}}_i \\ \dot{\omega}_i \end{bmatrix} = \begin{bmatrix} \mathbf{f}_i \\ \mathbf{n}_i - \tilde{\omega}_i \mathbf{J}_i \omega_i \end{bmatrix} \tag{1}$$

or,

$$\mathbf{M}_i \dot{\mathbf{v}}_i = \mathbf{g}_i$$

where m_i and $\mathbf{J}_i$ are the mass and inertia tensor, and $\mathbf{f}_i$ and $\mathbf{n}_i$ are the sum of forces and moments acting on the body. The rotational equations of motion can be expressed in terms of either the xyz or the $\xi\eta\zeta_i$ components of the vectors. If the inertia tensor with respect to the $\xi\eta\zeta_i$ coordinate system is denoted as the constant matrix ${}^i\mathbf{J}_i$, then $\mathbf{J}_i = \mathbf{A}_i\, {}^i\mathbf{J}_i\, \mathbf{A}_i^T$.

4. Point Coordinates

The position and orientation of a rigid body in a nonmoving xyz coordinate system can be described by the position of several points on the body. It will be shown that the most general case requires four points. These points will be referred to as the *primary points*.

Other points on the body will be called the *secondary* or *nonprimary* points, where their coordinates can be described in terms of the coordinates of the primary points.

4.1. REPRESENTING A BODY BY PRIMARY POINTS

A rigid body may be represented by two, three, or four primary points, as shown in Fig. 2. In such cases we need six, nine, or twelve Cartesian coordinates, respectively, to define the position of these points. We also need, respectively, one, three, or six constraints of the type

$$(\mathbf{r}^i - \mathbf{r}^j)^{\mathrm{T}} (\mathbf{r}^i - \mathbf{r}^j) - \ell^{i,j2} = 0 \qquad (2)$$

Such a constraint keeps the distance between i and j points on the same rigid body a constant. We will refer to these constraints as the *primary constraints* in order to distinguish them from the kinematic constraints associated with the kinematic joints. The Cartesian coordinates of these primary points are referred to as the *primary coordinates* of the body. One major *advantage* of using primary coordinates, instead of three translational and three (or four) rotational coordinates for a body, is the elimination of the rotational coordinates and the corresponding rotational transformation matrix.

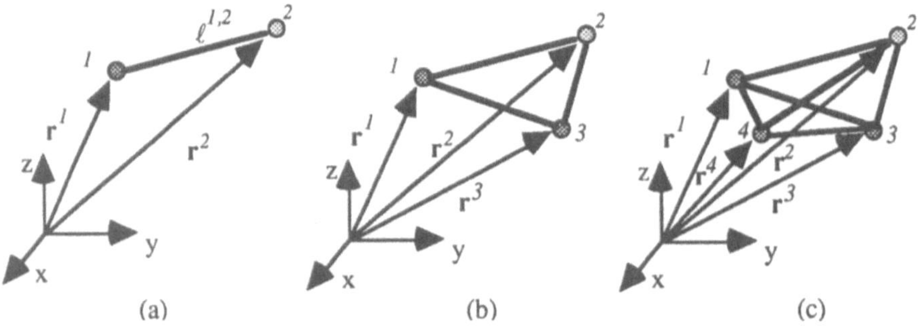

Figure 2. Locating a body by its primary points.

4.2. LOCATING A NONPRIMARY POINT

In order to describe the coordinates of a nonprimary point on a body in terms of the body primary coordinates, we need to examine the two-, three-, and four-point cases separately.

4.2.1. *Two Points:* Point A on the axis of the two primary points has distances a^1 and a^2 from the two primary points. A positive direction from point *1* to point *2* is defined as shown in Fig. 3(a). The coordinates of A can be expressed as $\mathbf{r}^A = (a^1\,\mathbf{r}^2 - a^2\,\mathbf{r}^1) / \ell^{1,2}$.

4.2.2. *Three Points.* We need to locate point A as a function of the coordinates of three primary points, as shown in Fig. 3(b). Assume that, at a given time (e.g., the initial time), $\mathbf{r}^1$, $\mathbf{r}^2$, $\mathbf{r}^3$, and $\mathbf{r}^A$ are known. (The following argument is also valid if the coordinates of these points are known in a local $\xi\eta\zeta_i$ coordinate system attached to the body.) The components of vectors $\mathbf{s}^{2,1}$, $\mathbf{s}^{3,1}$, and $\mathbf{s}^A$ are computed as $\mathbf{s}^{2,1} = \mathbf{r}^2 - \mathbf{r}^1$, $\mathbf{s}^{3,1} = \mathbf{r}^3 - \mathbf{r}^1$, and

$\mathbf{s}^A = \mathbf{r}^A - \mathbf{r}^1$. A vector $\mathbf{s}$ is defined perpendicular to $\mathbf{s}^{2,1}$ and $\mathbf{s}^{3,1}$ as $\mathbf{s} = \tilde{\mathbf{s}}^{2,1}\mathbf{s}^{3,1}$. Vector $\mathbf{s}^A$ can be described in terms of the components of $\mathbf{s}^{2,1}$, $\mathbf{s}^{3,1}$, and $\mathbf{s}$ as

$$\mathbf{s}^A = a^1\,\mathbf{s}^{2,1} + a^2\,\mathbf{s}^{3,1} + a^3\,\mathbf{s}$$

or,

$$\mathbf{s}^A = \mathbf{S}\;\mathbf{a}$$

where $\mathbf{S} \equiv \begin{bmatrix}\mathbf{s}^{2,1} & \mathbf{s}^{3,1} & \mathbf{s}\end{bmatrix}$ and $\mathbf{a} \equiv \begin{bmatrix}a^1 & a^2 & a^3\end{bmatrix}^{\mathrm{T}}$. Now the coefficient vector $\mathbf{a}$ is computed (only once) as $\mathbf{a} = \mathbf{S}^{-1}\mathbf{s}^A$. Then at any given time, since the coefficients are known, we can determine $\mathbf{r}^A$ for known $\mathbf{r}^1$, $\mathbf{r}^2$, and $\mathbf{r}^3$as

$$\mathbf{r}^A = \mathbf{r}^1 + a^1(\mathbf{r}^2 - \mathbf{r}^1) + a^2(\mathbf{r}^3 - \mathbf{r}^1) + a^3(\tilde{\mathbf{r}}^2 - \tilde{\mathbf{r}}^1)(\mathbf{r}^3 - \mathbf{r}^1)$$

4.2.3. *Four Points.* This is similar to the three-point case, i.e., to locate a point A, only three of the four points may be used. However, the fourth point can be used to obtain a third vector, $\mathbf{s}^{4,1}$, replacing vector $\mathbf{s}$ in the previous case. Note that the four primary points cannot be in the same plane.

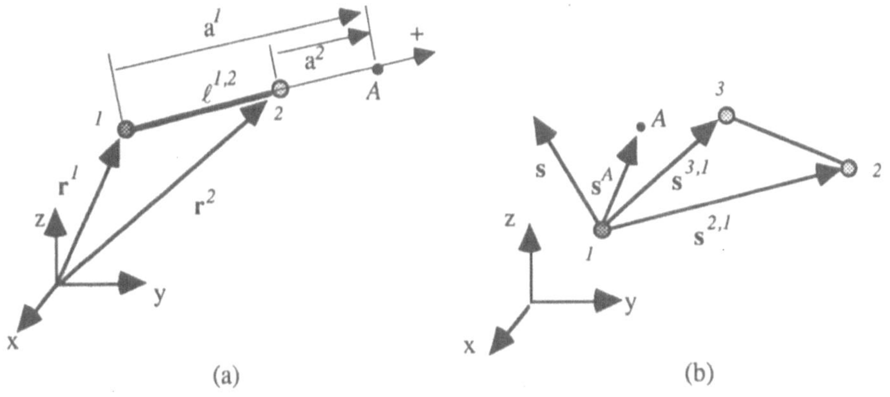

Figure 3. Locating a nonprimary point *A* in terms of the primary coordinates.

4.3. KINEMATIC JOINTS

Kinematic joints between rigid bodies are described in the form of algebraic constraints between the primary coordinates. For example if the primary point *k* on body *i* coincides with the primary point *l* on body *j* (e.g., a spherical joint), then we write a vector constraint as

$$\mathbf{r}_i^k - \mathbf{r}_j^l = \mathbf{0} \tag{3}$$

However, the idea behind the use of primary coordinates is to eliminate the need for defining some, if not all, of these simple constraints. This is achieved by allowing bodies to share primary points and, hence, reducing the total number of primary coordinates.

4.3.1. *Spherical Joint.* If two bodies are connected by a spherical joint, then one primary point is shared by the two bodies at the center of the joint, as shown in Fig. 4(a). In this case the two bodies are described by seven primary points, twelve primary constraints, and no constraints for the spherical joint.

4.3.2. *Revolute Joint.* Two primary points on the joint axis can be shared by the two bodies, as shown in Fig. 4(b). In this case we need six primary points and eleven primary constraints.

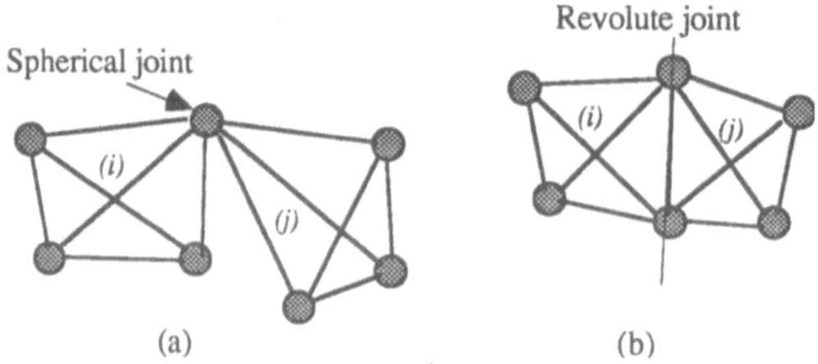

Figure 4. Shared primary points for bodies connected by spherical or revolute joints.

4.3.3. *Universal Joint.* Assume that vectors $\mathbf{s}_i$ and $\mathbf{s}_j$ are defined on the joint axes perpendicular to each other, as shown in Fig. 5(a). These vectors are also defined between the primary points on their respective bodies. One primary point is shared by the bodies at the intersect of the universal joint axes. Therefore, we need seven primary points, twelve primary constraints, and one additional constraint to keep vectors $\mathbf{s}_i$ and $\mathbf{s}_j$ perpendicular, i.e.,

$$\mathbf{s}_i^{\mathrm{T}}\mathbf{s}_j = 0 \tag{4}$$

4.3.4. *Cylindrical Joint.* For a cylindrical joint, as shown in Fig. 5(b), we need to keep three vectors, $\mathbf{s}_i$, $\mathbf{s}_j$, and $\mathbf{d}$ parallel since they are defined on the joint axis. Therefore, we will have four constraint equations,

$$\tilde{\mathbf{s}}_i\,\mathbf{s}_j = \mathbf{0} \tag{5.a}$$

$$\tilde{\mathbf{s}}_i\,\mathbf{d} = \mathbf{0} \tag{5.b}$$

A vector product, such as in Eq. 5.a or 5.b, yields only two independent algebraic equations, therefore we must select two of the three equations as our constraints. Normally due to the rotation of the bodies during an analysis, the choice of the two equations may change. In order to circumvent this issue completely, we recommend using the scalar product constraint twice, instead of a vector product. For example, if vectors $\mathbf{n}_i$ and $\mathbf{m}_i$ are defined perpendicular to vector $\mathbf{s}_i$, then, instead of Eqs. 5.a and 5.b, we may use

$$\begin{aligned} \mathbf{n}_i^{\mathrm{T}}\mathbf{s}_j = 0\,, \quad & \mathbf{m}_i^{\mathrm{T}}\mathbf{s}_j = 0 \\ \mathbf{n}_i^{\mathrm{T}}\mathbf{d} = 0\,, \quad & \mathbf{m}_i^{\mathrm{T}}\mathbf{d} = 0 \end{aligned} \tag{6}$$

4.3.5. *Prismatic Joint.* For a prismatic joint, in addition to Eqs. 5.a and 5.b we also need to eliminate the relative rotation between the two bodies. This can be accomplished by defining two vectors, such as $\mathbf{n}_i$ and $\mathbf{n}_j$, perpendicular to each other and also perpendicular to the joint axis. Therefore, the fifth constraint equation becomes,

$$\mathbf{n}_i^{\mathrm{T}}\mathbf{n}_j = 0$$

We can follow a similar procedure to describe the necessary constraints for other types of kinematic joints between bodies that are defined by primary coordinates.

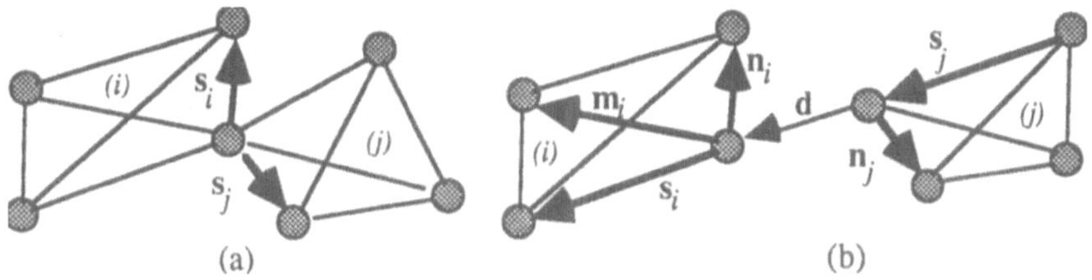

Figure 5. Primary points and vectors for describing kinematic constraints representing universal and cylindrical/prismatic joints.

4.4. KINEMATICS OF A MULTIBODY SYSTEM

For a multibody system with b rigid bodies interconnected by kinematic joints, assume that we have defined p primary points. Therefore, we need three 3 x p vectors of primary coordinates, velocities, and accelerations,

$$\mathbf{r} = \begin{bmatrix} \mathbf{r}^1 \\ \mathbf{r}^2 \\ \vdots \\ \mathbf{r}^p \end{bmatrix}, \quad \dot{\mathbf{r}} = \begin{bmatrix} \dot{\mathbf{r}}^1 \\ \dot{\mathbf{r}}^2 \\ \vdots \\ \dot{\mathbf{r}}^p \end{bmatrix}, \quad \ddot{\mathbf{r}} = \begin{bmatrix} \ddot{\mathbf{r}}^1 \\ \ddot{\mathbf{r}}^2 \\ \vdots \\ \ddot{\mathbf{r}}^p \end{bmatrix}$$

The primary coordinates are dependent through the primary and the kinematic joint constraints. Assume that there are m independent constraints of the form,

$$\mathbf{\Phi}(\mathbf{r}) = \mathbf{0} \tag{7}$$

Note that most of these constraints, if not all, are either linear or quadratic in terms of the primary coordinates. The first and second time derivatives of these constraints yield the velocity and acceleration constraints,

$$\dot{\mathbf{\Phi}} \equiv \mathbf{D}\dot{\mathbf{r}} = \mathbf{0} \tag{8}$$
$$\ddot{\mathbf{\Phi}} \equiv \mathbf{D}\ddot{\mathbf{r}} + \dot{\mathbf{D}}\dot{\mathbf{r}} = \mathbf{0} \tag{9}$$

where $\mathbf{D} \equiv \mathbf{\Phi}_{\mathbf{r}} \equiv \partial\mathbf{\Phi}/\partial\mathbf{r}$ is the Jacobian of the constraints and due to the linear or quadratic nature of the constraints it exhibits a very simple form.

4.5. MASS DISTRIBUTION

Normally the inertial characteristics of a rigid body are described by its mass and inertia tensor. Since we are defining a rigid body by several points, we need to distribute the body mass to these points, while preserving the inertial characteristics of the rigid body. We first demonstrate a mass distribution technique for the four-point case, then we specialize that to the three- and two-point cases.

4.5.1. *Four Primary Points*. The mass distributions to points must satisfy the total mass, the first moment, and the second moment conditions. These conditions provide ten algebraic equations; therefore, we can have up to ten unknowns to solve for. Four of the unknowns can be the mass of the four primary points, and the other six unknowns can be six coordinates associated with the position of the four primary points. For example, two of the primary points can have known positions, but the other two can be placed in such positions that our ten equations are satisfied. Although this process is quite practical, it may not be desirable due to several reasons: If the position of some of the primary points are considered as the unknowns, then the resulting algebraic equations become nonlinear, which may not be an attractive feature. We also want to have the freedom of positioning the primary points on the bodies in accordance with the joints that connect the bodies in order to reduce the number of primary coordinates. For these reasons, an alternative technique is proposed.

In addition to the four primary points, we introduce six secondary points with unknown masses. This makes the total number of unknown masses equal to the number of equations, i.e., ten. The ten equations are linear in terms of the unknown masses. However, this requires the position of the secondary points to be described in terms of the position of the primary coordinates. There are infinite possibilities for positioning the six secondary points. One such possibility, which is also quite simple, is shown in Fig. 6(a). Each secondary point is located between two primary points. We number the primary points *1*, ..., *4* and the secondary points *5*, ..., *10*. Hence, the position of the secondary points can be described as

$$\mathbf{s}_i^5 = (\mathbf{s}_i^1 + \mathbf{s}_i^2)/2 \qquad (10)$$
$$\vdots$$

Note that a vector $\mathbf{s}_i$ locates a point relative to the body mass center.

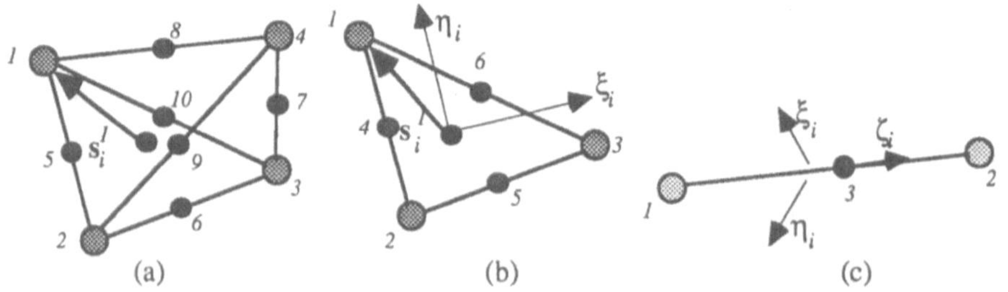

Figure 6. Primary and secondary points for mass distribution.

The ten equations are written as

$$\sum_{j=1}^{10} m^j = m_i \tag{11}$$

$$\sum_{j=1}^{10} {}^i\mathbf{s}_i^j \; m^j = \mathbf{0} \tag{12}$$

$$-\sum_{j=1}^{10} {}^i\tilde{\mathbf{s}}_i^j \; {}^i\tilde{\mathbf{s}}_i^j \; m^j = {}^i\mathbf{J}_i \tag{13}$$

The first expression yields one equation, the second expression yields three equations, and the third expression yields six equations. These equations can be solved for the ten unknown masses.

4.5.2. *Three Primary Points.* This is a special case of the four-primary-point formulation. In order to satisfy the ten equations, the three primary points *(1, 2, 3)*, the three secondary points *(4, 5, 6)*, and the body mass center must all be in the same plane. As shown in Fig. 6(b), we assume that these six points and the mass center are in the $\xi\eta_i$ plane, i.e., $\zeta_i^j = 0$; $j = 1, ..., 6$. Equations 11-13 can be used here again, but for six masses instead of ten. Some of these ten equations are automatically satisfied: the third equation in Eq. 12 and two equations associated with the ζ_i products of inertia in the Eq. 13. One necessary condition is that the moment of inertia about the ζ_i axis to be equal to the sum of the moments of inertia about the other two axes, i.e., $j^{(\zeta\zeta)} = j^{(\xi\xi)} + j^{(\eta\eta)}$. This condition automatically satisfies another equation in Eq. 13. Then we are left with six equations and six unknown masses:

$$\sum_{j=1}^{6} m^j = m_i\,, \qquad \sum_{j=1}^{6} \xi_i^j\, m^j = 0, \qquad \sum_{j=1}^{6} \eta_i^j\, m^j = 0,$$

$$\sum_{j=1}^{6} \eta_i^{j2}\, m^j = j^{(\xi\xi)}, \qquad \sum_{j=1}^{6} \xi_i^{j2}\, m^j = j^{(\eta\eta)}, \qquad -\sum_{j=1}^{6} \xi_i^j\, \eta_i^j\, m^j = j^{(\xi\eta)}$$

4.5.3. *Two Primary Points.* Primary points *1* and *2* form a line that passes through the mass center as shown in Fig. 6(c). One secondary point (point *3*) along this line is defined. Since the ξ_i and η_i coordinates for all three points are zero, the ten equations yield the following necessary conditions: all three products of inertia must be zero ($j^{(\xi\eta)} = j^{(\eta\zeta)} = j^{(\zeta\xi)} = 0$); the moment of inertia about the ζ_i axis must be zero ($j^{(\zeta\zeta)} = 0$); and the moments of inertia about the other two axes must be equal ($j^{(\xi\xi)} = j^{(\eta\eta)}$). Then we have three equations in three unknown masses:

$$m^1 + m^2 + m^3 = m_i$$

$$m^1\,\zeta_i^1 + m^2\,\zeta_i^2 + m^3\,\zeta_i^3 = 0$$

$$m^1\,\zeta_i^{12} + m^2\,\zeta_i^{22} + m^3\,\zeta_i^{32} = j^{(\xi\xi)}$$

One likely situation is that the two primary points have equal distances from the mass center and then the secondary point is positioned at the mass center itself. If the length between the two primary points is denoted as ℓ, then the three masses are found to be

$m^1 = m^2 = 2j^{(\xi\xi)} / \ell^2$ and $m^3 = m_i - 4j^{(\xi\xi)} / \ell^2$. Furthermore, if $j^{(\xi\xi)} = m_i\ell^2 / 12$, then $m^1 = m^2 = m_i / 6$ and $m^3 = 2m_i / 3$.

4.6. FORCE DISTRIBUTION

A force or a moment acting on a rigid body must be resolved into one or more forces acting on the primary points. The resultant force and moment associated with this force distribution must be equivalent to the original force and/or moment.

4.6.1. *Four Primary Points (Force).* Consider a single force acting at point *P* as shown in Fig. 7(a). Point *P* is positioned from the mass center by vector $\mathbf{s}_i^P$. We need to find an equivalent set of four forces acting on the four particles. We may assume that the four forces are all parallel to the original force $\mathbf{f}_i^P$, i.e., $\mathbf{f}^j = \alpha^j\ \mathbf{f}_i^P;\ j = 1, ..., 4$, where α^j are four unknown coefficients. We need to satisfy the following conditions:

$$\sum_{j=1}^{4} \mathbf{f}^j = \mathbf{f}_i^P \tag{14}$$

$$\sum_{j=1}^{4} \tilde{\mathbf{s}}_i^j\ \mathbf{f}^j = \tilde{\mathbf{s}}_i^P\ \mathbf{f}_i^P \tag{15}$$

Since these equations must be valid for any $\mathbf{f}_i^P$, we get four equations in four unknowns:

$$\begin{bmatrix} \xi_i^1 & \xi_i^2 & \xi_i^3 & \xi_i^4 \\ \eta_i^1 & \eta_i^2 & \eta_i^3 & \eta_i^4 \\ \zeta_i^1 & \zeta_i^2 & \zeta_i^3 & \zeta_i^4 \\ 1 & 1 & 1 & 1 \end{bmatrix} \begin{bmatrix} \alpha^1 \\ \alpha^2 \\ \alpha^3 \\ \alpha^4 \end{bmatrix} = \begin{bmatrix} \xi_i^P \\ \eta_i^P \\ \zeta_i^P \\ 1 \end{bmatrix}$$

Note that the unknown coefficients are a function of the position of point *P* and not a function of the magnitude or the direction of the applied force. Therefore, as long as the point of application of the force does not change, the coefficients are solved for only once.

4.6.2. *Three Primary Points (Force).* Since the three primary points, the body mass center, and the point of application of the force must all be in the same plane, as shown in Fig. 7(b), we can write the following three equations in three unknown coefficients:

$$\begin{bmatrix} \xi_i^1 & \xi_i^2 & \xi_i^3 \\ \eta_i^1 & \eta_i^2 & \eta_i^3 \\ 1 & 1 & 1 \end{bmatrix} \begin{bmatrix} \alpha^1 \\ \alpha^2 \\ \alpha^3 \end{bmatrix} = \begin{bmatrix} \xi_i^P \\ \eta_i^P \\ 1 \end{bmatrix}$$

4.6.3. *Two Primary Points (Force).* As shown in Fig. 7(c), the two primary points and point *P* form a straight line. The equations for the three-point case are further simplified to:

$$\begin{bmatrix} \zeta_i^1 & \zeta_i^2 \\ 1 & 1 \end{bmatrix} \begin{bmatrix} \alpha^1 \\ \alpha^2 \end{bmatrix} = \begin{bmatrix} \zeta_i^P \\ 1 \end{bmatrix}$$

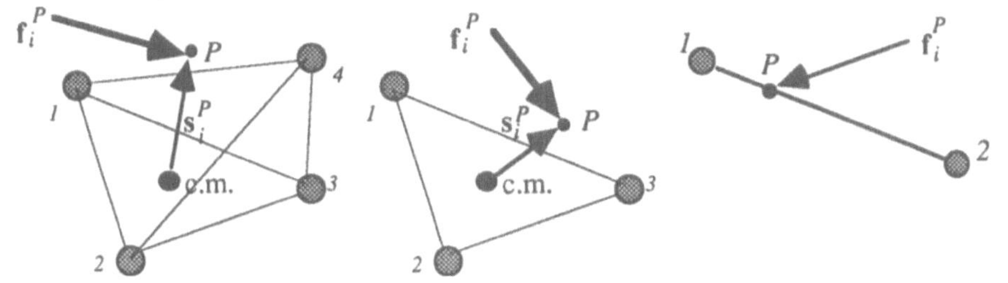

Figure 7. A force acting on a body represented by primary points.

4.6.4. *Four Primary Points (Moment).* Assume a pure moment acts on a body. There is more than one way to find a set of forces acting on the primary points equivalent to the applied moment. One solution is to assume six forces with known directions but unknown magnitudes, as shown in Fig. 8. The six forces must satisfy the following equations:

$$\sum_{j=1}^{6} \mathbf{f}^j = \mathbf{0} \tag{16}$$

$$\sum_{j=1}^{6} \tilde{\mathbf{s}}_i^j \, \mathbf{f}^j = \mathbf{n}_i \tag{17}$$

If we describe unit vectors $\mathbf{u}^{i,j}$ along the axes, we can solve the following six equations for six unknown magnitudes:

$$\begin{bmatrix} \mathbf{u}^{2,1} & \mathbf{u}^{3,1} & \mathbf{u}^{4,1} & \mathbf{u}^{3,2} & \mathbf{u}^{4,2} & \mathbf{u}^{4,3} \\ \tilde{\mathbf{s}}^1\mathbf{u}^{2,1} & \tilde{\mathbf{s}}^1\mathbf{u}^{3,1} & \tilde{\mathbf{s}}^1\mathbf{u}^{4,1} & \tilde{\mathbf{s}}^2\mathbf{u}^{3,2} & \tilde{\mathbf{s}}^2\mathbf{u}^{4,2} & \tilde{\mathbf{s}}^3\mathbf{u}^{4,3} \end{bmatrix} \begin{bmatrix} \alpha^1 \\ \alpha^2 \\ \alpha^3 \\ \alpha^4 \\ \alpha^5 \\ \alpha^6 \end{bmatrix} = \begin{bmatrix} \mathbf{0} \\ \mathbf{n} \end{bmatrix}$$

We note that the solution to these equations is a function of the magnitude and the direction of the applied moment $\mathbf{n}$.

Another method for obtaining a solution to this problem is to consider four parallel forces acting on the four primary points as

$$\mathbf{f}^j = \alpha^j \mathbf{u} \, , \; j = 1, \ldots, 4$$

where **u** is a unit vector defined perpendicular to the moment **n**. Applying the conditions given by Eqs. 16 and 17, but written for four forces instead of six, we can derive the following equations:

$$\sum_{j=1}^{4} \alpha^j = 0$$

$$\sum_{j=1}^{4} \mathbf{s}_i^j \; \alpha^j = \tilde{\mathbf{u}} \; \mathbf{n}_i$$

These four equations can be solved for the four unknown coefficients.

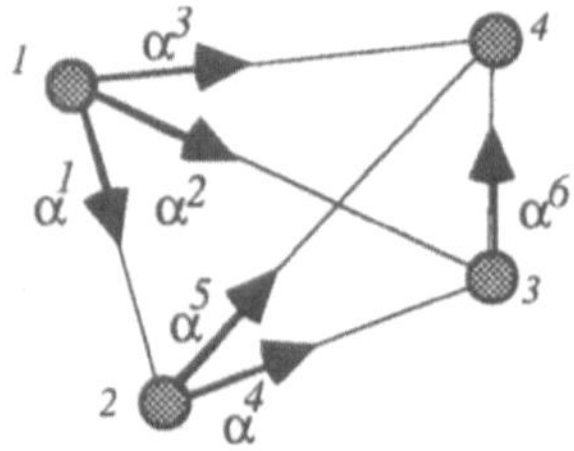

Figure 8. Representing a moment by six forces.

4.7. EQUATIONS OF MOTION AND INERTIA MATRIX

In order to derive the equations of motion for a rigid body using the primary coordinate representation, we need the equations of motion for a system of particles with the application of Lagrange multiplier technique. We first show the process for a two-primary-point case, then we repeat the process for three- and four-point cases. In each case we assume that the masses of the primary and secondary points have already been determined.

4.7.1. *Two Primary Points.* Assume that two forces, $\mathbf{f}^1$ and $\mathbf{f}^2$, act on two primary points, as shown in Fig. 9. Between the two primary and one secondary points, the following constraints exist:

$$(\mathbf{r}^1 - \mathbf{r}^2)^{\mathrm{T}} \, (\mathbf{r}^1 - \mathbf{r}^2) = \ell^{1,2^2} \quad (18.a)$$

$$\mathbf{r}^1 + \mathbf{r}^2 - 2\mathbf{r}^3 = 0 \quad (18.b)$$

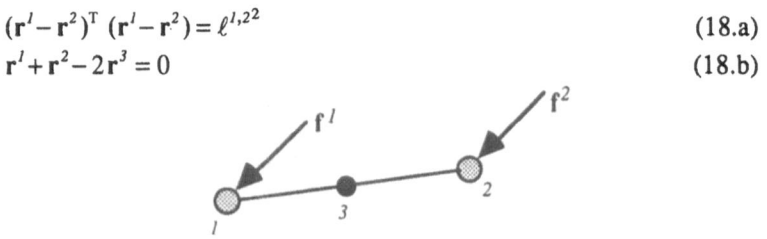

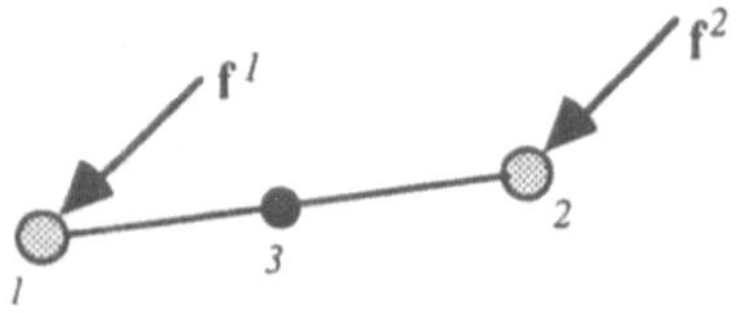

Figure 9. Two forces acting on a body represented by two primary and one secondary points.

The first and second time derivatives of these constraints yield the velocity and acceleration constraints as:

$$(\mathbf{r}^1 - \mathbf{r}^2)^T (\dot{\mathbf{r}}^1 - \dot{\mathbf{r}}^2) = 0 \tag{19.a}$$
$$\dot{\mathbf{r}}^1 + \dot{\mathbf{r}}^2 - 2\dot{\mathbf{r}}^3 = 0 \tag{19.b}$$

$$(\mathbf{r}^1 - \mathbf{r}^2)^T (\ddot{\mathbf{r}}^1 - \ddot{\mathbf{r}}^2) = -(\dot{\mathbf{r}}^1 - \dot{\mathbf{r}}^2)^T (\dot{\mathbf{r}}^1 - \dot{\mathbf{r}}^2) \tag{20.a}$$
$$\ddot{\mathbf{r}}^1 + \ddot{\mathbf{r}}^2 - 2\ddot{\mathbf{r}}^3 = 0 \tag{20.b}$$

Using the Lagrange multiplier technique, the equations of motion for these three constrained particles are written as

$$\mathrm{m}^1 \ddot{\mathbf{r}}^1 - (\mathbf{r}^1 - \mathbf{r}^2)\lambda^1 - \lambda^2 = \mathbf{f}^1 \tag{21.a}$$
$$\mathrm{m}^2 \ddot{\mathbf{r}}^2 + (\mathbf{r}^1 - \mathbf{r}^2)\lambda^1 - \lambda^2 = \mathbf{f}^2 \tag{21.b}$$
$$\mathrm{m}^3 \ddot{\mathbf{r}}^3 + 2\lambda^2 = 0 \tag{21.c}$$

where λ^1 contains one multiplier associated with the first constraint equation and λ^2 contains three multipliers associated with the second set of constraints.

In order to eliminate the Lagrange multipliers λ^2 and the acceleration vector $\ddot{\mathbf{r}}^3$ associated with the secondary point, we add Eqs. 21.a and 21.c, add Eqs. 21.b and 21.c, then we use Eq. 20.b. This yields the equations of motion for a rigid body in terms of the primary point accelerations as:

$$\begin{bmatrix} (\mathrm{m}^1 + \frac{\mathrm{m}^3}{4})\mathbf{I} & \frac{\mathrm{m}^3}{4}\mathbf{I} \\ \frac{\mathrm{m}^3}{4}\mathbf{I} & (\mathrm{m}^2 + \frac{\mathrm{m}^3}{4})\mathbf{I} \end{bmatrix} \begin{bmatrix} \ddot{\mathbf{r}}^1 \\ \ddot{\mathbf{r}}^2 \end{bmatrix} - \begin{bmatrix} \mathbf{r}^1 - \mathbf{r}^2 \\ \mathbf{r}^2 - \mathbf{r}^1 \end{bmatrix} \lambda^1 = \begin{bmatrix} \mathbf{f}^1 \\ \mathbf{f}^2 \end{bmatrix} \tag{22}$$

The complete set of equations of motion contains Eqs. 18.a, 19.a, 20.a, and 22. Note that the mass matrix is 6 x 6.

For the special case with $\mathrm{m}^1 = \mathrm{m}^2 = 2\mathrm{j}^{(\xi\xi)} / \ell^{1,2^2}$ and $\mathrm{m}^3 = \mathrm{m}_i - 4\mathrm{j}^{(\xi\xi)} / \ell^{1,2^2}$, and the special case with $\mathrm{m}^1 = \mathrm{m}^2 = \mathrm{m}_i / 6$ and $\mathrm{m}^3 = 2\mathrm{m}_i / 3$, the mass matrices become, respectively,

$$\begin{bmatrix} (\frac{\mathrm{m}_i}{4} + \frac{\mathrm{j}^{(\xi\xi)}}{\ell^{1,2^2}})\mathbf{I} & (\frac{\mathrm{m}_i}{4} - \frac{\mathrm{j}^{(\xi\xi)}}{\ell^{1,2^2}})\mathbf{I} \\ (\frac{\mathrm{m}_i}{4} - \frac{\mathrm{j}^{(\xi\xi)}}{\ell^{1,2^2}})\mathbf{I} & (\frac{\mathrm{m}_i}{4} + \frac{\mathrm{j}^{(\xi\xi)}}{\ell^{1,2^2}})\mathbf{I} \end{bmatrix}, \quad \begin{bmatrix} \frac{\mathrm{m}_i}{3}\mathbf{I} & \frac{\mathrm{m}_i}{6}\mathbf{I} \\ \frac{\mathrm{m}_i}{6}\mathbf{I} & \frac{\mathrm{m}_i}{3}\mathbf{I} \end{bmatrix}$$

4.7.2. *Three Primary Points.* We repeat a process similar to the two-point case by writing the necessary constraint equations and the equations of motion for three primary and three secondary points (particles). Then the Lagrange multipliers and the acceleration vectors associated with the secondary points are eliminated to obtain

$$\begin{bmatrix} m_{11}\mathbf{I} & m_{12}\mathbf{I} & m_{13}\mathbf{I} \\ m_{21}\mathbf{I} & m_{22}\mathbf{I} & m_{23}\mathbf{I} \\ m_{31}\mathbf{I} & m_{32}\mathbf{I} & m_{33}\mathbf{I} \end{bmatrix} \begin{bmatrix} \ddot{\mathbf{r}}^1 \\ \ddot{\mathbf{r}}^2 \\ \ddot{\mathbf{r}}^3 \end{bmatrix} - \begin{bmatrix} \mathbf{r}^1-\mathbf{r}^2 & 0 & \mathbf{r}^1-\mathbf{r}^3 \\ \mathbf{r}^2-\mathbf{r}^1 & \mathbf{r}^2-\mathbf{r}^3 & 0 \\ 0 & \mathbf{r}^3-\mathbf{r}^2 & \mathbf{r}^3-\mathbf{r}^1 \end{bmatrix} \begin{bmatrix} \lambda^1 \\ \lambda^2 \\ \lambda^3 \end{bmatrix} = \begin{bmatrix} \mathbf{f}^1 \\ \mathbf{f}^2 \\ \mathbf{f}^3 \end{bmatrix} \tag{23}$$

where

$$m_{11} = \mathrm{m}^1 + \frac{\mathrm{m}^4+\mathrm{m}^6}{4}, \quad m_{22} = \mathrm{m}^2 + \frac{\mathrm{m}^4+\mathrm{m}^5}{4}, \quad m_{33} = \mathrm{m}^3 + \frac{\mathrm{m}^5+\mathrm{m}^6}{4}$$

$$m_{21} = m_{12} = \frac{\mathrm{m}^4}{4}, \quad m_{31} = m_{13} = \frac{\mathrm{m}^6}{4}, \quad m_{32} = m_{23} = \frac{\mathrm{m}^5}{4}$$

In addition to these equations, we must consider three primary constraints and their first and second time derivatives.

4.7.3. *Four Primary Points.* A process similar to the previous cases is performed for four primary and six secondary points. Following the elimination of the Lagrange multipliers, the constraints, and the accelerations associated with the secondary points, the equations of motion are found to be:

$$\begin{bmatrix} m_{11}\mathbf{I} & m_{12}\mathbf{I} & m_{13}\mathbf{I} & m_{14}\mathbf{I} \\ m_{21}\mathbf{I} & m_{22}\mathbf{I} & m_{23}\mathbf{I} & m_{24}\mathbf{I} \\ m_{31}\mathbf{I} & m_{32}\mathbf{I} & m_{33}\mathbf{I} & m_{34}\mathbf{I} \\ m_{41}\mathbf{I} & m_{42}\mathbf{I} & m_{43}\mathbf{I} & m_{44}\mathbf{I} \end{bmatrix} \begin{bmatrix} \ddot{\mathbf{r}}^1 \\ \ddot{\mathbf{r}}^2 \\ \ddot{\mathbf{r}}^3 \\ \ddot{\mathbf{r}}^4 \end{bmatrix} - \begin{bmatrix} -\mathbf{s}^{2,1} & -\mathbf{s}^{3,1} & -\mathbf{s}^{4,1} & 0 & 0 & 0 \\ \mathbf{s}^{2,1} & 0 & 0 & -\mathbf{s}^{3,2} & -\mathbf{s}^{4,2} & 0 \\ 0 & \mathbf{s}^{3,1} & 0 & \mathbf{s}^{3,2} & 0 & -\mathbf{s}^{4,3} \\ 0 & 0 & \mathbf{s}^{4,1} & 0 & \mathbf{s}^{4,2} & \mathbf{s}^{4,3} \end{bmatrix} \begin{bmatrix} \lambda^1 \\ \lambda^2 \\ \lambda^3 \\ \lambda^4 \\ \lambda^5 \\ \lambda^6 \end{bmatrix} = \begin{bmatrix} \mathbf{f}^1 \\ \mathbf{f}^2 \\ \mathbf{f}^3 \\ \mathbf{f}^4 \end{bmatrix} \tag{24}$$

where $\mathbf{s}^{i,j} = \mathbf{r}^i - \mathbf{r}^j$ and

$$m_{11} = \mathrm{m}^1 + \frac{\mathrm{m}^5+\mathrm{m}^8+\mathrm{m}^{10}}{4} \qquad m_{12} = m_{21} = \frac{\mathrm{m}^5}{4} \qquad m_{14} = m_{41} = \frac{\mathrm{m}^8}{4}$$

$$m_{22} = \mathrm{m}^2 + \frac{\mathrm{m}^5+\mathrm{m}^6+\mathrm{m}^9}{4} \qquad m_{13} = m_{31} = \frac{\mathrm{m}^{10}}{4} \qquad m_{24} = 42 = \frac{\mathrm{m}^9}{4}$$

$$m_{33} = \mathrm{m}^3 + \frac{\mathrm{m}^6+\mathrm{m}^7+\mathrm{m}^{10}}{4} \qquad m_{23} = m_{32} = \frac{\mathrm{m}^6}{4} \qquad m_{34} = m_{43} = \frac{\mathrm{m}^7}{4}$$

$$m_{44} = \mathrm{m}^4 + \frac{\mathrm{m}^7+\mathrm{m}^8+\mathrm{m}^9}{4}$$

In addition to these equations, we must consider six primary constraints and their first and second derivatives. Note that the sum of all 16 components of the mass matrix is equal to the mass of the body.

4.8. DYNAMICS OF A MULTIBODY SYSTEM

The primary coordinate representation of rigid bodies allows us to easily determine the equations of motion of a system of multibodies interconnected by kinematic joints. Here we demonstrate the process by using a simple example. Assume that two bodies are connected by a spherical joint, as shown in Fig. 10. The system is represented by seven primary points. Since the primary point number *4* is shared by the two bodies, its mass receives contribution from both bodies. Similarly, the applied force on this point receives contribution from the forces that act on both bodies. The mass matrix and the force vector are written as:

$$\mathbf{M} = \begin{bmatrix} m_{11}\mathbf{I} & m_{12}\mathbf{I} & m_{13}\mathbf{I} & m_{14}\mathbf{I} & & & \\ m_{21}\mathbf{I} & m_{22}\mathbf{I} & m_{23}\mathbf{I} & m_{24}\mathbf{I} & & & \\ m_{31}\mathbf{I} & m_{32}\mathbf{I} & m_{33}\mathbf{I} & m_{34}\mathbf{I} & & & \\ m_{41}\mathbf{I} & m_{42}\mathbf{I} & m_{43}\mathbf{I} & m_{44}\mathbf{I} & m_{45}\mathbf{I} & m_{46}\mathbf{I} & m_{47}\mathbf{I} \\ & & & m_{54}\mathbf{I} & m_{55}\mathbf{I} & m_{56}\mathbf{I} & m_{57}\mathbf{I} \\ & & & m_{64}\mathbf{I} & m_{64}\mathbf{I} & m_{64}\mathbf{I} & m_{64}\mathbf{I} \\ & & & m_{74}\mathbf{I} & m_{75}\mathbf{I} & m_{76}\mathbf{I} & m_{77}\mathbf{I} \end{bmatrix}, \quad \mathbf{f} = \begin{bmatrix} \mathbf{f}^1 \\ \mathbf{f}^2 \\ \mathbf{f}^3 \\ \mathbf{f}^4 \\ \mathbf{f}^5 \\ \mathbf{f}^6 \\ \mathbf{f}^7 \end{bmatrix}$$

where $m_{44} = m_{44_i} + m_{44_j}$ and $\mathbf{f}^4 = \mathbf{f}_i^4 + \mathbf{f}_j^4$.

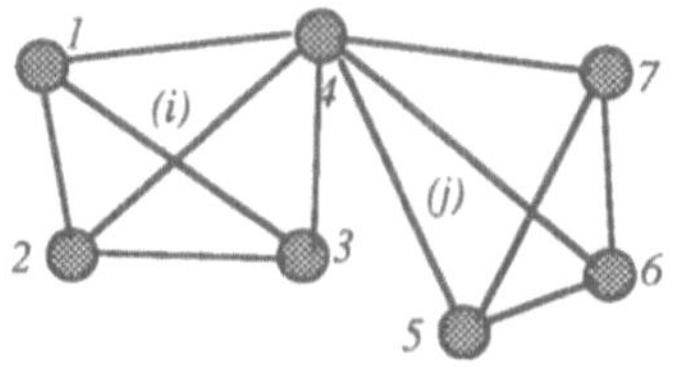

Figure 10. Two bodies connected by a spherical joint.

This example shows how the mass and the force of shared primary points are constructed. In general, the equations of motion for a multibody system are written as

$$\mathbf{\Phi}(\mathbf{r}) = \mathbf{0} \tag{7}$$

$$\dot{\mathbf{\Phi}} \equiv \mathbf{D}\dot{\mathbf{r}} = \mathbf{0} \tag{8}$$

$$\ddot{\mathbf{\Phi}} \equiv \mathbf{D}\ddot{\mathbf{r}} + \dot{\mathbf{D}}\dot{\mathbf{r}} = \mathbf{0} \tag{9}$$

$$\mathbf{M}\ddot{\mathbf{r}} - \mathbf{D}^{\mathrm{T}}\lambda = \mathbf{f} \tag{25}$$

where $\mathbf{\Phi}(\mathbf{r}) = \mathbf{0}$ contains all of the primary and the kinematic joint constraints, and vector λ contains the multipliers associated with all of these constraints.

5. Joint Coordinate Formulation for Open Loop Systems

The constrained equations of motion expressed by Eqs. 7-9 and 25 can be converted to a smaller set of equations in terms of a set of relative coordinates known as the *joint coordinates*. For multibody systems with open kinematic loops, this conversion yields a set of ordinary differential equations equal to the number of degrees of freedom of the system.

In a multibody system with open kinematic loops, we define the necessary joint coordinates for each kinematic joint in the system. For example, revolute and prismatic joints require one joint coordinate each, universal and cylindrical joints require two joint coordinates each, and for a spherical joint we need three joint coordinates. If the system is not connected to the ground by any kinematic joints, then we select one body as a *floating body*, and we assume it is attached to the ground by a joint, called a *floating joint*, exhibiting six degrees of freedom. The ground is considered as the *base* or the *root*. The body furthest from the base is called a *leaf*. An open-loop system only has one base but it may have more than one leaf.

The joints and their corresponding joint coordinates are numbered in any desired order. However, here for convenience, a joint carries the index number of the connecting body in the direction of the leaf. In order to clarify the notation and some of the definitions, assume that two bodies in a branch are connected by a joint as shown in Fig. 11. Joint i is called the incoming joint for body i and the exit joint for body i-. Therefore, it is said that joint i belongs to body i, not body i-. In an open-loop system, each body has only one incoming joint, but it may have none, one, or more exit joints. Each body has a *reference point*, which is a point defined by its joint. For example, the center of a spherical joint or any point on the axis of a revolute joint is the reference point for its body. Again, for convenience, the primary point which is selected as the reference point for a body carries the same index (number) as the body (and its joint), with the possible exception of the floating body.

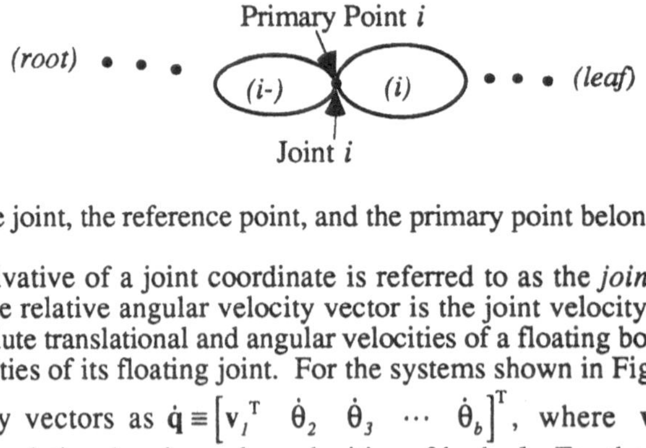

Figure 11. The joint, the reference point, and the primary point belonging to body i.

The time derivative of a joint coordinate is referred to as the *joint velocity*. For a spherical joint, the relative angular velocity vector is the joint velocity between the two bodies. The absolute translational and angular velocities of a floating body are considered as the joint velocities of its floating joint. For the systems shown in Fig. 12(a) we define the joint velocity vectors as $\dot{\mathbf{q}} \equiv \left[\mathbf{v}_1^{\mathrm{T}} \ \ \dot{\theta}_2 \ \ \dot{\theta}_3 \ \ \cdots \ \ \dot{\theta}_b\right]^{\mathrm{T}}$, where $\mathbf{v}_1$ is a 6-vector containing the translational and angular velocities of body *1*. For the system shown in Fig. 12(b), the vector of joint velocities is defined as $\dot{\mathbf{q}} \equiv \left[\dot{\theta}_1 \ \ \dot{\theta}_2 \ \ \cdots \ \ \dot{\theta}_b\right]^{\mathrm{T}}$. The dimension of $\dot{\mathbf{q}}$ in both cases is equal to the number of degrees of freedom of the system.

For multibody systems where the bodies are described by primary points, it can be shown that there exists a simple transformation between the joint velocity vector and the vector of point velocities as

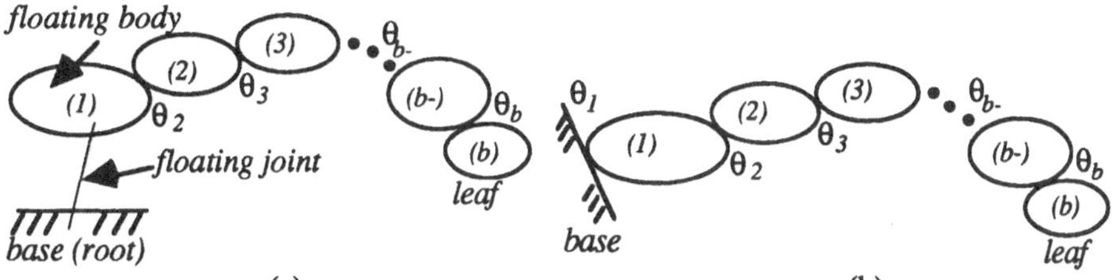

(a) (b)

Figure 12. Two single-branch open-loop systems with and without a *floating joint*..

$$\dot{\mathbf{r}} = \mathbf{B}\dot{\mathbf{q}} \tag{26}$$

Matrix **B** is called the *velocity transformation matrix*, and it is a function of the primary coordinates. We can show that this matrix is orthogonal to the Jacobian matrix **D** by substituting Eq. 26 into Eq. 8 to obtain $\mathbf{D}\mathbf{B}\dot{\mathbf{q}} = \mathbf{0}$. Since $\dot{\mathbf{q}}$ is a vector of independent velocities, then

$$\mathbf{DB} = \mathbf{0} \tag{27}$$

The time derivative of Eq. 26 gives the transformation formula for the accelerations:

$$\ddot{\mathbf{r}} = \mathbf{B}\ddot{\mathbf{q}} + \dot{\mathbf{B}}\dot{\mathbf{q}} \tag{28}$$

Substituting this equation into Eq. 25, premultiplying both sides by $\mathbf{B}^{\mathrm{T}}$, and then taking advantage of Eq. 27 yield

$$\mathcal{M}\ddot{\mathbf{q}} = \mathcal{f} \tag{29}$$

where

$$\mathcal{M} = \mathbf{B}^{\mathrm{T}}\mathbf{M}\mathbf{B} \tag{30}$$

$$\mathcal{f} = \mathbf{B}^{\mathrm{T}}(\mathbf{f} - \mathbf{M}\dot{\mathbf{B}}\dot{\mathbf{q}}) \tag{31}$$

Equation 29 represents the equations of motion for an open-loop multibody system.

5.1. VELOCITY TRANSFORMATION MATRIX

Systematic construction of the velocity transformation matrix can be demonstrated with an example.

Example 1: Assume that in the single-branch open-loop system shown in Fig. 13(a) we have a floating body, two revolute, one translational, and one spherical joint. As shown in Fig. 13(b), we need to use fifteen primary points to represent the five bodies in this system. Since body *1* is selected to be connected to the ground by a floating joint, a reference frame $\xi\eta\zeta_1$ is attached to its mass center, and the translational and angular velocities of this frame relative to a nonmoving reference frame are considered as the floating joint velocity vector. For convenience, we denote the mass center of the base as point *0*. We have numbered the primary points such that points *0, 2, 3, 4* and *5* are the reference points for bodies *1, 2, 3, 4,* and *5*, respectively. We can write the following velocity equations:

$$\begin{aligned}
&\dot{\mathbf{r}}^{1} = \dot{\mathbf{r}}^{0} - \tilde{\mathbf{d}}^{1,0}\,\omega_1\,, \quad \dot{\mathbf{r}}^{2} = \dot{\mathbf{r}}^{0} - \tilde{\mathbf{d}}^{2,0}\,\omega_1\,, \quad \dot{\mathbf{r}}^{6} = \dot{\mathbf{r}}^{0} - \tilde{\mathbf{d}}^{6,0}\,\omega_1\,,\\
&\dot{\mathbf{r}}^{7} = \dot{\mathbf{r}}^{0} - \tilde{\mathbf{d}}^{7,0}\,\omega_1\,, \quad \dot{\mathbf{r}}^{8} = \dot{\mathbf{r}}^{2} - \tilde{\mathbf{d}}^{8,2}\,\omega_2\,, \quad \dot{\mathbf{r}}^{9} = \dot{\mathbf{r}}^{2} - \tilde{\mathbf{d}}^{9,2}\,\omega_2\,,\\
&\dot{\mathbf{r}}^{4} = \dot{\mathbf{r}}^{3} - \tilde{\mathbf{d}}^{4,3}\,\omega_3\,, \quad \dot{\mathbf{r}}^{10} = \dot{\mathbf{r}}^{3} - \tilde{\mathbf{d}}^{10,3}\,\omega_3\,, \quad \dot{\mathbf{r}}^{11} = \dot{\mathbf{r}}^{3} - \tilde{\mathbf{d}}^{11,3}\,\omega_3\,,\\
&\dot{\mathbf{r}}^{5} = \dot{\mathbf{r}}^{4} - \tilde{\mathbf{d}}^{5,4}\,\omega_4\,, \quad \dot{\mathbf{r}}^{12} = \dot{\mathbf{r}}^{4} - \tilde{\mathbf{d}}^{12,4}\,\omega_4\,, \quad \dot{\mathbf{r}}^{13} = \dot{\mathbf{r}}^{4} - \tilde{\mathbf{d}}^{13,4}\,\omega_4\,,\\
&\dot{\mathbf{r}}^{14} = \dot{\mathbf{r}}^{5} - \tilde{\mathbf{d}}^{14,5}\,\omega_5\,, \quad \dot{\mathbf{r}}^{15} = \dot{\mathbf{r}}^{5} - \tilde{\mathbf{d}}^{15,5}\,\omega_5\,.
\end{aligned} \tag{a}$$

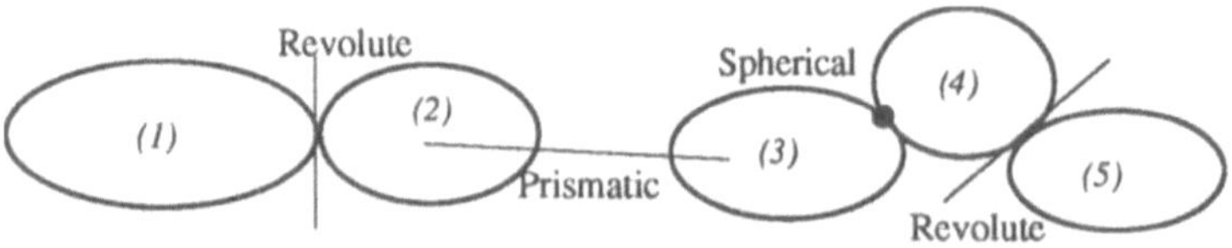

(a)

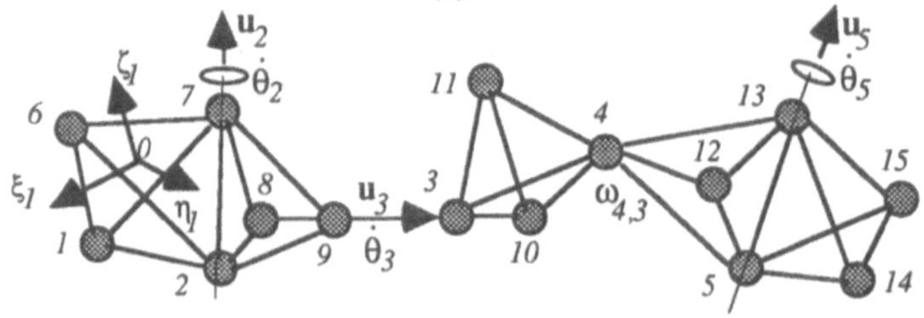

(b)

Figure 13. An open-loop system and its representation by primary points.

We note the following relationships:

$$\begin{aligned}
&\dot{\mathbf{r}}^{0} = \dot{\mathbf{r}}_1\\
&\dot{\mathbf{r}}^{3} = \dot{\mathbf{r}}^{9} - \tilde{\mathbf{u}}_3\,\omega_2 + \mathbf{u}_3\,\dot{\theta}_3\\
&\omega_2 = \omega_1 + \mathbf{u}_2\,\dot{\theta}_2\\
&\omega_3 = \omega_2\\
&\omega_4 = \omega_3 + \omega_{4,3}\\
&\omega_5 = \omega_4 + \mathbf{u}_5\,\dot{\theta}_5
\end{aligned} \tag{b}$$

We now substitute Eqs. (b) into Eqs. (a) in a forward process from the base toward the leaf. We also simplify the equations by using relationships such as $\mathbf{d}^{i,k} + \mathbf{d}^{k,j} = \mathbf{d}^{i,j}$. Then, the following velocity transformation equation is obtained in matrix form:

$$
\begin{bmatrix} \dot{\mathbf{r}}^1 \\ \dot{\mathbf{r}}^6 \\ \dot{\mathbf{r}}^2 \\ \dot{\mathbf{r}}^7 \\ \dot{\mathbf{r}}^8 \\ \dot{\mathbf{r}}^9 \\ \dot{\mathbf{r}}^3 \\ \dot{\mathbf{r}}^{10} \\ \dot{\mathbf{r}}^{11} \\ \dot{\mathbf{r}}^4 \\ \dot{\mathbf{r}}^{12} \\ \dot{\mathbf{r}}^5 \\ \dot{\mathbf{r}}^{13} \\ \dot{\mathbf{r}}^{14} \\ \dot{\mathbf{r}}^{15} \end{bmatrix}
=
\begin{bmatrix}
\mathbf{I} & -\tilde{\mathbf{d}}^{1,0} & & & & \\
\mathbf{I} & -\tilde{\mathbf{d}}^{6,0} & & & & \\
\mathbf{I} & -\tilde{\mathbf{d}}^{2,0} & & & & \\
\mathbf{I} & -\tilde{\mathbf{d}}^{7,0} & & & & \\
\mathbf{I} & -\tilde{\mathbf{d}}^{8,0} & -\tilde{\mathbf{d}}^{8,2}\mathbf{u}_2 & & & \\
\mathbf{I} & -\tilde{\mathbf{d}}^{9,0} & -\tilde{\mathbf{d}}^{9,2}\mathbf{u}_2 & & & \\
\mathbf{I} & -\tilde{\mathbf{d}}^{3,0} & -\tilde{\mathbf{d}}^{3,2}\mathbf{u}_2 & \mathbf{u}_3 & & \\
\mathbf{I} & -\tilde{\mathbf{d}}^{10,0} & -\tilde{\mathbf{d}}^{10,2}\mathbf{u}_2 & \mathbf{u}_3 & & \\
\mathbf{I} & -\tilde{\mathbf{d}}^{11,0} & -\tilde{\mathbf{d}}^{11,2}\mathbf{u}_2 & \mathbf{u}_3 & & \\
\mathbf{I} & -\tilde{\mathbf{d}}^{4,0} & -\tilde{\mathbf{d}}^{4,2}\mathbf{u}_2 & \mathbf{u}_3 & & \\
\mathbf{I} & -\tilde{\mathbf{d}}^{12,0} & -\tilde{\mathbf{d}}^{12,2}\mathbf{u}_2 & \mathbf{u}_3 & -\tilde{\mathbf{d}}^{12,4} & \\
\mathbf{I} & -\tilde{\mathbf{d}}^{5,0} & -\tilde{\mathbf{d}}^{5,2}\mathbf{u}_2 & \mathbf{u}_3 & -\tilde{\mathbf{d}}^{5,4} & \\
\mathbf{I} & -\tilde{\mathbf{d}}^{13,0} & -\tilde{\mathbf{d}}^{13,2}\mathbf{u}_2 & \mathbf{u}_3 & -\tilde{\mathbf{d}}^{13,4} & \\
\mathbf{I} & -\tilde{\mathbf{d}}^{14,0} & -\tilde{\mathbf{d}}^{14,2}\mathbf{u}_2 & \mathbf{u}_3 & -\tilde{\mathbf{d}}^{14,4} & -\tilde{\mathbf{d}}^{14,5}\mathbf{u}_5 \\
\mathbf{I} & -\tilde{\mathbf{d}}^{15,0} & -\tilde{\mathbf{d}}^{15,2}\mathbf{u}_2 & \mathbf{u}_3 & -\tilde{\mathbf{d}}^{15,4} & -\tilde{\mathbf{d}}^{15,5}\mathbf{u}_5
\end{bmatrix}
\begin{bmatrix} \dot{\mathbf{r}}_1 \\ \omega_1 \\ \dot{\theta}_2 \\ \dot{\theta}_3 \\ \omega_{4,3} \\ \dot{\theta}_5 \end{bmatrix}
$$

In this equation we have the velocity of the primary points on the left-hand side and the joint velocities on the right-hand side. The coefficient matrix of the joint velocity vector is the velocity transformation matrix **B**. Instead of listing the velocities of the primary points in ascending order, we have grouped them by bodies in order to demonstrate the triangular form of this matrix. From the structure of this matrix, we note that the matrix can be partitioned into submatrices (block matrices), which are associated with different types of kinematic joints. Table 1 provides the block matrices for several kinematic joints. The last column in this table provides the time derivatives of these block matrices, since they are needed for evaluating the time derivative of matrix **B**. Block matrices for other types of joints can be constructed from the elementary block matrices, as shown in Table 2. Note that the components of $\mathbf{d}^{i,j}$ and $\dot{\mathbf{d}}^{i,j}$ can be computed as $\mathbf{d}^{i,j} = \mathbf{r}^i - \mathbf{r}^j$, $\dot{\mathbf{d}}^{i,j} = \dot{\mathbf{r}}^i - \dot{\mathbf{r}}^j$.

Table 1. Elementary Block Matrices

Joint type	*Matrix size*	*Identifier*	*Entries*	*Time derivative*
Floating	3 x 6	$\mathbf{F}^{i,j}$	$\begin{bmatrix}\mathbf{I} & -\tilde{\mathbf{d}}^{i,j}\end{bmatrix}$	$\begin{bmatrix}\mathbf{0} & -\dot{\tilde{\mathbf{d}}}^{i,j}\end{bmatrix}$
Revolute	3 x 1	$\mathbf{R}^{i,j}$	$\begin{bmatrix}-\tilde{\mathbf{d}}^{i,j}\mathbf{u}_j\end{bmatrix}$	$\begin{bmatrix}-(\dot{\tilde{\mathbf{d}}}^{i,j}+\tilde{\mathbf{d}}^{i,j}\tilde{\omega}_j)\mathbf{u}_j\end{bmatrix}$
Prismatic	3 x 1	$\mathbf{P}^{i,j}$	$\begin{bmatrix}\mathbf{u}_j\end{bmatrix}$	$\begin{bmatrix}\tilde{\omega}_j\mathbf{u}_j\end{bmatrix}$
Spherical	3 x 3	$\mathbf{S}^{i,j}$	$\begin{bmatrix}-\tilde{\mathbf{d}}^{i,j}\end{bmatrix}$	$\begin{bmatrix}-\dot{\tilde{\mathbf{d}}}^{i,j}\end{bmatrix}$

Table 2. Composite Block Matrices

Joint type	*Matrix size*	*Identifier*	*Entries*
Universal	3 x 2	$\mathbf{U}^{i,j}$	$\left[-\tilde{\mathbf{d}}^{i,j}\mathbf{u}_j^{(1)} \quad -\tilde{\mathbf{d}}^{i,j}\mathbf{u}_j^{(2)}\right]$
Cylindrical	3 x 2	$\mathbf{C}^{i,j}$	$\left[-\tilde{\mathbf{d}}^{i,j}\mathbf{u}_j \quad \mathbf{u}_j\right]$

A velocity transformation matrix can be constructed directly from the topology of the multibody system and the block matrix entries of Table 1 as

$$
\mathbf{B} = \begin{bmatrix}
\mathbf{F}^{1,0} & & & & \\
\mathbf{F}^{6,0} & & & & \\
\mathbf{F}^{2,0} & & & & \\
\mathbf{F}^{7,0} & & & & \\
\mathbf{F}^{8,0} & \mathbf{R}^{8,2} & & & \\
\mathbf{F}^{9,0} & \mathbf{R}^{9,2} & & & \\
\mathbf{F}^{3,0} & \mathbf{R}^{3,2} & \mathbf{P}^{3,3} & & \\
\mathbf{F}^{10,0} & \mathbf{R}^{10,2} & \mathbf{P}^{10,3} & & \\
\mathbf{F}^{11,0} & \mathbf{R}^{11,2} & \mathbf{P}^{11,3} & & \\
\mathbf{F}^{4,0} & \mathbf{R}^{4,2} & \mathbf{P}^{4,3} & \mathbf{S}^{4,4} & \\
\mathbf{F}^{12,0} & \mathbf{R}^{12,2} & \mathbf{P}^{12,3} & \mathbf{S}^{12,4} & \\
\mathbf{F}^{11,0} & \mathbf{R}^{11,2} & \mathbf{P}^{11,2} & \mathbf{S}^{11,2} & \\
\mathbf{F}^{13,0} & \mathbf{R}^{13,2} & \mathbf{P}^{13,3} & \mathbf{S}^{13,4} & \\
\mathbf{F}^{14,0} & \mathbf{R}^{14,2} & \mathbf{P}^{14,3} & \mathbf{S}^{14,4} & \mathbf{R}^{14,5} \\
\mathbf{F}^{15,0} & \mathbf{R}^{15,2} & \mathbf{P}^{15,3} & \mathbf{S}^{15,4} & \mathbf{R}^{15,5}
\end{bmatrix}
$$

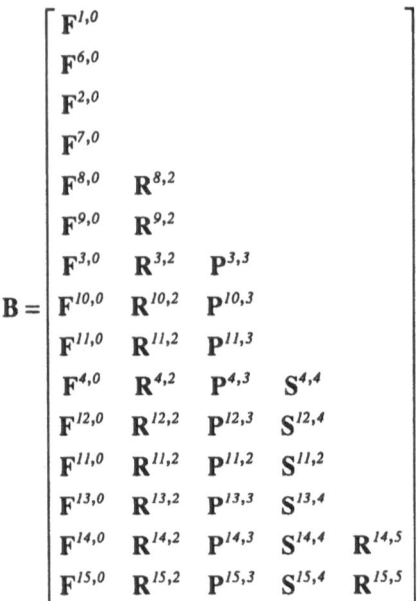

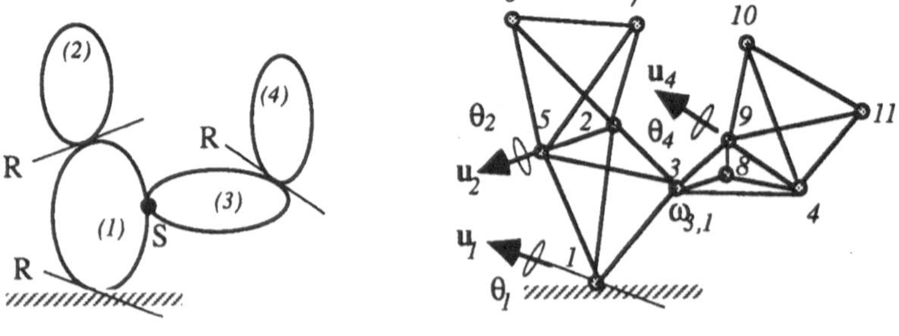

(a) (b)

Figure 14. A multi-branch system and its primary point representation.

Example 2: The multibody system shown in Fig. 14(a) has three revolute and one spherical joint. The primary point representation of the system is shown in Fig. 14(b). Since the incoming joint for body *1* is fixed to the ground, we need not consider the primary point *1* in the vector of velocities since its velocity remains zero. From the topology of the system, the velocity relations, as well as the **B** matrix, are written as

$$\begin{bmatrix} \dot{\mathbf{r}}^2 \\ \dot{\mathbf{r}}^3 \\ \dot{\mathbf{r}}^4 \\ \dot{\mathbf{r}}^5 \\ \dot{\mathbf{r}}^6 \\ \dot{\mathbf{r}}^7 \\ \dot{\mathbf{r}}^8 \\ \dot{\mathbf{r}}^9 \\ \dot{\mathbf{r}}^{10} \\ \dot{\mathbf{r}}^{11} \end{bmatrix} = \begin{bmatrix} \mathbf{R}^{2,1} & & & \\ \mathbf{R}^{3,1} & & & \\ \mathbf{R}^{4,1} & & \mathbf{S}^{4,3} & \\ \mathbf{R}^{5,1} & & & \\ \mathbf{R}^{6,1} & \mathbf{R}^{6,2} & & \\ \mathbf{R}^{7,1} & \mathbf{R}^{7,2} & & \\ \mathbf{R}^{8,1} & & \mathbf{S}^{8,3} & \\ \mathbf{R}^{9,1} & & \mathbf{S}^{9,3} & \\ \mathbf{R}^{10,1} & & \mathbf{S}^{10,3} & \mathbf{R}^{10,4} \\ \mathbf{R}^{11,1} & & \mathbf{S}^{11,3} & \mathbf{R}^{11,4} \end{bmatrix} \begin{bmatrix} \dot{\theta}_1 \\ \dot{\theta}_2 \\ \omega_{3,1} \\ \dot{\theta}_3 \end{bmatrix}$$

5.2. INTEGRATION OF THE EQUATIONS OF MOTION

The equations of motion for open-loop multibody systems represent a set of nonlinear ordinary differential equations that can be put in the standard form

$$\dot{\mathbf{y}} = \mathbf{f}(\mathbf{y}, t) \tag{32}$$

where $\mathbf{y}$ and $\dot{\mathbf{y}}$ arrays contain the joint coordinates, velocities, and accelerations as:

$$\dot{\mathbf{y}} \equiv \begin{bmatrix} \dot{\mathbf{q}} \\ \ddot{\mathbf{q}} \end{bmatrix}, \quad \mathbf{y} \equiv \begin{bmatrix} \mathbf{q} \\ \dot{\mathbf{q}} \end{bmatrix} \tag{33}$$

The numerical solution of the equations of motion requires a numerical integration process that predicts the elements of $\mathbf{y}$ at any time step t. The solution of the equations of motion must determine and return the elements of $\dot{\mathbf{y}}$ to the integration algorithm. The elements of $\dot{\mathbf{y}}$ can be obtained from the elements of $\mathbf{y}$ by implementing the following steps:

1. The contents of $\mathbf{y}$ are known; i.e., $\mathbf{q}$ and $\dot{\mathbf{q}}$.
2. In a forward process, moving from the base towards the leaves, compute the primary coordinates $\mathbf{r}$ (refer to section 5.2.1).
3. Evaluate matrix $\mathbf{B}$.
4. In a forward process, moving from the base towards the leaves, or by using matrix $\mathbf{B}$, compute the primary velocities $\dot{\mathbf{r}}$.
5. Evaluate matrix $\dot{\mathbf{B}}$.
6. Evaluate the mass matrix and force vector, $\mathbf{M}$ and $\mathbf{f}$ (refer to Eq. 25).
7. Evaluate the mass matrix and force vector, $\boldsymbol{M}$ and $\boldsymbol{f}$ (refer to Eqs. 30 and 31)

8. Solve the equations of motion for $\ddot{\mathbf{q}}$ (refer to Eq. 29).
9. Construct $\dot{\mathbf{y}}$ array and return the contents to the integration algorithm.

5.2.1. *Forward Process.* In Step 2 of the preceding algorithm, we need to compute the primary coordinates **r** from the joint coordinates **q**. For this purpose, we move from the base toward a leaf and calculate the primary coordinates. At each step we assume that the primary coordinates of body *i-* and the joint coordinate(s) *i* are known, then we compute the primary coordinates of body *i*. For each type of joint, we may derive closed form expressions to perform the required computation recursively. As an example, consider a revolute joint between bodies *i-* and *i* as shown in Fig. 15. Primary points *1, 2,* and *3* on body *i-* have known coordinates $\mathbf{r}^1$, $\mathbf{r}^2$, and $\mathbf{r}^3$. The joint coordinate θ_i is also known. We want to compute the coordinates of the primary point *4* on body *i*. It is assumed that the primary points *2* and *3* are defined on the joint axis.

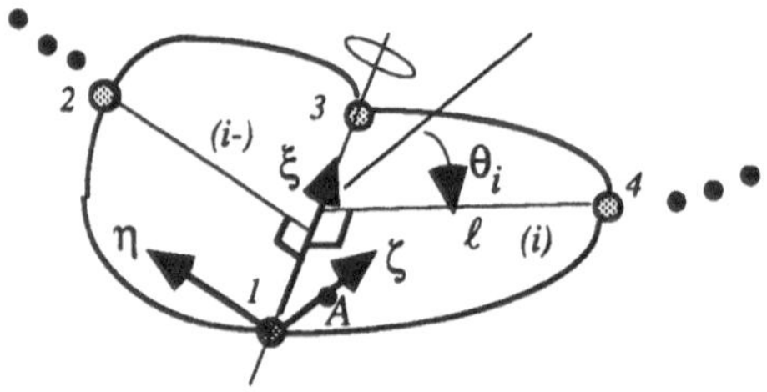

Figure 15. Two bodies represented by primary points and connected by a revolute joint.

We define a $\xi\eta\zeta$ coordinate system on body *i-* such that the ξ-axis is along the joint axis, the η-axis is in the plane of the three points on body *i-*, and the ζ-axis is orthogonal to the other two axes. The coordinates of these three points and point *A* defined on ζ-axis are assumed to be known in the $\xi\eta\zeta$ coordinate system as:

$$^{i-}\mathbf{s}^{2,1} = \begin{bmatrix}\xi^2 & \eta^2 & 0\end{bmatrix}^{\mathrm{T}}, \quad ^{i-}\mathbf{s}^{3,1} = \begin{bmatrix}\xi^3 & 0 & 0\end{bmatrix}^{\mathrm{T}}, \quad ^{i-}\mathbf{s}^{A,1} = \begin{bmatrix}0 & 0 & \xi^3\eta^2\end{bmatrix}^{\mathrm{T}}$$

The coordinates of point *4* in the $\xi\eta\zeta$ coordinate system can be expressed as a function of θ^i as $^{i-}\mathbf{s}^{4,1} = \begin{bmatrix}\xi^4 & -\ell\sin\theta_i & \ell\cos\theta_i\end{bmatrix}^{\mathrm{T}}$, where θ_i is defined as shown. We let

$$^{i-}\mathbf{s}^{4,1} = a^1\, {}^{i-}\mathbf{s}^{3,1} + a^2\, {}^{i-}\mathbf{s}^{2,1} + a^3\, {}^{i-}\mathbf{s}^{A,1}$$

We define $\mathbf{S} \equiv \begin{bmatrix}\mathbf{s}^{3,1} & \mathbf{s}^{2,1} & \mathbf{s}^{A,1}\end{bmatrix}$ and $\mathbf{a} \equiv \begin{bmatrix}a^1 & a^2 & a^3\end{bmatrix}^{\mathrm{T}}$, then $^{i-}\mathbf{s}^{4,1} = \mathbf{S}\mathbf{a}$, or $\mathbf{a} = \mathbf{S}^{-1}\, {}^{i-}\mathbf{s}^{4,1}$. This yields

$$\begin{bmatrix}a^1\\ a^2\\ a^3\end{bmatrix} = \frac{1}{\eta^2\xi^3}\begin{bmatrix}\eta^2 & -\xi^2 & 0\\ 0 & \xi^3 & 0\\ 0 & 0 & 1\end{bmatrix}\begin{bmatrix}\xi^4\\ -\ell\sin\theta_i\\ \ell\cos\theta_i\end{bmatrix}$$

These coefficients are evaluated as a function of the local constant coordinates of the primary points and the variable joint coordinate θ_i. Then the primary coordinates $\mathbf{r}^4$ can be computed as:

$$\mathbf{r}^4 = \mathbf{r}^1 + a^1(\mathbf{r}^3 - \mathbf{r}^1) + a^2(\mathbf{r}^2 - \mathbf{r}^1) + a^3(\tilde{\mathbf{r}}^3 - \tilde{\mathbf{r}}^1)(\mathbf{r}^2 - \mathbf{r}^1)$$

6. Joint Coordinate Formulation for Closed-Loop Systems

For multibody systems with closed kinematic loops, the equations of motion in terms of a set of joint coordinates can be determined in several ways. We first derive these equations as a set of differential-algebraic equations, then we reduce them to a minimal set of ordinary-differential equations. We note that a closed-loop system may contain one or more closed loops. A closed loop can be eliminated from a system by removing one joint, which is called a *cut joint*. By cutting as many joints as the number of closed loops, an open-loop system, which is called a *reduced system*, is obtained. This process, in most cases, yields additional branches, and hence new leaves are formed. The cutting process of a closed loop is shown in Fig. 16. All matrices and vectors associated with a cut joint carry a right superscript or subscript $\otimes$.

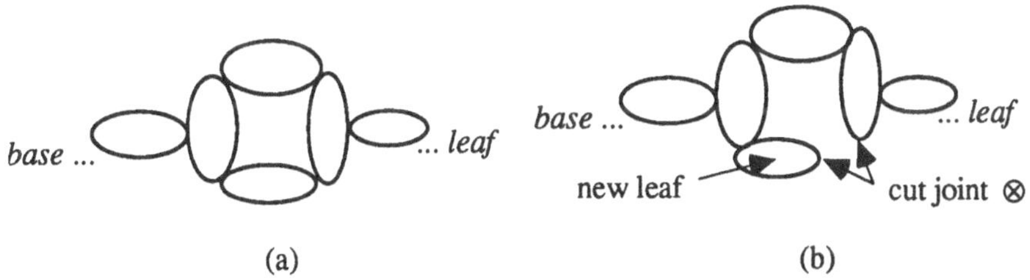

Figure 16. A closed-loop system and its reduced open-loop representation.

6.1. DIFFERENTIAL-ALGEBRAIC EQUATIONS OF MOTION

For the reduced system we define a vector of joint coordinates $\mathbf{q}$, and then we determine its corresponding matrix $\mathbf{B}$. We note that for the cut joint(s), we do not define any joint coordinates. Furthermore, for the reduced system we write the equations of motion as described in Eq. 29. Now the cut joint is put back in order to obtain the original closed loop system. In the original system the joint coordinates that are within a closed loop are no longer independent. The dependency of these joint coordinates can be described by constraint equations between the primary points of the two bodies that share the cut joint. These constraints are written as

$$\mathbf{\Phi}^{\otimes}(\mathbf{r}) = \mathbf{0} \tag{34}$$

where $\mathbf{r}$ only contains some of the primary coordinates of the connecting bodies. The time derivative of these constraints provides the velocity constraints,

$$\dot{\mathbf{\Phi}}^{\otimes} \equiv \mathbf{D}^{\otimes}\dot{\mathbf{r}} = \mathbf{D}^{\otimes}\mathbf{B}\dot{\mathbf{q}} = \mathbf{C}\dot{\mathbf{q}} = \mathbf{0} \tag{35}$$

where $\mathbf{D}^{\otimes}$ is the Jacobian of the constraints in Eq. 34 and $\mathbf{C} \equiv \mathbf{D}^{\otimes}\mathbf{B}$ is the coefficient matrix of the joint velocities in Eq. 35. Some of the constraints in Eq. 34, depending on the nature of the closed loop, may be redundant. Hence, some of the rows of $\mathbf{C}$ may also be redundant and must therefore be eliminated. The time derivative of Eq. 35 yields the acceleration constraints,

$$\ddot{\mathbf{\Phi}}^{\otimes} \equiv \mathbf{C}\ddot{\mathbf{q}} + \dot{\mathbf{C}}\dot{\mathbf{q}} = \mathbf{0} \tag{36}$$

where $\dot{\mathbf{C}} = \mathbf{D}^{\otimes}\dot{\mathbf{B}} + \dot{\mathbf{D}}^{\otimes}\mathbf{B}$. Due to these constraints, with the use of Lagrange multipliers, Eq. 29 is modified as

$$\boldsymbol{M}\ddot{\mathbf{q}} - \mathbf{C}^{\mathrm{T}}\,\mathbf{v} = \boldsymbol{f} \tag{37}$$

where $\mathbf{v}$ contains a set of Lagrange multipliers associated with the constraints of Eq. 34. Equations 34-37 form a set of differential-algebraic equations describing the dynamics of a multibody system with closed loops.

6.1.1. *Evaluation of Matrix* **C.** The elements of **C** can be found using different techniques. The product $\mathbf{D}^{\otimes}\mathbf{B}$ can be evaluated numerically since the elements of both matrices can be computed numerically. However, since the elements of $\mathbf{D}^{\otimes}$ and **B** are also available in closed form, the elements of **C** may also be found in closed form.

The elements of $\mathbf{D}^{\otimes}$ can be expressed in closed form for most common kinematic joints that may end up as cut joints. It is shown in section 4.3 that we can construct the necessary constraint equations describing various kinematic joints by combining some of the following constraints (refer to Fig. 17):

$$\mathbf{r}_i^k - \mathbf{r}_j^l = \mathbf{0} \tag{3}$$

$$\mathbf{s}_i^{\mathrm{T}}\mathbf{s}_j = 0 \tag{4}$$

$$\mathbf{s}_i^{\mathrm{T}}\mathbf{d} = 0 \tag{6}$$

where $\mathbf{s}_i = \mathbf{r}_i^m - \mathbf{r}_i^k$, $\mathbf{s}_j = \mathbf{r}_j^n - \mathbf{r}_j^l$, $\mathbf{d} = \mathbf{r}_j^l - \mathbf{r}_i^k$. The entries of the Jacobian matrix $\mathbf{D}^{\otimes}$ associated with these constraints are shown in Table 3. The columns are associated with the primary points that appear in the constraints.

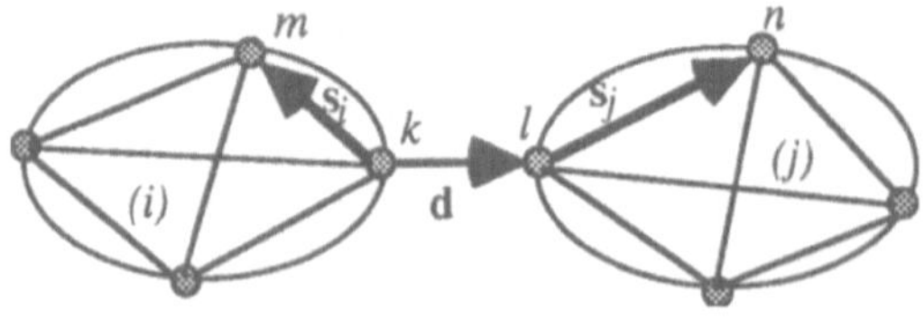

Figure 17. A cut joint (constraints) between two bodies.

Table 3. Entries of the Jacobian matrix $\mathbf{D}^{\otimes}$

Point ⇒ Cut joint ⇓	k	m	l	n
$\mathbf{r}_i^k - \mathbf{r}_j^l = \mathbf{0}$	$\mathbf{I}$		$-\mathbf{I}$	
$\mathbf{s}_i{}^{\mathrm{T}}\mathbf{s}_j = 0$	$-\mathbf{s}_j{}^{\mathrm{T}}$	$\mathbf{s}_j{}^{\mathrm{T}}$	$-\mathbf{s}_i{}^{\mathrm{T}}$	$\mathbf{s}_i{}^{\mathrm{T}}$
$\mathbf{s}_i{}^{\mathrm{T}}\mathbf{d} = 0$	$-(\mathbf{d}+\mathbf{s}_i)^{\mathrm{T}}$	$\mathbf{d}^{\mathrm{T}}$	$\mathbf{s}_i{}^{\mathrm{T}}$	

The entries of the **C** matrix for different cut joints (cut constraints) can be found in closed form by inspecting the product of the entries of Tables 1 and 3. The results of such inspection are shown in Table 4 for different joint coordinates. Figure 18 shows the indices and vectors used in this table. The entries of the table are stated for a joint in the branch associated with body i. For a joint in the branch associated with body j, the sign of the entry must be reversed.

Table 4. Entries of the **C** matrix

B ⇒ Cut joint ⇓	Floating	Revolute	Prismatic	Spherical
$\mathbf{r}_i^k - \mathbf{r}_j^l = \mathbf{0}$	$\mathbf{0}$	$-\tilde{\mathbf{d}}^{\otimes,r}\mathbf{u}_r$	$\mathbf{u}_r$	$-\tilde{\mathbf{d}}^{\otimes,r}$
$\mathbf{s}_i{}^{\mathrm{T}}\mathbf{s}_j = 0$	$\mathbf{0}$	$\mathbf{s}_i{}^{\mathrm{T}}\tilde{\mathbf{s}}_j\mathbf{u}_r$	0	$\mathbf{s}_i{}^{\mathrm{T}}\tilde{\mathbf{s}}_j$
$\mathbf{s}_i{}^{\mathrm{T}}\mathbf{d} = 0$	$\mathbf{0}$	$\mathbf{s}_i{}^{\mathrm{T}}\tilde{\mathbf{d}}^{\otimes,r}\mathbf{u}_r$	$-\mathbf{s}_i{}^{\mathrm{T}}\mathbf{u}_r$	$\mathbf{s}_i{}^{\mathrm{T}}\tilde{\mathbf{d}}^{\otimes,r}$

Note: $\mathbf{d}^{\otimes,r} = \mathbf{d}^{k,r} = \mathbf{d}^{l,r}$

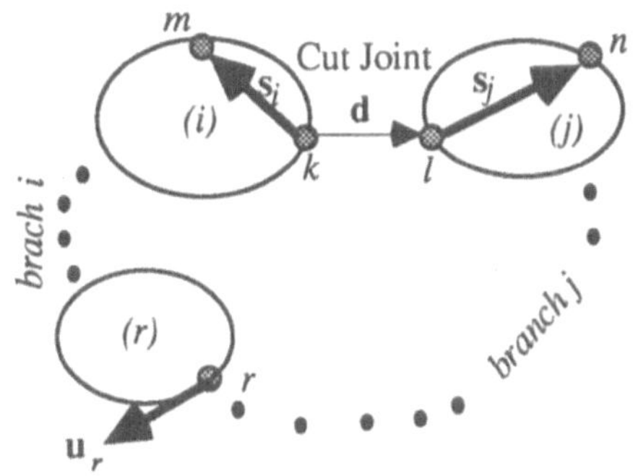

Figure 18. The indices and vectors used in generating matrix **C**.

Example 3: Consider the closed-loop system shown in Fig. 19 containing two revolute, one spherical, and one universal joint. The system is not attached to the ground, therefore body *1* is considered as the floating body. If the universal joint is selected to be the cut joint, we need the following constraints from Eqs. 3 and 4:

$$\mathbf{r}_3^5 - \mathbf{r}_4^7 = \mathbf{0}$$
$$\mathbf{s}_3^{\mathrm{T}}\mathbf{s}_4 = 0$$

We define the vector of joint velocities as $\dot{\mathbf{q}} = \left[\mathbf{v}_1^{\mathrm{T}},\ \dot{\theta}_2,\ \omega_{3,2}^{\mathrm{T}},\ \dot{\theta}_4\right]^{\mathrm{T}}$. Then Table 3 yields the Jacobian matrix **C** as

$$\mathbf{C} = \begin{bmatrix} \mathbf{0} & -\tilde{\mathbf{d}}^{5,2}\mathbf{u}_2 & -\tilde{\mathbf{d}}^{5,3} & \tilde{\mathbf{d}}^{7,4}\mathbf{u}_4 \\ \mathbf{0} & \mathbf{s}_3^{\mathrm{T}}\tilde{\mathbf{s}}_4\mathbf{u}_2 & \mathbf{s}_3^{\mathrm{T}}\tilde{\mathbf{s}}_4 & \mathbf{s}_4^{\mathrm{T}}\tilde{\mathbf{s}}_3\mathbf{u}_4 \end{bmatrix}$$

The elements in the column(s) associated with the velocity vector of body *1* are zero since body *1* is a floating body. The actual Jacobian is a 4 x 5 matrix, i.e., the closed loop exhibits one degree of freedom.

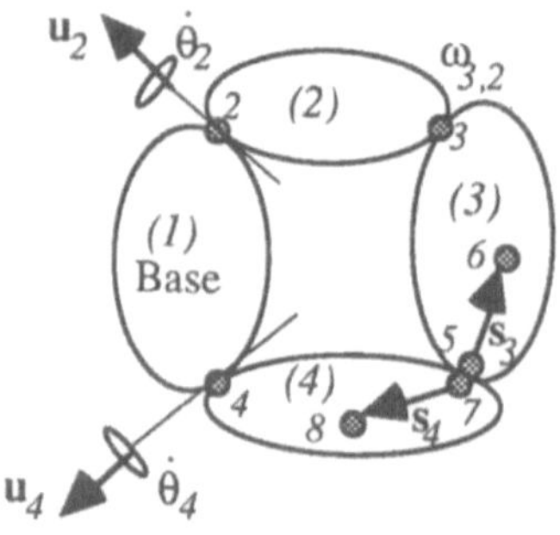

Figure 19. A closed-loop system.

6.1.2. *Multiple Loops*. A multibody system with more than one closed loop yields a **C** matrix containing submatrices that are either uncoupled or loosely coupled. For the two uncoupled loops shown in Fig. 20(a), the two submatrices of **C** are also uncoupled,

$$\mathbf{C} = \begin{bmatrix} \mathbf{0} & [\mathbf{C}_1] & \mathbf{0} & \mathbf{0} & \mathbf{0} \\ \mathbf{0} & \mathbf{0} & \mathbf{0} & [\mathbf{C}_2] & \mathbf{0} \end{bmatrix}$$

In this case, the submatrices can be treated separately during the analysis. For the two coupled loops shown in Fig. 20(b), the submatrices of **C** are coupled,

$$\mathbf{C} = \begin{bmatrix} \mathbf{0} & [& \mathbf{C}_1 &] & \mathbf{0} & \mathbf{0} \\ \mathbf{0} & \mathbf{0} & [& \mathbf{C}_2 &] & \mathbf{0} \end{bmatrix}$$

When the submatrices of **C** are uncoupled, their corresponding constraints can be treated independently during an analysis.

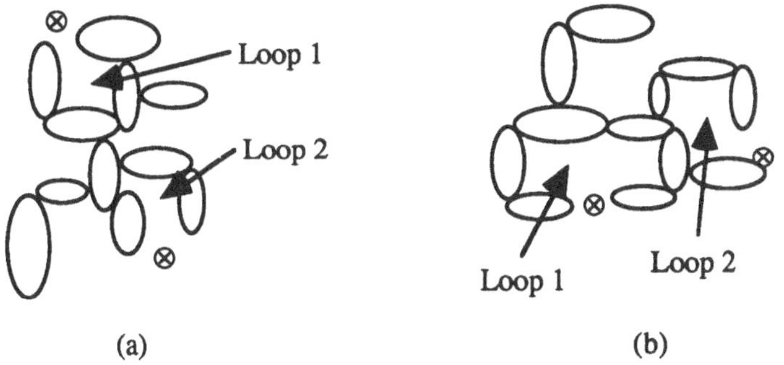

Figure 20. Examples of uncoupled and coupled closed-loop systems.

6.1.3. *Integration of the Equations of Motion* . The mixed differential-algebraic equations of motion for closed-loop multibody systems, presented in Eqs. 34-37, can be solved for the dynamic response of the system by using a process similar to the algorithm of section 5.2. The $\mathbf{y}$ and $\dot{\mathbf{y}}$ arrays are defined as in Eqs. 32 and 33, and the first seven steps of the algorithm remain unchanged. The final steps of the new algorithm are:

•
•
•

8. Evaluate matrix **C**.
9. Solve the equations of motion for $\ddot{\mathbf{q}}$ and $\mathbf{v}$ (refer to Eqs. 36 and 37).
10. Construct $\dot{\mathbf{y}}$ array and return the contents to the integration algorithm.

This is a simple algorithm which does not account for possible constraint violation due to numerical errors.

6.2. DIFFERENTIAL EQUATIONS OF MOTION

The differential-algebraic equations of motion for a closed loop system can be converted to a set of ordinary-differential equations. For this purpose, within each closed loop we select a set of independent joint velocities, equal to the number of degrees of freedom associated with the loop, to form a vector of independent joint velocities $\dot{\mathbf{q}}^{(i)}$. A closed-loop velocity transformation matrix **E** is defined as

$$\dot{\mathbf{q}} = \mathbf{E}\dot{\mathbf{q}}^{(i)} \tag{38}$$

One characteristic of **E** is that it is orthogonal to the Jacobian **C**. This can be shown by substituting Eq. 38 into Eq. 35 to obtain $\mathbf{C}\mathbf{E}\dot{\mathbf{q}}^{(i)} = \mathbf{0}$. Since $\dot{\mathbf{q}}^{(i)}$ contains independent velocities, then

$$\mathbf{C}\mathbf{E} = \mathbf{0} \tag{39}$$

The time derivative of Eq. 38 yields the acceleration transformation formula as:

$$\ddot{\mathbf{q}} = \mathbf{E}\ddot{\mathbf{q}}^{(i)} + \dot{\mathbf{E}}\dot{\mathbf{q}}^{(i)} \tag{40}$$

Now substituting Eq. 40 into Eq. 37, premultiplying both sides by $\mathbf{E}^T$, then using Eq. 39 yields

$$\bar{M}\ddot{\mathbf{q}}^{(i)} = \bar{f} \tag{41}$$

where

$$\bar{M} = \mathbf{E}^T M \mathbf{E} \tag{42}$$

$$\bar{f} = \mathbf{E}^T(f - M\dot{\mathbf{E}}\dot{\mathbf{q}}^{(i)}) \tag{43}$$

Equation 41 provides a set of nonlinear ordinary-differential equations of motion equal to the number of degrees of freedom of the system.

6.2.1. *Evaluation of Matrix* **E**. Matrix **E** can be obtained from matrix **C** using the constraints of Eq. 35. By partitioning the vector of joint velocities into two dependent and independent sets, and respectively partitioning **C** into two submatrices, Eq. 35 can be written as

$$\mathbf{C}^{(i)}\dot{\mathbf{q}}^{(i)} + \mathbf{C}^{(d)}\dot{\mathbf{q}}^{(d)} = \mathbf{0}$$

This yields $\dot{\mathbf{q}}^{(d)} = -\mathbf{C}^{(d)^{-1}}\mathbf{C}^{(i)}\dot{\mathbf{q}}^{(i)}$ or

$$\dot{\mathbf{q}} \equiv \begin{bmatrix} \dot{\mathbf{q}}^{(i)} \\ \dot{\mathbf{q}}^{(d)} \end{bmatrix} = \begin{bmatrix} \mathbf{I} \\ -\mathbf{C}^{(d)^{-1}}\mathbf{C}^{(i)} \end{bmatrix} \dot{\mathbf{q}}^{(i)}$$

This provides a closed form formula for matrix **E** as a function of the submatrices of **C** as

$$\mathbf{E} = \begin{bmatrix} \mathbf{I} \\ -\mathbf{C}^{(d)^{-1}}\mathbf{C}^{(i)} \end{bmatrix} \tag{44}$$

In most practical applications, matrix **C** is quite small in dimensions. Therefore one way to obtain **E** is to evaluate it numerically.

6.2.2. *Evaluation of* $\dot{\mathbf{E}}\dot{\mathbf{q}}^{(i)}$. The acceleration constraints of Eq. 36 can be written as $\mathbf{C}\mathbf{E}\ddot{\mathbf{q}}^{(i)} - \mathbf{C}\dot{\mathbf{E}}\dot{\mathbf{q}}^{(i)} + \dot{\mathbf{C}}\dot{\mathbf{q}} = \mathbf{0}$. The first term in this equation is zero since **CE** = **0**. The identities in **E** result in 0's in $\dot{\mathbf{E}}$. Therefore, we have

$$\dot{\mathbf{E}}\dot{\mathbf{q}}^{(i)} = \begin{bmatrix} \mathbf{0} \\ \mathbf{C}^{(d)^{-1}}\dot{\mathbf{C}}\dot{\mathbf{q}} \end{bmatrix} \tag{45}$$

This term is needed for evaluating the generalized force vector described by Eq. 43.

6.3. INTEGRATION OF THE EQUATIONS OF MOTION

The equations of motion expressed by Eq. 41 can be put into the standard form of Eq. 32, where $\mathbf{y}$ and $\dot{\mathbf{y}}$ arrays contain the independent joint coordinates, velocities, and accelerations as,

$$\dot{\mathbf{y}} \equiv \begin{bmatrix} \dot{\mathbf{q}}^{(i)} \\ \ddot{\mathbf{q}}^{(i)} \end{bmatrix}, \quad \mathbf{y} \equiv \begin{bmatrix} \mathbf{q}^{(i)} \\ \dot{\mathbf{q}}^{(i)} \end{bmatrix}$$

The process of numerical solution of these equations, in general, is similar to that presented in section 5.2. However, the intermediate steps that are required in evaluating the vectors and matrices for Eq. 41 are more extensive than those of Eq. 29 or 37.

7. Conclusions

In this paper we have presented a computational procedure for generating the equations of motion for rigid-multibody systems. In this procedure, the bodies are described as a collection of constrained particles. This description results into a large set of mixed differential-algebraic equations of motion. A velocity transformation process is then employed to transform these equations to either a much smaller set of mixed differential-algebraic equations or even a minimal set of ordinary-differential equations.

The idea of describing a body as a collection of several particles at first may appear to be awkward, however, this description has certain interesting features which for some applications may be found useful. Two of the most useful features of this methodology are (a) the elimination of the use of rotational coordinates for bodies and (b) a natural extension to the finite element representation for a deformable body.

The elimination of rotational coordinates leads to possible savings in computation time when this procedure is compared against the body-joint coordinate formulation. It has been determined that numerical computations associated with rotational transformation matrices and their corresponding coordinate transformations between reference frames is time consuming and, therefore, if these computations are avoided, more efficient codes may be developed. The elimination of rotational coordinates can also be found very beneficial in design sensitivity analysis of multibody systems. In most procedures for design sensitivity analysis, leading to an optimal design process, the derivatives of certain functions with respect to a set of design parameters are required. Analytical evaluation of these derivatives are much simpler if the rotational coordinates are not present and if we only deal with translational coordinates.

Some practical applications of multibody dynamics require one or more bodies in the system to be described as deformable in order to obtain a more realistic dynamic response. Deformable bodies are normally modeled by the finite element technique. Assume that the deformable body is connected to a rigid body described by a set of particles. Then, one or more particles of the rigid body can coincide with one or more nodes of the deformable body in order to describe the kinematic joint between the two bodies. This is a much simpler process that when the rigid body is described by a set of translational and rotational coordinates.

In summary, the methodologies presented in this paper have many interesting characteristics that may be found useful in some applications. These methodologies can be combined with other methods to develop even more efficient, accurate, and flexible procedures. It should be noted that there is no single multibody formulation to be considered as "the best formulation" for general multibody dynamics. Each formulation has its own unique or common features and, therefore, selected features should be adopted to our advantage.

References

1. Nikravesh, P. E. (1988) *Computer-Aided Analysis of Mechanical Systems*, Prentice-Hall.
2. Jerkovsky, W. (1978) "The Structure of Multibody Dynamics Equations," *J. Guidance and Control*, Vol. 1, No. 3, pp. 173-182.
3. Kim, S. S. and Vanderploeg, M. J. (1986) "A General and Efficient Method for Dynamic Analysis of Mechanical Systems Using Velocity Transformations," *ASME J. Mech., Trans., and Auto. in Design*, Vol. 108, No. 2, pp. 176-182.
4. Nikravesh, P. E. and Gim, G. (1993) "Systematic Construction of the Equations of Motion for Multibody Systems Containing Closed Kinematic Loops," *J. of Mechanical Design,* Vol. 115, No. 1, pp. 143-149.
5. Serna, M. A., Aviles, R. and Garcia de Jalon, J. (1982) "Dynamic Analysis of Plane Mechanisms with Lower Pairs in Basic Coordinates," *Mechanisms and Machine Theory*, Vol. 7, No. 6, pp. 397-403.
6. Garcia de Jalon, J., Unda, J., Avello, A. and Jimenez, J. M. (1986) "Dynamic Analysis of Three-Dimensional Mechanisms in Natural Coordinates," *ASME* Design Engineering Technical conference, Columbus, OH, Paper No. 86-DET-137.
7. Affifi, H. A. (1992) "A Multi-Rigid-Body Formulation Based on Particles Dynamics and Velocity Transformation," Research Report, Department of Aerospace and Mechanical Engineering, University of Arizona.

DYNAMICS OF MULTIBODY SYSTEMS WITH MINIMAL COORDINATES

M. HILLER and A. KECSKEMÉTHY
Fachgebiet Mechatronik, Universität Duisburg, Lotharstraße 1, 47057 Duisburg, Germany

ABSTRACT: Discussed in this contribution is a particular approach for tackling the problem of formulating the equations of motion of minimal order for complex mechanical systems. The obective is to arrive at a system of pure differential equations, which is robust and for which efficient integration techniques exist. This is achieved by a special treatment of the kinematics, which are formulated by consideration of closed-form solutions for the subsystems where this is possible and reducing the generation of dynamical equations to a repetitive evaluation of the kinematics. Discussed in the paper are the necessary techniques for solving the arising subproblems, from methods for finding closed-form solutions in single- and multi-loop systems to the incorporation of non-holonomic constraints. Also, some remarks are given concerning the implementation of the methods based on modern programming paradigms, such as object-oriented programming and symbolical formula manipulation. The key concepts are illustrated by several examples, among which are actual research objects and applications in cooperation with industry.

1. Introduction

The dynamics of multibody systems is a field of research since now more then 30 years. Since its beginnings, several outstanding approaches have emerged which, in part, have reached commercial maturity. Early publications in this field are the papers of Hooker and Margulies [22] and Roberson and Wittenburg [36], and some of the most reknown "turn-key" program packages are IMP [39], DAMN [7], ADAMS [32], DADS [41], NEWEUL [26], COMPAMM [10], MESA VERDE [43], MECANO [11] and SIMPACK [37]. These program packages are usually endowed with easy-to-use graphical interfaces and universal schemes for automatic equation generation by means of which the engineer is freed from the burden of establishing the model equations of the system. The German Research Council (DFG) has just finished a nationwide five-year research project devoted to dynamics of multibody systems, gathering some of the best contemporary multibody formalisms in a new general purpose program ([38]). Why thus a further approach?

The reason for the development of the present approach lies in the fact that, in industry, for complex systems and corresponding applications still a large amount of modelling is performed by hand, because engineers find that (a) such programs can be "tuned" to run faster (e.g. for hardware-in-the-loop applications), (b) hand-tailored program modules implementing non-standard solution techniques can be more easily incorporated, and (c) these programs are better understood. As it turns out, for many practical systems huge parts of the underlying equations can be solved either in closed form or in some simplified manner, and incorporation of these solutions into the general procedure is (today) only possible with the help of human intelligence.

The objective of this paper is to show some techniques and systematics for helping the designer to establish where closed form solutions may arise in the system, and how to incorporate them into the dynamics-generation procedure. This knowledge

M. F. O. S. Pereira and J. A. C. Ambrósio (eds.),
Computer-Aided Analysis of Rigid and Flexible Mechanical Systems, 61–100.

can be used for generating the equations of motion of minimal order for quite general systems, and thus to include very efficient computer-models of mechanical systems into more sophisticated programming environments. It is also intended to present the main results of the work done by the author and his co-workers in the recent years.

The rest of the paper is organized as follows: After giving an overview of the basic steps of the overall procedure based on the example of the modelling of the dynamics of an upper-class passenger car in Section **2**, the idea of reducing the dynamics to a repetitive evaluation of the kinematics is presented in Section **3**. At the heart of the approach is the problem of appropriately solving the kinematics of the single loop, which is discussed in Section **4**. Here, the method of the "Characteristic Pair of Joints" for establishing appropriate closure conditions is elaborated, and a short overview of an approach for determining the polynomial of minimal order for the general case is given. Also, a scheme for the automatic generation of closed-form solutions, where possible, is described. Section **5** addresses the problem of the dissection of a general multi-loop system into modules which can be used for incorporating the results of the previous section into the general-case problem. This is achieved by regarding the independent multibody loops as individual "kinematical transformers" which are then coupled together by linear equations to make up the original general multiple loop system. Section **6** covers the treatment of non-holonomic constraints, showing as an example of a complex application a combined wheeled and legged vehicle. Finally, in Section **7** some remarks concerning implementation-specific issues are given.

2. Basic Modelling Steps in the Minimal Coordinate Approach

Various techniques exist for transforming a real technical system into a mechanical model depending on the kind of investigation to be carried out. As an example, consider the simulation of the dynamics of a passenger car with active components such as anti-lock-systems (ABS) and drive-slide-control (ASR) (see Fig. 1 and Fig. 2). Here, eigenfrequencies of the system up to 25 Hz have to be taken into account, which is accomplished by representing the complete vehicle as a multibody system including the full nonlinear kinematics of the wheel suspensions, the suspension of the engine together with the powertrain, as well as dynamic tire models. In addition, elasticities at particular hinges have to be taken into account.

The corresponding multibody system is characterized by a complex topology with many kinematical loops (see Fig. 3). The model consists of 30 rigid bodies, constrained by 40 joints (12 Hooke, 18 spherical, 10 revolute) as well as 3 virtual joints guiding the chassis with respect to the inertial system and the front and rear suspension carriers with respect to the carrier. From these, a total of 13 multibody loops and 25 degrees of freedom arise. Furthermore, a power train with four degrees of freedom as well as tire models of different degrees of complexity are incorporated into the mechanical model.

To obtain the dynamical equations of motion in minimal coordinates, the global kinematics of the system, describing the motion of all bodies in the system with respect to the inertial frame, is required. For this purpose, the following modelling steps are introduced:

1. Decomposition of the *global* kinematics into *relative* and *absolute* kinematics by introducing joint coordinates. Thus the originally large set of implicit equations in absolute coordinates is separated into a small set of implicit or partly implicit

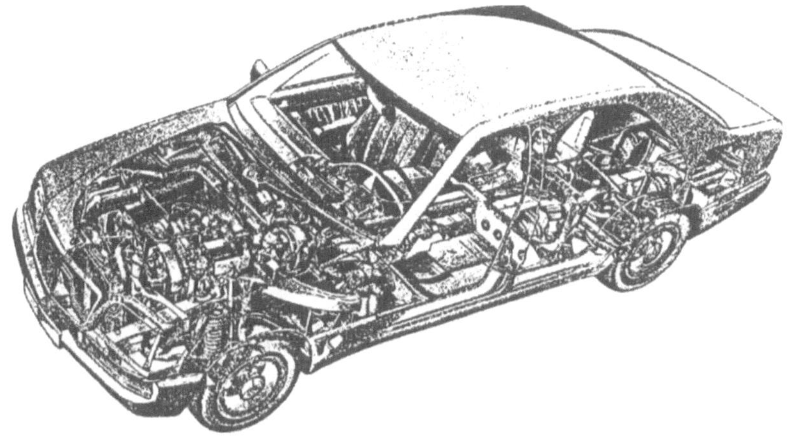

Figure 1: Upper-class passenger car.

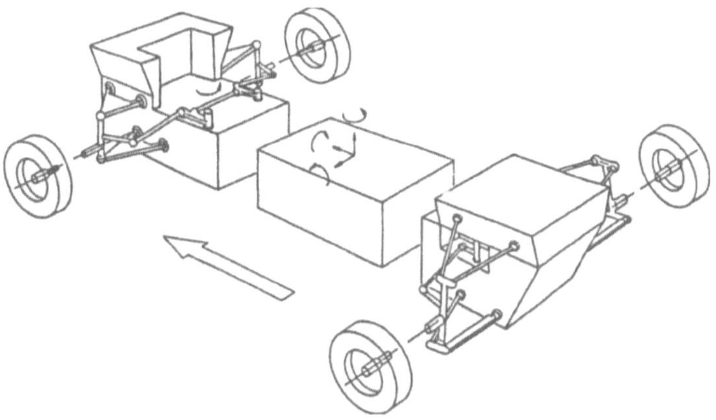

Figure 2: Model of the passenger car.

equations governing the relative kinematics and a set of *explicit* equations for the absolute kinematics (see Section **3**).

2. Decomposition of the equations for the relative kinematics into components corresponding to individual kinematical loops or subsystems of several kinematical loops. These components are then treated as "kinematical transformers" (see Sections **4**, **5** and **6**).

3. Determination of closed-form solutions for individual components, if possible. For example, the five kinematical loops contained in the front-axle of the vehicle shown in Fig. 3 can be explicitly solved as individual loops and also as a complete subsystem (see Sections **4**, **5** and **6**).

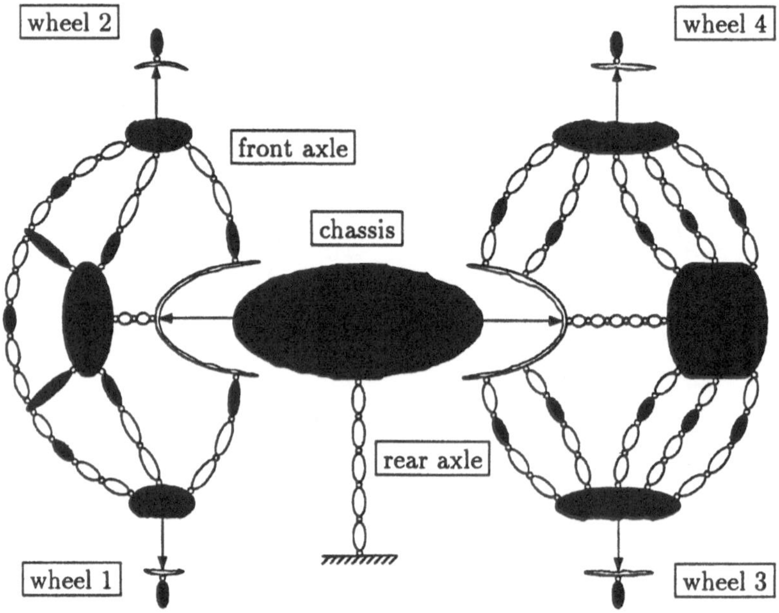

Figure 3: Topology of the passenger car.

4. Assembly of the solution of the individual components to obtain the global kinematics of the complete multibody system (see Fig. 4). The use of object-oriented programming techniques permits this assembly process to be implemented in a simple and intuitive manner (see Section **7**).

5. Generation of the overall dynamical equations.

3. Equations of Motion for Complex Multibody Systems with Minimal Coordinates

3.1. DYNAMICS

The equations of motion for general multibody systems can be stated starting from D'ALEMBERT's principle. This principle applied to a scleronomic, holonomic or non-holonomic multibody system consisting of n_B rigid bodies reads (see also [14])

$$\sum_{i=1}^{n_B} \Big[(m_i \boldsymbol{a}_{S_i} - \boldsymbol{f}_i^e) \cdot \delta \boldsymbol{s}_i + (\boldsymbol{\Theta}_{S_i} \dot{\boldsymbol{\omega}}_i + \boldsymbol{\omega}_i \times \boldsymbol{\Theta}_{S_i} \boldsymbol{\omega}_i - \boldsymbol{\tau}_{S_i}^e) \cdot \delta \boldsymbol{\phi}_i \Big] = 0 \qquad (1)$$

where, for body $\mathcal{B}_i$,

m_i	-	mass,
$\boldsymbol{\Theta}_i$	-	tensor of mass-inertia,
$\boldsymbol{f}_i$	-	resultant vector of applied forces,
$\boldsymbol{\tau}_{S_i}$	-	resultant vector of applied moments at center of gravity,
$\boldsymbol{a}_{S_i} = \ddot{\boldsymbol{s}}_i$	-	vector of acceleration of center of gravity,
$\boldsymbol{\omega}_i$	-	vector of angular velocity,
$\dot{\boldsymbol{\omega}}_i$	-	vector of angular acceleration,
$\delta \boldsymbol{s}_i$	-	vector of virtual displacement of center of gravity,
$\delta \boldsymbol{\phi}_i$	-	vector of virtual rotation.

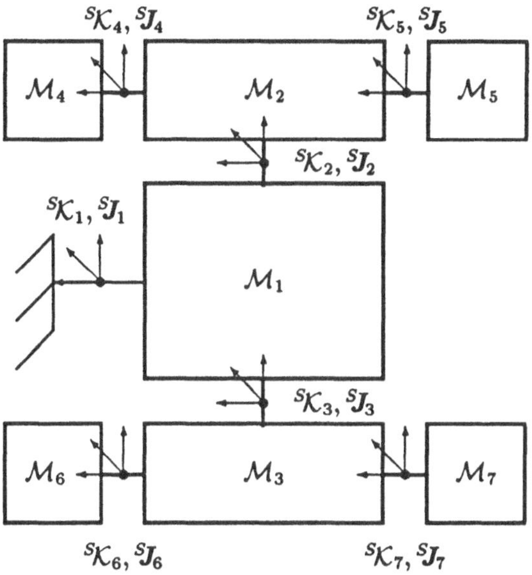

Figure 4: Assembly of the components of the passenger cars.

All vectors appearing in Eq. (1) are tensors of valence one, thus Eq. (1) is stated in a component-independent form. Note that the constraint forces do not appear in Eq. (1) because the virtual displacements $\delta \boldsymbol{s}_i$ and $\delta \boldsymbol{\phi}_i$ are assumed to be compatible with *all* the constraints imposed on the system. For the general case where the virtual displacements $\delta \boldsymbol{s}_i$ and $\delta \boldsymbol{\phi}_i$ are not independent, one introduces f independent generalized coordinates $\underline{q} = [q_1, \ldots, q_f]^{\mathrm{T}}$ with corresponding independent virtual displacements $\delta \underline{q} = [\delta q_1, \ldots, \delta q_f]^{\mathrm{T}}$, f being the number of degrees of freedom of the multibody system. The dependent virtual displacements are related to the independent virtual displacements by linear, homogeneous transformations which can be obtained directly from the velocity transformations

$$\boldsymbol{v}_i = \boldsymbol{J}_{s\,i}\,\dot{\underline{q}} \quad , \quad \boldsymbol{\omega}_i = \boldsymbol{J}_{\omega\,i}\,\dot{\underline{q}} \tag{2}$$

by substituting velocities with virtual displacements:

$$\delta \boldsymbol{s}_i = \boldsymbol{J}_{s\,i}\,\delta \underline{q} \quad , \quad \delta \boldsymbol{\phi}_i = \boldsymbol{J}_{\omega\,i}\,\delta \underline{q} \ . \tag{3}$$

The $3 \times f$ transformation matrices $\boldsymbol{J}_{s\,i}$ and $\boldsymbol{J}_{\omega\,i}$ are still to be determined.

The relationships between the dependent and independent accelerations, are derived from Eq. (2) as

$$\boldsymbol{a}_i = \boldsymbol{J}_{s\,i}\,\ddot{\underline{q}} + \dot{\boldsymbol{J}}_{S\,i}\,\dot{\underline{q}} \quad , \quad \dot{\boldsymbol{\omega}}_i = \boldsymbol{J}_{\omega\,i}\,\ddot{\underline{q}} + \dot{\boldsymbol{J}}_{\omega\,i}\,\dot{\underline{q}} \ . \tag{4}$$

Insertion of these transformations into D'ALEMBERT's principle yields, due to the independency of virtual displacements $\delta q_1, \ldots, \delta q_f$, the equations of motion of minimal order

$$\boldsymbol{M}\,\ddot{\underline{q}} + \underline{b} = \underline{Q} \ , \tag{5}$$

where the $f \times f$ generalized mass-matrix $\boldsymbol{M}$, the $f \times 1$ matrix of generalized centrifugal and coriolis forces $\underline{b}$ and the $f \times 1$ matrix of generalized applied forces $\underline{Q}$ read, respectively,

$$\boldsymbol{M}(\underline{q}) = \sum_{i=1}^{n_B} \left[m_i \, \boldsymbol{J}_{s_i}^{\mathrm{T}} \, \boldsymbol{J}_{s_i} + \boldsymbol{J}_{\omega_i}^{\mathrm{T}} \Theta_i \boldsymbol{J}_{\omega_i} \right] , \tag{6}$$

$$\underline{b}(\underline{q}, \dot{\underline{q}}) = \sum_{i=1}^{n_B} \left[m_i \, \boldsymbol{J}_{s_i}^{\mathrm{T}} \, \dot{\boldsymbol{J}}_{s_i} \dot{\underline{q}} + \boldsymbol{J}_{\omega_i}^{\mathrm{T}} \left(\Theta_i \, \dot{\boldsymbol{J}}_{\omega_i} \dot{\underline{q}} + \omega_i \times \Theta_i \, \omega_i \right) \right] , \tag{7}$$

$$\underline{Q}(\underline{q}, \dot{\underline{q}}) = \sum_{i=1}^{n_B} \left[\boldsymbol{J}_{s_i}^{\mathrm{T}} \, \boldsymbol{f}_i + \boldsymbol{J}_{\omega_i}^{\mathrm{T}} \, \boldsymbol{\tau}_i \right] . \tag{8}$$

Once the transformation matrices $\boldsymbol{J}_{s_i}$ and $\boldsymbol{J}_{\omega_i}$ are established the problem of stating the equations of motion of minimal order is solved. The difficulty for complex multibody systems is that the transformations Eq. (2) to Eq. (4) are mostly hard to obtain. Although the position coordinates of the bodies are known to be analytic functions of the generalized coordinates

$$\left. \begin{array}{rcl} \boldsymbol{s}_i &=& \boldsymbol{s}_i(q_1, \ldots, q_f) \\ \boldsymbol{R}_i &=& \boldsymbol{R}_i(q_1, \ldots, q_f) \end{array} \right\} \; i = 1, 2, \; \ldots \; , n_B \; , \tag{9}$$

where $\boldsymbol{R}_i$ denotes the tensor measuring the rotation of body $\mathcal{B}_i$ with respect to the inertial frame, these functions are generally not known explicitly. Thus defining $\boldsymbol{J}_{s_i}$ and $\boldsymbol{J}_{\omega_i}$ by analytical differentiation, as in

$$\boldsymbol{J}_{s_i} = \frac{\partial \boldsymbol{s}_i}{\partial \underline{q}} \tag{10}$$

leads to very long formulas which cannot be applied to complex multibody systems. For this reason, most of the present methods avoid this kind of formulation by using LAGRANGE-multipliers. This is equivalent to "transfering" some — or all — of the constraints from the kinematics to the dynamics.

In this paper, the method applied is to produce suitable expressions for the transformations in Eqs. (2) and (4) using "kinematical differentials". Suppose that an effective formulation of the *global kinematics* exists for a given multibody system, which provides the relationships between position, velocity and acceleration of all bodies for given values of the generalized coordinates and their time derivatives, as shown in Fig. 5. Such a full solution of the kinematics, from which the quantities needed for the transformations in Eqs. (3) and (4) can be determined easily, is described in [15]. Obviously, these relationships can be evaluated for any set of values

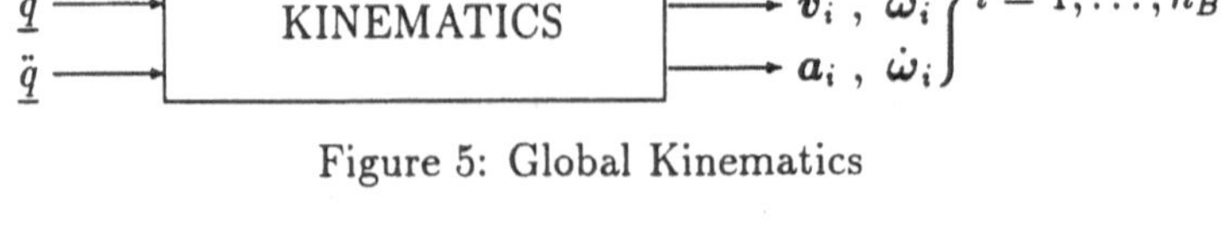

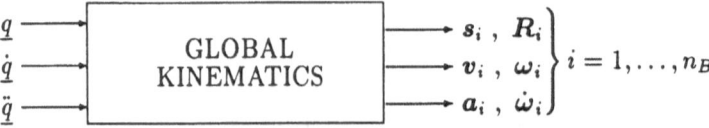

Figure 5: Global Kinematics

of the generalized velocities $\dot{q}_1, \ldots, \dot{q}_f$. In particular, evaluating the kinematics for a fixed position and a special set of generalized velocities

$$\hat{\dot{q}}_i^{(j)} = \begin{cases} 1 & \text{if } i = j \\ 0 & \text{otherwise} \end{cases} \qquad \text{for } j = 1, \ldots, f \ , \tag{11}$$

yields *pseudo*-velocities $\hat{\boldsymbol{v}}_i^{(j)}$ and $\hat{\boldsymbol{\omega}}_i^{(j)}$, which, compared with Eq. (2), give just the j-th column of the transformation matrices $\boldsymbol{J}_{si}$ and $\boldsymbol{J}_{\omega i}$, respectively. Similarly, evaluation of the kinematics for fixed position and fixed velocity, but with the special generalized accelerations

$$\hat{\ddot{\underline{q}}} = 0 \ , \tag{12}$$

yields the *pseudo*-accelerations $\hat{\boldsymbol{a}}_i$ and $\hat{\dot{\boldsymbol{\omega}}}_i$ which, compared with Eq. (4), equal the terms $\dot{\boldsymbol{J}}_{si}\,\dot{\underline{q}}$ and $\dot{\boldsymbol{J}}_{\omega i}\,\dot{\underline{q}}$, respectively.

With these quantities, the differential relationships of Eq. (3) and Eq. (4) can be re-stated as

$$\left.\begin{aligned} \boldsymbol{s}_i &= \sum_{j=1}^{f} \hat{\boldsymbol{v}}_i^{(j)} \delta q_j \quad ; \quad \boldsymbol{a}_i = \sum_{j=1}^{f} \hat{\boldsymbol{v}}_i^{(j)} \ddot{q}_j + \hat{\boldsymbol{a}}_i \ , \\ \delta \boldsymbol{\phi}_i &= \sum_{j=1}^{f} \hat{\boldsymbol{\omega}}_i^{(j)} \delta q_j \quad ; \quad \dot{\boldsymbol{\omega}}_i = \sum_{j=1}^{f} \hat{\boldsymbol{\omega}}_i^{(j)} \ddot{q}_j + \hat{\dot{\boldsymbol{\omega}}}_i \ , \end{aligned}\right\} \tag{13}$$

and the coefficients for the matrices of the equations of motion of minimal order read:

$$\left.\begin{aligned} M_{jk} &= \sum_{i=1}^{n_B} \left[m_i \, \hat{\boldsymbol{v}}_i^{(j)} \cdot \hat{\boldsymbol{v}}_i^{(k)} + \hat{\boldsymbol{\omega}}_i^{(j)} \cdot \boldsymbol{\Theta}_i \, \hat{\boldsymbol{\omega}}_i^{(k)} \right] \\ \underline{b}_j &= \sum_{i=1}^{n_B} \left[m_i \, \hat{\boldsymbol{v}}_i^{(j)} \cdot \hat{\boldsymbol{a}}_i + \hat{\boldsymbol{\omega}}_i^{(j)} \cdot \left(\boldsymbol{\Theta}_i \, \hat{\dot{\boldsymbol{\omega}}}_i + \boldsymbol{\omega}_i \times \boldsymbol{\Theta}_i \, \boldsymbol{\omega}_i \right) \right] \\ \underline{Q}_j &= \sum_{i=1}^{n_B} \left[\hat{\boldsymbol{v}}_i^{(j)} \cdot \boldsymbol{f}_i + \hat{\boldsymbol{\omega}}_i^{(j)} \cdot \boldsymbol{\tau}_i \right] \end{aligned}\right\} . \tag{14}$$

It should be noted that by Eq. (14) the problem of stating the equations of motion of minimal order for arbitrary multibody systems has been reduced to a purely kinematical problem. Moreover, the coefficients of the equations of motion are written in terms of "physical" quantities, i.e. tensors, which are independent to coordinate transformations. Thus, one is still free to choose an appropriate coordinate system for the evaluation of each individual term. Together with the process of determining the partial derivatives by a simple re-evaluation of the kinematics — denominated *kinematical differentials* — Eq. (14) gives a very simple approach for stating the equations of motion of minimal order which is also very easily implemented. The key to the effectiveness of the method is a particular formulation of the kinematics, which will be discussed next.

3.2. KINEMATICS

For a multibody system with independent coordinates $\underline{q}$, position, velocity and acceleration of all bodies have to be expressed as a function of $\underline{q}, \dot{\underline{q}}, \ddot{\underline{q}}$. For systems with complex topology, such as those containing kinematical loops, it is suitable to divide the calculation into the two steps (see Fig. 6)

- *relative kinematics*, where all dependent joint coordinates $\underline{\beta}$ and its derivatives $\dot{\underline{\beta}}, \ddot{\underline{\beta}}$ are expressed as functions of $\underline{q}\,,\,\dot{\underline{q}}\,,\,\ddot{\underline{q}}$.
- *absolute kinematics*, in which, by a forward kinematics procedure, kinematical quantities $\boldsymbol{s}_i, \boldsymbol{R}_i, \boldsymbol{v}_i, \boldsymbol{\omega}_i, \boldsymbol{a}_i, \dot{\boldsymbol{\omega}}_i$ are calculated for all bodies as a function of $\underline{\beta}\,,\,\dot{\underline{\beta}}\,,\,\ddot{\underline{\beta}}$.

Both steps are combined to obtain the *global kinematics* described in the preceding section (see Fig. 5).

Figure 6: Kinematics of multibody systems with closed loops.

In the following, several particular aspects which arise in multibody systems with closed kinematical loops will be discussed. Consider a spatial multibody system, consisting of n_B *moving* rigid bodies (i.e. without counting the inertial frame), and n_G *elementary* joints, i.e. revolute or prismatic joints with a single degree of freedom. According to Eq. (9), the position and orientation of all bodies can be stated as nonlinear analytic functions of $f = 6\,n_B - 5\,n_G$ generalized coordinates $q_1, \ldots, q_f$. An alternative, more compact representation of Eq. (9) is

$$\underline{p}_i = \underline{p}_i(\underline{q}) \tag{15}$$

where $\underline{p}_i$ is a 6–tuple holding three translational and three rotational parameters, or a $(6+\epsilon)$–tuple, ϵ being the number of redundant rotational parameters. For example

$$\underline{p}_i = [x_i, y_i, z_i; \phi_i, \theta_i, \psi_i]^T \tag{16}$$

with coordinates x_i,y_i,z_i for the reference point of body i and EULER-angles ϕ_i,θ_i,ψ_i (or another set of rotational parameters) for its orientation. The $n_p = 6\,n_B$ position coordinates of all bodies can be put together in an array

$$\underline{p} = [\underline{p}_1^T, \ldots, \underline{p}_{n_B}^T]^T. \tag{17}$$

For a general multiloop system, the functions in Eq. (15) will not be known a priori in closed form. Instead, they are defined implicitly by a system of $6\,n_B$ nonlinear equations

$$y_i(\underline{p}; \underline{q}) = 0 \; ; \; i = 1, \ldots, 6\,n_B\,, \tag{18}$$

which consist of the constraints at the joints, as well as additional equations defining the generalized coordinates in terms of the position parameters of the bodies. This form represents a large "sparse" system of relatively simple equations. An alternative formulation of Eq. (18) is obtained after introducing the joint coordinates $\underline{\beta} = [\beta_1, \ldots, \beta_{n_\beta}]^T$ as auxiliary variables and then splitting Eq. (18) into two subsystems:

$$\underline{p} = \underline{p}(\underline{\beta}) \, , \tag{19}$$

$$g_i(\underline{\beta}; \underline{q}) = 0 \quad ; \; i = 1, \ldots, n_\beta \, . \tag{20}$$

The first subsystem represents the *forward kinematics*, i.e. the process by which one obtains the absolute motion of the bodies for known relative motion at the joints. This is a task which can always be stated recursively in closed form and will not be discussed here. The second subsystem, which defines the functions $\underline{\beta}(\underline{q})$ implicitly, represents the *relative kinematics* and consists of a reduced system of constraint equations, together with f equations describing the choice of the generalized coordinates (Fig. 6). Note that this choice can be formulated as simple one-one correspondences to particular joint coordinates, resulting in f of the equations out of Eq. (20) being mostly trivial. The remaining $r_\beta = n_\beta - f$ nonlinear equations represent the "core" of the reduced system of constraint equations.

In recent years, several methods to state the constraint equations of the "core" of the reduced system have been developed. Among others, one approach is based on the following concepts:

- *The characteristic pair of joints* to state the six constraint equations of a single multibody loop in a mostly recursive form ([21]). If a fully recursive formulation is possible, this solution can be found automatically, as shown in [23].
- The concepts of *kinematical transformer* and *block diagram*. Here, the individual kinematical loop is treated as a transmission element, and a complex multibody system consisting of many closed loops – usually interconnected by linear equations in the joint variables – is represented as a block diagram. This can also be regarded as an oriented graph which visualizes the kinematical flow in the system ([14]).
- The concept of the *kinematical differentials* for an efficient evaluation of the time derivatives required for the kinematics as well as for the dynamics. This has already been described in the previous section.

4. Kinematics of Single Multibody Loops

4.1. STATING LOOP CLOSURE CONDITIONS

Considering a single multibody loop L_i consisting of $n_B(L_i)$ bodies and $n_G(L_i) = n_B(L_i)$ elementary joints, one introduces the $n_\beta(L_i) = n_G(L_i)$ (relative) joint coordinates $\underline{\beta}(L_i) = [\beta_1^{L_i}, \ldots, \beta_{n_\beta}^{L_i}]^{\mathrm{T}}$ as auxiliary variables. For these coordinates there exist, in the general case, six independent constraint equations, which arise from the closure conditions of the loop. Special configurations, where the equations become linearly dependent, shall not be considered here. It is then always possible to define six joint coordinates as functions of the other $f(L_i) = n_G(L_i) - 6$ independent joint coordinates. This formulation is independent of the overall motion of the loop and is thus a "local" property of the loop. Correspondingly, the number $f(L_i)$ of independent coordinates is called the local degree of freedom of the loop, and the process of solving the constraint equations is called *relative kinematics*.

In principle, any approach can be adopted for the formulation and solution of the constraint equations, e.g. the well–known method of Denavit and Hartenberg [8], where the closure conditions are derived from the component representation of a closed chain of 4×4 transformation matrices. But the drawback of such approaches is that they are not very effective because the solutions must be found numerically.

There are some principal methods for stating geometric closure conditions in kinematical loops which are independent of a particular mathematical formulation.

4.1.1. Disconnection of the Multibody-Loop at a Body. The two halves of the disconnected body have to perform the same motion relative to an arbitrary reference frame. The main advantage of this approach is the universal mathematical formulation for multibody-loops with arbitrary joints and geometric dimensions. However, closed form solutions for the six unknown joint coordinates can be only obtained by performing algebraic eliminations from six carefully chosen independent closure equations (see Section **4.4**).

4.1.2. Disconnection of the multibody-loop at a joint. Disconnection of the loop at a joint with f_G dependent joint coordinates gives an implicit "core" system of $6 - f_G$ closure equations in which the $fjoint$ joint coordinates do not appear. Thus, $fjoint$ unknowns are immediately eliminated without particular algebraic manipulations being necessary.

4.1.3. The "Characteristic Pair of Joints". Here, the kinematical loop is disconnected at two joints $\mathcal{G}_a$ and $\mathcal{G}_b$ having $f_{\mathcal{G}_a}$ and $f_{\mathcal{G}_b}$ degrees of freedom, respectively (Fig. 7). One obtains two open chains which will be designated as the "upper segment" and the "lower segment" of the multibody-loop (see also [15, 20, 44]). Then, the six closure equations can be split up into two subsystems. An implicit "core" system

$$\underline{g}_{char}\left(\underline{\beta}_{char}, \underline{q}\right) = \underline{0} \tag{21}$$

with

$$h = 6 - (f_{\mathcal{G}_a} + f_{\mathcal{G}_b}) \tag{22}$$

equations gives the h dependent joint coordinates $\underline{\beta}_{char}$ not belonging to the characteristic pair of joints. Thus a maximum number of $h = 4$ equations occurs if the joints $\mathcal{G}_a$ and $\mathcal{G}_b$ are revolute or prismatic joints each having one degree of freedom, i.e. $f_{\mathcal{G}_a} = f_{\mathcal{G}_b} = 1$. These equations have to be numerically solved. However, in the most advantageous case the number of dependent joint coordinates in the characteristic pair of joints is five and the implicit core system consists of only one equation which can be explicitly solved.

There are $6 - h$ additional equations

$$\underline{g}_{comp}\left(\underline{\beta}_{char}, \underline{\beta}_{comp}, \underline{q}\right) = \underline{0} \tag{23}$$

to determine the $6 - h$ "complementary" joint coordinates $\underline{\beta}_{comp}$ belonging to the pair of joints $\mathcal{G}_a$ and $\mathcal{G}_b$. It can always be explicitly solved as with the evaluation of the implicit core system (21) the relative position of the bodies within both segments is known.

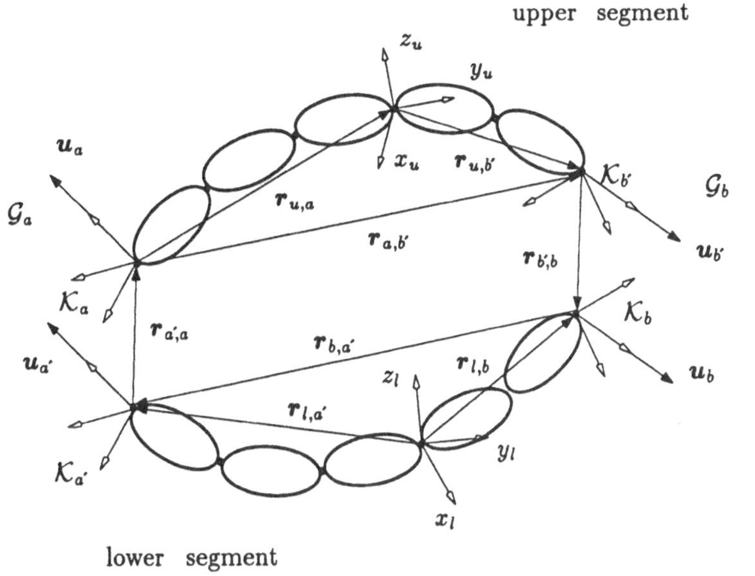

Figure 7: The characteristic pair of joints

4.2. THE METHOD OF THE "CHARACTERISTIC PAIR OF JOINTS"

4.2.1. The Characteristic Loop Closure Parameters. The loop closure conditions in Eq. (21) can be expressed by certain distances and angles – the "characteristic loop closure parameters" – measured between the reference frames $\mathcal{K}_a$ and $\mathcal{K}_{b'}$ on the upper segment and $\mathcal{K}_b$ and $\mathcal{K}_{a'}$ on the lower segment. These loop closure parameters can be confined to five fundamental types expressing simple geometric relations between points, axes or planes of the joints $\mathcal{G}_a$ and $\mathcal{G}_b$. In the following, they are visualized by corresponding characteristic pairs of joints.

(I) *Distance between two points* (Fig. 8a). A characteristic pair of a spherical joint (S) and a "reduced" spherical joint (universal joint, S_R) with altogether $f_{\mathcal{G}_a} + f_{\mathcal{G}_b} = 5$ degrees of freedom gives, according to Eq. (22), $h = 1$ closure parameter g_I . This is the square of the distance d of the centers $\mathcal{O}_a$ and $\mathcal{O}_b$ of the joints which can be expressed both in the upper segment (left index u) and in the lower segment (left index l):

$$ {}_u g_I = \boldsymbol{r}_{a,b'} \cdot \boldsymbol{r}_{a,b'} \;\; = {}^u\underline{r}^{\mathrm{T}}_{a,b'}\, {}^u\underline{r}_{a,b'} \quad , \quad \boldsymbol{r}_{a,b'} = \boldsymbol{r}_{u,b'} - \boldsymbol{r}_{u,a} \;\; , \tag{24} $$

$$ {}_l g_I = \boldsymbol{r}_{b,a'} \cdot \boldsymbol{r}_{b,a'} \;\; = {}^l\underline{r}^{\mathrm{T}}_{b,a'}\, {}^l\underline{r}_{b,a'} \quad , \quad \boldsymbol{r}_{b,a'} = \boldsymbol{r}_{l,a'} - \boldsymbol{r}_{l,b} \;\; . \tag{25} $$

Here, the dot product is written both in the coordinate-free vector notation and – for the numerical evaluation – in matrix notation using the components of the vectors in reference frames $\mathcal{K}_u$ and $\mathcal{K}_l$ fixed to the upper and lower segments (left upper indices u and l), respectively.

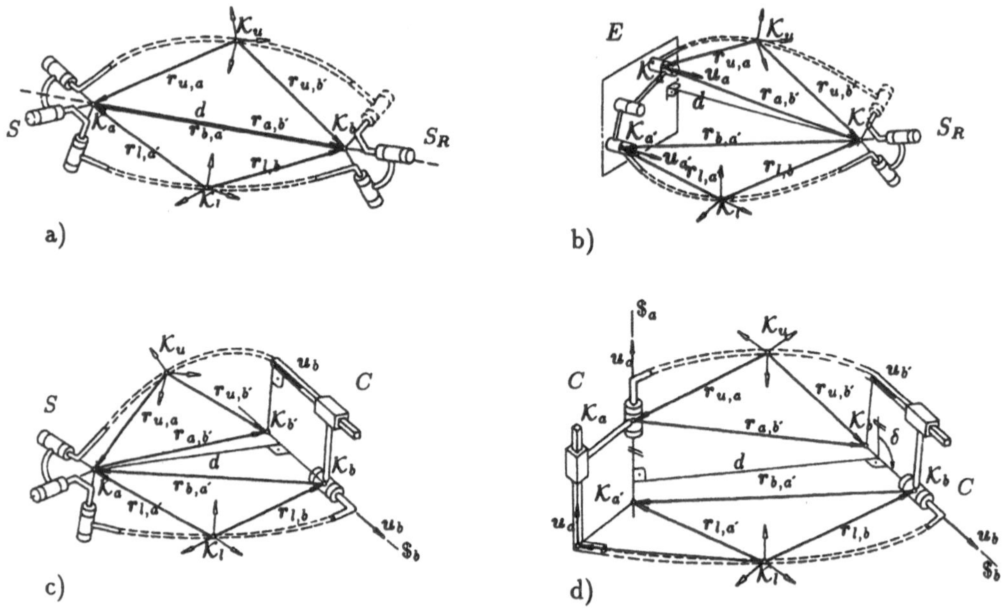

Figure 8: Implicit loop closure conditions

(II) *Distance of a Point from a Plane* (Fig. 8b). The pair of a planar joint (E) and a "reduced" spherical joint (S_R) has $f_{\mathcal{G}_a} + f_{\mathcal{G}_b} = 5$ degrees of freedom and requires $h = 1$ implicit closure parameter g_{II} . This is the distance d of the center $\mathcal{O}_b$ of the reduced spherical joint from the plane Π_a of the planar joint:

$$ {}_u g_{II} = \boldsymbol{r}_{a,b'} \cdot \boldsymbol{u}_a \quad = {}^u\underline{r}^{\mathrm{T}}_{a,b'}\; {}^u\underline{u}_a \quad , \tag{26} $$

$$ {}_l g_{II} = \boldsymbol{r}_{b,a'} \cdot \boldsymbol{u}_{a'} \quad = -{}^l\underline{r}^{\mathrm{T}}_{b,a'}\; {}^l\underline{u}_a \quad . \tag{27} $$

(III) *Distance of a Point from a Line* (Fig. 8c). For a pair of a spherical joint (S) and a cylindric joint (C), $f_{\mathcal{G}_a} + f_{\mathcal{G}_b} = 5$, the characteristic closure parameter g_{III} is the square of the distance d of the center $\mathcal{O}_a$ of the spherical joint from the joint axis $\$_b$ of the cylindric joint:

$$ {}_u g_{III} = \boldsymbol{n}_{a,b'} \cdot \boldsymbol{n}_{a,b'} \quad = {}^u\underline{n}^{\mathrm{T}}_{a,b'}\; {}^u\underline{n}_{a,b'} \quad , \quad \boldsymbol{n}_{a,b'} = \boldsymbol{r}_{a,b'} \times \boldsymbol{u}_{b'} \quad , \tag{28} $$

$$ {}_l g_{III} = \boldsymbol{n}_{b,a'} \cdot \boldsymbol{n}_{b,a'} \quad = {}^l\underline{n}^{\mathrm{T}}_{b,a'}\; {}^l\underline{n}_{b,a'} \quad , \quad \boldsymbol{n}_{b,a'} = \boldsymbol{r}_{b,a'} \times \boldsymbol{u}_{b} \quad . \tag{29} $$

(IV) and V) *Dual Angle between Two Axes* (Fig. 8d.) The pair of two cylindric joints (C), $f_{\mathcal{G}_a} + f_{\mathcal{G}_b} = 4$, requires $h = 2$ closure parameters g_{IV} and g_V. The first of them is the cosine of angle α between the two axis directions:

$$ {}_u g_{IV} = \boldsymbol{u}_a \cdot \boldsymbol{u}_{b'} \quad = {}^u\underline{u}^{\mathrm{T}}_{a}\; {}^u\underline{u}_{b'} \quad , \tag{30} $$

$$ {}_l g_{IV} = \boldsymbol{u}_b \cdot \boldsymbol{u}_{a'} \quad = -{}^l\underline{u}^{\mathrm{T}}_{b}\; {}^l\underline{u}_{a'} \quad . \tag{31} $$

The further closure parameter g_V is the expression $d \sin\alpha$ with the shortest distance d between the lines $\$_a$ and $\$_b$ along the common perpendicular:

$$ {}_u g_{V} = \boldsymbol{n}_{a,b'} \cdot \boldsymbol{r}_{a,b'} \quad = {}^u\underline{n}^{\mathrm{T}}_{a,b'}\; {}^u\underline{r}_{a,b'} \quad , \quad \boldsymbol{n}_{a,b'} = \boldsymbol{u}_{a} \times \boldsymbol{u}_{b'} \quad , \tag{32} $$

$$ {}_I g_V = \boldsymbol{n}_{b,a'} \cdot \boldsymbol{r}_{b,a'} \quad = {}^I\underline{n}^{\mathrm{T}}_{b,a'}\, {}^I\underline{r}_{b,a'} \quad , \quad \vec{n}_{b,a'} = \vec{u}_b \times \vec{u}_{a'} \ . \tag{33} $$

4.2.2. Systematics of the Loop Closure Equations. Generally, the loop closure equations can be stated in four steps which are summarized in the following.

Joint $\mathcal{G}_a$		Joint $\mathcal{G}_b$			loop closure parameter				
type	$f_{\mathcal{G}_a}$	type	$f_{\mathcal{G}_b}$	h	I	II	III	IV	V
S	3	S_R	2	1	1				
E	3	S_R	2	1		1			
S	3	E_R	2	1		1			
E	3	E_P	2	1				1	
S	3	C	2	1			1		
E	3	C	2	1				1	
C	2	C	2	2				1	1
S	3	R	1	2	1	1			
E	3	R	1	2		1		1	
C	2	R	1	3			1	1	1
R	1	R	1	4	1	2		1	
S	3	P	1	2		2			
E	3	P	1	2				2	
C	2	P	1	3				2	1
R	1	P	1	4		2		2	
P	1	P	1	4				3	1

P: Prismatic joint S: Spherical joint C: Cylindric joint E: Planar joint

S_R: Two revolute joints with concurrent axes

E_R: Two revolute joints with parallel axes or one revolute and one prismatic joint with orthogonal axes

E_P: One revolute and one prismatic joint with orthogonal axes or two prismatic joints with non–parallel axes

Table 1: Implicit loop closure conditions

Step 1 *Choice of the Characteristic Pair of Joints.* Two kinematic pairs having joint coordinates belonging to the dependent coordinates $\underline{\beta}$ of the multibody-loop are chosen as a characteristic pair of joints. The possible combinations are shown in Table 1. The two joints are chosen in such manner that the number h of implicit equations is as low as possible. If there are different numbers $f_{\mathcal{G}_a}$ and $f_{\mathcal{G}_b}$ of joints coordinates, the joint with the higher number of joint coordinates is designated by $\mathcal{G}_a$. With the maximum number of joint coordinates in the characteristic pair of joints being $h = 5$, the highest number $f_{\mathcal{G}_a}$ and $f_{\mathcal{G}_b}$ are three and two, respectively.

Step 2 *Implicit Loop Closure Equations.* The h implicit loop closure equations (21) have the general form:

$$ \underline{g}_{char}\left(\underline{\beta}_{char}, \underline{q}\right) = \begin{bmatrix} g_{1,char} \\ \vdots \\ g_{h,char} \end{bmatrix} = \begin{bmatrix} {}_I g_1 - {}_u g_1 \\ \vdots \quad \vdots \\ {}_I g_h - {}_u g_h \end{bmatrix} = \underline{0} \ . \tag{34} $$

Here, ${}_u g_i$ and ${}_o g_i$ are closure parameters chosen from the five elementary types of Eqs. (24) to (33). The number h and the type of the implicit equations depend on the type of the two joints $\mathcal{G}_a$ and $\mathcal{G}_b$ and are shown in Table 1. Having evaluated Eq. (34) the joint coordinates within the upper and the lower segment are known. Then, the direct transitions between the reference frames *Framea* and $\mathcal{K}_{b'}$ on the upper segment and $\mathcal{K}_b$ and $\mathcal{K}_{a'}$ on the lower segment can be expressed as:

$$ {}^a\boldsymbol{R}_{b'} = {}^a\boldsymbol{R}_u\, {}^u\boldsymbol{R}_{b'} \quad , \quad {}^{b'}\underline{r}_{a,b'} = {}^{b'}\boldsymbol{R}_u \left({}^u\underline{r}_{u,b'} - {}^u\underline{r}_{u,a} \right) \quad , \tag{35} $$

$$ {}^b\boldsymbol{R}_{a'} = {}^b\boldsymbol{R}_l\, {}^l\boldsymbol{R}_{a'} \quad , \quad {}^b\underline{r}_{b,a'} = {}^b\boldsymbol{R}_l \left({}^l\underline{r}_{l,a'} - {}^l\underline{r}_{l,b} \right) \quad . \tag{36} $$

To determine the remaining joint coordinates of joints $\mathcal{G}_a$ and $\mathcal{G}_b$ the relative position of the two segments has to be considered which has been not yet used up to this step.

Step 3 *Joint Coordinates of Joint* $\mathcal{G}_b$. The two segments are built together in such manner that the corresponding geometric elements of the two joints $\mathcal{G}_a$ and $\mathcal{G}_b$ on the upper and the lower segment do coincide. With respect to the different joint types one obtains constraint equations for at most two joint coordinates of the joint $\mathcal{G}_b$ ([44]). For these, the already computed transitions (35) and (36) and the constant dimensions of the bodies in joint $\mathcal{G}_b$ are used. One obtains the matrices ${}^{b'}\boldsymbol{R}_b$ and ${}^b\underline{r}_{b',b}$ describing the transition from $\mathcal{K}_{b'}$ to $\mathcal{K}_b$. Together with Eqs. (35) and (36) the relative position of the reference frames $\mathcal{K}_{a'}$ and $\mathcal{K}_a$ of joint $\mathcal{K}_a$ can be expressed:

$$ {}^a\boldsymbol{R}_{a'} = {}^a\boldsymbol{R}_{b'}\, {}^{b'}\boldsymbol{R}_b\, {}^b\boldsymbol{R}_{a'} \quad , \tag{37} $$

$$ {}^a\underline{r}_{a,a'} = {}^a\boldsymbol{R}_{b'} \left({}^{b'}\underline{r}_{a,b'} + {}^{b'}\underline{r}_{b',b} + {}^{b'}\boldsymbol{R}_b\, {}^b\underline{r}_{b,a'} \right) \quad . \tag{38} $$

Step 4 *Joint Coordinates of Joint* $\mathcal{G}_a$. With known relative position of reference frames $\mathcal{K}_{a'}$ and $\mathcal{K}_a$ the joint coordinates of $\mathcal{G}_a$ can be determined using constraint equations which depend only on the type of joint $\mathcal{G}_a$ ([44]).

The cases which lead to closed–form solutions represent the class of *recursively solvable* multibody loops. Such cases are encountered very often in technical applications of complex multibody systems and are thus of great interest. For a more detailed description of this method the reader is refered to the more elaborate expositions of [12], [21] and [44] (see Section **4.4**).

4.3. METHOD OF THE "MINIMAL POLYNOMIAL EQUATION"

This approach can be regarded as an extension of the above mentioned method of the "characteristic pair of joints". The idea is to state the constraint equations for all possible joint configurations in a recursive form, whereby for the first unknown joint coordinate a polynomial of minimal order has to be derived. In the case of only rotational coordinates, the polynomial is stated in $\tan\frac{\beta_1}{2}$ and the order of the polynomial is dependent on the geometry as well as on the type and combination of joint coordinates. If the kinematic loop consists of six arbitrarily arranged revolute joints, the minimal order of the polynomial is 16 (this represents the "worst case"). The order of the polynomial equation depends on the type of the unknown joints as well as on their arrangement, and may vary between 16 and 2. All possible combinations together with the order of the polynomial equation (which is equal to the number of possible configurations) are shown in Table 2. In comparison to the method of the "characteristic pair of joints", some additional geometrical conditions have to be stated. This method was first developed by Lee and Liang [27] and has been elaborated in [28]. Further improvements are given in [34].

Order of polynomial	Mechanisms	
2	R-3C	RCCC
	2R-2S	RSSR
	2R-P-2C	RCPCR, RRCPC RCPRC, RCRPC, RPCRC RRPCC, RPCCR RPRCC
	3R-2P-C	RPCPRR RPRPCR, RRPRPC RPRPRC RCPPRR, RRRPPC, RPPCRR RRPPRC, RCRPPR, RPPRRC RPRCPR, RPRRPC
	4R-3P	RRPPPRR, RPPPRRR RPRPPRR, RRPRPPR RPPRRPR RPRPRPR
4	4R-S	RRSRR, RSRRR
	4R-E	RRERR, RERRR
	3R-2C	RCRCR, RRCRC
8	3R-2C	RCRRC RCCRR,RRRCC
	4R-P-C	RRRPCR, RRCPRR RRRRPC, RPCRRR RCRPRR, RRCRPR, RRRPRC RCRRPR, RRPRRC
	5R-2P	RRPRPRR, RRRPRPR RPRRPRR, RPRRRPR RRRRPPR, RRRPPRR
16	5R-C	RRRRCR, RRRCRRR, RRRRRC
	6R-P	RRRPRRR, RRRRPRR, RRRRRPR
	7R	RRRRRRR

P: Prismatic joint S: Spherical joint C: Cylindric joint E: Planar joint

Table 2: Types of single-loop mechanisms.

The principal idea behind the method of minimal polynomial equations is to state an appropriate set of closure equations from which the unknown joint angles can be algebraically eliminated in such a way that the degree of the final polynomial equation becomes not higher than 16. The method developed by Lee and Liang uses a two–step elimination of four unknown angles from a set of 14 loop closure equations containing five unknown joint angles. The complete elimination process described in ([29]) gives the following calculation scheme for the arbitrarily arranged six unknown joint angles ($\underline{p}$ is a vector containing constant parameters):

$$\sum_{i=1}^{16} d_i(\underline{p}) \tan^i \frac{\beta_1}{2} = 0 \tag{39}$$

$$\beta_2 = \beta_2\left(\underline{p}, \beta_1\right) \tag{40}$$

$$\beta_6 = \beta_6\left(\underline{p}, \beta_1\right) \tag{41}$$

$$\beta_4 = \beta_4\left(\underline{p}, \beta_1, \beta_2, \beta_6\right) \tag{42}$$

$$\beta_3 = \beta_3\left(\underline{p}, \beta_1, \beta_2, \beta_6, \beta_4\right) \tag{43}$$

$$\beta_5 = \beta_5\left(\underline{p}, \beta_1, \beta_2, \beta_6, \beta_4, \beta_3\right) \tag{44}$$

In the general case, Eq. (39) has to be solved iteratively, while Eqs. (40) to (44) can be resolved in closed form, either uniquely or yielding two solutions, respectively.

4.4. AUTOMATIC GENERATION OF CLOSED-FORM SOLUTIONS

As shown in Table 2, closed-form solutions are always possible if the order of the minimal polynomial equations equals two. By a suitable combination of the geometrically intuitive approach of the characteristic pair of joints with algebraic techniques known from robotics ([33]), it is possible to derive an algorithm for the automatic determination of closed-form solutions of the inverse kinematics problem for loops in which such solutions exist. Such an approach was recently developed by Kecskeméthy, and published in [23]. In the sequel, the main ideas of this approach are described.

One considers a single closed multibody loop modelled as a sequence of homogeneous transformations A_i, $i = 1, \ldots, n$ (Fig. 9).

Recall that a homogeneous transformation A_i models the motion from a reference frame $\mathcal{K}_{i-1}$ to a reference frame $\mathcal{K}_i$, expressed as the transformation of point coordinates defined with respect to $\mathcal{K}_i$ to corresponding coordinates with respect to $\mathcal{K}_{i-1}$. Such a transformation matrix has the structure

$$\mathrm{A}_i = \begin{bmatrix} {}^{i-1}\boldsymbol{R}_i & {}^{i-1}_{i-1}\underline{r}_i \\ 0 & 1 \end{bmatrix} = \begin{bmatrix} \rho_{11} & \rho_{12} & \rho_{13} & r_1 \\ \rho_{21} & \rho_{22} & \rho_{23} & r_2 \\ \rho_{31} & \rho_{32} & \rho_{33} & r_3 \\ 0 & 0 & 0 & 1 \end{bmatrix}, \tag{45}$$

where ${}^{i-1}\boldsymbol{R}_i$ is the orthogonal 3×3 matrix of rotation transforming vector components from $\mathcal{K}_i$ to $\mathcal{K}_{i-1}$ and ${}^{i-1}_{i-1}\underline{r}_i$ is the radius vector connecting the origin $\mathcal{O}_{i-1}$ of $\mathcal{K}_{i-1}$ to the origin $\mathcal{O}_i$ of $\mathcal{K}_i$ in the decomposition with respect to $\mathcal{K}_{i-1}$ (the indices i and $i-1$ have been left out in coefficient-wise notation for better clarity).

The closure of the loop is achieved by stating $\mathcal{K}_n \equiv \mathcal{K}_0$, or, equivalently

$$\mathrm{A}_1 \mathrm{A}_2 \cdots \mathrm{A}_n = \boldsymbol{I}_4 . \tag{46}$$

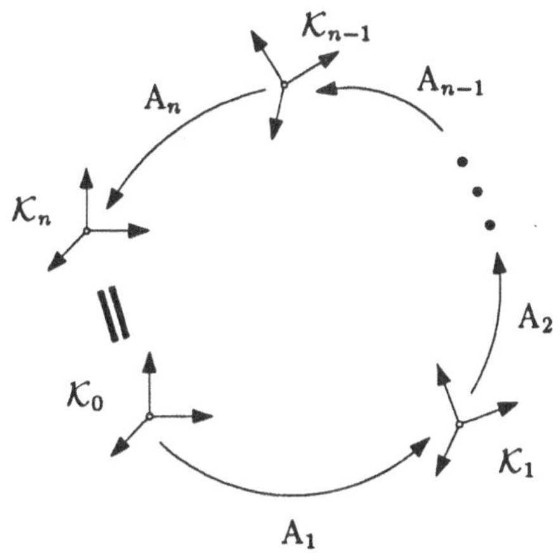

Figure 9: Basic structure of a loop

Eq. (46) contains twelfe non-trivial scalar equations to be fulfilled for the loop to stay closed. Out of these, six are dependent because of the orthogonality condition of the rotation matrix. However, just striking out six equations is not feasible, because then (a) not all uniqueness conditions of the solutions can be fulfilled, and (b) closed-form solutions will not become evident.

Actually, in order to find closed-form solutions, even more equations have to be taken into account by considering also alternative forms of the closure condition Eq. (46), such as

$$A_{i_{j+1}} \cdots A_{i_k} = A_{i_j}^{-1} \cdots A_{i_1}^{-1} A_{i_n}^{-1} \cdots A_{i_{k+1}}^{-1} \tag{47}$$

where $1 < j < k < n$ and $i_1, \ldots, i_n$ is a cyclic permutation of $1, \ldots, n$. These equations state that any two possible branches within the loop starting at an arbitrary reference frame $\mathcal{K}_{i_j}$ and terminating at another arbitrary reference frame $\mathcal{K}_{i_{k+1}}$ must yield the same transformation.

The key issue of an algorithm for finding closed-form solution is thus to pick out from Equations (46) and (47) a set of six scalar equations

$$\left.\begin{array}{rcl} f_1(\beta_1) & = & 0 \\ f_2(\beta_2\,;\,\beta_1) & = & 0 \\ & \vdots & \\ f_6(\beta_6\,;\,\beta_5,\ldots,\beta_1) & = & 0 \end{array}\right\}, \tag{48}$$

plus some additional equations needed for unique solutions, where possible, with functions f_i containing exactly one unknown more than the preceding equations and being mostly of order two in the corresponding unknown variable β_i (or $\tan\frac{\beta_i}{2}$ in the case of a revolute joint).

This problem can be solved by resorting to the geometrical properties of the transformation matrices A_i, in particular, to their invariance properties. We introduce as objects of interest the three coordinate planes and the origin of the reference systems, respectively. These geometrical objects have the following representations as homogeneous vectors

$$\overset{\text{H}}{\underline{e}}_1 = \begin{bmatrix} 1 \\ 0 \\ 0 \\ 0 \end{bmatrix}, \quad \overset{\text{H}}{\underline{e}}_2 = \begin{bmatrix} 0 \\ 1 \\ 0 \\ 0 \end{bmatrix}, \quad \overset{\text{H}}{\underline{e}}_3 = \begin{bmatrix} 0 \\ 0 \\ 1 \\ 0 \end{bmatrix}, \quad \overset{\text{H}}{\underline{o}} = \begin{bmatrix} 0 \\ 0 \\ 0 \\ 1 \end{bmatrix}, \tag{49}$$

where the symbol 'H' is stacked above homogeneous vectors for better clarity (but will be dropped later when there is no risk of confusion between homogeneous and euclidian vectors) and the vectors $\overset{\text{H}}{\underline{e}}_i$ are particular instances of unit vectors $\overset{\text{H}}{\underline{u}} = [\underline{u}^T, 0]^T$ representing the points at infinity of the projective space P_n ([5]). Note that the following properties hold

$$\mathrm{A}^{-1}\overset{\text{H}}{\underline{u}} = \mathrm{A}^T\overset{\text{H}}{\underline{u}}, \tag{50}$$

$$\overset{\text{H}}{\underline{u}}{}^T\mathrm{A} = \left(\mathrm{A}^T\overset{\text{H}}{\underline{u}}\right)^T = \left(\mathrm{A}^{-1}\overset{\text{H}}{\underline{u}}\right)^T, \tag{51}$$

$$\|\underline{\xi}\|_{\text{H}} = \sqrt{\|\underline{\xi}\|^2 - 1}, \tag{52}$$

where $\|\cdot\|_{\text{H}}$ denotes the "homogeneous norm"

$$\|\mathrm{A}\overset{\text{H}}{\underline{o}}\|_{\text{H}} = \|\mathrm{A}^{-1}\overset{\text{H}}{\underline{o}}\|_{\text{H}}. \tag{53}$$

The transformations A_i can be divided into *elementary, general* and *trivial transformations*. The elementary transformations

$$\mathtt{Rot}[\underline{e}_1, \Theta] = \begin{pmatrix} 1 & 0 & 0 & 0 \\ 0 & \cos\Theta & -\sin\Theta & 0 \\ 0 & \sin\Theta & \cos\Theta & 0 \\ 0 & 0 & 0 & 1 \end{pmatrix}, \quad \mathtt{Trans}[\underline{e}_1, s] = \begin{pmatrix} 1 & 0 & 0 & s \\ 0 & 1 & 0 & 0 \\ 0 & 0 & 1 & 0 \\ 0 & 0 & 0 & 1 \end{pmatrix}$$

$$\mathtt{Rot}[\underline{e}_2, \Theta] = \begin{pmatrix} \cos\Theta & 0 & \sin\Theta & 0 \\ 0 & 1 & 0 & 0 \\ -\sin\Theta & 0 & \cos\Theta & 0 \\ 0 & 0 & 0 & 1 \end{pmatrix}, \quad \mathtt{Trans}[\underline{e}_2, s] = \begin{pmatrix} 1 & 0 & 0 & 0 \\ 0 & 1 & 0 & s \\ 0 & 0 & 1 & 0 \\ 0 & 0 & 0 & 1 \end{pmatrix}$$

$$\mathtt{Rot}[\underline{e}_3, \Theta] = \begin{pmatrix} \cos\Theta & -\sin\Theta & 0 & 0 \\ \sin\Theta & \cos\Theta & 0 & 0 \\ 0 & 0 & 1 & 0 \\ 0 & 0 & 0 & 1 \end{pmatrix}, \quad \mathtt{Trans}[\underline{e}_3, s] = \begin{pmatrix} 1 & 0 & 0 & 0 \\ 0 & 1 & 0 & 0 \\ 0 & 0 & 1 & s \\ 0 & 0 & 0 & 1 \end{pmatrix}$$

are denoted as $\mathrm{A}_E(\underline{e}_i\,;\,\beta)$, translational and rotational transformations being distinguished by a boolean variable σ which is 1 in the first case and 0 in the second, and for which $\overline{\sigma} = 1 - \sigma$. Recall that these transformations form a basis for the group of rigid rotations, so that any rigid-body motion can be decomposed into a sequence of six such transformations. For the general transformations, the sub-types "general translational" (A_T), "general rotational" (A_R) and "general spatial" (A_G) motion

$$A_T = \begin{bmatrix} I_3 & \underline{r} \\ 0 & 1 \end{bmatrix} , A_R = \begin{bmatrix} R & \underline{0} \\ 0 & 1 \end{bmatrix} , A_G = \begin{bmatrix} R & \underline{r} \\ 0 & 1 \end{bmatrix} \tag{54}$$

are taken into account. Finally, there exist 24 "trivial transformations" which just interchange the coordinate axes (and thus involve no numerical computations). These can be collected with the notation

$$A_P = \mathtt{Perm}[i_1, i_2, i_3] \quad , \tag{55}$$

where the coefficients of the rotation part of A_P fulfill the relationships

$$\rho_{jk}(\mathtt{Perm}[i_1, i_2, i_3]) = \begin{cases} +1 & \text{for} \quad i_j = k \\ -1 & \text{for} \quad i_j = -k \\ 0 & \text{otherwise} \end{cases} \quad . \tag{56}$$

We now assume that the sequence of transformations within the loop is such that the unknown variables only appear in *elementary transformations*, and that all trivial transformations have been eliminated. This can be always achieved by reducing composite transformations, such as the four-parametric DENAVIT-HARTENBERG-form, into elementary transformations, and shifting trivial transformations to the left or to the right ([23]). The key idea of the algorithm is to regard the characteristic measurements within the loop as particular projection operations, and to search the chain for sequences of transformations which leave any of the geometric elements mentioned above invariant. This is explained in the sequel.

4.4.1. Projection Operators. Projection operators represent the basic means of obtaining scalar equations from the general closure condition Eq. (47). As general criteria, projection operators should not be any linear combination of the invariants of a 4×4 matrix, and should yield as simple expressions as possible in order to allow the closed-form resolution for any unknowns contained in the projected matrix. In order to achieve this, projection operators are lead back to the basic geometrical measurements used in kinematics. Currently, five such basic measurements are being used[1]: (I) the quadratic distance between two points g_{PP}, (II) the distance of a point to a plane g_{EP}, (III) the cosine of the angle between two planes (or orientations) g_{EE}, (IV) the distance (along the common perpendicular) between two lines g_{LL} and (V) the quadratic distance of a point to a line g_{LP}, see e. g. [44] or [14]. In order to carry out these measurements using homogeneous transformations, two reference frames $\mathcal{K}$ (fixed) and $\mathcal{K}'$ (moved) are introduced, such that the points, lines or planes mentioned above correspond with either an origin $\underline{o}$, a coordinate axis $\mathcal{L}_i$ or a coordinate plane Π_i, respectively. Denote by A the homogeneous matrix relating coordinates with respect to $\mathcal{K}'$ to coordinates with respect to $\mathcal{K}$. Then, the

[1]A sixth measurement type was introduced in [27], which can be interpreted as the projection of the orientation vector of a line to the reflection of the orientation vector of a second line about the plane perpendicular to a vector connecting one reference point on each line, see [3]. The incorporation of this measurement type in the general theory is subject of future research.

measurements mentioned above define the following projection operations

$$g_{PP}(A) = \|\mathrm{A}\,\overset{\mathrm{H}}{\underline{o}}\,\|^2_{\mathrm{H}} \tag{57}$$

$$g_{EP}(A\,;\,\underline{e}_i) = \overset{\mathrm{H}^{\mathrm{T}}}{\underline{e}_i}\,\mathrm{A}\,\overset{\mathrm{H}}{\underline{o}}\ \ (= r_i(A)\ ) \tag{58}$$

$$g_{EE}(A\,;\,\underline{e}_i,\underline{e}'_j) = \overset{\mathrm{H}^{\mathrm{T}}}{\underline{e}_i}\,\mathrm{A}\,\overset{\mathrm{H}'}{\underline{e}_j}\ \ (= \rho_{ij}(A)\ ) \tag{59}$$

$$g_{LL}(A\,;\,\underline{e}_i,\underline{e}'_j) = \underline{e}_i^{\mathrm{T}}\left[\mathtt{Rot[\,A\,]}\,\underline{e}'_j \times \mathtt{Trans[\,A\,]}\right]\ \ (= \epsilon_{ikl}\cdot\rho_{kj}(A)\cdot r_l(A)\ ) \tag{60}$$

$$g_{LP}(A\,;\,\underline{e}_i) = \|\mathrm{A}\,\overset{\mathrm{H}}{\underline{o}}\,\|^2_{\mathrm{H}} - (\overset{\mathrm{H}^{\mathrm{T}}}{\underline{e}_i}\,\mathrm{A}\,\overset{\mathrm{H}}{\underline{o}})^2 \tag{61}$$

$\mathtt{Rot[\,A\,]}$ and $\mathtt{Trans[\,A\,]}$ are operators which extract the rotational and translational part of a homogeneous transformation matrix A , respectively.

Note that for the projections g_{EP} or g_{LP} the corresponding plane or line is chosen from the *fixed* frame. This is consistent with the property that premultiplication of homogeneous matrices is only meaningful for orientation vectors. Resolution of the projected matrix with respect to an unknown joint-coordinate β contained in A is obtained by decomposing

$$\mathrm{A} = \mathrm{A}_\ell\,\mathrm{A}_E(\underline{e}_\nu\,;\,\beta)\,\mathrm{A}_r = \begin{bmatrix} \boldsymbol{R}_\ell & \underline{r}_\ell \\ 0 & 1 \end{bmatrix}\begin{bmatrix} \boldsymbol{R}_E(\underline{e}_\nu\,;\,\beta) & \underline{r}_E(\underline{e}_\nu\,;\,\beta) \\ 0 & 1 \end{bmatrix}\begin{bmatrix} \boldsymbol{R}_r & \underline{r}_r \\ 0 & 1 \end{bmatrix}, \tag{62}$$

where $\mathrm{A}_E(\underline{e}_\nu\,;\,\beta)$ is an elementary transformation with

$$\boldsymbol{R}_E(\underline{e}_\nu\,;\,\beta) = \boldsymbol{I}_3 + \overline{\sigma}\left(\sin\beta\,\tilde{\underline{e}}_\nu + (1-\cos\beta)\,\tilde{\underline{e}}_\nu^2\right) = \boldsymbol{I}_3 + \overline{\sigma}\,T(\underline{e}_\nu\,;\,\beta) \tag{63}$$

$$\underline{r}_E(\underline{e}_\nu\,;\,\beta) = \sigma\,\underline{e}_\nu\,\beta \tag{64}$$

and $\tilde{\underline{v}}$ denotes the anti-symmetrical matrix

$$\tilde{\underline{v}} = \begin{pmatrix} 0 & -v_3 & v_2 \\ v_3 & 0 & -v_1 \\ -v_2 & v_1 & 0 \end{pmatrix}.$$

Note that all projection types can be described by a *general projection operator* $\pi\,(\underline{\xi}_L,\underline{\xi}_R\,;\,\mathrm{A}\,)$, where $\underline{\xi}_L,\underline{\xi}_R \in \{\underline{o},\mathcal{L}_i,\Pi_i\}$ denote a "left" and a "right" geometrical element, between which the measurement is carried out. Table 3 shows the actual projections which are performed in each case.

4.4.2. Use of Isotropy Groups for Elimination of Unknowns. If one can find a sequence of transformations which leaves a geometric element invariant, it is clear that any projection involving that geometric elements will be independent of the variables contained in that sequence. The characteristic properties of such sequences of transformations can be quickly recollected as those of particular subgroups of the group of general rigid motion, namely the *isotropy groups* of the geometric elements in question.

Let G be a group acting on a smooth manifold M. For each $x \in M$, the *isotropy group* is defined to be $G^x = \{\ g \in G : g\cdot x = x\ \}$ (cf. [31]). The isotropy groups of interest in the formulation of constraints are the subgroups of rigid motion which leave the origin, the coordinate planes and the coordinate axes invariant, respectively, i.e. the sets

$$\hat{\mathrm{A}}^{\mathrm{iso}(\xi)} = \left\{\ \mathrm{A} : \mathrm{A}\,\underline{\xi} = \underline{\xi}\ \right\}\ ,\ \underline{\xi} \in \{\,\underline{o},\,\Pi_i\,,\,\mathcal{L}_i\,\}\,. \tag{65}$$

		$\underline{\xi}_R$		
		$\underline{o}$	Π_j	$\mathcal{L}_j$
$\underline{\xi}_L$	$\underline{o}$	$g_{PP}(\mathrm{A})$	$g_{EP}(\mathrm{A}^{-1}\,;\,\underline{u}_j)$	$g_{LP}(\mathrm{A}^{-1};\,\underline{u}_j)$
	Π_i	$g_{EP}(\mathrm{A}\,;\,\underline{u}_i)$	$g_{EE}(\mathrm{A}\,;\,\underline{u}_i\,,\,\underline{u}_j)$	none
	$\mathcal{L}_i$	$g_{LP}(\mathrm{A}\,;\,\underline{u}_i)$	none	$g_{LL}(\mathrm{A}\,;\,\underline{u}_i\,,\,\underline{u}_j)$

Table 3: Definition of the general projection operator $\pi\,(\underline{\xi}_L, \underline{\xi}_R\,;\,\mathrm{A})$

Denote by $\hat{\mathrm{A}}^{(\xi)} \in \hat{\mathrm{A}}^{\mathrm{iso}(\xi)}$ a particular element of an isotropy group, where the superscript (ξ) will be dropped when not needed. The following three (group) properties are an immediate consequence of the definition of isotropy groups

(IG1) $\boldsymbol{I}_4 \in \hat{\mathrm{A}}^{\mathrm{iso}(\xi)}$

(IG2) $\hat{\mathrm{A}}^{(\xi)} \in \hat{\mathrm{A}}^{\mathrm{iso}(\xi)} \;\Leftrightarrow\; \hat{\mathrm{A}}^{(\xi)^{-1}} \in \hat{\mathrm{A}}^{\mathrm{iso}(\xi)}$

(IG3) $\hat{\mathrm{A}}_1^{(\xi)} \in \hat{\mathrm{A}}^{\mathrm{iso}(\xi)} \;\wedge\; \hat{\mathrm{A}}_2^{(\xi)} \in \hat{\mathrm{A}}^{\mathrm{iso}(\xi)} \;\Rightarrow\; \hat{\mathrm{A}}_1^{(\xi)}\,\hat{\mathrm{A}}_2^{(\xi)} \in \hat{\mathrm{A}}^{\mathrm{iso}(\xi)}$.

Particularly because of property (IG3), the detection of subsequences of transformations which leave a geometric element invariant is very simple, namely just a gathering of adjacent transformations with this property. Suppose that one has found two non-overlapping subsequences $\hat{\mathrm{A}}_A$ and $\hat{\mathrm{A}}_B$ with respective invariant geometric elements $\underline{\xi}_A$ and $\underline{\xi}_B$, and that both of these contain as much elements and as much unknowns as possible among all possible sequences of transformations, whereby $\hat{\mathrm{A}}_B$ contains a smaller or equal number of unknowns than $\hat{\mathrm{A}}_A$. After appropriate cyclic permutation, the closure condition can be transformed into the unique form

$$\mathrm{A}_I\,\hat{\mathrm{A}}_B\,\mathrm{A}_{II}\,\hat{\mathrm{A}}_A = \boldsymbol{I}_4 \quad , \qquad \underline{\text{(Closure Type 0)}} \tag{66}$$

which can be immediately transformed into

$$\hat{\mathrm{A}}_B\,\mathrm{A}_{II}\,\hat{\mathrm{A}}_A = \mathrm{A}_I^{-1} \quad . \tag{67}$$

Then, because $\hat{\mathrm{A}}_A$ leaves $\underline{\xi}_A$ invariant, and $\hat{\mathrm{A}}_B$ leaves $\underline{\xi}_B$ invariant, the projection operator $\pi\,(\underline{\xi}_B\,,\,\underline{\xi}_A\,;\,\hat{\mathrm{A}}_B\,\mathrm{A}_{II}\,\hat{\mathrm{A}}_A)$ will be invariant of both $\hat{\mathrm{A}}_A$ *and* $\hat{\mathrm{A}}_B$. Thus, applying this projection on Eq. (67) yields the scalar equation

$$\pi\,(\underline{\xi}_B\,,\,\underline{\xi}_A\,;\,\mathrm{A}_{II}) = \pi\,(\underline{\xi}_B\,,\,\underline{\xi}_A\,;\,\mathrm{A}_I^{-1}) \tag{68}$$

which is independent of all variables contained in $\hat{\mathrm{A}}_A$ as well as in $\hat{\mathrm{A}}_B$. Clearly, if all but one of the current unknowns are contained in $\hat{\mathrm{A}}_A$ or $\hat{\mathrm{A}}_B$, then either A_I or A_{II} contains exactly one unknown variable, and after decomposing according to Eq. (62) and applying the projection operator selected by $\underline{\xi}_A$ and $\underline{\xi}_B$, a resolvable scalar equation results. Eq. (67) corresponds to the division of the loop in four sub-chains (Figure 10), of which $\hat{A}_B$ and $\hat{A}_A$ have no influence on the projected equation. This division can also be used for the efficient formulation of the Jacobian of the loop. It is conjectured that this division is optimal in this sense.

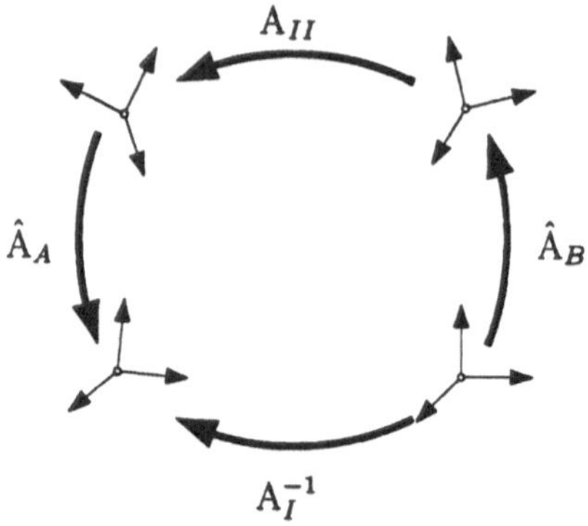

Figure 10: Qualitative structure of a loop for the "optimal" division

In principle, after finding a first closure condition with the properties discussed above, the remaining unknowns can be resolved in a straight-forward manner using existing algorithms. However, by a slight modification of expectable closure conditions, the same basic steps can be applied for all unknowns in the loop and even in the case of some overconstrained mechanisms, which are ***not paradoxical*** in the sense defined in [4]. Then, at any stage of the analysis, besides the closure condition denominated Type 0 above, only one of the following situations can arise:

$$\hat{A}_A = I_4 \qquad \text{(Closure Type 1)} \qquad (69)$$

$$\hat{A}_B\,\hat{A}_A = I_4 \qquad \text{(Closure Type 2)} \qquad (70)$$

$$A_I\,\overline{A}_B\,A_{II}\,\hat{A}_A = I_4 \qquad \text{(Closure Type 3)} \qquad (71)$$

$$A_I\,\overline{A}_A = I_4 \qquad \text{(Closure Type 4)} \qquad (72)$$

Clearly, in the first of these cases the whole sequence of transformations in the loop is element of the isotropy group of some geometric element $\underline{\xi}_A$. Thus, any projection carried out with this element yields a scalar equation which is identically fulfilled. As there are three independent projections which can be carried out with one geometric element, a closure condition of Type 1 reduces the number of independent constraint equations of the loop by three. For example, in the plane four-bar mechanism all transformations share as invariant element the plane perpendicular to the rotation axes, so the number of independent constraint equations is here only three. Similarly, in the closure condition of Type 2 the chain is decomposed into two sequences which contain together all unknowns and are elements of the isotropy groups of geometric elements $\underline{\xi}_A$ and $\underline{\xi}_B$, respectively. Thus the projection operator selected by these two elements yields again a scalar equation which is independent of any unknown variables. Such a situation arises for example in the case of a Cardan shaft, where the six rotational joints can be grouped into two sets of respectively three intersecting axes with the intersection points as corresponding invariant geometric elements. Note that for these two types of closure conditions, the sequences $\hat{A}_A$ or $\hat{A}_B$ may be bordered with additional transformations containing no unknowns without changing the basic results. Such bordering transformations have been intentionally left out for better clarity.

The closure conditions of Type 3 and Type 4 do not occur in the initial analysis of the loop, but in the process of eliminating unknowns contained in the sequences $\hat{A}_B$ and $\hat{A}_A$. The present algorithmic approach for obtaining scalar equations for these unknowns is as follows: after having produced the projection pertaining to $\underline{\xi}_B$ and $\underline{\xi}_A$, the invariance property associated with $\underline{\xi}_B$ is removed from the elements of $\hat{A}_B$ together with the current resolved unknown. Then, a new closure condition is searched by applying the same criteria as above. Eventually, no more invariant properties remain besides those in $\hat{A}_A$, but there is still an unknown in the remaining transformations. This is the situation in the closure condition of Type 3, where $\overline{A}_B$ contains the remaining unknown in a form similar to Eq. (62). Then, a projection as defined in Eq. (68) is carried out, but this time $\underline{\xi}_B$ is taken as a geometric element which actually *is* transformed by $\overline{A}_B$, thus yielding a scalar equation which contains this unknown. In the case that $\overline{A}_B$ is a rotation, two elements corresponding to both coordinate planes parallel to the rotational axis, and thus a uniquely solvable pair of equations is obtained. After performing this step, the invariance properties of $\underline{\xi}_A$ are artificially removed from the elements in $\hat{A}_A$, and the whole process described above is repeated, until eventually one last unknown is pending, but no invariant element remains whatever. This is the situation of closure condition Type 4, where $\overline{A}_A$ holds the remaining unknown. Then, again a similar projection as Eq. (68) is carried out, but this time $\underline{\xi}_B$ and $\underline{\xi}_A$ are taken as geometrical elements which are actually transformed by $\overline{A}_A$, yielding as in the case of the closure condition of Type 3 a unique solution.

A description of an implementation of this method has been given in [23]. Further, an application of the generation of closed-form solutions to the automatic programming of high-speed processors, like the "CORDIC" (COordinate DIgital Computer) has been worked out in [35].

5. Kinematics of Multiloop Systems

5.1. THE CONCEPT OF THE "KINEMATICAL TRANSFORMER"

For the following it will suffice to note that the relative kinematics of a multibody loop can be reduced to a system of equations which yield six dependent joint coordinates as functions of $f\ (L_i)$ independent joint coordinates as shown in the previous section. After the appropriate formulation of the constraint equations these functions can be regarded as producing a nonlinear transmission behaviour between independent "input" variables and dependent "output" variables. This is represented in Fig. 11 by a "black box" termed *kinematical transformer*, where for better clarity the loop index L_i is dropped and independent joint coordinates are denoted $\underline{q}$. With these transmission elements, the constraint equations of general multibody systems can be systematically partitioned into smaller subsystems, allowing one to find, where applicable, closed form solutions for the constraints of systems with multiple multibody loops by resorting to the methods developed so far for single loop systems.

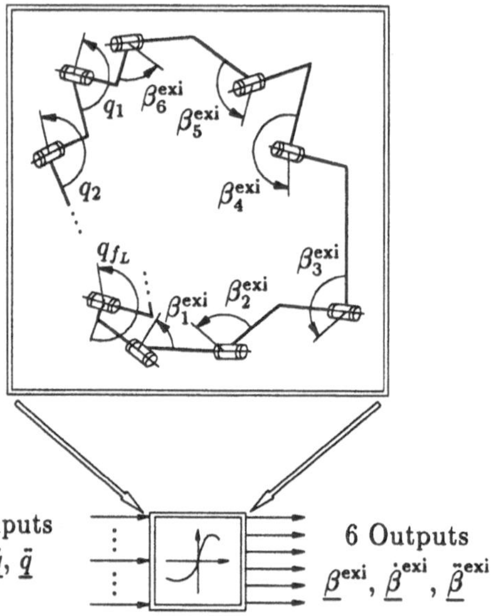

Figure 11: The general *kinematical transformer*

In a general multiple-loop multibody system, it is possible to define a independent set of multibody loops, which correspond to a cycle basis of the associated graph of interconnection. To take advantage of this property, one regards the multibody system now as a multiloop system, where, after introducing appropriate joint coordinates and formulating the constraint equations of each multibody loop *individually*, a set of "*kinematical transformers*" is obtained. Clearly, the complete set of joint coordinates introduced to describe each individual multibody loop as a kinematical transformer leads to a redundant set of joint coordinates. Thus additional conditions have to be formulated to make the extended set of variables $\underline{\beta}$ consistent. These consistency conditions will define the interconnection of the individual transmission elements, as shown in the following.

Consider a joint $\mathcal{G}_i$, contained in $n_L(\mathcal{G}_i)$ loops, and connecting $n_B(\mathcal{G}_i)$ bodies, see Fig. 12. The relative position of the bodies $b_1^{\mathcal{G}_i}, \ldots, b_{n_B(\mathcal{G}_i)}^{\mathcal{G}_i}$ connected by the joint can be described by a unique joint coordinate μ_j for each body b_j, as measured from one of the bodies, say body b_1. Within each loop L_k incident with joint $\mathcal{G}_i$, an additional joint coordinate $\beta_k^{\mathcal{G}_i}$, describing the relative position between two bodies $b_{b_1(k)}$ and $b_{b2(k)}$, is introduced. With these, the following relationships hold at the joint:

$$\mu_1 = 0 \ , \tag{73}$$

$$\mu_{b_1(L_k)} - \mu_{b_2(L_k)} = \beta_k^{\mathcal{G}_i} - \alpha_k^{\mathcal{G}_i}; \quad k = 1, \ldots, n_L(\mathcal{G}_i) \ , \tag{74}$$

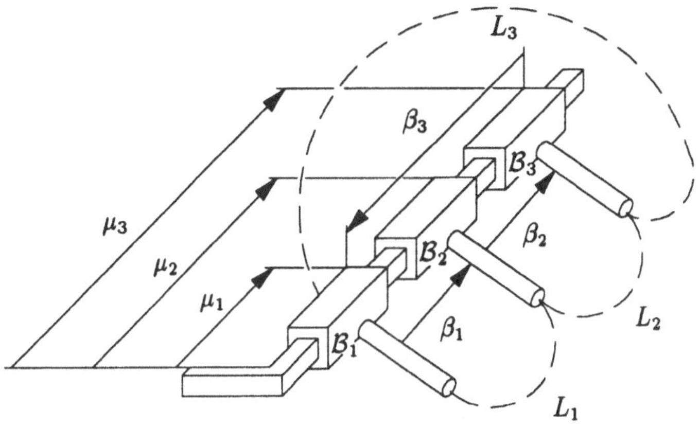

Figure 12: A joint connecting several bodies

where in $\alpha_k^{\mathcal{G}_i}$ any arising constants are collected. This is a system of $n_L(\mathcal{G}_i)+1$ linear equations which can be put into the compact matrix notation

$$P^{\mathcal{G}_i}\underline{\mu} = \underline{\beta}^{\mathcal{G}_i} + \underline{\alpha}^{\mathcal{G}_i} \quad , \tag{75}$$

with

$$\underline{\mu} = \begin{bmatrix} \mu_1 \\ \vdots \\ \mu_{n_B(\mathcal{G}_i)} \end{bmatrix} ; \underline{\beta}^{\mathcal{G}_i} = \begin{bmatrix} 0 \\ \beta_1^{\mathcal{G}_i} \\ \vdots \\ \beta_{n_L(\mathcal{G}_i)}^{\mathcal{G}_i} \end{bmatrix} ; \underline{\alpha}^{\mathcal{G}_i} = \begin{bmatrix} 0 \\ \alpha_1^{\mathcal{G}_i} \\ \vdots \\ \alpha_{n_L(\mathcal{G}_i)}^{\mathcal{G}_i} \end{bmatrix} . \tag{76}$$

Clearly, for the $(n_L(\mathcal{G}_i)+1) \times n_B(\mathcal{G}_i)$ matrix $P^{\mathcal{G}_i}$ having general rank $r^{\mathcal{G}_i}$, with

$$r^{\mathcal{G}_i} \leq n_B(\mathcal{G}_i) \quad , \quad r^{\mathcal{G}_i} \leq n_L(G)+1 \quad , \tag{77}$$

Eq. (75) only has a solution if $\underline{\beta}^G$ satisfies the $n_L(G)+1-r^{\mathcal{G}_i}$ linear equations

$$H^{\mathcal{G}_i}\underline{\beta}^{\mathcal{G}_i} + \underline{\alpha}^{\mathcal{G}_i} = 0 \quad , \tag{78}$$

where $H^{\mathcal{G}_i}$ represents the $(n_L(G)+1-r^{\mathcal{G}_i}) \times n_L(G)$ matrix which is the orthogonal complement of $P^{\mathcal{G}_i}$, and which, after appropriate column pivoting of matrix $P^{\mathcal{G}_i}$, contains only zeros and ± 1.

A case of particular interest is obtained if the matrix $P^{\mathcal{G}_i}$ has full column rank,

$$r^{\mathcal{G}_i} = n_B(\mathcal{G}_i) \quad . \tag{79}$$

In this case the number of independent linear equations in Eq. (78) is

$$p(\mathcal{G}_i) = n_L(\mathcal{G}_i) - n_B(\mathcal{G}_i) + 1 \quad , \tag{80}$$

and the number of interconnections between the loops can be established very easily at each joint. If Eq. (79) holds for all joints $\mathcal{G}_i$, the multibody system is said to be *completely loop-connected*. This is the case for a *2-connected* graph, i. e. one in which at least *two* joints have to be removed in order for the mechanism to fall apart in two or more parts. Complex multibody systems, or at least subsystems of complex multibody systems, typically belong to this category.

Eq. (78) describes the ***independent linear*** relationships which hold between the joint coordinates of the individual multibody loops. Clearly, they are defined uniquely at each joint and involve only signed sums of joint coordinates. Thus they can be represented by ***summing junctions*** connecting the individual kinematical transformers in a block diagram designated ***kinematical network***, see Fig. 13 below.

It is easy to show that the sum of the local degrees of freedom of the multibody loops $f(L_i)$, reduced by the sum of the number of interconnection equations at each joint $p(\mathcal{G}_i)$, results in exactly the number of degrees of freedom of the corresponding multibody system. Thus the "assembled" kinematical transformers thus represent an isomorphism of the relative kinematics of complex multibody systems to a much simpler representation. From this representation, a further optimization of the system of constraint equations is possible, as shown next.

5.3. DETERMINATION OF EQUATION ORDERING

In the block diagram of kinematical transformers, the orientation of the edges represents the sequence in which the individual equations are to be solved. An aspect of particular interest is to find an ordering of the constraint equations such that the individual blocks are recursively solvable. This aim, which involves also the choice of generalized coordinates for the overall system, can be formulated as an orientation problem in the block diagram. The conditions for recursive solution are:

1. The number of external inputs is equal to the number of degrees of freedom of the system.
2. The number of inputs for each multibody loop L_i is equal to the local degree of freedom $f(L_i)$ of the loop.
3. Each summing junction has exactly one output.
4. There are no closed circuits.
5. The local kinematics of the transformers are recursively solvable.

The analysis of complex multibody systems shows that for the majority of technical applications conditions (1) through (5) can be accomplished. These systems are termed *recursively solvable systems*. Systems for which not all conditions can be fulfilled are called *non-recursively solvable systems*. The most common reason for the appearance of a non-recursively solvable system is that conditions (1) through (4) can not be accomplished. The cases for which condition (5) is violated are very rare and shall not be regarded here.

A method for finding an orientation of the edges of the block diagram which fulfills conditions (1) to (4) in the case of *recursively solvalbe* systems is proposed in [13] and [24]. The equation ordering can be found very easily in this case by considering the degree of coupling of the elements (i.e. the number of edges connecting them to other elements) as compared with the number of *allowable* inputs: starting from unoriented edges, one subsequently orients the edges of those elements (summing junctions or transformers), whose number of unoriented edges is not greater than the number of allowable inputs, as inputs. After orienting all edges, the block diagram now also contains the *solution flow* for the relative kinematics, which represents the required ordering of the constraint equations.

As an example of a recursively solvable system, a planar mechanism consisting of four interconnected planar four-bar loops is considered, see Fig. 13. The redundant set of relative coordinates includes for each loop four variables. Three of these can be solved as functions of the fourth in closed form, yielding corresponding kinematical transformers. There are three linear assembly equations occuring in the joints A, B, C. A sequence of elements for which unoriented edges can be oriented as described above is: L_4, C, L_3, L_2, B, A, L_1. This sequence yields a "solution flow" which obviously is recursive. Thus the constraint equations of this system are solvable in *closed form*.

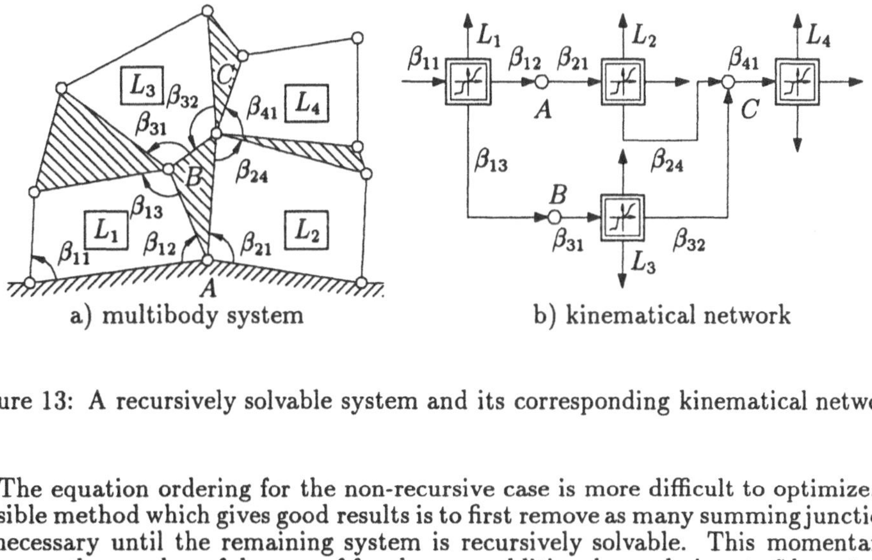

Figure 13: A recursively solvable system and its corresponding kinematical network

The equation ordering for the non-recursive case is more difficult to optimize. A possible method which gives good results is to first remove as many summing junctions as necessary until the remaining system is recursively solvable. This momentarily increases the number of degrees of freedom, so additional pseudo-inputs $\tilde{\underline{q}}$ have to be introduced. The linear equations which correspond to the removed junctions then define implicitly the functions $\tilde{\underline{q}}(\underline{q})$, which can be solved numerically.

An example of a non-recursively solvable system is shown in Fig. 14. The planar mechanism consists of five independent multibody loops which are again four-bar mechanisms. There are four linear assembly equations at the joints A, B, C, D. From the corresponding block diagram it is clear that the algorithm described above can not start, because there is no element which has an allowable number of inputs greater than the number of connections. This situation changes when the summing junction D is temporarily removed and an additional input $\tilde{q}$ is provided. In this case the system is recursively solvable. The original system is achieved by re-considering the closure condition at junction D, which yields the implicit equation for the determination of the function $\tilde{q}(q)$.

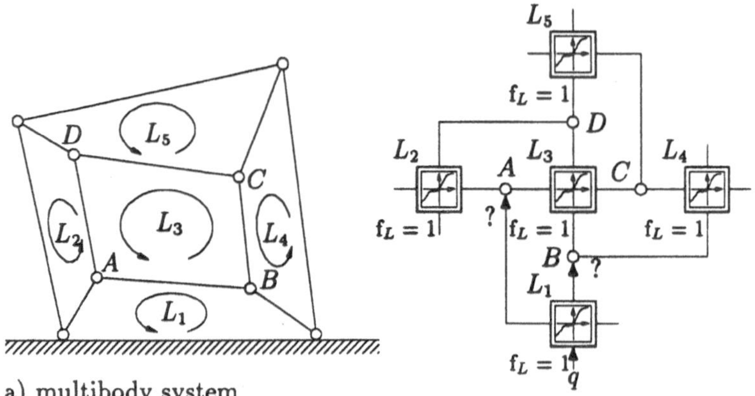

a) multibody system

b) kinematical network

Figure 14: A non-recursively solvable system and its kinematical network

In technical applications only very few, if any, of these additional implicit equations occur. As a consequence, the number of implicity defined functions is substantially reduced, and the kinematics, as well as the dynamics, can be solved very efficiently.

6. Nonholonomic Constraints

Instead of a general discussion of nonholonomic constraints, which is beyond the scope of this paper, a typical technical example of a system with nonholonomic constraints shall be discussed. It is the roboTRAC — a combined wheeled and legged vehicle — which can be modelled as a complex nonholonomic multibody system with kinematical loops and a time-varying structure (see Fig. 15). Several investigations ([17, 18, 19]) deal with the formulation of the kinematics and dynamics as well as the generation of walking patterns for this mobile mechanical platform. Here, the main results are summarized.

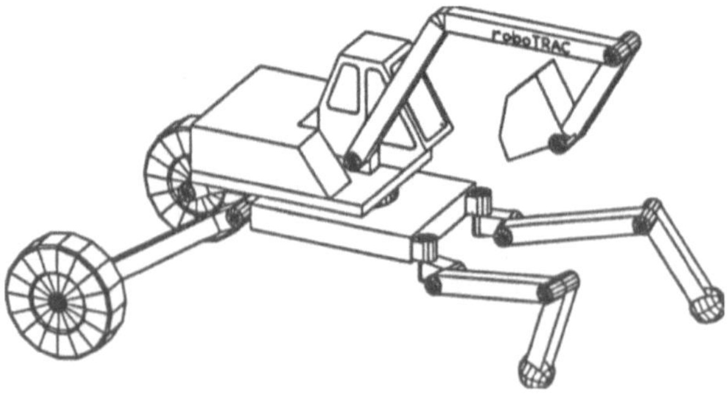

Figure 15: The roboTRAC ([42]).

6.1. KINEMATICAL MODEL

In this chapter, two different approaches for solving the nonholonomic kinematics of the roboTRAC, which have already been presented in detail in [17, 19], will be compared. For the investigation, the following assumptions concerning the kinematical model shown in Fig. 15 are made:

- The free motion of the carriage is described by six coordinates represented by three prismatic joints (3P) for the translational motion and three revolute joints (3R) for the rotational motion connecting the carriage to the inertial frame.
- The wheels are substituted by skids having the same influence on the kinematical behaviour of the system as the wheels.
- Every foot tip is connected to the environment by three prismatic joints (3P). This is a suitable approach to control the walking motion of the roboTRAC.

Without taking into account the nonholonomic constraints arising from the skids, the number of degrees of freedom of the complete holonomic system is

$$f_h = 12 \quad .$$

The configuration space thus has the dimension dim $\mathcal{C} = 12$ ([30]). Due to the two nonholonomic skid conditions, the number of degrees of freedom of the complete system is reduced to

$$f = 10 \quad ,$$

which is equal to the dimension dim $\mathcal{P}$ of the phase space.

6.2. CONSTRAINT EQUATIONS OF NONHOLONOMIC SYSTEMS

The nonholonomic constraints in a nonholonomic system are represented by non-integrable linear relations between velocities. Let the nonholonomic multibody system have $h = r_\beta$ holonomic constraints and n nonholonomic constraints. Then, the dimension of the configuration space $\mathcal{C}$ is $m = n_\beta - h$ while the dimension of the phase space $\mathcal{P}$ is $f = m - n$.

The holonomic constraints can be stated as a set of nonlinear algebraic equations

$$g_i\left(\underline{\beta}\right) = 0 \quad , \quad i = 1, \ldots, h \quad , \tag{81}$$

where $\underline{\beta}$ is the $n_\beta \times 1$ vector of the joint coordinates. Differentiating Eq. (81) with respect to time yields

$$\dot{\underline{g}} = \frac{\partial \underline{g}}{\partial \underline{\beta}} \dot{\underline{\beta}} = \boldsymbol{J}_h \dot{\underline{\beta}} = \boldsymbol{0} \quad , \tag{82}$$

with $\boldsymbol{J}_h$ as the $h \times n_\beta$ Jacobian of the holonomic system.

Additionally, the nonholonomic constraints can be expressed as

$$\boldsymbol{J}_n \dot{\underline{\beta}} = \boldsymbol{0} \quad , \tag{83}$$

where $\boldsymbol{J}_n$ is the $n \times n_\beta$ Jacobian corresponding to the nonholonomic constraints. Two procedures to solve the kinematics of a nonholonomic system are possible (Fig. 16).

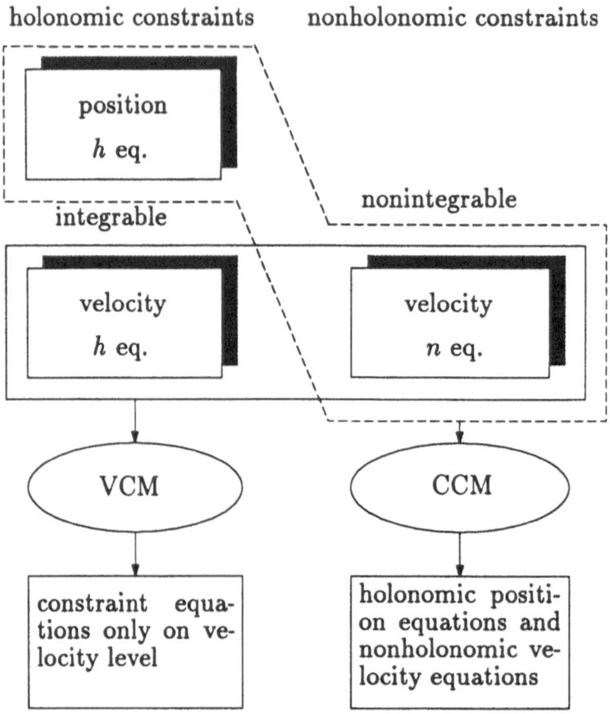

Figure 16: Constraints of nonholonomic systems

On the one hand, all $n+h$ constraint equations can be stated on velocity level as a set of linear equations; the corresponding position can be obtained from numerical integration of all joint velocities. This method will be named ***velocity constraint method*** (VCM). On the other hand, all h holonomic constraint equations can be stated on position level as a set of nonlinear equations, while the n nonholonomic constraints are stated as a set of linear equations on velocity level. The corresponding n position coordinates can be obtained from numerical integration. This method will be named ***combined constraint method*** (CCM).

6.3. KINEMATICS OF THE roboTRAC

In this section, only a short overview of both methods will be given; a more detailed description can be found in [17] and [19].

6.3.1. Velocity Constraint Method. Considering Fig. 17, the following velocity-level equations can be formulated for the holonomic constraints:

- The velocities of the foot reference points T_i, $i = 1, 2$ relative to the inertial frame $\mathcal{K}_I$, must vanish:

$$\boldsymbol{v}_{Ti} = \mathbf{0} \quad , \quad i = 1, 2 \quad . \tag{84}$$

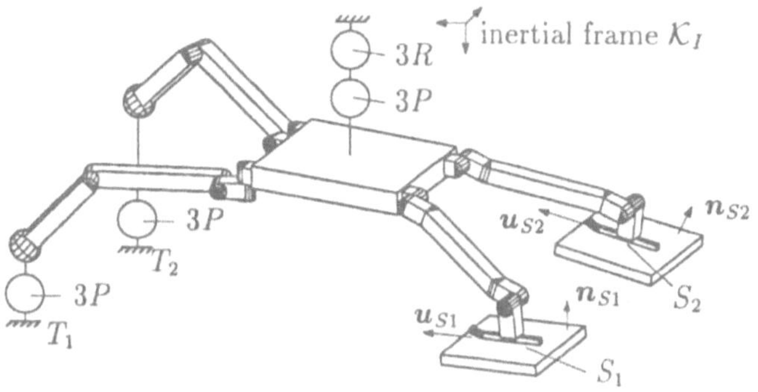

Figure 17: Kinematical model of the roboTRAC

- The projection of the velocity vectors $\boldsymbol{v}_{Si}$, $i = 1, 2$ of the skids onto the ground normal vectors $\boldsymbol{n}_{Si}$, $i = 1, 2$ must also vanish:

$$\boldsymbol{v}_{Si} \cdot \boldsymbol{n}_{Si} = 0 \quad , \quad i = 1, 2 \quad . \tag{85}$$

- The component of the angular velocity $\boldsymbol{\omega}_{Si}$, $i = 1, 2$ of the skids normal to the plane spanned by the ground normal vector and the longitudinal direction of the skid has to be zero:

$$\boldsymbol{\omega}_{Si} \cdot (\boldsymbol{u}_{Si} \times \boldsymbol{n}_{Si}) = 0 \quad , \quad i = 1, 2 \quad . \tag{86}$$

Furthermore, the nonholonomic constraints implying that the instantaneous motion of the skids must be parallel to their longitudinal axis have to be considered:

$$\boldsymbol{v}_{Si} \cdot (\boldsymbol{u}_{Si} \times \boldsymbol{n}_{Si}) = 0 \quad , \quad i = 1, 2 \quad . \tag{87}$$

Evaluating Eqs. (84) – (87) using forward kinematics yields

$$\boldsymbol{J}_{\beta} \dot{\underline{\beta}} = 0 \quad , \tag{88}$$

where $\boldsymbol{J}_{\beta}$ represents the (12×22) Jacobian of all constraint equations and $\dot{\underline{\beta}}$ is a (22×1) vector containing all joint velocities.

After choosing $f = 10$ velocities out of $\dot{\underline{\beta}}$ as independent velocities and assembling them into the vector of the pseudo-velocities $\dot{\underline{\pi}}$, the remaining 12 velocities $\dot{\underline{\beta}}_{dep}$ can be obtained by solving Eq. (88). To get the joint coordinates of the complete system at each time step, all 22 joint velocities have to be integrated numerically (Fig. 18).

6.3.2. Combined Constraint Method. In this method, first a holonomic system with $f = 12$ degrees of freedom is exhaustively modelled for efficient kinematics. The corresponding constraint equations on position level can be derived using the *characteristic pair of joints*. The solution flow based on the principle of *kinematical transformers* becomes obvious from Fig. 19. In this figure, L_A and L_B represent the kinematical loops between the inertial frame and the foot points T_1, T_2, while L_C and L_D represent the kinematical loops between the inertial frame and the skid reference points S_1, S_2, respectively.

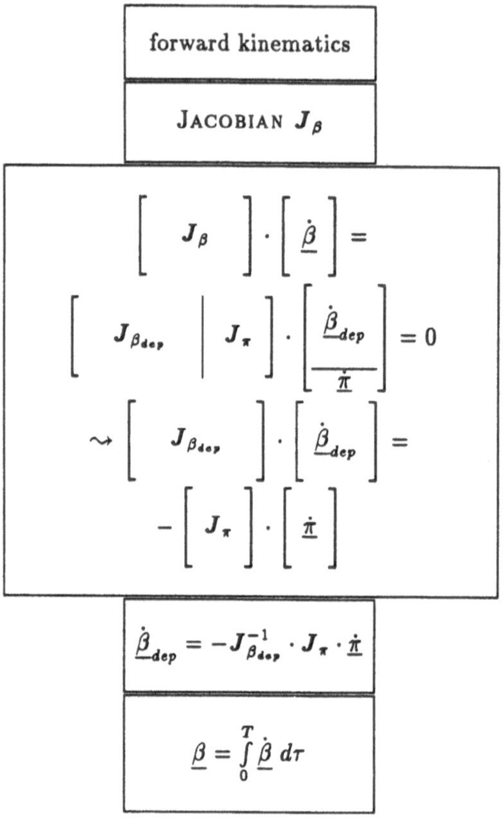

Figure 18: Solution steps for a nonholonomic problem

Regarding the nonholonomic system in a second step, the degrees of freedom are reduced from twelve to ten by the two nonholonomic skid conditions. Introducing the independent pseudo–velocities

$$\dot{\pi}_i = \dot{q}_i \quad , \quad i = 1, \ldots, 10 \quad ,$$

the two dependent velocities $\dot{q}_{11}$ and $\dot{q}_{12}$ can be determined from Eq. (87).

For this, the linear nonholonomic constraints can be expressed in terms of the pseudo velocities $\dot{\underline{\pi}}$ and the unknown velocities $\dot{q}_{11}$ and $\dot{q}_{12}$.

$$\dot{\underline{g}} = \boldsymbol{A} \begin{bmatrix} \dot{q}_{11} \\ \dot{q}_{12} \end{bmatrix} + \begin{bmatrix} c_1(\dot{\pi}_1, \ldots, \dot{\pi}_{10}) \\ c_2(\dot{\pi}_1, \ldots, \dot{\pi}_{10}) \end{bmatrix} = 0 \quad . \tag{89}$$

Eq. (89) can be represented by the following block diagram:

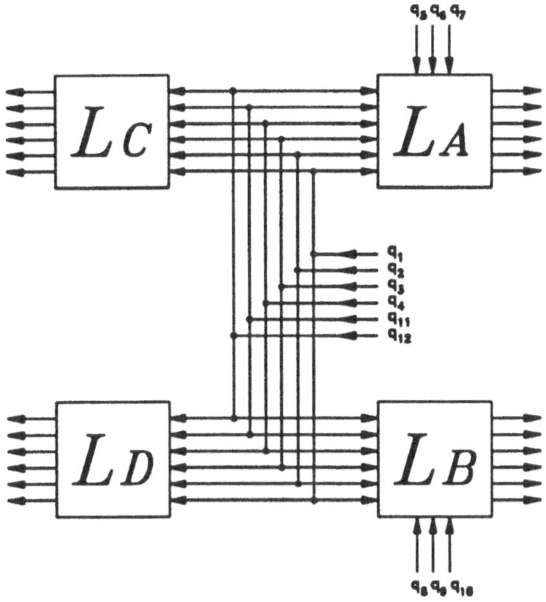

Figure 19: Block diagram of the holonomic system

$$\xrightarrow[\dot{q}]{q} \boxed{\begin{array}{c} \text{relative} \\ \text{kinematics} \\ \text{(holonomic)} \end{array}} \xrightarrow[\underline{\dot{\beta}}]{\underline{\beta}} \boxed{\begin{array}{c} \text{nonholonomic} \\ \text{constraints} \end{array}} \xrightarrow[\underline{\dot{g}}]{}$$

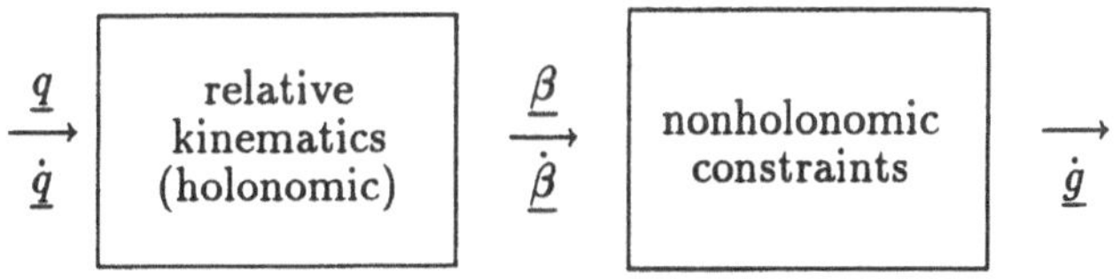

The still unknown matrix A and vector $\underline{c}$ can be determined by evaluating Eq. (89) with particular inputs as described in Table 4.

Now, the unknown velocities can be calculated as

$$\begin{bmatrix} \dot{q}_{11} \\ \dot{q}_{12} \end{bmatrix} = -A^{-1}c \quad . \tag{90}$$

To get the position corresponding to each time step, Eq. (90) has to be integrated numerically.

6.3.3. <u>Comparison of Methods</u> Once the possibilities of solving the complex kinematics of the roboTRAC have been presented, it is interesting to compare the number of mathematical operations and CPU-time needed for the different methods. For this comparison, a HP Apollo Workstation Series 400 with MC 68030 processor has been used. Table 5 contains the number of mathematical operations as well as the CPU time needed for calculating the right hand side of the differential equations. The optimizations involved in VCM(O) and CCM(O) are described in [40]. Finally, it has to be emphasized that the formulation and implementation of the combined constraint method (CCM) is more complicated and time consuming.

$\dot{q}_{11} = \dot{q}_{12} = 0$ $\dot{\underline{\pi}}$ as given $\rightsquigarrow$	$\dot{\underline{g}} = \underline{c}$
$\dot{q}_{11} = 1,\ \dot{q}_{12} = 0$ $\dot{\underline{\pi}} = 0$ $\rightsquigarrow$	$\dot{\underline{g}}^{(11)} \hat{=} 1^{st}$ column of $\boldsymbol{A}$
$\dot{q}_{11} = 0,\ \dot{q}_{12} = 1$ $\dot{\underline{\pi}} = 0$ $\rightsquigarrow$	$\dot{\underline{g}}^{(12)} \hat{=} 2^{nd}$ column of $\boldsymbol{A}$

Table 4: Determination of $\boldsymbol{A}$ and $\underline{c}$

method	VCM	VCM(O)	CCM	CCM(O)
multiplications	1277	740	983	598
additions	980	594	716	475
trigonometrical functions	26	26	48	44
CPU-time for the right hand side [s]	0.0154	0.0115	0.0157	0.0098

VCM - Velocity constraint method
CCM - Combined constraint method
VCM(O) - Optimized variant of VCM
CCM(O) - Optimized variant of CCM

Table 5: Comparison of methods.

6.4. KINEMATICAL CONTROL OF AN EXPERIMENTAL SETUP

Up to now, simulation results can be visualized on the computer by means of diagrams or computer animation (Fig. 20). Another possibility is to transform the simulation results into the motion of a scaled model of the mechanical system. For this purpose, a controller is based on the micro-processor INTEL 8039 has been developed. It is fed from a digital computer by the parallel interface. The output of the controller is connected to a multiplexer distributing the signals to eight servos, which represent the actuators of the mechanical model of the roboTRAC (Fig. 21).

7. Implementation Issues

The concepts described in the previous sections offer a high potential for program optimization and modularity. However, they are difficult to realize, because (a) their efficiency depends to a large extent on a thorough modelling of the sub-components, (b) the components contain a great deal of data and sub-functions which must be administered within the running program, and (c) the solution techniques must be

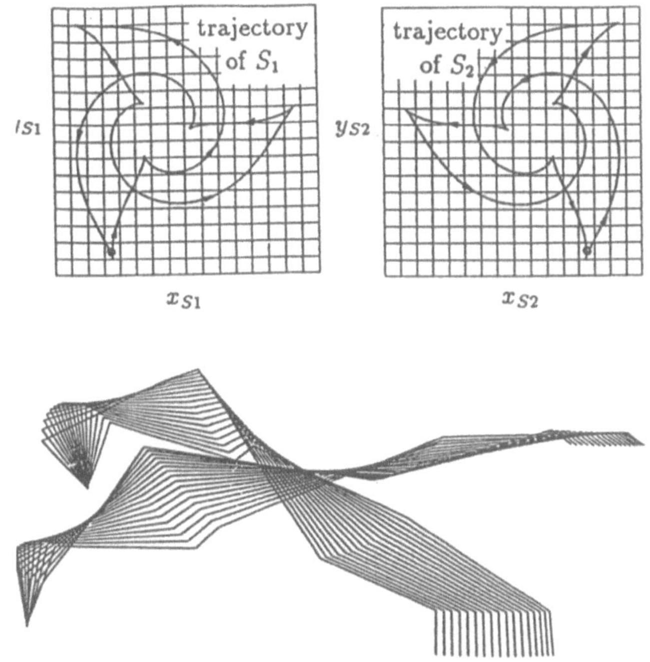

Figure 20: Usual visualization methods.

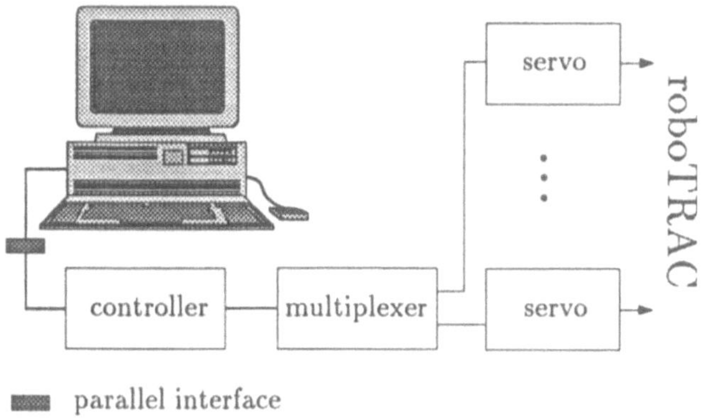

Figure 21: Connection between computer and model.

carefully adapted to the structure of the equations. To cope with these problems, a series of techniques have been developed, which shall be briefly described in the following.

7.1. SYMBOLIC FORMULA MANIPULATION

The resolution schemes described in Section **4.4.** and Section **5.** lend themselves for symbolical formula manipulation. Specifically, an implementation of the methods of Section **4.4.** has already been carried out upon the symbolic programming language *Mathematica*. Such an implementation can be used to obtain symbolical expressions of the resulting scalar equations. For example, the session for the inverse kinematics of a planar four-bar mechanism looks like this:

```
In[2]:= PlaneFourBarMechanism

Out[2]= {DHTransform[beta1, 0, 0, 1] . DHTransform[beta2, 0, 0, r] .
>         DHTransform[beta3, 0, 0, d] . DHTransform[beta4, 0, 0, s],
>         {beta2, beta3, beta4}}

In[3]:= GenerateConstraints[PlaneFourBarMechanism]

Out[3]= {ESF[1, Null, 0, 0, 0],
>         ESF[1, Null, 0, 0, 0],
>         ESF[1, -Null, 0, 0, 0],
>         ESF[0, -beta2, -2 r ADH1[1, 4], 2 r s,
              2     2     2                2
>          -d   + r   + s   + ADH1[1, 4]  ],
>         ESF[0, beta3, d, 0, -ADH3[1, 4], -ADH3[2, 4]],
>         ESF[0, beta4, Sin[beta3], Cos[beta3], -ADH3[2, 1], -ADH3[1, 1]]}

In[4]:= GetSubstitutions[%]

Out[4]= {ADH1[1, 4] -> -1 + s,
>         ADH3[1, 1] -> Cos[beta1] Cos[beta2] - Sin[beta1] Sin[beta2],
>         ADH3[1, 4] -> -r + ADH1[1, 4] Cos[beta2] + s Sin[beta2],
>         ADH3[2, 1] -> -(Cos[beta2] Sin[beta1]) - Cos[beta1] Sin[beta2],
>         ADH3[2, 4] -> s Cos[beta2] - ADH1[1, 4] Sin[beta2]}
```

Here, the notation `DHTransform[beta1, 0, 0, 1]` is a short-cut for specifying the four DENAVIT-HARTENBERG-parameters of a general transformation, and `ESF[sigma, beta, A, B, C]` is a shortcut for specifying a scalar equations of order 2 for the unknown variable `beta`, which is either rotational (`sigma` $= 0$) or translational (`sigma` $= 1$). The lines `ESF[1, Null, 0, 0, 0]` denote identically fulfilled constraint equations, which are typical for special cases of over-constrained, but movable mechanisms.

7.2. OBJECT-ORIENTED PROGRAMMING

The problem of integrating the concepts in a running program can be effectively tackled by resorting to the paradigm of object-oriented programming. Several approaches have emerged in the last years for carrying out such a modelling, three of which are of particular interest:

- Using object-oriented programming for program structurization. Excellent experience where made endowing traditional FORTRAN-programs with an object-oriented shell. By this, the problems of variable passing in huge programs could be avoided, and the efficient implementations already performed earlier could be integrated into large programs as modules with literally no side-effects (see e.g. [16]).

- Using object-oriented programming for algebraic manipulations. At this level, the most frequent operations involving scalars, vectors, and matrices, are implemented at an abstract level, supplying the user with simple interfaces for the definition and evaluation of composite functions and their derivatives ([1]).
- Using object-oriented programming for mechanical modelling. This type of modelling aims at describing the physical interrelationships within the mechanical system at such an abstract level, that generic objects can be identified whose actions can be described without resorting to any particular representations. A particular modeling in this direction is the treatment of mechanical components as "kinetostatical transmission elements", which transmit motion, forces and inertia properties along the multibody system ([25]).

7.3. SPECIAL SOLUTION TECHNIQUES

The method described in Section **3** for formulating the equations of motion of multibody systems in minimal coordinates is particularly efficient for systems with complex topology including multiple kinematical loops, but other solution techniques, which are often based on other types of coordinates and different mechanical principles, may be more suitable for particular system topologies. Taking advantage of object-oriented programming methods, it is quite realistic to apply alternate solution techniques within the same program environment. Several such specialized solution techniques have indeed been implemented and will be briefly described in the following:

Recursive methods such as those of [9, 6] were formulated for the forward-dynamics problem and are particularly efficient for tree-structured systems, achieving an $O(N)$ operation count, where N is the number of bodies. By a reinterpretation in the context of a differential-geometric approach it was possible to implement a variant of this method as an option in the group's software package (see [24]).

Hybrid methods which combine features of the minimal-coordinate method, recursive methods and absolute-coordinate methods can be applied due to the open and modular nature of the object-oriented programmign environment. Such methods permit formulations tailored to a particular mechanical system (see [24, 1]).

Direct methods for differential-algebraic equations are a prerequisite for the use of absolute-coordinate methods or the hybrid methods mentioned above and permit numerical integration of the equations of motion without resorting to such devices as coordinate-partitioning or constraint stabilization. In the form given in [2], such methods are easily integrated into existing multibody codes.

8. Conclusions

The approach discussed in this paper shows that it is possible to design specialized methods for the formulation of the equations of motion of minimal order for general multibody systems by solving the kinematics efficiently. This is achieved by regarding the individual kinematical loop as the main building brick of the modelling, and developing appropriate solution schemes for the solutions of its local kinematics. Subsequently, the kinematics can be incorporated in the general dynamics procedure, supplying the algorithm with the necessitated terms. The advantages of the approach lie in its possibility of yielding compact, efficient code for systems of virtually any complexity. This is demonstrated by examples ranging from combined wheeled and legged vehicles to complete passenger cars. Also, its implementation gets increasingly simpler by using modern programming techniques, such as object-oriented programming and symbolical formula manipulation. These features, together with new possibilities arising with the advent of faster hardware, make it feasible to use the approach e. g. for incorporating complex models of mechanical systems in hardware-in-the-loop applications.

References

1 Martin Anantharaman. Flexible multibody systems — An object-oriented approach. In M. Pereira and J.A.C. Ambrósio, editors, *Proceedings of the NATO-Advanced Study Institute on Computer Aided Analysis of Rigid and Flexible Mechanical Systems*, pages 383–402, Tróia, Portugal, 27 June - 9 July 1993.

2 Martin Anantharaman and Manfred Hiller. Numerical simulation of mechanical systems using methods for differential-algebraic equations. *International Journal of Numerical Methods in Engineering*, 32:1531–1542, 1991.

3 J. Angeles and K.E. Zanganeh. The semigraphical determination of all real inverse kinematic solutions of general six-revolute manipulators. In *Proceedings of the RoManSy*, 1992.

4 Jorge Angeles. *Rational Kinematics.* Springer Tracts in Natural Philosophy 34. Springer-Verlag, New York, 1988.

5 O. Bottema and B. Roth. *Theoretical Kinematics.* Applied Mathematics and Mechanics 24. North-Holland Publishing Company, Amsterdam, Oxford, New York, 1970.

6 H. Brandl, R. Johanni, and M. Otter. A very efficient algorithm for the simulation of robots and similar multibody systems without inversion of the masss matrix. In *IFAC/IFIP/IMACS Symposium on Robotics*, Wien, December 1986.

7 M.A. Chace and D.A. Smith. DAM—Digital computer program for the dynamic analysis of generalized mechanical systems. *SAE Paper No. 710244*, January 1971.

8 J. Denavit and R.S. Hartenberg. A kinematic notation for lower-pair mechanisms based on matrices. *Transactions of the ASME, Journal of Applied Mechanics*, pages 215–221, June 1955.

9 R. Featherstone. Position and velocity transformations between robot end-effector coordinates and joint angles. *The International Journal of Robotics Research*, 2(2):35–45, 1983.

10 J. Garcia de Jalón, J. Unda, A. Avello, and J.M. Jiménez. Dynamic analysis of three-dimensional mechanisms in "natural coordinates". ASME-Paper 86-Det-137, 1986.

11 M. Geradin and A. Cardona. Kinematics and dynamics of rigid and flexible mechanisms using finite elements and quaternion algebra. *Computational Mechanics*, 4:115–135, 1989.

12 M. Hiller. *Empfindlichkeitsanalyse zur Erfassung von Fertigungstoleranzen.* VDI-Berichte 596. VDI-Verlag, 1986.

13 M. Hiller and M. Anantharaman. Systematische Strukturierung der Bindungsgleichungen mehrschleifiger Mechanismen. *Zeitschrift für angewandte Mathematik und Mechanik*, 69:T303–T305, 1989.

14 M. Hiller and A. Kecskeméthy. Equations of motion of complex multibody systems using kinematical differentials. *Transactions of the Canadian Society of Mechanical Engineers*, 13(4):113–121, 1989.

15 M. Hiller, A. Kecskeméthy, and C. Woernle. A loop-based kinematical analysis of spatial mechanisms. ASME-Paper 86-DET-184, New York, 1986.

16 M. Hiller and V. Pichler. Vehicle dynamics simulation using object-oriented programming. In *Third Pan American Congress of Applied Mechanics — PACAM III, January 4–8*, São Paulo (Brazil), 1993.

17 M. Hiller and Th. Schmitz. Kinematics and dynamics of the combined legged and wheeled vehicle 'RoboTRAC'. In *CSME Mech. Eng. Forum*, pages 387–392, Toronto, June 1990.

18 M. Hiller and Th. Schmitz. Robotrac — An Example of a Mechatronic System. In P.A. MacConaill, P. Drews, and K.-H. Robrock, editors, *Mechatronics & Robotics, I*, Advances in Design and Manufacturing, pages 31–41, Amsterdam, 1991. Espirit CIM-Europe, IOS Press.

19 M. Hiller, G. Schweitzer, and C. Woernle. Kinematical control of the combined wheeled and legged vehicle RoboTRAC. In *8th CISM-IFToMM Symposium Ro.man.sy*, Cracow, July 1990. Hermes Press, Paris.

20 M. Hiller and C. Woernle. A systematic approach for solving the inverse kinematic problem of robot manipulators. In E. Bautista, J Garcia-Lomas, and Navarro A., editors, *Proceedings 7th World Congress Th. Mach. Mech.*, pages 1135–1139, Sevilla, September 1987. IFTOMM, Pergamon Press.

21 M. Hiller and C. Woernle. The characteristic pair of joints — an effective approach for the solution of the inverse kinematics problem for robots. In *Proceedings of the International Conference on Robotics and Automation*, Philadelphia, April 1988. IEEE.

22 W.W. Hooker and G. Margulies. The dynamical attitude equations for an n-body satellite. *The Journal of Astronautical Sciences*, XII(4):123–128, 1965.

23 A. Kecskeméthy and M. Hiller. Automatic closed-form kinematics-solutions for recursive single-loop chains. In *Flexible Mechanisms, Dynamics, and Analysis, Proc. of the 22nd Biennal ASME-Mechanisms Conference, Scottsdale (USA)*, pages 387–393, September 1992.

24 Andrés Kecskeméthy. *Objektorientierte Modellierung der Dynamik von Mehrkörpersystemen mit Hilfe von Übertragungselementen.* PhD thesis, Universität - GH - Duisburg, 1993.

25 Andrés Kecskeméthy. Sparse-matrix generation of Jacobians for the object-oriented modelling of multibody dynamics. In M.S. Pereira and J.A.C. Ambrósio, editors, *Proceedings of the NATO-Advanced Study Institute on Computer Aided Analysis of Rigid and Flexible Mechanical Systems*, pages 71–90, Tróia, Portugal, 27 June - 9 July 1993.

26 Edwin Kreuzer. *Symbolische Berechnung der Bewegungsgleichungen von Mehrkörpersystemen.* Fortschrittberichte der VDI Reihe 11 Nr. 32. VDI-Verlag, Düsseldorf, 1979.

27 H.Y. Lee and C.G. Liang. A new vector-theory for the analysis of spatial mechanisms. *Mechanism and Machine Theory*, 23:209–217, 1988.

28 H.Y. Lee, C. Woernle, and M. Hiller. A complete solution for the inverse kinematic problem of the general 6R robot manipulator. *Transactions of the ASME, Journal of Mechanical Design*, 113(4):481–486, 1991.

29 H.J. Li. *Ein Verfahren zur vollständigen Lösung der Rückwärtstransformation für Industrieroboter mit allgemeiner Geometrie*. PhD thesis, Universität - GH - Duisburg, 1991.

30 J.I. Neimark and N.A. Fufaev. *Dynamics of Nonholonomic Systems*. Providence, Rhode Island: American Mathematical Society, 1972.

31 Peter J. Olver. *Applications of Lie Groups to Differential Equations*. Graduate Texts in Mathematics 107. Springer-Verlag, 1986.

32 N. Orlandea, M.A. Chace, and D.A. Calahan. A sparsity-oriented approach to the dynamic analysis and design of mechanical design of mechanical systems — Parts 1 & 2. *Transactions of the ASME, Journal of Engineering for Industry*, pages 773–784, August 1979.

33 Richard P. Paul. *Robot Manipulators: Mathematics, Programming, and Control.* The MIT Press Series in Artificial Intelligence. The MIT Press, Cambridge (Massachusetts), London (England), 1986.

34 M. Raghavan and B. Roth. Kinematic analysis of a 6R manipulator of general geometry. In H. Miura and S. Arimoto, editors, *Proceedings of the 5th International Symposium on Robotics Research*, Cambridge, 1990. MIT Press.

35 Wolfgang Risse. Konzeption und Entwicklung eines Emulators für CORDIC-Felder in kinematischen Anwendungen. Master's thesis, Universität Duisburg, Fachgebiet Mechatronik, 1992.

36 R.E. Roberson and J. Wittenburg. A dynamical formalism for an arbitrary number of interconnected rigid bodies. with reference to the problem of satellite attitude control. In *Proceedings of the 3rd IFAC Congress 1966*, pages 46D.2–46D.9, London, 1968.

37 Wolfgang Rulka. SIMPACK — A computer program for simulation of large-motion multibody systems. In Werner Schiehlen, editor, *Multibody Systems Handbook*, pages 265–284. Springer-Verlag, Berlin, Heidelberg, New York, 1990.

38 Werner Schiehlen, editor. *Advanced Multibody System Dynamics*, Solid Mechanics and its Applications, Dordrecht, Boston, London, 1993. Kluwer Academic Publishers.

39 P. N. Sheth and J. J. Uicker Jr.. IMP (Integrated Mechanisms Program), A computer-aided design analysis system for mechanisms and linkage. *Transactions of the ASME, Journal of Engineering for Industry*, pages 454–464, May 1972.

40 S. Vogel. Modellierung der Kinematik des Schreitroboters roboTRAC auf Geschwindigkeits– bzw. Lageebene zur Entwicklung von Schreitstrategien. Technical report, Universität–GH–Duisburg, Fachgebiet Mechanik, 1991.

41 R.A. Wehage and E.J. Haug. Generalized coordinate partitioning for dimension reduction in analysis of constrained dynamic systems. *Transactions of the ASME, Journal of Mechanical Design*, 104:247–255, 1982.

42 M. Werder. Cross-country vehicle. US Patent No. 4 779 691, October 1988.

43 J. Wittenburg and U. Wolz. MESA VERDE: Ein Computerprogramm zur Simulation der nichtlinearen Dynamik von Vielkörpersystemen. *Robotersysteme*, 1:7–18, 1985.

44 C. Woernle. *Ein systematisches Verfahren zur Aufstellung der geometrischen Schließbedingungen in kinematischen Schleifen mit Anwendung bei der Rückwärtstransformation für Industrieroboter*. Fortschrittberichte VDI Reihe 18 Nr. 59. VDI Verlag, Düsseldorf, 1988.

SYMBOLIC COMPUTATIONS IN MULTIBODY SYSTEMS

W. SCHIEHLEN
Institute B of Mechanics, University of Stuttgart
D-70550 Stuttgart, Germany

ABSTRACT: Symbolic formula manipulation has proven to be an efficient tool in the dynamical analysis of multibody systems. A multibody system data base is introduced and its implementation using a CAD-3D-software is shown. Starting from the data base the equations of motion are generated by a coordinate partitioning approach combined with the projection criterion. For the symbolical-numerical solution inverse kinematics algorithms are applied. The simulation results are visualized by computer animation. A four-bar mechanism and a crank-slider mechanism serve as examples. Two approaches for the dynamical analysis of flexible multibody systems are presented. Further, the optimization of multibody systems is treated using an actively controlled vehicle suspension as an example.

1. Introduction

An integrated approach for modeling, generation of symbolical equations of motion, simulation and visualization of multibody systems is described. A general object-oriented data model for all multibody formalisms is presented. With respect to existing CAD-interfaces, different solid model design methods and various visualization demands, the data model allows multibody modeling with a direct interface to a data base. Some software tools like an integrated Newton-Euler formalism are able to use immediately the parametrized multibody system data base. For multibody systems with closed kinematic loops a set of ordinary differential equations is formulated automatically which can be solved with explicit multistep integration algorithms. This is achieved by different minimal sets of generalized coordinates being specified by a coordinate partitioning approach during the numerical integration. The basic steps and the extreme flexibility of this automated mechanical design and simulation process is demonstrated for mechanisms.

Machines, mechanisms, road vehicles and spacecrafts can be modeled properly as multibody systems for the design and the dynamical analysis. The complexity of

M. F. O. S. Pereira and J. A. C. Ambrósio (eds.),
Computer-Aided Analysis of Rigid and Flexible Mechanical Systems, 101–136.

the dynamical equations called for the development of computer-aided formalisms a quarter of a center ago. The theoretical background is today available from a number of textbooks authored e.g. by Roberson and Schwertassek [1], Nikravesh [2], Haug [3] and Shabana [4]. The state-of-the-art is also presented at a series of IUTAM/IAVSD symposia, documented in the corresponding proceedings, see, e.g. Kortüm and Schiehlen [5], Bianchi and Schiehlen [6], Kortüm and Sharp [7].

In addition, a number of commercially distributed computer codes was developed, a summary of which is given in the Multibody System Handbook [8]. The computer codes available shows different capabilities: some of them generate only the equations of motion in numerical or symbolical form, respectively, some of them provide numerical integration and simulation, too. Moreover, there are also extensive software systems on the market which offer additionally graphical data input, animation of body motions and automated signal data analysis.

2. Multibody systems data model

Modeling of a mechanical system by the method of multibody systems is characterized by a composition of rigid bodies, joints, springs, dampers, and servomotors, see Figure 1. Force elements like springs, dampers, and servomotors acting in discrete nodal points result in applied forces and torques on the rigid bodies. Joints with different properties connecting the various bodies constrain their motion, they are often identified as constraint elements.

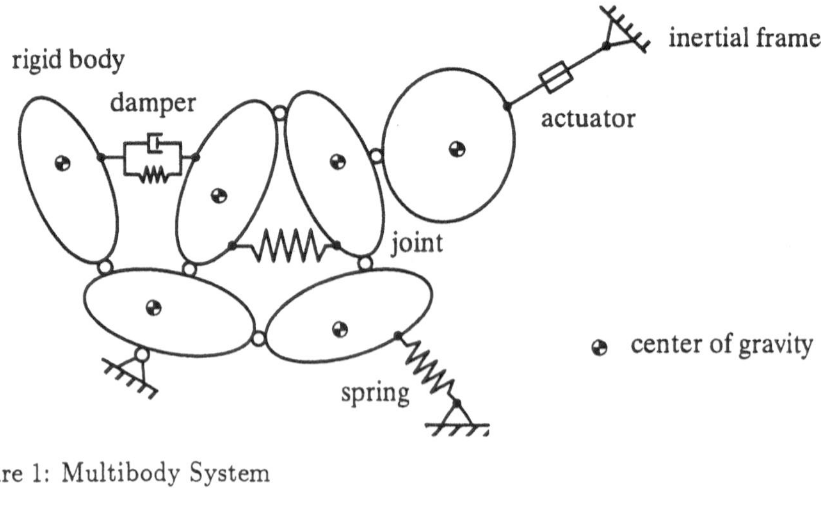

Figure 1: Multibody System

For the generation of the equations of motion computer programs may be used. Well known multibody system computer codes producing exclusively numerical data are ADAMS, Orlandea [9], and DADS, Haug [10]. To the contrary, computer programs like SD-FAST, Rosenthal and Sherman [11] and NEWEUL, Kreuzer [12] provide the explicit symbolical expressions for the system equations.

Nowadays CAD-systems are widely embedded in the industrial design and construction process, while a general application of three-dimensional CAD-systems is still rare. They support an analytically and topologically complete modeling, a collision detection, and the calculation of surface and volume properties closely related to the geometric representation of solid models.

Some couplings of solid modelers with multibody simulation software are realized for the numerical computer code ADAMS, e.g. for the CAD-system ARIES [13]. A CAD-3D-system independent approach is included in the program package RASNA and is described by Hollar and Rosenthal [14].

A system dynamics analysis requires as basic parameters mass, center of gravity, and moments of inertia of each body related to the geometry model and modeling method of the CAD-system used. A modular software concept demands an exchange of complete or single object data between the CAD-system and the multibody formalism. Therefore, a general interface to multibody computer codes is demanded to serve as a compatible and comfortable CAD-post processor, taking the different algorithms and implementations of multibody computer codes into account. The commercially available multibody modeling software tools within CAD-systems are mostly dedicated to a particular multibody dynamics computer code. Often, no options are supplied for a parametric multibody system description or the modeling is restricted to either robot, mechanism or vehicle dynamics. This variety of systems, each with different model data and the growing problems in the exchange of data, requires the development and production of cheaper and more reliable software products.

Consequently this leads to a database concept for the CAD-3D-modeling of multibody systems, see Figure 2:

- Collect the necessary data describing uniquely a multibody model for the different multibody programs.
- Examine the different geometry models of CAD-systems for solids and extract the relevant data for multibody systems.
- Define a geometry model for the representation of multibody elements.
- Design data types and operations and construct a software interface for a code-independent modeling of multibody systems.

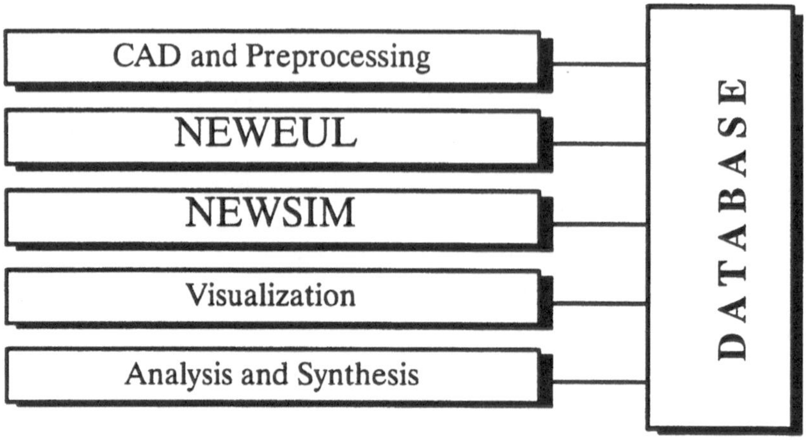

Figure 2: Modules within the database concept

A dynamic simulation environment for multibody systems represents in practice a large, sophisticated software system. Therefore, an important step is the definition of an abstract data model on a conceptual level. A first effort to develop a generalized data model for multibody systems including symbolical parameters and a postprocessing of CAD-data is described by Otter, Hocke, Daberkow, and Leister [15]. Each of the bodies is described by body-fixed reference frames. Further body-fixed frames, related joints and force elements are described. Additional symbolical parameters are defined for the position and orientation of the frames with respect to each other as well as the mass properties of the bodies. Consequently, for symbolical as well as numerical formalisms a generalized data base relies upon the basic modeling elements frame, body, joint, and force and is further adapted and extended with respect to the geometry models in CAD-3D and graphics systems.

A property of a solid in a CAD-3D system can be derived from a face normal specifying the inner and outer parts of an object, while the coincidence of the vertices of adjoining faces is not guaranteed. The geometric modeling by parametrized shapes is appropriate for geometric objects, whose shape is uniquely defined by a restriced number of parameters. Examples of parametrized shapes with an equivalent wire representation are shown in Figure 3.

For the global properties volume, surface area, moment of inertia, and center of gravity of solid models, integrals have to be evaluated like

$$I = \int_{Solid} f^V dV \tag{1}$$

see e.g. Mortenson [16], where $f^V = f^V(x, y, z)$ denotes a scalar property function.

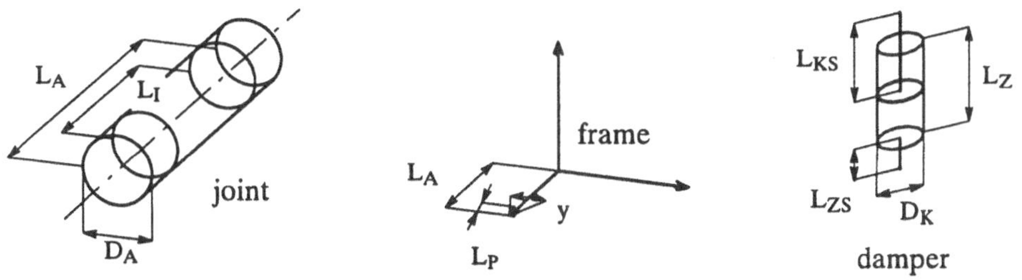

Figure 3: Parametrized wire representations of multibody elements

While Constructive Solid Geometry suggests the calculation of mass properties by the following recursively applied formulas

$$\begin{aligned}
\int_{Solid1 \cup Solid2} f^V dV &= \int_{Solid1} f^V dV + \int_{Solid2} f^V dV - \int_{Solid1 \cap Solid2} f^V dV, \\
\int_{Solid1 - Solid2} f^V dV &= \int_{Solid1} f^V dV - \int_{Solid1 \cap Solid2} f^V dV, \qquad (2)
\end{aligned}$$

boundary representations allow the evaluation via surface integrals.

The examination of different geometry models yield the following results:

- Mass property calculation modules for multibody systems do not depend on the model geometry (CSG or B-Rep). These results can be related directly with the input entities needed for the rigid bodies.
- A planar face model derived from the geometric entities of the solid body yield the graphic data for the description of the body's shape necessary for visualization.
- The parametrized shapes are well suited to serve as a geometry model for multibody modeling elements like frame, joint, and force.

The object-oriented data model conceptually developed by Otter et al. [15] results in classes defined for the elements *part*, *frame*, *body*, *interact*, *joint*, *force*, *global*, and *param* and additional operations valid for these classes.

An object of class *body*, e.g. Figure 4, comprises alle time-invariant data of a rigid body. It is obvious that the components inertia matrix and mass of an object of class body are supplied by their numerical values, too. A location of the center of gravity different from the body-fixed reference frame is taken into consideration by reference to an equivalent object of class *frame*.

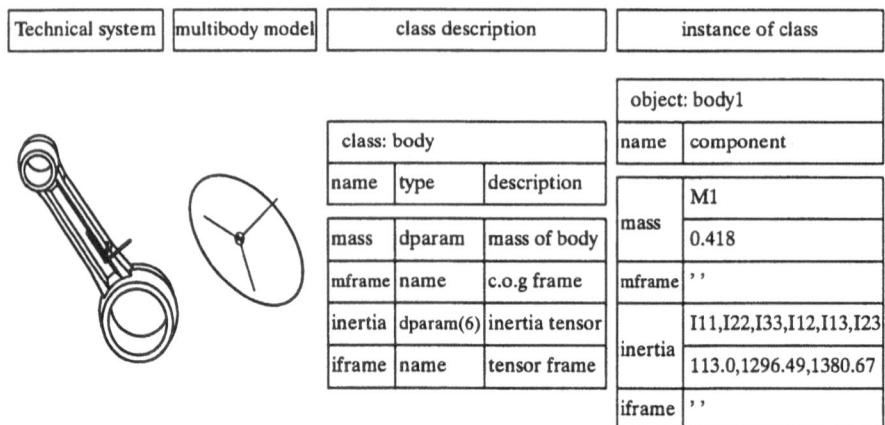

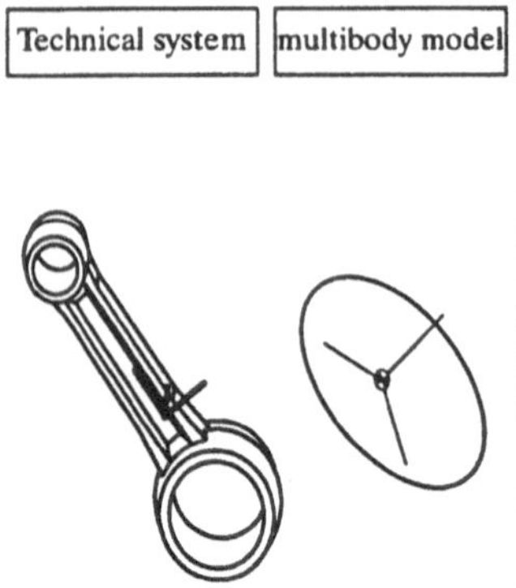

class: body		
name	type	description
mass	dparam	mass of body
mframe	name	c.o.g frame
inertia	dparam(6)	inertia tensor
iframe	name	tensor frame

object: body1	
name	component
mass	M1
	0.418
mframe	' '
inertia	I11,I22,I33,I12,I13,I23
	113.0,1296.49,1380.67
iframe	' '

Figure 4: Object of class *body* with its data model

Coupling elements of a multibody system are collected in class *interact*. Interactions are valid between two objects of class *frame* on different objects of class *part*.

Due to object-oriented software techniques, the definition of abstract data types in classes furthermore demands a description of the operations valid on the objects. These operations are designed for a practical, interactive multibody modeling process, e.g. in a CAD-3D-system. For all classes the basic operations 'create', 'delete', 'modify', and 'list' are defined, more complex operations take the relationships between objects of a multibody system into account.

Further classes are required for the graphical representation, like the actual frame axis length, its color or visibility, which depend on the actual multibody size and modeling state. An equivalent geometry data model for multibody elements well suitable for machine, robot and vehicle dynamics requires a unique spatial representation of the multibody elements, their function and physical quantity, see Daberkow [17]. From Figure 3 it is obvious that spatial parametrized shapes satisfy a graphic representation for objects of class *frame*, *joint*, and *force*. The definition of the geometry 3D classes *g3frame*, *g3joint* and *g3force* and operations for the geometry data model is equivalent to the multibody data model and includes classes comprising color, projection and viewpoint data.

3. Implementation and CAD-3D-realization

The implementation of the object-oriented data model in the data base system RSYST [18] allows storage and modification of multibody system objects. To realize fast access and interactive graphic visualization, the implementation of the object-oriented classes and operations within the CAD-3D-system is performed by means of data types and routines, which result in a system-independent modeling kernel library for multibody systems, see Daberkow [17]. This high level library DAMOS-C (DAta MOdel Standard implemented in C) supplies interfaces for modeling, input, and output as well as for the graphic representation. This open interface allows the integration in the commercially available CAD-3D- system SIGRAPH [19] and a new developed graphics-system.

The integration scheme in Figure 5 shows the interfaces to the CAD-3D software moduls of SIGRAPH. An extension of the CAD command language supplies additional commands which are necessary for the execution of multibody modeling operations. The CAD-3D-system menu is completed by special multibody system icons. To assure the graphic display of the modeling elements, the parametrized shapes are modeled via the 3D-wireframe entities of the CAD-graphic subsystem. A multibody command language of RSYST serves as a multibody system neutral file to store the multibody objects, see Otter et al. [20]

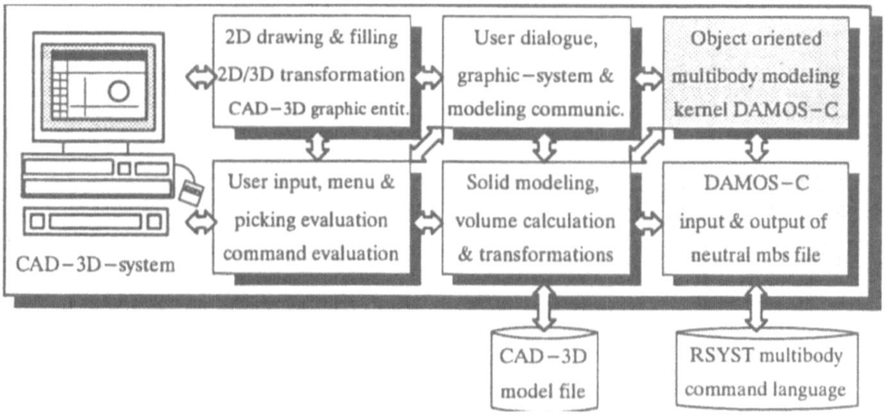

Figure 5: Integration of the multibody modeling kernel

The solid model design of a crank slider mechanism is performed by volume oriented techniques in PARASOLID from a disassembled model, Figure 6. All bodies of the crank-slider mechanism of a single four stroke engine are shown in Figure 6. Each body is supplied with adequate density attributes.

The first multibody modeling step is the initialization. Here, an appropriate solid is chosen as the inertial body of the multibody system, see Figure 6. In the next step arbitrary solids are interactively chosen to have the properties of a multibody part. Each object of class *body* retrieves its mass and inertia components from the mass property calculation modul of PARASOLID. To visualize the multibody part property, the equivalent solids are supplied by reference frames, located in the center of gravity.

By default, the orientation of further created joint and force definition frames is parallel to the specified reference frame. The position of these frames is defined by the CAD-3D-picking commands performed by the user. Figure 6 shows these modeling steps and the graphic representation of the objects. Joint definition frames are located along the unit normals of those faces, which form bearing surfaces or bearing bores of a solid.

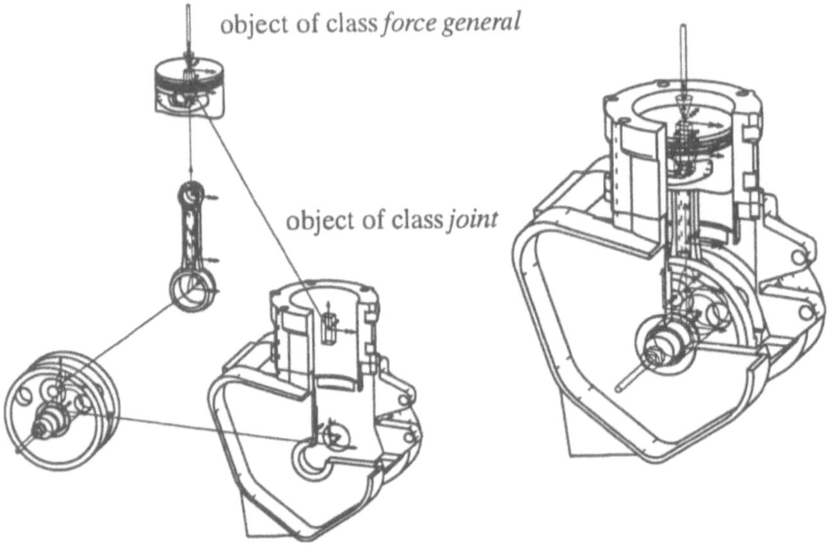

Figure 6: Disassembled and assembled mechanism with joint and force objects

A planar system modeled for spatial analysis demands a proper constraint selection. Redundant constraints remain if a mechanism is supplied with joints of class *revolute* and *translational*, making the determination of reaction forces impossible. Consequently, for an analysis modified joints have to be chosen. Objects of class *revolute* are visualized by the parametrized shapes and the wireframe entities. The connection between the objects of class *part* by the object of class *interact* is visualized by a 3D-line entity between the interacted frames.

The multibody modeling kernel library implemented in the CAD-3D-system supports an assembling of arbitrary pairs of class *part*. Figure 6 shows the assembling of individual solids over the equivalent objects of class joint. By modifying the **rangle** component of arbitrary objects of class *joint*, an initial multibody configuration is adjusted interactively, providing therefore an initial estimate for closed loop systems. Finally, an object of class *force general* is added to the piston part.

4. Generation of equations of motion starting from the database

The generation of equations of motion and the embedding of these equations to simulation software is especially in case of large multibody models very time consuming and prone of errors. Starting from the description of the multibody system stored on the database, the modul NEWEUL, Kreuzer and Leister [21], generates symbolic equations of motions and all information necessary for the automatic simulation. The modul NEWSIM, Leister [22], uses in the next step the compiled symbolical equations of motion for the simulation. Using the object-oriented datamodel the modules NEWEUL and NEWSIM are tools of a modular software package of the multibody system approach, see Figure 7.

In a first step the information stored in the database has to be extracted. In a modular concept the generation of equations of motion and the simulation have to be separated. The datamodel includes all the information neccessary for the generation of the equations of motion and, an adapted version of NEWEUL can be used as module in the database concept. Based upon a Newton-Euler formalism the symbolical equations of motion are generated using d'Alembert's or Jourdain's principle to eliminate the reactions forces and torques, see Ref. [23]. By means of a special, for the multibody system approach developed formulamanipulator, it is possible to generate the equations of motion with minimal costs of computation time, see Kreuzer [12]. The symbolical equations of motion can be used on the one hand in the simulation environment NEWSIM and on the other hand in any general purpose simulation environment, e.g. ACSL [24] or DSSIM [25].

At first, from the objects *interact* and *joint* the topology of the multibody system is computed. Additionally from the object *joint* the generalized coordinates are determined. The kinematical description of multibody systems is done by the definition of frames relatively to any arbitrary frame. These frames define rigid bodies, joints, auxiliary frames, and reference frames, too. Additionally the mass-geometric properties and the applied forces and moments are neccessary. These data can be found in the objects *interact* and *force*, see Figure 7.

The modul NEWSIM serves for the numerical simulation of the generated symbolic equations of motion. It is easy to study the influence of parameters or to optimize the dynamical behaviour with respect to some specified criteria. NEWSIM has the possibility to treat additional differential or differential-algebraic equations. For

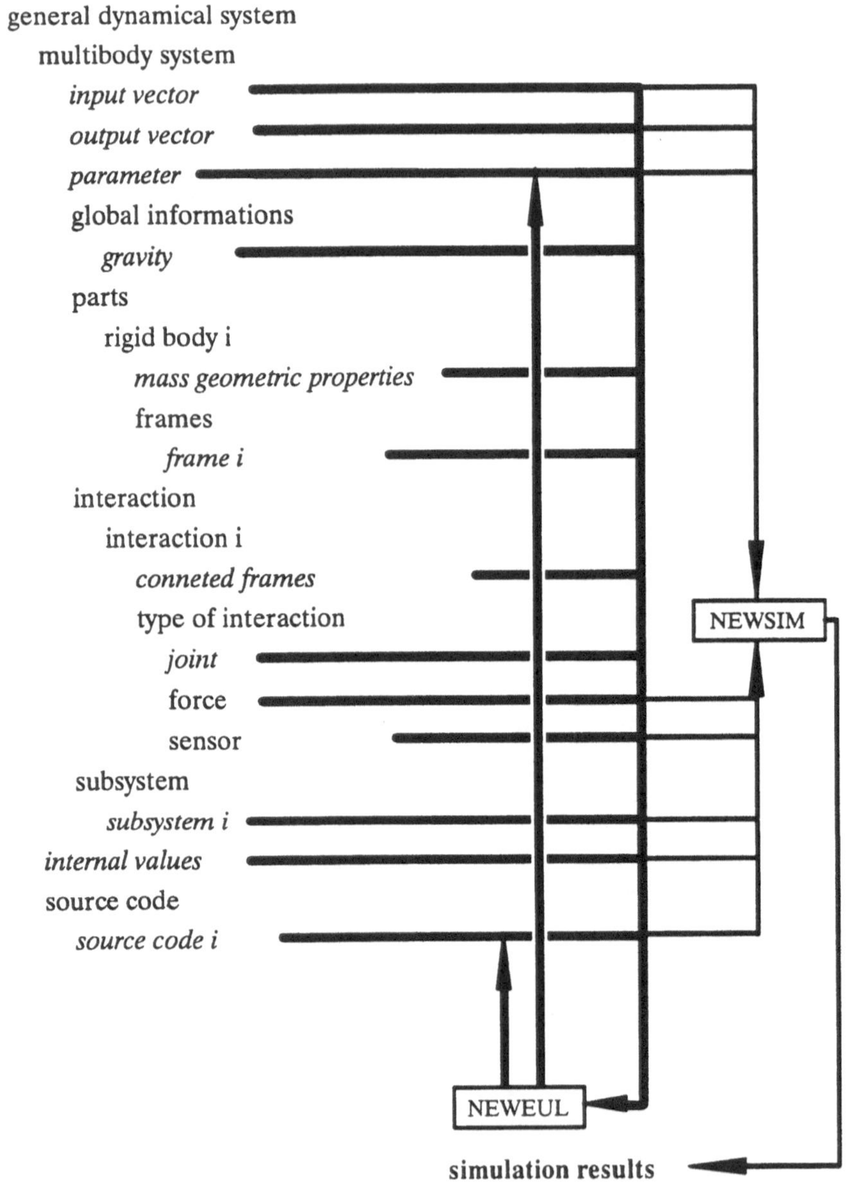
general dynamical system
multibody system
input vector
output vector
parameter
global informations
gravity
parts
rigid body i
mass geometric properties
frames
frame i
interaction
interaction i
conneted frames
type of interaction
joint
force
sensor
subsystem
subsystem i
internal values
source code
source code i
NEWSIM
NEWEUL
simulation results

Figure 7: Dataflow of the datamodel

integration in the time-domain different integration schemes are e.g. Runge-Kutta methods, Adams-methods, BDF-methods. For multibody systems including closed loops a modified Adams-Bashforth-Moulton method is implemented, see Leister [26]. All neccesary routines for the automatic simulation software are generated by NEWEUL, Figure 8. After the compilation and binding step the problem-specific programm takes all parameters and options from the datafile. This program reads all options, initial conditions, fixed system parameters like masses, moments of inertia, geometric data, stiffness constants, and further data from the input file and solves the equations of motion of the problem.

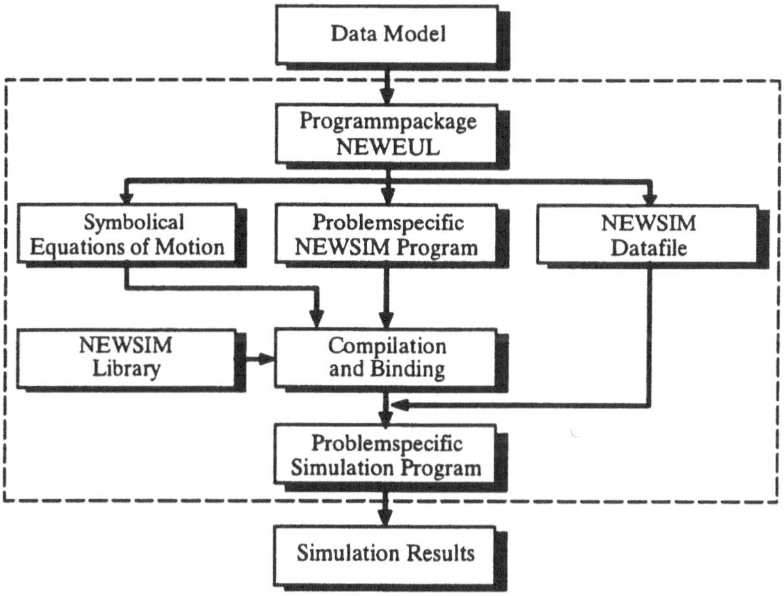

Figure 8: Simulation of the dynamic behaviour with NEWEUL and NEWSIM

5. Formalism for multibody systems using coordinate partitioning

Modeling dynamical systems by the method of multibody systems results in either ordinary differential equations (ODEs) using minimal coordinates or coupled differential and algebraic equations using cartesian and redundant coordinates (DAEs). Often ODEs are integrated numerically by explicit multistep integration algorithms whereas DAEs have to be integrated by implicit or halfimplicit methods. Numerical experiments have shown, Leister [26], that the integration algorithms for ODEs seems to be more efficient than algorithms for DAEs. Thus, it is advantageous to describe multibody systems by a minimal number of pure differential equations, the so-called state space form.

Consider a mechanical system modelled by e generalized coordinates $\boldsymbol{x} = [x_1, \ldots, x_e]^T$ and subject to q holonomic constraints represented by at least twice differentiable functions $\boldsymbol{\Phi}(\boldsymbol{x}, t) = [\Phi_1(\boldsymbol{x}, t), \ldots, \Phi_q(\boldsymbol{x}, t)]^T$. The governing equations of constrained motion of the system can be written in the following DAE form, see e.g. Ref. [23]

$$\boldsymbol{M}(\boldsymbol{x}, t)\, \ddot{\boldsymbol{x}} = \boldsymbol{h}(\dot{\boldsymbol{x}}, \boldsymbol{x}, t) + \boldsymbol{Q}(\boldsymbol{x}, t)\, \boldsymbol{g} , \tag{3}$$

$$\boldsymbol{\Phi}(\boldsymbol{x}, t) = \boldsymbol{0} , \tag{4}$$

where $\boldsymbol{M}$ is the $e \times e$ symmetric positive-definite mass matrix; $\boldsymbol{h}$ represents the components of applied forces on the system and the gyroscopic terms; $\boldsymbol{Q}^T = \boldsymbol{C} = \partial\boldsymbol{\Phi}/\partial\boldsymbol{x}$ is the $q \times e$ constraint matrix; and $\boldsymbol{g} = [\lambda_1, \ldots, \lambda_q]^T$ conserves Lagrange multipliers or generalized constraint forces, respectively. The constraint equations (2) can be differentiated to:

$$\dot{\boldsymbol{\Phi}} = \boldsymbol{C}(\boldsymbol{x}, t)\, \dot{\boldsymbol{x}} + \boldsymbol{a}(\boldsymbol{x}, t) = \boldsymbol{0}, \tag{5}$$

$$\ddot{\boldsymbol{\Phi}} = \boldsymbol{C}(\boldsymbol{x}, t)\, \ddot{\boldsymbol{x}} + \boldsymbol{b}(\dot{\boldsymbol{x}}, \boldsymbol{x}, t) = \boldsymbol{0}, \tag{6}$$

where $\boldsymbol{a} = \partial\boldsymbol{\Phi}/\partial t$, and $\boldsymbol{b} = \dot{\boldsymbol{C}}\dot{\boldsymbol{x}} + \dot{\boldsymbol{a}}$.

The coordinate partitioning method makes use of the fact that only $f = e - q$ from the e *initial* coordinates $\boldsymbol{x}$ are *independent*, denoted $\boldsymbol{y} = [y_1, \ldots, y_f]^T$; the others are refered to as the *dependent* coordinates in the meaning of this method, $\boldsymbol{x}_D = [x_{D1}, \ldots, x_{Dq}]^T$. Thus, according to the symbolic partition,

$$\boldsymbol{x} = \begin{bmatrix} \boldsymbol{y}^T & \boldsymbol{x}_D^T \end{bmatrix}^T , \tag{7}$$

the constraint equations (5) can be rewritten as

$$\boldsymbol{C}_I(\boldsymbol{x}, t)\, \dot{\boldsymbol{y}} + \boldsymbol{C}_D(\boldsymbol{x}, t)\, \dot{\boldsymbol{x}}_D + \boldsymbol{a}(\boldsymbol{x}, t) = \boldsymbol{0}. \tag{8}$$

For clarity in the mathematical formulation, this symbolic notation, partitioned relative to independent and dependent coordinates, will be used through the whole paper. In numerical algorithms, however, it is usually more convenient to complete this task by assigning appropriate *addresses* to the entries of matrices and vectors being partitioned.

If constraints (4) are independent, $rank(\boldsymbol{C}) = q$, there exist at least one set of $\dot{\boldsymbol{x}}_D$ such that the corresponding square submatrix $\boldsymbol{C}_D$ is nonsingular, $det(\boldsymbol{C}_D) \neq 0$.

This enables one to express $\dot{\boldsymbol{x}}_D$ as linear combinations of $\dot{\boldsymbol{y}}$, and then $\ddot{\boldsymbol{x}}_D$ as linear combination of $\ddot{\boldsymbol{y}}$, i.e.:

$$\dot{\boldsymbol{x}}_D = -\boldsymbol{C}_D^{-1}(\boldsymbol{C}_I\dot{\boldsymbol{y}} + \boldsymbol{a}) = \boldsymbol{A}(\boldsymbol{x},t)\,\dot{\boldsymbol{y}} + \boldsymbol{\eta}(\boldsymbol{x},t), \tag{9}$$

$$\ddot{\boldsymbol{x}}_D = \boldsymbol{A}(\boldsymbol{x},t)\ddot{\boldsymbol{y}} + \boldsymbol{\xi}(\dot{\boldsymbol{y}},\boldsymbol{x},t). \tag{10}$$

Using (9) and (10), the following interdependences between the initial and independent velocities and accelerations can be introduced:

$$\dot{\boldsymbol{x}} = \begin{bmatrix} \dot{\boldsymbol{y}} \\ \dot{\boldsymbol{x}}_D \end{bmatrix} = \begin{bmatrix} \boldsymbol{I}^{(f)} \\ \boldsymbol{A} \end{bmatrix} \dot{\boldsymbol{y}} + \begin{bmatrix} \boldsymbol{0} \\ \boldsymbol{\eta} \end{bmatrix} = \boldsymbol{D}^T\dot{\boldsymbol{y}} + \begin{bmatrix} \boldsymbol{0} \\ \boldsymbol{\eta} \end{bmatrix}, \tag{11}$$

$$\ddot{\boldsymbol{x}} = \boldsymbol{D}^T\ddot{\boldsymbol{y}} + \begin{bmatrix} \boldsymbol{0} \\ \boldsymbol{\xi} \end{bmatrix}, \tag{12}$$

where $\boldsymbol{I}^{(f)}$ denotes the $f \times f$ identity matrix; and $\boldsymbol{0}$ is the f-dimensional null vector.

The $f \times e$ matrix $\boldsymbol{D}(\boldsymbol{x},t)$ is a priori of maximal rank, and is an orthogonal complement matrix to the constraint matrix $\boldsymbol{C}$ in the e-space of the system's configuration, i.e. $\boldsymbol{D}\boldsymbol{C}^T = \boldsymbol{0}$. Thus, the columns of $\boldsymbol{D}^T$ are (contravariant) components of vectors $\underline{\boldsymbol{d}}_j (j = 1, \ldots, f)$ which span the *tangent* subspace in the e-space. On the other hand, the columns of $\boldsymbol{C}^T$ are (covariant) components of constraint vectors $\underline{\boldsymbol{c}}_i$ $(i = 1, \ldots, q)$ which span the *orthogonal* (or *constrained*) subspace. The tangent and orthogonal subspaces complement each other in the e-space, and $\boldsymbol{D}\boldsymbol{C}^T = \boldsymbol{0}$ expresses the orthogonality conditions $\underline{\boldsymbol{d}}_j \cdot \underline{\boldsymbol{c}}_i = 0$ $(j = 1(1)f;\ i = 1(1)q)$. Since of linear independence, $\underline{\boldsymbol{c}}_1, \ldots, \underline{\boldsymbol{c}}_q, \underline{\boldsymbol{d}}_1, \ldots, \underline{\boldsymbol{d}}_f$ span a new base $\boldsymbol{e}_{cd} = \begin{bmatrix} \boldsymbol{e}_c^T & \boldsymbol{e}_d^T \end{bmatrix}^T$ in the e-space, where $\boldsymbol{e}_c = \begin{bmatrix} \underline{\boldsymbol{c}}_1, \ldots, \underline{\boldsymbol{c}}_q \end{bmatrix}^T$ and $\boldsymbol{e}_d = \begin{bmatrix} \underline{\boldsymbol{d}}_1, \ldots, \underline{\boldsymbol{d}}_f \end{bmatrix}^T$ are the base vectors of orthogonal and tangent subspaces, respectively. The transformation formula between the (covariant) bases $\boldsymbol{e}_{cd}$ and $\boldsymbol{e}_x$ is

$$\boldsymbol{e}_{cd} = \begin{bmatrix} \boldsymbol{e}_c \\ \boldsymbol{e}_d \end{bmatrix} = \begin{bmatrix} \boldsymbol{C}\boldsymbol{M}^{-1} \\ \boldsymbol{D} \end{bmatrix} \boldsymbol{e}_x = \boldsymbol{T}_{cd}\,\boldsymbol{e}_x\,, \tag{13}$$

where $\boldsymbol{e}_x = [\underline{\boldsymbol{k}}_1, \ldots, \underline{\boldsymbol{k}}_e]^T$ are the base vectors spanning the directions of $\boldsymbol{x}$. The appearance of $\boldsymbol{M}^{-1}$ in the upper part of $\boldsymbol{T}_{cd}$ comes evident after a little inspection. Since $\boldsymbol{C}^T$ contains covariant components of the base vectors of the orthogonal subspace, the transformation between the covariant base vectors $\boldsymbol{e}_c$ and $\boldsymbol{e}_x$ requires the $\boldsymbol{C}\boldsymbol{M}^{-1}$ formula. On the other hand, $\boldsymbol{D}^T$ contains contravariant components, and the transformation between $\boldsymbol{e}_d$ and $\boldsymbol{e}_x$ is defined by matrix $\boldsymbol{D}$. For details refer to Blajer [27].

Using the above definitions, the dynamic equations (3) can be projected into the base $\boldsymbol{e}_{cd}$, which is equivalent to the left-sided premultiplication of these equations by $\boldsymbol{T}_{cd}$. The tangential projection (into $\boldsymbol{e}_d$ base), after considering (11) and (12), leads to the minimal set of constraint reaction-free (or canonical) dynamic equations in independent coordinates

$$M_d(x,t)\,\ddot{y} = h_d(\dot{y},x,t)\,, \tag{14}$$

where

$$M_d = DMD^T, \tag{15}$$
$$h_d = D\left(h - M\left[0^T\ \xi^T\right]^T\right). \tag{16}$$

As M is the metric tensor matrix of base e_x, the metric tensor matrix of base e_{cd} can be written as

$$M_{cd} = T_{cd}MT_{cd}^T = \begin{bmatrix} CM^{-1}C^T & 0 \\ 0^T & DMD^T \end{bmatrix} = \begin{bmatrix} M_c & 0 \\ 0^T & M_d \end{bmatrix}, \tag{17}$$

where $M_c = CM^{-1}C^T$ and $M_d = DMD^T$ are the metric tensor matrices of bases e_c and e_d, respectively; and 0 is the $q \times f$ null matrix. The above relation, which will be of use in the following, indicates that the orthogonal and tangent subspaces really complement each other in the e-space.

By appending $\dot{y} = v$ to (14), $2f$ first-order differential equations in v and y follow. However, since M_d and h_d depend on all initial coordinates x, the constraint equation (4) have to be solved at each step of integration for x_D as function of the current values of y, and this process is usually computationally very expensive. To overcome this problem related to the method described, the projection criterion will be presented for a proper choice of the independent coordinates and a symbolical inverse kinematics approach will be proposed.

6. Projective criterion for coordinate partitioning

The projective criterion for coordinate partitioning proposed in this paper deals with a system's configuration space which is not a Cartesian one but an e-dimensional Riemannian space. The norm of a vector in such a space has thus to be redefined according to the vector space algebra. The aspects of contravariant/covariant vector representations are of importance for this definition and for the further base transformations in the e-space, see e.g. Blajer [27]. The transformation matrix T_{cd} defined in (13) is the mapping of the covariant representations k_i^* of vectors $\underline{k}_i = k_i^{*T} e_x^*$ $(i = 1(1)e)$,

$$k_i^* = [0,\ldots,0,1,0,\ldots,0]^T\,, \tag{18}$$

into e^*_{cd} base, i.e.

$$k_i^{*(cd)} = T_{cd} k_i^* = \begin{bmatrix} CM^{-1} \\ D \end{bmatrix} k_i^* . \tag{19}$$

The vector $\underline{k}_i$ defined this way can be interpreted as a *unit* vector along $\dot{x}_i$ direction, $\underline{\dot{x}} \cdot \underline{k}_i = \dot{x}^T k_i^* = \dot{x}_i$, and this elucidate its (covariant) representation in (18). Then, it comes from (19) that the i-th column of CM^{-1} is the (covariant) representation of $\underline{k}_i$ in e_c^* base, whereas the i-th column of D is the (covariant) representation of $\underline{k}_i$ in e_d^* base. Denoting these representations by $k_i^{*(c)}$ and $k_i^{*(d)}$, respectively, it can be written that:

$$\begin{aligned} CM^{-1} &= \left[k_1^{*(c)}\ k_2^{*(c)} \cdots k_e^{*(c)}\right]_{(q\times e)}, \\ D &= \left[k_1^{*(d)}\ k_2^{*(d)} \cdots k_e^{*(d)}\right]_{(f\times e)}, \end{aligned} \tag{20}$$

i.e. $k_i^{*(c)}$ and $k_i^{*(d)}$ are the i-th columns of CM^{-1} and D, respectively.

Using generalized scalar products, $|\underline{k}_i|^2$, $|\underline{k}_i^{(c)}|^2$ and $|\underline{k}_i^{(d)}|^2$ can be written as follows:

$$\begin{aligned} |\underline{k}_i|^2 &= k_i^{*T} M^{-1} k_i^* = M^{-1}(i,i), \\ |\underline{k}_i^{(c)}|^2 &= \left(k_i^{*(c)}\right)^T M_c^{-1} k_i^{*(c)}, \\ |\underline{k}_i^{(d)}|^2 &= \left(k_i^{*(d)}\right)^T M_d^{-1} k_i^{*(d)}, \end{aligned} \tag{21}$$

where $M^{-1}(i,i)$ is the iith entry of M^{-1}; and M_c and M_d are defined in (15). Basing on (21), the following generalized formulation of the projective criterion for coordinate partitioning can be introduced:

$$\cos^2 \alpha_i = \frac{|\underline{k}_i^{(d)}|^2}{|\underline{k}_i|^2} = \frac{(k_i^{*(d)})^T M_d^{-1} k_i^{*(d)}}{M^{-1}(i,i)}, \tag{22a}$$

$$\cos^2 \beta_i = \frac{|\underline{k}_i^{(c)}|^2}{|\underline{k}_i|^2} = \frac{(k_i^{*(c)})^T M_c^{-1} k_i^{*(c)}}{M^{-1}(i,i)} . \tag{22b}$$

The bigger $\cos^2 \alpha_i$ (the smaller $\cos^2 \beta_i$) the closer is $\underline{k}_i$ to the tangent hyperplane and the better x_i as an independent coordinate.

In fact two formulae for the reported criterion have been introduced, (22a) and (22b). Respectively, they express the squared cosines (generalized to the e-spaces) of angels between the vector $\underline{\boldsymbol{k}}_i$ and its projections $\underline{\boldsymbol{k}}_i^{(d)}$ and $\underline{\boldsymbol{k}}_i^{(c)}$ into the tangent and orthogonal subspaces. The matrix $\boldsymbol{M}_d$ used in (22a) is actually the mass matrix of the minimal-dimension dynamic equations (14), and thus is available (more or less explicitly) in its inverted form at each instant of the system motion simulation. The matrix $\boldsymbol{M}_c = \boldsymbol{C}\boldsymbol{M}^{-1}\boldsymbol{C}^T$ used in (22b) has to be formulated and inverted individually. Therefore, the formulation (22a) is recommendable for the reported formulation.

For the current set $\boldsymbol{y}$, the reported criterion can be applied occasionally to check or redefine the choice for $\boldsymbol{y}$ as related those components of $\boldsymbol{x}$ whose corresponding $\cos^2 \alpha_i$ $(i = 1, \ldots, e)$ have the biggest values.

7. Application of inverse kinematics algorithms

The essential shortcoming of the coordinate partitioning method is the necessity of inverting $\boldsymbol{C}_D$ in order to determine $\boldsymbol{A} = -\boldsymbol{C}_D^{-1}\boldsymbol{C}_I$, $\boldsymbol{\eta} = \boldsymbol{C}_D^{-1}\boldsymbol{a}$, and $\boldsymbol{\xi} = \boldsymbol{C}_D^{-1}\boldsymbol{b}$, required for the formulation of equations (12) or (17). During the simulation process $\boldsymbol{C}_D$ has to be inverted at each step of integration, and this may bring some inefficiency in computations.

In this section advantages are emphasized that may arise in the coordinate partitioning approach to the dynamic analysis of constrained mechanical systems by adapting special algorithms of inverse kinematics developed in the field of robotics, and of remarkable importance is a technique developed by Woernle [30]. According to this technique, the kinematic chains are parted into two open chains so that to select relations with a reduced number of unknowns. Then, setting some coordinates to be *frozen* (independent), the recursive relations for the other (dependent) coordinates as function of the *frozen* ones are found without introducing the constraint equations in the form (4), see also Eppinger and Kreuzer [31], and Blajer, Schiehlen and Schirm [32], and Schiehlen and Blajer [33]. These recursive relations are denoted symbolically as

$$\boldsymbol{x}_D = \boldsymbol{x}_D(\boldsymbol{y}, t)\,, \tag{23}$$

and are recognized also as *closing conditions*, Ref. [34]. In fact, (23) are often quite complex, and the amount of labour required for their derivation depends greatly on the skill of the investigator in using the inverse kinematics procedures. Nevertheless, this initial work pays in the further analysis. The (recursive) relations for (9) and (10) are usually not so laborious to be obtained analytically. They can also be derived by using computer symbolical formalisms like NEWEUL [8], [21].

The application of inverse kinematics algorithms benefits in analytical (though recursive) formulae for $\boldsymbol{x}_D$, $\boldsymbol{A}$, $\boldsymbol{\eta}$ and $\boldsymbol{\xi}$. This accelerates usually the numerical formulation of the tangent dynamic equations (14), and the final governing equations of motion can be written in the following simplified $2f$-order form:

$$\hat{\boldsymbol{M}}_d(\boldsymbol{y})\,\dot{\boldsymbol{v}} = \hat{\boldsymbol{h}}_d(\boldsymbol{v},\boldsymbol{y},t)\,, \tag{24a}$$

$$\dot{\boldsymbol{y}} = \boldsymbol{v} \tag{24b}$$

where $\hat{\boldsymbol{M}}_d$ and $\hat{\boldsymbol{h}}_d$ correspond to $\boldsymbol{M}_d$ and $\boldsymbol{h}_d$ defined in (15) and (16) after substituting $\boldsymbol{x} = [\boldsymbol{y}^T \quad \boldsymbol{x}_D^T(\boldsymbol{y},t)]^T$ and $\dot{\boldsymbol{y}} = [\boldsymbol{v}^T \quad (\boldsymbol{A}(\boldsymbol{y},t)\boldsymbol{v} + \boldsymbol{\eta}(\boldsymbol{y},t))^T]^T$, where $\boldsymbol{x}_D(\boldsymbol{y},t)$, $\boldsymbol{A}(\boldsymbol{y},t)$, $\boldsymbol{\eta}(\boldsymbol{y},t)$ and $\boldsymbol{\xi}(\boldsymbol{v},\boldsymbol{y},t)$ represent the recursive formulae from the inverse kinematics. Note that the closing conditions (23) replace the constraint equations (4), i.e. it can be written

$$\hat{\boldsymbol{\Phi}}(\boldsymbol{x},t) = -\boldsymbol{x}_D(\boldsymbol{y},t) + \boldsymbol{x}_D = \boldsymbol{0}\,. \tag{25}$$

Thus, the solution of (24) is released from the problem of constraint violation. Note also that, as all the entries of $\dot{\boldsymbol{x}}$ and $\boldsymbol{x}$ are determined at each step of integration of (24), an eventual transition from one set of $\dot{\boldsymbol{y}}$ to another will not yield any inconsistency in the initial value problem of accordingly reformulated governing equations. Obviously, an appropriate number of recursive formulae (23), (9) and (10) for different possible (or all) sets of $\boldsymbol{y}$ from $\boldsymbol{x}$ has to be prepared in advance.

Consider the planar four-bar linkage shown in Fig.9. In order to build an open-loop system, each of the joints O_i $(i = 1(1)4)$ can be cut. The coordinates $\boldsymbol{x}$ of

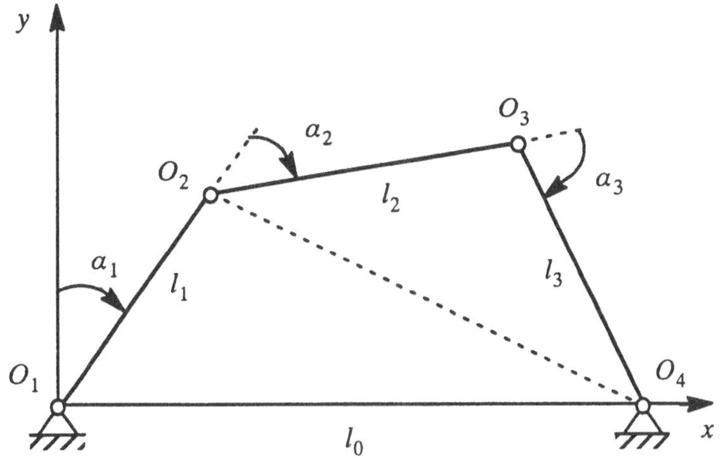

Figure 9: Four-bar mechanism

the *unconstrained* system can also be defined differently. In the example case, the mechanism was cut in joint O_4 and the relative coordinates $\boldsymbol{x} = [\alpha_1,\ \alpha_2,\ \alpha_3]^T$ have been chosen to represent a planar manipulator with the end-effector fixed in point O_4. The dynamic equations of the system, corresponding to (14), will not be reported here.

The constraints of the system can be expressed either implicitly by constraint equations (4) or explicitly by closing conditions (23), which yields respective formulations of matrix $\boldsymbol{D}$ defining the tangent subspace. In the following an application of inverse kinematics algorithms leading to recursive relations for the closing conditions will be demonstrated. The subsequent derivation for the simple example bases on the approach given by Woernle [30]. According to the approach, the mechanism is separated, by cutting in joints O_2 and O_4, into the *lower* and *upper* segments, and the closure condition is

$$\vec{\boldsymbol{r}}_{lower} \circ \vec{\boldsymbol{r}}_{lower} - \vec{\boldsymbol{r}}_{upper} \circ \vec{\boldsymbol{r}}_{upper} = 0 \ , \tag{26}$$

where $\vec{\boldsymbol{r}}_{lower} := O_1O_4 - O_1O_2$, $\vec{\boldsymbol{r}}_{upper} := O_3O_2 - O_3O_4$. Due to the used segmentation, (26) depends on α_1 and α_3 (does not depend on α_2), and only one of these coordinates can be chosen for an independent one in the subsequent derivation (the choice $\boldsymbol{y} = [\alpha_2]$ would require a different segmentation). Here, the relations (23), (11) and (12) are reported only for $\boldsymbol{y} = [\alpha_1]$.

Solving (26), one obtaines

$$\alpha_3 = \pm \arccos\left(\frac{l_0^2 + l_1^2 - l_2^2 - l_3^2 - 2l_0l_1 \sin\alpha_1}{2l_2l_3}\right) \ . \tag{27}$$

The complementary relation for α_2 is obtained then as suggested in [30] from

$$\begin{aligned} \sin\alpha_2 &= \frac{l_0l_2\cos\alpha_1 + l_0l_3\cos(\alpha_1+\alpha_3) + l_1l_3\sin\alpha_3}{l_0^2 + l_1^2 - 2l_0l_1\sin\alpha_1} \ , \\ \cos\alpha_2 &= \frac{l_0l_2\sin\alpha_1 + l_0l_3\sin(\alpha_1+\alpha_3) - l_1l_3\cos\alpha_3 - l_1l_2}{l_0^2 + l_1^2 - 2l_0l_1\sin\alpha_1} \ . \end{aligned} \tag{28}$$

The relations (27) and (28) express recursively $\alpha_3(\alpha_1)$ and $\alpha_2(\alpha_1)$ as defined in (23). Differentiation of these closing conditions leads to:

$$\begin{aligned} \dot{\alpha}_3 &= \frac{l_0l_1}{l_2l_3}\frac{\cos\alpha_1}{\sin\alpha_3}\dot{\alpha}_1 \ , \\ \dot{\alpha}_2 &= -\frac{l_0}{l_3}\frac{\cos(\alpha_1+\alpha_2)}{\sin\alpha_3}\dot{\alpha}_1 - \dot{\alpha}_3 \ , \end{aligned} \tag{29}$$

and therefore, the matrix $\boldsymbol{D}$ defined in (11) can be stated as

$$\boldsymbol{D} = \begin{bmatrix} 1 & -\dfrac{l_0 l_1 \cos\alpha_1 + l_0 l_2 \cos(\alpha_1 + \alpha_2)}{l_2 l_3 \sin\alpha_3} & \dfrac{l_0 l_1}{l_2 l_3} \cdot \dfrac{\cos\alpha_1}{\sin\alpha_3} \end{bmatrix} . \tag{30}$$

After differentiating (29) the vector $\boldsymbol{\xi}$ introduced in (12) can be formed as shown by Blajer, Schiehlen and Schirm [32].

Now the problem of the *best* independent coordinate choice is discussed. For particular linkage data, the results obtained by using the projective criterion are shown in Fig.10. The linkage geometry was set to assure that the choice $\boldsymbol{y} = [\alpha_1]$ never leads to a singularity, enabling one to plot the results throughout the whole range $\alpha_1 \in < 0, 2\pi >$. As seen, both α_1 and α_2 are acceptable choices for $\boldsymbol{y}$ in any linkage configuration, none of them, however, can be assigned to be the *best* independent coordinate over the hole range of α_1. It is evident, also, that the choice $\boldsymbol{y} = [\alpha_3]$ is the worst as leading to singularities at $\alpha_1 = \pi/2$ and $\alpha_1 = 3\pi/2$, and due to relatively small values of tangent projections.

The numerical simulation can now be carried out in the *best* independent coordinate according to the projective criterion, i.e. changes between the coordinates α_1 and α_2 are necessary to ensure the integration with the *best* coordinate. This change can be done without a loss in integration order and stepsize by using a modified Adams-Bashforth integration code.

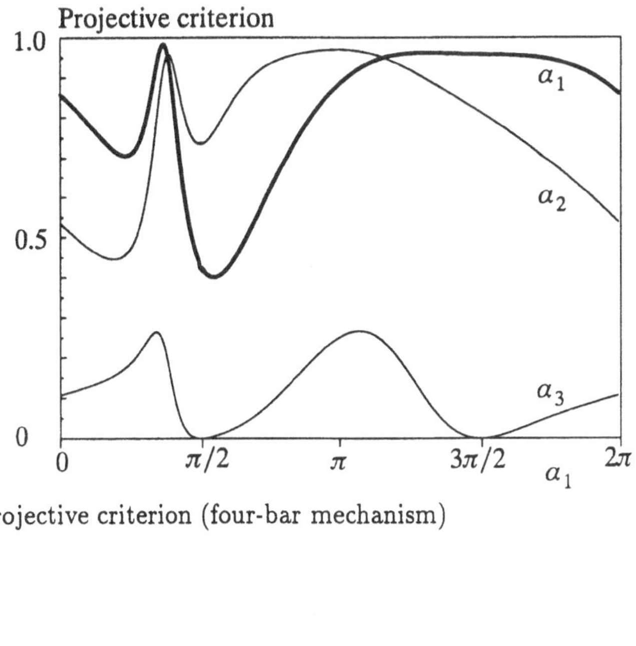

Figure 10: Projective criterion (four-bar mechanism)

8. Visualization of simulation results

A convenient verification a dynamic visualization of a multibody system simulation is obtained by a 3D computer graphics animation. Animation methods differ according to the geometry model, rendering algorithms and possible user interaction. The most sophisticated animation method is achieved by rendering algorithms like raytracing and radiosity. These rendering techniques result in realistic images, but suffer from time-consuming computations. During image display, no interactive modification of the view projection is possible. A raytraced image of the crank-slider mechanism is shown in Figure 11.

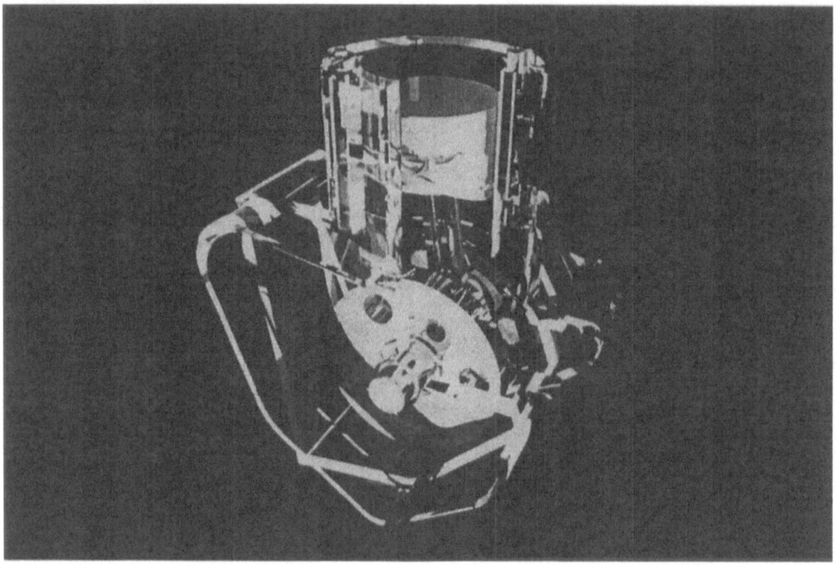

Figure 11: Raytracing of crank slider mechanism

Most CAD-3D-systems offer modules for the generation of images with hidden line and hidden surfaces removal and shaded surfaces. Often, the solid model and rendering algorithms yield sophisticated 2D drawings for documentation purposes, but allow a dynamic visualization only in a wireframe mode.

Consequently the unified approach to display a broad variety of simulation result for different initial conditions, visualization systems and applications is based on the planar face model. The visualization module VISANI for the interactive, high speed animation of arbitrary multibody systems is described by Daberkow [17]. As a result of the simulation, a time plot of the crankshaft bearing force of the mechanism under an applied piston gas force and an animated sequence is shown in Figure 12.

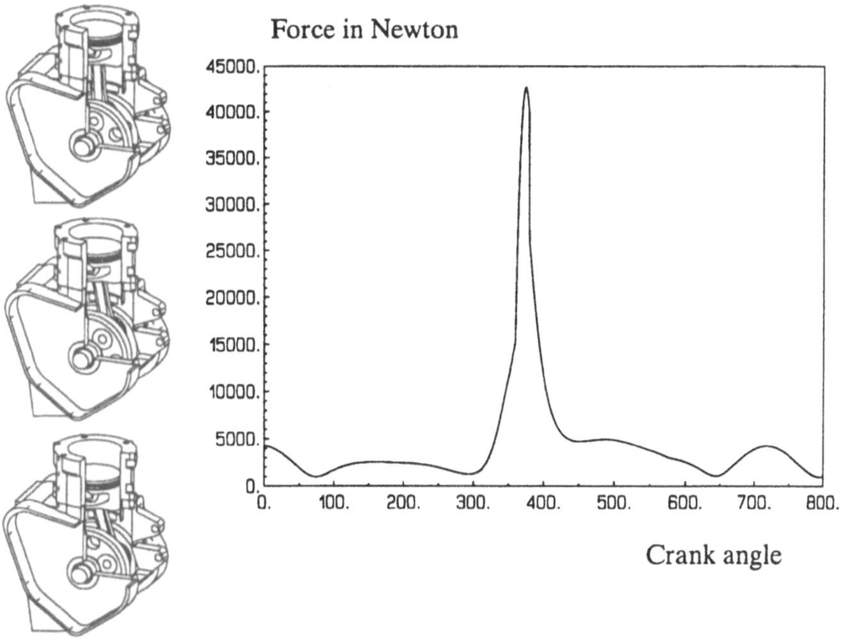

Figure 12: Time plot and animated sequences of the crank slider mechanism

9. Flexible multibody systems

A free rigid body i has $e_{is} = 6$ degrees of freedom which is also true for the reference frame of an elastic body in its undeformed configuration. The deformation of a flexible body depends on a displacement field $\boldsymbol{u} = \boldsymbol{u}(\boldsymbol{c}, t)$ characterizing the position of a material particle identified by the vector $\mathbf{c}$ in the undeformed configuration. Assuming a discretization of the elastic body, the displacement field is expressed as a linear combination of selected deformation modes

$$\boldsymbol{u}(\boldsymbol{c}, t) = \boldsymbol{\Phi}(\boldsymbol{c})\boldsymbol{q}(t) \tag{31}$$

where the vector $\boldsymbol{q}(t) = [q_1, \ldots, q_{e_{if}}]^T$ represents the e_{if} flexible coordinates of the body i and $\boldsymbol{\Phi}(\boldsymbol{c})$ means the corresponding shape function vector of the deformation modes. The flexible coordinates enlarge the number of degrees of freedom of body i to $e_i = e_{is} + e_{if}$. The same is true for the total multibody system with p bodies, $i = 1(1)p$. On the other hand, the q holonomic constraints reduce the number of degrees of freedom. As result, one obtains for the flexible system's degrees of freedom $f = e_s - q + e_f$. Then, the global $f \times 1$ position vector of generalized coordinates can be partitioned as

$$y_q = \left[y^T \; q^T \right]^T \tag{32}$$

where the $(e_s - q) \times 1$ vector $\boldsymbol{y}$ denotes the rigid body motion and the $e_f \times 1$ vector $\boldsymbol{q}$ describes the elastic coordinates of the system. Further, it is assumed that the elastic coordinates are small compared to the rigid body motion of the system.

The equations of motion of the flexible multibody systems are an extension of (24) and result in

$$\begin{bmatrix} \hat{M}_d(y) & \hat{M}_{de}(y_q) \\ \hat{M}_{de}^T(y_q) & \hat{M}_e \end{bmatrix} \begin{bmatrix} \ddot{y} \\ \ddot{q} \end{bmatrix} + \begin{bmatrix} 0 & 0 \\ 0 & \hat{K}_e \end{bmatrix} \begin{bmatrix} y \\ q \end{bmatrix} = \hat{h}_q(y_q, \dot{y}_q, t) \tag{33}$$

where the $e_s \times e_f$ matrix $\hat{M}_{de}$ represents the coupling between flexible and rigid body motion and the $e_f \times e_f$ matrices $\hat{M}_e$ and $\hat{K}_e$ characterizes the small structural vibrations of the system.

For a more detailed evaluation it requires a proper choice of the deformation modes characterized by the shape functions $\boldsymbol{\Phi}(\mathbf{c})$, see (31). An approach closely related to rigid body modeling uses the rigid-elastic superelement, Fig. 13, and was presented by Rauh [35].

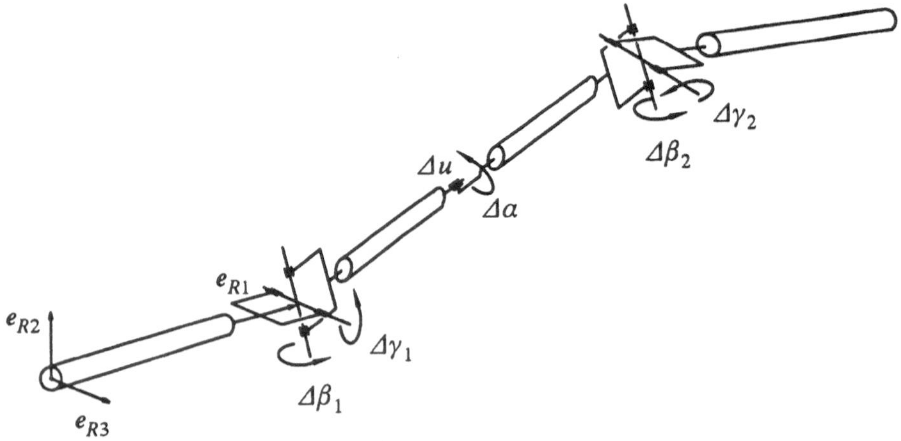

Figure 13: Rigid-elastic superelement for a flexible beam

The rigid-elastic superelement represents a piecewise constant shape function and the elastic properties result in discrete spring coefficients with respect to the elastic coordinates $\Delta\alpha$, $\Delta\beta_1$, $\Delta\beta_2$, $\Delta\gamma_1$, $\Delta\gamma_2$ and Δu. The advantage of this approach is that standard rigid body codes like NEWEUL can be applied for the generation of the equations of motion.

Another approach related to the finite element method uses standard finite elements, Fig. 14, and was presented by Bremer and Pfeiffer [36], Melzer [37] and Sorge [38]. It turns out that most of the volume integrals summarized in (33) are time-invariant

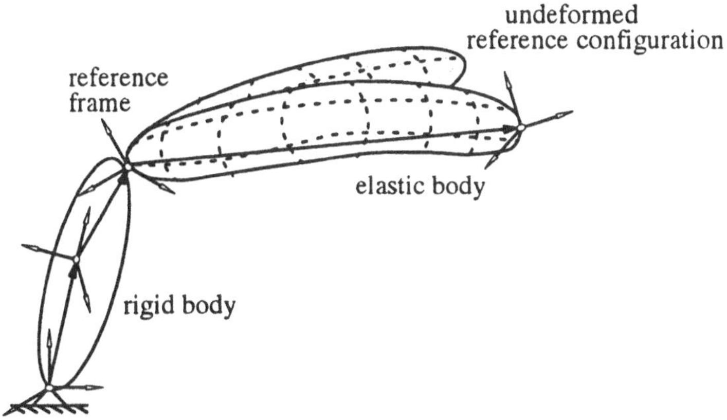

Figure 14: Finite element representation of an elastic body

and may be precomputed, see Melzer [37]. This means that flexible multibody systems with small elastic displacements can be analysed using symbolic computations. Compared to purely numerical computations by the standard finite element method, the symbolic computations are much more efficient.

10. Optimization of multibody systems

Optimization is considered as an integrated modeling and parameter evaluation process for multibody systems. Computer aided generation of symbolic equations of motion and sensitivity analysis helps to cut down the time for designing such systems. The method will be presented for vehicle suspension systems following Bestle, Eberhard and Schiehlen [39]

Designing a mechanical system by computer aided engineering involves several steps: (i) modeling, (ii) choosing design variables, (iii) defining performance criteria and constraints, (iv) solving the optimization problem resulting in optimal values for the design variables. Finally, the optimal system has to be realized, and there may occur some economic constraints requiring a reformulation of the design problem. This may cause changes in the model, in the choice of the design variables, and in definitions of performance criteria, or it requires additional constraints.

Multibody system models are well accepted for simulating the dynamic behavior of vehicle systems in a frequency domain less than $50Hz$. For low frequency motions the carbody, the driver, and the axles can be considered as rigid bodies connected by ideal links and coupled by force elements like springs, dampers, and active elements, Fig. 15.

Springs in a suspension system act on both the carbody and the axles. For linear springs the force proportional to the relative displacement reads as

$$F_c = c_F(z_P - z_Q) \tag{34}$$

where c_F is the stiffness coefficient. The force of a linear damper is proportional to the relative velocity:

$$F_d = d_F(\dot{z}_P - \dot{z}_Q) \,. \tag{35}$$

For improved modeling nonlinear characteristics like a cubic damper may be used, too:

$$F_d = d_F(\dot{z}_P - \dot{z}_Q)^3 \,. \tag{36}$$

For more effective damping of the carbody motions forces proportional to the absolute velocity of the carbody, i.e.

$$F_d = d_A \dot{z}_P \,, \tag{37}$$

may be better suited than a relative damper (35). Such a device is called a *skyhook damper* which has to be realized by an active element.

To some extent, the designer of a suspension system has a free choice of the values of the stiffness coefficients c_F, damping coefficients d_F, and control parameters d_A. Any of these parameters may be chosen as a design variable. But geometrical data like the height of the center of gravity of the carbody or the masses and moments of inertia of some of the bodies may be changed within given ranges, too. This will result in a different behavior due to changes in the inertia matrix and the generalized forces, respectively.

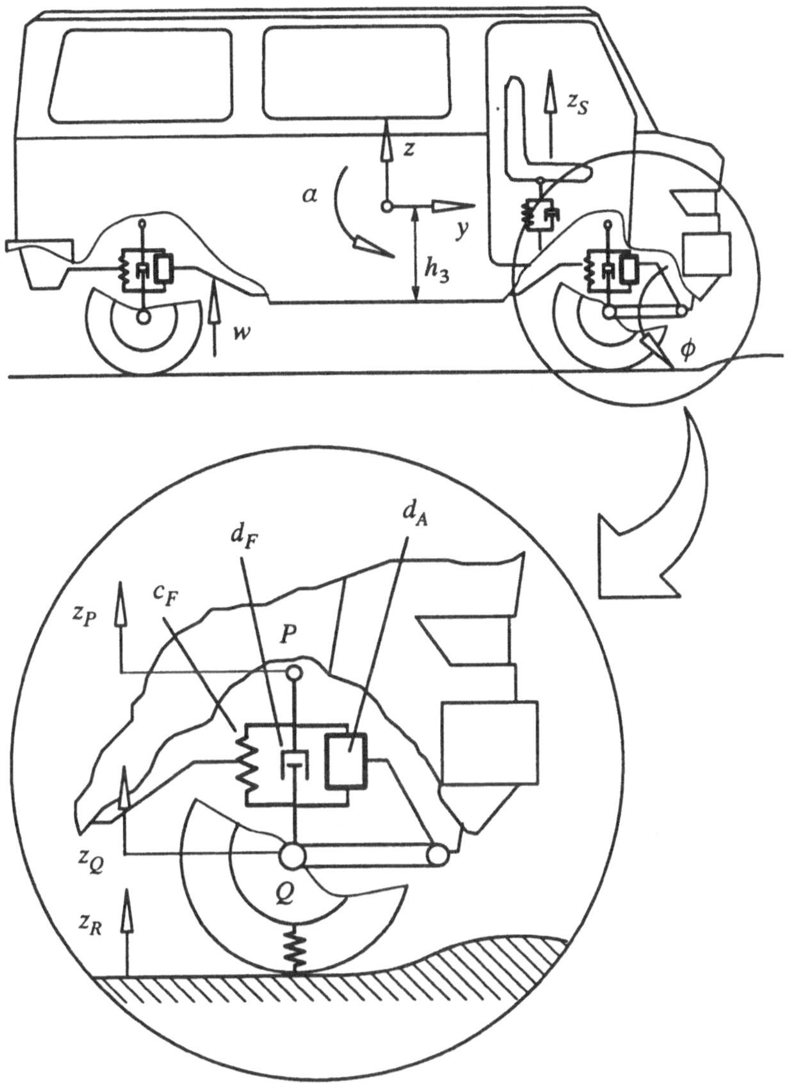

Figure 15: Vehicle modeled for plane motion

After summarizing all design variables in a $h \times 1$ vector $\boldsymbol{p}$, the equations of motion (24) have to be rewritten as

$$\boldsymbol{M}(\boldsymbol{y},\boldsymbol{p})\,\dot{\boldsymbol{v}} = \boldsymbol{h}(\boldsymbol{v},\boldsymbol{y},t,\boldsymbol{p}) \,, \tag{38a}$$

$$\dot{\boldsymbol{y}} = \boldsymbol{v} \tag{38b}$$

These equations are supplemented by initial conditions for the state coordinates $\boldsymbol{y}$ and $\boldsymbol{v}$ at some given time t^0:

$$\begin{aligned} \boldsymbol{y}^0 &:= \boldsymbol{y}(t^0): \quad \boldsymbol{\Phi}^0(t^0,\boldsymbol{y}^0,\boldsymbol{p}) = \boldsymbol{0}, \qquad det\frac{\partial \boldsymbol{\Phi}^0}{\partial \boldsymbol{y}^0} \neq 0, \\ \boldsymbol{v}^0 &:= \boldsymbol{v}(t^0): \quad \dot{\boldsymbol{\Phi}}^0(t^0,\boldsymbol{y}^0,\boldsymbol{v}^0,\boldsymbol{p}) = \boldsymbol{0}, \quad det\frac{\partial \dot{\boldsymbol{\Phi}}^0}{\partial \boldsymbol{v}^0} \neq 0, \end{aligned} \tag{39}$$

where the Jacobians of $\boldsymbol{\Phi}^0$ and $\dot{\boldsymbol{\Phi}}^0$ have to be regular for determining the initial state $\boldsymbol{y}^0$ and $\boldsymbol{v}^0$ uniquely.

The application of systematic methods for searching optimal values of the design variables requires the definition of performance criteria. An important function of suspension systems is to improve the riding comfort by isolating the carbody from roadway unevenness. The comfort of a vehicle can be evaluated by the acceleration $\ddot{z}_s$ acting on the driver. If the driving over a bump is considered as a test, accelerations may be penalized by the square of the time, too. An integral performance criterion can then be expressed by

$$\psi_C = \int_{t^0}^{t^1} t^2 \ddot{z}_s^2 dt \,. \tag{40}$$

Another important task of a suspension system is to provide safety which is related to the dynamic variation in the load between the wheels and the road. If the tire is considered as a linear spring, the load is proportional to the relative displacement between wheel and road surface, Fig. 15. A performance function evaluating the safety of a vehicle is

$$\psi_S = \int_{t^0}^{t^1} (z_Q - z_R)^2 dt \,. \tag{41}$$

Carbody isolation results in very soft suspensions, but this will also yield large relative motions between carbody and axles. For limiting the relative displacements, a criterion like

$$\psi_D = \int_{t^0}^{t^1} (\frac{z_P - z_Q}{s_0})^6 dt \,. \tag{42}$$

may be used where s_0 is a predefined amplitude following from the vehicle design, which should not be exceeded to much.

The final time t^1 in the criteria (40)-(42) may be defined by the user or given implicitly by some final state $\boldsymbol{y}^1 = \boldsymbol{y}^1(t^1)$, $\boldsymbol{v}^1 = \boldsymbol{v}^1(t^1)$:

$$\begin{aligned} t^1 \ &: \ H^1(t^1, \boldsymbol{y}^1, \boldsymbol{v}^1, \boldsymbol{p}) = 0, \\ &\dot{H}^1 := \frac{dH^1}{dt^1} = \frac{\partial H^1}{\partial t^1} + \left(\frac{\partial H^1}{\partial \boldsymbol{y}^1}\right)^T \boldsymbol{v}^1 + \left(\frac{\partial H^1}{\partial \boldsymbol{v}^1}\right)^T \dot{\boldsymbol{v}}^1 \neq 0\,. \end{aligned} \tag{43}$$

E.g., for the vehicle excited by a bump the final time has to be chosen sufficiently large to guarantee substantial decrease of vibrations.

The problem as stated can be reduced to a nonlinear programming problem. If some value is assigned to each design variable, the state $\boldsymbol{y}(t)$, $\boldsymbol{v}(t)$ is completely determined by the equations of motion (38) and initial conditions (39). Further, the final time is given by equation (43). Then, the criteria (40)-(42) are functions of the design variables only. In engineering problems, there are always some restrictions on the design variables. If there are only lower and upper bounds, the feasible design space is given by

$$P = \left(\boldsymbol{p} \in R^h | p_i^l \le p_i \le p_i^u,\ i = 1(1)h\right)\,. \tag{44}$$

Minimizing each of the criteria (40)-(42) individually represents a nonlinear programming problem which may be solved by any general purpose optimization algorithm. In general, the optimal points in the design space, i.e.

$$\begin{aligned} \boldsymbol{p}_C \ &: \ \psi_C^* := \psi_C(\boldsymbol{p}_C) = \min_{\boldsymbol{p}\in P} \psi_C(\boldsymbol{p}), \\ \boldsymbol{p}_S \ &: \ \psi_S^* := \psi_S(\boldsymbol{p}_S) = \min_{\boldsymbol{p}\in P} \psi_S(\boldsymbol{p}), \\ \boldsymbol{p}_D \ &: \ \psi_D^* := \psi_D(\boldsymbol{p}_D) = \min_{\boldsymbol{p}\in P} \psi_D(\boldsymbol{p}), \end{aligned} \tag{45}$$

will be different from each other. A feasible design where all three criteria have simultaneously minimal values does not exist in general. E.g., high riding comfort requires a very soft suspension whereas low relative displacement between carbody and axles can only be achieved by a stiff suspension. Multicriteria optimization theory, see e.g. Osyczka [40] offers optimal designs in such conflicting situations, too.

An often used method is the weighted objectives method. Instead of the individual criteria a scalar weighted-sum criterion has to be minimized:

$$\psi = w_C \frac{\psi_C}{\psi_C^*} + w_S \frac{\psi_S}{\psi_S^*} + w_D \frac{\psi_D}{\psi_D^*}, \quad w_C + w_S + w_D = 1. \tag{46}$$

The numerical procedure for solving such problems is an iterative process starting from an user-defined design and finding a better one step by step. The application of such procedures for the optimization of vehicle systems is very time consuming. In each step the criterion (46) has to be evaluated which requires the numerical integration of the equations of motion. Therefore, rapidly converging optimization algorithms like the sequential quadratic programming (SQP) method is recommended, see Fletcher [41].

Advanced optimization algorithms like SQP are based not only on function evaluations but also on gradient information of the objective and constraint functions. Since the gradients cannot be computed analytically for engineering systems, some algorithms use the possibility of computing gradients by numerical differentiation, e.g. with the forward difference formula:

$$\nabla\psi_k \approx \frac{\psi(\boldsymbol{p} + \Delta p_k \boldsymbol{e}_k) - \psi(\boldsymbol{p})}{\Delta p_k}, \quad k = 1(1)h\ , \tag{47}$$

where $\Delta p_k \boldsymbol{e}_k$ is a small perturbation of the design variable p_k. But applying difference formulas to the optimization of multibody systems will yield poor results due to the rather large error in evaluating ψ by numerical integration. Therefore, it is advisable to use sensitivity analysis methods for generating analytical information on the gradients.

For the sensitivity analysis of integral type functions like equations (40)-(42) two methods have been developed as reported by Haug [42]: the direct differentiation method and the adjoint variable method. From a computational point of view the latter is preferable and will be discussed here. For a general criterion

$$\psi(\boldsymbol{p}) = G^1(t^1, \boldsymbol{y}^1, \boldsymbol{v}^1, \boldsymbol{p}) + \int_{t^0}^{t^1} F(t, \boldsymbol{y}, \boldsymbol{v}, \dot{\boldsymbol{v}}, \boldsymbol{p}) dt \tag{48}$$

the gradient can be computed as

$$\begin{aligned} \nabla\psi = \frac{d\psi}{dp} &= \frac{\partial G^1}{\partial p} - \tau^1 \frac{\partial H^1}{\partial p} - \left(\frac{\partial \boldsymbol{\Phi}^0}{\partial p}\right)^T \boldsymbol{\zeta}^0 - \left(\frac{\partial \dot{\boldsymbol{\Phi}}^0}{\partial p}\right)^T \boldsymbol{\eta}^0 \\ &+ \int_{t^0}^{t^1} \left[\frac{\partial F}{\partial p} - \left(\frac{\partial(\boldsymbol{M}\dot{\boldsymbol{v}} - \boldsymbol{h})}{\partial p}\right)^T (\boldsymbol{\nu} - \boldsymbol{\xi})\right] dt \end{aligned} \tag{49}$$

where the adjoint variables τ^1, $\boldsymbol{\mu}(t)$, $\boldsymbol{\nu}(t)$, $\boldsymbol{\xi}(t)$, $\boldsymbol{\zeta}^0$, $\boldsymbol{\eta}^0$ have to be obtained from algebraic and differential equations, see Bestle and Eberhard [43]. The variable τ^1 and the final values of $\boldsymbol{\mu}(t^1)$ and $\boldsymbol{\nu}(t^1)$ are given by

$$\begin{aligned} \tau^1 &= \frac{\dot{G}^1 + F^1}{\dot{H}^1}, \\ \boldsymbol{\mu}(t^1) &= \frac{\partial G^1}{\partial \boldsymbol{y}^1} - \tau^1 \frac{\partial H^1}{\partial \boldsymbol{y}^1}, \\ \boldsymbol{M}^1 \boldsymbol{\nu}(t^1) &= \frac{\partial G^1}{\partial \boldsymbol{v}^1} - \tau^1 \frac{\partial H^1}{\partial \boldsymbol{v}^1}. \end{aligned} \tag{50}$$

The functions $\boldsymbol{\mu}(t)$ and $\boldsymbol{\nu}(t)$ have to be computed by numerical backward integration of the adjoint differential equations

$$\begin{aligned} \dot{\boldsymbol{\mu}} &= \left(\frac{\partial(\boldsymbol{M}\dot{\boldsymbol{v}} - \boldsymbol{h})}{\partial \boldsymbol{y}}\right)^T (\boldsymbol{\nu} + \boldsymbol{\xi}) - \frac{\partial F}{\partial \boldsymbol{y}}, \\ \boldsymbol{M}\dot{\boldsymbol{\nu}} &= -\boldsymbol{\mu} - \dot{\boldsymbol{M}}\boldsymbol{\nu} - \left(\frac{\partial \boldsymbol{h}}{\partial \boldsymbol{v}}\right)^T (\boldsymbol{\nu} + \boldsymbol{\xi}) - \frac{\partial F}{\partial \boldsymbol{v}} \end{aligned} \tag{51}$$

where the auxiliary variables $\boldsymbol{\xi}(t)$ have to be obtained simultaneously from the linear algebraic equations

$$\boldsymbol{M}\boldsymbol{\xi} = \frac{\partial F}{\partial \boldsymbol{v}} \tag{52}$$

Finally, $\boldsymbol{\eta}^0$ and $\boldsymbol{\zeta}^0$ are available successively from the linear algebraic equations

$$\left(\frac{\partial \dot{\boldsymbol{\Phi}}^0}{\partial \boldsymbol{v}^0}\right)^T \boldsymbol{\eta}^0 = \boldsymbol{M}^0 \boldsymbol{\nu}(t^0), \qquad \left(\frac{\partial \boldsymbol{\Phi}^0}{\partial \boldsymbol{y}^0}\right)^T \boldsymbol{\zeta}^0 = \boldsymbol{\mu}(t^0) - \left(\frac{\partial \dot{\boldsymbol{\Phi}}^0}{\partial \boldsymbol{y}^0}\right)^T \boldsymbol{\eta}^0. \tag{53}$$

Further studies have shown the adjoint variable method to be suitable for symbolic computations, too. In particular, the partial derivatives with respect to the design variables and the state variables are computed automatically by the programmable formula manipulation package MAPLE [44]. The resulting equations of motion and the adjoint differential equations are then solved by numerical integration. Because of the complexity of these equations and the dependence of the adjoint equations on state variables it is advantageous to use multistep integration algorithms and a corresponding interpolation scheme, as shown by Bestle and Eberhard [43].

The planar model of a vehicle, Fig. 15, consists of 4 rigid bodies. The 6×1 vector $\boldsymbol{y}$ summarizing the generalized coordinates reads as

$$\boldsymbol{y} = [y,\ z,\ \alpha,\ \phi,\ w,\ z_S]^T\ . \tag{54}$$

The dynamic behavior will be described by a set of twelve first order differential equations of motion. The vehicle is assumed to drive with a constant velocity of $20m/s$ over a sinusoidal bump of height $0.1m$ and length $3m$, often found in residential areas as "sleeping policemen". The vehicle suspension is modeled by force elements (34)-(37) in parallel configuration. If the control parameter d_A is zero, the suspension is called passive, otherwise it is active. The dampers may be linear (34) or have a progressive characteristic (35).

In the following, the damping and stiffness coefficients of the front and rear suspension, i.e. d_F, d_R, c_F, c_R, the height of the center of gravity h_3, and in case of active suspensions the control parameter d_A are chosen as design variables. They can be summarized in the 6×1 vector of design variables

$$\boldsymbol{p} = [d_F,\ d_R,\ c_F,\ c_R,\ h_3,\ d_A]^T\ . \tag{55}$$

Optimization of the vehicle has been performed using criterion (46) with different sets of weighting factors w_c, w_s, w_d. Fig.16 shows the vertical acceleration of the driver for the initial and the optimized design with respect to criterion (40) using the planar car model with progressive dampers.

For both active and passive suspensions optimization leads to an improved riding comfort. In particular, the active suspension reduces the maximum acceleration as well as long term vibrations.

Fig. 17 shows the riding comfort of an optimized design using criteria (40)-(42) one after the other and linear dampers. Considering the relative displacement leads to high maximal accelerations while optimization of the riding safety yields low but poorly damped accelerations. The analysis of conflicting optimization criteria shows that the improvement of one criterion worsens the other criteria. Thus, only a multi-criteria approach will give an engineering trade-off, Fig. 18. E.g., if a weighted-sum criterion is used instead of the comfort criterion (40) only, the improvement in riding comfort is not as high as in Fig. 18. However, the final result shows a strong improvement of the riding comfort compared to the initial situation without much pay-off in riding safety or relative displacement. It will be a matter of engineering intuition to make a good choice on the weighting coefficients w_c, w_s, w_d.

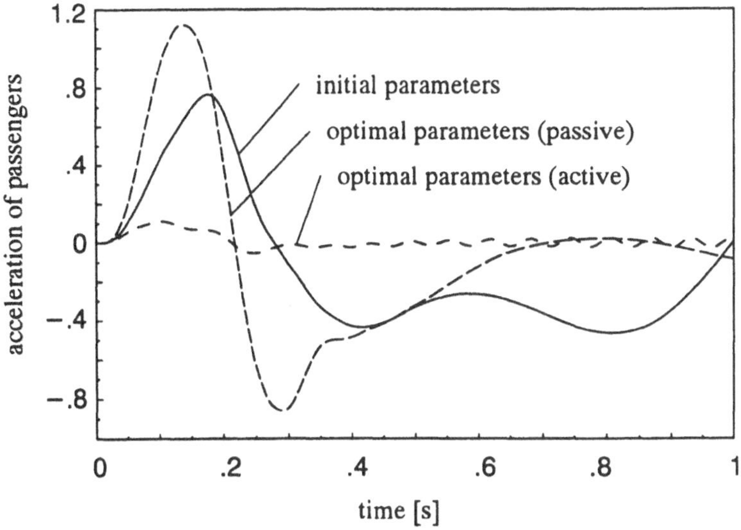

Figure 16: Optimization of ride comfort, progressive damper

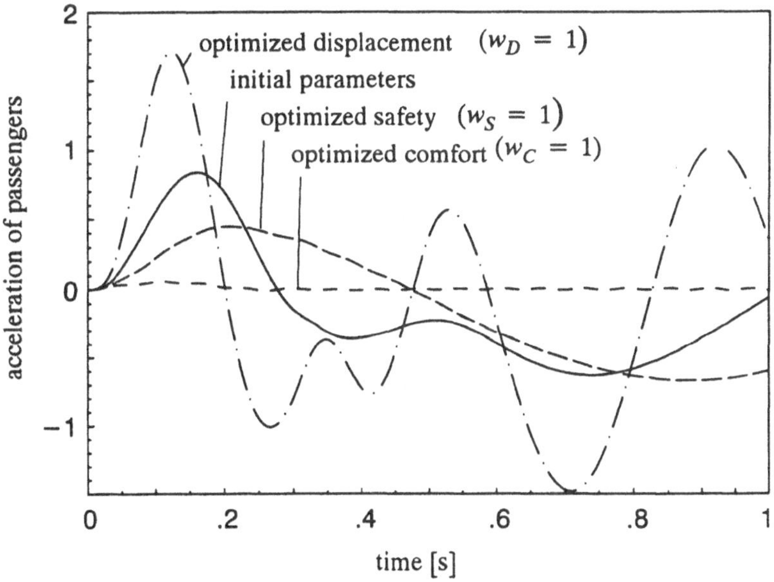

Figure 17: Optimization by single criteria, linear damper

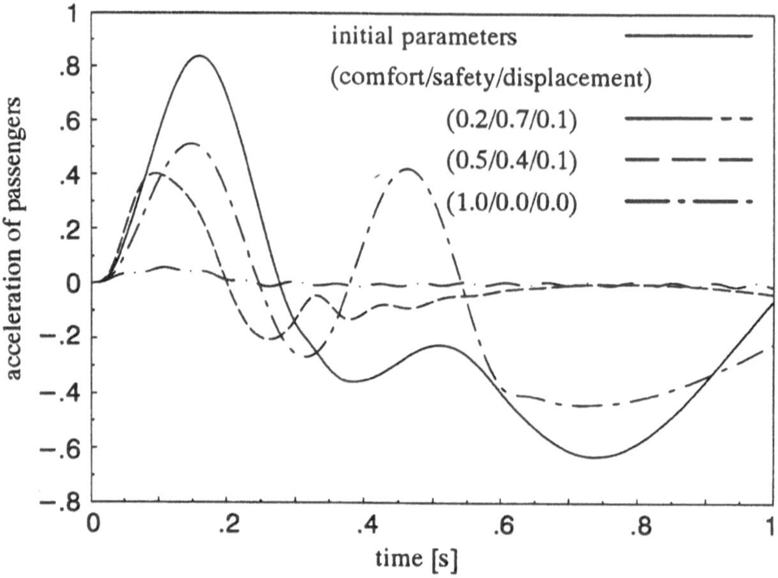

Figure 18: Optimization by mixed criteria,linear damper

11. Conclusion

In this paper an integrated modeling, simulation, visualization, and optimization of multibody system dynamics is introduced. A unified general data model including the graphic description is presented. To support the preceding CAD-3D-modeling stage, a unified spatial graphic representation for multibody elements is designed. Object-oriented classes and operations are then implemented in a system independent multibody modeling kernel library and integrated in a commercial CAD-3D system. From the multibody model data base, an integrated Newton-Euler formalism generates a set of symbolical ordinary differential equations, which are solved by explicit multistep integration algorithms. Thereby, a minimal set of generalized coordinates is specified during numerical integration without restart of the integration algorithm, using the projection method within the coordinate partitioning approach. The visualization of the crank slider mechanism demonstrates that this integrated approach fits the criteria of a modular, automated design and simulation environment. It is shown that symbolic computations are also very efficient in the dynamical analysis of flexible multibody systems. Further, the optimization of design parameters is introduced and the adjoint variable method is presented. For the gradient computation a combination of NEWEUL and MAPLE generated symbolical equations of motion and of adjoint variables. The multicriteria optimization overcomes the conflicts typical for engineering problems. As an example the optimization of a vehicle's suspension system is presented.

References

1 Roberson, R.E., Schwertassek, R. (1988): *Dynamics of Multibody Systems.* Springer, Berlin.

2 Nikravesh, P.E. (1988): *Computer-aided Analysis of Mechanical Systems.* Prentice-Hall, New Jersey.

3 Haug, E.J. (1989): *Computer-aided Kinematics and Dynamics of Mechanical Systems.* Allyn and Bacon, Boston.

4 Shabana, A. (1989): *Dynamics of Multibody Systems.* Wiley, New York.

5 Kortüm, W.; Schiehlen, W. (1985): General purpose vehicle system dynamics software based on multibody formalisms. *Vehicle System Dynamics* **14**, 229-263.

6 Bianchi, G.; Schiehlen, W. (eds.) (1986): *Dynamics of Multibody Systems.* Springer-Verlag, Berlin.

7 Kortüm, W.; Sharp, R. S. (1991): A report on the state-of-affairs on application of multibody computer codes to vehicle system dynamics. *Vehicle System Dynamics* **20**, 177-184.

8 Schiehlen, W. (ed) (1990): *Multibody Systems Handbook.* Springer–Verlag, Berlin.

9 Orlandea, N. (1973): *Node-Analogous Sparsity-Oriented Methods for Simulation of Mechanical Systems.* Ph.D. dissertation, University of Michigan.

10 Haug, E.J. (1989): *Computer Aided Kinematics and Dynamics of Mechanical Systems.* Allyn and Bacon, Boston.

11 Rosenthal, D.E., Sherman, M.A. (1986): High performance multibody simulation via symbolic equation manipulation and Kane's method. *Journal of Astronautical Sciences* **34**, 223-239.

12 Kreuzer, E. (1979): *Symbolische Berechnung der Bewegungsgleichungen von Mehrkörpersystemen*, Ph.D. Dissertation, Stuttgart.

13 ARIES Conceptstation Software Simulation Mechanism Reference. Aries Technology Inc., Lowell, MA, 1990.

14 Hollar, M.G.; Rosenthal, D.E. (1991): *Concurrent Design and Analysis of Mechanisms.* Rasna Corporation, San Jose.

15 Otter, M.; Hocke, M.; Daberkow, A.; Leister, G. (1990): *Ein objektorientiertes Datenmodell zur Beschreibung von Mehrkörpersystemen unter Verwendung von RSYST.* Institut B für Mechanik, IB-16, Stuttgart.

16 Mortenson, M.E. (1985): *Geometric Modeling.* John Wiley, New York.

17 Daberkow, A. (1992): *Zur CAD-gestützten Modellierung von Mehrkörpersystemen.* Ph.D. Dissertation, Stuttgart.

18 Rühle, R. (1973): RSYST, ein integriertes Modulsystem mit Datenbasis zur automatischen Berechnung von Kernreaktoren. IKE 4-12 1973, Stuttgart.

19 SIGRAPH-CAD-3D. SIEMENS NIXDORF AG, München, 1992.

20 Otter, M.; Hocke, M.; Daberkow, A.; Leister, G. (1993): An object oriented data-model for multibody systems. In: *Advanced Multibody System Dynamics.* Kluwer, Dordrecht, 19-48.

21 Kreuzer, E.; Leister, G. (1991): Programmsystem NEWEUL'90, Anleitung AN-24, Institut B für Mechanik, Stuttgart.

22 Leister, G. (1991): Programmpaket NEWSIM. Anleitung AN-25, Institut B für Mechanik, Stuttgart.

23 Schiehlen, W. (1986): *Technische Dynamik.* Teubner Verlag, Stuttgart.

24 ACSL-Advanced Continuous Simulation Language Reference Manual. Inc. Concord/Mass.: Mitchell u. Gauthier Assoc., 1987.

25 Otter, M.; Gaus, N. (1991): ANDECS-DSSIM: Modular Dynamic Simulation With Database Integration. User's Guide, Version 2.1. Oberpfaffenhofen.

26 Leister, G. (1992): *Beschreibung und Simulation von Mehrkörpersystemen mit geschlossenen kinematischen Schleifen.* Ph.D. Dissertation, Stuttgart.

27 Blajer, W. (1991): Contribution to the projection method of obtaining equations of motion, *Mechanics Research Communications* **18**, 293-301.

28 Baumgarte, J. (1972): Stabilization of constraints and integrals of motion in dynamical systems, *Computational Methods in Applied Mechanics and Engineering*, **1**,1-16.

29 Ostermeyer, G.-P. (1990): On Baumgarte stabilization for differential algebraic equations. In: *Real-Time Integraion Methods for Mechanical System Simulation*, NATO ASI Series, Vol. F69, Springer-Verlag, Berlin-Heidelberg, 193-207.

30 Woernle, C. (1988): *Ein systematisches Verfahren zur Aufstellung der geometrischen Schließbegingungen in kinematischen Schleifen mit Anwendung bei der Rückwärtstransformation für Industrieroboter*, Ph.D. Dissertation, Stuttgart.

31 Eppinger, M.; Kreuzer, E. (1990): Evaluation of methods for solving the inverse kinematics of manipulators, In: Proc. of the 8-th CISM-IFToMM Symp. on Theory and Practice on Robots and Manipulators, Cracow.

32 Blajer, W.; Schiehlen, W.; Schirm, W. (1993): Dynamic analysis of constrained multibody systems using inverse kinematics. *Mechanism and Maschine* **28**(3), 397-405.

33 Schiehlen, W.; Blajer, W. (1992): Closing conditions and reaction forces of multibody systems. *Zeitschrift für Angewandte Mathematik und Mechanik* (ZAMM) **72**, T45-T47.

34 Schiehlen, W. (1991): Computational aspects in multibody system dynamics, *Computer Methods in Applied Mechanics and Engineering* **30**, 569-582.

35 Rauh, J. (1988): *Ein Beitrag zur Modellierung elastischer Balkensysteme*, Ph.D. Dissertation, Stuttgart.

36 Bremer, H.; Pfeiffer, F. (1992): *Elastische Mehrkörpersysteme*, Teubner, Stuttgart.

37 Melzer, F. (1993): Symbolic computations in flexible multibody systems. In: Computer Aided Analysis of Rigid and Flexible Mechanical Systems (NATO ASI, Tróia, 27 June - 7 July 1993). Pereira, M.S. (ed.). To appear.

38 Sorge, K. (1992): *Mehrkörpersysteme mit starr-elastischen Subsystemen*. Ph.D. Thesis, München.

39 Bestle, D., Eberhard, P. and Schiehlen, W.: Optimizing of an actively controlled vehicle system. In: *Proceedings of the IUTAM Symposium Optimal Control of Mechanical Systems* (Moscow, Russia, 20-24 April 1992). Chernousko, F.L. (ed.). To appear.

40 Osyczka, A. (1984): *Multicriteria Optimization in Engineering.* Ellis Horwood, New York.

41 Fletcher, R. (1987): *Practical Methods of Optimization.* Wiley, Chichester.

42 Haug, E. J. (1987): Design Sensitivity Analysis of Dynamic Systems. In: *Computer Aided Design: Structural and Mechanical Systems*, Mota-Soares, C.A. (ed.), Springer, Berlin.

43 Bestle, D. and Eberhard, P. (1992): Analyzing and Optimizing Multibody Systems. *Mechanics of Structures and Machines* **20**, 67-92.

44 Char, B.W. et. al. (1990): *MAPLE-Reference Manual.* Waterloo Maple Publ., Waterloo.

ON-LINE DYNAMIC ANALYSIS OF MECHANICAL SYSTEMS

THOMAS R. KANE
Professor of Applied Mechanics
Stanford University
Stanford, CA 94305
USA

ABSTRACT. By working with a symbol manipulation computer program created specifically for this purpose, a dynamicist can use a personal computer to analyze motions of mechanical systems in a highly efficient manner. The theory underlying the computer program is discussed, and illustrative examples are presented.

1. Introduction

The behavior of a mechanical system possessing a finite number of degrees of freedom is governed, in general, by a set of coupled, nonlinear, ordinary differential equations. Since solutions of such equations only rarely can be found in closed form, it was a rather thankless task to formulate such equations prior to the advent of computers. Not surprisingly, the subject of equation formulation methodology thus received scant attention until computers made it possible to obtain with little effort numerical solutions of nonlinear differential equations; and then it became apparent that the task of formulating equations of motion could become very burdensome, especially in connection with many problems of practical interest. To overcome this difficulty, dynamicists began to create computer programs that could **formulate** equations of motion, as well as solve them numerically; and many such "multibody programs", as they have come to be called, are widely used today.

Powerful and useful as they may be, all multibody programs suffer to varying degrees from one major defect, which is that they impose restrictions on the way a system is modeled mathematically. Most notably, the dependent variables employed tend to be fixed once and for all, which means that they can be quite unsuitable in some situations, leading to computationally inefficient solutions. Additionally, the class of systems accommodated by a given multibody program usually is quite limited. For example, only systems with a certain topology may be analyzable, say a tree-structure; or Coulomb friction forces may be inadmissible, etc. To say it simply, the analyst resorting to the use of a multibody program frequently sacrifices freedom for convenience.

M. F. O. S. Pereira and J. A. C. Ambrósio (eds.),
Computer-Aided Analysis of Rigid and Flexible Mechanical Systems, 137–157.

A relatively recent development in the field of computing makes it possible at present to formulate equations of motion with the aid of a computer while retaining all of the freedom enjoyed by an analyst deriving equations of motion "by hand." This development is the creation of symbol manipulation languages and their incorporation in programs specifically designed for dynamic analysis. With the aid of such a program, one can perform analytical work with a personal computer, doing so in an interactive fashion, that is, by typing a command that causes the computer to perform an analytical task and to report the result almost instantaneously, whereupon one issues the next command, and so on. It is the purpose of this paper to describe this process in detail, to explore the rationale underlying the process, to discuss a number of related issues, and to present some illustrative examples.

The sequel is arranged as follows. Section 2 deals with on-line mathematical analyses not involving any mechanics-related commands. In Sec. 3, problems of kinematics are addressed. Inertia calculations are the subject of Sec. 4, and issues that arise when one attempts to employ a symbol manipulator to formulate equations of motion are explored in Sec. 5. The on-line determination of forces is considered in Sec. 6, and concluding remarks appear in Sec. 7.

2. On-line Symbolic Mathematics

Computer programs capable of carrying out symbolic manipulations have existed for a long time, and have been used to perform tasks such as, for example, expanding $(A+B+C)^7$ to obtain

$$\begin{aligned} A^7 &+ B^7+C^7+7(AB^6+AC^6+BA^6+BC^6+CA^6+CB^6) \\ &+ 21(A^2B^5+A^2C^5+A^5B^2+A^5C^2+B^2C^5+B^5C^2) \\ &+ 35(A^3B^4+A^3C^4+A^4B^3+A^4C^3+B^3C^4+B^4C^3) \\ &+ 42(ABC^5+ACB^5+BCA^5)+105(AB^2C^4+AB^4C^2 \\ &+ BA^2C^4+BA^4C^2+CA^2B^4+CA^4B^2)+140(AB^3C^3 \\ &+ BA^3C^3+CA^3B^3)+210(A^2B^2C^3+A^2B^3C^2+A^3B^2C^2) \end{aligned}$$

Originally, this necessitated creating a program and then executing it on a mainframe computer, but more recently it has become possible to type an instruction such as, for example,

```
EXPAND((A+B+C)^7)
```

on a personal computer and then see the result displayed on the computer screen within a short time after pressing the ENTER key. To illustrate some capabilities of a particular program of this kind, called AUTOLEV, let us consider a number of specific examples.

Suppose that F_1 and F_2, the first a function of x, the second a function of x and y, are given by

$$F_1 = 3x - 5\cos(x), \quad F_2 = e^{xy}$$

and we need the derivative of F_1 with respect to x and the partial derivative of F_2 with respect to y. Activation of AUTOLEV causes the line number (1) to appear on the screen, whereupon one types VARIABLES X,Y, so that the screen now appears as follows:

```
(1) VARIABLES X,Y
(2)
```

Next, typing F1 = 3*X-5*COS(X) and pressing ENTER causes the following to appear on the screen:

```
  (1) VARIABLES X,Y
  (2) F1=3*X-5*COS(X)
>>(3) F1 = 3*X-5*COS(X)
```

Line (3) is an "echo" of the assignment statement entered in line (2). No such echo was produced by line (1) because this line serves as a declaration rather than as an assignment statement. Proceeding similarly, one can enter

```
(4) F2=EXP(X*Y)
```

to which the program responds with

```
>>(5) F2 = EXP(X*Y)
```

and now the desired derivatives are found by typing

```
(6) DF1_DX=D(F1,X)
```

which produces

```
>>(7) DF1_DX = 3 + 5*SIN(X)
```

while entering

```
(8) DF2_DY=D(F2,Y)
```

leads to

```
>>(9) DF2_DY =  X*EXP(X*Y)
```

The left-hand sides of lines (6) and (8) are names chosen by the analyst, whereas D(F1,X) and D(F2,Y) are instructions issued in the language of the program.

The lines

```
  (1) CONSTANTS A,B,C,D,E,F
  (2) U=[A,B,C]
>>(3) U = [A, B, C]
  (4) V=[D;E;F]
>>(5) V = [D, E, F]
  (6) W=U*V
```

cause AUTOLEV to treat U and V as a 1×3 row matrix and a 3×1 column matrix, respectively, and to report the inner product of U and V as the 1×1 matrix W in line (7):

```
>>(7) W = [A*D + B*E + D*F]
```

If a 3×3 matrix R is introduced as

```
  (8) R = [A,B,C;D,B,E;F,A,D]
>>(9) R = [A, B, C; D, B, E; F, A, D]
```

then the inverse of R, arbitrarily called S by the user, is generated in response to

```
  (10) S = INVERT(R)
```

which yields

```
>>(11) S[1,1] = -(A*E-B*D)/(B*D*(A-D)-B*F*(C-E)-A*(A*E-C*D))
>>(12) S[1,2] = (A*C-B*D)/(B*D*(A-D)-B*F*(C-E)-A*(A*E-C*D))
>>(13) S[1,3] = -B*(C-E)/(B*D*(A-D)-B*F*(C-E)-A*(A*E-C*D))
>>(14) S[2,1] = -(D^2-E*F)/(B*D*(A-D)-B*F*(C-E)-A*(A*E-C*D))
>>(15) S[2,2] = (A*D-C*F)/(B*D*(A-D)-B*F*(C-E)-A*(A*E-C*D))
>>(16) S[2,3] = -(A*E-C*D)/(B*D*(A-D)-B*F*(C-E)-A*(A*E-C*D))
>>(17) S[3,1] = (A*D-B*F)/(B*D*(A-D)-B*F*(C-E)-A*(A*E-C*D))
>>(18) S[3,2] = -(A^2-B*F)/(B*D*(A-D)-B*F*(C-E)-A*(A*E-C*D))
>>(19) S[3,3] = B*(A-D)/(B*D*(A-D)-B*F*(C-E)-A*(A*E-C*D))
```

Representative vector operations, so important in dynamics, can be performed by introducing a reference frame, say N, with the declaration

```
  (20) FRAMES N
```

which causes AUTOLEV to regard `N1>`, `N2>`, and `N3>` as a right-handed set of mutually perpendicular unit vectors normally written $\mathbf{N}_1$, $\mathbf{N}_2$, $\mathbf{N}_3$. Hence, vectors $\mathbf{V} = A\mathbf{N}_1 + B\mathbf{N}_2 + C\mathbf{N}_3$ and $\mathbf{W} = D\mathbf{N}_1 + E\mathbf{N}_2 + F\mathbf{N}_3$ can be entered as

```
  (3) V>=A*N1>+B*N2>+C*N3>
>>(4) V> = A*N1> + B*N2> + C*N3>
  (5) W>=D*N1>+E*N2>+F*N3>
>>(6) W> = D*N1> + E*N2> + F*N3>
```

and **X**, the cross-product of **V** and **W**, is then found by typing

```
(7) X>=CROSS(V>,W>)
```

which produces

```
>>(8) X> = (B*F-C*E)*N1> + (C*D-A*F)*N2> + (A*E-B*D)*N3>
```

Furthermore, since operations can be "nested", the scalar triple product of **V**, **W**, and **X** is obtained as

```
(9) TRIPLE=DOT(V>,CROSS(W>,X>))

>>(10) TRIPLE = A*(E*(A*E-B*D)+F*(A*F-C*D))
              - C*(D*(A*F-C*D)+E*(B*F-C*E))
              - B*(D*(A*E-B*D)-F*(B*F-C*E))
```

As a final illustration in the use of AUTOLEV for purely mathematical purposes, we consider the program's ability to simplify expressions such as, for example,

$$[\cos(A) + 5\cos^3(A) + 6\cos^3(A)\tan^2(A)]/\cos(A)$$

Simplification is accomplished by typing

```
(11) (COS(A)+5*COS(A)^3+6*COS(A)^3*TAN(A)^2)/COS(A)
```

and pressing ENTER, which produces

```
RESULT: 6 + SIN(A)^2
```

3. On-Line Kinematic Analysis

Certain problems of kinematics can be solved on-line simply by using the methods of analytic geometry. For example, Fig. 1 shows a manipulator formed by pin-connected elements A, B, C, and D. The lengths x_1, x_2, and x_3 of A, B, and C, respectively, are presumed to be variable, whereas LD, the length of D is fixed. D represents a load-carrying platform, and the system may be regarded as driven either by motors at points G, H, and I, which can cause the angles QA, QB, and QC to take on desired values, or by rack-and-pinion drives that cause the lengths of A, B, and C to acquire assigned values.

For given values of the angles QA, QB, QC and the lengths LD, LE, LF, it is a relatively simple matter to find x_1, x_2, and x_3, for this can be accomplished by solving a set of linear equations. By way of contrast, to deal with what is sometimes called the "inverse kinematics" of the system, that is, to determine the values of QA, QB, QC, and QD corresponding to given values of x_1, x_2, x_3, LD, LE, and LF, one must solve a set of coupled, nonlinear equations, namely,

$$x_1\cos(QA) + LD\cos(QD) - x_2\cos(QB) - LF = 0$$

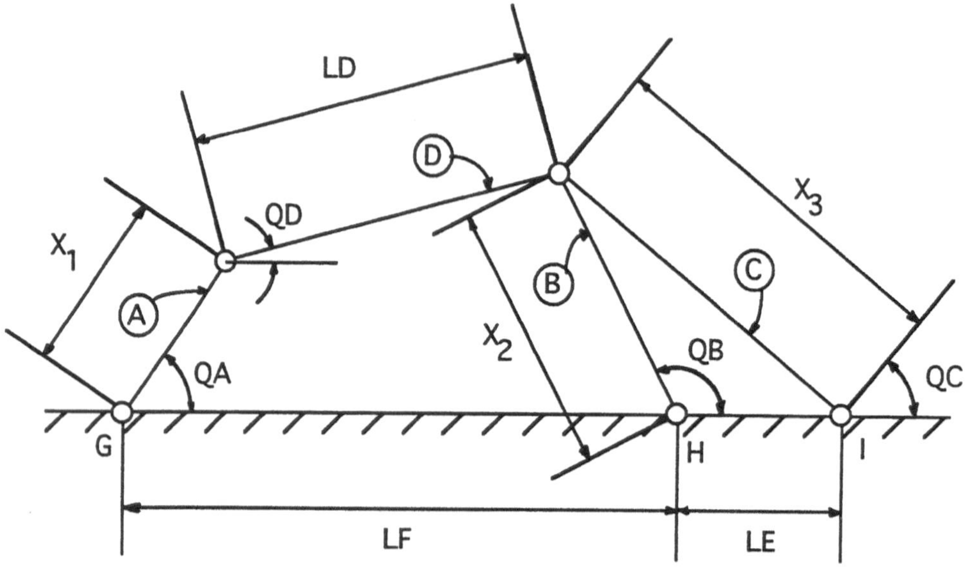

Figure 1: Manipulator

$$x_1 \sin(QA) + LD \sin(QD) - x_2 \sin(QB) = 0$$

$$x_2 \cos(QB) - x_3 \cos(QC) - LE = 0$$

$$x_2 \sin(QB) - x_3 \sin(QC) = 0$$

To this end, one can execute the AUTOLEV program

```
VARIABLES QA,QB,QC,QD
CONSTANTS X1,X2,X3,LD,LE,LF
Z[1]=X1*COS(QA)+LD*COS(QD)-X2*COS(QB)-LF
Z[2]=X1*SIN(QA)+LD*SIN(QD)-X2*SIN(QB)
Z[3]=X2*COS(QB)-X3*COS(QC)-LE
Z[4]=X2*SIN(QB)-X3*SIN(QC)
NONLINEAR(Z,QA,QB,QC,QD)
```

which causes AUTOLEV to create a FORTRAN program called NONLIN.FOR and to prompt one to enter in a file called NONLIN.IN the values of the given quantities, as well as guesses for the values of QA, QB, QC, and QD. Execution of the program NONLIN.FOR then produces QA, QB, QC, QD. For example, for $x_1 = 3.1$ m, $x_2 = 3.5$ m, $x_3 = 5.1$ m, $LD = 4.0$ m, $LE = 4.3$ m, $LF = 4.4$ m, one obtains $QA = 71.50$ deg, $QB = 80.96$ deg, $QC = 137.3$ deg, $QD = 74.21$ deg.

Vector analysis comes into play in dynamics in connection with angular velocities and angular accelerations of rigid bodies and velocities and accelerations of points. A device represented schematically by Fig. 2 can serve as a case in point. It consists of a rigid body A that rotates in a reference frame N with a constant angular speed $W1$ about an axis that is fixed in A and N parallel to a unit vector **A**1 and that supports an arm B which rotates relative to A with a constant angular speed $W2$ about an axis fixed in A and B and parallel to a unit vector **A**2. The magnitude of the acceleration of point D in N is to be expressed in terms of $W1$, $W2$, $L1$ and $L2$ for an instant at which the angle E in Fig. 2 is equal to 90 deg. An AUTOLEV solution of this problem begins with the declarations

```
CONSTANTS W1,W2,L1,L2
FRAMES N,A,B
POINTS C,D
```

Next, the velocity of C in N, the angular velocity of A in N, the angular velocity of B in N, and the position vector from C to D are entered by inspection as

```
V_C_N>=-L1*W1*A3>
W_A_N>=W1*A1>
W_B_N>=W_A_N>-W2*A3>
P_C_D>=L2*A1>
```

Corresponding equations can be produced equally easily by hand. But the next command,

```
ALF_B_N>=DT(W_B_N>,N)
```

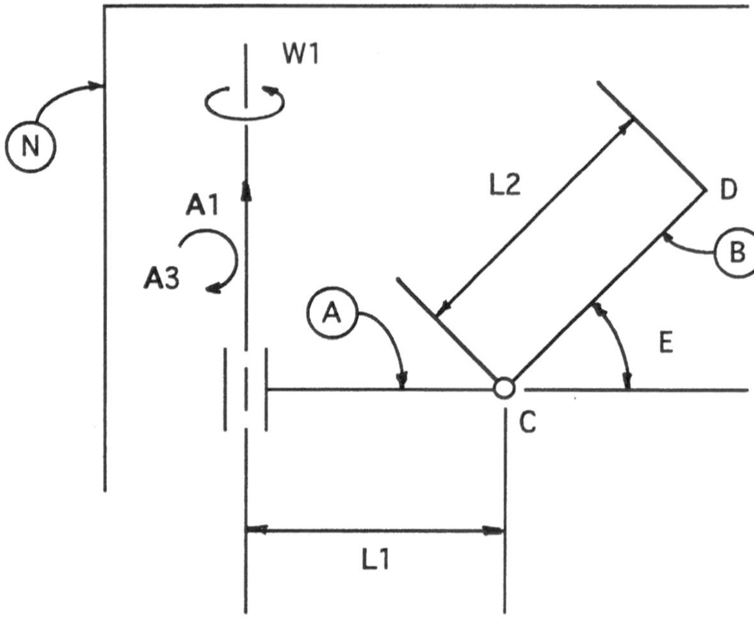

Figure 2: Acceleration Analysis

creates an expression for the angular acceleration of B in N by employing the program's ability to differentiate vectors in a specified reference frame, and this saves a considerable amount of hand labor. Moreover, once this line has been executed, significant savings in time and effort are realized when an expression for the acceleration of point D in N is created in response to the command

```
A2PTS(N,B,C,D)
```

This command, based on a familiar kinematical theorem, instructs AUTOLEV to find the acceleration of C in N by differentiating the velocity of C in N with respect to t in N and then adding to the resulting vector the vector produced with the command

```
CROSS(ALF_B_N>,P_C_D>)+CROSS(W_B_N>,CROSS(W_B_N>,P_C_D>))
```

AUTOLEV calls the result `A_D_N>`, so that all that remains to be done is to type

```
MAGNITUDE=MAG(A_D_N>)
```

which yields

```
>>(17) MAGNITUDE = (L1^2*W1^4+L2^2*W2^4+4*L2^2*W1^2*W2^2)^0.5
```

This example shows that the symbol manipulator under consideration "knows" certain kinematical theorems. In fact, it contains many commands based on theorems dealing with orientation angles, Euler parameters, direction cosines, etc.

4. On-Line Inertia Calculations

Figure 3 shows a solid, right-circular cone C of radius R and height H, as well as mutually perpendicular unit vectors **C**1, **C**2, **C**3. The moment of inertia of C about line $A-B$, to be denoted by IAB, is to be expressed in term of R, H, and the mass M of C, in terms of which $I1$, $I2$, $I3$, the central moments of inertia of C, are given by

$$I1 = I3 = 3M(4R^2 + H^2), \quad I2 = 3MR^2/10$$

We begin the analysis with the declarations

```
CONSTANTS R,H
BODIES C
MASS C,M
POINTS A,B
```

and enter $I1$ and $I2$ as

```
I1=3*M*(4*R^2+H^2)
I2=3*M*R^2/10
```

The central inertia dyadic of C is created with the line

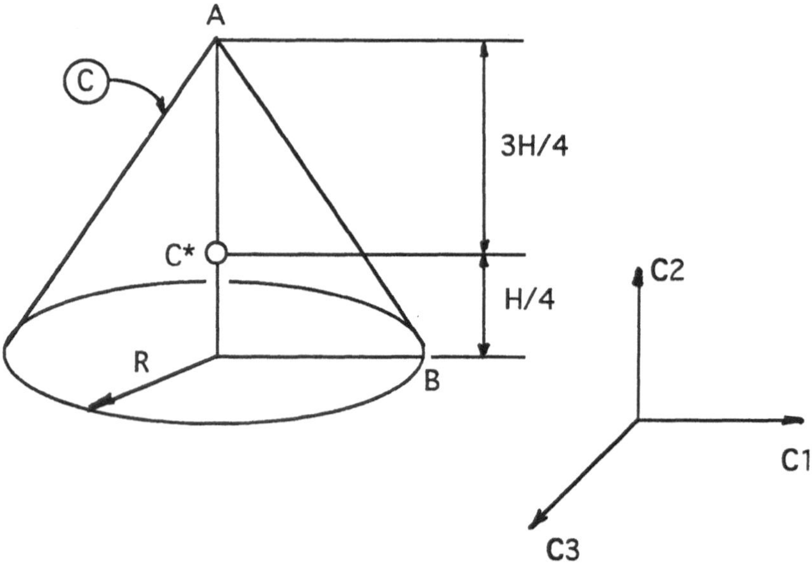

Figure 3: Cone

```
INERTIA C,I1,I2,I1,0,0,0
```

and the position vector from C^*, the mass center of C, to point A, by typing

```
P_CSTAR_A>=0.75*H*C2>
```

The command

```
PARALLEL(C,A)
```

thereupon causes AUTOLEV to construct the inertia dyadic of C for point A, which brings us into position to find IAB by pre– and post–multiplying this dyadic with a unit vector parallel to $A-B$. This unit vector, which we call AB>, is formed with the line

```
AB>=UNITVEC(R*C1>-H*C2>)
```

All that remains to be done, therefore, is to issue the command

```
IAB=EXPAND(DOT(AB>,DOT(I_C_A>>,AB>))
```

which produces

```
18) IAB = 3.8625*M*R^2*(H^2+3.1067961*R^2)/(H^2+R^2)
```

Should it be necessary to evaluate IAB for particular values of M, R, and H, say $M=1$, $R=2$, $H=3$, this is accomplished with the additional line

```
IABVALUE=EVALUATE(IAB,M=1,R=2,H=3)
```

which leads to the response

```
>>(20) IABVALUE = 25.465385
```

This example shows that AUTOLEV incorporates certain fundamental theorems associated with inertia concepts.

5. On-Line Formulation of Dynamical Equations of Motion

Dynamical equations of motion are created by expressing dynamical principles or formalisms, such as Newton's laws, the angular momentum principle, or Lagrange's equations, in terms of quantities characterizing the physical properties and behavior of a material system. Clearly, some of the on–line analysis capabilities already discussed can be used for this purpose; but, as we shall see, on–line equation of motion formulation can be performed most effectively only when certain issues specific to this process have been resolved optimally.

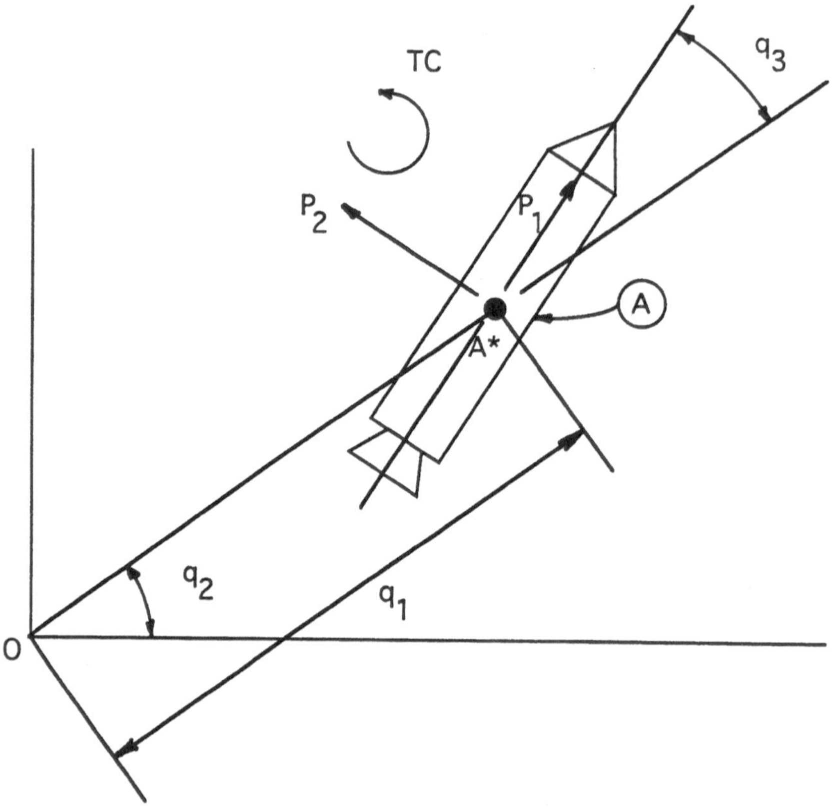

Figure 4: Rocket

Suppose that an analysis of rocket flight is to be undertaken by reference to the simple, planar model suggested by Fig. 4, where q_i $(i = 1,2,3)$ play the roles of generalized coordinates in the sense of Lagrange, while the set of all contact and distance forces acting on the rocket A is represented by the measure number TC of the torque of a couple and the measure numbers P_1 and P_2 of a force applied at A^*, the mass center of A. To formulate dynamical equations of motion with the aid of Lagrange's method, one can begin by introducing v_i $(i = 1,2,3)$ via the kinematical differential equations

$$dq_i/dt = v_i \quad (i = 1,2,3)$$

Next, after expressing the kinetic energy K of A as

$$K = \frac{1}{2}[M(v_1^2 + q_1^2 v_2^2) + I(v_2 + v_3)^2]$$

and the generalized forces F_i $(i = 1, 2, 3)$ as

$$F_1 = P_1 \cos(q_3) - P_2 \sin(q_3)$$

$$F_2 = q_1[P_1 \sin(q_3) + P_2 \cos(q_3)] + TC$$

$$F_3 = TC$$

one substitutes into Lagrange's equations. This necessitates various differentiations of K, which can be performed conveniently with a symbol manipulation computer program. For example, employing AUTOLEV, we enter the lines

```
CONSTANTS M,I
VARIABLES Q[3],V[3],P[2],TC
K=0.5*(M*(V1^2+Q1^2*V2^2)+I*(V2+V3)^2)
F1=P1*COS(Q3)-P2*SIN(Q3)
F2=Q1*(P1*SIN(Q3)+P2*COS(Q3))+TC
F3=TC
Q1'=V1
Q2'=V2
Q3'=V3
```

and then create the dynamical equations of motion with the commands

```
Z[1]=EXPAND(DT(D(K,V1))-D(K,Q1)-F1)
Z[2]=EXPAND(DT(D(K,V2))-D(K,Q2)-F2)
Z[3]=EXPAND(DT(D(K,V3))-D(K,Q3)-F3)
```

The resulting equations are

$$M(dv_1/dt - q_1 v_2^2) = P_1 \cos(q_3) - P_2 \sin(q_3)$$

$$(I + Mq_1^2)dv_2/dt + I dv_3/dt + 2Mq_1 v_1 v_2 = q_1[P_1 \sin(q_3) - P_2 \cos(q_3)] - TC$$

$$I(dv_2/dt + dv_3/dt) = TC$$

The relative complexity of these equations is attributable to the fact that v_i $(i = 1, 2, 3)$ are poor variables for dealing with the problem at hand. To see this, let us examine what happens when one uses as dependent variables, in addition to q_1, q_2, and q_3, generalized speeds u_1, u_2, and u_3 defined as follows: u_1 and u_2 are the measure numbers of the axial and transverse components of the inertial velocity of the mass center of A, and u_3 is an inertial angular speed of A. This leads to the kinematical differential equations

$$dq_1/dt = u_1 \cos(q_3) - u_2 \sin(q_3)$$

$$dq_2/dt = [u_1 \sin(q_3) + u_2 \cos(q_3)]/q_1$$

$$dq_3/dt = u_3 - [u_1 \sin(q_3) + u_2 \cos(q_3)]/q_1$$

and, resorting to the use of Newton–Euler methods by summing forces in the axial and transverse directions, one obtains the two dynamical equations

$$M(du_1/dt - u_2 u_3) = P_1$$

$$M(du_2/dt + u_3 u_1) = P_2$$

while taking moments about the mass center of A yields the remaining dynamical equation,

$$I du_3/dt = TC$$

Obviously, these equations are significantly simpler than their earlier counterparts. The reason we used Newton–Euler methods, rather than Lagrange's equations, to formulate them is that the latter are inapplicable in their usual form under the present circumstances, that is, when the kinetic energy is regarded as a function of q_r and u_r $(r = 1, 2, 3)$, rather than as a function of q_r and dq_r/dt $(r = 1, 2, 3)$. For this reason alone it is inadvisable to base a computer program intended to facilitate equation of motion formulation on Lagrange's equations. An additional fact that mitigates against the use of Lagrange's equations is that such use frequently necessitates the performing of operations leading to terms that ultimately cancel each other or, if they do not, give rise to computations that have no effect on final results. For purposes of computerized symbol manipulation, this is an important consideration, for it has serious implications regarding computer memory usage.

What about Newton–Euler methods? Does the fact that they serve well in connection with the rocket–flight example justify the inference that, in general, they constitute a sound basis for computerized equation of motion formulation? It becomes apparent that this is not the case when one attempts to formulate equations of motion for a system such as, for example, the one depicted in Fig. 1, that is, one containing "loops." Under these circumstances, the use of Newton–Euler methods for the formulation of equations of motion necessarily entails the introduction and subsequent elimination of certain constraint forces, processes which are sufficiently algorithmically subtle and computationally intensive to give rise to significant obstacles to the creation of an efficient on–line equation of motion formulation program.

As is explained in detail in [1], the equations

$$F_r + F_r^* = 0 \quad (r = 1, ..., p)$$

furnish an alternative to Lagrange's equations and Newton–Euler methods as a point of departure for the formulation of dynamical equations of motion of any mechanical system possessing p degrees of freedom. Given such a system, one begins the equation formulation process by performing a "first level" kinematic analysis, that is, by constructing expressions for angular velocities of rigid bodies, velocities of mass centers of these bodies, and velocities of particles. Next, one creates vectors representing active forces and torques, which brings one into position to form expressions for F_r and F_r^* by carrying out operations that can be programmed once and for all. This is the rationale underlying the program AUTOLEV. As a first illustrative example, we return to the rocket problem represented by Fig. 4.

The statement

```
NEWTONIAN N
```

serves to inform the program that a reference frame called N is to be regarded as Newtonian, that is, as one such that use of

$$F_r + F_r^* = 0 \quad (r = 1, ..., p)$$

leads to physically correct results. Next, the lines

```
VARIABLES U1,U2,U3,Q1,Q2,Q3
```

declare u_r and q_r $(r = 1, 2, 3)$ as time-dependent variables to be employed to characterize the motion and configuration of the rocket, whose mass and relevant inertia properties are brought into the program with the statements

```
BODIES A
MASS A,M
I_A_ASTAR=I*A3*A3
```

The aforementioned first level kinematic analysis is carried out simply by expressing the velocity of the mass center of A in N as

```
V_ASTAR_N>=U1*A1>+U2*A2>
```

and the angular velocity of A in N as

```
W_A_N>=U3*A3>
```

Forces and torques are entered with the lines

```
ACTIONS F1,F2,TOR
FORCE(ASTAR,F1*A1>+F2*A2>)
TORQUE(A,TOR*A3>)
```

The rest is accomplished by typing

```
ZERO=FR()+FRSTAR()
UNCOUPLE()
```

This yields the dynamical equations of motion by evoking the response

```
>>(20) U1' = U2*U3 + F1/M
>>(21) U2' = F2/M - U1*U3
>>(22) U3' = TOR/I
```

The four-bar linkage shown in Fig. 5 provides a second illustrative example, one involving a loop. A, B, and C are massless rods of length LA, LB, and LC, respectively; the revolute joints supporting A and C at P and O have vertical axes and are separated by a distance LD; PAB and PBC designate particles of mass M; and QA, QB, and QC are the angles between line $O - P$ and A, B, and C, respectively. Finally, a couple whose torque is indicated in Fig. 5 is applied to C. The objective of the analysis to be undertaken is to produce motion simulations, that is, plots of, say, QC as a function of time t.

As before, we begin with

```
NEWTONIAN N
```

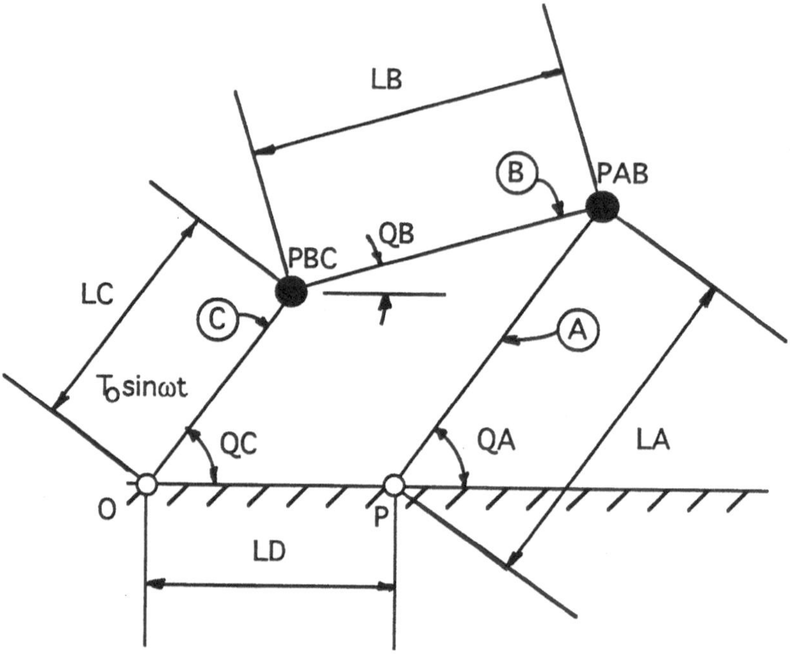

Figure 5: Four–bar Linkage

Although the system under consideration possesses only one degree of freedom, it is expedient (and permissible) to introduce three generalized speeds. Therefore, we continue by entering

```
VARIABLES U1,U2,U3,QA,QB,QC
CONSTANTS LA,LB,LC,LD
PARTICLES PAB,PBC
MASS PAB,M,PBC,M
POINTS PCO
```

The last line introduces a point named PCO, intended to be the point of C in contact with O.

Kinematical differential equations are created by typing

```
QA'=U1
QB'=U2
QC'=U3
```

and the angles QA, QB, QC are brought into the analysis with lines

```
FRAMES A,B,C
SIMPROT(N,A,3,QA)
SIMPROT(N,B,3,QB)
SIMPROT(N,C,3,QC)
```

which have the effect of creating direction cosine matrices relating unit vectors variously fixed in A, B, and C to unit vectors $\mathbf{N}_i$ $(i = 1,2,3)$ fixed in N. The direction cosine matrices relating the unit vectors variously fixed in A, B, and C to each other are generated with the commands

```
A_B=A_N*N_B
B_C=B_N*N_C
C_A=C_N*N_A
```

The angular velocities of A, B, and C in N and the velocities of PAB, PBC, and POC in N are entered as

```
W_A_N>=U1*A3>
W_B_N>=U2*B3>
W_C_N>=U3*C3>
```

and

```
V_PAB_N>=LA*U1*A2>
V_PBC_N>=V_PAB_N>-LB*U2*B2>
V_PCO_N>=V_PBC_N>-LC*U3*C2>
```

and two constraint equations are produced by noting that the velocity of POC in N must vanish, which requirement is enforced with the lines

```
CONSTRAIN[1]=DOT(V_PCO_N>,N1>)
CONSTRAIN[2]=DOT(V_PCO_N>,N2>)
CONSTRAIN()
```

The last line causes the program to solve the constraint equations, that is, to express $U2$ and $U3$ in terms of $U1$.

The couple applied to C is brought into the program with lines that characterize the torque of the couple, namely,

```
CONSTANTS TO,OMEGA
TORQUE(C,TO*SIN(OMEGA*T)*C3>)
```

and the equations of motion appear in response to the command

```
ZERO=FR()+FRSTAR()
```

Finally, to cause AUTOLEV to write a FORTRAN program for the numerical solution of the equations of motion, all one has to do is to type

```
CODE LINKAGE
```

To verify that the foregoing on-line operations can lead directly to tangible results, one can use the FORTRAN program LINKAGE.FOR created by AUTOLEV to determine the response of the linkage to the oscillatory torque applied to C if, for example, the system is initially at rest with $QA = 90$ deg, $QB = 0$, $QC = 60$ deg and $LB = 0.3$ m, $LC = 0.5$ m, $w = 0.6$ rad/s, and $T_o = 0.8$ Nm. Figure 6 is a plot of QC vs. t based on numerical results obtained with LINKAGE.FOR.

6. On-Line Determination of Forces

Frequently, the principal purpose of a dynamic analysis is to determine forces, sometimes called constraint forces, that come into play during the motion of a mechanical system. The motion itself may be known a priori, or it may have to be determined simultaneously with the unknown forces. In either case, it is disadvantageous to employ Lagrange's equations when these provide the desired information only with the aid of Lagrange multipliers, for this tends to complicate an analysis unnecessarily. Similarly, the use of Newton–Euler methods can necessitate the introduction and subsequent elimination of quantities not of interest in their own right. By way of contrast, the equation of motion formulation methodology discussed in Sec. 5 and underlying the program AUTOLEV furnishes the means for the selective determination of scalar unknowns associated with unknown forces. Suppose, for example, that it is desired not only to determine the motion of the linkage considered in the previous section, but also to evaluate for each instant of such a motion the magnitude of the force exerted on C by the bearing at point O. Only minor modifications of the earlier program are required to obtain the desired information. Specifically, following

```
ACTIONS TOR
```

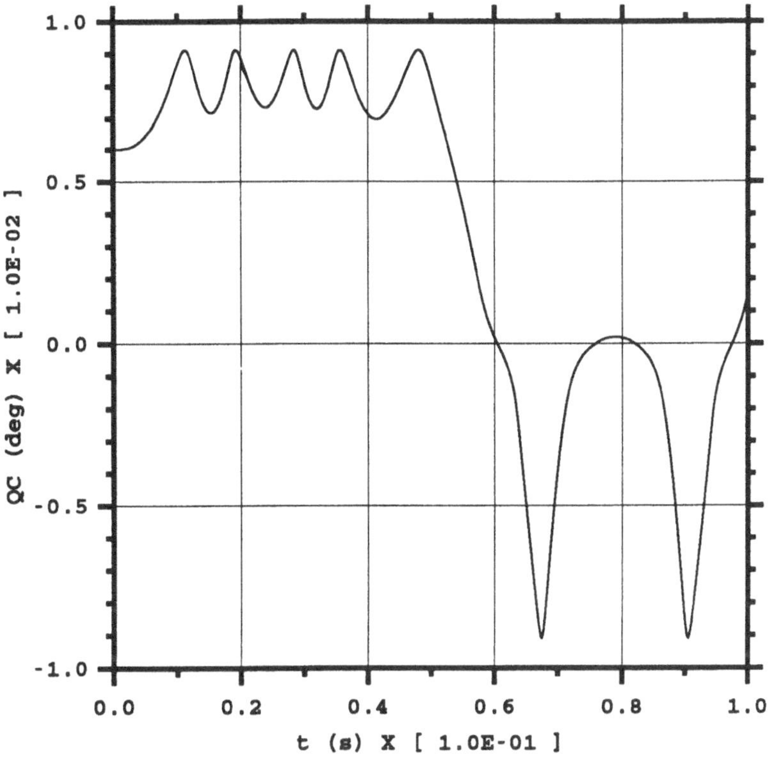

Figure 6: Linkage Response

the line

```
AUXACTIONS F1,F2
```

is added to introduce two scalars that will presently be used to characterize the force in question. The lines dealing with constraint conditions are modified to read

```
AUXCONSTRAIN[1]=DOT(V_PCO_N>,N1>)
AUXCONSTRAIN[2]=DOT(V_PCO_N>,N2>)
AUXCONSTRAIN()
```

and the force applied to C at O is accommodated with the statement

```
FORCE(PCO,F1*N1>+F2*N2>)
```

A new FORTRAN program is created by executing the new AUTOLEV program, and this brings one into position to generate the desired results. For instance, using the same data as before, one obtains the the force vs. t plot shown in Fig. 7.

7. Conclusion

While computers have played a major role in the solution of various problems of dynamics for many years, their use in the manner described in this paper is in its infancy. Experience gained to date suggests that the new approach can be highly effective not only in an industrial setting, but also in connection with the teaching of the subject of dynamics. Because a good symbol manipulation program allows one to carry out quickly and effortlessly many analytical tasks which, when performed by hand, are both arduous and time–consuming, such a program can enable one to devote more time and energy than would otherwise be available to the study of the fundamental concepts underlying a given analysis. Additionally, it is a simple matter to create a permanent, easily readable record of an on-line analysis, which can be of great value for purposes of checking and reviewing. It remains to be seen how widely the new approach will be adopted.

8. Reference

1 Kane, Thomas R. and Levinson, David A. (1985) *Dynamics: Theory and Applications*, McGraw–Hill, New York.

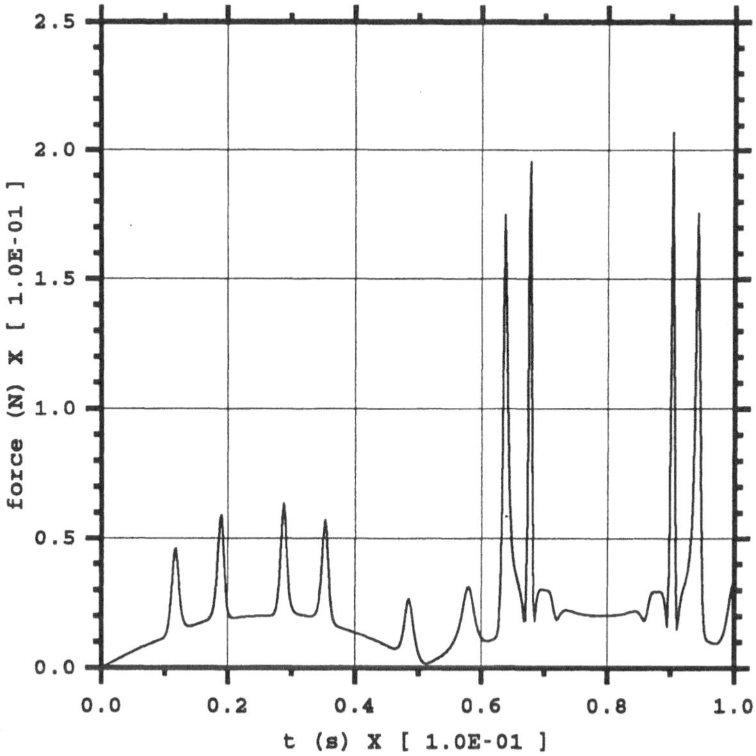

Figure 7: Force Magnitude

TOPOLOGICAL DESCRIPTION OF ARTICULATED SYSTEMS

J. WITTENBURG
Inst. of Technical Mechanics
University of Karlsruhe
D-76128 Karlsruhe, Germany

ABSTRACT. Articulated systems in the sense of this paper are systems composed of objects and of connections between objects. With, both, objects and connections time-varying physical quantities are associated which interact with one another. Three different kinds of articulated systems are presented. In electrical systems the objects are the nodes of a network and the connections are subsystems composed of impedances, voltage sources and current sources. In the first kind of mechanical systems presented the objects are bodies whereas the connections are joints characterized by kinematical constraints. Also in the second kind of mechanical systems presented the objects are bodies whereas the connections are springs and dampers. It is the purpose of this paper to point out that inspite of the apparent differences in nature of these systems there is a common characteristic, namely an interconnection topology. Matrices defined on directed graphs are shown to be appropriate parameters for describing this topology. Using such matrices equations can be formulated which are valid for any kind of system of the particular type under consideration. These formulations are a basis for the development of efficient general-purpose computer programs as well as for new analytical insight into system behaviour. The paper starts out with a chapter on directed graphs and on associated matrices. The remaining chapters are devoted to applications demonstrating the efficiency of these tools.

1. Directed Graphs and Associated Matrices

In this chapter connected, directed graphs are described. A graph consists of vertices labeled $i = 0, \ldots, n$ and of arcs labeled $a = 1, \ldots, m$. Fig.1 shows a graph with $n = 4$ and $m = 7$. A graph is connected if between any two vertices there is a connec-

M. F. O. S. Pereira and J. A. C. Ambrósio (eds.),
Computer-Aided Analysis of Rigid and Flexible Mechanical Systems, 159–196.

tion by (at least one) chain of arcs and vertices. Two vertices may be connected by more than one arc.

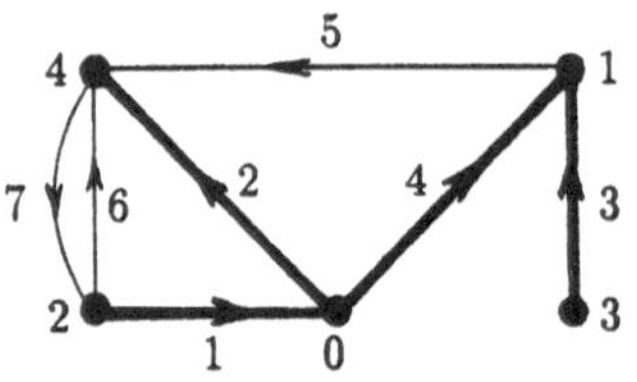

Figure 1. Directed graph.

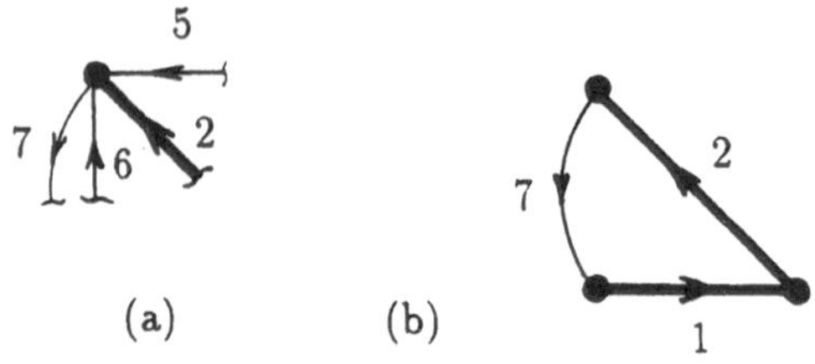

Figure 2. Cutset 2 (Fig.a) and circuit 7 (Fig.b) for the directed graph of Fig.1.

By giving each arc a sense of direction - indicated by an arrow - a directed graph is created. The sense of direction is necessary if it is desired to distinguish the two vertices connected by an arc. In applications to electrical networks vertices and arcs represent nodes and electrical connections, respectively. In this case the sense of direction allows to specify when a current through a connection is positive. The distinction between the two nodes at the ends of a connection allows an unambiguous definition of the voltage of one node relative to the other. In applications to articulated multibody systems vertices and arcs represent bodies and kinematical joints, respectively. In this case the distinction between two bodies coupled by a joint allows an unambiguous definition of velocity of one body relative to the other. It also allows to state on which of the two bodies an internal joint force is acting with positive sign and on which with negative sign.

A connected graph is called tree-structured if between any two vertices there exists a unique minimal chain of arcs and vertices connecting the two vertices. In a tree-structured, connected graph the identity $m = n$ holds. A connected graph with $m > n$ has circuits (also called loops). In such a graph $m - n$ arcs can be eliminated in such a way that a tree-structured graph - a so-called spanning tree of the original graph - remains. Normally, a graph with circuits has more than one spanning tree. An arbitrary spanning tree is chosen. In Fig.1 the arcs drawn in heavy lines represent a spanning tree. The arcs of the spanning tree are labeled $1, \ldots, n$ in an arbitrary order. The remaining arcs are called chords. They are labeled $n + 1, \ldots, m$ in an arbitrary order. In what follows indices i and j refer to vertices, a and b to arcs and c to chords.

For each arc $a = 1, \ldots, m$ let $i^+(a)$ and $i^-(a)$ be the labels of the two vertices at the starting point and at the terminating point of arc a, respectively. Thus $i^+(a)$ and $i^-(a)$ are the names of two integer functions with integer arguments. For the directed graph of Fig.1 the two functions are given in Table 1.

Table 1. Integer functions describing system topology.

a	1	2	3	4	5	6	7
$i^+(a)$	2	0	3	0	1	2	4
$i^-(a)$	0	4	1	1	4	4	2

The two functions fully describe the topology of a graph including the sense of direction of the arcs. In what follows other mathematical quantities - matrices - will be defined which serve the same purpose. It should be clear that these matrices can be traced back to the two integer functions. In formulating the dynamics of electrical networks and of articulated systems of rigid bodies matrices are very convenient as will be seen. They allow compact formulations. However, when it comes to programming such a formalism for numerical computations it may be advantageous to return to the integer functions in order to save computation time.

Each arc of the spanning tree uniquely defines a cutset of the complete graph. It consists of the arc itself and of the minimal set of chords which must be cut in order to split the graph in two subgraphs (s. Fig.2a). The cutset associated with arc a will also be called cutset a. A chord belonging to cutset a is said to be positively directed (in the cutset) if it points towards the same subgraph as arc a. Otherwise it is negatively directed.

Each chord uniquely defines a circuit. It consists of the chord itself and of the minimal set of tree arcs creating a circuit (s. Fig.2b). The circuit associated with chord c will also be called circuit c. A tree arc belonging to circuit c is said to be positively directed (in the circuit) if its sense of direction around the circuit is the same as that of chord c. Otherwise it is negatively directed.

1.1. MATRICES ASSOCIATED WITH GRAPHS

For a connected, directed graph and associated with a chosen spanning tree the following matrices are defined:

Incidence matrix $\mathbf{S}$ with elements

$$S_{ia} = \begin{cases} +1 & (i = i^+(a)) \\ -1 & (i = i^-(a)) \\ 0 & (\text{else}) \end{cases} \qquad (i = 1, \ldots, n;\ a = 1, \ldots, m), \tag{1}$$

Cutset matrix **P** with elements

$$P_{ab} = \begin{cases} +1 & \text{(arc } b \text{ belongs to cutset } a \text{ and is positively directed)} \\ -1 & \text{(arc } b \text{ belongs to cutset } a \text{ and is negatively directed)} \\ 0 & \text{(arc } b \text{ does not belong to cutset } a\text{)}, \end{cases}$$

$$(a = 1, \ldots, n;\ b = 1, \ldots, m),$$

Circuit matrix **K** with elements

$$K_{ca} = \begin{cases} +1 & \text{(arc } a \text{ belongs to circuit } c \text{ and is positively directed)} \\ -1 & \text{(arc } a \text{ belongs to circuit } c \text{ and is negatively directed)} \\ 0 & \text{(arc } a \text{ does not belong to circuit } c\text{)} \end{cases}$$

$$(c = n+1, \ldots, m;\ a = 1, \ldots, m),$$

Path matrix **T** of the spanning tree with elements

$$T_{ai} = \begin{cases} +1 & \text{(arc } a \text{ is on the path between vertices 0 and } i \text{ and is directed towards vertex 0)} \\ -1 & \text{(arc } a \text{ is on the path between vertices 0 and } i \text{ and is directed towards vertex i)} \\ 0 & \text{(arc } a \text{ is not on the path between vertices 0 and } i\text{)} \end{cases} \tag{2}$$

$$(a, i = 1, \ldots, n).$$

It should be noted that in the matrices **S** and **T** the vertex $i = 0$ is not represented. In applications to electrical networks vertex 0 represents ground. In applications to articulated multibody systems vertex 0 represents a carrier vehicle the motion of which is prescribed as a function of time (in simple cases inertial space). For the carrier vehicle no equations of motion are required. In Branin [1] and Wittenburg [2] a row $i = 0$ of the incidence matrix is defined. It is not necessary for the present paper.

Each of the matrices **S**, **P** and **K** is partitioned into a submatrix associated with the tree (columns $1, \ldots, n$; index t) and a submatrix associated with chords (columns $n+1, \ldots, m$; index c). From the definitions follows that $\mathbf{P}_t = \mathbf{I}$, $\mathbf{K}_c = \mathbf{I}$ where **I** is the unit matrix (of different dimensions in the two cases). Thus

$$\mathbf{S} = [\mathbf{S}_t \,\vdots\, \mathbf{S}_c], \qquad \mathbf{P} = [\,\mathbf{I} \,\vdots\, \mathbf{P}_c], \qquad \mathbf{K} = [\mathbf{K}_t \,\vdots\, \mathbf{I}\,]. \tag{3}$$

For the directed graph and its spanning tree shown in Fig.1 the four matrices are:

$$\mathbf{S} = \begin{bmatrix} 0 & 0 & -1 & -1 & +1 & 0 & 0 \\ +1 & 0 & 0 & 0 & 0 & +1 & -1 \\ 0 & 0 & +1 & 0 & 0 & 0 & 0 \\ 0 & -1 & 0 & 0 & -1 & -1 & +1 \end{bmatrix}, \qquad \mathbf{T} = \begin{bmatrix} 0 & +1 & 0 & 0 \\ 0 & 0 & 0 & -1 \\ 0 & 0 & +1 & 0 \\ -1 & 0 & +1 & 0 \end{bmatrix},$$

$$\mathbf{P} = \begin{bmatrix} +1 & 0 & 0 & 0 & 0 & +1 & -1 \\ 0 & +1 & 0 & 0 & +1 & +1 & -1 \\ 0 & 0 & +1 & 0 & 0 & 0 & 0 \\ 0 & 0 & 0 & +1 & -1 & 0 & 0 \end{bmatrix}, \qquad \mathbf{K} = \begin{bmatrix} 0 & -1 & 0 & +1 & +1 & 0 & 0 \\ -1 & -1 & 0 & 0 & 0 & +1 & 0 \\ +1 & +1 & 0 & 0 & 0 & 0 & +1 \end{bmatrix}.$$

The incidence matrix is most directly related to the two integer functions $i^+(a)$ and $i^-(a)$. In the case of the other matrices this relationship is not obvious. It is an interesting exercise to formulate an algorithm for calculating these matrices directly from the two functions without making use of the following theorems. Between the matrices **P**, **K**, **S** and **T** there exist relationships.

Theorem 1: $\mathbf{SK}^T = \mathbf{0}$.

Proof:

$$(\mathbf{SK}^T)_{ic} = \sum_{a=1}^{m} S_{ia}K_{ca} = \sum_{a=1}^{n} S_{ia}K_{ca} + \sum_{a=n+1}^{m} S_{ia}\underbrace{K_{ca}}_{\delta_{ca}} = \sum_{a=1}^{n} S_{ia}K_{ca} + S_{ic}$$

$(i = 1, \ldots, n;\ c = n+1, \ldots, m)$. Two cases must be distinguished. Case 1: Vertex i is incident with chord c ($S_{ic} \neq 0$). Then vertex i is incident with exactly one tree arc and the sum over a is equal to $-S_{ic}$ independent of the senses of direction of chord c and of this single tree arc. Case 2: Vertex i is not incident with chord c ($S_{ic} = 0$). Then vertex i is incident with 2 tree arcs and the sum over a is equal to zero independent of whether the two tree arcs belong to circuit c or not. End of proof. Theorem 1 represents an orthogonality relationship.

Theorem 2: $\mathbf{TS}_t = \mathbf{S}_t\mathbf{T} = \mathbf{I}$.

Proof:

$$(\mathbf{TS}_t)_{ab} = \sum_{i=1}^{n} T_{ai}S_{ib} = T_{a,i^+(b)} - T_{a,i^-(b)} \qquad (a, b = 1, \ldots, n).$$

Two cases must be distinguished. Case 1: $b = a$. Arc a is either pointing towards vertex 0 or away from it. If the former is true then $T_{a,i^+(a)} = 1$, $T_{a,i^-(a)} = 0$. If the latter is true then $T_{a,i^+(a)} = 0$, $T_{a,i^-(a)} = -1$. Thus, $(\mathbf{TS}_t)_{aa} = 1$. Case 2: $b \neq a$. Consider the path between vertex 0 and vertex $i^+(b)$ and the path between vertex 0 and vertex $i^-(b)$. Arc a belongs either to both paths or to none of them. In either case, $T_{a,i^+(b)} = T_{a,i^-(b)}$ and, hence, $(\mathbf{TS}_t)_{ab} = 0$. End of proof.

Theorem 3: $\quad \mathbf{TS}_c = \mathbf{P}_c$.

Proof:

$$(\mathbf{TS}_c)_{ac} = \sum_{i=1}^{n} T_{ai} S_{ic} = T_{a,i^+(c)} - T_{a,i^-(c)}$$

($a = 1, \ldots, n;\ c = n+1, \ldots, m$). Two cases must be distinguished. Case 1: Chord c belongs to cutset a. Then, independent of the senses of direction of arc a and of chord c, $T_{a,i^+(c)} - T_{a,i^-(c)} = P_{ac} \neq 0$ holds. Case 2: Chord c does not belong to cutset a. Then, $T_{a,i^+(c)} = T_{a,i^-(c)}$ and, hence, $T_{a,i^+(c)} - T_{a,i^-(c)} = 0 = P_{ac}$. End of proof.

Theorems 2 and 3 together establish the relationship

$$\mathbf{TS} = \mathbf{P}. \tag{4}$$

When this is postmultiplied by $\mathbf{K}^T$ one gets, due to Theorem 1, the equation

$$\mathbf{PK}^T = \mathbf{0} \tag{5}$$

which represents another orthogonality relationship. Using the partitioning of (3), the equation takes the form $\mathbf{K}_t^T + \mathbf{P}_c = \mathbf{0}$, i.e.

$$\mathbf{K}_t^T = -\mathbf{P}_c \tag{6}$$

Theorems 1 to 3 and Eqs.(4) to (6) show that a directed graph is fully described by its incidence matrix $\mathbf{S}$ and by the path matrix $\mathbf{T}$ for one of its spanning trees. The cutset matrix $\mathbf{P}$ and the circuit matrix $\mathbf{K}$ can be expessed one in terms of the other. Either one of them can be expressed in terms of the matrices $\mathbf{S}$ and $\mathbf{T}$.

The matrices $\mathbf{S}$, $\mathbf{P}$ and $\mathbf{K}$ were known to mathematicians for a long time (s. Busacker/Saaty [3] The path matrix $\mathbf{T}$ was first defined by Branin [1] for electrical networks and independently by Roberson/Wittenburg [4] for articulated multibody systems.

1.2. TREE-STRUCTURED GRAPHS

For graphs with tree structure a few more notations are introduced. For every arc $a = 1, \ldots, n$ a number σ_a is defined:

$$\sigma_a = \begin{cases} +1 & \text{(arc } a \text{ is directed towards vertex 0)} \\ -1 & \text{(arc } a \text{ is directed away from vertex 0)} \end{cases} \qquad (a = 1, \ldots, n). \quad (7)$$

For every pair of arcs a, b a set κ_{ab} of vertices is defined as follows. Cutting two arcs a and b $(a, b = 1, \ldots, n)$ results either in 3 subgraphs $(a \neq b)$ or in 2 subgraphs $(a = b)$. In the case $a \neq b$ κ_{ab} is the set of vertices of that subgraph which contains no vertex which is incident with arc b. In the case $a = b$ κ_{ab} is the set of vertices of that subgraph which does not contain vertex 0.

For two arcs a and $b \neq a$ the ordering relationship arc $b >$ arc a means that arc a is on the path from vertex 0 to vertex $i_{i+(b)}$ (and thus also on the path from vertex 0 to vertex $i_{i-(b)}$). It should be noted that for two arcs a and $b \neq a$ neither arc $b >$ arc a nor arc $a >$ arc b may be true, namely if the two arcs are located on different tree branches as seen from vertex 0.

For every vertex $i = 1, \ldots, n$ there is a unique path to vertex 0. On this path a single arc a is incident with vertex i (it is the only arc for which $S_{ia}T_{ai} \neq 0$). This arc is called the *inboard* arc of vertex i (inboard as seen from vertex 0). The vertex $\neq i$ with which the inboard arc is incident is called *inboard* vertex of vertex i.

The labeling of arcs and vertices of a tree-structured graph is called *regular* if for every vertex $i = 1, \ldots, n$ the label of the inboard arc equals i and, furthermore, the label of the inboard vertex is $< i$. A regular labeling is always possible in at least one way. Both, the incidence matrix $\mathbf{S}$ as well as the path matrix $\mathbf{T}$ of a regularly labeled tree-structured graph are upper triangular matrices with nonzero elements along the diagonal. If, in addition, the sense of direction of the arcs is such that each arc is pointing away from vertex 0 than all diagonal elements of both matrices are -1.

2. Electrical Networks

In this section classical methods of analysis for electrical networks are described. For reference see Branin [1]. Let the vertices and arcs of a connected graph represent the nodes and the physical connections, respectively, of an electrical d-c or a-c network. Each of its arcs $a = 1, \ldots, m$ has the form of Fig.3. It consists of an impedance Z_a, an ideal voltage source providing a voltage E_a^s and - in parallel arrangement - an ideal current source providing a current J_a^s. In a d-c network Z_a, E_a^s and J_a^s $(a = 1, \ldots, m)$ are real constants. In an a-c network these quantities are constant complex amplitudes provided all quantities are harmonically oscillating with one single excitation frequency. The same is true for all quantities introduced further below.

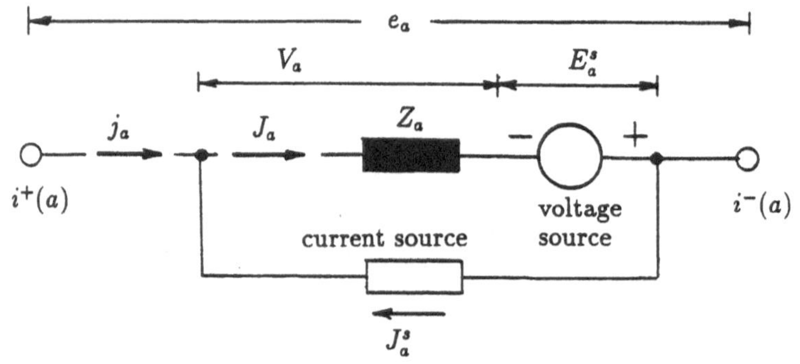

Figure 3. Arc a between the vertices $i^+(a)$ and $i^-(a)$ represents a system composed of an impedance Z_a, a voltage source and a current source. e_a, E_a^s and V_a are voltages and j_a, J_a, J_a^s are currents.

Unknowns are the the arc voltages e_a and the arc currents j_a $(a = 1, \ldots, m)$ shown in the figure. Let E_i $(i = 1, \ldots, n)$ be the node voltages against node 0. By definition e_a is the difference

$$e_a = E_{i^+(a)} - E_{i^-(a)} = \sum_{i=1}^{n} S_{ia} E_i \qquad (a = 1, \ldots, m). \tag{8}$$

In matrix form these equations read

$$\mathbf{e} = \mathbf{S}^T \mathbf{E}. \tag{9}$$

When this is premultiplied with $\mathbf{K}$, and use is made of Theorem 1 the equation

$$\mathbf{K}\mathbf{e} = \mathbf{0} \tag{10}$$

is obtained. It expresses Kirchhoff's circuit law for voltages. The equation is equivalent to (9). Kirchhoff's node law for currents reads

$$\sum_{a=1}^{m} S_{ia} j_a = 0 \qquad (i = 1, \ldots, n) \tag{11}$$

or in matrix form

$$\mathbf{Sj} = \mathbf{0} \quad \text{with} \quad \mathbf{j} = [j_1, \ldots, j_m]^T. \tag{12}$$

In (8) and (11) the incidence matrix is used for two different purposes. In (8) it is used for expressing a relative quantity e_a as the difference of two absolute quantities E_i or. We may also say that it is used for summing up quantities associated with all vertices which are incident with a given arc a. In this application the summation is carried out with respect to the first index of $\mathbf{S}$. This results in the appearance of $\mathbf{S}^T$ in (9). In (11) the incidence matrix is used for summing up quantities associated with all arcs which are incident with a given vertex i. In this application the summation is carried out with respect to the second index of $\mathbf{S}$. This results in the appearance of $\mathbf{S}$ in (12).

Let V_a and J_a be the voltage across and the current through the impedance Z_a, respectively. Ohm's generalized law for arc a reads

$$V_a = \sum_{b=1}^{m} Z_{ab} J_b \qquad (a, b = 1, \ldots, m)$$

with constant self- and transimpedances Z_{ab} $(a, b = 1, \ldots, m)$. The matrix form of these equations is

$$\mathbf{V} = \mathbf{ZJ} \qquad \Leftrightarrow \qquad \mathbf{J} = \mathbf{YV} \qquad (\mathbf{Y} = \mathbf{Z}^{-1}). \tag{13}$$

$\mathbf{Z}$ and $\mathbf{Y}$ are called impedance and atmittance matrix, respectively. From Fig.3 it follows that

$$V_a = E^s_a + e_a, \qquad J_a = J^s_a + j_a \quad (a = 1, \ldots, m)$$

or in matrix form

$$\mathbf{V} = \mathbf{E}^s + \mathbf{e}, \qquad \mathbf{J} = \mathbf{J}^s + \mathbf{j}. \tag{14}$$

The equations (9) to (14) allow the explicit solution for the unknowns $\mathbf{e}$ and $\mathbf{j}$ for given $\mathbf{E}^s$, $\mathbf{J}^s$ and $\mathbf{Z}$. As a first step the m arc currents are expressed in terms of $m - n$ circuit currents i_c $(c = n + 1, \ldots, m)$. If we put

$$\mathbf{j} = \mathbf{K}^T \mathbf{i}, \quad \text{with} \quad \mathbf{i} = [i_{n+1}, \ldots, i_m]^T, \tag{15}$$

then (12) is satisfied because of Theorem 1. The physical interpretation of the circuit currents is seen from Fig.4. i_c is positive if it is directed in the sense of chord c. In an arc a common to two circuits the arc current j_a is the sum or the difference of the circuit currents associated with these circuits. In mechanics there is an analogy to this: In the derivation of Bredt's formula for the torsion of a thin-walled bar with multi-cellular cross section the m shear flows in as many walls (arcs) are expessed in terms of $m - n$ circuit flows associated with as many cells (circuits) by the same relationship. For details see Pestel/Wittenburg [5]. Note the analogies between (10) and (12) and between (9) and (15). They show that the matrix **K** is used in two different ways as is the matrix **S**.

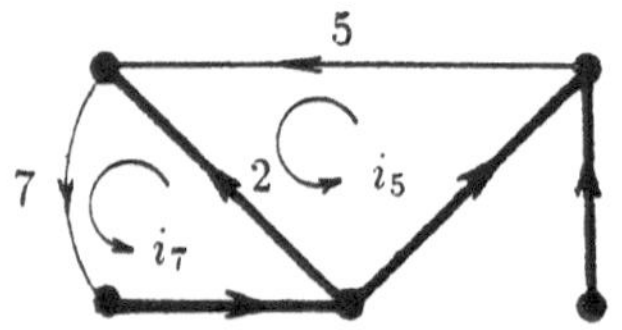

Figure 4. Relationship between arc currents j_a and circuit currents i_c : $j_2 = i_7 - i_5$.

Two methods of solution for **e** and **j** from Eqs.(9) to (14) are available.
Circuit Method: From (13) and (14) follows

$$\mathbf{Zj} - \mathbf{e} = \mathbf{E}^s - \mathbf{ZJ}^s \tag{16}$$

and with (10)

$$\mathbf{KZj} = \mathbf{K}(\mathbf{E}^s - \mathbf{ZJ}^s)$$

or with (15)

$$\mathbf{KZK}^T\mathbf{i} = \mathbf{K}(\mathbf{E}^s - \mathbf{ZJ}^s). \tag{17}$$

This is solved for **i**. (15) and (16) then yield **j** and **e**.

Node method: From (13) and (14) follows

$$\mathbf{Ye} = \mathbf{J}^s - \mathbf{YE}^s + \mathbf{j} \tag{18}$$

and with (12)

$$\mathbf{SYe} = \mathbf{S}(\mathbf{J}^s - \mathbf{YE}^s)$$

or with (9)

$$\mathbf{SYS}^T\mathbf{E} = \mathbf{S}(\mathbf{J}^s - \mathbf{YE}^s). \tag{19}$$

This is solved for **E**. (9) and (18) then yield **e** and **j**.

In electrical networks the number m of arcs is typically much larger than the number n of nodes. This is reflected by the fact that the path matrix **T** of a spanning tree does not play an explicite role in the above analysis.

3. Multibody Systems

In this chapter articulated mechanical systems, also called multibody systems, are investigated. The individual bodies are assumed to be rigid. The bodies are labeled $1, \ldots, n$. As a reference body inertial space is used. It is counted as body 0. For more detailed information see Wittenburg [2,6].

3.1. DYNAMICS. THE PRINCIPLE OF VIRTUAL POWER

Equations of motion will be obtained from the principle of virtual power. The Lagrangian form of the principle is

$$\sum_{i=1}^{n} \int_{m_i} \delta\dot{\vec{r}} \cdot (\ddot{\vec{r}}dm - d\vec{F}) = 0. \tag{20}$$

$\vec{r}$ is the radius vector of a mass element dm in inertial space; $\ddot{\vec{r}}$ is its absolute acceleration; $\delta\dot{\vec{r}}$ is a virtual change of its velocity under the condition that time t as well as the location of the system is held constant; $d\vec{F}$ is the total force applied to dm. Constraint forces caused by ideal kinematical constraints do not contribute to virtual power because for any pair of constraint forces $\vec{F}_1 = +\vec{F}$ and $\vec{F}_2 = -\vec{F}$ (actio = reactio) the term $\delta\dot{\vec{r}}_1 \cdot \vec{F}_1 + \delta\dot{\vec{r}}_2 \cdot \vec{F}_2$ is equal to zero. Note: It is often said that a single constraint force such as $\vec{F}_1$ has no virtual power. This is not correct since $\delta\dot{\vec{r}}_1$ and $\vec{F}_1$ need not be orthogonal.

For a system of rigid bodies (20) can be given the form

$$\sum_{i=1}^{n} \left[\delta\dot{\vec{r}}_i \cdot (m_i\ddot{\vec{r}}_i - \vec{F}_i) + \delta\vec{\omega}_i \cdot (J_i\dot{\vec{\omega}}_i + \underbrace{\vec{\omega}_i \times J_i\vec{\omega}_i - \vec{M}_i}_{-\vec{M}_i^*})\right] = 0 \tag{21}$$

($\vec{r}_i$ radius vector of body i center of mass; $\vec{\omega}_i$ absolute angular velocity of body i; m_i, J_i mass and central inertia tensor, respectively, of body i; $\vec{F}_i$ resultant external force on body i; $\vec{M}_i$ resultant external torque on body i about its center of mass). In matrix form the equation reads

$$\delta\dot{\vec{\mathbf{r}}}^T \cdot (\mathbf{M}\ddot{\vec{\mathbf{r}}} - \vec{\mathbf{F}}) + \delta\vec{\omega}^T \cdot (\mathbf{J}\dot{\vec{\omega}} - \vec{\mathbf{M}}^*) = 0 \tag{22}$$

M and **J** are diagonal $(n \times n)$-matrices of masses and inertia tensors, respectively. All other matrices are column matrices of vectors, for example $\delta\vec{\omega} = [\delta\vec{\omega}_1, \ldots, \delta\vec{\omega}_n]^T$. In writing (22) the notion of matrix and of matrix product is generalized. In addition to normal matrices with scalar elements also matrices are used the elements of which are vectors or tensors. Vector in this context is an intrinsic quantity and not a triple of scalar coordinates in some particular reference base. The same applies to tensors. In (22) the matrix products $\mathbf{M}\ddot{\vec{\mathbf{r}}}$, $\mathbf{J}\dot{\vec{\omega}}$ and $\delta\dot{\vec{\mathbf{r}}}^T \cdot \vec{\mathbf{F}}$ occur. Later also the cross product of vector matrices will occur. All these products are defined as natural generalisations of the usual matrix product. This means that every element of the product of two matrices is a sum of products (defined in different ways) of elements of the two matrices.

If a multibody system is not kinematically constrained to inertial space (as, for example, a multibody satellite in orbit) then Newton's equation for the composite system center of mass can be decoupled from the remaining equations. It reads

$$m_{tot}\ddot{\vec{r}}_C = \sum_{i=1}^{n} \vec{F}_i, \qquad m_{tot} = \sum_{i=1}^{n} m_i. \tag{23}$$

The remaining equations are obtained from (21) by substituting $\vec{r}_i = \vec{r}_C + \vec{R}_i$ $(i = 1, \ldots, n)$ with $\vec{R}_i$ being the vector from the composite system center of mass to the body i center of mass. This results in

$$\delta\dot{\vec{\mathbf{R}}}^T \cdot (\mathbf{M}\ddot{\vec{\mathbf{R}}} - \vec{\mathbf{F}}) + \delta\vec{\omega}^T \cdot (\mathbf{J}\dot{\vec{\omega}} - \vec{\mathbf{M}}^*) = 0. \tag{24}$$

The relationship between $\vec{R}_i$ and $\vec{r}_i$ can be written in the form

$$\vec{R}_i = \vec{r}_i - \vec{r}_C = \vec{r}_i - \frac{1}{m_{tot}} \sum_{j=1}^{n} m_j\vec{r}_j \qquad (i = 1, \ldots, n) \tag{25}$$

or in matrix form

$$\vec{\mathbf{R}} = \mu^T \vec{\mathbf{r}} \tag{26}$$

where μ is the matrix with elements

$$\mu_{ij} = \delta_{ij} - \frac{m_i}{m_{tot}} \qquad (i,j = 1,\ldots,n). \tag{27}$$

This matrix is singular ($\vec{\mathbf{r}}$ cannot be recovered from $\vec{\mathbf{R}}$). It has the properties

$$\mu\mu = \mu, \qquad \mu\mathbf{M}\mu^T = \mathbf{M}\mu^T. \tag{28}$$

(22) and (24) constitute the dynamics part of the formulation of equations of motion for multibody systems. What remains to be done is kinematics.

3.2. KINEMATICS. OUTLINE OF THE METHOD

Let $\mathbf{q} = [q_1,\ldots,q_N]^T$ be an arbitrary set of generalized coordinates which are suitable for specifying the location and orientation of a multibody system. These coordinates may be either joint variables or variables of position relative to inertial space or a combination of the two. At this point it is also not necessary to know whether $\mathbf{q}$ represents a minimal set of variables equal in number to the total number of degrees of freedom of the system or whether there exist constraint equations between the variables. Holonomic, scleronomic constraint equations have the general form

$$f_i(q_1,\ldots,q_N) = 0 \qquad (i = 1,\ldots,\nu). \tag{29}$$

For more general constraint equations see Wittenburg [2]. In any case the radius vector of the body i center of mass ($i = 1,\ldots,n$) can be expressed as some more or less complicated nonlinear function $\vec{r}_i(q_1,\ldots,q_N)$ of the chosen variables. It follows that equations of the following forms exist:

$$\delta\dot{\vec{\mathbf{r}}} = \vec{\mathbf{a}}_1\delta\dot{\mathbf{q}}, \qquad \ddot{\vec{\mathbf{r}}} = \vec{\mathbf{a}}_1\ddot{\mathbf{q}} + \vec{\mathbf{b}}_1 \tag{30}$$

where $\vec{\mathbf{a}}_1$ is an $(n \times N)$-matrix of as yet unknown vectors which depend on $q_1,\ldots,q_N$. $\vec{\mathbf{b}}_1$ is a column matrix of vectors which depends on $q_1,\ldots,q_N$ and also on $\dot{q}_1,\ldots,\dot{q}_N$. In analogy to (30) there exist relationships of the forms

$$\delta\vec{\omega} = \vec{\mathbf{a}}_2\delta\dot{\mathbf{q}}, \qquad \dot{\vec{\omega}} = \vec{\mathbf{a}}_2\ddot{\mathbf{q}} + \vec{\mathbf{b}}_2 \tag{31}$$

with other matrices $\vec{\mathbf{a}}_2$ and $\vec{\mathbf{b}}_2$.

Substitution of (30) and (31) into (22) results in the equation

$$\delta\dot{\mathbf{q}}^T(\mathbf{A}\ddot{\mathbf{q}} - \mathbf{B}) = 0 \tag{32}$$

with

$$\left.\begin{aligned} \mathbf{A} &= \vec{\mathbf{a}}_1^T \cdot \mathbf{M}\vec{\mathbf{a}}_1 + \vec{\mathbf{a}}_2^T \cdot \mathbf{J}\vec{\mathbf{a}}_2, \\ \mathbf{B} &= \vec{\mathbf{a}}_1^T \cdot (\vec{\mathbf{F}} - \mathbf{M}\vec{\mathbf{b}}_1) + \vec{\mathbf{a}}_2^T \cdot (\vec{\mathbf{M}}^* - \mathbf{J}\vec{\mathbf{b}}_2). \end{aligned}\right\} \tag{33}$$

If the variables $q_1, \ldots, q_N$ are independent then it follows that

$$\mathbf{A}\ddot{\mathbf{q}} = \mathbf{B}. \tag{34}$$

For a system not kinematically constrained to inertial space (22) is replaced by (24). In view of (26) this has the effect that (34) remains valid if in (33) $\vec{\mathbf{a}}_1$ and $\vec{\mathbf{b}}_1$ are replaced by $\mu^T\vec{\mathbf{a}}_1$ and $\mu^T\vec{\mathbf{b}}_1$, respectively.

(34) is a minimal set of differential equations governing the multibody system under consideration. The matrix $\mathbf{A}$ is symmetric. It is also positive definite since it represents the coefficient matrix of the kinetic energy expression. It depends on $q_1, \ldots, q_N$. The column matrix $\mathbf{B}$ depends on $q_1, \ldots, q_N$ and, in addition, on $\dot{q}_1, \ldots, \dot{q}_N$. Both matrices have scalar elements.

Next, we consider the case that the chosen variables are subject to ν constraint equations of the form (29). It follows that

$$\delta\dot{f}_i = \sum_{j=1}^{N} \frac{\partial f_i}{\partial q_j} \delta\dot{q}_j, \qquad \ddot{f}_i = \sum_{j=1}^{N} \frac{\partial f_i}{\partial q_j} \ddot{q}_j + \sum_{j=1}^{N}\sum_{k=1}^{N} \frac{\partial^2 f_i}{\partial q_j \partial q_k} \dot{q}_j\dot{q}_k$$

($i = 1, \ldots, \nu$). Both sets of equations are solved for dependent quantities in terms of independent quantities. The number of independent quantities equals the rank of the Jacobian with elements $\partial f_i/\partial q_j$ ($i = 1, \ldots, \nu$; $j = 1, \ldots, N$). The result of this procedure are expressions of the form

$$\delta\dot{\mathbf{q}} = \mathbf{G}\delta\dot{\mathbf{q}}^*, \qquad \ddot{\mathbf{q}} = \mathbf{G}\ddot{\mathbf{q}}^* + \mathbf{H}. \tag{35}$$

Here, $\mathbf{q}$ is the column matrix of all N variables; $\mathbf{q}^*$ is the column matrix of all independent variables; $\mathbf{G}$ is a rectangular scalar matrix which depends on $q_1, \ldots, q_N$. The column matrix $\mathbf{H}$ depends on $q_1, \ldots, q_N$ and, in addition, on $\dot{q}_1, \ldots, \dot{q}_N$. Substitution of (35) into (32) results in the equation

$$\delta\dot{\mathrm{q}}^{*T}(\mathbf{A}^*\ddot{\mathrm{q}}^* - \mathbf{B}^*) = 0 \tag{36}$$

with

$$\mathbf{A}^* = \mathbf{G}^T\mathbf{A}\mathbf{G}, \qquad \mathbf{B}^* = \mathbf{G}^T(\mathbf{B} - \mathbf{J}\mathbf{H}). \tag{37}$$

From (36) follows

$$\mathbf{A}^*\ddot{\mathrm{q}}^* = \mathbf{B}^*. \tag{38}$$

This represents a minimal set of equations of motion. The matrices $\mathbf{A}^*$ and $\mathbf{B}^*$ depend on all variables $q_1, \ldots, q_N$. This means that in the course of numerical integration every time $\mathbf{A}^*$ and $\mathbf{B}^*$ are evaluated the nonlinear constraint equations (29) must be solved for the dependent variables. For two reasons this is not a very time-consuming task. 1) The Jacobian $\mathbf{G}$ is available so that a Newton-Raphson method can be applied. 2) The previous solution is always a good approximation for the actual solution.

As a conclusion of this chapter it can be stated that a minimal set of equations of motion for a multibody system is explicitly available as soon as the 4 kinematical matrices $\vec{\mathbf{a}}_1$, $\vec{\mathbf{b}}_1$ and $\vec{\mathbf{a}}_2$, $\vec{\mathbf{b}}_2$ from (30) and (31) are known in terms of the chosen variables. In the case of dependent variables also the constraint equations (29) must be known together with the matrices $\mathbf{G}$ and $\mathbf{H}$ deduced from them. The following chapters are devoted to the development of these expressions.

3.3. KINEMATICS FOR JOINT VARIABLES

Kinematics is not concerned with forces. Consequently, neither passive force elements (springs or dampers) nor active force elements (actuators) are of interest in this chapter. Only holonomical constraints caused by joints will be taken into consideration. Between any pair of bodies there can be at most one joint which, thus, encompasses all physical connections between the two bodies. A joint cannot be shared by more than one pair of bodies. This assumption implies that joints seemingly interconnecting more than two bodies are counted as several distinct joints.

In what follows the notion of joint is very general. If a body of a multibody system is not kinematically constrained at all then it is necessary to describe its location and angular orientation with respect to some suitably chosen reference body by 6 generalized coordinates. In this case the body and the chosen reference body are said to be connected by a 6-d.o.f-joint. Thus, joints can have any number of degrees of freedom $1 \leq f \leq 6$.

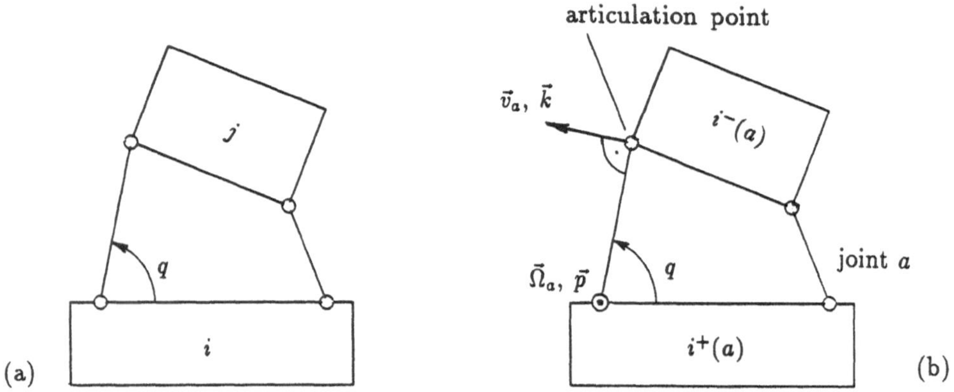

Figure 5a. The kinematical loop of 4 bodies represents a 2-body system with a single 1-d.o.f.-joint if the rods are treated as massless bodies.
Figure 5b. Kinematics of a 1-d.o.f. joint.

Another usage of the term joint is explained as follows. Consider the situation that two bodies i and j of large mass (or moment of inertia) are kinematically interconnected by a leight-weight mechanism which consists of one or more individual bodies. The entire subsystem consisting of the mechanism and of bodies i and j constitutes a multibody system of its own which, normally, has kinematical loops. If the inertia properties of the interconnecting mechanism are negligible compared with the two bodies i and j than the only purpose of the mechanism is to impose kinematical constraints upon the motion of body i relative to body j. In this case the whole mechanism represents a single joint between bodies i and j. Fig.5a shows an example. The system consists of 4 bodies forming a kinematical loop. If the two rods are treated as massless bodies then the system consists of only two bodies i and j with a 1-d.o.f.-joint for which the angle called q can be used as a generalized coordinate. After these introductory remarks let the vertices $0, \ldots, n$ of a graph represent the bodies, and let the arcs $1, \ldots, m$ represent the joints of the system. Circuits in the graph correspond to kinematical loops in the mechanical system. In a system with tree structure the total number of degrees of freedom equals the sum of the numbers of degrees of freedom of the indivual joints. Examples are living mechanisms (the human body in a phase of motion with only one foot on the ground) and some industrial robots. Most mechanisms in machines have a total number of degrees of freedom equal to 1. This is achieved by a sufficiently large number of kinematical loops.

3.3.1. *Systems With Tree Structure.* In this section the kinematics of tree-structured multibody systems is presented. It is particularly simple. Each arc (i.e each joint) is given a sense of direction. The resulting tree-structured, directed graph has an incidence matrix **S** and a path matrix **T** which is the inverse of **S** (see Theorem 2).

As in Chapter 1 $i^+(a)$ and $i^-(a)$ are the labels of the two vertices incident with arc a (i.e. of the two bodies connected by joint a). If $1 \le f_a \le 6$ is the number of degrees of freedom in joint a then f_a generalized joint variables are chosen for describing the position and angular orientation of body $i^-(a)$ relative to body $i^+(a)$. Fig.5b shows an example with a single degree of freedom in the joint. Let N be the total number of generalized coordinates thus defined for all joints together. As in Sec.2.2. we define $\mathbf{q} = [q_1, \ldots, q_N]^T$. These variables are not subject to constraint equations of the form (29).

Kinematics of Individual Joints. The kinematics of body $i^-(a)$ relative to body $i^+(a)$ is described in terms of the variables for a single joint a and of their first and second time derivatives. Six kinematical quantities are required. Three of them are the position vector, the velocity $\vec{v}_a$ and the acceleration $\vec{a}_a$ of a single point of body $i^-(a)$ relative to body $i^+(a)$. The other three are the angular orientation, the angular velocity $\vec{\Omega}_a$ and the angular acceleration $\vec{\epsilon}_a$ of body $i^-(a)$ relative to body $i^+(a)$. The single point on body $i^-(a)$ can be chosen arbitrarily. In what follows it is called the articulation point for joint a. Its radius vector in a reference frame fixed on body $i^-(a)$ is called $\vec{c}_{i^-(a),a}$ with the first index identifying the body and the second identifying the joint. Fig.6 shows the two bodies with body-fixed reference frames $\vec{\mathbf{e}}^{i^+(a)}$ and $\vec{\mathbf{e}}^{i^-(a)}$ attached at the body centers of mass and the vectors $\vec{c}_{i^+(a),a}$ and $\vec{c}_{i^-(a),a}$. $\vec{c}_{i^+(a),a}$ is the position vector of the articulation point relative to body $i^+(a)$. In general, it is a nonlinear function of the joint variables. The angular orientation of body $i^-(a)$ relative to body $i^+(a)$ is described by the direction cosine matrix $\mathbf{A}_a$ relating the reference frame $\vec{\mathbf{e}}^{i^-(a)}$ to the reference frame $\vec{\mathbf{e}}^{i^+(a)}$. Also $\mathbf{A}_a$ is normally a function of the joint variables. The six kinematical quantities associated with joint a are

$$\vec{c}_{i^+(a),a}(q_1, \ldots, q_N), \qquad \mathbf{A}_a(q_1, \ldots, q_N),$$

$$\vec{v}_a = \sum_{j=1}^{N} \vec{k}_{aj}\dot{q}_j, \qquad \vec{\Omega}_a = \sum_{j=1}^{N} \vec{p}_{aj}\dot{q}_j,$$

$$\vec{a}_a = \sum_{j=1}^{N} \vec{k}_{aj}\ddot{q}_j + \vec{a}_{a0}, \qquad \vec{\epsilon}_a = \sum_{j=1}^{N} \vec{p}_{aj}\ddot{q}_j + \vec{\epsilon}_{a0} \quad (a = 1, \ldots, n). \tag{39}$$

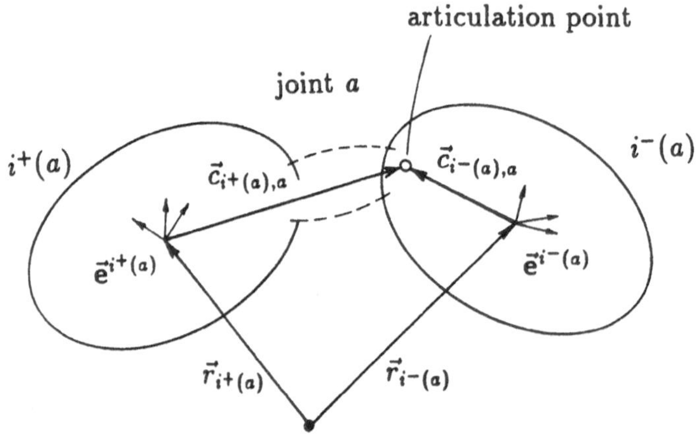

Figure 6. Two bodies coupled by a joint a of unspecified nature. The articulation point is fixed on body $i^-(a)$ and - normally - moving relative to body $i^+(a)$. Radius vectors $\vec{r}_{i^+(a)}$ and $\vec{r}_{i^-(a)}$ of the body centers of mass in inertial space.

To the sums in (39) only the variables associated with joint a contribute nonzero terms. For all other variables the vectors $\vec{k}_{aj}$ and $\vec{p}_{aj}$ are zero. The vectors $\vec{k}_{aj}$ and $\vec{p}_{aj}$ associated with the variables of joint a have simple physical interpretations. As an example the system in Fig.5b is considered. The joint between the two bodies has a single coordinate q. This means that $\vec{v}_a$ and $\vec{\Omega}_a$ have the simple forms

$$\vec{v}_a = \vec{k}(q)\dot{q}, \qquad \vec{\Omega}_a = \vec{p}(q)\dot{q}$$

with a single vector $\vec{k}$ and $\vec{p}$, respectively. These vectors are easily obtained from a kinematics analysis of the four-bar mechanism. Their directions are shown in the figure. The quantities $\vec{v}_{a0}$ and $\vec{\epsilon}_{a0}$ are abbreviations for quantities depending on zero- and first-order derivatives of joint variables. As a preparation for what follows the four vector equations in (39) are written in matrix form:

$$\left.\begin{array}{ll} \vec{\mathbf{v}} = \vec{\mathbf{k}}^T\dot{\mathbf{q}}, & \vec{\mathbf{\Omega}} = \vec{\mathbf{p}}^T\dot{\mathbf{q}}, \\ \vec{\mathbf{a}} = \vec{\mathbf{k}}^T\ddot{\mathbf{q}} + \vec{\mathbf{a}}_0, & \vec{\epsilon} = \vec{\mathbf{p}}^T\ddot{\mathbf{q}} + \vec{\epsilon}_0. \end{array}\right\} \tag{40}$$

The matrices $\vec{\mathbf{k}}$ and $\vec{\mathbf{p}}$ have block-diagonal structure.

Kinematics of Absolute Motion. For the principle of virtual power in (22) expressions for $\ddot{\vec{\mathbf{r}}}$ and $\dot{\vec{\omega}}$ are required in terms of joint variables. By definition

$$\vec{\Omega}_a = \vec{\omega}_{i^-(a)} - \vec{\omega}_{i^+(a)} = -\sum_{i=1}^{n} S_{ia}\vec{\omega}_i \qquad (a = 1, \ldots, n)$$

or in matrix form

$$\vec{\boldsymbol{\Omega}} = -\mathbf{S}^T\vec{\omega}.$$

From Theorem 2 and from (40) follows

$$\vec{\omega} = -\mathbf{T}^T\vec{\boldsymbol{\Omega}} = -(\vec{\mathbf{p}}\mathbf{T})^T\dot{\mathbf{q}}. \tag{41}$$

For angular accelerations the analogous equations read

$$\dot{\vec{\Omega}}_a = \vec{\epsilon}_a + \underbrace{\vec{\omega}_{i^+(a)} \times \vec{\Omega}_a}_{\vec{f}_a} = -\sum_{i=1}^{n} S_{ia}\dot{\vec{\omega}}_i \qquad (a = 1, \ldots, n),$$

$$\dot{\vec{\omega}} = -(\vec{\mathbf{p}}\mathbf{T})^T\ddot{\mathbf{q}} - \mathbf{T}^T\vec{\mathbf{f}}$$

with $\vec{\mathbf{f}} = [\vec{f}_1, \ldots, \vec{f}_n]^T$. This equation has the desired form of (31). Two of the four matrices $\vec{\mathbf{a}}_1$, $\vec{\mathbf{b}}_1$, $\vec{\mathbf{a}}_2$ and $\vec{\mathbf{b}}_2$ have, thus, been found:

$$\vec{\mathbf{a}}_2 = -(\vec{\mathbf{p}}\mathbf{T})^T, \qquad \vec{\mathbf{b}}_2 = -\mathbf{T}^T\vec{\mathbf{f}}. \tag{42}$$

The other two matrices are developed from an expression for $\vec{\mathbf{r}}$. Fig.6 yields

$$\vec{c}_{i^+(a),a} - \vec{c}_{i^-(a),a} = \vec{r}_{i^-(a)} - \vec{r}_{i^+(a)} \tag{43}$$

or

$$\sum_{i=1}^{n} S_{ia}\vec{c}_{ia} = -\sum_{i=1}^{n} S_{ia}\vec{r}_i \qquad (a = 1, \ldots, n). \tag{44}$$

Remark: The expression on the left-hand side is correct only if the articulation point for the joint located on body 0 is fixed on body 0. This is the case if the arc associated with this joint is given the direction towards body 0. For the general case see Wittenburg [2].

After this digression we consider Eq.(44) again. The left-hand side expression suggests the definition of a weighted incidence matrix $\vec{\mathbf{C}}$ with the elements

$$\vec{C}_{ia} = S_{ia}\vec{c}_{ia} \qquad (i, a = 1, \dots, n). \tag{45}$$

The matrix form of the equations then reads

$$\vec{\mathbf{C}}^T \mathbf{1} = -\mathbf{S}^T \vec{\mathbf{r}} \tag{46}$$

where $\mathbf{1} = [1\ 1\ \dots\ 1\,]^T$ is a summation operator. From Theorem 2 follows

$$\vec{\mathbf{r}} = -\mathbf{T}^T \vec{\mathbf{C}}^T \mathbf{1}. \tag{47}$$

For (30) we need an expression for $\ddot{\vec{\mathbf{r}}}$. It is most easily found from the second time derivative of (43):

$$\ddot{\vec{r}}_{i^-(a)} - \ddot{\vec{r}}_{i^+(a)} = -\vec{c}_{i^+(a),a} \times \dot{\vec{\omega}}_{i^+(a)} + \vec{c}_{i^-(a),a} \times \dot{\vec{\omega}}_{i^-(a)} + \vec{a}_a + \vec{h}_a \qquad (a = 1, \dots, n),$$

$$\vec{h}_a = \vec{\omega}_{i^+(a)} \times (\vec{\omega}_{i^+(a)} \times \vec{c}_{i^+(a),a}) - \vec{\omega}_{i^-(a)} \times (\vec{\omega}_{i^-(a)} \times \vec{c}_{i^-(a),a}) + 2\vec{\omega}_{i^+(a)} \times \vec{v}_a$$

$(a = 1, \dots, n)$ or

$$-\sum_{i=1}^{n} S_{ia}\ddot{\vec{\mathbf{r}}}_i = -\sum_{i=1}^{n} \vec{C}_{ia} \times \dot{\vec{\omega}}_i + \vec{a}_a + \vec{h}_a \qquad (a = 1, \dots, n)$$

or in matrix form

$$-\mathbf{S}^T \ddot{\vec{\mathbf{r}}} = -\vec{\mathbf{C}}^T \times \dot{\vec{\omega}} + \vec{\mathbf{a}} + \vec{\mathbf{h}}.$$

From this follows

$$\ddot{\vec{\mathbf{r}}} = \mathbf{T}^T \vec{\mathbf{C}}^T \times \dot{\vec{\omega}} - \mathbf{T}^T \vec{\mathbf{a}} - \mathbf{T}^T \vec{\mathbf{h}}.$$

For $\dot{\vec{\omega}}$ and $\vec{\mathbf{a}}$ the expressions from (42) and (40) are substituted. This yields

$$\ddot{\vec{\mathbf{r}}} = -\mathbf{T}^T \left(\vec{\mathbf{C}}^T \mathbf{T}^T \times \vec{\mathbf{p}}^{\,T} + \vec{\mathbf{k}}^T \right) \ddot{\mathbf{q}} - \mathbf{T}^T \left(\vec{\mathbf{C}}^T \mathbf{T}^T \times \vec{\mathbf{f}} + \vec{\mathbf{a}}_0 + \vec{\mathbf{h}} \right)$$

whence follows

$$\vec{\mathbf{a}}_1 = -\mathbf{T}^T \left(\vec{\mathbf{C}}^T \mathbf{T}^T \times \vec{\mathbf{p}}^{\,T} + \vec{\mathbf{k}}^T \right), \quad \vec{\mathbf{b}}_1 = -\mathbf{T}^T \left(\vec{\mathbf{C}}^T \mathbf{T}^T \times \vec{\mathbf{f}} + \vec{\mathbf{a}}_0 + \vec{\mathbf{h}} \right). \tag{48}$$

These matrices together with $\vec{\mathbf{a}}_2$ and $\vec{\mathbf{b}}_2$ from (42) are substituted into (33). This then yields the equations of motion (34) for a system with tree structure.

The equations are easily programmed. Input data are

- the number n of bodies and joints
- the integer functions $i^+(a)$ and $i^-(a)$ $(a = 1, \ldots, n)$
- masses and inertia tensors for all bodies
- for each joint $a = 1, \ldots, n$ the parameters shown in (39):
  the body-fixed vector $\vec{c}_{i^-(a),a}$
  the quantities $\vec{c}_{i^+(a),a}(q_1, \ldots, q_N)$ and $\mathbf{A}_a(q_1, \ldots, q_N)$
  the vectors $\vec{k}_{aj}$, $\vec{p}_{aj}$, $\vec{a}_{a0}$, $\vec{\epsilon}_{a0}$
- forces $\vec{F}_i$ and torques $\vec{M}_i$ $(i = 1, \ldots, n)$ acting on the bodies.

Up to this point vectors and tensors have not been decomposed in reference frames. For each vector and tensor the scalar coordinates are given as input data in a reference frame in which the coordinates are most easily available ($\vec{c}_{i^-(a),a}$ in $\vec{\mathbf{e}}^{i^-(a)}$, weight in $\vec{\mathbf{e}}^0$ etc.). The label of the chosen reference frame is part of the input data. From the integer functions $i^+(a)$ and $i^-(a)$ the program constructs the matrices $\mathbf{S}$ and $\mathbf{T}$. The latter contains the information how to transform via a chain of joints and with the help of the matrices $\mathbf{A}_a$ vectors from given reference frames into a common reference frame. These transformations are carried out automatically. In this way, all the matrices required for calculating $\mathbf{A}$ and $\mathbf{B}$ of (34) are constructed from the input data. The structure of such a program is simple. The computer code MESA VERDE programed by Wolz [7] is one such realisation. It produces expressions in alphanumerical form. In addition to the matrices $\mathbf{A}$ and $\mathbf{B}$ it produces expressions which enable the user of the program to calculate various kinematical quantities as well as constraint forces and torques. MESA VERDE is not restricted to tree-structured systems.

Most recently Weber [8] developed an object-oriented code written in $C++$ which enables a user to program multibody formalisms of more general nature in alphanumerical form. Equations such as (33) occupy a single line of code.

At the beginning of chapter 1 it was said that the matrices defined for graphs allow compact formulations. This has been demonstrated. It was said also that in programming such formalisms computation time can be saved by going back to the integer functions $i^+(a)$ and $i^-(a)$. This is obvious since the matrices defined in the chapter on kinematics are sparse. Moreover, the location of nonzero elements in the matrices is determined by the integer functions. For tree-structured systems a regular labeling of bodies and joints is useful (a program may have a module for relabeling a non-regularly labeled graph so that the program user is free in giving labels to bodies and joints). An important aspect of programming is the use of recursive formulas. This subject has been treated by many authors (see, for example, Featherstone [9] and Kim/Haug [10]). How closely it is related to the present formalism can be seen from Eq.(41), for example: $\vec{\omega} = -\mathbf{T}^T\vec{\Omega}$. Whenever $\mathbf{T}^T$ appears as a leading factor a recursive calculation along branches of the tree from body 0 in the outward direction is indicated.

3.3.2. *Systems With Kinematical Loops.* A kinematical loop is a loop formed of rigid bodies and of joints which individually have less than 6 degrees of freedom each. The joint variables of a kinematical loop are subject to kinematical constraints. In the case of holonomic, scleronomic constraints the constraint equations have the form of Eq.(29). In Sec.3.2 a method was explained which leads to a minimal set of equations of motion in the presence of constraint equations. In explaining this method it was assumed that initially a system with tree structure exists which is then subjected to constraint equations. From this follows that in the analysis of systems with kinematical loops one must solve the problems

- how to generate the tree-structured system to start with
- how to formulate the set of constraint equations (29)
- how to determine the number of independent variables
- how to select these variables and
- how to calculate the matrices $\mathbf{G}$ and $\mathbf{H}$ of (35).

These purely kinematical problems will not be studied in detail in this paper. For generating a tree-structured system two different methods are described. The first method consists in *eliminating joints* in sufficient number in order to arrive at a system with tree structure. Elimination of joints also means elimination of joint variables. The column matrix $\mathbf{q}$ of joint variables for the resulting tree-structured system does not contain the variables of the eliminated joints. Constraint equations must be formulated for the remaining variables contained in $\mathbf{q}$.

Another method called *duplication of bodies* arrives at a tree-structured system by the following procedure first described by Lilov/Chirikov [11]. See also Wolz [7] and Weber [8]. One single body in a loop is replaced by two identical twin-bodies 1 and 2. The original body has two joints which belong to the loop under consideration. One of these joints is eliminated on body 1 and the other is eliminated on body 2. The bodies 1 and 2 are then allowed to drift apart. This results in a tree-structured system with one additional body and with the original set of joints. Constraint equations express conditions that the bodies do not drift apart.

For a given kinematical loop both methods allow the formulation of constraint equations. In practical applications in machines and mechanisms certain types of kinematical loops are typical. The simplest possible example is the parallelogram mechanism. The planar four-bar mechanism is another example for which constraint equations can be formulated in such a way that dependent variables are explicitly expressed in terms of independent variables. Also the matrices **G** and **H** can be stated explicitly. For many more loops the constraint equations are found in the literature either in explicite or in implicite form. Woernle [12] describes an efficient way for formulating constraint equations. His basic idea is to cut a pair of joints in a loop. The method is referenced elsewhere in this book. A general-purpose computer program for multibody dynamics should have not only a library of joints but also a library of kinematical loops in which the relevant mathematical expressions are available.

With a computer program that generates the kinematical quantities described in Sec. 3.3.1 constraint equations for kinematical loops can be formulated fully automatically if the method of body duplication is used. Conditions for two bodies 1 and 2 not to drift apart are formulated on the level of position variables, on the level of velocities and on the level of accelerations (see Wolz [7] and Weber[8]).
Level of Position Variables: The two body centers of mass must coincide. This is the condition $\vec{r}_1 = \vec{r}_2$. Both vectors are elements of the matrix $\vec{\mathbf{r}}$ in Eq.(47). Both are functions of the joint variables belonging to the loop under consideration. Also the angular orientation of the two bodies must be the same. This is the condition that the product of the direction cosine matrices $\mathbf{A}_a$ or $\mathbf{A}_a^T$ (depending on the sense of direction of the arcs) of the joints belonging to the loop must be the unit matrix. Of the 9 equations represented by this condition anly 3 need be considered.
Acceleration Level: The conditions

$$\ddot{\vec{r}}_1 = \ddot{\vec{r}}_2, \qquad \dot{\vec{\omega}}_1 = \dot{\vec{\omega}}_2$$

together with the equations

$$\ddot{\vec{\mathbf{r}}} = \vec{\mathbf{a}}_1\ddot{\mathbf{q}} + \vec{\mathbf{b}}_1, \qquad \dot{\vec{\omega}} = \vec{\mathbf{a}}_2\ddot{\mathbf{q}} + \vec{\mathbf{b}}_2$$

establish 6 inhomogeneous linear equations for $\ddot{\mathbf{q}}$. The rank of the coefficient matrix is the number of independent acceleration variables $\ddot{\mathbf{q}}^*$. The chosen dependent acceleration variables are expressed in terms of the independent ones. Together with the identity $\ddot{\mathbf{q}}^* = \ddot{\mathbf{q}}^*$ this establishes (35):

$$\ddot{\mathbf{q}} = \mathbf{G}\ddot{\mathbf{q}}^* + \mathbf{H}.$$

On the velocity level the equation is $\dot{\mathbf{q}} = \mathbf{G}\dot{\mathbf{q}}^*$. It is obvious how one has to proceed if the number of kinematical loops in a system and, thus, the number of duplicated bodies is larger than 1.

3.4. CONSTRAINT FORCES AND TORQUES IN JOINTS

In a technical joint two bodies are in contact in points of a more or less complicated surface. Constraint forces are, therefore, distributed forces. The engineer responsable for the design of the joint must be interested in the distribution of the constraint force. The dynamics analyst who assumes rigid surfaces in the joint can only calculate an equivalent force system which consists of a single force and a single torque. The torque depends upon the choice of the point at which the force is thought to be acting. It is natural to use for each joint a the articulation point located on body $i^-(a)$. Let $\vec{X}_a$ and $\vec{Y}_a$ be the constraint force and the constraint torque, respectively, thus defined for joint a. For unambiguity we specify that $+\vec{X}_a$ and $+\vec{Y}_a$ are acting on body $i^+(a)$ and $-\vec{X}_a$ and $-\vec{Y}_a$ on body $i^-(a)$.

The resultant of all constraint forces applied to body i of a multibody system with bodies $i = 1, \ldots, n$ and joints $a = 1, \ldots, m$ is the expression

$$\sum_{a=1}^{m} S_{ia}\vec{X}_a \qquad (i = 1, \ldots, n).$$

Newton's and Euler's theorems for isolated bodies, thus, have the forms

$$m_i\ddot{\vec{r}}_i = \vec{F}_i + \sum_{a=1}^{m} S_{ia}\vec{X}_a \qquad (i = 1, \ldots, n),$$

$$J_i\dot{\vec{\omega}}_i + \vec{\omega}_i \times J_i\vec{\omega}_i = \vec{M}_i + \sum_{a=1}^{m} S_{ia}\vec{c}_{ia} \times \vec{X}_a + \sum_{a=1}^{m} S_{ia}\vec{Y}_a \qquad (i = 1, \ldots, n).$$

All quantities in these equations have the same definitions as in (21). In matrix form the equations read

$$\mathbf{M}\ddot{\vec{\mathbf{r}}} = \vec{\mathbf{F}} + \mathbf{S}\vec{\mathbf{X}}, \qquad \mathbf{J}\dot{\vec{\omega}} = \vec{\mathbf{M}}^* + \vec{\mathbf{C}} \times \vec{\mathbf{X}} + \mathbf{S}\vec{\mathbf{Y}}. \tag{49}$$

All quantities in these equations have the same definitions as in (22) and (45). If the multibody system has tree structure then $\mathbf{S}$ has the inverse $\mathbf{T}$. $\vec{\mathbf{X}}$ and $\vec{\mathbf{Y}}$ are then found explicitly by premultiplying both equations by $\mathbf{T}$. The constraint forces and torques are calculated in parallel with the numerical integration of the equations of motion so that all kinematical quantities are known as functions of time.

If the multibody system has kinematical loops then the number m of constraint forces and torques is larger than n. Eqs.(49) are not sufficient. In this case additional information is contained in the equations

$$\vec{\mathbf{k}} \cdot \vec{\mathbf{X}} = \mathbf{0}, \qquad \vec{\mathbf{p}} \cdot \vec{\mathbf{Y}} = \mathbf{0}. \tag{50}$$

Here, $\vec{\mathbf{k}}$ and $\vec{\mathbf{p}}$ are the matrices defined by Eq.(40). The first equation expresses the fact that the constraint force $\vec{X}_a$ is orthogonal to all possible directions $\vec{k}_{aj}$ of relative translation in joint a. The second equation is the equivalent for rotation. Eqs.(50) are satisfied also in tree-structured systems. In that case, however, they do not give information which is not yet contained in (49). For systems with kinematical loops (49) and (50) together determine the constraint forces and torques provided the system is dynamically determinate. The term dynamically determinate is a generalisation of the term statically determinate used in statics of rigid-body systems. As a matter of fact, in statics of multibody systems constraint forces and torques in supports and joints are determined from Eqs.(49) and (50) under the conditions $\ddot{\vec{\mathbf{r}}} \equiv \vec{\mathbf{0}}$ and $\vec{\omega} \equiv \vec{\mathbf{0}}$. As in statics also in dynamics systems can be indeterminate. For a determinate system only some of the equations (50) are required.

Good machine design uses dynamically determinate systems in order to prevent constraint forces resulting from inaccurate manifacturing or from temperature changes. A dynamics analyst interested in formulating equations of motion for a mechanism tends to assume fewer degrees of freedom in joints than the designer actually provided for the said reasons. The mechanism thus simulated has soluable equations of motion. But it may be dynamically indeterminate. This is the case if the rank of the system of Eqs.(49) and (50) is smaller than m.

3.5. AUGMENTED BODIES

The previous chapters were devoted to formulations of multibody kinematics and dynamics with the aim of developing general-purpose computer programs for industrial applications. The present chapter deals with more theoretical aspects of analytical mechanics. For a restricted class of multibody systems equations of motion are developed in a form suitable for non-numerical investigations. Subject of investigation are tree-structured multibody systems without kinematical constraints to inertial space and with spherical joints only. Starting point is the principle of virtual power in the form (24):

$$\delta\dot{\vec{\mathbf{R}}}^T \cdot (\mathbf{M}\ddot{\vec{\mathbf{R}}} - \vec{\mathbf{F}}) + \delta\vec{\omega}^T \cdot (\mathbf{J}\dot{\vec{\omega}} - \vec{\mathbf{M}}^*) = 0. \tag{51}$$

For $\vec{\mathbf{R}}$ Eqs.(26) and (47) yield

$$\vec{\mathbf{R}} = -(\vec{\mathbf{C}}\mathbf{T}\mu)^T\mathbf{1}. \tag{52}$$

As articulation point of a spherical joint the center of the sphere is chosen. This has the consequence that not only the vector $\vec{c}_{i^-(a),a}$ is fixed on body $i^-(a)$ but also the vector $\vec{c}_{i^+(a),a}$ is fixed on body $i^+(a)$.

A comment is necessary on the joint connnecting the system to body 0 (inertial space). This joint has 6 degrees of freedom. Since Newton's equation for the translation of the composite system center of mass has been decoupled already ((23)) one is left with 3 degrees of freedom. This allows to interprete the joint as a fictitious spherical joint between body 0 and an arbitrarily chosen body at an arbitrary point on this body. In what follows this joint will be labeled joint 1. It is located at the center of mass of body 1. From this follows that

$$\vec{c}_{11} = \vec{0}. \tag{53}$$

The elements of the matrix $\vec{\mathbf{C}}\mathbf{T}\mu$ in (52) allow an interesting physical interpretation. First, the matrix $\vec{\mathbf{C}}\mathbf{T}$ is investigated.

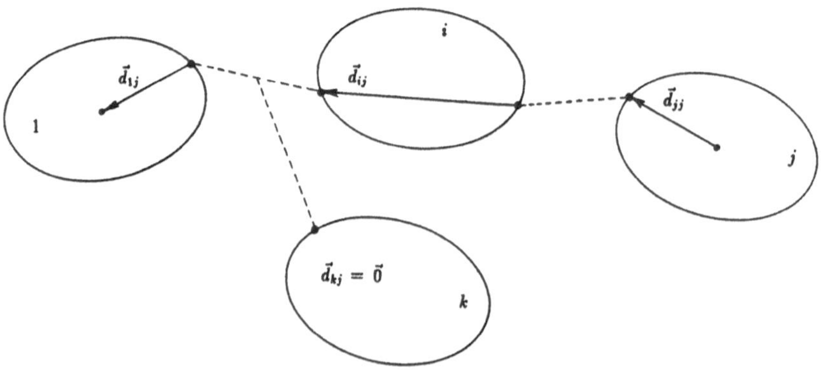

Figure 7. The vectors $\vec{d}_{ij}$ on different bodies i for a given body j.

It has the elements

$$\vec{d}_{ij} = (\vec{\mathbf{C}}\mathbf{T})_{ij} = \sum_{a=1}^{n} S_{ia} T_{aj} \vec{c}_{ia} \qquad (i, a = 1, \dots, n). \tag{54}$$

For given vertices i and j the product $S_{ia}T_{aj}$ is nonzero only for those arcs which are on the path from body 0 to body j ($T_{aj} \neq 0$) and which, in addition, are incident with vertex i ($S_{ia} \neq 0$). No arc satisfies both conditions if the vertices i and j are located on different branches of the tree as seen from body 0. Both conditions are satisfied for a single arc if $i = j$ and for exactly 2 arcs if vertex i is on the path from body 0 to body j (but $i \neq j$). Fig.7 illustrates the vectors $\vec{d}_{ij}$ in the three cases. $\vec{d}_{1j}$ on body 1 is as shown because of (53). Having interpreted the elements of $\vec{\mathbf{C}}\mathbf{T}$ we now turn to the matrix $\vec{\mathbf{C}}\mathbf{T}\mu$. We begin by introducing the concept of augmented bodies. For each body $i = 1, \dots, n$ an augmented body is constructed as follows. To the tip of each vector $\vec{c}_{ia}$ fixed on body i a point mass is attached which is equal to the sum of the masses of all bodies (except body 0) which are connected with body i either directly or indirectly via the respective joint a. From this definition follows that every augmented body is a rigid body of mass m_{tot}. The augmented body i has a center of mass which, in general, does not coincide with the center of mass of the original body i. In Fig.8 body i is depicted with its original center of mass C_i and

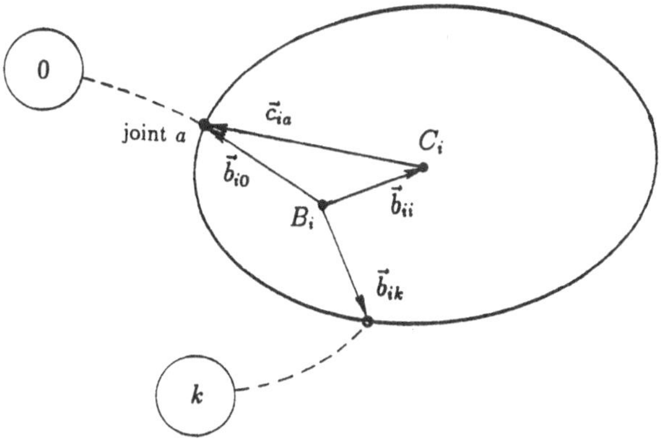

Figure 8. Vectors $\vec{b}_{ij}$ on the augmented body i for different bodies j.

with the new center of mass B_i. Also shown are body 0 and another body k. On the augmented body i vectors $\vec{b}_{ij}$ ($j = 0, \dots, n$) are defined. They point from B_i to C_i in the case $i = j$ and from B_i to the tip of the vector $\vec{c}_{ia}$ leading either directly or indirectly to body j in the case $j \neq i$. As examples the vectors $\vec{b}_{ii}$, $\vec{b}_{i0}$ and $\vec{b}_{ik}$ are shown in the figure. The vectors $\vec{b}_{ij}$ with $j \neq i$ play the same role for the augmented bodies that is played by the vectors $\vec{c}_{ia}$ for the original bodies. One should notice, however, the difference in notation. The second index of $\vec{c}_{ia}$ is that of an arc (joint) whereas the second index of $\vec{b}_{ij}$ is that of a vertex (body). All vectors $\vec{b}_{ij}$ $(i = 1, \ldots, n;\ j = 0, \ldots, n)$ are defined. However, the number of different vectors is smaller than the number of combinations of indices i, j. From the definition follows that

$$\sum_{j=1}^{n} \vec{b}_{ij} m_j = \vec{0} \qquad (i = 1, \ldots, n). \tag{55}$$

Between the vectors $\vec{b}_{ij}$ and $\vec{d}_{ij}$ defined in (54) there exist the relationships

$$\vec{d}_{ij} = \vec{b}_{i0} - \vec{b}_{ij} \qquad (i, j = 1, \ldots, n) \tag{56}$$

as is easily verified for the different cases shown in Fig.7. After these preparations the elements of the matrix $\vec{\mathbf{C}}\mathbf{T}\mu$ are found to be

$$\begin{aligned} (\vec{\mathbf{C}}\mathbf{T}\mu)_{ij} &= \sum_{k=1}^{n} \vec{d}_{ik} \mu_{kj} = \sum_{k=1}^{n} (\vec{b}_{i0} - \vec{b}_{ij})(\delta_{kj} - m_k/m_{tot}) \\ &= \vec{b}_{i0}(1-1) - \vec{b}_{ij} + \frac{1}{m_{tot}} \sum_{k=1}^{n} \vec{b}_{ik} m_k \\ &= -\vec{b}_{ij} \qquad (i, j = 1, \ldots, n) \end{aligned} \tag{57}$$

because of (55).

(52) now yields

$$\delta \dot{\vec{\mathbf{R}}}^T = \mathbf{1}^T \left[\delta\vec{\omega}_i \times \vec{b}_{ij} \right] = \delta\vec{\omega}^T \times \left[\ \vec{b}_{ij}\ \right], \qquad \ddot{\vec{\mathbf{R}}} = \left[\ \ddot{\vec{b}}_{ij}\ \right]^T \mathbf{1}.$$

When this is substituted into (51) the equation is obtained:

$$\delta\vec{\omega}^T \cdot \left\{ \left[\ \vec{b}_{ij}\ \right] \times \left(\mathbf{M} \left[\ \ddot{\vec{b}}_{ij}\ \right]^T \mathbf{1} - \vec{\mathbf{F}} \right) + \mathbf{J}\dot{\vec{\omega}} - \vec{\mathbf{M}}^* \right\} = 0$$

and since the angular velocities are unconstrained

$$\left[\ \vec{b}_{ij}\ \right] \times \left(\mathbf{M} \left[\ \ddot{\vec{b}}_{ij}\ \right]^T \mathbf{1} - \vec{\mathbf{F}} \right) + \mathbf{J}\dot{\vec{\omega}} - \vec{\mathbf{M}}^* = \vec{0}. \tag{58}$$

This is a set of n vector differential equations of motion. Let $\vec{\mathbf{G}}$ with elements $\vec{g}_{ij}$ $(i,j=1,\ldots,n)$ be an abbreviation for the matrix

$$\vec{\mathbf{G}} = \left[\ \vec{b}_{ij}\ \right] \times \mathbf{M}\left[\ \ddot{\vec{b}}_{ij}\ \right]^T.$$

Then the equations of motion are

$$\sum_{j=1}^{n} \vec{g}_{ij} - \sum_{j=1}^{n} \vec{b}_{ij} \times \vec{F}_j + J_i\dot{\vec{\omega}}_i + \vec{\omega}_i \times J_i\vec{\omega}_i - \vec{M}_i = \vec{0} \qquad (i=1,\ldots,n). \qquad (59)$$

Here, the original meaning of $\vec{M}_i^*$ shown in (21) has been re-introduced. For further simplifications we investigate the vectors

$$\vec{g}_{ij} = \left[\left[\ \vec{b}_{ij}\ \right] \times \mathbf{M}\left[\ \ddot{\vec{b}}_{ij}\ \right]^T\right]_{ij} = \sum_{k=1}^{n} m_k\vec{b}_{ik} \times \ddot{\vec{b}}_{jk} \qquad (i,j=1,\ldots,n).$$

In what follows only the case $i \neq j$ is considered. Let the graph be divided into two subgraphs by drawing a line through an arbitrary arc on the path between vertex i and vertex j. Let the set of labels of all vertices of the subgraph containing the vertex i be denoted by I and the set of labels of all vertices of the subgraph containing the vertex j by II. Then, for all labels k belonging to I (abbreviated $k \in I$) the identity $\vec{b}_{jk} = \vec{b}_{ji}$ holds and for all labels $k \in II$ the identity $\vec{b}_{ik} = \vec{b}_{ij}$. With this, $\vec{g}_{ij}$ becomes

$$\vec{g}_{ij} = \left(\sum_{k\in I} m_k\vec{b}_{ik}\right) \times \ddot{\vec{b}}_{ji} + \vec{b}_{ij} \times \sum_{k\in II} m_k\ddot{\vec{b}}_{jk} \qquad (i \neq j).$$

With the help of (55) the term in brackets can be given the form

$$\sum_{k\in I} m_k\vec{b}_{ik} = \sum_{k=1}^{n} m_k\vec{b}_{ik} - \sum_{k\in II} m_k\vec{b}_{ik} = -\vec{b}_{ij}\sum_{k\in II} m_k.$$

In a similar manner

$$\sum_{k\in II} m_k\ddot{\vec{b}}_{jk} = -\ddot{\vec{b}}_{ji}\sum_{k\in I} m_k.$$

With this, $\vec{g}_{ij}$ takes the form

$$\vec{g}_{ij} = -\Big(\sum_{k\in I} m_k + \sum_{k\in II} m_k\Big)\vec{b}_{ij} \times \ddot{\vec{b}}_{ji} = -m_{tot}\vec{b}_{ij} \times \ddot{\vec{b}}_{ji} \qquad (i,j=1,\ldots,n;\ i \neq j).$$

Substitution into (59) yields the equations

$$J_i\dot{\vec{\omega}}_i + \vec{\omega}_i \times J_i\vec{\omega}_i + \sum_{k=1}^{n} \vec{b}_{ik} \times \ddot{\vec{b}}_{ik} m_k = \vec{M}_i + \sum_{j=1}^{n} \vec{b}_{ij} \times \vec{F}_j + m_{tot} \sum_{j \neq i} \vec{b}_{ij} \times \ddot{\vec{b}}_{ji} \quad (60)$$

$(i = 1, \ldots, n)$. This can be further simplified if it is recognized that originally

$$J_i\dot{\vec{\omega}}_i + \vec{\omega}_i \times J_i\vec{\omega}_i = \frac{d}{dt}\int_{m_i} \vec{\rho} \times \dot{\vec{\rho}}\, dm = \int_{m_i} \vec{\rho} \times \ddot{\vec{\rho}}\, dm$$

where $\vec{\rho}$ is the vector from the the body i center of mass C_i to the mass element dm. It follows that the expression on the left-hand side of (60) can be written in the form $K_i\dot{\vec{\omega}}_i + \vec{\omega}_i \times K_i\vec{\omega}_i$ where K_i is the inertia tensor of the augmented body with respect to its center of mass B_i. Thus, the equations of motion are obtained in their final form:

$$K_i\dot{\vec{\omega}}_i + \vec{\omega}_i \times K_i\vec{\omega}_i = \vec{M}_i + \sum_{j=1}^{n} \vec{b}_{ij} \times \vec{F}_j + m_{tot} \sum_{j \neq i} \vec{b}_{ij} \times \ddot{\vec{b}}_{ji} \quad (61)$$

$$\text{with} \quad \ddot{\vec{b}}_{ji} = \dot{\vec{\omega}}_j \times \vec{b}_{ji} + \vec{\omega}_j \times (\vec{\omega}_j \times \vec{b}_{ji}) \qquad (i, j = 1, \ldots, n). \quad (62)$$

Each of these equations has the form of Euler's equations of motion for a single rigid body. Here, the rigid body is the augmented body i of mass m_{tot}. The equations are coupled through the term $\ddot{\vec{b}}_{ji}$. They cannot be uncoupled. The right-hand side expression contains the torques $\vec{b}_{ij} \times (m_{tot}\ddot{\vec{b}}_{ji} + \vec{F}_j)$ $(j \neq i)$. The force $m_{tot}\ddot{\vec{b}}_{ji} + \vec{F}_j$ can be interpreted as follows. Let the augmented body j $(j \neq i)$ be suspended as a pendulum in inertial space at its joint leading toward body i and let it be subject to its external force $\vec{F}_j$. When it is then given its actual angular velocity and acceleration it exerts on the suspension point the force $m_{tot}\ddot{\vec{b}}_{ji} + \vec{F}_j$. The equations (61) were published by Roberson/Wittenburg [4]. They allow non-numerical investigations of a number of problems which represent generalisations of classical problems of rigid body dynamics. For details see Wittenburg [13] and Wittenburg/Lilov [14]

4. Linear Vibrations of Chains of Bodies

In this chapter systems of bodies are investigated which are interconnected not by joints with kinematical constraints but by linear springs and dampers. Moreover, each body has a single degree of freedom only. Fig.9 depicts an example with $n = 4$ bodies. Body 0 is inertial space. Altogether $m^S = 9$ springs and $m^D = 3$ dampers interconnect the bodies $0, \ldots, n$. The bodies can translate along the x-axis. The following formulations apply also to systems with rotatory instead of translatory oscillators.

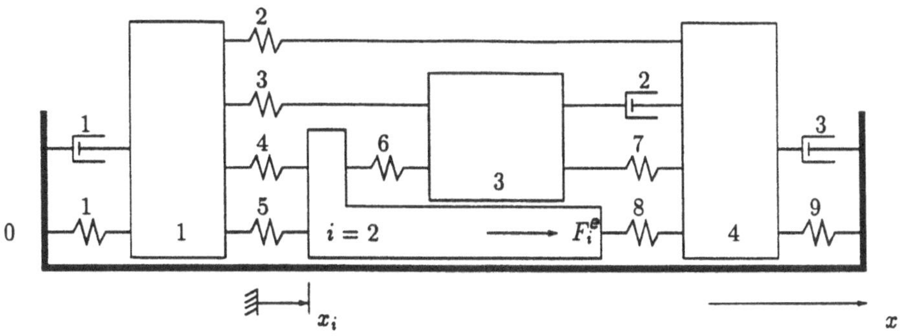

Figure 9. A chain of bodies with connecting springs and dampers.

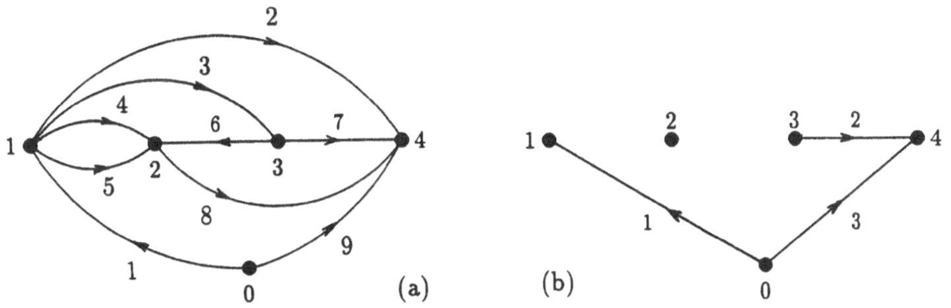

Figure 10ab. The spring graph (Fig.a) and the damper graph (Fig.b) of the system in Fig.9.

Figs.10a,b show two graphs with vertices $i = 0, \ldots, n$. The arcs in Fig.10a represent the springs, and the arcs in Fig.10b represent the dampers. To every arc a sense of direction is given in an arbitrary way. The spring graph happens to be connected but the damper graph is not. This does not cause problems because only the incidence matrices of the graphs are needed. The definition (1) of the incidence matrix remains valid if the graph is not connected. For the graphs shown in Figs.10a,b the two matrices are

$$\mathbf{S}^{\mathrm{S}} = \begin{bmatrix} -1 & 1 & 1 & 1 & 1 & 0 & 0 & 0 & 0 \\ 0 & 0 & 0 & -1 & -1 & -1 & 0 & 1 & 0 \\ 0 & 0 & -1 & 0 & 0 & 1 & 1 & 0 & 0 \\ 0 & -1 & 0 & 0 & 0 & 0 & -1 & -1 & -1 \end{bmatrix}, \qquad \mathbf{S}^{\mathrm{D}} = \begin{bmatrix} -1 & 0 & 0 \\ 0 & 0 & 0 \\ 0 & 1 & 0 \\ 0 & -1 & -1 \end{bmatrix}.$$

As generalized coordinates $q_1, \ldots, q_n$ relative as well as absolute displacements of bodies are accepted. The arbitrarily chosen coordinates are represented by the arcs of still another directed graph called the coordinate graph. Also this graph has vertices $i = 0, \ldots, n$. Definition: The coordinate q_a $(a = 1, \ldots, n)$ is the displacement along the x-axis of body $i^-(a)$ relative to body $i^+(a)$. Fig.11a depicts a coordinate graph in which q_1 and q_2 are absolute displacements of the bodies 1 and 3, respectively, whereas q_3, q_4 and q_5 are relative displacements. If n absolute displacements are chosen as coordinates then the coordinate graph has the form shown in Fig.11b. The coordinate graph for any suitably chosen set of coordinates is connected and tree-structured. If it would consist of unconnected subgraphs then the coordinates would not be suitable for describing the location relative to one another of the sets of bodies associated with these subgraphs. If the graph would have a circuit then the coordinates associated with the arcs of this circuit would not be independent. Thus, for the coordinate graph there exist an incidence matrix $\mathbf{S}^{\mathrm{C}}$ and a path matrix $\mathbf{T}^{\mathrm{C}}$ which is the inverse of $\mathbf{S}^{\mathrm{C}}$.

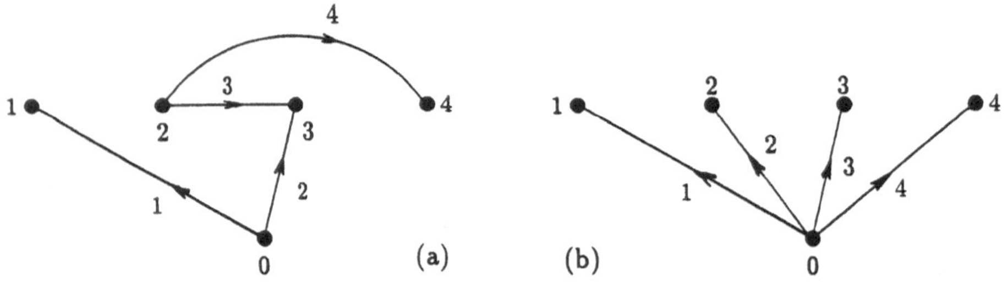

Figure 11ab. Two coordinate graphs for the system in Fig.9.

For the coordinate graph shown in Fig.11a the two matrices are

$$\mathbf{S}^{\mathrm{C}} = \begin{bmatrix} -1 & 0 & 0 & 0 \\ 0 & 0 & 1 & 1 \\ 0 & -1 & -1 & 0 \\ 0 & 0 & 0 & -1 \end{bmatrix}, \qquad \mathbf{T}^{\mathrm{C}} = \begin{bmatrix} -1 & 0 & 0 & 0 \\ 0 & -1 & -1 & -1 \\ 0 & 1 & 0 & 1 \\ 0 & 0 & 0 & -1 \end{bmatrix}.$$

4.1. EQUATIONS OF MOTION

We begin by formulating Newton's equation of motion for the isolated bodies $i = 1,\ldots,n$. Let x_i $(i = 1,\ldots,n)$ be the absolute displacement of body i in the positive x-direction. The displacements are zero when the system is in a state of rest in the absence of external forces. In this state the springs may be prestressed. We also define $x_0 \equiv 0$. The chosen generalized coordinates were defined to be

$$q_a = x_{i^-(a)} - x_{i^+(a)} = -\sum_{i=1}^{n} S^{\mathrm{C}}_{ia} x_i \qquad (a = 1,\ldots,n). \tag{63}$$

These equations are combined in the matrix equation $\mathbf{q} = -\mathbf{S}^{\mathrm{C}^T}\mathbf{x}$. From Theorem 2 follows

$$\mathbf{q} = -\mathbf{S}^{\mathrm{C}^T}\mathbf{x} \qquad \Leftrightarrow \qquad \mathbf{x} = -\mathbf{T}^{\mathrm{C}^T}\mathbf{q}. \tag{64}$$

Newton's equation for body i reads

$$m_i \ddot{x}_i = F^e_i + R^{\mathrm{S}}_i + R^{\mathrm{D}}_i \qquad (i = 1,\ldots,n). \tag{65}$$

F^e_i is the external force acting on body i. It is assumed to be given. R^{S}_i is the resultant of all spring forces on body i, and R^{D}_i is the resultant of all damper forces. In what follows spring forces and, therefore, the spring graph will be considered. The spring a $(a = 1,\ldots,m^{\mathrm{S}})$ with spring constant k_a connects the bodies $i^+(a)$ and $i^-(a)$. In analogy to the displacements q_a defined in (63) for the coordinate graph we define for the spring graph the quantities

$$\Delta\ell_a = x_{i^-(a)} - x_{i^+(a)} = -\sum_{j=1}^{n} S^{\mathrm{S}}_{ja} x_j \qquad (a = 1,\ldots,m^{\mathrm{S}}).$$

$\Delta\ell_a$ is the change of length of spring a compared with its length in the equilibrium configuration of the forcefree system. Independent of the sense of direction of arc a body $i^+(a)$ is subject to the force $+F_a = k_a \Delta\ell_a$ and body $i^-(a)$ is subject to the force $-F_a$ (in addition to forces already acting in the equilibrium position due to prestressing of springs). From the definition of the incidence matrix it follows that the resultant spring force R^{S}_i on body i has the form

$$R^{\mathrm{S}}_i = \sum_{a=1}^{m^{\mathrm{S}}} S^{\mathrm{S}}_{ia} F_a = -\sum_{a=1}^{m^{\mathrm{S}}} S^{\mathrm{S}}_{ia} k_a \sum_{j=1}^{n} S^{\mathrm{S}}_{ja} x_j \qquad (i = 1,\ldots,n).$$

In this equation the incidence matrix is used again in the two different ways that were explained in the context of (8) and (11). For the resultant damper force R_i^{D} on body i the equivalent expression is obtained with elements of the incidence matrix $\mathbf{S}^{\text{D}}$ for the damper graph and with damping constants d_a ($a = 1, \ldots, m^{\text{D}}$) instead of spring constants k_a. With these expressions (65) takes the form

$$m_i \ddot{x}_i + \sum_{a=1}^{m^{\text{D}}} S_{ia}^{\text{D}} d_a \sum_{j=1}^{n} S_{ja}^{\text{D}} \dot{x}_j + \sum_{a=1}^{m^{\text{S}}} S_{ia}^{\text{S}} k_a \sum_{j=1}^{n} S_{ja}^{\text{S}} x_j = F_i^e$$

($i = 1, \ldots, n$). All n equations are combined in the matrix equation

$$\mathbf{M}\ddot{\mathbf{x}} + \mathbf{S}^{\text{D}}\mathbf{D}\mathbf{S}^{\text{D}T}\dot{\mathbf{x}} + \mathbf{S}^{\text{S}}\mathbf{D}\mathbf{S}^{\text{S}T}\mathbf{x} = \mathbf{F}^e. \tag{66}$$

In this $\mathbf{M}$ is the diagonal matrix of the n masses, $\mathbf{D}$ is the diagonal matrix of the m^{D} damper constants and $\mathbf{K}$ is the diagonal matrix of the m^{S} spring constants.

For $\mathbf{x}$ the expression in the right-hand side equation (64) is substituted. The equation is then premultiplied by $\mathbf{T}^{\text{C}}$. This results in the following equation of motion for the variables $\mathbf{q}$ with symmetric mass, damping and stiffness matrices:

$$\begin{aligned} &\mathbf{T}^{\text{C}}\mathbf{M}\mathbf{T}^{\text{C}T}\ddot{\mathbf{q}} + (\mathbf{T}^{\text{C}}\mathbf{S}^{\text{D}})\mathbf{D}(\mathbf{T}^{\text{C}}\mathbf{S}^{\text{D}})^T\dot{\mathbf{q}} \\ &\qquad + (\mathbf{T}^{\text{C}}\mathbf{S}^{\text{S}})\mathbf{K}(\mathbf{T}^{\text{C}}\mathbf{S}^{\text{S}})^T\mathbf{q} = -\mathbf{T}^{\text{C}}\mathbf{F}^e. \end{aligned} \tag{67}$$

In this formulation the physical parameters, the parameters describing the system topology and the parameter $\mathbf{T}^{\text{C}}$ describing the choice of coordinates appear separately in the equations.

4.2. THE SPECIAL CASE OF FREE CHAINS OF BODIES

A chain of bodies is called free if it has no spring and damper connections to body 0. In this case all spring and damper forces are internal forces. This means that the absolute acceleration $\ddot{x}_C$ of the composite system center of mass is not influenced by springs or dampers. It is obtained by summing up all n Eqs.(66):

$$\ddot{x}_C = \frac{1}{m_{tot}} \sum_{i=1}^{n} F_i^e. \tag{68}$$

Since $\ddot{x}_C$ is independent of the variables $\mathbf{q}$ it must be possible to extract from Eqs.(66) $n-1$ independent equations of motion for as many relative displacements. In the special case of a free chain with $n = 2$ bodies the procedure is known. One must multiply the first equation by m_2, the second by m_1 and then take the difference. This results in the well-known equation

$$\frac{m_1 m_2}{m_1 + m_2}\,\ddot{q} + d\dot{q} + kq = \frac{m_1 F_2^e - m_2 F_1^e}{m_1 + m_2} \tag{69}$$

with the reduced mass $m_1 m_2/(m_1 + m_2)$ und with the relative displacement $q = x_2 - x_1$. In what follows the procedure will be shown for the general case $n \geq 2$. The properties of the resulting $n-1$ equations will be discussed.

Auxiliary coordinates $\mathbf{z}$ are defined by the equations $z_i = x_i - x_C$ $(i = 1, \ldots, n)$. This is the scalar form of (25). Therefore, (26) is valid:

$$\mathbf{z} = \mu^T \mathbf{x}. \tag{70}$$

The principle of virtual power in the formulation (24) for a system without constraints to inertial space is reduced to the scalar equation

$$\delta\dot{\mathbf{z}}^T(\mathbf{M}\ddot{\mathbf{z}} - \mathbf{F}) = 0. \tag{71}$$

From (70) follows $\ddot{\mathbf{z}} = \mu^T\ddot{\mathbf{x}}$ und $\delta\dot{\mathbf{z}}^T = \delta\dot{\mathbf{x}}^T\mu$ and from (66)

$$\mathbf{F} = \mathbf{F}^e - \mathbf{S}^D\mathbf{D}\mathbf{S}^{D^T}\dot{\mathbf{x}} - \mathbf{S}^S\mathbf{K}\mathbf{S}^{S^T}\mathbf{x}.$$

Since the elements of $\delta\dot{\mathbf{x}}$ are independent, one obtains the equation

$$\mu\mathbf{M}\mu^T\ddot{\mathbf{x}} + \mu\mathbf{S}^D\mathbf{D}\mathbf{S}^{D^T}\dot{\mathbf{x}} + \mu\mathbf{S}^S\mathbf{K}\mathbf{S}^{S^T}\mathbf{x} = \mu\mathbf{F}^e. \tag{72}$$

The damping matrix and the stiffness matrix only appear to be unsymmetric. In the case of a free chain of bodies the identities hold: $\mu\mathbf{S}^S = \mathbf{S}^S$ and $\mu\mathbf{S}^D = \mathbf{S}^D$. Proof for $\mathbf{S}^S$: When no spring is connected to body 0 then every column of $\mathbf{S}^S$ contains exactly one element $+1$ and one element -1 so that $\sum_{i=1}^n S_{ia}^S = 0$ $(a = 1, \ldots, n)$. From this follows

$$(\mu\mathbf{S}^S)_{ia} = \sum_{j=1}^n (\delta_{ij} - \frac{m_i}{m_{tot}}) S_{ja}^S = S_{ia}^S.$$

End of proof. Hence (72) is identical with

$$\mu\mathbf{M}\mu^T\ddot{\mathbf{x}} + \mathbf{S}^D\mathbf{D}\mathbf{S}^{D^T}\dot{\mathbf{x}} + \mathbf{S}^S\mathbf{K}\mathbf{S}^{S^T}\mathbf{x} = \mu\mathbf{F}^e. \tag{73}$$

This equation can be obtained in a much simpler way. Just premultiply (66) with μ and make use of the identities $\mu\mathbf{M} = \mu\mathbf{M}\mu^T$, $\mu\mathbf{S}^{\mathrm{D}} = \mathbf{S}^{\mathrm{D}}$ and $\mu\mathbf{S}^{\mathrm{S}} = \mathbf{S}^{\mathrm{S}}$.

For $\mathbf{x}$ again the expression from (64) is substituted. Then the equation is premultiplied by $\mathbf{T}^{\mathrm{C}}$. This results in the equations of motion

$$(\mathbf{T}^{\mathrm{C}}\mu)\mathbf{M}(\mathbf{T}^{\mathrm{C}}\mu)^T\ddot{\mathbf{q}} + (\mathbf{T}^{\mathrm{C}}\mathbf{S}^{\mathrm{D}})\mathbf{D}(\mathbf{T}^{\mathrm{C}}\mathbf{S}^{\mathrm{D}})^T\dot{\mathbf{q}} \\ + (\mathbf{T}^{\mathrm{C}}\mathbf{S}^{\mathrm{S}})\mathbf{K}(\mathbf{T}^{\mathrm{C}}\mathbf{S}^{\mathrm{S}})^T\mathbf{q} = -\mathbf{T}^{\mathrm{C}}\mu\mathbf{F}^e. \tag{74}$$

The damping matrix and the stiffness matrix are the same as in (67). Only the mass matrix and the right-hand side term are different. In what follows a physical interpretation is given to the elements of the mass matrix. This is done with the help of the numbers σ_i $(i = 1, \ldots, n)$, of the sets κ_{ab} $(a, b = 1, \ldots, n)$ of vertices and of the ordering relationship arc $a >$ arc b (see (7) and the text following this equation). Let m_{ab} be the total mass of all bodies which are associated with the vertices in the set κ_{ab} $(m_0 = 0)$. Examples: For the coordinate graph of Fig.11a one has $\sigma_1 = \sigma_2 = -1$, $\sigma_3 = +1$, $m_{13} = m_1$, $m_{31} = m_2 + m_5$ und $m_{22} = m_2 + m_3 + m_4 + m_5$. From the definitions follow the relationships (in what follows $\mathbf{T}$ is written instead of $\mathbf{T}^{\mathrm{C}}$) $\sum_{i=1}^n T_{ai}m_i = \sigma_a m_{aa}$,

$$\sum_{i=1}^n T_{ai}T_{bi}m_i = \begin{cases} m_{aa} & a = b \\ m_{aa} = \sigma_a\sigma_b m_{ab} & \text{arc } a > \text{arc } b \\ m_{bb} = \sigma_a\sigma_b m_{ba} & \text{arc } b > \text{arc } a \\ 0 & \text{sonst,} \end{cases}$$

$$(\mathbf{T}\mu)_{ai} = \sum_{\ell=1}^n T_{a\ell}\left(\delta_{\ell i} - \frac{m_\ell}{m_{tot}}\right) = T_{ai} - \frac{\sigma_a m_{aa}}{m_{tot}}. \tag{75}$$

With this the elements of the mass matrix can be given the form

$$\begin{aligned} \left[(\mathbf{T}^{\mathrm{C}}\mu)\mathbf{M}(\mathbf{T}^{\mathrm{C}}\mu)^T\right]_{ab} &= \sum_{i=1}^n (\mathbf{T}\mu)_{ai}(\mathbf{T}\mu)_{bi}m_i \\ &= \sum_{i=1}^n \left(T_{ai} - \frac{\sigma_a m_{aa}}{m_{tot}}\right)\left(T_{bi} - \frac{\sigma_b m_{bb}}{m_{tot}}\right)m_i \\ &= \sum_{i=1}^n T_{ai}T_{bi}m_i - \sigma_a\sigma_b\frac{m_{aa}m_{bb}}{m_{tot}} \\ &= \begin{cases} m_{aa}(m_{tot} - m_{aa})/m_{tot} & a = b \\ +\sigma_a\sigma_b m_{ab}m_{ba}/m_{tot} & \begin{cases} \text{arc } a > \text{arc } b \\ \text{or} \\ \text{arc } b > \text{arc } a \end{cases} \\ -\sigma_a\sigma_b m_{ab}m_{ba}/m_{tot} & \text{else.} \end{cases} \end{aligned} \tag{76}$$

Thus, the elements of the mass matrix are reduced masses in a generalized sense. An example: For the coordinate graph of Fig.11a one finds $\left[(\mathbf{T}^c\mu)\mathbf{M}(\mathbf{T}^c\mu)^T\right]_{13} = -m_1(m_2+m_5)/m_{tot}$.

Let the coordinates $\mathbf{q}$ be chosen such that only q_n is an absolute displacement (of an arbitrary body). All other coordinates are relative displacements. Then, only arc n is incident with vertex 0 in the coordinate graph. Then $m_{nn} = m_{tot}$ and $m_{na} = 0$ $(a \neq n)$. From this follows that all elements in the nth row and in the nth column of the mass matrix are equal to zero.

Each of the remaining terms in (73) or in the equivalent Eq.(72) has the form $\mathbf{T}\mu\mathbf{X}$ with some column matrix $\mathbf{X}$. The nth element is the sum $\sum_{i=1}^{n}(\mathbf{T}\mu)_{ni}X_i$. With the special choice of coordinates it follows from (75) that $(\mathbf{T}\mu)_{ni} = 0$ $(i = 1, \ldots, n)$ since $T_{ni} = \sigma_n$ $(i = 1, \ldots, n)$ and $m_{nn} = m_{tot}$. This proves that the nth equation of (74) is the identity $0 = 0$. Thus, we have $n-1$ independent equations for relative displacements $q_1, \ldots, q_{n-1}$. In the special case $n = 2$ the system of $n-1$ equations has the form (69).

References

1. Branin, F.H. (1962) 'The relation between Kron's method and the classical methods of network analysis',The Matrix and Tensor Quarterly 12,69-105.
2. Wittenburg, J. (1977) Dynamics of systems of rigid bodies, Teubner, Stuttgart
3. Busacker, R.G. and Saaty, T.L. (1965) Finite Graphs and networks, Mc Graw-Hill, New York
4. Roberson, R.E. and Wittenburg, J. (1968) 'A dynamical formalism for an arbitrary number of interconnected rigid bodies. With reference to the problem of satellite attitude control' 3rd IFAC Congr. London 1966, Proc. 46D.2-46D.9
5. Pestel, E. and Wittenburg, J. (1992) Technische Mechanik Bd.2: Festigkeitslehre (2.Aufl.),B.I.-Verlag, Mannheim
6. Haug, E.J. (ed.) (1984) Computer aided analysis and optimization of mechanical system dynamics, NATO ASI ser. F, Springer, Berlin
7. Wolz, U. (1986) 'Dynamik von Mehrkoerpersystemen - Theorie und symbolische Programmierung' (Diss. Karlsruhe), Fortschritt-Berichte VDI, Reihe 11,75, VDI-Verlag, Düsseldorf
8. Weber, B. (1993) 'Symbolische Programmierung in der Mehrkörperdynamik' (Diss. Karlsruhe), Preprint Nr.93/3, Inst. f. Wiss. Rechnen u. Math. Modellbildung, Universität Karlsruhe
9. Featherstone, R. (1983) 'The calculation of robot dynamics using articulated-body inertias' Robotics Research 2(1),13-30

10. Kim, S.S. and Haug, E.J. (1988) ' A recursive formulation for flexible multibody dynamics. Part I' Comp.Meth.in Appl.Mech.and Eng. 71, 293-314
11. Lilov, L. and Chirikov, V. (1981) 'On the dynamics equations of systems of interconnected bodies' J. Appl.Math.Mech.(PMM) 45,383-390
12. Woernle, C. (1988) 'Ein systematisches Verfahren zur Aufstellung der geometrischen Schliessbedingungen in kinematischen Schleifen mit Anwendung bei der Rueckwaertstransformation fuer Industrieroboter' (Diss. Stuttgart)
13. Wittenburg, J. (1974) 'Permanente Drehungen zweier durch ein Kugelgelenk gekoppelter, starrer Koerper' Acta Mech.19,215-226
14. Wittenburg, J. and Lilov, L. (1975) 'Relative equilibrium positions and their stability for a multi-body satellite in a circular orbit' Ing.-Arch.44, 269-279

PART II
Dynamics of Flexible Mechanical Systems

FLEXIBILITY IN MULTIBODY DYNAMICS WITH APPLICATIONS TO CRASHWORTHINESS

JORGE A. C. AMBRÓSIO and MANUEL SEABRA PEREIRA
Instituto Superior Técnico
Technical University of Lisbon
1096 Lisboa, Portugal

ABSTRACT. Formulations based on multibody dynamics for the analysis of crashworthiness and impact of vehicles and structural systems are reviewed in this paper. A methodology to incorporate the elastodynamics effects, suitable to describe the elastic deformations of flexible bodies, is discussed. The limitations of this methodology for crash impact are overcome in a more general formulation where the deformation of the flexible (or partially flexible) bodies is described using an updated Lagrangian formulation. This allows for geometric and material nonlinear behavior of the multibody components. A major drawback of this nonlinear formulation is the inability to describe zones of concentrated deformation due to local instabilities. For this purpose the plastic hinge concept, where the structural plastic deformation is modelled by nonlinear joint-spring set-up, is used. The validity of this model is assessed by carrying out an experimental test where a hollow steel extruded beam collide with a rigid block. By predicting where and when failure is likely to occur using a flexible model, the present technique provides an efficient tool to access the crashworthiness design of a broad class of impact excited structural configurations with general kinematic constraints. Finally these methodologies are applied to model the rollover of a truck in order to illustrate their capabilities.

1. Introduction

During the last twenty five years computer aided analysis of crashworthiness and structural impact has received a large attention and is now emerging as a powerful methodology which can be successfully applied in practical and industrial situations. In this paper, multibody dynamics based methodologies, applicable to crashworthiness are reviewed.

Several approaches using experiments [1-4] and/or numerical simulations have been adopted in the past. Different numerical formulations with varying degree of complexity and accuracy have been proposed using spring-mass models [5-9], finite difference methods [10-12], and finite element methodologies [13-16]. Hybrid approaches [5,17,18] utilizing data obtained from quasi-static crushing of different segments of the colliding structure have also been developed. In these methods the generalized non-linear load-displacement characteristics are kinematically coupled to the global structural system to obtain the overall dynamic response of the structure.

In some cases the experimental load-deformation characteristics can be adjusted to take into consideration strain rate effects [19]. The access to such experimental data allows an insight to complex phenomena such as wrinkling, friction and failure of different connection elements which are, in many occasions difficult, if not impossible, to obtain in general purpose nonlinear finite element computer codes.

In standard finite element formulations the large displacements and deformations of the gross motion are not generally taken into consideration. However recent efforts in the

M. F. O. S. Pereira and J. A. C. Ambrósio (eds.),
Computer-Aided Analysis of Rigid and Flexible Mechanical Systems, 199–232.

field of nonlinear structural dynamics have contributed for the development of well known commercially available codes such as PAM-CRASH [20], DYNA-3D [21], DYCAST [13] and WHAMS-3D [22] . These programs are now able to simulate with improved accuracy several different structural impact phenomena such as large localized deformations, structural instabilities, transient vibrations, stress wave propagations and eventually structural collapse due to material damage and loads causing stresses above the ultimate strength. These codes, however, require large computer resources and normally involve time consuming modelling data preparation which make them rather unsuitable as a design tool during the initial design stages.

In crashworthiness and impact analysis of structural mechanical systems, the elasto-dynamic effects play an important role on the system behavior. During the impact period the deformation of the components interfere with the motion of the system which results in a strong coupling between the structural flexibilities and the gross motion of the different components. For this purpose several researchers have suggested procedures that successfully introduce the elasto-dynamic effects into multibody dynamic formulations [23-25]. However, there are some unsolved difficulties related with the complexity of the models obtained

The problems associated with the introduction of flexibility effects in a multibody system are related with the complex geometries of the flexible bodies and it is not always obvious how to develop proper and judicious simplified truss type models to adequately represent integrated beam and sheet metal structural components. Basically this area deals with the understanding of the failure and collapse mechanisms based on experimental results which makes possible to tune accurate and cost effective simplified analytical techniques.

In many impact situations, the individual structural members are overloaded principally in bending giving rise to plastic deformations in highly localized regions, called plastic hinges. These hinges occur at points of maximum bending moments, at load application points, at joints and in locally weak areas. If the levels of plastic deformation are large, a plastic hinge allows relative rotation between the parts of the structure and it becomes reasonable to model these phenomena within the framework of rigid-flexible body dynamics formulations. For complex cross sections and joints the plastic behavior is more complex involving local buckling and eventually fracture which can only be accurately predicted by simple and cheap tests on localized parts of the structure [26].

In this paper a multibody dynamic formulation for systems with linear and nonlinear structural deformations is reviewed and the plastic-hinge modelling approach as applied to a rigid-flexible multibody system is presented. These flexibility effects, which may be important during the impact period, can be taken in consideration with the present formulation. The example of a rotating beam is presented to illustrate the effect of geometric nonlinear deformations on the system components. A colliding beam example is analyzed and a corresponding experimental test is carried out to assess the validity of the proposed formulation. Finally, these methodologies are applied to the rollover and crashworthiness of an utility truck.

2. Multibody Dynamics Using Joint Coordinates

A multibody system is a collection of rigid and flexible bodies joined together by kinematic joints and force elements as depicted in Figure 1. For the ith body in the system $\mathbf{q}_i$ denotes a vector of coordinates which contains the Cartesian translational coordinates $\mathbf{r}_i$, a set of rotational coordinates $\mathbf{p}_i$, and a set of nodal coordinates $\mathbf{q}'_f$, $\mathbf{u}'$ or δ' (if body i is flexible). A vector of velocities for a rigid body i is defined as $\mathbf{v}_i$, which contains a 3-vector of translational velocities $\dot{\mathbf{r}}_i$ and a 3-vector of angular velocities ω_i (defined in the XYZ coordinate system). If body i is flexible then the vector of velocities $\mathbf{v}_i$ contains $\dot{\mathbf{r}}_i$, ω_i (defined in the $\xi\eta\zeta_i$ coordinate system) and a vector of nodal velocities $\dot{\mathbf{q}}'_f$ or $\dot{\delta}'$. The vector of accelerations for the body is denoted by $\dot{\mathbf{v}}_i$ and it is simply the time derivative of $\mathbf{v}_i$. For a multibody system containing *nb* bodies, the vectors of coordinates, velocities, and accelerations are $\mathbf{q}$, $\mathbf{v}$ and $\dot{\mathbf{v}}$ which contain the elements of $\mathbf{q}_i$, $\mathbf{v}_i$ and $\dot{\mathbf{v}}_i$, respectively, for i=1, ...,*nb*.

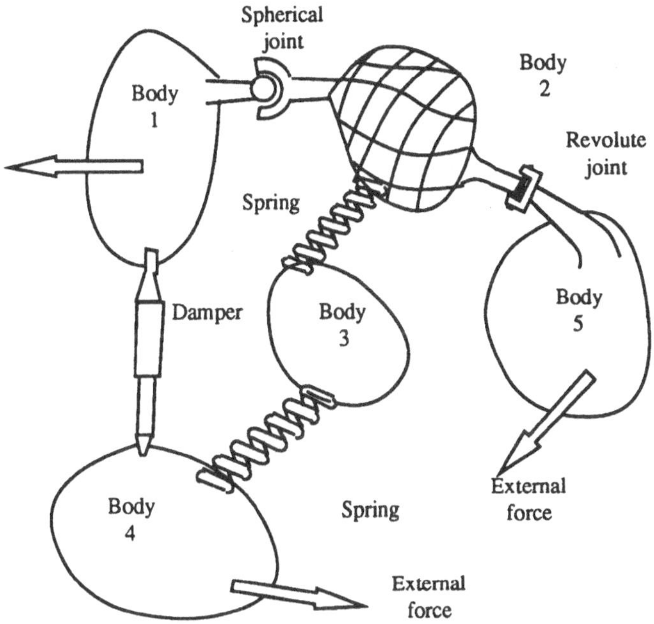

Figure 1 Schematic representation of a multibody system

Let the kinematic joints between rigid bodies be described by *mr* independent constraints as

$$\Phi(\mathbf{q}) = \mathbf{0} \tag{1}$$

The first and second derivatives of the constraints yield the kinematic velocity and acceleration equations.

$$\dot{\Phi} \equiv \mathbf{D}\mathbf{v} = \mathbf{0} \tag{2}$$
$$\ddot{\Phi} \equiv \dot{\mathbf{D}}\mathbf{v} + \mathbf{D}\dot{\mathbf{v}} = \mathbf{0} \tag{3}$$

where **D** is the Jacobian matrix of the constraints. The equation of motion for the system of rigid bodies are written (see reference [27])

$$\mathbf{M}\dot{\mathbf{v}} - \mathbf{D}^T\boldsymbol{\lambda} = \mathbf{g} \qquad (4)$$

where **M** is the inertia matrix, λ is a vector of Lagrange multipliers, and $\mathbf{g} = \mathbf{g}(\mathbf{q}, \mathbf{v})$ contains the forces and moments that act on the bodies, and the gyroscopic terms.

The constrained equations of motion expressed by equations (1) to (4) can be converted to a smaller set of equations in terms of a set of coordinates known as joint coordinates. Such transformation is briefly discussed here (for more details refer to reference [27]). The relative configurations of two adjacent bodies are described by a set of relative coordinates, equal to the number of relative degrees of freedom between the bodies. The vector of joint coordinates for a system of rigid bodies is denoted by β and it contains all the joint coordinates and the absolute coordinates of the floating base bodies. The vector of joint velocities, defined as $\dot{\beta}$, is the time derivative of β, being its relation with **v** is given by [27]

$$\mathbf{v} = \mathbf{B}\dot{\beta} \qquad (5)$$

where matrix **B** is the velocity transformation matrix and can be shown to be orthogonal to the Jacobian matrix **D**. The transformation of the accelerations is obtained by deriving equation (5) with respect to time. This is written as

$$\dot{\mathbf{v}} = \dot{\mathbf{B}}\dot{\beta} + \mathbf{B}\ddot{\beta} \qquad (6)$$

Substituting equation (6) into equation (4), premultiplying by $\mathbf{B}^T$, and using the orthogonality condition between **B** and **D** yield

$$\mathcal{M}\ddot{\beta} = f \qquad (7)$$

where

$$\mathcal{M} = \mathbf{B}^T\mathbf{M}\mathbf{B} \qquad (8)$$

$$f = \mathbf{B}^T\left(\mathbf{g} - \mathbf{M}\dot{\mathbf{B}}\dot{\beta}\right) \qquad (9)$$

Equation (7) represents the generalized equation of motion for an open-loop system of rigid bodies. This equation, containing the minimum number of second-order differential equations, can be used instead of the mixed set of differential-algebraic equations given by equations (1) through (4). Closed-loops can be analyzed as open-loops provided that the loop is cut at a joint and the corresponding kinematic constraint is not eliminated from the equations of motion. In a second step, the methodology described here can be used to eliminate the explicit use of the kinematic constraint that represents the cut joint. However, it is computationally more advantageous not to preform this second step of velocity transformations in most of the problems. In reference [28] the equations of motion of a system containing closed kinematic loops are presented and discussed.

3. Flexible Multibody Dynamics

For the crashworthiness and impactanalysis, using a multibody formalism, the description of the flexibility of its components may be necessary. The behavior of systems subjected to impact is characterized by zones of large deformations and by zones where only elastic deformations take place. For the purpose of describing this behavior, linear and nonlinear formulations of multibody systems are reviewed in this section.

3.1. LINEAR DEFORMATIONS

It has been shown [25,29] that the configuration of a deformable body in a multibody system can be described by a set of global reference coordinates $\mathbf{q}_r^i$ and local elastic coordinates $\mathbf{u}^i$ which are defined using the finite element methodology. As shown in figure 2, the position of a flexible body in the non-moving reference frame XYZ is specified by the spatial location $\mathbf{r}^i$ of a body fixed frame $\xi\eta\zeta$ and a set of angular orientation coordinates ϕ^i, thus the coordinates describing the gross motion of the body are $\mathbf{q}_r^{iT} = [\mathbf{r}^{iT}, \phi^{iT}]$.

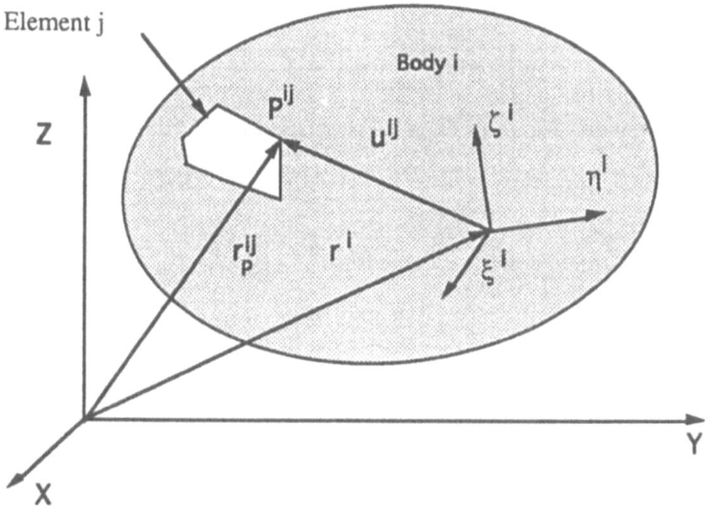

Figure 2. Reference generalized coordinates

Let $\mathbf{q}^i = [\,\mathbf{q}_r^{iT}, \mathbf{u}^{iT}]^T$ be the vector of generalized coordinates of body *i*. Assuming all coordinates to be independent, the Lagrange equations of motion for this flexible body can be written in the form

$$\frac{d}{dt}\left(\frac{\partial T^i}{\partial \dot{\mathbf{q}}^i}\right) - \left(\frac{\partial T^i}{\partial \mathbf{q}^i}\right) + \left(\frac{\partial U^i}{\partial \mathbf{q}^i}\right) - \mathbf{g}^i = \mathbf{0} \qquad (10)$$

Using the finite element method to describe the flexibility of body i, the kinetic energy T^i is a function of $\dot{\mathbf{q}}^i$ and $\mathbf{q}^i$, and the elastic energy U^i is function of $\mathbf{q}^i$. The equations of motion (10) for body i take the form [24,25,29,30]

$$\mathbf{M}^i(\mathbf{q}^i)\,\ddot{\mathbf{q}}^i + \mathbf{K}^i\,\mathbf{q}^i = \mathbf{g}^i(\dot{\mathbf{q}}^i,\mathbf{q}^i,t) + \mathbf{s}^i(\dot{\mathbf{q}}^i,\mathbf{q}^i) \tag{11}$$

where $\mathbf{M}^i$, $\mathbf{K}^i$ are the mass and stiffness matrices of body i, respectively, $\mathbf{g}^i$ is the vector of generalized forces of body i, $\mathbf{s}^i$ is a vector containing velocity quadratic terms and other acceleration independent terms. In a less compact form, equation (11) is written as:

$$\begin{bmatrix} \mathbf{M}_{rr} & \mathbf{M}_{r\phi} & \mathbf{M}_{rf} \\ \mathbf{M}_{\phi r} & \mathbf{M}_{\phi\phi} & \mathbf{M}_{\phi f} \\ \mathbf{M}_{fr} & \mathbf{M}_{f\phi} & \mathbf{M}_{ff} \end{bmatrix} \begin{bmatrix} \ddot{\mathbf{r}} \\ \dot{\omega}' \\ \ddot{\mathbf{u}}' \end{bmatrix} + \begin{bmatrix} \mathbf{0} & \mathbf{0} & \mathbf{0} \\ \mathbf{0} & \mathbf{0} & \mathbf{0} \\ \mathbf{0} & \mathbf{0} & \mathbf{K} \end{bmatrix} \begin{bmatrix} \mathbf{0} \\ \mathbf{0} \\ \mathbf{u}' \end{bmatrix} = \begin{bmatrix} \mathbf{g}_r \\ \mathbf{g}'_\phi \\ \mathbf{g}'_f \end{bmatrix} - \begin{bmatrix} \mathbf{s}_r \\ \mathbf{s}'_\phi \\ \mathbf{s}'_f \end{bmatrix} \tag{12}$$

In this equation the submatrices $\mathbf{M}_{rr}$, $\mathbf{M}_{\phi r}$, $\mathbf{M}_{r\phi}$ and $\mathbf{M}_{\phi\phi}$, associated with the gross motion of the body-fixed coordinate frame, and $\mathbf{M}_{ff}$, the standard finite element mass matrix, are time invariant. Assuming small linear elastic deformations for the flexible body, the stiffness matrix $\mathbf{K}$ is also constant. The remaining terms of the mass matrix are time variant and must be calculated every time step. The mean axis conditions can be applied to equation (12) resulting in a constant mass matrix where the inertia coupling between rigid and flexible degrees of freedom disapears [29,30]. Another methodology to transform the mass matrix $\mathbf{M}^i$ into a diagonal constant matrix is discussed next section.

3.2. EQUATIONS OF MOTION FOR CONSTRAINED FLEXIBLE BODY

Consider now a mechanical system with nb bodies connected by joints which are described by m holonomic constraints in the form

$$\mathbf{\Phi}(\mathbf{q},t) = \mathbf{0} \tag{13}$$

where $\mathbf{\Phi}(\mathbf{q},t) = [\Phi_1(\mathbf{q},t)^T, \ldots, \Phi_m(\mathbf{q},t)^T]^T$. These equations express the dependency between the generalized cartesian coordinates $\mathbf{q}$.

Consider, for example, two bodies i and j connected through a revolute joint in a common point k, as illustrated in figure 3. The vectorial equation which forces point k to be coincident in both bodies at all times is written in the form

$$\mathbf{r}^i + \mathbf{A}^i\mathbf{b}'^i - \mathbf{r}^j - \mathbf{A}^j\mathbf{b}'^j = \mathbf{0} \tag{14}$$

where $\mathbf{b}'^i$, $\mathbf{b}'^j$ are position vectors of point k in bodies i and j respectively; $\mathbf{A}^i$, $\mathbf{A}^j$ are transformation matrices from the body coordinate systems to the global inertia frame.

This joint has two algebraic constraint equations. If both bodies are flexible $\mathbf{b}'^i$, $\mathbf{b}'^j$ depend on the generalized elastic coordinates, implying that these vectors have to be calculated at each time for the current deformation state. Then

$$\mathbf{r}^i + \mathbf{A}^i(\mathbf{b}'^i_0 + \delta^i_k) - \mathbf{r}^j - \mathbf{A}^j(\mathbf{b}'^j_0 + \delta^j_k) = \mathbf{0} \tag{15}$$

where $\mathbf{b'}_0{}^i$, $\mathbf{b'}_0{}^j$ correspond to the position vectors of point k in the undeformable state, $\delta_k{}^i$, $\delta_k{}^j$ are the flexible displacements of the connection node (point k) of bodies *i* and *j*, respectively.

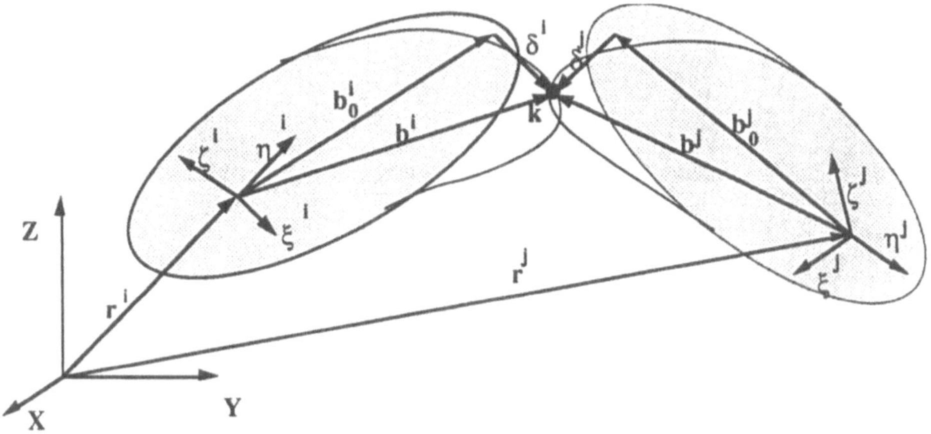

Figure 3. Revolute Joint

The holonomic kinematic constraints are introduced in the variational form of the equations of motion of body *i* using Lagrange multipliers. After substituting all energy expressions in these equations, the dynamic equations of motion for a flexible body are written in a compact form as:

$$\mathbf{M}^i\left(\mathbf{q}^i\right)\ddot{\mathbf{q}}^i + \mathbf{D}^T\lambda = \mathbf{g}^i\left(\dot{\mathbf{q}}^i,\mathbf{q}^i,t\right) + \mathbf{s}^i\left(\dot{\mathbf{q}}^i,\mathbf{q}^i\right) - \mathbf{K}^i\,\mathbf{q}^i \tag{16}$$

where $\mathbf{D} = \left(\partial\Phi/\partial\mathbf{q}^i\right)$ is the Jacobian matrix for the constraints.

Once these equations have been obtained for each body, it is necessary to assemble the equations for all bodies of the mechanical system. For the equations obtained, the angular acceleration of the body fixed coordinate frame must be transformed to global components such that the accelerations in equation (11) are consistent with the rigid body accelerations used in the transformations of the joint coordinate method, expressed by equations (5). The constraint equations and their corresponding Lagrange multipliers can be eliminated from the equations of motion by using velocity transformations. The interested reader is directed to references [31,32] for a more detailed discussion on the use of the joint co-ordinate method with flexible bodies.

3.3. GEOMETRIC AND MATERIAL NONLINEAR DEFORMATIONS

The description of the deformation of the flexible body *i* presented before is not suitable, by itself, to applications where the nonlinear deformations play a major role in the dynamics of the multibody system. This is the case of applications involving the impact and crashworthiness of vehicles. In order to overcome these limitations, a more general formulation of a flexible body was proposed by Ambrósio and Nikravesh [33]. In this methodology an updated Lagrangian formulation is used to describe the kinetics of the flexible body. Moreover, the finite element method is used to represent the flexible body.

The kinetic energy, deformation energy and external forces are calculated using the updated Lagrangian formulation. Using equation (10) for each finite element and assembling their contributions leads to the flexible body equations of motion written as:

$$\begin{bmatrix} \mathbf{M}_{rr} & \mathbf{M}_{r\phi} & \mathbf{M}_{rf} \\ \mathbf{M}_{\phi r} & \mathbf{M}_{\phi\phi} & \mathbf{M}_{\phi f} \\ \mathbf{M}_{fr} & \mathbf{M}_{f\phi} & \mathbf{M}_{ff} \end{bmatrix} \begin{bmatrix} \ddot{\mathbf{r}} \\ \dot{\omega}' \\ \ddot{\mathbf{u}}' \end{bmatrix} = \begin{bmatrix} \mathbf{g}_r \\ \mathbf{g}'_\phi \\ \mathbf{g}'_f \end{bmatrix} - \begin{bmatrix} \mathbf{s}_r \\ \mathbf{s}'_\phi \\ \mathbf{s}'_f \end{bmatrix} - \begin{bmatrix} \mathbf{0} \\ \mathbf{0} \\ {}^{t}_{t'}\mathbf{F} \end{bmatrix} - \begin{bmatrix} \mathbf{0} & \mathbf{0} & \mathbf{0} \\ \mathbf{0} & \mathbf{0} & \mathbf{0} \\ \mathbf{0} & \mathbf{0} & {}^{t}_{t'}\mathbf{K}_L + {}^{t}_{t'}\mathbf{K}_{NL} \end{bmatrix} \begin{bmatrix} \mathbf{0} \\ \mathbf{0} \\ \mathbf{u}' \end{bmatrix} \tag{17}$$

In this equations the left superscripts are referred to the configuration in which an event occurs while a left subscripts refer the configuration in which the event is measured. The configurations in which an event can occur or be measured are: the current configuration; the last known equilibrium configuration; and the initial configuration, respectively denoted by $t+\Delta t$, t and 0. For instance, vector ${}^{t}_{t'}\mathbf{F}$ denotes the nodal forces equivalent to the actual state of stress that occur in the reference configuration t and are measured in a corotated configuration t'. Vector $\mathbf{u}'$ denotes the increments of displacements from the updated configuration to the current configuration due to the incremental nature of this formulation.

In equation (17) the mass matrix is equal to the mass matrix calculated for equation (12), the right-hand side is composed by a vector of externally applied generalized forces, a vector of gyroscopic forces, and internal forces due to the deformation of the flexible body. The vector of the external applied generalized forces is evaluated over the updated configuration and it is written as:

$${}_{t'}\mathbf{g} = \begin{bmatrix} \int_{{}^{t'}A} {}^{t+\Delta t}_{t'}\mathbf{f}_s \, {}^{t'}da + \int_{{}^{t'}V} {}^{0}\rho \, {}^{t+\Delta t}_{t'}\mathbf{f}_b \, {}^{t'}dv \\ \int_{{}^{t'}A} \tilde{\mathbf{b}}'\mathbf{A}^T \, {}^{t+\Delta t}_{t'}\mathbf{f}_s \, {}^{t'}da + \int_{{}^{t'}V} {}^{0}\rho\tilde{\mathbf{b}}'\mathbf{A}^T \, {}^{t+\Delta t}_{t'}\mathbf{f}_b \, {}^{t'}dv \\ \int_{{}^{t'}A} \mathbf{N}^T\mathbf{A}^T \, {}^{t+\Delta t}_{t'}\mathbf{f}_s \, {}^{t'}da + \int_{{}^{t'}V} {}^{0}\rho\mathbf{N}^T\mathbf{A}^T \, {}^{t+\Delta t}_{t'}\mathbf{f}_b \, {}^{t'}dv \end{bmatrix} \tag{18}$$

where $\mathbf{A}$ is the transformation matrix from the body fixed coordinate system to the inertial frame, $\mathbf{N}$ is the matrix of shape functions, ${}^{t+\Delta t}_{t'}\mathbf{f}_b$ and ${}^{t+\Delta t}_{t'}\mathbf{f}_s$ are the body and surface forces respectively. The vector of gyroscopic forces is written as:

$$\mathbf{s} = \begin{bmatrix} \mathbf{A}\tilde{\omega}'\tilde{\omega}' \int_{{}^{0}V} {}^{0}\rho\mathbf{b}' \, {}^{0}dv \\ \int_{{}^{0}V} {}^{0}\rho\tilde{\mathbf{b}}'\tilde{\omega}'\tilde{\omega}\mathbf{b}' \, {}^{0}dv \\ \int_{{}^{0}V} {}^{0}\rho\mathbf{N}^T\tilde{\omega}'\tilde{\omega}\mathbf{b}' \, {}^{0}dv \end{bmatrix} + 2 \begin{bmatrix} \mathbf{A}\tilde{\omega}' \int_{{}^{0}V} {}^{0}\rho\mathbf{N} \, {}^{0}dv \\ \int_{{}^{0}V} {}^{0}\rho\tilde{\mathbf{b}}'\tilde{\omega}\mathbf{N} \, {}^{0}dv \\ \int_{{}^{0}V} {}^{0}\rho\mathbf{N}^T\tilde{\omega}\mathbf{N} \, {}^{0}dv \end{bmatrix} \dot{\mathbf{u}}' \tag{19}$$

In equation (17) matrices ${}_{t'}^{t}\mathbf{K}_L$ and ${}_{t'}^{t}\mathbf{K}_{NL}$ are respectively the linear and nonlinear stiffness matrices, and ${}_{t'}^{t}\mathbf{F}$ denotes the vector of equivalent nodal forces due to the actual state of stress. These quantities are given by:

$$ {}_{t'}^{t}\mathbf{K}_L = \int_{{}^{t'}V} {}_{t'}^{t}\mathbf{B}_L^T \; {}_{t'}\mathbf{C} \; {}_{t'}^{t}\mathbf{B}_L \; {}^{t'}dv \tag{20} $$

$$ {}_{t'}^{t}\mathbf{K}_{NL} = \int_{{}^{t'}V} {}_{t'}^{t}\mathbf{B}_{NL}^T \; {}_{t'}^{t}\boldsymbol{\tau}' \; {}_{t'}^{t}\mathbf{B}_{NL} \; {}^{t'}dv \tag{21} $$

$$ {}_{t'}^{t}\mathbf{F} = \int_{{}^{t'}V} {}_{t'}^{t}\mathbf{B}_L^T \; {}_{t'}^{t}\hat{\boldsymbol{\tau}}' \; {}^{t'}dv \tag{22} $$

In these equations ${}_{t'}^{t}\mathbf{B}_L^T$ and ${}_{t'}^{t}\mathbf{B}_{NL}^T$ denote the linear and nonlinear strain matrices, respectively, and ${}_{t'}^{t}\boldsymbol{\tau}'$ is the Cauchy stress tensor for the updated configuration. It should be noted that the reference to the linearity of the stiffness matrices ${}_{t'}^{t}\mathbf{K}_L$ and ${}_{t'}^{t}\mathbf{K}_{NL}$ is related to their relation with the displacements. If the constitutive tensor ${}_{t'}\mathbf{C}$ is not constant then both ${}_{t'}^{t}\mathbf{K}_L$ and ${}_{t'}^{t}\mathbf{K}_{NL}$ are not linear.

A multibody system may experience elasto-plastic deformations of one or more of its components. For these problems, an elasto-plastic constitutive tensor ${}_{t'}\mathbf{C}$ must be used in the equation (20). Isotropic hardening and isothermal conditions can be assumed for the description of this tensor. The material yield condition is written as:

$$ f\left({}^{t'}\tau, {}^{t'}\kappa\right) = 0 \tag{23} $$

where ${}^{t'}\tau$ is the Cauchy stress tensor and ${}^{t'}\kappa$ is the hardening parameter (which is a function of the state of strain). Yielding occurs when equation (23) is satisfied. Any further strain increment will be partially elastic and partially plastic. These strain increments are related with the total strain increment by

$$ d\,{}_{t'}\mathbf{e} = d\,{}_{t'}\mathbf{e}^P + d\,{}_{t'}\mathbf{e}^E \tag{24} $$

Furthermore, let associated plasticity be assumed. In these conditions Zienckiewicz [34] shows that the form of the elasto-plastic constitutive tensor is given by

$$ {}_{t'}\mathbf{C} = {}_{t'}\mathbf{C}^E - {}_{t'}\mathbf{C}^E \frac{\partial f}{\partial\, {}^{t'}\tau} \left(\frac{\partial f}{\partial\, {}^{t'}\tau}\right)^T {}_{t'}\mathbf{C}^E \left[H + \left(\frac{\partial f}{\partial\, {}^{t'}\tau}\right)^T {}_{t'}\mathbf{C}^E \frac{\partial f}{\partial\, {}^{t'}\tau} \right]^{-1} \tag{25} $$

where $\mathbf{C}^E$ is the elastic constitutive tensor. The parameter H is the slope of the plot of the stress versus plastic strain for the uniaxial test if the Huber-Von Mises surface is used in equation (23).

3.4. PARTIALLY FLEXIBLE BODY

Equation (17) describes thoroughly the motion of a flexible body. However the form of this equation is not efficient for numerical implementation because not only all the quantities of the right-hand side are not constant but also the mass matrix is variant. A simpler form of the equations of motion for a flexible body is obtained if a lumped mass formulation is used and the accelerations $\ddot{\mathbf{u}}'$ are substituted by a vector of nodal accelerations relative to the nonmoving reference frame $\ddot{\mathbf{q}}_f'$ [23].

The vectors of nodal accelerations can be partitioned into translational and angular accelerations as:

$$\ddot{\mathbf{u}}' = \begin{bmatrix} \ddot{\delta}' \\ \ddot{\theta}' \end{bmatrix} \quad ; \quad \ddot{\mathbf{q}}_f' = \begin{bmatrix} \ddot{\mathbf{d}}' \\ \alpha' \end{bmatrix}$$

The relation between the relative and absolute nodal accelerations for a node k is described by:

$$\begin{bmatrix} \ddot{\delta}' \\ \ddot{\theta}' \end{bmatrix}_k = \begin{bmatrix} \ddot{\mathbf{d}}' \\ \alpha' \end{bmatrix}_k - \begin{bmatrix} \mathbf{A}^T & -(\tilde{\mathbf{x}}^k + \tilde{\delta}^k)' \\ \mathbf{0} & \mathbf{I} \end{bmatrix} \begin{bmatrix} \ddot{\mathbf{r}} \\ \dot{\omega}' \end{bmatrix} - \begin{bmatrix} \tilde{\omega}'\tilde{\omega}'(\mathbf{x}^k + \delta^k)' + 2\tilde{\omega}'(\dot{\delta}^k)' \\ \tilde{\omega}'(\dot{\theta}^k)' \end{bmatrix} \tag{26}$$

where $\mathbf{x}^k$ is the position of node k in the reference configuration. Equation (26) is evaluated for all nodes of body i and substituted into equation (17) yielding

$$\sum_{k=1}^{n} (m\ddot{\mathbf{d}}')_k = \mathbf{g}_r \tag{27a}$$

$$\sum_{k=1}^{n} \left[m(\tilde{\mathbf{x}} + \tilde{\delta})' \ddot{\mathbf{d}}' \right]_k = \mathbf{g}_\theta' \tag{27b}$$

$$\mathbf{M}_{ff}\ddot{\mathbf{q}}_f' = \mathbf{g}_f' - {}_{t'}^{t}\mathbf{F} - ({}_{t'}^{t}\mathbf{K}_L + {}_{t'}^{t}\mathbf{K}_{NL})\mathbf{u}' \tag{27c}$$

Equations (27a) and (27b) are the equations of motion for the center of mass of a system of particles [35]. Equation (27c) is the equation of motion for the nodes of the flexible body, expressed in the body fixed coordinate system. Note that due to the use of the lumped mass formulation the mass matrix $\mathbf{M}_{ff}$ is diagonal and written as:

$$\mathbf{M}_{ff} = \mathrm{Diag}(m_1\mathbf{I}, \mathbf{0}, \cdots, m_k\mathbf{I}, \mathbf{0}, \cdots, m_n\mathbf{I}, \mathbf{0})$$

where m_k is the lumped mass of node k, and $\mathbf{I}$ and $\mathbf{0}$ are 3x3 identity and null matrices associated with the translational and rotational degrees of freedom, respectively.

If the origin of the body fixed coordinate system is coincident with the center of mass of the flexible body, equations (27a) and (27b) are the equations of motion of the origin of the $\xi\eta\zeta$ referencial. Very often it is useful to locate the origin in some other point of the flexible body. For this purpose let it be assumed that the flexible body has one rigid part and one flexible part. Let the body fixed coordinate frame be attached to the center of mass of the rigid part as shown in figure 4. The flexible part is attached to the rigid part by the nodes that belong to boundary ψ. The body-fixed coordinate frame is the same for the rigid and flexible domains.

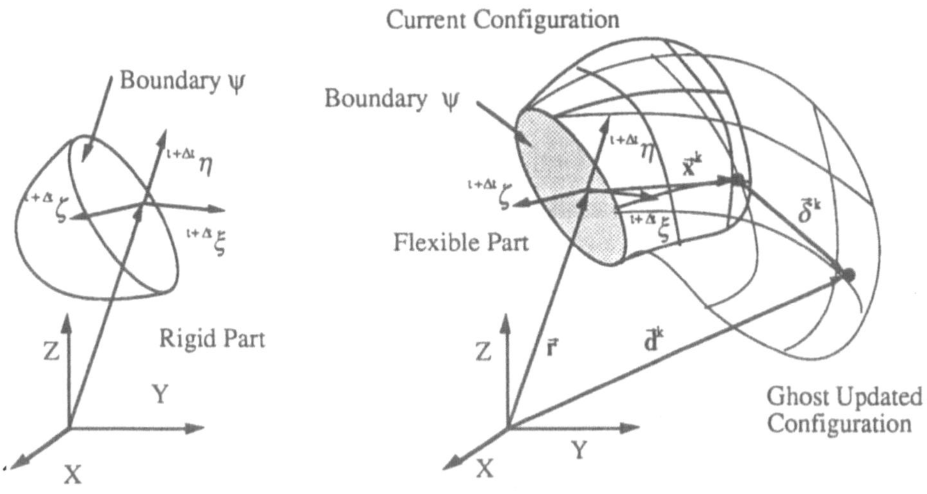

Figure 4 Flexible body with a rigid part

The Newton-Euler equations of motion for a rigid body are written as

$$m\ddot{\mathbf{r}} = \mathbf{f} \tag{28a}$$

$$\mathbf{J}'\dot{\omega}' = \mathbf{n}' - \tilde{\omega}'\mathbf{J}\omega' \tag{28b}$$

where **f** and **n** are the external forces and moments applied over the center of mass of the rigid part of the body. Equations (28a) and (28b) can be used instead of (27a) and (27b) provided that proper kinematic constraints are introduced between the flexible and rigid parts of the body. These kinematic constraints, that only affect the nodes in the boundary ψ, are described by:

$$\delta' = \dot{\delta}' = \ddot{\delta}' = \theta' = \dot{\theta}' = \ddot{\theta}' = \mathbf{0} \tag{29}$$

The constraint equations can be applied to the equation (26) for each of the boundary nodes using the Lagrange multiplier technique. At a latter step these multipliers are eliminated from the equations of motion resulting in the dynamic equations [23]

$$\begin{bmatrix} m + \overline{\mathbf{A}}\underline{\mathbf{M}}^{*}\overline{\mathbf{A}}^{T} & -\overline{\mathbf{A}}\underline{\mathbf{M}}^{*}\mathbf{S} & \mathbf{0} \\ -\left(\overline{\mathbf{A}}\underline{\mathbf{M}}^{*}\mathbf{S}\right)^{T} & \mathbf{J}' + \mathbf{S}^{T}\underline{\mathbf{M}}^{*}\mathbf{S} & \mathbf{0} \\ \mathbf{0} & \mathbf{0} & \mathbf{M}_{ff} \end{bmatrix} \begin{bmatrix} \ddot{\mathbf{r}} \\ \dot{\omega}' \\ \ddot{\mathbf{q}}'_{f} \end{bmatrix} = \begin{bmatrix} \mathbf{f} + \overline{\mathbf{A}}\mathbf{C}'_{\delta} \\ \mathbf{n}' - \tilde{\omega}'\mathbf{J}'\omega' - \mathbf{S}^{T}\mathbf{C}'_{\delta} - \overline{\mathbf{I}}^{T}\mathbf{C}'_{\theta} \\ \mathbf{g}'_{f} - {}^{t}_{t'}\mathbf{F} - \left({}^{t}_{t'}\mathbf{K}_{L} + {}^{t}_{t'}\mathbf{K}_{NL}\right)\mathbf{u}' \end{bmatrix} \tag{30}$$

where $\underline{\mathbf{M}}^{*}$ is a diagonal mass matrix containing the mass of the $\underline{n}$ boundary nodes. Matrices $\overline{\mathbf{A}}^{\mathrm{T}}, \mathbf{S}$ and $\overline{\mathbf{I}}$ are made from (3x3) matrices as:

$$\overline{\mathbf{A}}^{\mathrm{T}} = \begin{bmatrix} \mathbf{A}^{\mathrm{T}} \\ \mathbf{A}^{\mathrm{T}} \\ \vdots \\ \mathbf{A}^{\mathrm{T}} \end{bmatrix} \quad ; \quad \mathbf{S} = \begin{bmatrix} \left(\tilde{\mathbf{x}}^{1} + \tilde{\delta}^{1}\right)' \\ \left(\tilde{\mathbf{x}}^{2} + \tilde{\delta}^{2}\right)' \\ \vdots \\ \left(\tilde{\mathbf{x}}^{\underline{n}} + \tilde{\delta}^{\underline{n}}\right)' \end{bmatrix} \quad ; \quad \overline{\mathbf{I}} = \begin{bmatrix} \mathbf{I} \\ \mathbf{I} \\ \vdots \\ \mathbf{I} \end{bmatrix}$$

Vectors $\mathbf{C}'_{\delta}$ and $\mathbf{C}'_{\theta}$ represent respectively the reaction force and moment of the flexible part of the body over the rigid part. These reaction force/moments are written as

$$\mathbf{C}'_{\delta} = \mathbf{g}'_{\underline{\delta}} - {}_{t'}^{\,t}\mathbf{F}_{\underline{\delta}} - \left({}_{t'}^{\,t}\mathbf{K}_{\mathrm{L}} + {}_{t'}^{\,t}\mathbf{K}_{\mathrm{NL}}\right)_{\underline{\delta}\delta}\delta' - \left({}_{t'}^{\,t}\mathbf{K}_{\mathrm{L}} + {}_{t'}^{\,t}\mathbf{K}_{\mathrm{NL}}\right)_{\underline{\delta}\theta}\theta' \tag{31}$$

$$\mathbf{C}'_{\theta} = -\mathbf{g}'_{\underline{\theta}} + {}_{t'}^{\,t}\mathbf{F}_{\underline{\theta}} + \left({}_{t'}^{\,t}\mathbf{K}_{\mathrm{L}} + {}_{t'}^{\,t}\mathbf{K}_{\mathrm{NL}}\right)_{\underline{\theta}\delta}\delta' + \left({}_{t'}^{\,t}\mathbf{K}_{\mathrm{L}} + {}_{t'}^{\,t}\mathbf{K}_{\mathrm{NL}}\right)_{\underline{\theta}\theta}\theta' \tag{32}$$

In these equations the subscripts δ and θ refer to the partition of the vectors and matrices with respect to the translational and rotational nodal degrees of freedom. The underlined subscripts are referred to the boundary nodes between the rigid and flexible parts.

Equation (30) is the finite element equation of motion for a flexible body. When finite elements with rotational degrees of freedom are used to discretize the flexible body, some null elements appear in the diagonal of the mass submatrix $\mathbf{M}_{\mathrm{ff}}$. Therefore equation (30) cannot be solved explicitly for the accelerations. Three approaches can be used to solve this problem. In the first approach rotational inertias obtained by lumping the off-diagonal terms of the consistent mass matrix $\mathbf{M}_{\mathrm{ff}}$ are used to replace the null coefficients. In the second approach a static condensation of the nodal rotational degrees of freedom is used. In a third approach the modal superposition technique is used to eliminate the explicit use of nodal rotations In what follows, any reference to the use of equation (30) implies the use of the first approach. The second and third approaches are discussed next.

3.5. STATIC CONDENSATION OF NODAL ROTATIONS

In order to use the static condensation of the rotational degrees of freedom let the nodal equations of motion be partitioned into translational and rotational degrees of freedom. The relation between the translational degrees of freedom and the rotational coordinates is described by

$$\theta' = {}_{t'}^{\,t}\mathbf{K}_{\theta\theta}^{-1}\left(\mathbf{g}'_{\theta} - {}_{t'}^{\,t}\mathbf{F}_{\theta} - {}_{t'}^{\,t}\mathbf{K}_{\theta\delta}\delta'\right) \tag{33}$$

Applying equation (33) to equation (30) results in the equations of motion of the reduced system, i.e., without the explicit use of the rotational degrees of freedom. These equations are written as:

$$\begin{bmatrix} m+\overline{\mathbf{A}}\underline{\mathbf{M}}^{*}\overline{\mathbf{A}}^{T} & -\overline{\mathbf{A}}\underline{\mathbf{M}}^{*}\mathbf{S} & \mathbf{0} \\ -(\overline{\mathbf{A}}\underline{\mathbf{M}}^{*}\mathbf{S})^{T} & \mathbf{J}'+\mathbf{S}^{T}\underline{\mathbf{M}}^{*}\mathbf{S} & \mathbf{0} \\ \mathbf{0} & \mathbf{0} & \mathbf{M}_{\delta\delta} \end{bmatrix} \begin{bmatrix} \ddot{\mathbf{r}} \\ \dot{\omega}' \\ \ddot{\delta}' \end{bmatrix} = \begin{bmatrix} \mathbf{f}+\overline{\mathbf{A}}\mathbf{C}'_{\delta} \\ \mathbf{n}'-\tilde{\omega}'\mathbf{J}'\omega'-\mathbf{S}^{T}\mathbf{C}'_{\delta}-\overline{\mathbf{I}}^{T}\mathbf{C}'_{\theta} \\ \mathbf{g}'_{\delta}-{}^{t}_{t'}\mathbf{F}_{\delta}-{}^{t}_{t'}\mathbf{K}_{\delta\theta}\,{}^{t}_{t'}\mathbf{K}^{-1}_{\theta\theta}\left(\mathbf{g}'_{\theta}-{}^{t}_{t'}\mathbf{F}_{\theta}\right)-\left({}^{t}_{t'}\mathbf{K}_{\delta\delta}-{}^{t}_{t'}\mathbf{K}_{\delta\theta}\,{}^{t}_{t'}\mathbf{K}^{-1}_{\theta\theta}\,{}^{t}_{t'}\mathbf{K}_{\theta\delta}\right)\delta' \end{bmatrix} \tag{34}$$

By a proper choice for the location and orientation of the body fixed coordinate system in the rigid part of the flexible body, the mass matrix in equations (30) and (34) is turned into a diagonal invariant matrix. For this purpose the position of its origin must be coincident with the center of the system of masses composed by the rigid part and the boundary nodes of the flexible part of the body. Furthermore the coordinate system must be aligned with the principal directions of inertia of the rigid part plus boundary nodes.

3.6. MODAL SUPERPOSITION TECHNIQUE

In order to achieve computational efficiency in the solution of the flexible body equations of motion, the modal superposition technique has been widely used [24,29]. This method is well suited to reduce the number of degrees of freedom of a flexible body when the mass and stiffness matrix are time invariants and the frequency contents of the external applied forces are of the same order as the lower natural frequencies of the flexible body. This procedure can still be applied for cases where the stiffness matrix shows some level of nonlinearity. Assume that the stiffness matrix is decomposed into an invariant matrix and a displacement dependent matrix. For cases where the material constitutive tensor is constant (linear elastic material) the constant stiffness matrix is ${}^{t}_{t'}\mathbf{K}_{L}$ while the displacement dependent matrix is ${}^{t}_{t'}\mathbf{K}_{NL}$. Moreover, assume that the first two rows of equation (30) or equation (34) have been solved for $\ddot{\mathbf{r}}$ and $\dot{\omega}'$.

Substituting the relation between the global nodal accelerations and the nodal accelerations relative to the body fixed coordinate system, given by equation (26), into the third row of equation (30) gives

$$\mathbf{M}_{ff}\ddot{\mathbf{u}}'+{}^{t}_{t'}\mathbf{K}_{L}\,{}^{t}\mathbf{u}'=\mathbf{g}'_{f}-{}^{t}_{t'}\mathbf{F}-{}^{t}_{t'}\mathbf{K}_{NL}\mathbf{u}'-\mathbf{f}_{\varepsilon} \tag{35}$$

where vector $\mathbf{f}_{\varepsilon}$ represents the inertia forces due to the substitution of global nodal accelerations by local accelerations. This vector is written as:

$$\mathbf{f}_{\varepsilon}=\begin{bmatrix} \mathbf{M}^{*}\left(\overline{\mathbf{A}}^{T}\ddot{\mathbf{r}}-\mathbf{S}\dot{\omega}'-\mathbf{W}_{2}-2\mathbf{W}_{1}\dot{\delta}'\right) \\ \mathbf{0} \end{bmatrix} \tag{36}$$

Here $\mathbf{W}_1$ and $\mathbf{W}_2$ are represented by

$$\mathbf{W}_{1}=\mathrm{Diag}(\tilde{\omega}',\tilde{\omega}',\cdots,\tilde{\omega}')$$

$$\mathbf{W}_2 = \begin{bmatrix} \tilde{\omega}'\tilde{\omega}'\left(\mathbf{x}^1 + \delta^1\right)' \\ \tilde{\omega}'\tilde{\omega}'\left(\mathbf{x}^2 + \delta^2\right)' \\ \vdots \\ \tilde{\omega}'\tilde{\omega}'\left(\mathbf{x}^n + \delta^n\right)' \end{bmatrix}$$

The solution of the eigenproblem, posed by equating the right-hand side of equation (35) to zero, is a set of natural frequencies and corresponding modes of vibration for the flexible body. The nodal displacements can be expressed as a linear combination of the modes of vibration, i.e.

$$^{i}\mathbf{u}' = \mathbf{X}\mathbf{z} \tag{37}$$

where $\mathbf{X}$ is the modal matrix. The number of modes of vibration envolved in equation (34) is n_m which is normally much smaller than the number of nodal degrees of freedom of the flexible body. Once the modes of vibration are not time dependent, the modal accelerations and velocities are given by

$$\ddot{\mathbf{u}}' = \mathbf{X}\ddot{\mathbf{z}} \tag{38}$$
$$\dot{\mathbf{u}}' = \mathbf{X}\dot{\mathbf{z}} \tag{39}$$

Equations (37) through (39) are now substituted into equation (35) and the result pre-multiplied by $\mathbf{X}^T$. Using the property of orthonormality of the modal matrix with the mass matrix it is found that

$$\ddot{\mathbf{w}} = \mathbf{X}^T\left(\mathbf{g}_f' - {}^{i}_{i'}\mathbf{F} - {}^{i}_{i'}\mathbf{K}_{NL}\mathbf{u}' - \mathbf{f}_\varepsilon\right) - \Lambda\mathbf{w} \tag{40}$$

where Λ is a diagonal matrix with the squares of the natural frequencies. Equation (40) is the modal equation of motion for the flexible body. The complete set of equations of motion for the flexible body is composed by the two first rows of equation (30) and equation (40).

3.7. APPLICATION TO A ROTATING BEAM

The problem of a canteliver beam attached to a rigid hub, which is spun up from rest to a constant angular speed, is analyzed here. This problem, first proposed by Kane et al. [36] is studied in order to show the performance of the methodologies presented, namely to show the difference between the application of the different types of coordinates used to describe the deformations of the flexible body.

The canteliver beam, with a lenght of 10 meter and annular cross section is presented in figure 5. The angular speed of the hub is a function of time prescribed as:

$$\omega(t) = \begin{cases} \dfrac{6}{15}\left[t - \dfrac{15}{2\pi}\sin\left(\dfrac{2\pi t}{15}\right)\right] & \text{rad/s} \quad 0 \le t \le 15 \\ 6 & \text{rad/s} \quad t \ge 15 \end{cases}$$

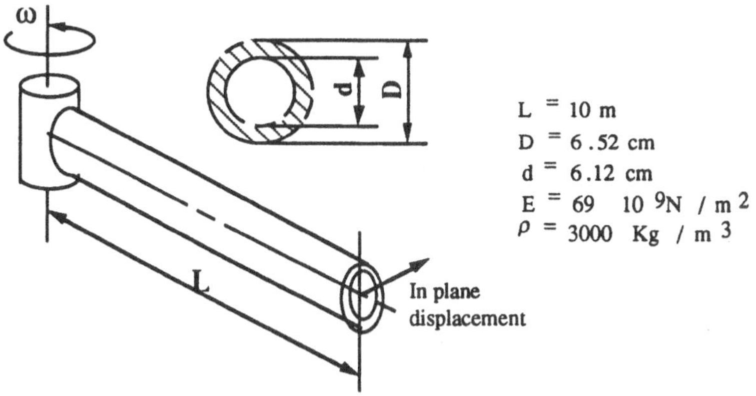

Figure 5 Rotating beam

The results presented in figure 6 show that if a linear behavior is assumed for the beam, i.e., the geometric stiffness is neglected and the deformations are small, the tip displacement of the beam with respect to the body fixed coordinates becomes infinite after 7 seconds of simulation. If equation (34) is used to represent the flexible body, the results are similar to those obtained by Kane et al., and the tip displacement increases while the angular acceleration of the hub is increasing. The tip of the beam ends up oscillating about its undeformed position after the angular speed becomes constant.

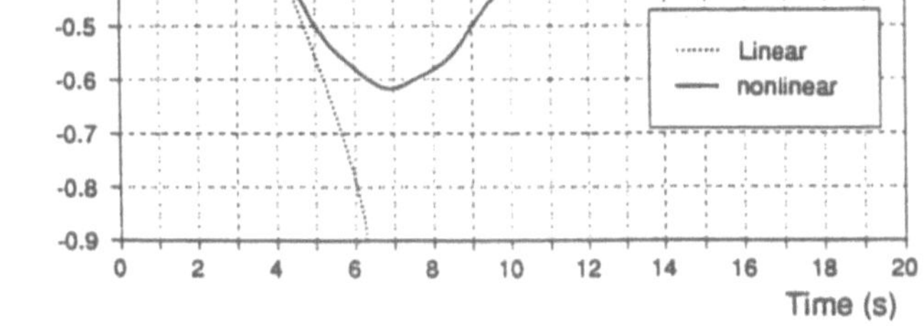

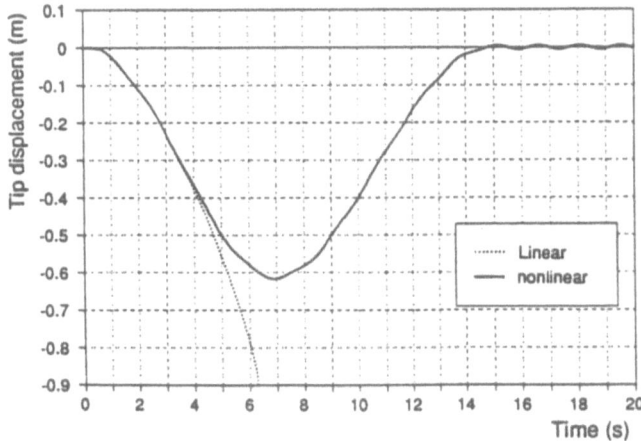

Figure 6 In plane displacement of the tip of the rotating beam
with respect to the underformed position

The different types of equations of motion are referred here as: equations with no coordinate reduction (30); statically condensed equations (34); and modal equations of motion (40). When a linear behavior for the beam was assumed all forms of the equations of motion provided the same behavior. When the beam was allowed to show a geometric nonlinear behavior the solution of the equations of motion based in the modal superposition had an error of 10% relative to the results obtained with the equations of motion with static condensation or with no coordinate reduction.

4. Concentrated Deformations - Plastic Hinges

The plastic hinge concept has been previously developed by utilizing generalized spring elements to represent constitutive characteristics of localized plastic deformation of beams. Bending plastic deformation at an attachment node has been modelled by revolute joints, as shown in figure 7.

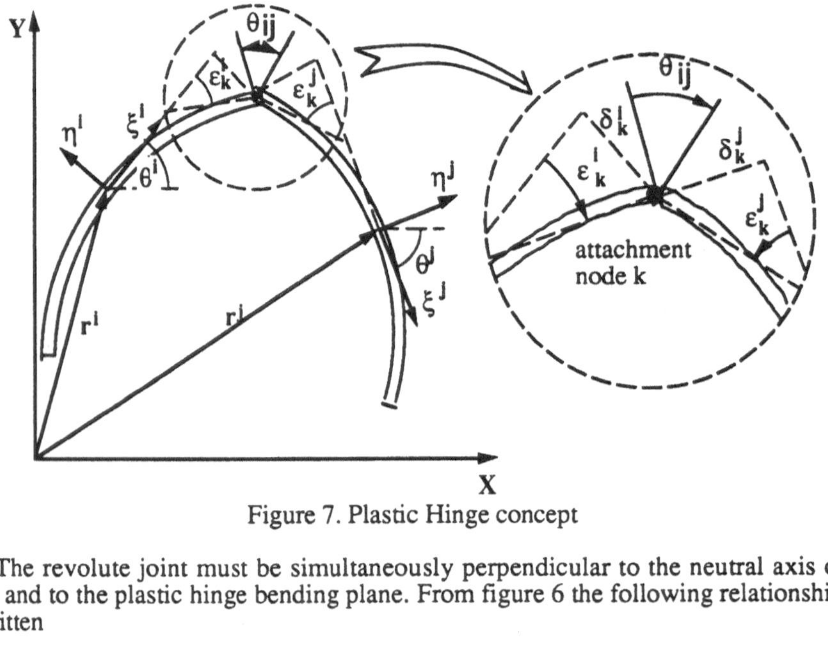

Figure 7. Plastic Hinge concept

The revolute joint must be simultaneously perpendicular to the neutral axis of the beam and to the plastic hinge bending plane. From figure 6 the following relationship can be written

$$\theta^{ij} = \theta^{i} + \varepsilon_k^{i} - \theta^{j} - \varepsilon_k^{j} \tag{41}$$

which shows the dependency of the plastic hinge angle on the rigid body positions and on the elastic rotations of body *i* and body *j* at the attachment node k. The angle values are directly obtained as relative coordinates from the integration process and correspond to the relative degree of freedom, θ_{ij}, of the revolute joint under consideration.

Figure 8 illustrates a typical torque-angle constitutive relationship which has been obtained in an earlier work [37] for the case of a steel tubular cross section based on a kinematic folding model [38]. This model was modified to take into account elastic plastic material properties including strain hardening.

The plastic hinge modelling technique generally requires that the spring data must be multiplied by a dynamic correction factor. Currently, the formula suggested by Winmer [39]

$$P_d / P_s = 1 + 0.07\, V_0^{.82} \tag{42}$$

has been used where P_d and P_s are the dynamic and static forces, respectively, and V_0 is the impact velocity. This equation may not be valid for a wide range of cross sections. Its only advantage resides in its generality for accounting strain rate effects. The user, however, may be able to incorporate other correction factors.

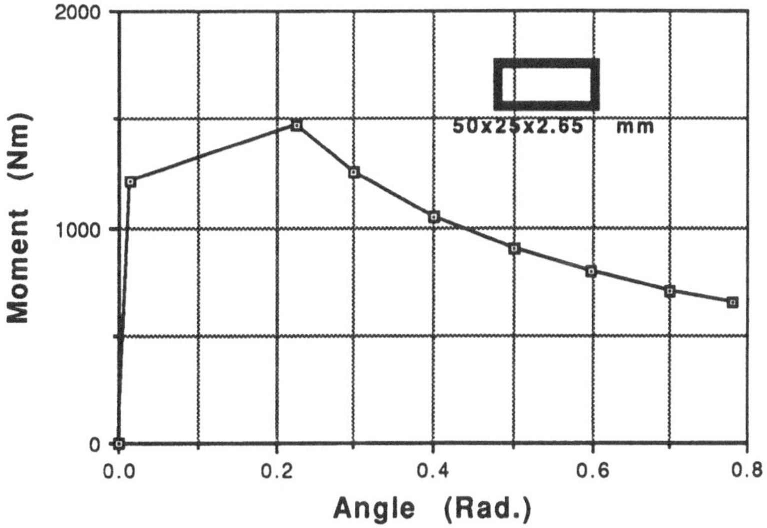

Figure 8. Plastic Hinge constitutive relationship

4.1. EXAMPLE OF AN IMPACTING BEAM

The formulations for rigid and flexible body dynamics including the plastic hinge model have been implemented in a computer program. In order to verify the proposed analytical technique a comparison with an experimental test [40,41] was carried out.

The experimental set up for this test case and the test specimen are shown in figure 9. The test bar is an E36 steel box-beam tube, 1 m long, with a 50x25 mm hollow rectangular section and 2.65 mm thick. One of the bar's ends is articulated to the ground structure with a revolute joint which is realized by means of a coupling sleeve allowing the rotation of the bar around an horizontal axis. A ballasting mass of 5.95 kg is attached to the other end of the bar. This mass is made of an XC38 steel cube 90.5mm long.

The test procedure is similar to the pendulum ram impact test. The bar is accelerated by means of a fast cylinder which actuates until an angular velocity of 11.85 rad/s is reached. After stabilization of the velocity, the bar collides with a rigid block. The edge of this rigid block is transverse to the beam longitudinal axis and located at a distance of 0.5 m away from the axis of the revolute joint. During the impact, accelerations have been measured using accelerometers which were implanted on the ballasting mass. A post impact observation clearly indicates the existence of a localized plastic bending zone which supports the consideration of a plastic hinge and a final value of the permanent bending angle was measured and was found to be 22°.

Figure 9 Experimental configuration

A series of computer simulations have been performed for the analysis of the impacting beam referred to above. Different rigid and rigid-flexible models have been considered which are shown schematically in figure 10.

Two rigid models with two and four bodies and two and four revolute joints respectively; and two flexible models with the same topology of the rigid body cases. Plastic hinges have been assumed in all intermediate revolute joints. Each flexible body was discretized with one finite element across and with extreme nodes located in the revolute joints and in the ballasting mass. A translational penalty spring with a very high stiffness was included to represent the unilateral contact between the beam and the impacted edge. This spring actuates only in the compression stage. During the initial stages, before contact, and in the final rebound phase this spring element is stretched and no force is considered.

Figure 11 illustrates a sequence of computer generated positions for the two rigid body model during the simulation time. The predicted gross motion shows a similar trend when compared with high speed photographs. An assessment of the accuracy of the theoretical models is made on the basis of the results obtained for the permanent bending angles. This is justified as the final configuration of the beam is strongly dependant on the dynamics of the problem and also on the mechanisms of energy absorption. The measured and calculated final bending angles are summarized in the table 1. An excellent agreement can be observed when comparing the experimental value with the different numerical simulations.

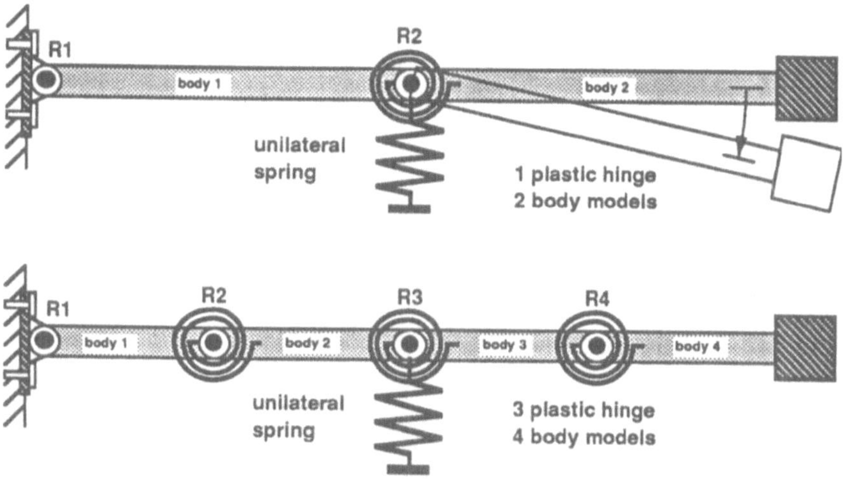

Figure 10. Multibody models

The flexible formulation yields lower values of the bending angle when compared to the rigid body simulations. For the flexible cases a small amount of the initial kinetic energy is absorbed in the excitation of linear elastic structural vibrations, thus reducing the energy available for plastic bending. The higher values obtained in the rigid body simulations are acceptable since, in these cases, the initial kinetic energy is totally absorbed in the plastic hinge mechanism.

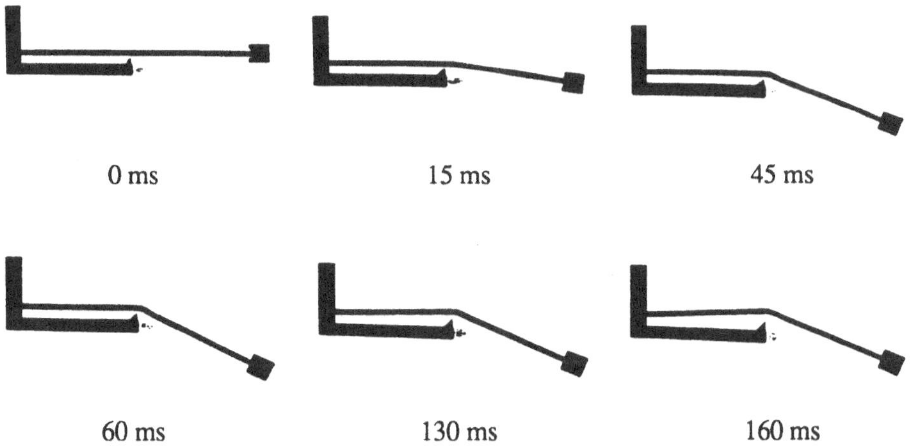

Figure 11. Evolution of the bending angle. Rigid body model

Table 1. Comparison of Results

	Permanent bending angles	Discrepancy (in %)
Experimental test	22.0°	-
RIGID		
1 hinge	23.0°	4.7
3 hinges	22.2°	1.0
FLEXIBLE		
1 hinge	21.6°	1.8
3 hinges	22.1°	0.4

5. Kinetostatic Method

For some applications of the crashworthiness analysis, the mass of the deformable part of the flexible body is relatively small compared to that of the rigid part. This may happen when the deformable part is a layer around the rigid body or a flexible appendage. This concept is illustrated in figure 12 for the rollover of a vehicle where the flexible part is designated by χ.

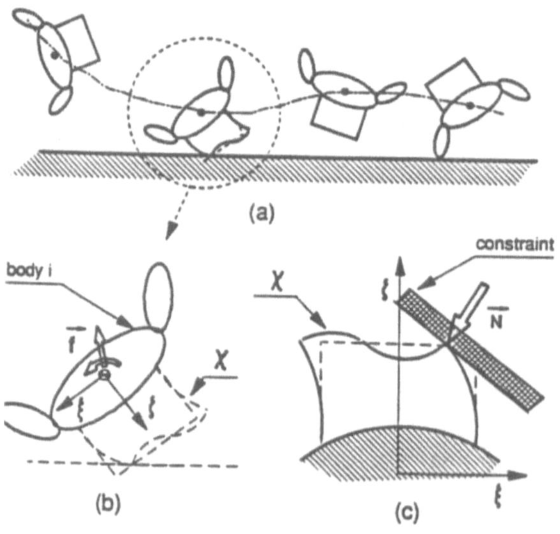

Figure 12 Schematic representation of the rollover of a vehicle with a rollbar cage: (a) Global displacement of the vehicle; (b) Deformation of the rollbar cage during impact; (c) Force/Moment reaction over the chassis due to structural deformations.

In this method, the following assumptions are made: the mass and moments of inertia of the structure can be neglected when compared with that of the rigid body; the mass and moments of inertia of the rigid part include that of the flexible part; the deformation of the flexible part does not change the inertial characteristics of the body.

For simplicity of notation, the deformable part of the flexible body is here designated by "the structure" while the rigid part of the same body is referred to as "rigid body *i*". It is assumed that the material constitutive law for the structure is linear elastic; the deformations are small; no other force besides the reaction forces from impact is applied over the structure.

The equations of motion for a rigid/flexible body are given by equations (30). The rigid and flexible equations are numerically uncoupled. In the right hand side, the action/reaction forces between the flexible and the rigid part are accounted for. Using the assumption of a massless structure for the flexible equations of motion, all of the nodal masses are eliminated. In order to maintain the total mass of the rigid/flexible body, the inertia of the structure is added to that of the rigid body *i*. Equation (30) becomes:

$$\begin{bmatrix} \mathbf{m} & \mathbf{0} & \mathbf{0} \\ \mathbf{0} & \mathbf{J}' & \mathbf{0} \\ \mathbf{0} & \mathbf{0} & \mathbf{0} \end{bmatrix} \begin{bmatrix} \ddot{\mathbf{r}} \\ \dot{\omega}' \\ \ddot{\mathbf{q}}'_f \end{bmatrix} = \begin{bmatrix} \mathbf{f} + \overline{\mathbf{A}}\mathbf{C}'_\delta \\ \mathbf{n}' - \tilde{\omega}'\mathbf{J}'\omega' - \mathbf{S}^T\mathbf{C}'_\delta - \overline{\mathbf{I}}^T\mathbf{C}'_\theta \\ \mathbf{g}'_f - {}^t_{t'}\mathbf{F} - \left({}^t_{t'}\mathbf{K}_L + {}^t_{t'}\mathbf{K}_{NL}\right)\mathbf{u}' \end{bmatrix} \tag{43}$$

where the mass **m** and the inertia tensor **J**' contain the inertial properties of the rigid body *i* and structure. The first two rows of equation (43) are the equations of motion for the rigid body. Vectors $\mathbf{c}_\delta'$ and $\mathbf{c}_\theta'$ represent the reaction forces and moments and are given by equations (31) and (32). The last row of equation (43) is the static equilibrium equation for the structure.

Assuming that the structure is linear elastic, i.e., the constitutive equation is linear and the deformations are small, equation (43) can be partitioned into:

$$\begin{bmatrix} \mathbf{m} & \\ & \mathbf{J}' \end{bmatrix} \begin{bmatrix} \ddot{\mathbf{r}} \\ \dot{\omega}' \end{bmatrix} = \begin{bmatrix} \mathbf{f} + \overline{\mathbf{A}}\mathbf{C}'_\delta \\ \mathbf{n}' - \tilde{\omega}'\mathbf{J}'\omega' - \mathbf{S}^T\mathbf{C}'_\delta - \overline{\mathbf{I}}^T\mathbf{C}'_\theta \end{bmatrix} \tag{44a}$$

$$\mathbf{K}\,\mathbf{u}' = \mathbf{g}'_f \tag{44b}$$

Comparing equation (44b) with equation (27c) it is observed that the nonlinear stiffness matrix ${}^t_{t'}\mathbf{K}_{NL}$ vanishes due to the assumption of the small deformations. The linear stiffness matrix ${}^t_{t'}\mathbf{K}_L$ becomes constant due to the assumption of the linear elastic material law and it is denoted here by **K**. In this sense **u**' is a vector of total nodal displacements rather than a vector of displacement increments. This implies that the vector of the equivalent nodal forces ${}^t_{t'}\mathbf{F}$ vanishes. The vector of the applied nodal forces $\mathbf{g}_f'$ still contains the external applied forces over the nodes and the forces due to the force elements that are connected to the structure. For the purpose of deriving an expression for the nodal forces due to the impact with an obstacle, let it be assumed, for the moment, that all forces applied over the structure (vector $\mathbf{g}_f'$) are only due to the impact, i.e., $\mathbf{g}_f'$ are the reaction forces of the obstacle over the structure plus the contact friction forces.

When the structure impacts another object, for example a rigid nonmoving obstacle, it undergoes some deformations as shown in figure 12(b). In turn, the effect of the deformation of the structure over the attached rigid body is described by applying a force/moment on body i, as depicted in figure 12(c). This force/moment, denoted by **f**, is simply the resultant of the reaction forces of the structure over the rigid body. Referring to equation (44a), the reaction force/moment is given by

$$\mathbf{f} = \begin{bmatrix} \overline{\mathbf{A}}\mathbf{C}'_\delta \\ -\mathbf{S}^T\mathbf{C}'_\delta - \overline{\mathbf{I}}^T\mathbf{C}'_\theta \end{bmatrix} \tag{45}$$

where vectors $\mathbf{c}_\delta'$ and $\mathbf{c}_\theta'$ depend upon the nodal displacements of the structure. The objective is to calculate the vector of nodal displacements $\mathbf{u}'$ in an efficient manner, so that the reaction forces **f** can be obtained.

Let the jth node of the finite element representation of the structure come in contact with the surface of an obstacle. The surface of the obstacle is defined by the global coordinates of a point Q_j, denoted by $\mathbf{d}_j^Q$, and a normal unit vector $\mathbf{n}_j$ as shown in figure 13. The subscript j is used for point Q and vector **n** in order to indicate that this surface contacts node j.

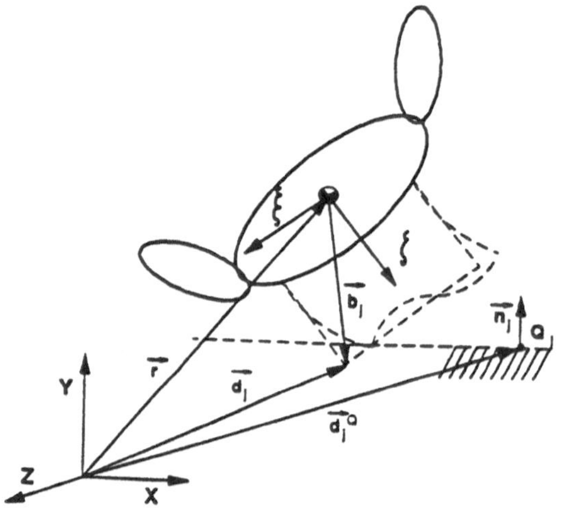

Figure 13 Definition of the position constraint posed by the rigid surface

At any given instant, the global coordinates of the undeformed position of node j, denoted by vector $\mathbf{d}_j$, can be calculated from the global configuration of body i. The "apparent penetration" α_j of node j into the contacting surface, as illustrated in figure 14, canm be calculated as

$$\alpha_j = \mathbf{n}_j^T\left(\mathbf{d}_j^Q - \mathbf{d}_j\right) \tag{46}$$

In reality, the structure deforms in such a way that node $\mathbf{d}_j$ remains on the surface of the obstacle. Denoting by δ_j the vector of nodal displacements of node j, the projection of this vector onto the normal to the surface must be equal to the apparent penetration α_j, i.e.

$$\mathbf{n}_j^T \delta_j = \alpha_j \tag{47}$$

Equation (46) is substituted into equation (47) to yield one constraint equation for node *j* as:

$$\mathbf{n}_j^T \left(\delta_j + \mathbf{d}_j + \mathbf{d}_j^Q \right) = 0 \tag{48}$$

In order to apply this constraint to the static equilibrium equations of the structure, equation (44b), the nodal constraint equation (48) must be written in terms of the nodal displacements with respect to the body fixed coordinate system, i.e.,

$$\mathbf{n}_j^T \left[\mathbf{A} \left(\delta_j' + \mathbf{b}_j' \right) + \mathbf{r} - \mathbf{d}_j^Q \right] = 0 \tag{49}$$

This equation is rearranged as:

$$\mathbf{n}_j^T \mathbf{A}\ \delta_j' = -\mathbf{n}_j^T \left(\mathbf{A}\ \mathbf{b}_j' + \mathbf{r} - \mathbf{d}_j^Q \right) \tag{50}$$

This equation, which is another form of equation (47), is the constraint equation on the displacement of node *j*. If more than one node simultaneously contact one or more obstacles, equation (50) is written for each node and the resulting set of constraints become

$$\mathbf{G}\mathbf{u}' = \alpha \tag{51}$$

where the rows of matrix **G** contain the components of vectors normal to the obstacle, **u**' is the vector of nodal displacements for all of the nodes, and vector α contains α_j's for all of the contacting nodes.

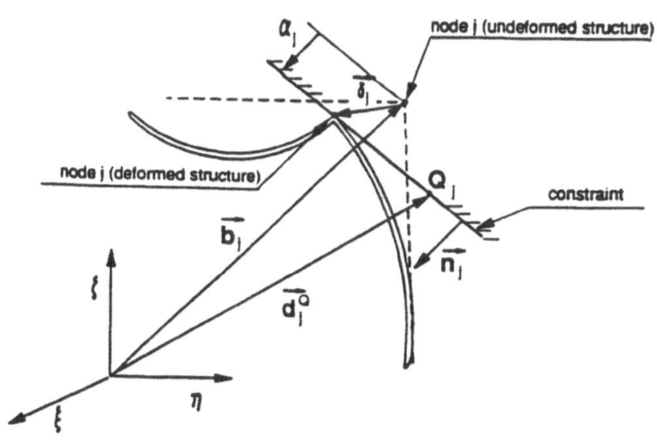

Figure 14 Contact between node *j* on the structure and a rigid surface

The reaction force at node *j* is denoted by $\mathbf{N}_j$. It may be assumed that there is also a friction force acting on the structure at that node. This friction force $\mathbf{g}'_j$,as shown in figure 15, can be expressed by

$$\mathbf{g}'_j = -\mu \left|\mathbf{N}_j\right| \mathbf{A}^T \mathbf{v}_j \tag{52}$$

where μ is the friction coefficient and $\mathbf{v}_j$ is a unit vector in the direction of the velocity of node *j* projected on the constraint surface. It must be noted that this force is valid only if there is sliding of node *j*. If friction force given by equation (52) is less than the product of dynamic friction coefficient by the normal reaction force due to the ground normal reaction, then stiction occurs. This case is not considered here.

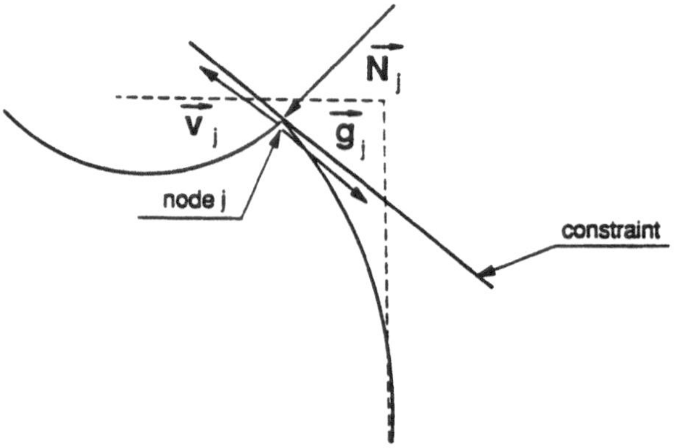

Figure 15 Reaction and friction forces at node *j*

These constraints are introduced in the static equilibrium equation of the structure using the Lagrange multiplier technique. Denoting by σ the Lagrange multipliers associated with the constraints of the contacting nodes, the system of constrained equilibrium equations is

$$\begin{bmatrix} \mathbf{K} & \mathbf{G}^T \\ \mathbf{G} & \mathbf{0} \end{bmatrix} \begin{bmatrix} \mathbf{u}' \\ \sigma \end{bmatrix} = \begin{bmatrix} \mathbf{g}'_f \\ \alpha \end{bmatrix} \tag{53}$$

Note that the term $\mathbf{G}^T\sigma$ in equation (53) represents the constraint forces due to the impact. For the *j*th node, the quantity $\mathbf{N}_j$ is:

$$\mathbf{N}_j = \mathbf{G}_j^T \sigma_j = \mathbf{n}_j \sigma_j \tag{54}$$

which is exactly the reaction force normal to the constraint surface at node *j*. Substituting this equation into equation (52) and observing that $\mathbf{n}_j$ is a unit vector, yields

$$\mathbf{g}'_j = -\mu \left|\sigma_j\right| \mathbf{A}^T \mathbf{v}_j \tag{55}$$

The nodal constraints are unilateral; i.e., σ_j does not change sign and it is a positive quantity as long as there is contact. Therefore, equation (55) is written as

$$\mathbf{g}'_j = \mathbf{H}_j^T \sigma_j \tag{56}$$

where $\mathbf{H}_j^T = -\mu\, \mathbf{A}^T \mathbf{v}_j$. If more than one node is in contact with the obstacle, and friction forces are the only external forces on the structure (excluding the reaction forces), then equation (56) is evaluated for all those nodes. This yields:

$$\mathbf{g}'_f = \mathbf{H}^T \sigma \tag{57}$$

where matrix $\mathbf{H}^T$ contains all $\mathbf{H}_j^T$'s. Substituting equation (57) into equation (53) results

$$\begin{bmatrix} \mathbf{K} & (\mathbf{G}-\mathbf{H})^T \\ \mathbf{G} & \mathbf{0} \end{bmatrix} \begin{bmatrix} \mathbf{u}' \\ \sigma \end{bmatrix} = \begin{bmatrix} \mathbf{0} \\ \alpha \end{bmatrix} \tag{58}$$

This equation is solved to find σ as

$$\sigma = \left[\mathbf{G}\mathbf{K}^{-1}(\mathbf{G}-\mathbf{H})^T\right]^{-1} \alpha \tag{59}$$

The dimension of matrix $[\mathbf{G}\ \mathbf{K}^{-1}\ (\mathbf{G} - \mathbf{H})]$ is k xk where k is the number of nodes in contact. Therefore, this is usually a very small matrix, and in most cases, a scalar. This means that the inversion of this matrix is not computationally expensive. The stiffness matrix **K** needs to be inverted only once as long as its elements are not changed. This is the case when only small linear elastic deformations are considered.

After evaluating the Lagrange multipliers σ from equation (59), equations (53) and (54) are used to calculate the reaction and friction forces at every contacting node. Since the structure is in static equilibrium, the set of reaction force/moments **f** as given by equations (45) is equivalent to the set of forces **N** and $\mathbf{g}_f$ as if they are directly applied to the rigid body i. For a typical contact node j, $\mathbf{N}_j$ and $\mathbf{g}_f$ act on body i at point j which is considered as an extension of body i. These forces cause a moment on body i due to the moment arm, which is a vector locating point j drelative to the origin of body i.

During a simulation, as long as none of the nodes in the structure is in contact with any obstacles, **f** is a null vector. This means that the dynamic analysis proceeds as a multi-rigid body system. In order to detect if a particular node j contacts or penetrates a surface at a certain time step, the term α_j is calculated from equation (46). A positive α_j means no contact and a negative α_j indicates a penetration. When penetration is detected, the corresponding reaction force/moment is calculated and included in the vector of forces. This reaction force/moment is updated as long as the node is in contact with the obstacle. When more than one node is detected to be in penetration, the sign of all Lagrange multipliers in vector σ must be verified. If any of these multipliers turns out to be negative, its corresponding constraint must be removed and equation (59) must be solved again. This situation can be described by referring to figure 16. As illustrated in figure 16(a), the undeformed configuration of two of the nodes are in "apparent penetration", i.e., negative values for α_j's are obtained from equation (45). However, in reality, if the deformation of the structure is considered, only one node may be in contact with the surface as shown in figure 16(b). If two nodal constraints are enforced, then an incorrect structural deformation is obtained. The negative Lagrange multiplier indicates which nodal constraint is enforced incorrectly and consequently must be removed.

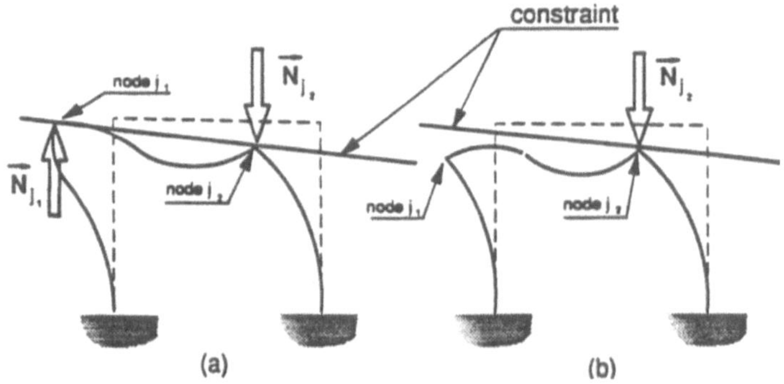

Figure 16 An apparent penetration of two nodes may yield: (a) a positive and a negative reaction force for two incorrectly enforced contacts; (b) removal of the contact constraint on one node yields correct deformation of the structure

6. Application to an Utility Truck Rollover

The vehicle simulated here is an utility truck. Originally this vehicle did not have any protection in case of a rollover. In order to provide that extra protection for passengers, a rollbar cage was attached to the chassis of the truck. This study involves determining the safety of the vehicle while the bars undergo structural deformations during the rollover.

The model of the vehicle, excluding the rollbar cage, consists of the main chassis, the complete suspension system, and four wheels, as shown in figure 17. The front wheels are connected to the main chassis by unequal A-arms (double wishbones). The rear wheels are connected to the main chassis by semi-trailing arms. Suspension springs, shock absorbers, and jounce stops are modeled by point-to-point spring-damper elements with nonlinear characteristics, as presented in figure 18. Tire characteristics including traction, braking and lateral forces due to steering were considered depending on such factors as normal force, slip angle and camber angle. The vehicle model consists of fifteen joint coordinates, equal to the number of degrees of freedom of the system. Six degrees of freedom correspond to the main chassis, four to the four suspension systems, four to the rolling wheels, and one to the steering.

The rollbar cage is a flexible frame mounted over the chassis to protect the passengers in case of a rollover. The rollbars are made of 1025-1030 steel with a yield strength of 30,000 psi. The cross-sectional area of each bar is annular with an inside radius of 2.14 cm and an outside radius of 2.54 cm. Two models for the rollbar cage are considered here: a model based on the plastic hinge technique, and; a finite element model.

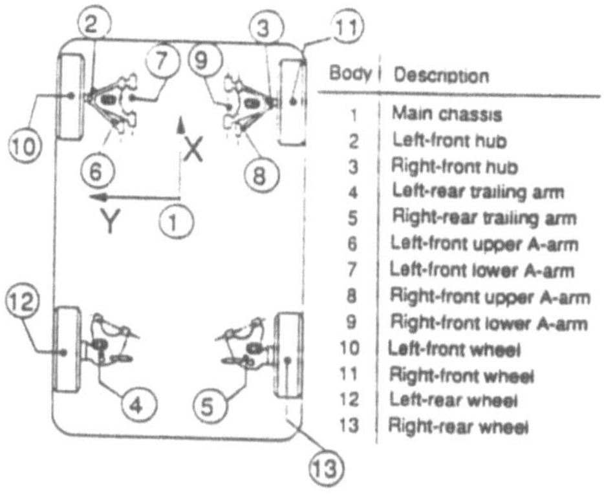
Body
Description
1 Main chassis
2 Left-front hub
3 Right-front hub
4 Left-rear trailing arm
5 Right-rear trailing arm
6 Left-front upper A-arm
7 Left-front lower A-arm
8 Right-front upper A-arm
9 Right-front lower A-arm
10 Left-front wheel
11 Right-front wheel
12 Left-rear wheel
13 Right-rear wheel
X
Y

Figure 17 Schematic model of the utility truck

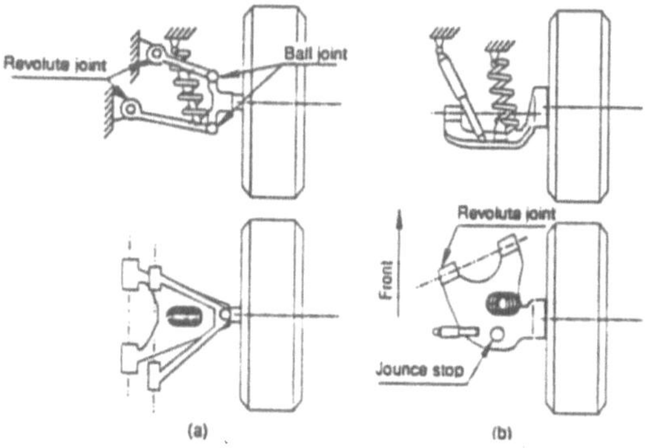

Figure 18 Front and rear suspension systems of the utility truck

6.1. PLASTIC HINGE MODEL

Within the framework of rigid body dynamics it is possible to simplify the plastic hinge model by assuming: (i) Point particles with only three translational degrees of freedom to model the rollbar cage beam elements in a lumped manner; (ii) These particles are assumed to be connected to one another and to the main chassis by massless links; (iii) The relative angular orientation of these massless links are monitored during the roll over analysis and moments are applied accordingly using some constitutive law obtained from analytical models or in suitable experimental tests. This model is schematically shown in figure 19.

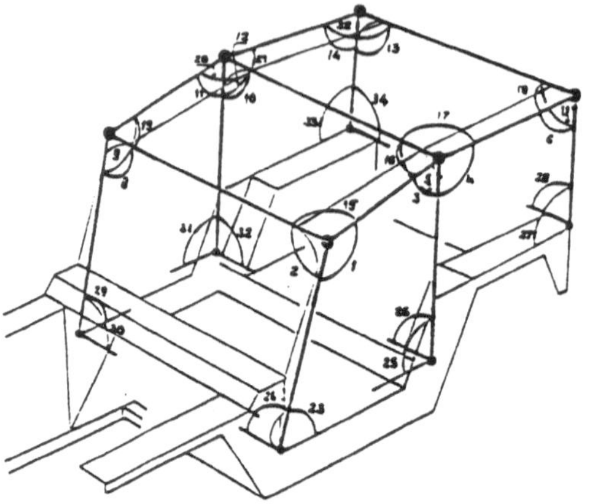

Figure 19 Representation of the rollbar cage using plastic hinges

A simple structure is considered in figure 20 to illustrate the present plastic hinge model as compared to the model described previously. In the model (a) moments are applied in the revolute joints according to some suitable constitutive law $M_i = M_i(\theta_i)$. In model (b), the resistive moments will be represented by forces actuating on the particles.

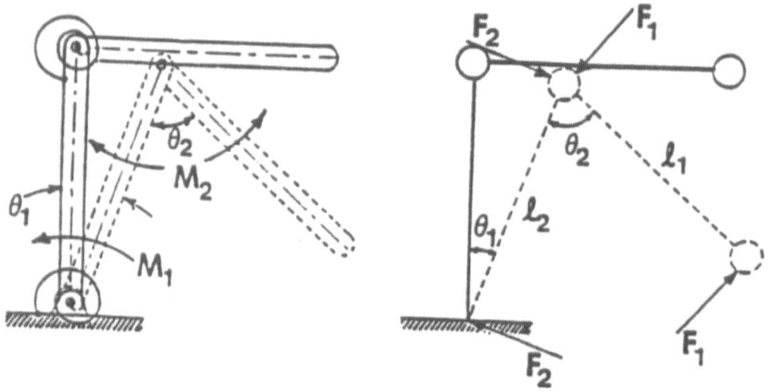

Figure 20 Plastic hinge model used for the rollbar cage

For example, for the relative angular motion θ_2, these forces can be calculated using the following simple vector relationships,

$$F_1 = \frac{M_2(\theta_2)}{\ell_1}; \quad F_2 = \frac{M_1(\theta_1)}{\ell_2}$$

A rollover test for the truck was performed by placing the vehicle over a cart moving at a speed of 30 m.p.h. and impacting a water filled decelerator system, thereby throwing the truck off the cart. The initial roll angle was 23 degrees, and the height of release was 30 cm as shown in figure 21. In order to maintain the total kinetic energy of the vehicle approximately the same as in the experimental test, an initial speed for the truck at the time of departure is assumed to be 25 m.p.h. plus a angular velocity of 1.5 rad/s in the roll direction._Several accelerometers were attached to the vehicle to record the responses.

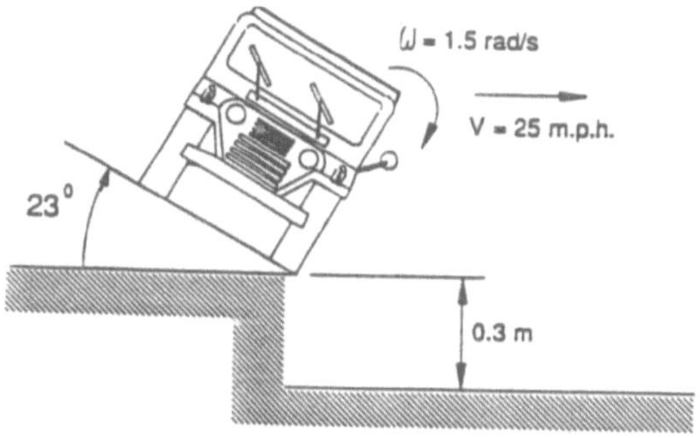

Figure 21 Initial conditions for the truck before rollover

The rollover simulation was performed from the instant the vehicle departed the cart. A friction model describing the interaction between the vehicle and the concrete ground was developed. The vehicle experienced a 247° roll and came to a stop on its side after approximately 4 seconds. The recorded and simulated accelerations at the center of mass of the main chassis were compared in the lateral (y) and in the vertical (z) directions and are sumarized in table 2 for peak accelerations. Both the test and simulation showed that the rolbar cage did not collapse, and a maximum plastic deformation of 4 cm was observed.

Table 2. Comparison of peak accelerations

	Experiment		Simulation	
	Acceleration (g)	Time (sec)	Acceleration (g)	Time (sec)
y_{min}	-7.5	0.43	-7-7	0.44
y_{max}	6.5	1.80	6.3	1.78
z_{min}	-15.0	0.43	-15.0	0.44
z_{max}	----*	----*	5.2	2.24

* - Not available due to accelerometer damage

6.2. FINITE ELEMENT MODEL

The finite element model of the cage is composed of 13 beam elements and 12 nodes. To simulate the attachment of the cage to the chassis 6 of the nodes are fixed to body 1, as shown in figure 22. This leads to a finite element model with 36 degrees of freedom.

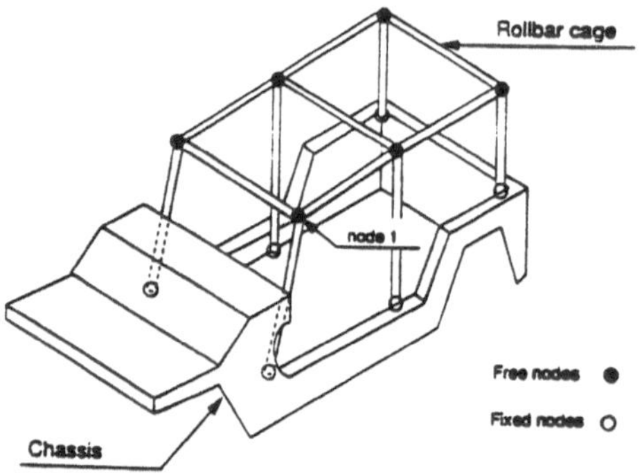

Figure 22 Rollbar cage and chassis

In order to evaluate the performance of the methodologies described in this paper, the simulations are made assuming no friction between the contacting rollbar cage nodes and the ground. Moreover, an unilateral constraint between contact nodes and ground is introduced whenever a node of the rollbar cage iniciates its contact with the ground. This constraint is removed when the sign of the Lagrange multiplier associated with this constraint changes.

Figure 23 shows the vertical displacement of the center of the chassis using complete coupling between the gross motion and the deformation of the flexible bodies of the system. In the kinetostatic method only a linear elastic material behavior is considered. For comparison the first simulation was run with a similar material behavior for the rollbar cage. A second simulation was performed using a material with the yield strength referred to above and a tangential plastic modulus given as $E^T=E/10$.

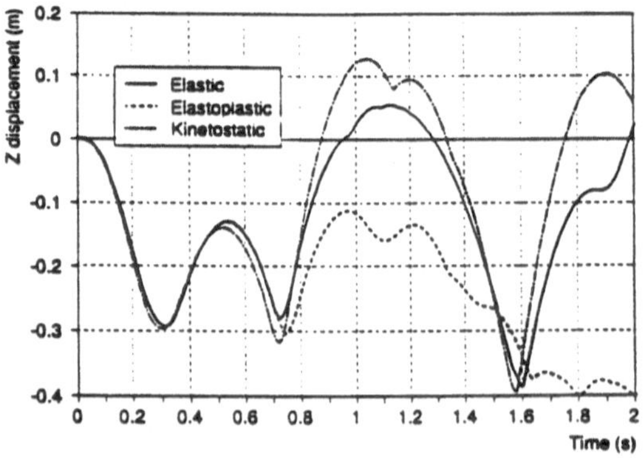

Figure 23 Vertical displacement of the center of mass of the chassis

These simulations show that both methods predict very similar behaviors within the first 1.5 seg. when a linear elastic behavior is assumed for the rollbar cage. The height of the vertical displacement of the chassis for the current method is lower than for the kinetostatic method because: some of the energy of the system is dissipated in the vibration of the cage; there is inertia coupling between the vehicle gross motion and the cage deformations. This dissipation of energy is clear from figure 24 where the lateral deflection of node 1 relative top the center of mass of the chassis is presented. In this figure the damping of the vibration of the rollbar cage is shown. Due to the extreme nonlinearity of the problem, small initial deviations between the results yield quite different motions after the initial period. When an elasto-plastic behavior is allowed the motion of the vehicle is, as expected, completely different from that of the kinetostatic method.

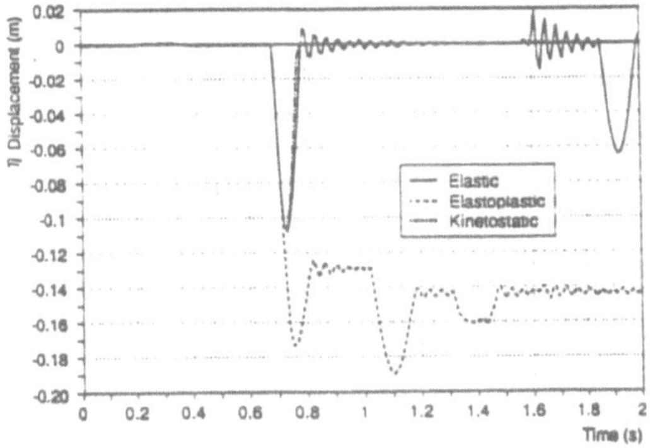

Figure 24 Lateral deflection for node 1 of the rollbar cage

7. Conclusions

A general formulations for the dynamic analysis of rigid and flexible multibody systems suitable for crashworthiness and impact were reviewed in this paper. Based on an updated Lagrangian formulation, the geometric and material nonlinear deformation of a general flexible body was described and introduced into the multibody system description. The equations obtained in this form were highly nonlinear and inefficient for computational purposes. A simpler form of these equations was obtained using a lumped mass formulation and referring the nodal accelerations to the inertial coordinate frame. Though a constant diagonal mass matrix was obtained, the numerical performance of these equations was still not adequate, if the methodology is to be used as a design tool.

Zones of localized instabilities are difficult to model in a simple form using the finite element method. A plastic hinge model and finite elements have been combined to provide an hybrid model for obtaining the dynamic plastic response of structural systems under impact conditions. The numerical method is quite general and the impact responses of stress, acceleration, velocity, position and structural generalized displacements of any part of the structure can be calculated at any instant of time. It has been shown that this procedure is adequate to predict the behavior of a rotating beam impacting a fixed edge. It was found that good agreement exists between the theoretical/numerical results and an experimental test which was carried out to validate the proposed methodology.

For situations where only a localized part of the flexible body deforms as a result of an impact it may happen that the mass of such region is neglectable when compared with the mass of the undeformed part. In this case the flexible part of the body may be assumed massless. This assumption led to the development of a kinetostatic model for crash-impact. The main advantage of this model is that the calculations of the deformation of the flexible bodies are only carried on while contact between ground and body lasts. This methodology seems promising, but some work needs still to be carried on to allow the structural behavior of the impacting bodies to be nonlinear.

The computing times for the flexible models are at least one order of magnitude larger then the times required for the equivalent rigid body models (plastic-hinge and kinetostatic). This is due to the increase in degrees of freedom of the problem and the need to drastically reduce the integration time steps in order to accurately integrate the high frequency content resulting from the linear elastic structural vibrations of the flexible bodies.

The methodologies described in this paper, are quite general and can be applied in the impact analysis of complex structures undergoing gross motion as, for example, the case of vehicle crash situations.

Acknowledgments

The authors acknowledge the support of: AGARD - Advisory Group for Aerospace and Research Development, project P77; JNICT - Junta Nacional de Investigação Cientifica e Tecnologica, Project STRD/C/TPR/569/92. Also the collaboration and support of Prof. Parviz Nikravesh was greatly appreciated.

References

1. R. A. Wilson (1970) "A Review of Vehicle Impact Testing; How it Began and What is Being Done", *SAE Trans.*, SAE Paper No. 700403
2. G. H. Tidbury (1984) "Future trends in Simulation of Crashworthiness", *Proc. International Conference on Vehicle Simulation*", I. Mech.Eng. paper C179/84.
3. I. D. Nielson (1984) "Trends in the Design of Car Front and side Structures To Meet Future Safety Needs:, *Proc. International Conference on Vehicle Simulation*", I. Mech.Eng. paper C180/84.
4. W. Johnson and A. G. Mamalis (1978) *Crashworthiness of Vehicle*", Mechanical engineering Publications Ltd.,London, England,.
5. R. I. Mori (1968) "Analytical Approach to Automobile Collision", *Automobile Engineering Conference*, SAE Paper No. 680016.
6. M. Tani and R. I. Mori (1970), "A Study of Automobile Crashworthiness", SAE Paper No. 700175.
7. M. M. Kamal (1970) "Analysis and Simulation of Vehicle to Barrier Impact", *International Automobile Safety Conference*, SAE Paper No. 700414.
8. K. H. Lin (1973) "A Rear-end Barrier Impact simulation Model for Unibody Passenger Cars", *SAE Trans.* **82**: Paper 730156.
9. C. J. Dressler and R. E. Schorry (1979) "High Speed Impact and Aggressivity Analysis of the CALSPAN/Chrysler Research Safety Vehicles (RSV)", in *3rd International Conference on Vehicle Structural Mechanics*, SAE Paper No. 790993.
10. S. J. Fenves and A. Gonzalez-Caro (1971) "Network Topological Formulation of Analysis and Design of Rigid Plastic Framework Structures", *International Journal for Numerical Methods in Engineering*, **3**, 425-441.

11. C. M. Ni (1973) "Impact Response of Curved Box Beam-columns with Large Global and Local Deformations", *AIAA Dynamics Specialists Conference,* Paper No. 73-410.
12. C. M. Ni (1974) "A Numerical Method for Non-linear Impact Analysis of Planar Frames", *Proc. International Conference on Computational Methods in Non-linear Mechanics*, Austin, Texas, September 23-25.
13. A. B. Pifko and R. Winter (1981) "Theory and Applications of Finite Element Analysis to Structural Crash Simulation", *J. Computers and Structures,* **13**, 277-285.
14. R. J. Hayduk, R. Winter, A. B. Pifko and E.L. Fesanella (1983) "Application of the Non-linear Finite Element Computer Program DYCAST to Aircraft Crash Analysis", *Structural Crashworthiness,* Eds. N. Jones and T. Wierzbicky, Butterworths, London, England, 283-307.
15. E. Haug, F. Arnaudea, J. Du Bois and A. Rouvay (1983) "Static and Dynamic Finite Element Analysis of Structural Crashworthiness in the Automotive and Aerospace Industries", *Structural Crashworthiness*, Eds. N. Jones and T. Wierzbicky, Butterworths, London, England, 175-217.
16. P. Du Bois and J. F. Chedmall (1987) "Automotive Crashworthiness Performance on a Supercomputer", *SAE Trans.*, SAE paper No. 870565.
17. M. S. Pereira, P. Nikravesh, G. Gim and J.A.C. Ambrósio (1987),"Dynamic Analysis of Roll-over and Impact of Vehicles", *XVIII Bus and Coach Experts Meeting*, Budapest, Hungria.
18. T. B. Sato (1966) "Dynamical Considerations in Automobile Collisions", *J. of the Society of Automotive Engineers of Japan,* **20**,(5).
19. Y. Ohkubo, T. Akamatsu and K. Shirasawa (1974) "Mean Crushing Strength of Closed Heat Section Members", SAE paper No. 740040.
20. E. Haug (1989) "The PAMCRASH Code as an Efficient tool for Crashworthiness Simulation and Design", *Second European Cars/trucks Simulation Symposium, Schliersee*, Germany, May 22-24.
21. J. O. Halquist (1982) *Theoretical Manual for DYNA-3D*, Lawrence Livermore Laboratory.
22. T. Belytschko and J. M. Kenedy (1986) *WHAMS-3D, An Explicit 3D Finite Element Program*, KBS2 Inc. P.O. Box 453, Willow Springs, IL 60480.
23. J.A.C. Ambrósio, P.E. Nikravesh and M. S. Pereira (1990), "Crashworthiness Analysis of a Truck", *J. Mathematical Computer Modelling*, **14**, 959-964.
24. A. Shabana and R. Wehage (1983) "Variable Degree of Freedom Component Mode Analysis of Inertia Variant Flexible Mechanical Systems", *ASME J. of Mech. Trans. and Auto. Design*, **105**, 371-378.
25. P.E. Nikravesh, J.A.C. Ambrósio and M. S. Pereira (1990) "Rollover Simulation and Crashworthiness Analysis of Trucks", *Journal of Forensic Engineering*, **2**(3) 387-401.
26. J. C. Brown (1990) "The Design and Type Approval of Coach Structures for Roll-over using the CRASH-D Program", *Int. J. of Vehicle Design*, **11**(4/5), 361-373.
27. P. E. Nikravesh (1990) "Systematic Reduction of Multibody Equations of Motion To a Minimal Set", *Int. J. Non-Linear Mechanics*, **25**, 143-151.
28. P. E. Nikravesh and G.H. Gim (1989) "Systematic Construction of The Equations of Motion For Multibody Systems Containing Closed Kinematic Loops", in *Proc. of the ASME Design Automation Conference*, Montreal, Canada, Sept. 17-20.
29. M. S. Pereira, and P. L. Proença (1991) "Dynamic Analysis of Spatial Flexible Multibody Systems Using Joint Coordinates", *Int. J. Num. Meth. in Engng.*, **32**, 1799-1812.

30. A. A. Shabana (1989) *Dynamics of Multibody Systems*, John Wiley & Sons, New York.
31. J.A.C. Ambrosio (1991) *Elastic-Plastic Large Deformation of Flexible Multibody Systems In Crash Analysis*, University of Arizona.
32. P.E. Nikravesh, and J.A.C. Ambrósio (1991) "Systematic Construction of Equations of Motion for Rigid-Flexible Multibody Systems Containing Open and Closed Kinematic Loops", *Int. J. for Nume. Methods in Engng.*, **32**, 1749-1766.
33. J.A.C. Ambrosio, and P.E. Nikravesh (1992) "Elastic-Plastic Deformation In Multibody Dynamics", *Nonlinear Dynamics*,**3**, 85-104,.
34. O.C. Zienckiewicz (1977) *The Finite Element Method*, McGraw-Hill.
35. D.T. Greenwood (1965) *Principles of Dynamics*, Prentice-Hall.
36. T.R. Kane, R.R. Ryan, and A.K. Banerjee (1987) "Comprehensive Theory For The Dynamics of a General Beam Attached to a Moving Rigid Base", *J. of Guidance, Control and Dynamics*, **10**, 139-151.
37. Anceau, J. H., Drazetic, P. and Ravalard, I.(1992) "Plastic Hinges Behaviour in the Multibody Systems", *Mécanique Matériaux Électricité*, nº 444, France.
38. D. Kecman (1983), "Bending Collapse of Rectangular and Square section Tubes", *Int. J. of Mech. Sci.*, **25**(9-10), 623-636.
39. Winmer, A. (1977) Einfluss der Belastungsgeshwindigheit auf das Festigkeits und Verformungsverhaten am Beispiel von Kraftfarhzengen", *ATZ* **77**(10),281-286.
40. M.S. Pereira, P. Drazetic, and Y. Ravalard (1993) "An Hybrid Method For Impact Analysis of Rigid-Flexible Structural Systems Undergoing Gross Motion", submitted to *J. Impact Engng.*
41. F. Dacheux, P. Drazetic, A. Marrisol, and Y. Ravalard (1991) "Impact Behavior of Structures", in *Proc. of Int. Conf. of Nonlinear Engineering Computations*, Swansea, U.K., September 16-20.

FINITE ELEMENT MODELING CONCEPTS IN MULTIBODY DYNAMICS

M. GÉRADIN[1], A. CARDONA[2], D.B.DOAN[1] and J.DUYSENS[1]
[1]*LTAS, Univ. of Liège*
Rue E. Solvay 21, B-4000 Liège, BELGIUM
[2]*INTEC (CONICET/UNL)*
Güemes 3450, 3000 Santa Fé, ARGENTINA

ABSTRACT : The paper describes a finite element formulation of flexible multibody systems. The discretized equations of motion are formulated using the augmented lagrangian approach and are solved in an implicit manner. Symbolic computation is utilized to develop the element models. Flexible members are treated in two ways: either in a fully nonlinear manner using a geometrically exact beam model, or through the substructuring concept. Two complex joint models are presented: a cam pair with double curvature and a flexible slider. Dry friction effects are taken into account using a regularization procedure.

1. Introduction

The computer approach to flexible multibody systems presented in this survey paper results from a research project started at the Aerospace Laboratory (LTAS) of the University of Liège since 1984 under the direction of the first author. It has significantly progressed from 1986 to 1989 thanks to the contribution of A. CARDONA who prepared and presented his PhD thesis [1] on the subject at the University of Liège in 1989. The resulting software (MECANO, a specific module of the general finite element software SAMCEF) has now become an industrial product but its development still remains a subject of intense industrial research at LTAS. The other two co-authors are members of the LTAS research team who have later contributed to the project on specific aspects such as joint modelling [5] and automatic software generation through symbolic computation [11].

Our objective has been to generalize the concept of finite element to articulated systems, starting from the methodology adopted in nonlinear structural dynamics codes based on *implicit* time integration.

Indeed, the evaluation of the existing mechanism analysis softwares which were available commercially when this project was started and which still are on the market today

M. F. O. S. Pereira and J. A. C. Ambrósio (eds.),
Computer-Aided Analysis of Rigid and Flexible Mechanical Systems, 233–284.

revealed that most of them have been designed for systems made of rigid bodies and therefore do not perform efficiently when flexibility effects in the members have to be taken into account.

The finite element approach to flexible multibody systems may be regarded as a particular case of the *cartesian coordinate approach*. An essential difference, however, remains in the manner in which the kinematics of flexible motion is described. When dealing with flexible bodies, it is usual to assume that global motion is decomposed into a rigid-body motion to which is superimposed a small deformation. The main limitation of this decomposition is that linear elasticity is necessarily assumed in the rigid body frame and therefore important nonlinear effects such as geometric stiffening may be missed in the resulting analysis.

The finite element methodology described here represents a full departure from traditional approaches in the sense that the total motion (including thus rigid body motion and elastic deformation) is directly referred to the inertial frame.

The following advantages result from this assumption:

- The representation of inertia forces is greatly simplified;
- the stiffness properties of each elastic member may be described in a quite rigorous manner, including the geometric stiffening effects.

The general principles of this finite element approach are described in section 2. Starting from an adequate parametrization of finite rotations and displacements we compute, according to figure 1.1, appropriate measures of strain and relative motion in terms of which the structural matrices of the elements are developed. They are built from the augmented lagrangian description of the constraints, assumed holonomic at this stage for sake of simplicity.

Due to the stiff character of the motion equations obtained in differential-algebraic form after discretization, the method of solution adopted is of implicit type (based on Newton-Raphson iteration) and therefore the motion equations have to be developed in linearized form. Efficient time integration is dealt with separately in a companion paper [10].

The concept of finite element has been applied up to now to develop a quite extensive library of rigid and flexible joint and member elements which cannot all be described in the present contribution. Section 3 deals with flexibility effects in the members and is itself divided into two main parts: subsection 3.1 presents the formalism adopted to develop a 3-D elastic beam element, while subsection 3.2 deals with the specific problem of *substructuring*, the objective being to model the structural behavior of flexible components of arbitrary

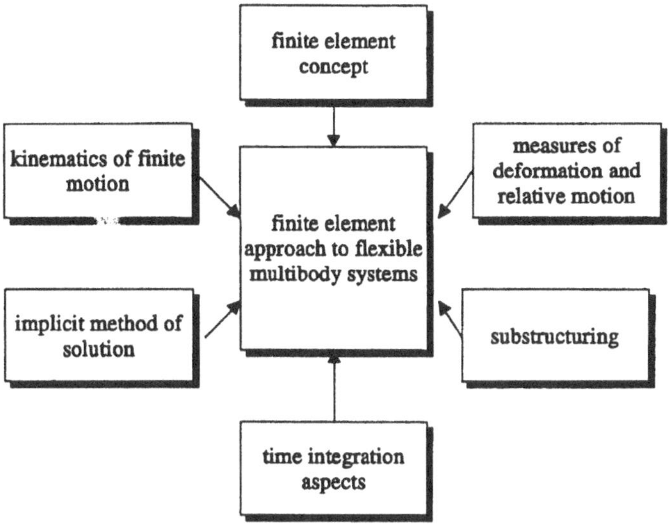

figure 1.1 : Principle of the finite element approach to multibody systems.

shape starting from a *component mode representation* obtained standard dynamic linear analysis. Section 4 is devoted to the finite element description of joints, and starts with the extension to non-holonomic constraints of the concepts presented in section 2. The very simple case of a hinge joint is treated next in order to show that automatic element generation can be performed through symbolic computation. Finally, subsections 4.3 and 4.4 present two joints of very complex nature: a cam pair with double curvature and a flexible slider element. In both cases, dry friction is taken into account using a regularization procedure.

Section 6 describes three applications which demonstrate the validity of the concepts presented.

2. General Concepts

2.1. FINITE ROTATION DESCRIPTION

Numerous techniques exist to represent a finite rotation in space which have each their respective advantages and drawbacks. The main criteria to be considered for selecting an appropriate formalism are [2] the number of parameters involved (3 or 4), their physical meaning, their algebraic properties, the existence of singularities and the form taken by the associated composition law for successive rotations.

According to these criteria, the system of parameters that we have selected is the set of 3 parameters formed by the cartesian components of the rotation vector

$$\boldsymbol{\Psi} = \mathbf{n}\, \Psi \tag{2.1.1}$$

where $\mathbf{n}$ represents the instantaneous rotation axis, and Ψ is the rotation amplitude about it.

Let us recall that the exponential form

$$\mathbf{R} = \mathbf{1} + \widetilde{\boldsymbol{\Psi}} + \frac{1}{2!}\widetilde{\boldsymbol{\Psi}}^2 + \cdots = \exp(\widetilde{\boldsymbol{\Psi}}) \tag{2.1.2}$$

allows constructing the rotation operator $\mathbf{R}$ in terms of the vector (2.1.1), where $\widetilde{\boldsymbol{\Psi}}$ is the skew-symmetric matrix made of the components of $\boldsymbol{\Psi}$ ($\widetilde{\Psi}_{ij} = -\epsilon_{ijk}\Psi_k$) . If one denotes by $\widetilde{\boldsymbol{\Theta}}$ the material rotation increment, i.e. expressed in a referential frame attached to the moving and/or deforming body, the incremental rotation is then expressed by the matrix

$$\delta\mathbf{R} = \mathbf{R}\, \delta\widetilde{\boldsymbol{\Theta}} \tag{2.1.3}$$

and the material rotation increments are themselves related to the finite rotation parameters by a linear relationship of type

$$\delta\boldsymbol{\Theta} = \mathbf{T}(\boldsymbol{\Psi})\, \delta\boldsymbol{\Psi} \tag{2.1.4}$$

with the matrix $\mathbf{T}(\boldsymbol{\Psi})$ given by [2]

$$\mathbf{T}(\boldsymbol{\Psi}) = \frac{\sin\|\boldsymbol{\Psi}\|}{\|\boldsymbol{\Psi}\|}\mathbf{I} + \left(1 - \frac{\sin\|\boldsymbol{\Psi}\|}{\|\boldsymbol{\Psi}\|}\right)\mathbf{n}\mathbf{n}^T - \frac{1}{2}\left(\frac{\sin\|\boldsymbol{\Psi}\|/2}{\|\boldsymbol{\Psi}\|/2}\right)^2 \widetilde{\boldsymbol{\Psi}} \tag{2.1.5}$$

Equation (2.1.4), which forms the basis of the adopted formalism, allows computing the angular velocities with a similar relationship. Their time derivative provides also the expression of angular accelerations

$$\begin{aligned} \boldsymbol{\Omega} &= \mathbf{T}(\boldsymbol{\Psi})\, \dot{\boldsymbol{\Psi}} \\ \mathbf{A} &= \mathbf{T}(\boldsymbol{\Psi})\, \ddot{\boldsymbol{\Psi}} + \dot{\mathbf{T}}(\boldsymbol{\Psi})\, \dot{\boldsymbol{\Psi}} \end{aligned} \tag{2.1.6}$$

The elements of the cartesian rotation vector allow to represent rotations of any magnitude. However, equation (2.1.5) shows that matrix $\mathbf{T}(\boldsymbol{\Psi})$ becomes singular when $\|\boldsymbol{\Psi}\| \rightarrow (2k\pi,\ k = 1, 2, \ldots)$ and therefore the parametrization presents differentiability holes. This inconvenience can be overcome by restricting the rotation vector to the range

$$\|\boldsymbol{\Psi}\| \leq \pi \tag{2.1.7}$$

Whenever the rotation violates condition (2.1.7), the rotational vector is modified according to [3]

$$\boldsymbol{\Psi}^* = \left(1 - \frac{2\pi}{||\boldsymbol{\Psi}||}\right)\boldsymbol{\Psi} \tag{2.1.8}$$

It is easy to verify that $\boldsymbol{\Psi}^*$ verifies the conditions

$$\mathbf{R}(\boldsymbol{\Psi}) = \mathbf{R}(\boldsymbol{\Psi}^*) \qquad \text{and} \qquad ||\boldsymbol{\Psi}^*|| \leq \pi \tag{2.1.9}$$

2.2. THE CONCEPT OF FINITE ELEMENT IN MULTIBODY DYNAMICS

The concept of finite element model may be adopted in a most general sense to represent any type of functionality appearing in the description of a multibody system : rigid or elastic member, mechanical joint, mechanism of interaction either between members or between a member and the external world.

In all cases, adequate kinematic description and parametrization of finite motion allows to define appropriate measures of deformation. Rigid elements are then characterized by the condition of zero deformation, while flexible elements are derived from a virtual work expression and the assumption of a constitutive law. This very general reasoning allows to construct a finite element library specialized to multibody analysis in terms of which most mechanical interactions may easily be described. The element library available in MECANO [4] includes rigid and elastic bodies, different types of rigid and deformable joints, active elements, element describing various interaction modes such as dissipation ; it also allows to customize the library through the concept of user element.

The global description of the finite element model of any multibody system can be made using the following definitions and notations :

- $\mathbf{q}$ is a global set of degrees of freedom (DOF) describing the absolute positions and orientations of the representative points of the system ;
- $\mathbf{q}_e$ denotes the DOF set of a given element, and $\mathbf{L}_e$ is a boolean operator such that the relationship between elemental and global DOF

$$\mathbf{q}_e = \mathbf{L}_e\mathbf{q} \tag{2.2.1}$$

implicitly contains the topological description of the system.

The kinematic constraints may express joint constraints, behavior restrictions or driving constraints. They are always defined at the element level and take the most general (nonholonomic, rheonomic) form

$$\boldsymbol{\Phi}(\mathbf{q}_e, \dot{\mathbf{q}}_e, t) = 0 \tag{2.2.2}$$

They are introduced through the definition of a set $\boldsymbol{\lambda}$ of lagrangian multipliers.

Each element is also characterized by its strain and kinetic energies, so that the total internal energy of the system and its kinetic energy are computed through summation on individual elements

$$\mathcal{W}_{int} = \sum_e \mathcal{W}_e(\mathbf{q}_e) \quad \text{and} \quad \mathcal{K} = \sum_e \mathcal{K}_e(\mathbf{q}_e, \dot{\mathbf{q}}_e, t) \tag{2.2.3}$$

Likewise, the global dissipation forces result from elemental contributions to the virtual work of friction forces

$$\delta\mathcal{W}_{fr} = \sum_e \mathbf{L}_e^T \mathbf{d}_{fr,e}(\mathbf{q}_e, \dot{\mathbf{q}}_e)\delta\mathbf{q}_e \tag{2.2.4}$$

Finally, the external virtual work is directly written in terms of the external forces themselves

$$\delta\mathcal{W}_{ext} = -\mathbf{g}_{ext}(\mathbf{q}, \dot{\mathbf{q}}, t)\delta\mathbf{q} \tag{2.2.5}$$

The system equations of motion are deduced from the variational equation

$$\delta \int_{t_1}^{t_2} (\mathcal{K} - \mathcal{W} - \boldsymbol{\lambda}^T \boldsymbol{\Phi})dt = 0 \tag{2.2.6}$$

where $\delta\mathcal{W} = \delta\mathcal{W}_{int} + \delta\mathcal{W}_{ext}$.

In the holonomic case (the non-holonomic case will be considered in section 4.1), they take the form of the system of differential- algebraic equations (DAE)

$$\begin{cases} \mathbf{M}\ddot{\mathbf{q}} + \mathbf{B}^T\boldsymbol{\lambda} = \mathbf{g}(\mathbf{q}, \dot{\mathbf{q}}, t) \\ \boldsymbol{\Phi}(\mathbf{q}, t) = 0 \end{cases} \tag{2.2.7}$$

where $\mathbf{M}$ is a symmetric, positive definite mass matrix obtained from the assembling of the element contributions ; it is generally configuration-dependent ; the term $\mathbf{M}\ddot{\mathbf{q}}$ contains the relative inertia forces ; $\mathbf{g}(\mathbf{q}, \dot{\mathbf{q}}, t)$ is the sum of internal, external and complementary inertia forces ; $\mathbf{B} = \left[\frac{\partial \boldsymbol{\Phi}}{\partial \mathbf{q}}\right]$ is the gradient matrix of the kinematic constraints.

Let us finally mention that the numerical conditioning of equation system (2.2.7) may be significantly improved in view of its numerical solution by making use of the augmented lagrangian method [5]. It consists of adding to the variational equation (2.2.6) a penalty term in the constraints which reinforces the positive definite character of the functional. The modified functional takes the form

$$\delta \int_{t_1}^{t_2} (\mathcal{K} - \mathcal{W} - k\lambda^T \boldsymbol{\Phi} - p\boldsymbol{\Phi}\boldsymbol{\Phi}^T)dt = 0 \tag{2.2.8}$$

where k is a scaling factor on the constraints and p is a penalty term. The modified equations of motion are then

$$\begin{cases} \mathbf{M}\ddot{\mathbf{q}} + \mathbf{B}^T(k\boldsymbol{\lambda} + p\boldsymbol{\Phi}) = \mathbf{g}(\mathbf{q}, \dot{\mathbf{q}}, t) \\ k\boldsymbol{\Phi}(\mathbf{q}, t) = 0 \end{cases} \tag{2.2.9}$$

The solution of (2.2.9) obviously coincides with that of (2.2.7) since the term involving the constraints vanishes when the latter are verified.

2.3. IMPLICIT METHOD OF SOLUTION

The choice of an implicit method of solution allows to imbed any kind of analysis in the same formalism. In particular, the kinematic analysis of the system results from the determination of a succession of configurations with zero strain energy and a quasi-static analysis corresponds to the succession of equilibrium configurations obtained by omitting the kinetic energy of the system.

2.3.1 Linearization of Motion Equations. The implicit solution of the dynamic case relies upon linearization of the DAE equations (2.2.9) and proceeds as follows. Let us assume that $(\mathbf{q}^\star, \dot{\mathbf{q}}^\star, \ddot{\mathbf{q}}^\star, \boldsymbol{\lambda}^\star)$ represents an approximate solution of system (2.2.9) at time t. A corrected solution is obtained in the form

$$(\mathbf{q}^\star + \Delta\mathbf{q}, \dot{\mathbf{q}}^\star + \Delta\dot{\mathbf{q}}, \ddot{\mathbf{q}}^\star + \Delta\ddot{\mathbf{q}}, \boldsymbol{\lambda}^\star + \Delta\boldsymbol{\lambda}) \tag{2.3.1}$$

from the solution of the incremental equations

$$\begin{cases} \mathbf{M}\Delta\ddot{\mathbf{q}} + \mathbf{C}^t\Delta\dot{\mathbf{q}} + \mathbf{S}^t\Delta\mathbf{q} + k\mathbf{B}^T\Delta\boldsymbol{\lambda} = \mathbf{r}^* + O(\Delta^2) \\ k\mathbf{B}\Delta\mathbf{q} = -k\boldsymbol{\Phi}^* + O(\Delta^2) \end{cases} \tag{2.3.2}$$

where $\mathbf{r}$ is the residual vector of dynamic equilibrium

$$\mathbf{r} = \mathbf{g}(\mathbf{q}, \dot{\mathbf{q}}, t) - \mathbf{M}\ddot{\mathbf{q}} - \mathbf{B}^T(k\boldsymbol{\lambda} + p\boldsymbol{\Phi}) \tag{2.3.3}$$

and where the tangent stiffness and damping matrices $\mathbf{S}^t$ and $\mathbf{C}^t$ are computed from

$$\mathbf{S}^t = -\frac{\partial \mathbf{g}}{\partial \mathbf{q}} + \frac{\partial}{\partial \mathbf{q}}\left[\mathbf{B}^T\left(k\boldsymbol{\lambda} + p\boldsymbol{\Phi}\right)\right] \qquad \mathbf{C}^t = -\frac{\partial \mathbf{g}}{\partial \dot{\mathbf{q}}} \tag{2.3.4}$$

2.3.2 Time Integration. Time integration of the second-order DAE equations (2.3.2) is performed using an integration scheme of Newmark type [6]. The motivations of this choice are

- the filtering of high frequencies brought into the model by elasticity,
- the low dependence of algorithm stability with time step size,
- the software simplification brought by one-step, second-order time integration,
- the use of existing software architecture for structural dynamics,
- the accumulated experience in implicit nonlinear structural dynamics with Newmark type methods.

Newmark's integration scheme consists of a simultaneous interpolation of displacements and velocities, implicit in accelerations

$$\begin{aligned} \dot{\mathbf{q}}_{n+1} &= \dot{\mathbf{q}}_n + (1-\gamma)h\ddot{\mathbf{q}}_n + \gamma h\ddot{\mathbf{q}}_{n+1} + \mathbf{e}'_n \\ \mathbf{q}_{n+1} &= \mathbf{q}_n + h\dot{\mathbf{q}}_n + (\frac{1}{2}-\beta)h^2\ddot{\mathbf{q}}_n + \beta h^2\ddot{\mathbf{q}}_{n+1} + \mathbf{e}_n \end{aligned} \tag{2.3.5}$$

with the local truncation error on displacements [7]

$$\mathbf{e}_n = (\beta - \frac{1}{6})h^3\mathbf{q}^{(3)}(\tau) + O(h^4\mathbf{q}^{(4)}) \tag{2.3.6}$$

The constants (β, γ) coefficients are integration parameters. The values

$$\beta = \frac{1}{4} \qquad \text{and} \qquad \gamma = \frac{1}{2} \tag{2.3.7}$$

provide unconditional stability with maximum accuracy for a linear system.

It can be shown [8,26] that the straightforward application of Newmark's method with the parameters (2.3.7) to the DAE system (2.3.2) leads to a weak instability of the method induced by the algebraic constraints. This instability can be controlled by adapting the asymptotic behavior of the algorithm. A detailed discussion of the numerical aspects (i.e. stability, accuracy, time-step control) associated to time integration of equations (2.3.2) using Newmark type methods is made in [9-10].

2.3.3 Effective incremental procedure. Special care has to be taken in the incrementation procedure of the rotational DOF since finite rotations are not additive quantities.

Let us split the set of kinematic unknowns into translation and rotation parameters

$$[\mathbf{q}^T \;\; \boldsymbol{\lambda}^T] \;\Rightarrow\; [\mathbf{d}^T \;\; \boldsymbol{\Psi}^T \;\; \boldsymbol{\lambda}^T] \tag{2.3.8}$$

Displacements and lagrangian multipliers are incremented in the usual manner

$$\mathbf{d}(t) = \mathbf{d}_n + \Delta\mathbf{d}\,; \quad \boldsymbol{\lambda}(t) = \boldsymbol{\lambda}_n + \Delta\boldsymbol{\lambda} \tag{2.3.9}$$

while for rotations, we determine the incremental rotation necessary to carry from the previous configuration to the current one

$$\mathbf{R}(t) = \mathbf{R}_n \; \mathbf{R}_{inc}(t) \tag{2.3.10}$$

or, in terms of rotational vectors

$$exp(\tilde{\boldsymbol{\Psi}}) \;=\; exp(\tilde{\boldsymbol{\Psi}}_n) \; exp(\tilde{\boldsymbol{\Psi}}_{inc}) \tag{2.3.11}$$

where $\boldsymbol{\Psi}(t)$ is the rotational vector describing the actual rotation $\mathbf{R}(t)$, $\boldsymbol{\Psi}_n$ is the rotational vector of the reference configuration $\mathbf{R}_n$, and $\boldsymbol{\Psi}_{inc}(t)$ is the rotational vector of the incremental rotation $\mathbf{R}_{inc}(t)$.

This approach can be seen as an updated Lagrangian point of view for the rotation part of the system. The reference rotation is fixed to the previous step, so that the expressions for the variations of angular displacements, velocities and accelerations are simply obtained in terms of the incremental rotation by replacing $\boldsymbol{\Psi}$ by $\boldsymbol{\Psi}_{inc}(t)$ into equations (2.1.6)

$$\begin{aligned} \delta\boldsymbol{\Theta} &= \mathbf{T}(\boldsymbol{\Psi}_{inc}) \; \delta\boldsymbol{\Psi}_{inc} \\ \boldsymbol{\Omega} &= \mathbf{T}(\boldsymbol{\Psi}_{inc}) \; \dot{\boldsymbol{\Psi}}_{inc} \\ \mathbf{A} &= \mathbf{T}(\boldsymbol{\Psi}_{inc}) \; \ddot{\boldsymbol{\Psi}}_{inc} \; + \; \dot{\mathbf{T}}(\boldsymbol{\Psi}_{inc}) \; \dot{\boldsymbol{\Psi}}_{inc} \end{aligned} \tag{2.3.12}$$

By noting that $\mathbf{T}(0) \;=\; \mathbf{I}$, we get the starting values for the integration of the rotation parameters

$$\boldsymbol{\Psi}_{inc\;n} = \mathbf{0} \qquad \dot{\boldsymbol{\Psi}}_{inc\;n} = \boldsymbol{\Omega}_n \qquad \ddot{\boldsymbol{\Psi}}_{inc\;n} = \mathbf{A}_n \tag{2.3.13}$$

The same predictors and correctors may then be written on displacement and rotation variables and on lagrangian multipliers

$$\begin{aligned} \ddot{\mathbf{q}}^0_{n+1} &= 0 \\ \dot{\mathbf{q}}^0_{n+1} &= \dot{\mathbf{q}}_n + (1-\gamma)h\ddot{\mathbf{q}}_n \\ \mathbf{q}^0_{n+1} &= \mathbf{q}_n + h\dot{\mathbf{q}}_n + (\frac{1}{2} - \beta)h^2\ddot{\mathbf{q}}_n \\ \boldsymbol{\lambda}^0_{n+1} &= \boldsymbol{\lambda}_n \end{aligned} \tag{2.3.14}$$

and

$$\begin{aligned} \ddot{\mathbf{q}}^{i+1}_{n+1} &= \ddot{\mathbf{q}}^{i}_{n+1} + \frac{1}{\beta h^2}\Delta\mathbf{q} \\ \dot{\mathbf{q}}^{i+1}_{n+1} &= \dot{\mathbf{q}}^{i}_{n+1} + \frac{1}{\gamma h}\Delta\mathbf{q} \\ \mathbf{q}^{i+1}_{n+1} &= \mathbf{q}^{i}_{n+1} + \Delta\mathbf{q} \\ \boldsymbol{\lambda}^{i+1}_{n+1} &= \boldsymbol{\lambda}^{i}_{n+1} + \Delta\boldsymbol{\lambda} \end{aligned} \tag{2.3.15}$$

where the displacement and lagrangian multiplier increments are solutions of the tangent linear system

$$\begin{bmatrix} \mathbf{S}^t + \frac{1}{\gamma h}\mathbf{C}^t + \frac{1}{\beta h^2}\mathbf{M} & k\mathbf{B}^T \\ k\mathbf{B} & \mathbf{0} \end{bmatrix} \begin{bmatrix} \Delta\mathbf{q} \\ \Delta\boldsymbol{\lambda} \end{bmatrix} = \begin{bmatrix} \mathbf{r} \\ -k\boldsymbol{\Phi}^* \end{bmatrix} \tag{2.3.16}$$

Corrected values for $\mathbf{R}_{n+1}$, $\boldsymbol{\Omega}_{n+1}$, $\mathbf{A}_{n+1}$ are computed from

$$\begin{aligned} \mathbf{R}_{n+1} &= \mathbf{R}_n\mathbf{R}(\Psi_{inc,n+1}) \\ \boldsymbol{\Omega}_{n+1} &= \mathbf{T}(\Psi_{inc,n+1})\dot{\Psi}_{inc,n+1} \\ \mathbf{A}_{n+1} &= \mathbf{T}(\Psi_{inc,n+1})\dot{\Psi}_{inc,n+1} + \dot{\mathbf{T}}(\Psi_{inc,n+1})\dot{\Psi}_{inc,n+1} \end{aligned} \tag{2.3.17}$$

Iteration is pursued until the system reaches equilibrium state, which is characterized by the vanishing of the virtual work expressions

$$\delta\mathbf{q}^T\mathbf{r} = 0 \qquad \text{and} \qquad \delta\boldsymbol{\lambda}^T\boldsymbol{\Phi} = 0 \tag{2.3.18}$$

In practice, eqns (2.3.18) are considered to be satisfied whenever the inequalities

$$||\mathbf{r}|| < \epsilon \quad \text{and} \quad ||\boldsymbol{\Phi}|| < \eta \tag{2.3.19}$$

are satisfied with given tolerances ϵ and η.

2.4. GENERATION OF FE MODELS THROUGH SYMBOLIC COMPUTATION

The computation by hand of the rather complex mathematical expressions resulting from the present FE formulation has several drawbacks, namely

- the long and sometimes tedious programming phase (checking, validation...);
- the obtention of a Fortran source code which is not necessarily optimized.

Besides, this complexity can represent a real obstacle to the development of more elaborate elements.

An alternate approach for developing such a code consists in using computer algebra in a first step to generate automatically and to simplify all the cumbersome mathematical expressions that have to be evaluated [11]. The main advantages of this approach are :

- the obtention of a more reliable generated software (automatic generation of the mathematical expressions minimizes the risk of errors);
- the possibility to simplify the symbolic expressions generated through computer algebra system and thus, to minimize the number of arithmetic operations before generating an optimized Fortran source code;
- the increased facility to extend the capabilities of the software through automatic generation of new elements of which the manual development would be too combersome;

- a better efficiency of the developer who is relieved from performing complex algebraic developments.

A symbolic program has been developed upstream from the MECANO software [12] in order to automatically develop the FE models. The computer algebra system under which this symbolic macro-procedure is written is MAPLE [13]: it has been selected mainly for the power of its built-in programming tools (linear algebra package, differentiation facility, advanced programming language...).

This symbolic macro-procedure is divided into 3 mains parts:

- the symbolic treatment of the finite rotations;
- the symbolic computation of basic expressions such as kinetic and potential energies, virtual work and kinematic constraints;
- the automatic derivation of the tangent FE iteration matrices (mass, stiffness, gyroscopic and damping matrices).

Figure 2.4.1 summarizes the succesive steps of the procedure.

2.4.1 Symbolic treatment of the finite rotations. It consists essentially in the obtention of the symbolic forms of the rotation operator, the matrix $\mathbf{T}(\Psi)$ and the angular velocities and accelerations in terms of the rotation parameters (2.1.1).

2.4.2 Fully automatic derivation of the tangent FE matrices. In order to describe the capabilities of that part of the symbolic program, the development of a rigid body element is briefly decribed.

Starting from the symbolic expression of the kinetic energy, the procedure automatically computes the symbolic expressions of the material inertia forces, the mass matrix, the centrifugal stiffness matrix and the gyroscopic matrix.

The complete tangent FE iteration matrices obtained in this way are directly written in terms of generalized coordinates.

Figure 2.4.2 presents the symbolic expression obtained for the rotational part of the mass matrix. Matrix $\mathbf{M}_r$ is in fact the matrix product $\mathbf{T}^T\mathbf{J}\mathbf{T}$ where $\mathbf{T}$ is the linear operator (2.1.5) and $\mathbf{J}$ is the inertia tensor of the body, assumed here diagonal for sake of simplicity:

$$\mathbf{J} = diag(J_{11}\ J_{22}\ J_{33}) \tag{2.4.1}$$

2.4.3 Symbolic treatment of the kinematic constraints. The constraints expressing the indeformability of the rigid body element are treated by the Lagrange multiplier technique.

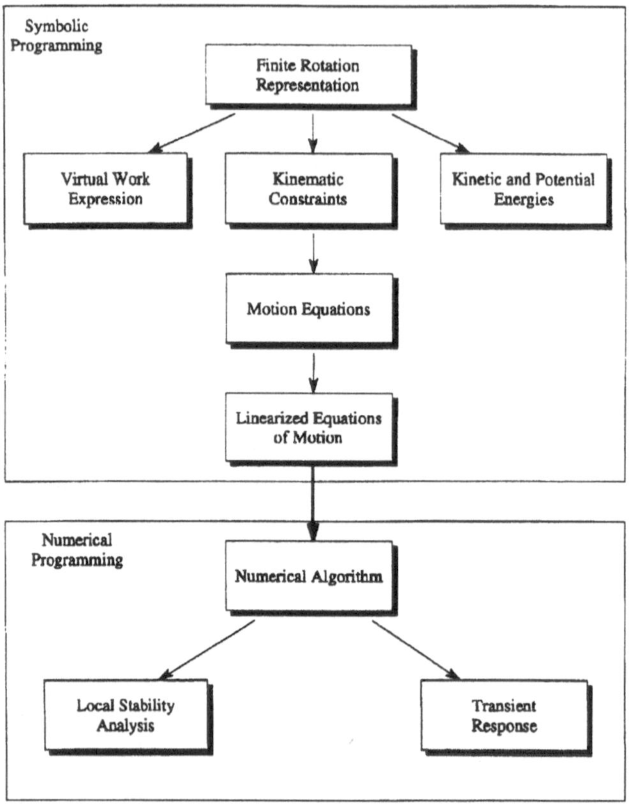

Figure 2.4.1 : Generation of a flexible multibody dynamics software throught symbolic programming

Their contribution to the FE iteration matrix implies the evaluation of the jacobian matrix of the constraints **B** and of their second derivatives (cf. 2.3.4). The latter evaluation can be very tedious when manually completed. It is not essential for computing a transient response but it can be shown to be essential for stability analysis.

A significant part of the symbolic procedure is devoted to the treatment of these kinematic constraints. Starting from their symbolic expression (written in vectorial form), the symbolic procedure automatically computes the first derivative of the kinematic constraints, the symbolic expression of the jacobian matrix **B** (the result being directly written in terms of the nodal parameters), the symbolic expression of the second derivatives and their contribution (in term of the nodal parameters) to the Hessian matrix (2.3.4).

The capabilities of this part of the symbolic procedure devoted to the treatment of the

```
> eval(Mt);
```

$$\left[\frac{\%13^2\, j11}{\%1^3}+\frac{\%10^2\, j22}{\%1^3}+\frac{\%6^2\, j33}{\%1^3},\ \frac{\%20\, j11\, \%13}{\%1^3}+\frac{\%19\, j22\, \%10}{\%1^3}+\frac{\%17\, j33\, \%6}{\%1^3},\ \frac{\%23\, j11\, \%13}{\%1^3}+\frac{\%22\, j22\, \%10}{\%1^3}+\frac{\%21\, j33\, \%6}{\%1^3}\right]$$

$$\left[\frac{\%20\, j11\, \%13}{\%1^3}+\frac{\%19\, j22\, \%10}{\%1^3}+\frac{\%17\, j33\, \%6}{\%1^3},\ \frac{\%20^2\, j11}{\%1^3}+\frac{\%19^2\, j22}{\%1^3}+\frac{\%17^2\, j33}{\%1^3},\ \frac{\%23\, j11\, \%20}{\%1^3}+\frac{\%22\, j22\, \%19}{\%1^3}+\frac{\%21\, j33\, \%17}{\%1^3}\right]$$

$$\left[\frac{\%23\, j11\, \%13}{\%1^3}+\frac{\%22\, j22\, \%10}{\%1^3}+\frac{\%21\, j33\, \%6}{\%1^3},\ \frac{\%23\, j11\, \%20}{\%1^3}+\frac{\%22\, j22\, \%19}{\%1^3}+\frac{\%21\, j33\, \%17}{\%1^3},\ \frac{\%23^2\, j11}{\%1^3}+\frac{\%22^2\, j22}{\%1^3}+\frac{\%21^2\, j33}{\%1^3}\right]$$

$\%1 := \psi_{[1]}^2+\psi_{[2]}^2+\psi_{[3]}^2$

$\%2 := \cos\left(\frac{1}{2}\sqrt{\%1}\right)$

$\%3 := \sqrt{\%1}\, \psi_{[2]}\, \%2^2$

$\%4 := \psi_{[1]}\, \psi_{[3]} \sin\left(\sqrt{\%1}\right)$

$\%5 := \sqrt{\%1}\, \psi_{[1]}\, \psi_{[3]}$

$\%6 := \%5 - \%4 + 2\sqrt{\%1}\, \psi_{[2]} - 2\, \%3$

$\%7 := \sqrt{\%1}\, \psi_{[3]}\, \%2^2$

$\%8 := \psi_{[1]}\, \psi_{[2]} \sin\left(\sqrt{\%1}\right)$

$\%9 := \sqrt{\%1}\, \psi_{[1]}\, \psi_{[2]}$

$\%10 := \%9 - \%8 - 2\sqrt{\%1}\, \psi_{[3]} + 2\, \%7$

$\%11 := \sin\left(\sqrt{\%1}\right) \psi_{[3]}^2$

$\%12 := \sin\left(\sqrt{\%1}\right) \psi_{[2]}^2$

$\%13 := \%12 + \%11 + \psi_{[1]}^2 \sqrt{\%1}$

$\%14 := \sqrt{\%1}\, \psi_{[2]}\, \psi_{[3]}$

$\%15 := \psi_{[2]}\, \psi_{[3]} \sin\left(\sqrt{\%1}\right)$

$\%16 := \sqrt{\%1}\, \psi_{[1]}\, \%2^2$

$\%17 := -2\sqrt{\%1}\, \psi_{[1]} + 2\, \%16 - \%15 + \%14$

$\%18 := \sin\left(\sqrt{\%1}\right) \psi_{[1]}^2$

$\%19 := \%18 + \%11 + \psi_{[2]}^2 \sqrt{\%1}$

$\%20 := \%9 - \%8 + 2\sqrt{\%1}\, \psi_{[3]} - 2\, \%7$

$\%21 := \%18 + \%12 + \psi_{[3]}^2 \sqrt{\%1}$

$\%22 := 2\sqrt{\%1}\, \psi_{[1]} - 2\, \%16 - \%15 + \%14$

$\%23 := -\%4 + \%5 - 2\sqrt{\%1}\, \psi_{[2]} + 2\, \%3$

Figure 2.4.2 : Automatic derivation of the tangent FE matrices.

kinematic constraints are illustrated in section 4.2 where the symbolic generation of a hinge joint element is presented.

3. Finite Element Representation of Elastic Components

3.1. BEAM REPRESENTATION OF ELASTIC MEMBERS

The appropriate description of flexible members requires in many cases the use of a beam formalism which incorporates properly the geometric nonlinear effects such as geometric stiffening. It is therefore essential to rely upon a true nonlinear beam theory [3,14-16].

3.1.1 Kinematic hypotheses. The behavior hypotheses adopted are the following:

(i) the beam is rectilinear,

(ii) the beam cross sections remain plane after deformation,

(iii) the shear deformation is allowed,

(iv) the rotational kinetic energy of cross sections is taken into account.

The kinematic assumptions (i) and (ii) can be summarized by the following equation

$$\mathbf{x} = \mathbf{x}_0 + X_\alpha \mathbf{t}_\alpha, \qquad \alpha = 2,3 \tag{3.1.1}$$

where $\mathbf{x}_0(t)$ represents the position of the beam neutral axis in the global reference frame and X_α $(\alpha = 1,2)$ are the material coordinates of a point on the cross section (figure 3.1.1). The base vectors $\mathbf{t}_\alpha$ are attached to the beam cross section and therefore, give the instantaneous orientation of the material frame $\mathbf{R}$. The current orientation of the base vectors is calculated in terms of the current rotation operator $\mathbf{R}(s)$

$$\mathbf{t}_j \;=\; \mathbf{R}\mathbf{E}_j \qquad (j = 1,2,3) \tag{3.1.2}$$

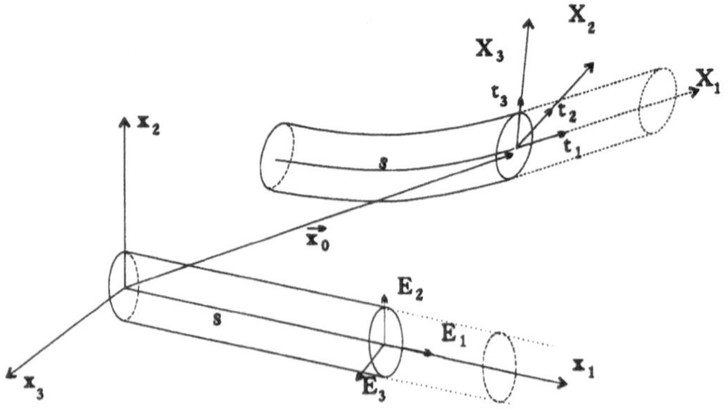

Figure 3.1.1 : Modeling of the flexible beam.

3.1.2 Material Measures of Beam Deformation. The measures of beam deformation are obtained from the comparison of displacement gradients in reference and current configurations.

The displacement gradients in reference configuration are computed from the reference position of a material point

$$\boldsymbol{\xi} = s\mathbf{E}_1 + X_2\mathbf{E}_2 + X_3\mathbf{E}_3 \quad \rightarrow \quad \frac{d\boldsymbol{\xi}}{ds} = \mathbf{E}_1 \tag{3.1.3}$$

The displacement gradients in the deformed configuration are obtained through derivation of (3.1.1)

$$\frac{d\mathbf{x}}{ds} = \frac{d\mathbf{x}_0}{ds} + X_2\frac{d\mathbf{t}_2}{ds} + X_3\frac{d\mathbf{t}_3}{ds} \tag{3.1.4}$$

with the base vector variations along the beam axis

$$\frac{d\mathbf{t}_j}{ds} = \frac{d\mathbf{R}}{ds}\mathbf{E}_j = \frac{d\mathbf{R}}{ds}\mathbf{R}^T\mathbf{t}_j \tag{3.1.5}$$

Substituting then (3.1.5) into (3.1.4) provides the relationship

$$\frac{d\mathbf{x}}{ds} = \frac{d\mathbf{x}_0}{ds} + \frac{d\mathbf{R}}{ds}\mathbf{R}^T(X_2\mathbf{t}_2 + X_3\mathbf{t}_3) \tag{3.1.6}$$

which expresses the position gradients in spatial coordinates. The substraction of (3.1.3) from (3.1.6) after expressing both quantities in the material frame provides the material measure of beam deformation

$$\mathbf{E}(s,\mathbf{X}) = \mathbf{R}^T(\frac{d\mathbf{x}_0}{ds} - \mathbf{t}_1) + \mathbf{R}^T\frac{d\mathbf{R}}{ds}(X_2\mathbf{E}_2 + X_3\mathbf{E}_3) \tag{3.1.7}$$

The first term involves the material measure of centroidal line, or *axial strain*

$$\boldsymbol{\Gamma} = \mathbf{R}^T(\frac{d\mathbf{x}_0}{ds} - \mathbf{t}_1) \tag{3.1.8}$$

Its components may be interpreted as follows: Γ_1 is the extensional strain, and $(\Gamma_2, \ \Gamma_3)$ are the shear strains along axes $\mathbf{t}_2$ and $\mathbf{t}_3$.

The second term involves the *material measure of curvature*

$$\tilde{\mathbf{K}} = \mathbf{R}^T\frac{d\mathbf{R}}{ds} \tag{3.1.9}$$

K_1 is the torsional deformation, while K_2 and K_3 are the bending curvatures along axes $\mathbf{t}_2$ and $\mathbf{t}_3$. The variations of the deformation measures (3.1.8) and (3.1.9) are given respectively by

$$\delta\boldsymbol{\Gamma} = \mathbf{R}^T\frac{d(\delta\mathbf{x}_0)}{ds} + \left(\widetilde{\mathbf{R}^T\frac{d\mathbf{x}_0}{ds}}\right)\delta\boldsymbol{\Theta} \qquad \text{and} \qquad \delta\mathbf{K} = \frac{d\delta\boldsymbol{\Theta}}{ds} + \tilde{\mathbf{K}}\delta\boldsymbol{\Theta} \tag{3.1.10}$$

In view of expressing dynamic equilibrium, they can also be put in the inverse forms

$$\frac{d}{ds}(\delta \mathbf{x}_0) = \mathbf{R}\delta\mathbf{\Gamma} - \frac{d\mathbf{x}_0}{ds} \times \delta\boldsymbol{\theta} \qquad \text{and} \qquad \frac{d}{ds}(\delta\boldsymbol{\theta}) = \mathbf{R}\delta\mathbf{K} \tag{3.1.11}$$

where $\delta\boldsymbol{\theta} = \mathbf{R}\delta\boldsymbol{\Theta}$ is the spatial rotation increment.

3.1.3 Local Expression of Dynamic Equilibrium. The stresses acting on the beam cross section are evaluated in terms of the Lagrange stress vector$\boldsymbol{\sigma}$ defined as the stress resultant per unit of undeformed cross section (figure 3.1.2). The latter is resolved along the base vectors attached to the cross section

$$\boldsymbol{\sigma} = \sigma_{11}\mathbf{t}_1 + \sigma_{12}\mathbf{t}_2 + \sigma_{13}\mathbf{t}_3 \tag{3.1.12}$$

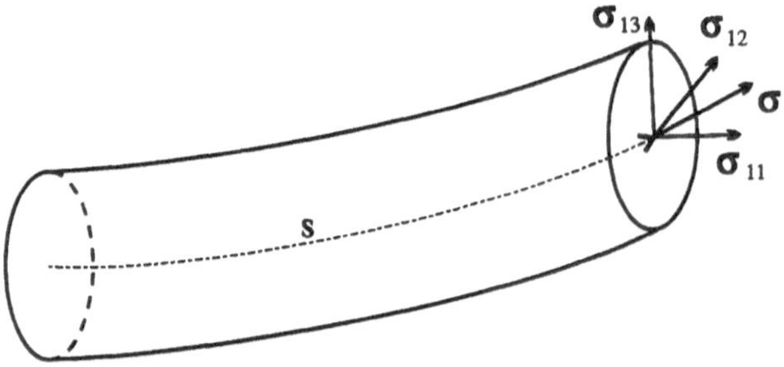

Figure 3.1.2 : Lagrangian stresses on beam cross section

Internal equilibrium is then expressed in terms of the following quantities: **b** the external force per unit of volume **b**, the specific mass ρ_0, the rotary inertia tensor of the cross setion expressed in spatial coordinates **I**, the spatial angular velocities and accelerations **a** and $\boldsymbol{\omega}$. Expressing translational equilibrium of a beam element of length ds provides the equation integrated over the cross section

$$\int_S [\frac{\partial\boldsymbol{\sigma}}{\partial s} + \mathbf{b} - \rho_0\ddot{\mathbf{x}}]ds = 0 \tag{3.1.13}$$

Similarly, rotational equilibrium can be expressed in the form

$$\int_S [\frac{d\mathbf{x}_0}{ds} \times \boldsymbol{\sigma} + (\mathbf{x} - \mathbf{x}_0) \times \frac{d\boldsymbol{\sigma}}{ds}]ds = \mathbf{I}\mathbf{a} + \boldsymbol{\omega} \times \mathbf{I}\omega - \int_S (\mathbf{x} - \mathbf{x}_0) \times \mathbf{b}ds \tag{3.1.14}$$

The dynamic equilibrium equations (3.1.13) and (3.1.14) can be expressed in terms of stress resultants obtained through integration over the cross section

$$\begin{aligned} \frac{d\mathbf{n}}{ds} &= -\overline{\mathbf{n}} + \mu\ddot{\mathbf{x}}_0 \\ \frac{d\mathbf{m}}{ds} + \frac{d\mathbf{x}_0}{ds} \times \mathbf{n} &= \mathbf{I}\mathbf{a} + \boldsymbol{\omega} \times \mathbf{I}\boldsymbol{\omega} - \overline{\mathbf{m}} \end{aligned} \tag{3.1.15}$$

where μ denotes the mass per unit length. The spatial measures of beam stress resultants and loads are defined by

$$\mathbf{n} = \int_S \boldsymbol{\sigma} dS = \text{the contact force on the cross section}$$

$$\mathbf{m} = \int_S (\mathbf{x} - \mathbf{x}_0) \times \boldsymbol{\sigma} dS = \text{the moment of stresses on the cross section}$$

$$\overline{\mathbf{n}} = \int_S \mathbf{b} dS = \text{the external force on the cross section}$$

$$\overline{\mathbf{m}} = \int_S (\mathbf{x} - \mathbf{x}_0) \times \mathbf{b} dS = \text{the external moment on the cross section}$$

One will also make use of the material counterparts of stress and load resultants

$$\mathbf{N} = \mathbf{R}^T \mathbf{n} \tag{3.1.16}$$

$$\mathbf{M} = \mathbf{R}^T \mathbf{m} \tag{3.1.17}$$

$$\overline{\mathbf{M}} = \mathbf{R}^T \overline{\mathbf{m}} \tag{3.1.18}$$

3.1.4 Weak Form of Dynamic Equilibrium. Let us start form the virtual work expression obtained through integration over the beam length of the equilibrium equations (3.1.13) and (3.1.14)

$$\int_0^\ell \left[\left(\delta \mathbf{x}_0^T \left(\frac{d\mathbf{n}}{ds} + \overline{\mathbf{n}} - \mu \ddot{\mathbf{x}}_0 \right) + \left(\delta \boldsymbol{\theta}^T \left(\frac{d\mathbf{m}}{ds} + \frac{d\mathbf{x}_0}{ds} \times \mathbf{n} - \mathbf{I}\mathbf{a} - \boldsymbol{\omega} \times \mathbf{I}\boldsymbol{\omega} + \overline{\mathbf{m}} \right) \right] ds = 0 \tag{3.1.19}$$

It can be integrated by parts in the form

$$\begin{aligned} \int_0^\ell [\mathbf{n}^T \delta \left(\frac{d\mathbf{x}_0}{ds} \right) + \left(\mathbf{m}^T \delta \frac{d\boldsymbol{\theta}}{ds} \right)] ds &= [\overline{\mathbf{n}}^T \delta \mathbf{x}_0 + \overline{\mathbf{m}}^T \delta \boldsymbol{\theta}]_0^\ell \\ &+ \int_0^\ell [(\delta \mathbf{x}_0^T (\overline{\mathbf{n}} - \mu \ddot{\mathbf{x}}_0) + \left(\delta \boldsymbol{\theta}^T (\overline{\mathbf{m}} + \frac{d\mathbf{x}_0}{ds} \times \mathbf{n} - \mathbf{I}\mathbf{a} - \boldsymbol{\omega} \times \mathbf{I}\boldsymbol{\omega} \right)] ds \end{aligned} \tag{3.1.20}$$

Let us next make use of expressions (3.1.10) for the variations of axial strains and curvatures: the left-hand side may then be expressed in terms of material strains and stresses, giving

$$\begin{aligned} \int_0^\ell [\mathbf{N}^T \delta \boldsymbol{\Gamma} + \mathbf{M}^T \delta \mathbf{K}] ds &= [\overline{\mathbf{n}}^T \delta \mathbf{x}_0 + \overline{\mathbf{m}}^T \delta \boldsymbol{\theta}]_0^\ell \\ &+ \int_0^\ell [(\delta \mathbf{x}_0^T (\overline{\mathbf{n}} - \mu \ddot{\mathbf{x}}_0) + \left(\delta \boldsymbol{\theta}^T (\overline{\mathbf{m}} - \mathbf{I}\mathbf{a} - \boldsymbol{\omega} \times \mathbf{I}\boldsymbol{\omega} \right)] ds \end{aligned} \tag{3.1.21}$$

Finally, equation (3.21) can be put in the form

$$\delta W_{int} = \delta W_{ext} - \delta W_{iner} \tag{3.1.22}$$

with

$$\delta W_{int} = \int_0^{\ell} [\mathbf{N}^T \delta \mathbf{\Gamma} + \mathbf{M}^T \delta \mathbf{K}] ds \tag{3.1.23}$$

and where the rotation terms of external and inertia forces are recasted in material coordinates, giving

$$\delta W_{ext} = \left[\overline{\mathbf{n}}^T \delta \mathbf{x}_0 + \overline{\mathbf{M}}^T \delta \mathbf{\Theta} \right]_0^{\ell} + \int_0^{\ell} [(\delta \mathbf{x}_0(\overline{\mathbf{n}}) + \left(\delta \mathbf{\Theta}^T (\overline{\mathbf{M}} \right)] ds \tag{3.1.24}$$

and

$$\delta W_{iner} = \int_0^{\ell} [(\delta \mathbf{x}_0^T (\mu \ddot{\mathbf{x}}_0) + \left(\delta \mathbf{\Theta}^T (\mathbf{J} \mathbf{A} + \mathbf{\Omega} \times \mathbf{J} \mathbf{\Omega} \right)] ds \tag{3.1.25}$$

with

$$\mathbf{J} = \mathbf{R}^T \mathbf{I} \mathbf{R} \tag{3.1.26}$$

3.1.5 Constitutive Equations. The material is assumed linear elastic, so that the internal stress resultants are related linearly to beam strain measures

$$\mathbf{\Sigma} = \mathbf{C} \mathbf{E} \tag{3.1.27}$$

with the diagonal matrix of elastic coefficients

$$\mathbf{C} = diag(EA, GA_2, GA_3, GJ, EI_2, EI_3) \tag{3.1.28}$$

EA is the axial stiffness, GJ is the torsional stiffness, and (GA_2, GA_3) and (EI_2, EI_3) denote respectively the shear and bending stiffnesses along transverse axes.

3.1.6 Finite Element Discretization. The finite element discretization of the virtual work expressions (3.1.22) is a lenghty process which we will only summarize here. A full derivation of the beam element together with numerical comparisons can be found in [3].

The discretization of eqns (3.1.22) is based on a linear interpolation of both displacements and rotation parameters

$$\mathbf{x}_0(s) = N_i(s) \mathbf{x}_{0i} \qquad \mathbf{\Psi}(s) = N_i(s) \mathbf{\Psi}_i \tag{3.1.29}$$

where $\mathbf{x}_{0i}, \mathbf{\Psi}_i$ are the nodal values of position and rotation parameters, collected in vector $\mathbf{q}_e$ of the element DOF, $N_i(s)$ is the linear interpolation function corresponding to node i, and summation is extended to the two nodes of the element.

The strain variations of element e can be expressed in terms of a configuration - dependent strain matrix $\mathbf{B}_e$

$$\delta \mathbf{E} = \mathbf{B}_e \, \delta \mathbf{q}_e \tag{3.1.30}$$

the strain matrix being of the form $\mathbf{B}_e = [\mathbf{B}_{(1)} \dots \mathbf{B}_{(n)}]$ with

$$\mathbf{B}_i = \begin{bmatrix} N_i'\mathbf{R}^T & N_i(\widetilde{\mathbf{R}^T\mathbf{x}_0'})\mathbf{T} \\ 0 & N_i'\mathbf{T} + N_i(\tilde{\mathbf{K}}\mathbf{T} + \mathbf{T}') \end{bmatrix} \tag{3.1.31}$$

where $(.)'$ denotes derivation with respect to s. The internal forces are such that

$$\delta W_{int} = \delta \mathbf{q}^T \mathbf{g}_{int} \tag{3.1.32}$$

and take thus the form

$$\mathbf{g}_{int} = \int_0^l \mathbf{B}^T \boldsymbol{\Sigma} ds \tag{3.1.33}$$

The stiffness matrix of the beam element results from the linearization of the internal forces according to eqn (2.3.4). It includes a material stiffness term and a geometric stiffness term

$$\mathbf{S}_e^t = \left(\frac{\partial \mathbf{g}_{int}}{\partial \mathbf{q}}\right)_e = (\mathbf{S}_{mat}^t + \mathbf{S}_\sigma^t)_e \tag{3.1.34}$$

with

$$\mathbf{S}_{mat}^t = \int_{[0,L_e]} \mathbf{B}_e^T \mathbf{C}_e \mathbf{B}_e \, ds \quad \text{and} \quad \mathbf{S}_\sigma^t = \int_{[0,L_e]} \frac{\partial \mathbf{B}_e}{\partial \mathbf{q}_e}^T \boldsymbol{\Sigma} \, ds \tag{3.1.35}$$

The inertia forces of the beam element likewise result from the discretization of

$$\delta W_{iner} = \delta \mathbf{q}^T \mathbf{g}_{iner} \tag{3.1.36}$$

They are expressed in the form

$$\mathbf{g}_{iner,e} = \mathbf{M}_e \ddot{\mathbf{q}}_e + \mathbf{h}_e(\mathbf{q}, \dot{\mathbf{q}}) \tag{3.1.37}$$

where the first term, which represents the relative inertia forces, is expressed in terms of the beam mass matrix

$$\mathbf{M}_e = \left(\frac{\partial \mathbf{g}_{iner}}{\partial \ddot{\mathbf{q}}}\right)_e \tag{3.1.38}$$

The contribution of nodes i and j takes the form

$$\mathbf{M}_{ij} = \int_0^l N_i(s) N_j(s) \begin{bmatrix} \mu \mathbf{I} & \mathbf{O} \\ \mathbf{O} & \mathbf{J} \end{bmatrix} ds \tag{3.1.39}$$

The second term of (3.1.37) represents the contribution of the centrifugal and complementary inertia forces. Its linearization according to (2.3.4) generates contributions to the tangent stiffness and damping matrices of eqn (2.3.2).

Let us finally mention that the integrals over the beam length are numerically computed by the Gauss integration rule, with only one Gauss point in order to avoid shear locking [17].

3.2. SUPERELEMENT REPRESENTATION OF COMPLEX MEMBERS [18]

Many cases occur where the deformation effects inside each body are small enough to consider that its elastic behavior remains linear in a local frame. Then, it may be said in some sense that the nonlinearities are limited to joint behavior. This fact allows the development of methods for modeling complex elastic mechanism members based on the linear expansion of the elastic displacements field in a basis of deformation modes of the body.

It is however important to stress out the limitations of the linearized approach. Situations may be identified where geometric stiffness effects are of paramount importance in at least some of the members of the system, in which case the linearized approach presented hereafter remains no longer valid and must be replaced by a fully nonlinear description of elastic body deformation.

This section describes one implementation of the component mode method for multibody analysis. In it, we seek to obtain a full independence between the vibration analysis module and the mechanism analysis one. The objective is to be able to use *any* existing linear finite element code of structural vibrations analysis to build the component mode model of the elastic member. In this way, we take advantage of the already developed capabilities of modeling complex structural members of many well established finite element programs for vibration analysis.

In the present formulation, flexible bodies are represented by a collection of fixed-boundary free vibration modes plus some "static correction" or "constraint" modes which account for local effects at the boundaries. The approach of fixed-boundary mode was chosen because it gives a perfect compatibility between bodies, a fact considered necessary to place appropriately the joints between them. The body is then linked to the rest of the system by the selected joints. The degrees of freedom of the superelement are the translations and the rotations at boundaries, plus a given number of internal mode amplitudes.

The inertia terms are computed from *a co-rotational approach* in which the consistent mass matrix provided by the linear analysis is used but the velocities interpolation is not kinematically coherent with the displacements one. this approach proved to be the best of all for the simplicity of formulation and easy interfacing of both modules. The sole information used from the vibration analysis module to build the superelement are the reduced stiffness and mass matrices.

3.2.1 Kinematics hypotheses. Let $\mathbf{x}$ be the position of an arbitrary point P of the flexible body; we write it in terms of variables in a local reference frame of the body:

$$\mathbf{x} = \mathbf{x}_0 + \mathbf{R}_0(\mathbf{X} + \mathbf{u}) \tag{3.2.1}$$

where $\mathbf{x}_0$ is the position of the local reference frame, $\mathbf{R}_0$ is the rotation of the local frame about the global one, $\mathbf{X}$ is the position of point P in the local frame and $\mathbf{u}$ is the elastic displacement of P measured in the local frame (see figure 3.2.1).

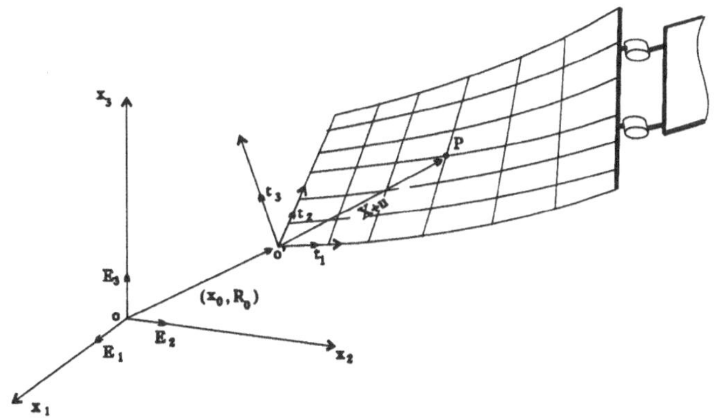

Figure 3.2.1 : Flexible body kinematics.

After time-differentiating equation (3.2.1), the virtual displacement, the velocity and the acceleration at point P result:

$$\begin{aligned} \delta\mathbf{x} &= \delta\mathbf{x}_0 + \mathbf{R}_0\delta\widetilde{\mathbf{\Theta}}_0(\mathbf{X} + \mathbf{u}) + \mathbf{R}_0\delta\mathbf{u} \\ \dot{\mathbf{x}} &= \dot{\mathbf{x}}_0 + \mathbf{R}_0\widetilde{\mathbf{\Omega}}_0(\mathbf{X} + \mathbf{u}) + \mathbf{R}_0\dot{\mathbf{u}} \\ \ddot{\mathbf{x}} &= \ddot{\mathbf{x}}_0 + \mathbf{R}_0(\widetilde{\mathbf{\Omega}}_0^2 + \widetilde{\mathbf{A}}_0)(\mathbf{X} + \mathbf{u}) + 2\mathbf{R}_0\widetilde{\mathbf{\Omega}}_0\dot{\mathbf{u}} + \mathbf{R}_0\ddot{\mathbf{u}} \end{aligned} \tag{3.2.2}$$

with $\mathbf{\Omega}_0, \mathbf{A}_0$ and $\delta\mathbf{\Theta}_0$ being the material angular velocities, accelerations and the variation of angular displacements of the local frame. Rotations can also be expressed as increments with respect to a reference value, giving

$$\mathbf{\Psi} = \mathbf{\Psi}_0 \circ \boldsymbol{\psi} \tag{3.2.3}$$

where the operation $\circ$ symbolizes the composition of rotations [2], and $\mathbf{\Psi}_0$ are the parameters of the current rotation of the local frame.

From equations (3.2.1,3.2.3), we can then compute the relative displacements and slopes inside the elastic body in terms of absolute positions and rotations:

$$\begin{bmatrix} \mathbf{u} \\ \boldsymbol{\psi} \end{bmatrix} = \begin{bmatrix} \mathbf{R}_0^T(\mathbf{x} - \mathbf{x}_0) - \mathbf{X} \\ (-\mathbf{\Psi}_0) \circ \mathbf{\Psi} \end{bmatrix} \tag{3.2.4}$$

Let us assume that the elastic displacements and slopes in the local reference frame of each body are small compared to the unity:

$$\frac{\|\mathbf{u}\|}{\|\mathbf{X}\|}\ ,\ \|\boldsymbol{\psi}\| \ll 1 \tag{3.2.5}$$

These requirements imply a geometric linearity condition in the local frame; that is to say, although the superelement as a whole undergoes finite rotations in the three-dimensional space, the displacements in a local frame remain small enough to assure the linearity of relations between local values of forces and displacements.

Let us also assume that the dynamic loading conditions are such that the local displacements and slopes can be accurately expanded in terms of a few global shape functions:

$$\begin{bmatrix} \mathbf{u} \\ \boldsymbol{\psi} \end{bmatrix} = \boldsymbol{\Phi}\mathbf{y} \tag{3.2.6}$$

Here, $\boldsymbol{\Phi}$ is the set of global shape functions and $\mathbf{y}$ are the generalized displacement amplitudes. The equations above express then a relation between local relative displacements of the body and global absolute positions, which can be conveniently used to formulate a discrete model of the body.

The relation between local and global variables can be employed to make a reduced model from a discrete one. The latter model, usually having a large number of degrees of freedom, could have been built by using, for instance, the finite element method. In this case, equations (3.2.1,3.2.3) can be rewritten at each node of the discretization in the following form:

$$\begin{bmatrix} \mathbf{x}_i \\ \boldsymbol{\Psi}_i \end{bmatrix} = \begin{bmatrix} \mathbf{x}_0 + \mathbf{R}_0(\mathbf{X}_i + \mathbf{u}_i) \\ \boldsymbol{\Psi}_0 \circ \boldsymbol{\psi}_i \end{bmatrix} = \begin{bmatrix} \mathbf{x}_0 + \mathbf{R}_0(\mathbf{X}_i + \boldsymbol{\Phi}_i\mathbf{y}) \\ \boldsymbol{\Psi}_0 \circ (\boldsymbol{\Phi}_i\mathbf{y}) \end{bmatrix} \tag{3.2.7}$$

where subindex i refers to magnitudes at node i, and index 0 denotes the same quantities computed at the origin of the reference frame.

If the Craig and Bampton component-mode method [19] is followed, the global shape functions are of two kinds:

$$\begin{bmatrix} \mathbf{u}_i \\ \boldsymbol{\psi}_i \end{bmatrix} = \boldsymbol{\Phi}_{B\,i}\mathbf{y}_B + \boldsymbol{\Phi}_{I\,i}\mathbf{y}_I \tag{3.2.8}$$

equation in which we distinguish between the boundary modes $\boldsymbol{\Phi}_B$ –obtained by the static condensation procedure– and the internal vibration modes $\boldsymbol{\Phi}_I$ –computed by fixing the boundaries. This particular choice of global shape functions permits to represent both the local deformation effects induced by the joints acting on the boundary degrees of freedom, and the global deformation effects induced by the dynamic behavior of the body itself.

The boundary generalized displacements $\mathbf{y}_B$ can be computed in terms of position and rotation values at the boundary:

$$\mathbf{y}_B = \begin{bmatrix} \mathbf{u}_B \\ \boldsymbol{\psi}_B \end{bmatrix} = \begin{bmatrix} \mathbf{R}_0^T(\mathbf{x}_B - \mathbf{x}_0) - \mathbf{X}_B \\ (-\boldsymbol{\Psi}_0) \circ (\boldsymbol{\Psi}_B) \end{bmatrix} \tag{3.2.9}$$

Then, equations (3.2.8,3.2.9) express a kinematics relation between the local relative displacements at the nodes of the discretized body, and global absolute positions and rotations. These global variables constitute the set of generalized displacements $\mathbf{q}$ of the superelement. It is formed by 6 degrees of freedom expressing position and orientation of the local frame, $3 \times (N_{B\,tras} + N_{B\,rot})$ degrees of freedom expressing positions (at $N_{B\,tras}$ nodes of the boundary) and orientations (at $N_{B\,rot}$ nodes of the boundary), and a certain number of internal modal amplitudes $\mathbf{y}_I$:

$$\mathbf{q}^T = [\mathbf{x}_0^T \quad \mathbf{\Psi}_0^T \quad \mathbf{x}_B^T \quad \mathbf{\Psi}_B^T \quad \mathbf{y}_I^T] \tag{3.2.10}$$

Positions and rotational vectors at the boundary connect the body to the rest of the multibody system.

In what follows, we treat $\mathbf{x}_B$ and $\mathbf{\Psi}_B$ as vectors with 3 components, but the reader should keep in mind that their actual dimension depends on the number of nodes (and degrees of freedom) retained at the boundary.

3.2.2 Computation of the strain energy. The energy of deformation of the body can be directly obtained by making the double discretization process on the continuum expression of the strain energy; i.e., if $\pi = \frac{1}{2}\int_V \sigma\epsilon \, dV$, the finite element method gives a first discrete equation as follows

$$\pi = \frac{1}{2}\mathbf{d}^T\mathbf{K}\mathbf{d} \tag{3.2.11}$$

with $\mathbf{d}$ the nodal displacements vector and $\mathbf{K}$ the stiffness matrix. The second discretization (expansion into the modal basis $\mathbf{d} = \mathbf{\Phi}\mathbf{y}$) gives the expression:

$$\pi = \frac{1}{2}\,\mathbf{y}^T\overline{\mathbf{K}}\mathbf{y} \tag{3.2.12}$$

where $\overline{\mathbf{S}}$ is the reduced stiffness matrix of the body:

$$\overline{\mathbf{K}} = \mathbf{\Phi}^T\mathbf{K}\mathbf{\Phi} = \begin{bmatrix} \overline{\mathbf{K}}_{BB} & \mathbf{0} \\ \mathbf{0} & \overline{\mathbf{K}}_{II} \end{bmatrix} \tag{3.2.13}$$

We note that in the latter equation we used the orthogonality relation which characterizes the constraint modes ($\mathbf{\Phi}_B^T\mathbf{K}\mathbf{\Phi}_I = \mathbf{0}$).

The variation of generalized displacements can be computed in terms of the variation of the superelement degrees of freedom $\mathbf{q}$ as follows:

$$\delta\mathbf{y} = \begin{bmatrix} \delta\mathbf{u}_B \\ \delta\boldsymbol{\psi}_B \\ \delta\mathbf{y}_I \end{bmatrix} = \begin{bmatrix} \mathbf{R}_0^T(\delta\mathbf{x}_B - \delta\mathbf{x}_0) + (\mathbf{X}_B + \mathbf{u}_B) \times \delta\mathbf{\Theta}_0 \\ \delta\mathbf{\Theta}_B - \delta\mathbf{\Theta}_0 \\ \delta\mathbf{y}_I \end{bmatrix} \tag{3.2.14}$$

By taking into account equation (3.2.4) which expresses the condition of small displacements and rotations in the local frame, the variation of displacements can be simplified to give:

$$\delta\mathbf{y} \simeq \begin{bmatrix} \mathbf{R}_0^T(\delta\mathbf{x}_B - \delta\mathbf{x}_0) + \widetilde{\mathbf{X}}_B\delta\mathbf{\Theta}_0 \\ \delta\mathbf{\Theta}_B - \delta\mathbf{\Theta}_0 \\ \delta\mathbf{y}_I \end{bmatrix} = \mathbf{Y}\delta\mathbf{q} \tag{3.2.15}$$

with the definition of the configuration dependent matrix

$$\mathbf{Y} = \begin{bmatrix} -\mathbf{R}_0^T & \widetilde{\mathbf{X}}_B & \mathbf{R}_0^T & \mathbf{0} & \mathbf{0} \\ \mathbf{0} & -\mathbf{I} & \mathbf{0} & \mathbf{I} & \mathbf{0} \\ \mathbf{0} & \mathbf{0} & \mathbf{0} & \mathbf{0} & \mathbf{I} \end{bmatrix} \tag{3.2.16}$$

and the vector of generalized coordinates

$$\delta\mathbf{q}^T = [\,\delta\mathbf{x}_0^T \;\; \delta\mathbf{\Theta}_0^T \;\; \delta\mathbf{x}_B^T \;\; \delta\mathbf{\Theta}_B^T \;\; \delta\mathbf{y}_I^T\,] \tag{3.2.17}$$

where $\delta\mathbf{\Theta}_0, \delta\mathbf{\Theta}_B$ are respectively the angular displacements variations at the reference frame and at the boundary nodes.

The internal forces vector of the superelement is then calculated as follows:

$$\delta\pi = \delta\mathbf{q}^T\,\mathbf{Y}^T\overline{\mathbf{K}}\mathbf{y} = \delta\mathbf{q}^T\,\mathbf{g}_{int} \tag{3.2.18}$$

By differentiating the internal forces and by neglecting the derivatives of $\mathbf{Y}$, we arrive at the expression of the stiffness matrix of the superelement:

$$\left[\frac{\partial\mathbf{g}_{int}}{\partial\mathbf{q}}\right]\Delta\mathbf{q} \simeq \mathbf{Y}^T\overline{\mathbf{K}}\mathbf{Y}\Delta\mathbf{q} = \overline{\mathbf{S}}^t\Delta\mathbf{q} \tag{3.2.19}$$

with

$$\overline{\mathbf{S}}^t = \mathbf{Y}^T\overline{\mathbf{K}}\mathbf{Y} \tag{3.2.20}$$

3.2.3 Co-rotational evaluation of the kinetic energy. The most convenient way to evaluate the kinetic energy of the superelement in a co-rotational manner

$$\mathcal{W} = \frac{1}{2}\int_V \dot{\mathbf{x}}\cdot\dot{\mathbf{x}}\,\rho\,dV = \frac{1}{2}\int_V (\mathbf{R}_0^T\dot{\mathbf{x}})\cdot(\mathbf{R}_0^T\dot{\mathbf{x}})\,\rho\,dV \tag{3.2.21}$$

Here, $\mathbf{R}_0$ gives the rotation of the reference frame attached to the elastic body at node 0.

Let us denote the co-rotational velocities by

$$\mathbf{v}(\mathbf{X}) = \mathbf{R}_0^T\dot{\mathbf{x}} \tag{3.2.22}$$

and interpolate them in terms of nodal velocities in the form

$$\mathbf{v}(\mathbf{X}) = \sum_{i=1}^{n} N_i(\mathbf{X})\ \mathbf{v}_i \tag{3.2.23}$$

where the summation extends to all nodes of the flexible member. Note that this interpolation is not consistent with the displacement interpolation used to build the strain energy expression and also that, for finite element models using rotational degrees of freedom, the same interpolation has to be made on the material angular velocities.

After performing the volume integral, the kinetic energy of the superelement can be written:

$$\mathcal{W} = \frac{1}{2}\sum_i\sum_j [\mathbf{v}_i^T \quad \boldsymbol{\Omega}_i^T] \int_V \mathbf{N}_i^T\mathbf{N}_j\, \rho\, dV \begin{bmatrix} \mathbf{v}_i \\ \boldsymbol{\Omega}_i \end{bmatrix} = \frac{1}{2}\sum_i\sum_j [\mathbf{v}_i^T \quad \boldsymbol{\Omega}_i^T]\ \mathbf{M}_{ij} \begin{bmatrix} \mathbf{v}_i \\ \boldsymbol{\Omega}_i \end{bmatrix} \tag{3.2.24}$$

where $\mathbf{M}_{ij}$ denotes the block of the mass matrix coupling nodes i and j.

The next step is to form a reduced model by following the component modes approach. A second stage discretization is made by assuming that the material velocities can be expressed in terms of a few global shape functions

$$\begin{bmatrix} \mathbf{v}_i \\ \boldsymbol{\Omega}_i \end{bmatrix} = \boldsymbol{\Phi}_i\dot{\mathbf{y}} = \Big[[\boldsymbol{\Phi}_{0\,u} \quad \boldsymbol{\Phi}_{0\,\Psi}]_i \quad [\boldsymbol{\Phi}_{B\,u} \quad \boldsymbol{\Phi}_{B\,\Psi}]_i \quad \boldsymbol{\Phi}_{I\,i} \Big] \begin{bmatrix} \mathbf{v}_0 \\ \boldsymbol{\Omega}_0 \\ \mathbf{v}_B \\ \boldsymbol{\Omega}_B \\ \dot{\mathbf{y}}_I \end{bmatrix} \tag{3.2.25}$$

where the part $[\mathbf{v}_0^T \quad \boldsymbol{\Omega}_0^T]$ corresponds to (material) velocities at the reference frame, $\dot{\mathbf{y}}_I$ are the time derivatives of the internal mode amplitudes, and the contribution of (material) velocities at the boundary nodes of the superelement can be written in the form:

$$[\mathbf{v}_B^T \quad \boldsymbol{\Omega}_B^T] = \Big[\Big((\mathbf{v}_1^T\,\boldsymbol{\Omega}_1^T)\ \ldots\ (\mathbf{v}_k^T\,\boldsymbol{\Omega}_k^T)\Big)\ \Big(\mathbf{v}_{k+1}^T\ \cdots\ \mathbf{v}_{k+\ell}^T\Big)\ \Big(\boldsymbol{\Omega}_{k+\ell+1}^T\ \cdots\ \boldsymbol{\Omega}_{k+\ell+m}^T\Big)\Big] \tag{3.2.26}$$

Note that at nodes $k+1, \ldots k+\ell$ of the boundary, only the translation degrees of freedom have been retained to form the superelement, while at nodes $k+\ell+1, \ldots k+\ell+m$ only the rotation terms are conserved. The sole restriction is that the three components of translation and/or rotation (if any) should be incorporated to the model. The translation material velocities are computed by projection over the reference frame, while material velocities are the true material velocities at the considered node.

Even when the flexible body suffers large rotations, the material velocities pattern does not change; then, this second stage discretization continues to be valid and we can still apply the same modal expansion. The total contribution of the superelement to the kinetic energy of the structure $\mathcal{W}$ can then be written as follows:

$$\mathcal{W} = \frac{1}{2}\,\dot{\mathbf{y}}^T\overline{\mathbf{M}}\dot{\mathbf{y}} = \frac{1}{2}\,[\mathbf{v}_0^T \quad \boldsymbol{\Omega}_0^T \quad \mathbf{v}_B^T \quad \boldsymbol{\Omega}_B^T \quad \dot{\mathbf{y}}_I^T] \begin{bmatrix} \overline{\mathbf{M}}^{(00)} & \overline{\mathbf{M}}^{(0B)} & \overline{\mathbf{M}}^{(0y)} \\ \overline{\mathbf{M}}^{(B0)} & \overline{\mathbf{M}}^{(BB)} & \overline{\mathbf{M}}^{(By)} \\ \overline{\mathbf{M}}^{(y0)} & \overline{\mathbf{M}}^{(yB)} & \mathbf{I} \end{bmatrix} \begin{bmatrix} \mathbf{v}_0 \\ \boldsymbol{\Omega}_0 \\ \mathbf{v}_B \\ \boldsymbol{\Omega}_B \\ \dot{\mathbf{y}}_I \end{bmatrix} \tag{3.2.27}$$

The constant mass matrix $\overline{\mathbf{M}}$ in (3.2.27) results from the projection of the element mass matrices over the modal basis

$$\overline{\mathbf{M}} = \sum_i \sum_j \boldsymbol{\Phi}_i^T \mathbf{M}_{ij} \boldsymbol{\Phi}_j \tag{3.2.28}$$

Inertia forces are computed by differentiating the kinetic energy. Its first variation is :

$$\delta\mathcal{W} = \delta\dot{\mathbf{y}}^T\, \overline{\mathbf{M}}\, \dot{\mathbf{y}} \tag{3.2.29}$$

The vector of variations of generalized velocities reads :

$$\delta\dot{\mathbf{y}}^T = [\,\delta\mathbf{v}_0^T \quad \delta\boldsymbol{\Omega}_0^T \quad \delta\mathbf{v}_B^T \quad \delta\boldsymbol{\Omega}_B^T \quad \delta\dot{\mathbf{y}}_I^T\,] \tag{3.2.30}$$

with the variations of material and angular velocities at nodes 0 and B computed from

$$\delta\mathbf{v} \;=\; \mathbf{R}_0^T\delta\dot{\mathbf{x}} - \delta\boldsymbol{\Theta}_0 \times (\mathbf{R}_0^T\dot{\mathbf{x}}) \qquad \text{and} \qquad \delta\boldsymbol{\Omega} \;=\; \delta\dot{\boldsymbol{\Theta}} + \boldsymbol{\Omega}_0 \times \delta\boldsymbol{\Theta} \tag{3.2.31}$$

By introducing the latter expressions into (3.2.29) and by integrating by parts, we get

$$\begin{aligned} \delta\mathcal{W} &= -\delta\mathbf{q}^T\ \mathbf{G}_{iner} \\ &= -\delta\mathbf{q}^T\ \left(\mathbf{P}\,\overline{\mathbf{M}}\,\mathbf{P}^T\,\ddot{\mathbf{q}} \;-\; \mathbf{P}\left(\overline{\mathbf{M}}\mathbf{W} \;+\; \mathbf{W}^T\overline{\mathbf{M}} \;+\; \mathbf{U}^T\overline{\mathbf{M}}\right)\mathbf{P}^T\,\dot{\mathbf{q}}\right) \end{aligned} \tag{3.2.32}$$

where variations of the generalized displacements at the global frame are given by (3.2.17) and where

$$\dot{\mathbf{q}}^T \;=\; [\dot{\mathbf{x}}_0^T \quad \boldsymbol{\Omega}_0^T \quad \dot{\mathbf{x}}_B^T \quad \boldsymbol{\Omega}_B^T \quad \dot{\mathbf{y}}_I^T] \tag{3.2.33}$$

$$\mathbf{P} = \begin{bmatrix} \mathbf{R}_0 & & & & \\ & \mathbf{I} & & & \\ & & \mathbf{R}_0 & & \\ & & & \mathbf{I} & \\ & & & & \mathbf{I} \end{bmatrix} \qquad \mathbf{W} = \begin{bmatrix} \widetilde{\boldsymbol{\Omega}}_0 & & & & \\ & 0 & & & \\ & & \widetilde{\boldsymbol{\Omega}}_0 & & \\ & & & 0 & \\ & & & & 0 \end{bmatrix} \qquad \mathbf{U} = \begin{bmatrix} 0 & \widetilde{\mathbf{v}}_0 & 0 & & \\ 0 & \widetilde{\boldsymbol{\Omega}}_0 & 0 & & \\ 0 & \widetilde{\mathbf{v}}_B & 0 & & \\ & & & \widetilde{\boldsymbol{\Omega}}_B & \\ & & & & 0 \end{bmatrix} \tag{3.2.34}$$

Then, by differentiating the inertia forces with respect to the generalized accelerations in the global frame $\ddot{\mathbf{q}}$ we get the superelement tangent mass matrix $\overline{\mathbf{M}}_{sup}$:

$$\overline{\mathbf{M}}_{sup} = \mathbf{P}\,\overline{\mathbf{M}}\,\mathbf{P}^T \tag{3.2.35}$$

The inertia forces also depend on the velocities $\dot{\mathbf{q}}$. In order to get full quadratic convergence rate, it will be necessary in some cases to compute the gyroscopic matrix of derivatives of the inertia forces with respect to velocities. This is a non symmetric matrix, which proved to be of value for improving convergence in several examples.

$$\overline{\mathbf{C}}_{sup} = \mathbf{P}\Big(\underbrace{\big[\overline{\mathbf{M}}\mathbf{U} - \mathbf{U}^T\overline{\mathbf{M}} - \mathbf{V}\big]}_{\text{antisym.}} - \underbrace{\big[\overline{\mathbf{M}}\mathbf{W} + \mathbf{W}^T\overline{\mathbf{M}}\big]}_{\text{sym.}}\Big)\mathbf{P}^T \tag{3.2.36}$$

The antisymmetric matrix $\mathbf{V}$ is defined by

$$\mathbf{V} = \begin{bmatrix} \mathbf{0} & \tilde{\mathbf{g}}_{u\,0} & & & \\ \tilde{\mathbf{g}}_{u\,0} & \tilde{\mathbf{g}}_{\Psi\,0} & \tilde{\mathbf{g}}_{u\,B} & & \\ & \tilde{\mathbf{g}}_{u\,B} & \mathbf{0} & & \\ & & & \tilde{\mathbf{g}}_{\Psi\,B} & \\ & & & & \mathbf{0} \end{bmatrix} \tag{3.2.37}$$

where vectors $\mathbf{g}_{u\,i}, \mathbf{g}_{\Psi\,i}$ are computed as follows:

$$[\mathbf{g}_{u\,0}^T \quad \mathbf{g}_{\Psi\,0}^T \quad \mathbf{g}_{u\,B}^T \quad \mathbf{g}_{\Psi\,B}^T \quad \mathbf{g}_I^T] = \dot{\mathbf{q}}^T\mathbf{P}\overline{\mathbf{M}} \tag{3.2.38}$$

We note that the pseudo-damping matrix $\overline{\mathbf{C}}_{sup}$ is formed by adding up two terms: a symmetric and an antisymmetric matrix which are clearly identified in equation (3.2.26). We finally remark that in this formulation, all contributions to the inertia terms (inertia forces $\mathbf{G}_{iner}$, mass matrix $\overline{\mathbf{M}}_{sup}$ and pseudo-damping matrix $\overline{\mathbf{C}}_{sup}$) are evaluated directly from the reduced mass matrix $\overline{\mathbf{M}}$, the projection over the modal basis of the finite element mass matrix. In this way, we can very easily interface the vibration analysis and the mechanism analysis modules.

4. Finite Element Representation of Kinematic Joints

4.1. GENERAL FORMULATION OF CONSTRAINTS [5]

The equations of motion of a dynamic system subjected to holonomic constraints have already been stated in the augmented lagrangian form (2.2.9). The role of the penalty term is essentially to add some positive curvature in the range space of $\left[\frac{\partial \mathbf{\Phi}}{\partial \mathbf{q}}\right]$ with a significant improvement of convergence. Besides, this term assures the positive definiteness of the displacements-associated submatrix of the Hessian matrix, so the only safeguard to be implemented against the appearance of null pivots during factorization is that terms associated to the Lagrange multipliers should be condensed after the degrees of freedom participating in the constraint equation.

The extension to systems with non-holonomic constraints is straightforward. The equations of motion in augmented lagrangian form now have for expression

$$\begin{cases} \mathbf{M}\ddot{\mathbf{q}} + \mathbf{Q}_h + \mathbf{Q}_{nh} = \mathbf{g}(\mathbf{q}, \dot{\mathbf{q}}, t) \\ k\mathbf{\Phi}_h(\mathbf{q}, t) = 0 \\ k\mathbf{\Phi}_{nh}(\mathbf{q}, \dot{\mathbf{q}}, t) = 0 \end{cases} \tag{4.1.1}$$

where $\mathbf{Q}_h$ and $\mathbf{Q}_{nh}$ denote respectively the constraint forces arising from holonomic and non-holonomic constraints

$$\begin{cases} \mathbf{Q}_h = \mathbf{B}_h^T(k\boldsymbol{\lambda} + p\mathbf{\Phi}_h) \\ \mathbf{Q}_{nh} = \mathbf{B}_{nh}^T(k\dot{\boldsymbol{\lambda}} + p\mathbf{\Phi}_{nh}) \end{cases} \tag{4.1.2}$$

with the jacobian matrices of holonomic and non-holonomic constraints

$$\mathbf{B}_h = \left[\frac{\partial \mathbf{\Phi}_h}{\partial \mathbf{q}}\right] \qquad \mathbf{B}_{nh} = \left[\frac{\partial \mathbf{\Phi}_{nh}}{\partial \dot{\mathbf{q}}}\right] \tag{4.1.3}$$

It is further assumed that the non-holonomic constraints are linear in the velocities, in which case they simplify into the form

$$\mathbf{\Phi}_{nh} = \mathbf{B}_{nh}(\mathbf{q})\dot{\mathbf{q}} + \mathbf{b}_{nh} \tag{4.1.4}$$

It is worthwhile noticing that the non holonomic constraints can also be seen as derived from a "pseudo-dissipation function" $\mathcal{D}$

$$\mathcal{D} = \frac{1}{2}p\|\mathbf{\Phi}_{nh}\|^2 + k\dot{\boldsymbol{\lambda}} \cdot \mathbf{\Phi}_{nh} \tag{4.1.5}$$

which generates the "dissipation forces":

$$\frac{\partial \mathcal{D}}{\partial \dot{\mathbf{q}}} = \mathbf{B}_{nh}^T(k\dot{\boldsymbol{\lambda}} + p\mathbf{\Phi}_{nh}) \qquad \frac{\partial \mathcal{D}}{\partial \dot{\boldsymbol{\lambda}}} = k\ \mathbf{\Phi}_{nh} \tag{4.1.6}$$

The non-holonomic constraints contribute then to the linearized equations of motion in the form

$$\begin{cases} \mathbf{M}\Delta\ddot{\mathbf{q}} + \mathbf{C}^t\Delta\dot{\mathbf{q}} + \mathbf{S}^t\Delta\mathbf{q} + k\mathbf{B}_h^T\Delta\boldsymbol{\lambda} + k\mathbf{B}_{nh}^T\Delta\dot{\boldsymbol{\lambda}} = \mathbf{r}^* + O(\Delta^2) \\ k\mathbf{B}_h\Delta\mathbf{q} = -k\mathbf{\Phi}_h^* + O(\Delta^2) \\ k\mathbf{B}_{nh}\Delta\dot{\mathbf{q}} = -k\mathbf{\Phi}_h^* + O(\Delta^2) \end{cases} \tag{4.1.7}$$

where $\mathbf{r}$ is the residual vector of dynamic equilibrium

$$\mathbf{r} = \mathbf{g}(\mathbf{q}, \dot{\mathbf{q}}, t) - \mathbf{M}\ddot{\mathbf{q}} - \mathbf{B}_h^T(k\boldsymbol{\lambda} + p\mathbf{\Phi}_h) - \mathbf{B}_{nh}^T(k\dot{\boldsymbol{\lambda}} + p\mathbf{\Phi}_{nh}) \tag{4.1.8}$$

and where the tangent stiffness and damping matrices $\mathbf{S}^t$ and $\mathbf{C}^t$ are computed from

$$\mathbf{S}^t = \frac{\partial}{\partial \mathbf{q}}\left[-\mathbf{g} + \mathbf{B}_h^T\ (k\boldsymbol{\lambda} + p\mathbf{\Phi}_h) + \mathbf{B}_{nh}^T\left(k\dot{\boldsymbol{\lambda}} + p\mathbf{\Phi}_{nh}\right)\right] \qquad \mathbf{C}^t = -\frac{\partial \mathbf{g}}{\partial \dot{\mathbf{q}}} + p\mathbf{B}_{nh}^T\mathbf{B}_{nh} \tag{4.1.9}$$

For sake of computational simplicity, the second-order derivative terms (such as the contribution of the non-holonomic constraints) may be omitted from the expression of the tangent matrix when when computing a nonlinear response through Newton-Raphson iteration. All terms have however to be evaluated to perform a linearized stability analysis.

4.2. SYMBOLIC GENERATION OF HINGE JOINT ELEMENT

As an example, let us consider the automatic generation of the vector and matrix quantities involved in the formulation of a hinge joint element. Let $\{\boldsymbol{\mu}_1, \boldsymbol{\mu}_2, \boldsymbol{\mu}_3\}$ and $\{\boldsymbol{\xi}_1, \boldsymbol{\xi}_2, \boldsymbol{\xi}_3\}$ be two triads of orthogonal unit vectors atached to nodes A and B respectively at the initial configuration. We suppose that we have already computed the configuration of the system at time t (reference configuration) and we want to compute the new situation at time $t + \Delta t$ (actual configuration). Let $\{\boldsymbol{\mu}'_1, \boldsymbol{\mu}'_2, \boldsymbol{\mu}'_3\}$ and $\{\boldsymbol{\xi}'_1, \boldsymbol{\xi}'_2, \boldsymbol{\xi}'_3\}$ be the triads obtained by mapping the initial ones into the reference configuration via the rotation operators at each node:

$$\boldsymbol{\mu}'_i = \mathbf{R}_{A\ ref}\ \boldsymbol{\mu}_i \tag{4.2.1}$$

$$\boldsymbol{\xi}'_i = \mathbf{R}_{B\ ref}\ \boldsymbol{\xi}_i \tag{4.2.2}$$

Finally, let $(\boldsymbol{\mu}''_1, \boldsymbol{\mu}''_2, \boldsymbol{\mu}''_2)$ and $(\boldsymbol{\xi}''_1, \boldsymbol{\xi}''_2, \boldsymbol{\xi}''_3)$ be the triads mapped into the actual configuration:

$$\boldsymbol{\mu}''_i = \mathbf{R}_{A\ ref}\mathbf{R}_{A\ inc}\ \boldsymbol{\mu}_i = \mathbf{R}_A \boldsymbol{\mu}_i \tag{4.2.3}$$

$$\boldsymbol{\xi}''_i = \mathbf{R}_{B\ ref}\mathbf{R}_{B\ inc}\ \boldsymbol{\xi}_i = \mathbf{R}_B \boldsymbol{\xi}_i \tag{4.2.4}$$

The rotation operators $\mathbf{R}_A$,$\mathbf{R}_{A\ ref}$,$\mathbf{R}_{A\ inc}$ give respectively the actual, reference and incremental rotations at node A.

The hinge joint is modeled by introducing six constraints: three imposing the equality of positions at the nodes, two fixing the rotations about two directions (figure 4.2.1). The last constraint introduces explicitly the joint angle α, and allows thus to apply a driving moment at the hinge.

$$sin(\theta - \alpha) = 0 \tag{4.2.5}$$

with

$$sin\theta = \boldsymbol{\mu}''_2.\boldsymbol{\xi}''_1 \quad , \quad cos\theta = \boldsymbol{\mu}''_1.\boldsymbol{\xi}''_1 \tag{4.2.6}$$

$$x_A = x_B$$
$$\mu_1'' \cdot \xi_3'' = 0$$
$$\mu_2'' \cdot \xi_3'' = 0$$

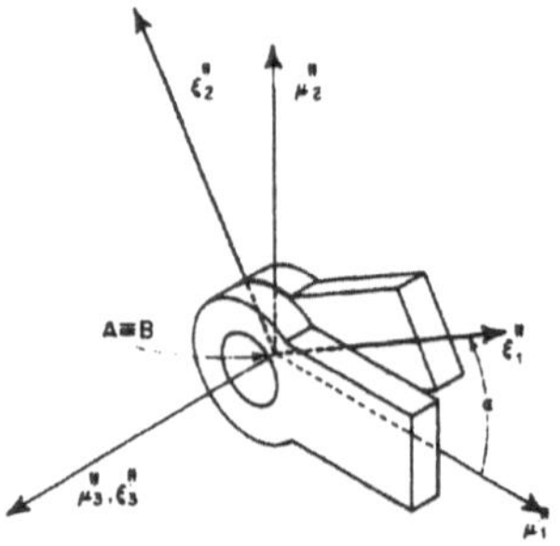

Figure 4.2.1 : Geometry of the hinge joint.

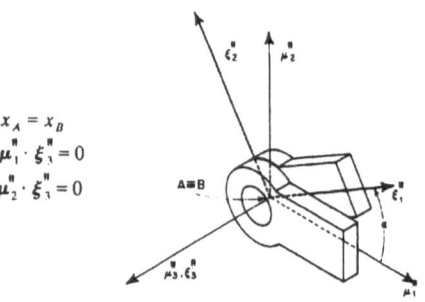

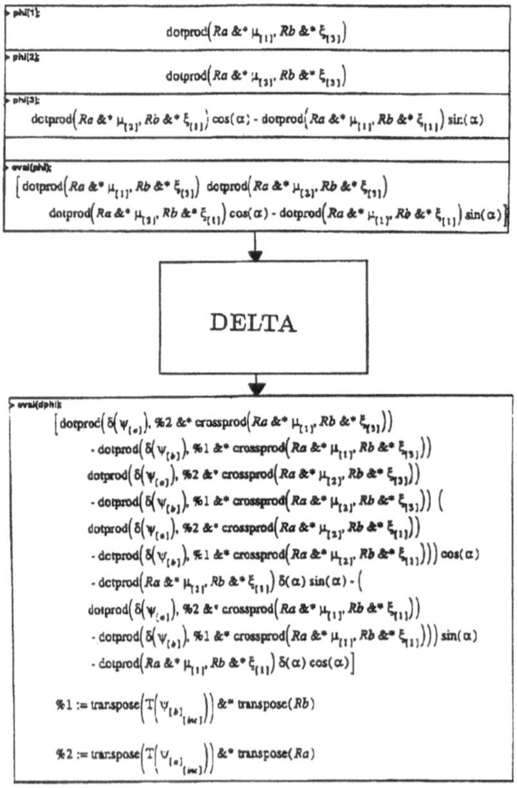

Figure 4.2.2 : The hinge joint : automatic differentiation of the kinematic constraints

The equality of positions is imposed by Boolean identifications of the corresponding degrees of freedom. The last three constraints are treated by the augmented Lagrangian procedure.

Starting from the symbolic vector expression of the last three constraints, the procedure DELTA (cf. figure 4.2.2) automatically computes the vector form of the constraint gradients. Afterwards, the procedure JOINT computes the symbolic expressions of matrix $\mathbf{B}$ and of the penalty term contribution to the Hessian matrix, both matrices being expressed in terms of the elements of matrices $\mathbf{R}$ and $\mathbf{T}$. The knowledge of the symbolic expression of $\mathbf{R}$ and $\mathbf{T}$ in terms of the rotation parameters (cf. section 2.4.1) permits to express matrix $\mathbf{B}$ in terms of these parameters, and through further differentiation the symbolic expression of the Hessian matrix term involving the second derivatives of the constraints is finally obtained.

4.3. FINITE ELEMENT DESCRIPTION OF A FLEXIBLE SLIDER [20]

Most applications in multibody dynamics literature concern relative motions between bodies linked together through rigid joints as hinges, cylindrical, prismatic joints, etc. Flexibility in joints have been sometimes considered, mostly for hinges. Flexible tracks can be seen as a special type of joint with applications such as modeling of erectable booms in space, dynamic simulation of landing gears and analysis of vehicle / flexible guideway interaction for ground transportation.

In this section we present a model of flexible straight slider joint. It can be described as a straight Bernoulli beam in a corotational frame over which slides a third node. A single track can be modeled by a series of elements aligned one after the other. The sliding node can pass from one joint to the next one that represents the track, thus allowing to refine the mesh up to achieving convergence. Dynamic friction effects between the sliding node and the track are also taken into account.

4.3.1 Kinematic Equations. Let us consider a straight flexible slider whose deformation is parameterized in terms of the position $\mathbf{x}_A, \mathbf{x}_B$ of its two extreme nodes A, B and of the orientation of two orthogonal triads $\mathbf{R}_A, \mathbf{R}_B$ attached to them. The element has a third sliding node C which freely moves along a rectilinear trajectory oriented parallel to the principal axis of the beam. The trajectory can be excentric from the beam axis. We will consider that, in a general case, the sliding node C has a second freedom: it can also move along a rectilinear trajectory $\mathbf{t}_S$ contained into the normal cross section of the beam (see figure 4.3.1).

Under these hypotheses, we can express the position of an arbitrary point on the sliding

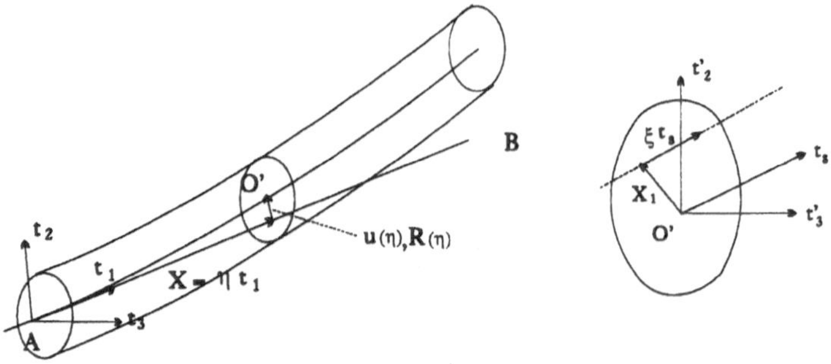

Figure 4.3.1 : Geometry of the slider element.

surface as:

$$\mathbf{x}(\eta, \xi) = \mathbf{x}_A + \mathbf{R}_A\left(\mathbf{X}(\eta) + \mathbf{u}(\eta) + \mathbf{R}(\eta)\left(\mathbf{X}_1 + \xi \mathbf{t}_S\right)\right) \tag{4.3.1}$$

where η, ξ are the coordinates along the principal axis $\mathbf{t}_1$ and along the secondary axis $\mathbf{t}_S$, $\mathbf{x}_A, \mathbf{R}_A$ are the position and rotation at node A, $\mathbf{X}(\eta) = \eta \mathbf{t}_1$ is the reference position of the cross section centroid in the reference frame, $\mathbf{u}(\eta)$ is the displacement of the centroid, $\mathbf{R}(\eta)$ is the rotation of the cross section with respect to the reference frame $\mathbf{R}_A$, and $\mathbf{X}_1$ gives the position of an arbitrary point of the secondary trajectory in the cross section. The reference frame $\mathbf{t}_i$ is oriented along the principal axes of the beam, but the secondary axis $\mathbf{t}_S$ does not necessarily coincide with the principal axis $\mathbf{t}_2$.

The cross section centroid position in the reference frame is

$$\mathbf{s}(\eta) = \mathbf{X}(\eta) + \mathbf{u}(\eta) \tag{4.3.2}$$

Then, the position of node B results

$$\mathbf{x}_B = \mathbf{x}_A + \mathbf{R}_A \mathbf{s}(L) \tag{4.3.3}$$

where L is the beam length.

We will assume that the beam behaves like a Bernoulli beam in the local reference frame. Then, the position and the angular displacement of the cross section at an arbitrary point of distance η from the origin of the reference frame is given by

$$\begin{bmatrix} \mathbf{s}(\eta) \\ \boldsymbol{\psi}(\eta) \end{bmatrix} = \mathbf{N}(\eta) \begin{bmatrix} \mathbf{s}_B \\ \boldsymbol{\psi}_B \end{bmatrix} = \mathbf{N}(\eta) \begin{bmatrix} \mathbf{R}_A^T(\mathbf{x}_B - \mathbf{x}_A) \\ (-\boldsymbol{\Psi}_A) \circ \boldsymbol{\Psi}_B \end{bmatrix} \tag{4.3.4}$$

where

$$\mathbf{N}(\eta)=\begin{bmatrix} \frac{\eta}{L} & 0 & 0 & 0 & 0 & 0 \\ 0 & (\frac{3\eta^2}{L^2}-\frac{2\eta^3}{L^3}) & 0 & 0 & 0 & (-\frac{\eta^2}{L}+\frac{\eta^3}{L^2}) \\ 0 & 0 & (\frac{3\eta^2}{L^2}-\frac{2\eta^3}{L^3}) & 0 & (\frac{\eta^2}{L}-\frac{\eta^3}{L^2}) & 0 \\ 0 & 0 & 0 & \frac{\eta}{L} & 0 & 0 \\ 0 & 0 & (-\frac{6\eta}{L^2}+\frac{6\eta^2}{L^3}) & 0 & (-\frac{2\eta}{L}+\frac{3\eta^2}{L^2}) & 0 \\ 0 & (\frac{6\eta}{L^2}-\frac{6\eta^2}{L^3}) & 0 & 0 & 0 & (-\frac{2\eta}{L}+\frac{3\eta^2}{L^2}) \end{bmatrix} \tag{4.3.5}$$

is the matrix of Hermite interpolation functions, $\mathbf{\Psi}_A$, $\mathbf{\Psi}_B$ are the rotational vectors at nodes A, B and $\circ$ symbolizes the composition of rotations.

Variations of positions and angular displacements in the local frame
From equation (4.3.4), we obtain

$$\begin{bmatrix} \delta\mathbf{s}(\eta) \\ \delta\boldsymbol{\psi}(\eta) \end{bmatrix} = \mathbf{N}(\eta)\begin{bmatrix} \delta\mathbf{s}_B \\ \delta\boldsymbol{\psi}_B \end{bmatrix} + \delta\eta \begin{bmatrix} \mathbf{s}'_B \\ \boldsymbol{\psi}'_B \end{bmatrix} \tag{4.3.6}$$

where

$$\begin{bmatrix} \mathbf{s}'_B \\ \boldsymbol{\psi}'_B \end{bmatrix} = \mathbf{N}'(\eta)\begin{bmatrix} \mathbf{s}_B \\ \boldsymbol{\psi}_B \end{bmatrix}$$

and where $\mathbf{N}'(\eta) = \frac{d\mathbf{N}}{d\eta}$ is the matrix of derivatives of the interpolation functions. We remark that we take variations with respect to η since this parameter is not fixed but expresses the degree of freedom of the sliding node along the principal axis of the beam.

By assuming that the displacements and rotations of the cross section in the reference frame are small, we can express the variation of positions and rotations at node B as follows

$$\begin{bmatrix} \delta\mathbf{s}_B \\ \delta\boldsymbol{\psi}_B \end{bmatrix} \simeq \begin{bmatrix} \mathbf{R}_A^T(\delta\mathbf{x}_B - \delta\mathbf{x}_A) + \widetilde{\mathbf{X}}_B\delta\mathbf{\Theta}_A \\ \delta\mathbf{\Theta}_B - \delta\mathbf{\Theta}_A \end{bmatrix} = \mathbf{B}\ \delta\mathbf{q}_1 \tag{4.3.7}$$

with

$$\mathbf{q}_1^T = [\mathbf{x}_A^T \quad \mathbf{\Theta}_A^T \quad \mathbf{x}_B^T \quad \mathbf{\Theta}_B^T] \qquad \text{and} \qquad \mathbf{B} = \begin{bmatrix} -\mathbf{R}_A^T & \widetilde{\mathbf{X}}_B & \mathbf{R}_A^T & \mathbf{0} \\ \mathbf{0} & -\mathbf{I} & \mathbf{0} & \mathbf{I} \end{bmatrix} \tag{4.3.8}$$

After replacing in equation (4.3.6), we obtain the variation of positions and angular displacements

$$\begin{bmatrix} \delta\mathbf{s}(\eta) \\ \delta\boldsymbol{\psi}(\eta) \end{bmatrix} = \mathbf{N}(\eta)\mathbf{B}\delta\mathbf{q}_1 + \begin{bmatrix} \mathbf{s}'_B \\ \boldsymbol{\psi}'_B \end{bmatrix}\delta\eta \tag{4.3.9}$$

4.3.2 Strain Energy. The internal strain energy of the element can be written in the form

$$U = \frac{1}{2}[\mathbf{s}_B \quad \boldsymbol{\psi}_B]\mathbf{K}^{(BB)}\begin{bmatrix} \mathbf{s}_B \\ \boldsymbol{\psi}_B \end{bmatrix} \tag{4.3.10}$$

where $\mathbf{K}^{(BB)}$ is the submatrix corresponding to node B of the standard stiffness matrix of a Bernoulli beam element.

The internal forces vector $\mathbf{g}_{int}$ is computed by differentiating the strain energy with respect to the nodal displacements

$$\delta U = \delta \mathbf{q}_1^T \, \mathbf{g}_{int} = \delta \mathbf{q}_1^T \mathbf{B}^T \mathbf{K}^{(BB)} \begin{bmatrix} \mathbf{s}_B \\ \boldsymbol{\psi}_B \end{bmatrix} \tag{4.3.11}$$

By further differentiation, and after neglecting the derivatives of $\mathbf{B}$, we obtain the tangent stiffness matrix

$$\mathbf{S}^t = \mathbf{B}^T \mathbf{K}^{(BB)} \mathbf{B} \tag{4.3.12}$$

Neglecting the derivatives of $\mathbf{B}$ is entirely compatible with the assumption of small displacements and rotations in the local frame.

4.3.3 Constraint Equations. Constraints are added to the system to impose node C to be permanently in contact with the sliding surface:

$$\mathbf{v} = \mathbf{x}_C - \mathbf{x}_A - \mathbf{R}_A \mathbf{s}(\eta) - \mathbf{R}_A \left(\mathbf{I} + \widetilde{\boldsymbol{\psi}}(\eta)\right)(\mathbf{X}_1 + \xi \mathbf{t}_S) = \mathbf{0} \tag{4.3.13}$$

In order to verify the sliding condition (4.3.13), the projection of the vector $\mathbf{v}$ along the principal axes of the beam is assumed to be zero:

$$\boldsymbol{\phi}_i = \mathbf{v}^T \mathbf{R}_A \mathbf{t}_i = \left(\mathbf{R}_A^T(\mathbf{x}_C - \mathbf{x}_A) - \mathbf{s} - (\mathbf{I} + \widetilde{\boldsymbol{\psi}})(\mathbf{X}_1 + \xi \mathbf{t}_S)\right)^T \mathbf{t}_i = 0 \tag{4.3.14}$$

Variations of the constraints $\boldsymbol{\phi}_i$ are then given by

$$\begin{aligned} \delta\boldsymbol{\phi}_i = {} & \delta \mathbf{x}_C^T \mathbf{R}_A \mathbf{t}_i - \delta \mathbf{x}_A^T \mathbf{R}_A \mathbf{t}_i + \delta \boldsymbol{\Theta}_A^T \left(\mathbf{t}_i \times \mathbf{R}_A^T(\mathbf{x}_C - \mathbf{x}_A)\right) \\ & - \underline{\left(\delta \mathbf{s}(\eta)^T \mathbf{t}_i + \delta \boldsymbol{\psi}(\eta)^T \left((\mathbf{X}_1 + \xi \mathbf{t}_S) \times \mathbf{t}_i\right)\right)} - \delta\xi \mathbf{t}_i^T (\mathbf{I} + \widetilde{\boldsymbol{\psi}}) \mathbf{t}_S \end{aligned} \tag{4.3.15}$$

After replacing the expressions for the variations of position and orientation (4.3.9), the underlined term on the right-hand-side can be written as

$$\delta \mathbf{s}(\eta)^T \mathbf{t}_i + \delta \boldsymbol{\psi}(\eta)^T (\mathbf{X}_1 + \xi \mathbf{t}_S) \times \mathbf{t}_i = \left(\mathbf{N}(\eta) \mathbf{B} \delta \mathbf{q}_1 + \delta\xi \begin{bmatrix} \mathbf{s}'_B \\ \boldsymbol{\psi}'_B \end{bmatrix}\right)^T \begin{bmatrix} \mathbf{t}_i \\ (\mathbf{X}_1 + \xi \mathbf{t}_S) \times \mathbf{t}_i \end{bmatrix} \tag{4.3.16}$$

By finally replacing the expression of $\mathbf{N}(\eta)$ into (4.3.16) and the latter one into (4.3.15), we obtain:

$$\begin{aligned} \delta\boldsymbol{\phi}_i = \delta \mathbf{q}^T \left[\frac{\partial \phi_i}{\partial \mathbf{q}}\right] = {} & \delta \mathbf{x}_A^T \mathbf{R}_A (\mathbf{f}_1 - \mathbf{t}_i) + \delta \boldsymbol{\Theta}_A^T \left(\mathbf{f}_2 + \widetilde{\mathbf{X}}_B \mathbf{f}_1 + \mathbf{t}_i \times \mathbf{R}_A^T (\mathbf{x}_C - \mathbf{x}_A)\right) \\ & - \delta \mathbf{x}_B^T \mathbf{R}_A \mathbf{f}_1 - \delta \boldsymbol{\Theta}_B^T \mathbf{f}_2 + \delta \mathbf{x}_C^T \mathbf{R}_A \mathbf{t}_i \\ & - \delta\eta \left(\mathbf{s}_B'^T \mathbf{t}_i + \boldsymbol{\psi}_B'^T (\mathbf{X}_1 + \xi \mathbf{t}_S) \times \mathbf{t}_i\right) - \delta\xi \left((\mathbf{I} + \widetilde{\boldsymbol{\psi}}) \mathbf{t}_S^T \mathbf{t}_i\right) \end{aligned} \tag{4.3.17}$$

where

$$\mathbf{f}_1 = \begin{bmatrix} \frac{\eta}{L} t_{i1} \\ (\frac{3\eta^2}{L^2} - \frac{2\eta^3}{L^3}) t_{i2} + (\frac{6\eta}{L^2} - \frac{6\eta^2}{L^3}) u_{i3} \\ (\frac{3\eta^2}{L^2} - \frac{2\eta^3}{L^3}) t_{i3} - (\frac{6\eta}{L^2} - \frac{6\eta^2}{L^3}) u_{i2} \end{bmatrix} \quad \text{and} \quad \mathbf{f}_2 = \begin{bmatrix} \frac{\eta}{L} u_{i1} \\ (\frac{\eta^2}{L} - \frac{\eta^3}{L^2}) t_{i3} + (-\frac{2\eta}{L} + \frac{3\eta^2}{L^2}) u_{i2} \\ (-\frac{\eta^2}{L} + \frac{\eta^3}{L^2}) t_{i2} + (-\frac{2\eta}{L} + \frac{3\eta^2}{L^2}) u_{i3} \end{bmatrix} \tag{4.3.18}$$

Here t_{i1}, t_{i2}, t_{i3} denote the first, second and third components of vector $\mathbf{t}_i$. The contributions to the internal forces vector and to the tangent stiffness matrix are then computed according to the procedure described in section (4.1).

4.3.4 Kinetic Energy. The procedure for computing the kinetic energy is the same as for the superelement (section 3.2.3). Using the co-rotational approach, the total contribution of the element to the kinetic energy of the structure $\mathcal{K}$ can be written as a double summation over the nodes of the element :

$$\mathcal{K} = \sum_{i,j=A,B} \mathcal{K}^{(ij)} \tag{4.3.19}$$

where $\mathcal{K}^{(ij)}$ is the contribution arising from nodes i and j :

$$\mathcal{K}^{(ij)} = \frac{1}{2} \begin{bmatrix} \mathbf{R}_i^T \dot{\mathbf{x}}_i \\ \boldsymbol{\Omega}_i \end{bmatrix}^T \mathbf{M}^{(ij)} \begin{bmatrix} \mathbf{R}_j^T \dot{\mathbf{x}}_j \\ \boldsymbol{\Omega}_j \end{bmatrix} = \frac{1}{2} \dot{\mathbf{q}}_i^T \mathbf{M}^{(ij)} \dot{\mathbf{q}}_j \tag{4.3.20}$$

Here, $\mathbf{M}^{(ij)}$ is the 6×6 submatrix corresponding to nodes i, j of a Bernoulli beam finite element mass matrix:

$$\mathbf{M} = \begin{bmatrix} \mathbf{M}^{(AA)} & \mathbf{M}^{(AB)} \\ \mathbf{M}^{(BA)} & \mathbf{M}^{(BB)} \end{bmatrix} \tag{4.3.21}$$

and $\dot{\mathbf{q}}_i$ is the 6-components vector of generalized velocities in the local frame to each node.

Inertia forces are computed by differentiating the kinetic energy. The first variation of the kinetic energy is :

$$\delta\mathcal{K}^{(ij)} = \delta\dot{\mathbf{q}}_i^T \mathbf{M}^{(ij)} \dot{\mathbf{q}}_j \tag{4.3.22}$$

while the variation of generalized velocities at node i reads :

$$\delta\dot{\mathbf{q}}_i = \begin{bmatrix} \mathbf{R}_i^T \delta\dot{\mathbf{x}}_i - \delta\boldsymbol{\Theta}_i \times (\mathbf{R}_i^T \dot{\mathbf{x}}_i) \\ \delta\dot{\boldsymbol{\Theta}}_i + \boldsymbol{\Omega}_i \times \delta\boldsymbol{\Theta}_i \end{bmatrix} \tag{4.3.23}$$

By introducing the latter expression into (4.3.22) and by integrating by parts – Hamilton's principle – we get

$$\delta\mathcal{K}^{(ij)} = \delta\dot{\mathbf{q}}_i^T \mathbf{g}_{iner,i} = -\begin{bmatrix} \delta\mathbf{x}_i \\ \delta\boldsymbol{\Theta}_i \end{bmatrix}^T \left(\begin{bmatrix} \mathbf{R}_i & \mathbf{0} \\ \mathbf{0} & \mathbf{I} \end{bmatrix} \mathbf{M}^{(ij)} \ddot{\mathbf{q}}_j + \begin{bmatrix} \mathbf{R}_i \widetilde{\boldsymbol{\Omega}}_i & \mathbf{0} \\ \widetilde{(\mathbf{R}_i^T \dot{\mathbf{x}}_i)} & \widetilde{\boldsymbol{\Omega}}_i \end{bmatrix} \mathbf{M}^{(ij)} \dot{\mathbf{q}}_j \right) \tag{4.3.24}$$

where the acceleration vector is

$$\ddot{\mathbf{q}}_j = \begin{bmatrix} \mathbf{R}_j^T \ddot{\mathbf{x}}_j + \boldsymbol{\Omega}_j \times (\mathbf{R}_j^T \dot{\mathbf{x}}_j) \\ \mathbf{A}_j \end{bmatrix} \tag{4.3.25}$$

Then, by differentiating the inertia forces with respect to the generalized accelerations, we get the element tangent mass matrix $\overline{\mathbf{M}}$:

$$\overline{\mathbf{M}}^{(ij)} = \begin{bmatrix} \mathbf{R}_i & \mathbf{0} \\ \mathbf{0} & \mathbf{I} \end{bmatrix} \mathbf{M}^{(ij)} \begin{bmatrix} \mathbf{R}_i & \mathbf{0} \\ \mathbf{0} & \mathbf{I} \end{bmatrix}^T \tag{4.3.26}$$

4.3.5 Dynamic Friction. During motion, Coulomb friction effects generates forces between node $\mathbf{x}_C$ and the sliding surface. These forces are directly proportional to the modulus of the normal contact force in the joint, through the friction coefficient :

$$\delta \mathbf{q}^T \mathbf{F}_f = -\delta\eta \; \mu_R(\dot{\eta}) \; ||\mathbf{F}_n|| \tag{4.3.27}$$

where the regularized friction coefficient $\mu_R(\dot{\eta})$ is

$$\mu_R(\dot{\eta}) = \begin{cases} \mu \left(2 - \left|\frac{\dot{\eta}}{\epsilon_v}\right|\right) \frac{\dot{\eta}}{\epsilon_v} & |\dot{\eta}| < \epsilon_v \\ \mu \frac{\dot{\eta}}{|\dot{\eta}|} & |\dot{\eta}| \geq \epsilon_v \end{cases} \tag{4.3.28}$$

and where ϵ_v is a small tolerance with dimension of speed, giving the velocity under which one considers the joint to be "stuck" , and μ is the Coulomb friction coefficient.

Differentiation of the friction force leads to the computation of the tangent damping matrix $\mathbf{C}_f$ and of the tangent stiffness matrix $\mathbf{S}_f$:

$$\left[\frac{\partial \mathbf{F}_f}{\partial \dot{\mathbf{q}}}\right] \Delta\dot{\mathbf{q}} = \mathbf{C}_f \Delta\dot{\mathbf{q}} = -\mu_R' ||\mathbf{F}_n|| \Delta\dot{\eta} \tag{4.3.29}$$

with

$$\mu_R'(\dot{\eta}) = \begin{cases} \frac{2\mu}{\epsilon_v}\left(1 - \left|\frac{\dot{\eta}}{\epsilon_v}\right|\right) & |\dot{\eta}| < \epsilon_v \\ 0 & |\dot{\eta}| \geq \epsilon_v \end{cases}$$

and

$$\left[\frac{\partial \mathbf{F}_f}{\partial \mathbf{q}}\right] \Delta\mathbf{q} = \mathbf{S}_f \Delta\mathbf{q} = -\mu_R(\dot{\eta}) \left[\frac{\partial ||\mathbf{F}_n||}{\partial \mathbf{q}}\right] \Delta\mathbf{q} \tag{4.3.30}$$

The contact force is given by the Lagrange multipliers vector $\boldsymbol{\lambda}$ (see equation (4.1.2) – the scaling factor k has been neglected here for simplicity) . The normal contact force $\mathbf{F}_n$ is obtained by eliminating all components of the total contact force into the longitudinal direction $\mathbf{t}_1$:

$$\mathbf{F}_n = (\mathbf{I} - \mathbf{t}_1 \mathbf{t}_1^T) \mathbf{R}^T \boldsymbol{\lambda} \tag{4.3.31}$$

where $\mathbf{R} = \exp(\widetilde{\boldsymbol{\phi}})$ is the rotation of the cross section at the contact point (node C) with respect to the reference frame $\mathbf{R}_A$. The modulus squared of the normal force is:

$$||\mathbf{F}_n||^2 = \boldsymbol{\lambda}^T \mathbf{R} (\mathbf{I} - \mathbf{t}_1 \mathbf{t}_1^T) \mathbf{R}^T \boldsymbol{\lambda} \tag{4.3.32}$$

where we have used the property

$$(\mathbf{I} - \mathbf{t}_1 \mathbf{t}_1^T)^k = (\mathbf{I} - \mathbf{t}_1 \mathbf{t}_1^T) \qquad \forall\, k \tag{4.3.33}$$

Differentiation of equation (4.3.32) gives

$$\left[\frac{\partial \|\mathbf{F}_n\|}{\partial \mathbf{q}}\right] \Delta \mathbf{q} = \frac{1}{\|\mathbf{F}_n\|}\Big((\mathbf{F}_n \times \boldsymbol{\lambda})^T \; \Delta \boldsymbol{\psi} + \mathbf{F}_n^T \mathbf{R}^T \Delta \boldsymbol{\lambda}\Big) \tag{4.3.34}$$

After replacing the expression of $\Delta\boldsymbol{\psi}$ (equation (4.3.9)) into (4.3.34), we obtain

$$\begin{aligned}\left[\frac{\partial \|\mathbf{F}_n\|}{\partial \mathbf{q}}\right] \Delta \mathbf{q} = \frac{1}{\|\mathbf{F}_n\|}\Big(&(\mathbf{F}_n \times \boldsymbol{\lambda})^T \mathbf{N}_\phi \mathbf{B} \; \Delta \mathbf{q}_1 \\ &+ (\mathbf{F}_n \times \boldsymbol{\lambda})^T \boldsymbol{\psi}'_B \Delta \eta + \mathbf{F}_n^T \mathbf{R}^T \Delta \boldsymbol{\lambda}\Big)\end{aligned} \tag{4.3.35}$$

and by replacing the latter into (4.3.30) we obtain the friction contribution to the tangent stiffness matrix :

$$\mathbf{S}_f = -\frac{\mu_R(\dot{\eta})}{\|\mathbf{F}_n\|}\begin{bmatrix} \mathbf{0} & \mathbf{0} & \mathbf{0} & \mathbf{0} & \mathbf{0} \\ \mathbf{0} & \mathbf{0} & \mathbf{0} & \mathbf{0} & \mathbf{0} \\ (\mathbf{F}_n \times \boldsymbol{\lambda})^T \mathbf{N}_\phi \mathbf{B} & \mathbf{0} & (\mathbf{F}_n \times \boldsymbol{\lambda})^T \boldsymbol{\psi}'_B & \mathbf{0} & \mathbf{F}_n^T \mathbf{R}^T \\ \mathbf{0} & \mathbf{0} & \mathbf{0} & \mathbf{0} & \mathbf{0} \\ \mathbf{0} & \mathbf{0} & \mathbf{0} & \mathbf{0} & \mathbf{0} \end{bmatrix} \tag{4.3.36}$$

We note that this matrix is non symmetric, as usual when modeling Coulomb friction.

4.4. FINITE ELEMENT DESCRIPTION OF A CAM ELEMENT

In this section we present a formulation of cam/follower systems in the context of our finite element approach to mechanism analysis. It takes into account full three-dimensional motion, elasticity of members, damping, friction dissipation and many other effects.

The following hypotheses and assumptions have been made to build the numerical model of the cam/follower subsystem:

- Cam and follower have relative planar motion. Contact should be of the punctual or linear type, at every time instant of the simulation. General surface to surface contact is, by now, not allowed.
- The contact surfaces are cylindrical with an almost arbitrary normal cross section. The normal section is described by a cubic spline defined in either cartesian or polar coordinates in a local frame of each body in contact.
- The contact surfaces have relative sliding motion, and intermittent contact is allowed.
- Friction forces due to sliding are taken into account.

The algorithm is quite general and does not make any assumption regarding which body is the cam and which the follower. This fact allows to consider complex situations in which the point of contact is not precisely defined in advance. Numerical examples are shown, demonstrating the potential of application of the approach.

4.4.1 Geometric description of the cam/follower pair. The cam to be modeled is a planar rigid body, with a external surface over which sliding of the follower (or of a second cam) can occur. The shape of the sliding surfaces of both components defines the particular behavior of the system (although usually, the follower is shaped in such way that mainly the form of the cam governs the system behavior). We do not make any distinction between cam and follower from the point of view of their geometric description, and both components are modeled in the same way.

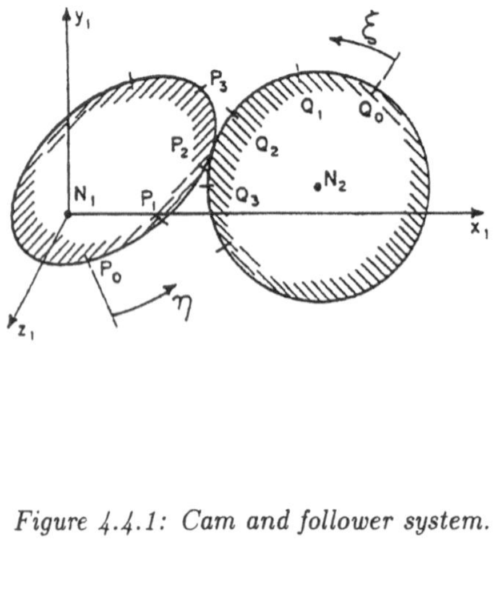

Figure 4.4.1: Cam and follower system.

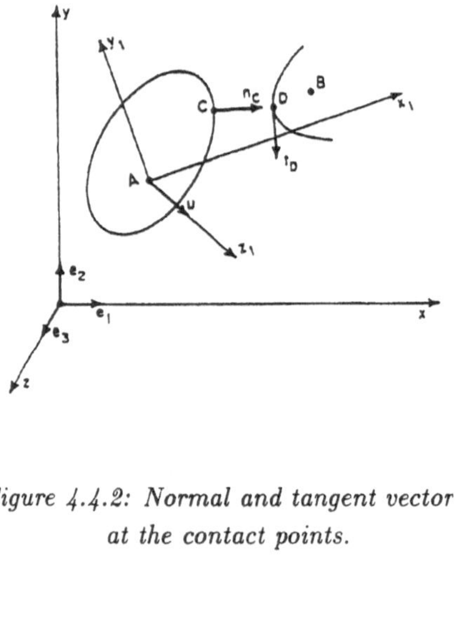

Figure 4.4.2: Normal and tangent vector at the contact points.

We will describe each sliding external surface (of cam and follower) as a closed curve $\mathbf{r}(\eta)$ in a local coordinate system $\{x_1, y_1\}$ to the plane of the considered body (see figure 4.4.1). η being the arc-length parameter along the curve. The curve $\mathbf{r}(\eta)$ is parameterized using a cubic spline interpolation [21].

4.4.2 Kinematics of the cam/follower pair. Let us consider a planar cam/follower pair, as shown in figure 4.4.2, such that its components can be described by spline functions $\mathbf{r}(\eta)$ and $\mathbf{s}(\xi)$ in their respective local coordinate systems. The cam and follower are shaped so that only point contact can occur between them, excluding any possibility of surface to surface contact.

Let A and B be two material points, one at each component, placed at the origin of the local coordinate systems $\{A; x_1, y_1\}$ and $\{B; x_1, y_1\}$. Let also C and D be two nodes located respectively over each external surface. The latter points are not fixed to the bodies, but slide over their external surfaces in such a way that they coincide with the (unique) contact point when the cam and follower come close together. Whenever the cams do not stay in contact, C and D are placed so as to be considered the natural candidates to come into contact.

The kinematics of the cam/follower pair can be completely defined in terms of the coordinates of nodes A, B, C and D, the rotation parameters at nodes A and B and the curvilinear coordinates η and ξ along the external surfaces [1,5]:

$$\mathbf{q}^T = [\mathbf{x}_A^T \quad \mathbf{\Psi}_A^T \quad \mathbf{x}_C^T \quad \mathbf{x}_B^T \quad \mathbf{\Psi}_B^T \quad \mathbf{x}_D^T \quad \eta \quad \xi] \tag{4.4.1}$$

where $\mathbf{\Psi}_A, \mathbf{\Psi}_B$ are the rotation parameters at nodes A and B, respectively, and $\mathbf{q}$ is the vector of generalized degrees of freedom of the joint.

In order to completely define the cam/follower behavior, we have to specify two sets of holonomic constraints and a contact law that can be defined through the use of a pseudo-potential. The equations of motion are derived afterwards following the augmented Lagrangian concept already described in section 4.1.

Since nodes C and D slide over the cams, their coordinates in the inertial frame are related to the coordinates of nodes A and B through the expressions

$$\mathbf{x}_C = \mathbf{x}_A + \mathbf{R}_A\mathbf{r}(\eta) \qquad\qquad \mathbf{x}_D = \mathbf{x}_B + \mathbf{R}_B\mathbf{s}(\xi) \tag{4.4.2}$$

where $\mathbf{R}_A$ and $\mathbf{R}_B$ represent the rotation operators at nodes A and B. They are parameterized in terms of the rotation parameters $\mathbf{\Psi}_A$ and $\mathbf{\Psi}_B$. These kinematic relationships constitute a first set of holonomic constraints to be satisfied. The following six constraints are used to impose them to the system:

$$\begin{aligned} [\Phi_1 \quad \Phi_2 \quad \Phi_3] &= [\mathbf{e}_1^T \quad \mathbf{e}_2^T \quad \mathbf{e}_3^T][\mathbf{x}_C - \mathbf{x}_A - \mathbf{R}_A\mathbf{r}(\eta)] \\ [\Phi_4 \quad \Phi_5 \quad \Phi_6] &= [\mathbf{e}_1^T \quad \mathbf{e}_2^T \quad \mathbf{e}_3^T][\mathbf{x}_D - \mathbf{x}_B - \mathbf{R}_B\mathbf{s}(\xi)] \end{aligned} \tag{4.4.3}$$

where $\mathbf{e}_1, \mathbf{e}_2, \mathbf{e}_3$ are the three base vectors of the cartesian inertial system (see figure 4.4.2).

In order to fix the curvilinear coordinates η and ξ along the cams surfaces, we have to introduce two additional constraints. These restrictions should express the fact that C and D are the natural candidates to come into contact. To this end, the two following constraints are proposed:

(i) The normal to the external surface of the first cam at the contact point (node C) should be perpendicular to the tangent to the external surface of the second cam at the same point (node D):

$$\Phi_7 = \mathbf{n}_C \cdot \mathbf{t}_D = 0 \tag{4.4.4}$$

where $\mathbf{n}_C$ is a unit vector normal to the first cam at C and where $\mathbf{t}_D$ is a unit vector tangent to the second cam at D.

(ii) Node D should always be placed over the normal to the first cam at C:

$$\Phi_8 = \mathbf{t}_C \cdot (\mathbf{x}_D - \mathbf{x}_C) = 0 \tag{4.4.5}$$

This restriction is naturally verified whenever the two cams are touching mutually (i.e. when nodes C and D coincide).

The satisfaction of these two constraints, (4.4.4) and (4.4.5), ensures that nodes C and D are coincident with the contact point when cam and follower come close together.

The tangent and normal vectors to the cam external surface are computed in terms of derivatives of the spline function describing the curve. The tangent vector $\mathbf{t}_C$ and the normal vector $\mathbf{n}_C$ to the first cam at node C are

$$\mathbf{t}_C = \mathbf{R}_A \mathbf{e}_{\mathbf{r}'} \qquad \mathbf{n}_C = \mathbf{R}_A(\mathbf{e}_{\mathbf{r}'} \times \mathbf{u}) \tag{4.4.6}$$

where $\mathbf{R}_A$ is the rotation of the cam expressed at node A, and $\mathbf{u}$ is a unit vector orthogonal to the plan of the joint (see figure 4.4.2). The tangent and normal vectors to the follower at D are similarly computed, giving:

$$\mathbf{t}_D = \mathbf{R}_B \mathbf{e}_{\mathbf{s}'} \qquad \mathbf{n}_D = \mathbf{R}_B(\mathbf{e}_{\mathbf{s}'} \times \mathbf{u}) \tag{4.4.7}$$

In order to calculate the contact forces, we should know the (normal) distance $q_{rel\,n}$ between cam and follower. This measure should be able to distinguish between states of penetration ($q_{rel\,n} < 0$) and of separation of the two bodies ($q_{rel\,n} > 0$). Therefore it is defined in the form:

$$q_{rel\,n} = \mathbf{n}_C \cdot (\mathbf{x}_D - \mathbf{x}_C) \tag{4.4.8}$$

We can impose afterwards the condition of non penetration through the definition of the pseudo-elastic potential:

$$\mathcal{V} = \frac{1}{2}\, k_{np}(q_{rel\,n})\; q_{rel\,n}^2 \qquad \text{with} \qquad k_{np}(q_{rel\,n}) = \begin{cases} k_{cont} & \text{if } q_{rel\,n} < 0 \\ 0 & \text{otherwise.} \end{cases} \tag{4.4.9}$$

During dynamic computations, spurious oscillations can be developed associated to the contact elastic potential. In order to damp out these oscillations, we include a small amount of dissipation derived from the following Rayleigh dissipation function:

$$\mathcal{D} = \frac{1}{2}\, c_{np}(q_{rel\ n})\ \dot{q}^2_{rel\ n} \qquad \text{with} \qquad c_{np}(q_{rel\ n}) = \begin{cases} c_{cont} & \text{if } q_{rel\ n} < 0 \\ 0 & \text{otherwise.} \end{cases} \tag{4.4.10}$$

The non penetration potential and dissipation functions so introduced give rise to the contact forces. The constants k_{cont} and c_{cont} should be chosen with caution, in order to avoid numerical ill-posed problems. We have obtained good results using as pseudo-elastic constant $k_{cont} = 1000 \times k$, where k is the scaling factor of constraints (see section 4.4.2).

Let κ_C and κ_D be the curvature of each cam at the contact points C and D. Curvatures can be very easily computed as follows:

$$\kappa_C = \frac{\mathbf{u} \times \mathbf{r}' \cdot \mathbf{r}''}{\|\mathbf{r}'\|^2} \qquad\qquad \kappa_D = \frac{\mathbf{u} \times \mathbf{s}' \cdot \mathbf{s}''}{\|\mathbf{s}'\|^2} \tag{4.4.11}$$

It is easy to verify that convex surfaces yield positive values of curvature. Then, verification of the following inequality ensures local mutual convexity of both surfaces, leading to contact unicity in a local sense:

$$\kappa_C + \kappa_D > 0 \tag{4.4.12}$$

Stringer conditions can be demanded, for instance by asking that

$$\min_{\eta} \kappa_C + \min_{\xi} \kappa_D > 0 \tag{4.4.13}$$

If the latter condition is verified, unicity of solution is assured for any relative position of both cams.

4.4.3 Computation of the element forces. In order to compute the constraint forces, we need the derivatives of constraints with respect to the generalized displacements vector of the joint (equation (4.1.2)).

After defining the non penetration potential $\mathcal{V}$ (4.4.9), we are able to compute the elastic contact forces as those forces conjugated to the variation of distance between cam and follower:

$$\delta\mathcal{V} = \delta q_{rel\ n}\ k_{np}(q_{rel\ n})\ q_{rel\ n} = \delta q_{rel\ n}\ \mathcal{F}_{elas} \tag{4.4.14}$$

with the elastic contact force $\mathcal{F}_{elas} = k_{np}(q_{rel\ n})\ q_{rel\ n}$.

When cam and follower are in contact, a friction force can arise between them owing to the eventual difference of tangential speeds. If we postulate a Coulomb mechanism of friction, this force can be considered directly proportional to the normal contact force and to the friction coefficient:

$$\delta q_{rel\ t}\ \mathcal{F}_{fr} = -\delta q_{rel\ t}\ \mu_R(\dot{q}_{rel\ t})\ |\mathcal{F}_n| \tag{4.4.15}$$

where $\delta q_{rel\ t}$ and $\dot{q}_{rel\ t}$ are the variation of relative tangential displacement and the relative tangential velocity between cams at the contact point; μ_R is a regularized friction coefficient:

$$\mu_R(\dot{q}_{rel\ t}) = \begin{cases} \mu\left(2 - \dfrac{|\dot{q}_{rel\ t}|}{\epsilon_v}\right)\dfrac{\dot{q}_{rel\ t}}{\epsilon_v} & \text{if } |\dot{q}_{rel\ t}| < \epsilon_v \\ \mu\dfrac{\dot{q}_{rel\ t}}{|\dot{q}_{rel\ t}|} & \text{if } |\dot{q}_{rel\ t}| \geq \epsilon_v \end{cases} \tag{4.4.16}$$

and $\mathcal{F}_n = \mathcal{F}_{elas} + \mathcal{F}_{diss}$ is the total contact force.

The relative speed at the point of contact can be computed in terms of velocities at nodes A and B:

$$\dot{\mathbf{q}}_{rel} = \dot{\mathbf{x}}_A + \mathbf{R}_A\tilde{\mathbf{\Omega}}_A\mathbf{r}(\eta) - \dot{\mathbf{x}}_B - \mathbf{R}_B\tilde{\mathbf{\Omega}}_B\mathbf{s}(\xi) \tag{4.4.17}$$

The tangential relative speed is obtained by projecting $\dot{\mathbf{q}}_{rel}$ over the tangential direction.

5. Numerical Examples

5.1. RETRACTION OF A THREE-LONGERON TRUSS

The system to be analyzed is a three longeron truss designed for use in a structural dynamics and control flight experiment. Each bay of the truss contains three longerons and diagonals. Interfaced between each bay is a triangle of batten members. The triangle of batten members may be envisioned as being in a horizontal plane, and has at each vertex a hinge body which connects two adjacent battens. Also attached to each hinge body are two longerons and two diagonals.

The truss is deployable, with two bays deploying at a time. During deployment, two batten triangles are held fixed while the intermediate one rotates about the z axis. The batten members connect rigidly to hinge bodies, while longerons and diagonals and hinged to them. To permit folding, the diagonals have mid-hinges along their length. The design is such that both fully deployed and folded configurations are nearly stress-free, while significant bending and twisting may occur during deployment.

Symmetry conditions have been used in order to limit the model to one bay. The model is made of 72 physical elements (51 beam elements, 6 rigid bodies and 15 hinge joints) and 7 additional constraints to impose the motion, giving a total of 421 DOF. The one bay model is shown in figure 9. Symmetry conditions with respect to the horizontal plane were imposed at triangle B. At triangle A, the equality of vertical positions at nodes A1, A2 and A3 was imposed, while rotations were keeped free. These boundary conditions are in accordance to those imposed at an experimental analysis of the mechanism [22]. The influence of other boundary conditions on efforts was determined and reported in [23]. Figure 5.1.1 displays

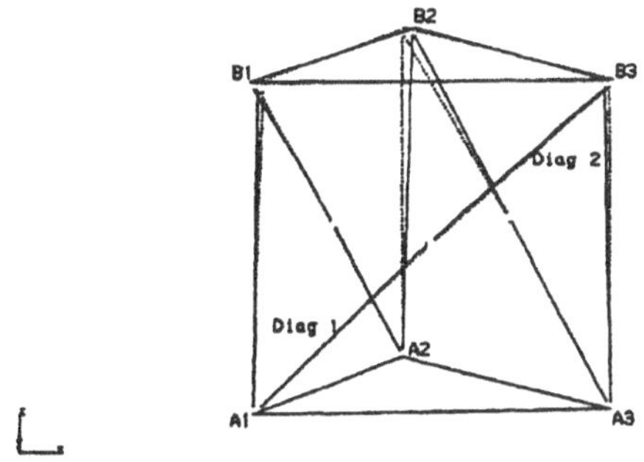

Figure 5.1.1 : One bay model of the truss.

the reference configuration (dotted line) and the initially stressed configuration obtained after assembling. Retraction is simulated in two phases:

a in order to unlock the mechanism, mid-diagonal hinge points are moved inwards and normally to lateral faces (times 0. to 2., see figure 5.1.2).

b the vertical displacement of the upper batten is then controlled up to complete retraction (figures 5.1.3 to 5.1.5). Figure 5.1.6 displays a vertical projection of the final configuration. Figure 5.1.7 provides information about the evolution of bending and torsion moments in longerons during retraction.

The kinematic analysis was made in 80 increments, with an average of 6.8 iterations per increment. The numerical model reproduces well the behavior of the experimental structure.

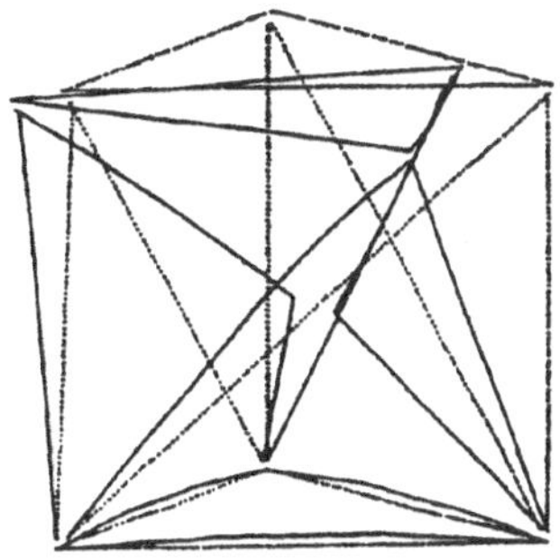

Figure 5.1.2 : Configuration at t=2.s.

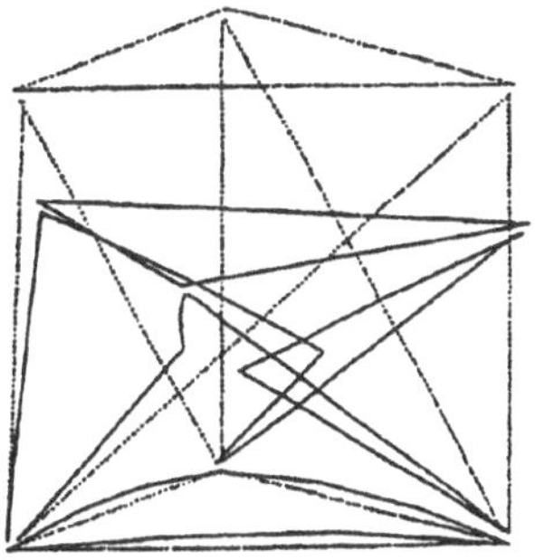

Figure 5.1.3 : Configuration at t=4.s.

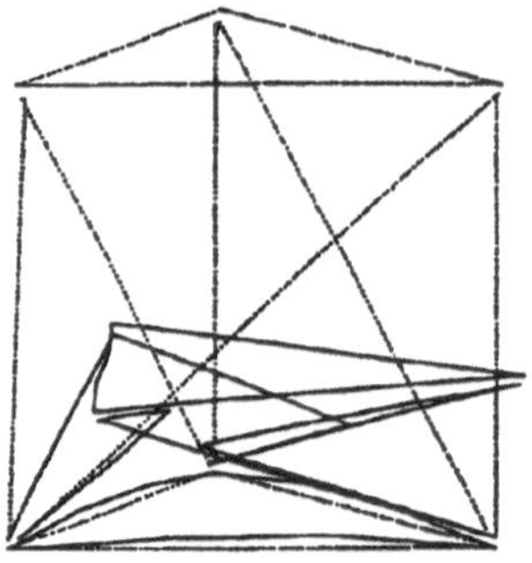

Figure 5.1.4 : Configuration at t=6.s.

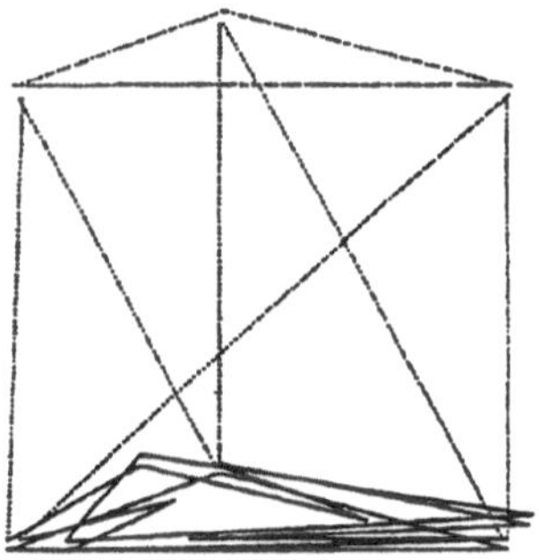

Figure 5.1.5 : Configuration at t=8.s.

5.2. LARGE FLEXIBLE SATELLITE ANTENNA

This example shows an aplication of the superelement concept to a practical case, the analysis of the deployment of a three-dimensional satellite antenna. The structure under consideration is made of five similar panels hinged together as shown by figure 5.2.1.

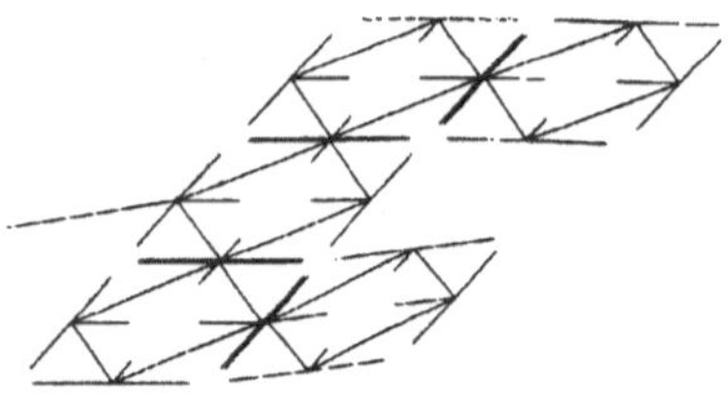

Figure 5.2.1 : geometry of 3-D satellite antenna.

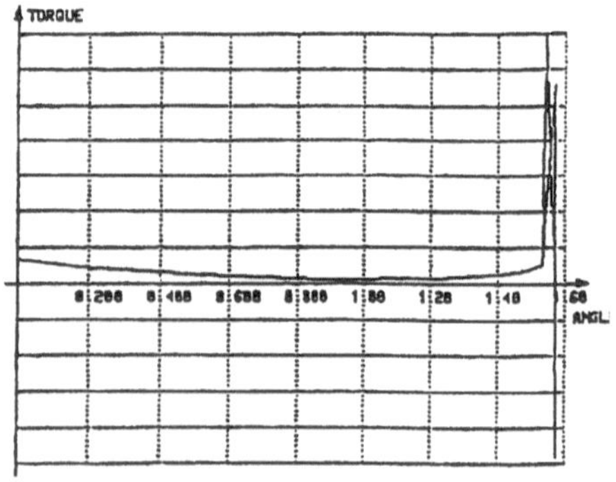

Figure 5.2.2 : torque/an gle law at the hinges.

The energy for deployment is provided by nonlinear springs acting at the hinges. with torque/rotation angle law as displayed in figure 5.2.2. The curve exhibits hysteresis in the vicinity of the locking angle, with an abrupt change of characteristics at this point (the horizontal scale was modified in the figure to allow better understanding of the phenomena). The first peak corresponds to the locking value of the torque, while the second one is generated by the hysteresis effect occuring at the locking/unlocking phase.

Each panel of the real structure is a stiffened sandwich plate made of composite material. It has thus been modeled as a sandwich flat shell with orthotropic stiffness properties and local reinforcements. An idea of the finite element model is given by figure 5.2.3 which shows a decomposition of the structure into four zones with different elastic properties. A complete description of the model is given in [24,25]. Each substructure has 584 DOF initially and is reduced to the four connecting nodes (the mid-side node along each panel edge) plus four internal vibration modes, giving a total of 28 DOF per panel.

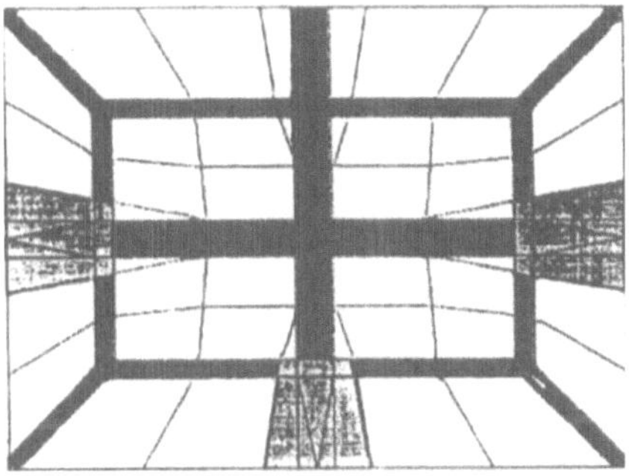

Figure 5.2.3 : finite element model and elastic properties of one panel.

The resulting mechanism model used to predict the dynamics during deployment has 242 DOF, with a quadratic mean bandwidth of 33. The time integration of the response was performed on a time interval of 41s.

Figure 5.2.4 shows a global view of the deployment process, while figure 5.2.5 displays the evolution of rotation angle at the hinges. As shown by figure 5.2.4. the structure is initially partially folded and complete deployment is achieved at time $T = 31s$. The rotation angle at the joints, on figure 5.2.5, increases regularly up to locking and then oscillates about this value. We observe that hinge 13 unlocks at time $T = 34$ due to the violent oscillations generated by locking at the other joints, and at time $t = 41$. it has not reached the full deployment state. We should point out, however, that since the stiffness characteristics of the joints possess extremely abrupt variations, the numerical simulations evidenced a nearly chaotic behavior – a strong dependence of results on the time integration parameters. Therefore. we are not able to assert if the unlocking at this joint is physically consistent or if it is purely numerical. and more complete tests should be performed in order to fully validate these results.

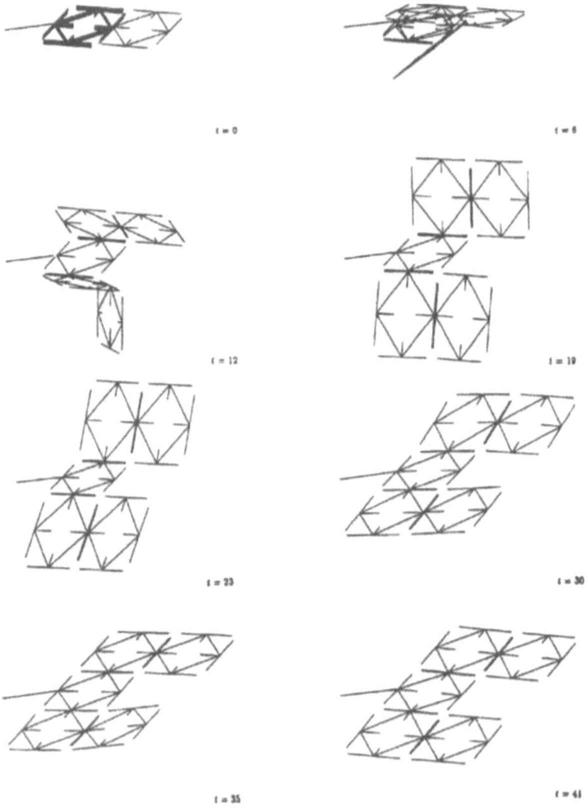

Figure 5.2.4 : Global view of the deployment process.

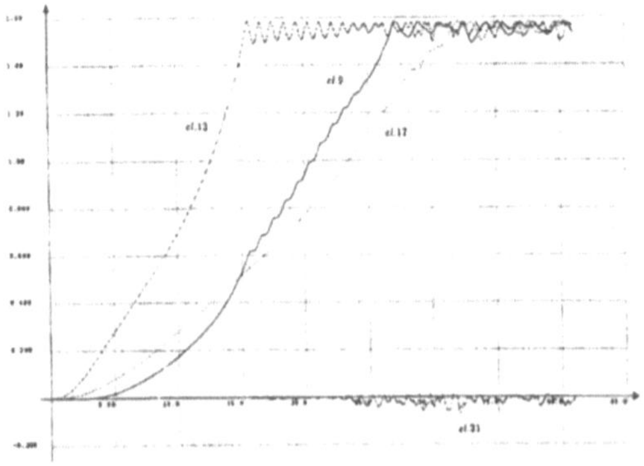

Figure 5.2.5 : Evolution of the rotation angle of the hinges.

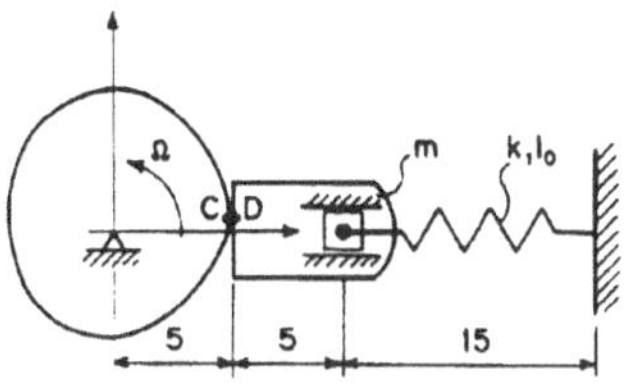

Figure 5.3.1 : Cam/follower/spring system.

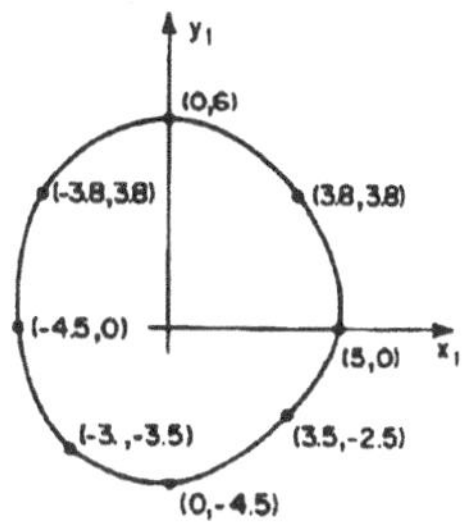

Figure 5.3.2 Cam geometry.

The example tested is that of figure 5.3.1: a cam/follower pair with a constant angular speed at the input shaft is analyzed, for several functioning conditions. The geometry of the cam was defined giving the eight points indicated in figure 5.3.2, and using a cubic spline interpolation between them. The spring constant k equals 500. , the unstressed length of the spring ℓ_0 is 22.5 and the mass of the follower m is 1. The angular speed at the input shaft is Ω = 1.75rev/s. All computations were made using a constant time step $h = 0.005$s.

Firstly, a purely kinetostatic analysis was made, for which we neglect all inertia forces. Follower displacements are plotted versus time in figure 5.3.3. Also shown is the input torque at the driving shaft necessary to statically balance at each time instant the force exerted by the follower spring (figure 5.3.4).

Secondly, a dynamic analysis was performed with the system submitted to the same input (constant speed at the driving shaft). Displacements of the follower are essentially the same as those of the kinetostatic analysis. However, the forces vary significantly: figure 5.3.5 displays the evolution in time of the contact forces between cam and follower for both kinetostatic analysis (continuous line) and dynamic analysis (dashed line).

It is worthwhile mentioning that in the dynamic analysis oscillations appear in the

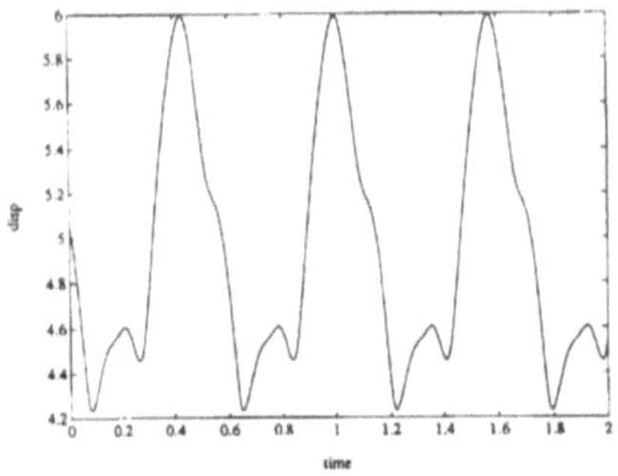

Figure 5.3.3: Displacements in time of the follower for kinetostatic analysis

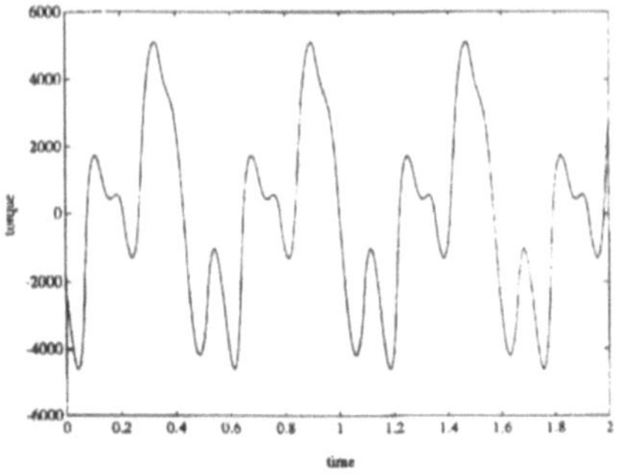

Figure 5.3.4 : Input torque in time at the driving shaft for the kinetostatic analysis

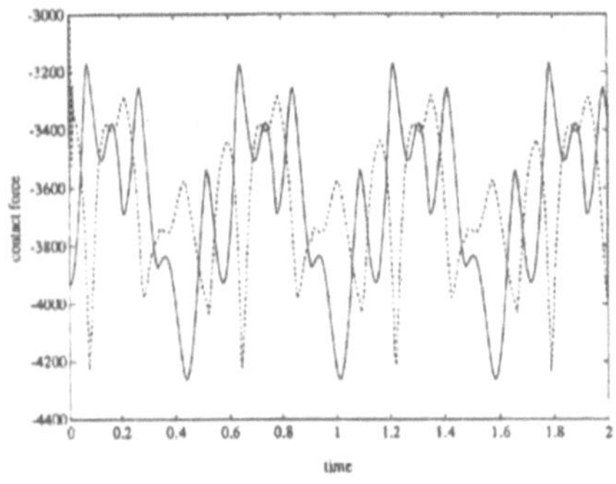

Figure 5.3.5: Contact forces between cam and follower for the kinetostatic analysis (continuous line) and for the dynamic analysis (dashed line).

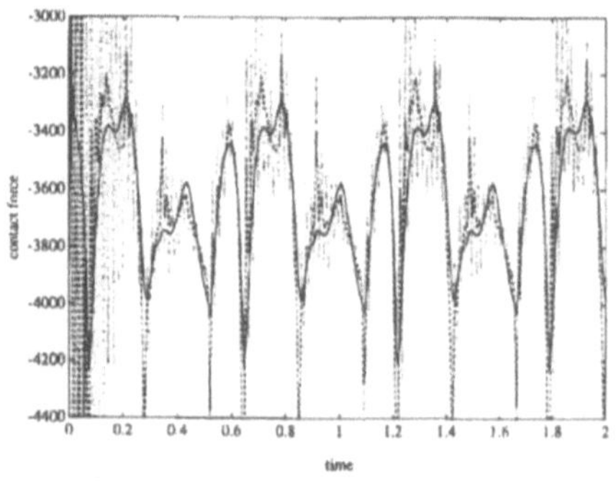

Figure 5.3.6: Contact forces between cam and follower for the dynamic analysis.
c_{cont} = 10000 *(continuous line);*
c_{cont} = 500 *(dashed line):*
c_{cont} = 0 *(points line)*

computed contact forces. These oscillations are of a purely numeric origin and are directly related to the values assigned to the contact stiffness k_{cont} and contact dissipation c_{cont}. Computations of figure 5.3.5 where obtained using a value of $c_{cont} = 10000$ for the contact dissipation constant. When decreasing this value, the force oscillations are magnified, as shown in figure 5.3.6. Clearly, the value of c_{cont} greatly influences the results. It should thus be chosen with caution by following a trial and error procedure.

The influence of friction is directly evidenced by the required computed torque to sustain motion. Figure 5.3.7 compares the required input torque for two different conditions: with friction $-\mu = 0.2-$ (continuous line) and without friction (dashed line). We can appreciate that the integral of the input torque over one period is null in the case of zero friction.

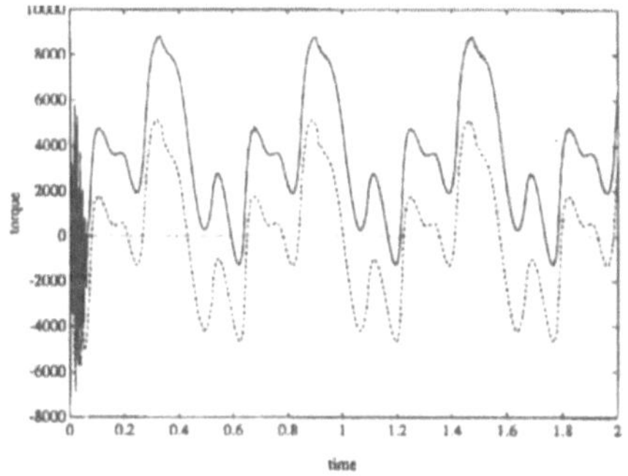

Figure 5.3.7: Input torque at the driving shaft. With friction (continuous line) and without friction (dashed line).

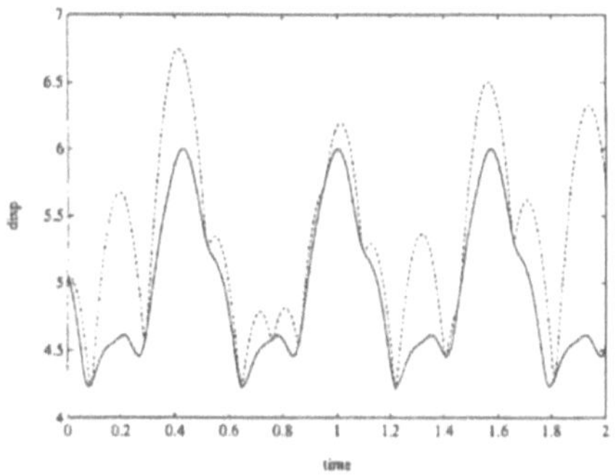

Figure 5.3.8: Displacements of node C (cam, continuous line) and of node D (follower, dashed line).

When increasing the mass of the follower, we can arrive to a situation in which continuous contact between bodies is not further assured. In figure 5.3.8 we plot the computed displacements of the (candidate) contact points of the cam (continuous line) and of the follower (dashed line) for this system. We see that the motion of the follower is almost chaotic, jumping continuously over the cam. Figure 5.3.9 displays the configurations in time of the system during the first revolution of the cam, for a case in which the follower mass is raised to $m = 15$.

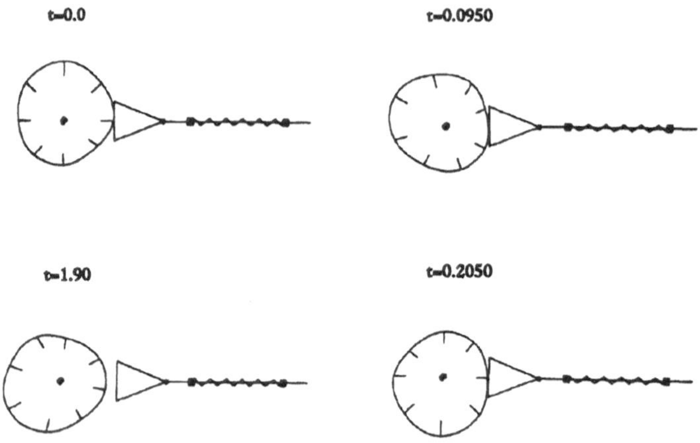

Figure 5.3.9: Evolution of the cam/follower system between times 0 and 0.665 s (first revolution).

6. references

1. CARDONA A., *An integrated approach to mechanism analysis*, PhD thesis, Université de Liège, Faculté des Sciences Appliquées (1989).

2. GERADIN M., PARK K.C. and CARDONA A., *On the representation of finite rotations in spatial kinematics*, LTAS report, University of Liège, Belgium (1988).

3. CARDONA A. and GERADIN M., *A beam finite element non linear theory with finite rotations*, Int. Jour. Num. Meth. Engng. Vol. 26, pp. 2403-2438 (1988).

4. CARDONA A., GERADIN M., GRANVILLE D. and RAEYMAEKERS V., *Module d'analyse de mécanismes flexibles MECANO - Manuel d'utilisation*, LTAS report, Université de Liège (1988).

5. CARDONA A., GERADIN M. and DOAN D.B., *Rigid and flexible joint modelling in multibody dynamics using finite elements*, Comp. Meth. Appl. Mech. Engng., Vol. 89, pp. 395-418 (1991).

6. NEWMARK N.M., *A method of computation for structural dynamics*, Jnl. Eng. Mch. Div. ASCE, No. 85 (EM3), proc. paper 2094, pp. 67-94 (1959).

7. GERADIN M., RIXEN D., *Théorie des Vibrations*, Masson Ed., Paris (1992).

8. FARHAT C.,GRIVELLI L. and GERADIN M., *On the spectral stability of time integration algorithms for a class of constrained dynamics problems*, AIAA paper 93-1306.

SDM 93 conference, La Jolla, CA, (AIAA, 1993).

9. CASSANO A. and CARDONA A., *A comparison between three variable-step algorithms for the integration of the equations of motion in structural dynamics*, Jnl of Latin American Research, Vol. 21, pp. 187-197 (1991).

10. CARDONA A. and GERADIN M., *Numerical integration of second order differential algebraic systems in flexible mechanism dynamics*, Computer Aided Analysis of Rigid and Flexible Mechanical Systems, NATO/ASI, Troia, Portugal (1993).

11. DUYSENS J. and GERADIN M., *Contribution of a symbolic calculation code to the elaboration of a finite element calculation code: a first application to the dynamic behavior of flexible mechanisms*, LTAS report VA 87, University of Liège, Belgium (1992).

12. DUYSENS J. and GERADIN M., *Flexible multibody dynamics analysis: a finite element approach aided by computer algebra*, Proc. Int. Workshop on Mechanism Design and Analysis (COMES'93), Clermont-Fd, France, 17-18 May 1993.

13. CHART B. et al., *MAPLE V - Language reference manual*, Springer-Verlag (1991).

14. SIMO J. C., *A finite strain beam formulation. The three-dimensional dynamic problem. Part I*, Comp. Meth. Appl. Mech. Engng., Vol. 49, pp. 55-70 (1985).

15. SIMO J. C. and VU-QUOC L., *A three-dimensional finite strain rod model. Part II: computational aspects*, Comp. Meth. Appl. Mech. Engng., Vol. 58, pp. 79-116 (1986).

16. PARK K.C., *Flexible beam dynamics: part I - formulation*, Center for Space Structures and Control, University of Colorado, Boulder (1986).

17. HUGHES T.J.R., *The finite element method : linear static and dynamic finite element analysis*, Prentice Hall, Englewood Cliffs, NJ (1987).

18. GERADIN M.and CARDONA A., *Substructuring techniques in flexible multibody systems*, Eight VPI & SU Symposium on Dynamics and Control of Large Space Structures, May 6-8, 1991.

19. CRAIG R. and BAMPTON M., *Coupling of substructures for dynamic analysis*, AIAA jnl, vol. 6, No. 7, pp.1313-1319 (1968).

20. CARDONA A. and GERADIN M., *Finite element modeling of flexible tracks*, Proc. Int. Conference: *Dynamics of flexible structures in space*, Cranfield, UK, 15-18 May 1990.

21. CARDONA A. and GERADIN M., *Kinematic and dynamic analysis of mechanisms*

with cams, Comp. Methods Appl. Mech. Eng., No. 103, pp. 115-134 (1993)

22. HOUSNER J., *Private communication* (1987).

23. GERADIN M. and CARDONA A., *Analysis of the retraction of a deployable three longeron truss*, LTAS report, University of Liège (1987).

24. GERADIN M., CARDONA A. and GRANVILLE D., *Deployment of large flexible space structures*, in space vehicle flight mechanics, Agard conf. proceedings, pp 28-1 - 28-11 (1989).

25. GRANVILLE D. and GERADIN M., *Calcul de déploiements d'antennes par MECANO*, LTAS report VF 62, Université de Liège (1989).

26. CARDONA A and GERADIN M., Time integration of the equations of motion in mechanisms analysis, Computers and Stuctures, Vol.33, No. 3, pp. 801-820 (1989).

KINEMATIC AND DYNAMIC SIMULATION OF RIGID AND FLEXIBLE SYSTEMS WITH FULLY CARTESIAN COORDINATES

J. GARCIA de JALON, J. CUADRADO, A. AVELLO and J.M. JIMENEZ
University of Navarre and CEIT
Manuel de Lardizábal 13
20009 San Sebastián
Spain

ABSTRACT. Multibody systems are quite often a complex combination or assembly of mechanical elements with very different mechanical behavior: rigid or flexible, linear or non-linear, etc. Sometimes it can be very difficult to carry out an efficient dynamic simulation with a single software package.

In practical applications, some bodies are so small and rigid that flexibility effects can be neglected safely, with the benefit of an improved numerical efficiency. In some studies, other bodies –such as the main hull of a car or a spacecraft– shall be considered as flexible and, because of their complex geometry and relatively high stiffness, finite elements and modal superposition techniques are the most suitable way to consider small elastic deformations, superimposed to large rigid body rotations and displacements. Finally, some bodies –as spatial booms or other very slender appendages– can be very flexible and experiment large (elastic) deformations and –probably– other second order or coupling effects, that can not be captured with linear methods, such as the standard mode superposition; in this case, large rotation theory of beams and shell finite elements is probably the most suitable solution.

This paper will describe a simple and efficient methodology that, by the use of a common set of variables, allows a unified study of multibody systems, where the three types of mechanical behavior described before coexist. This formulation is independent of the system topology, being able to deal with open and closed loops, and even with variable or changing topologies. The position variables used to simulate all these mechanical behaviors (rigid and elastic bodies, small and large deformations), are Cartesian coordinates of points, Cartesian components of unit vectors, joint coordinates (optionally) and modal coefficients (optionally). The use of a common set of Cartesian and global variables makes very easy the task of formulating the constraint equations. The resulting formulation is then very simple, general and efficient. An example of a complex mechanical system will be presented.

1. Introduction

In the last two decades a great deal of research has been done in computer simulation of complex multibody systems (MBS), most of them summarized in recent books by Nikravesh (1988), Roberson and Schwertassek (1988), Haug (1989), Shabana (1989), Huston (1990), Amirouche (1992) and García de Jalón and Bayo (1993). As a result of this research and of the necessity of practical solutions in the industry, several general-purpose computer programs have been developed (Schiehlen (1990)).

In the last few years, a great emphasis has been put on the efficiency of the methods of analysis. In kinematic simulation, interactivity is a very desirable capability of any program, and in the dynamic case large systems of non linear differential equations must

M. F. O. S. Pereira and J. A. C. Ambrósio (eds.),
Computer-Aided Analysis of Rigid and Flexible Mechanical Systems, 285–323.

be integrated as shortly as possible, even in real-time. Looking for this improved efficiency, some authors have developed symbolic methods for the derivation of the motion differential equations. When applicable, the symbolic formalisms have superior performance than the fully numerical formulations, but until now the latter remain the only general approach to the dynamic simulation of complex 3-D multibody systems.

This paper has as main objective to summarize some developments carried out by the team in the University of Navarre and CEIT (San Sebastián, Spain), in close collaboration with Prof. Bayo, in the University of California (Santa Barbara, USA)

1.1. CHOICE OF DEPENDENT COORDINATES

In a multibody system *independent coordinates* only determine actually the position of the input links. The position of the remaining links can be determined through the solution of the position problem, and the difficulty arises because this problem is non-linear and there are many possible solutions. This is the reason why it is necessary to use an extended set of coordinates, called *dependent coordinates*, that determine unambiguously the position and orientation of every body in the system.

Dependent coordinates are related by a system of nonlinear algebraic equations, the *constraint equations*, that play a very important role both in the kinematic and dynamic analysis of MBS. The kind of dependent coordinates used determines the number and complexity of the constraint equations, and thus the implementation effort and the computer time needed to solve practical cases.

It can be found in the bibliography that there are two main kinds of dependent coordinates: *relative* –or joint– coordinates and *reference point* –or Cartesian– coordinates. With relative coordinates the position of every body is determined with respect to the position of the previous one in the kinematic chain, using as many parameters as degrees of freedom of relative motion that are allowed by the pair that joins them. The number of relative coordinates is minimum among dependent coordinates, but they are more complicated to implement. With these coordinates the constraint equations arise from the closure of the independent kinematic loops. In open chain MBS relative coordinates are independent and so there are not constraint equations.

Reference point coordinates determine separately the position of each body through the Cartesian coordinates of a point and three or four parameters (usually Euler angles or Euler Parameters) to define its angular orientation. The number of reference point coordinates is higher but they are easier to manipulate. In this case the constraint equations arise from the kinematic joints that limit the relative motion of contiguous bodies.

Sometimes, it is interesting to use relative and reference point coordinates simultaneously. The resulting dependent coordinates are called *mixed coordinates*. Some relative coordinates, when selectively added to a full set of Cartesian coordinates, allows very simple implementation of actuator forces and/or torques, torsion springs, controls, etc.

Some authors, as Kim et Vanderploeg (1986a), use successively both systems of dependent coordinates, trying to gather the advantages of Cartesian coordinates (better and simpler user interface) and relative coordinates (easier control of relative motion and improved efficiency). This idea is very useful to improve the efficiency taking into account the MBS topology, as will be seen later on in this paper.

The MBS team at the University of Navarre and CEIT has introduced a new class of dependent coordinates, fully Cartesian, that they called *natural coordinates*. With these coordinates the position of a body is determined by the Cartesian coordinates of some points and the Cartesian components of some unit vectors rigidly attached to this body. At least two points and one non co-linear unit vector are necessary, in order to define completely the motion of the body, and no angular coordinates are needed. With natural coordinates the constraint equations arise mainly from the rigid body condition of the links, and secondarily from some kinematic joints.

The modeling of a three-dimensional mechanism with natural coordinates can be carried out following these general rules and recommendations:

1. The bodies must contain a sufficient number of points and unit vectors so that their motion is completely defined.
2. In each link, at least one point must be located on the axis of each joint of the link that has a preferred direction, such as revolute, cylindrical, prismatic or helical joints. A point shall be located on those joints in which there is a point common to the linked elements; this point can be shared.
3. A unit vector must be positioned at those joints having a rotational or translational axis, and should have the direction of the corresponding axis. Sometimes, the role performed by a unit vector can also be performed by a couple of basic points.
4. Some joints, such as the universal and gear joints, have their own particular requirements concerning the introduction of points and unit vectors.
5. Each unit vector is associated to a specific basic point, and the same single unit vector can be associated to several basic points. For example, on the robot's arm of figure 2, there are three rotational joints whose axes have the same direction; it is not necessary to enter three different unit vectors.
6. All points of interest, whose positions are to be considered as a primary unknown variable of the problem, can likewise be defined as basic points.

The *fully Cartesian* or *natural* coordinates have some interesting features, that are convenient to summarize at this stage.

1. Natural coordinates are composed of purely Cartesian variables and therefore they are easy to define and to represent geometrically.
2. Natural coordinates can be defined at the joints and then shared by contiguous bodies, contributing to define the position of both bodies and significantly simplifying the definition of joint constraint equations. At the same time, the total number of variables is kept moderate.
3. With other kinds of coordinates it is necessary to keep two sets of information: the variables that define the position and orientation of the reference frame attached to the moving body, and the local variables that define the body geometry (position and orientation of axis, etc.) with respect to the moving frame. With natural coordinates, a single set of variables define the geometry and the position of the body, directly in the global reference frame. It is only necessary to keep some constant values –distances, angles, etc.– that are independent of the reference frame.

4. With natural coordinates the constraint equations that arise from the rigid body and joint conditions are *quadratic* (or linear), so their Jacobian (matrix of partial derivatives) is a *linear* (or constant) function of the natural coordinates.
5. Natural coordinates can be complemented easily with relative angles and distances defined at the joints, to yield a *mixed* set of Cartesian and relative coordinates. Then, to drive an angle or a distance, and to define forces and/or torques in joints become rather straightforward. Relative coordinates also simplify the task of defining the constraint equations for some particular joints, such as the helical and gear joints.
6. In the constraint equations arising from natural coordinates, the design variables –lengths, angles, etc.– appear explicitly, not hidden by coordinate transformations. Thus, parametric and variational design, kinematic synthesis, sensitivity analysis and optimization may benefit from the use of these coordinates.

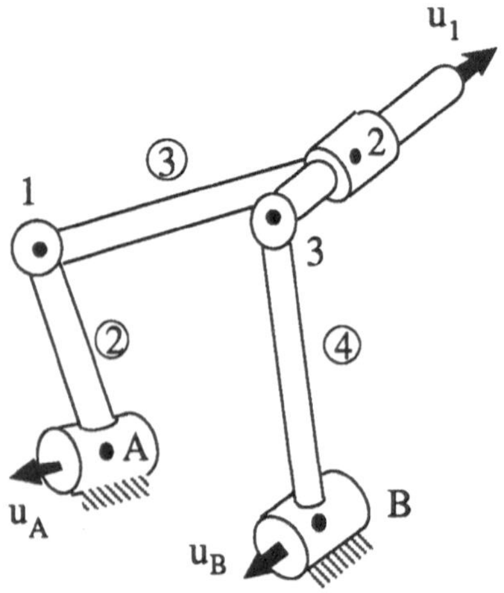

Figure 1. RSCR mechanism.

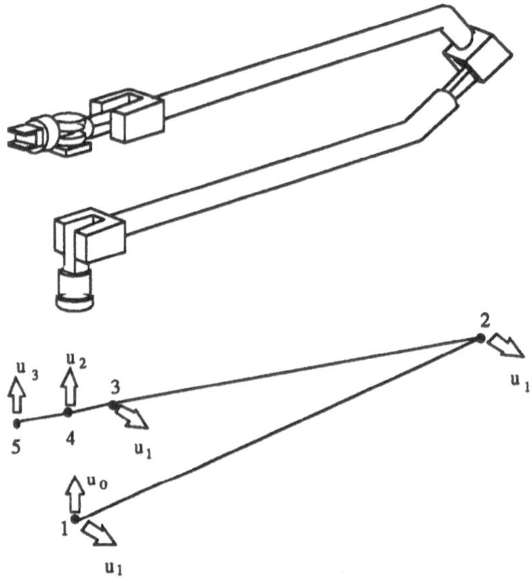

Figure 2. Space robot with 6 dof.

Figures 1 and 2 show an RSCR mechanism and a space robot modeled with natural coordinates. It can be seen how points and unit vectors are shared at the joints. For a full description of this coordinates and the corresponding constraint equations see García de Jalón and Bayo (1993).

1.2. GENERAL WAYS TO FORMULATE THE CONSTRAINT EQUATIONS

The fundamental topics of the formulation of the kinematic constraint equations will be addressed next. In the case of 3-D multibody systems, the constraint equations with natural coordinates originate in two ways:

1. from the rigid body condition of the elements, and,
2. from some of the kinematic joints that exist among them.

As an example, the constraint equations corresponding to the RSCR mechanism in figure 1 will be formulated next.

a) Rigid body conditions for body 2:

$$(x_1 - x_A)^2 + (y_1 - y_A)^2 + (z_1 - z_A)^2 - d_{1A}^2 = 0 \tag{1}$$

$$(x_1 - x_A)\, u_{Ax} + (y_1 - y_A)\, u_{Ay} + (z_1 - z_A)\, u_{Az} - d_{1A} \cos\alpha = 0 \tag{2}$$

b) Rigid body conditions for body 3:

$$(x_1 - x_2)^2 + (y_1 - y_2)^2 + (z_1 - z_2)^2 - d_{12}^2 = 0 \tag{3}$$

$$(x_1 - x_2)\, u_{1x} + (y_1 - y_2)\, u_{1y} + (z_1 - z_2)\, u_{1z} - d_{12} \cos\beta = 0 \tag{4}$$

$$u_{1x}^2 + u_{1y}^2 + u_{1z}^2 - 1 = 0 \tag{5}$$

c) Joint constraints for the cylindrical pair ($(\mathbf{r}_3 - \mathbf{r}_2) \times \mathbf{u}_1 = 0$; only two independent):

$$(y_3 - y_2)\, u_{1z} - (z_3 - z_2)\, u_{1y} = 0 \tag{7}$$

$$(z_3 - z_2)\, u_{1x} - (x_3 - x_2)\, u_{1z} = 0 \tag{8}$$

$$(x_3 - x_2)\, u_{1y} - (y_3 - y_2)\, u_{1x} = 0 \tag{9}$$

d) Rigid body conditions for body 4:

$$(x_3 - x_B)^2 + (y_3 - y_B)^2 + (z_3 - z_B)^2 - d_{3B}^2 = 0 \tag{10}$$

$$(x_3 - x_B)\, u_{1x} + (y_3 - y_B)\, u_{1y} + (z_3 - z_B)\, u_{1z} - d_{3B} \cos\gamma = 0 \tag{11}$$

$$(x_3 - x_B)\, u_{Bx} + (y_3 - y_B)\, u_{By} + (z_3 - z_B)\, u_{Bz} - d_{3B} \cos\phi = 0 \tag{12}$$

$$u_{1x}\, u_{Bx} + u_{1y}\, u_{By} + u_{1z}\, u_{Bz} - \cos\varphi = 0 \tag{13}$$

It is very easy to add relative coordinates, with the corresponding constraint equations. For instance, to add the distance (s) in the cylindrical pair it suffices to add the equation,

$$(x_3 - x_2)^2 + (y_3 - y_2)^2 + (z_3 - z_2)^2 - s^2 = 0 \tag{14}$$

Notice that the revolute and spherical joints do not introduce any constraint equations. It can be realized that all the equations (1)–(14) are quadratic. So, they will lead to linear Jacobians, very easy and cheap to evaluate.

If $\mathbf{q}$ is the vector of dependent coordinates, the equations (1)–(14) can be represented in the following compact form

$$\Phi(\mathbf{q}, t) = 0 \tag{15}$$

where time t can appear in the constraint equations through an externally driven coordinate, for instance, distance (s). A system of nonlinear equations similar to equations (1)–(14) can be developed for the robot in figure 2 or for any other system.

At this point it is convenient to divide the methods to solve the kinematic and/or dynamic equations in two groups: *global* and *topological* methods. In the global methods all the equations are set and then solved simultaneously, with no consideration of any particular characteristic of the system; they rely on efficient sparse matrix solvers. On the other hand, the topological methods try to take advantage of the system topology. For instance, the two systems in figures 1 and 2 are different: the RSCR is an closed loop chain and the manipulator is an open chain. The topological methods rely on the use of relative coordinates and on forward and backward recursive formulas.

2. Kinematic Analysis of Multi-Rigid-Body Systems

In this Section we will describe the kinematic analysis of rigid multibody systems. We will consider only the finite displacement problem. Both the global and topological methods will be consider, and a short subsection will be dedicated to the concept of *space of allowable motions*.

2.1. GLOBAL CONSTRAINT EQUATIONS

We start from a known position $\mathbf{q}$ in which all the constraint eqs. (15) are satisfied; then a finite increment is given to the input variables or degrees of freedom. In order to find the new position, it is necessary to solve the system of nonlinear constraint equations. This is carried out by the Newton–Raphson (N-R) method.

2.1.1. *Newton–Raphson and Modified Newton–Raphson Iteration.* The system of nonlinear eqs. (15) can be solved by the well known Newton–Raphson iteration formula

$$\mathbf{\Phi_q}(\mathbf{q}_i)\,(\mathbf{q}_{i+1} - \mathbf{q}_i) = -\,\mathbf{\Phi}(\mathbf{q}_i) \qquad (16)$$

This system of linear equations is sparse and can be solved with an appropriate routine found in a sparse linear algebra library. The *standard* N-R method requires a new factorization of the Jacobian for each iteration. The *modified* N-R iteration performs several forward and backward substitutions for each Jacobian factorization, allowing important efficiency improvements in many cases. However, if the increments in the input variables are large, the standard N-R shall be used because it is more robust and reliable.

Sometimes, it is possible to decrease the number of iterations of the N-R method by improving the position vector with which the iteration is started. This can be carried out by adding to the previous position a correction based on a velocity analysis computed with the last Jacobian available in factorized form. The cost of this improvement is a forward and a backward substitution.

2.1.2. *Redundant Constraints.* Practical difficulties in the solution of eqs. (15) and (16) can arise if there are redundant constraints, i.e. additional but compatible constraint equations that lead to a Jacobian matrix with more rows than columns.

There are two principal ways from which redundant constraint equations arise:

a) Due to convenience of implementation. For instance, if in the RSCR example in figure 1 all the eqs. (1)–(14) are kept, a redundant equation arise because only two of eqs. (7)–(9) corresponding to the cross product of vectors are independent.
b) In overconstrained multibody systems that are exceptions to the Grübler criterion, as for instance in spherical mechanisms and in many very important practical systems.

Redundant constraint equations can be detected and eliminated by a preprocessing of the constraint eqs. (15). The main disadvantage of this method is the need to repeat the dependent equation elimination process each time the multibody system changes its configuration, or –in some cases– after large changes in the position of the system. Thus this procedure is not suitable for real-time or interactive simulations, because there is no time to repeat the dependent equations elimination process. The second possibility is to solve systems (15) or (16) directly, with a procedure capable of directly tackling redundant constraints on a strictly standard form.

Let us assume that system (15), corresponding to a system with *n* coordinates and *f* degrees of freedom, has *m* nonlinear equations, of which only $(n-f)<m$ are independent. As a consequence, one may be tempted to think that the redundant equations in (15) just produce an excess of compatible equations in the linear system (16). If this were true no particular difficulties would appear during the solution, because there are a lot of ways and numerical routines to solve linear systems of equations with an excess of compatible equations. However, the problem is a little more complicated, because the redundant but compatible nonlinear equations (15) can induce an excess of non-compatible linear equations in (16) in the intermediate iterations. This does not happen, for instance, in velocity or acceleration analysis, because in these cases the Jacobian matrix is evaluated in the exact position, in which all constraint eqs. (15) are satisfied.

Sometimes this problem can be solved using Gaussian elimination with column pivoting and row scaling, because then, as long as $\mathbf{q}_i$ is approaching the true solution at which the constraint equations are fulfilled, the algorithm tends to disregard automatically the dependent equations. However, this procedure is not sufficiently robust and reliable.

A reliable algorithm to solve the redundant system of linear eqs. (16) is the *least-square* formulation. The normal equations corresponding to system (16) are,

$$\left[\Phi_{\mathbf{q}}^{T}\,\Phi_{\mathbf{q}}\right]_i(\mathbf{q}_{i+1}-\mathbf{q}_i)=-\left[\Phi_{\mathbf{q}}^{T}\right]_i\,\Phi(\mathbf{q}_i) \tag{17}$$

This algorithm converges on a very reliable way to the exact solution of all constraint equations. It can be argued that the solution of system (17) is less efficient than the solution of equation (16), mainly because of the product of matrices in the LHS. However, practical experience has shown that even for non redundant systems, eq. (17) can be more efficient than its counterpart (16). The reason is that in large MBS, the Jacobian tends to be very sparse, and then the product of matrices can be carried out very efficiently. System (17), although often less sparse than (16), has the advantage of being *symmetric*, with the possibility of saving storage and using simpler pivoting strategies.

Table 1 shows the results in CPU msec for the finite displacement problem of the spatial 6R robot shown in figure 2. The kinematic simulation consists of imposing an end-effector translation on an elliptic path contained in a plane perpendicular to the robot initial position plane. Three different conditions have been considered: standard N-R, modified N-R, and modified N-R with an improved initial approximation obtained from a velocity analysis. It can be concluded that in this case the improvements that result from using modified N-R and the velocity approximation are considerable.

Standard N-R	Modified N-R	Mod. N-R with vel. impr.
531	217	155

Table 1. CPU relative times for a finite displacement analysis.

2.1.3. *Improved Sparse Matrix Techniques*. The best way to improve the efficiency of global methods is to develop faster sparse matrix solvers, better suited for the size, sparsity pattern and characteristics of systems (15), (16) and (17). The best way to do that is to introduce the topology of the system into the sparse solver (see Duff et al. (1986)). Finally what one finds is a convergence of algorithms and procedures with the topological methods that will be considered later.

2.2. VELOCITY TRANSFORMATIONS: SPACE OF ALLOWABLE MOTIONS

Before entering the study of the dynamic problems, we will study in this Section the *possible* or *allowable* motions that the multibody system may have in accordance with the constraint equations. The study of these motions and the methods of expressing them is a kinematic problem, that has important implications in the formulation of the differential equations of motion.

The actual velocity vector $\dot{\mathbf{q}}$ of a constrained MBS, is a vector that belongs to a particular vector space that can be called the *space of allowable motions* (the term *motions* should actually be *velocities*). The study of this vector space and the ability to find a basis for it constitute very important points, for both kinematics and dynamics formulations. Many authors have been –explicitly or implicitly– referring to it (see Kamman and Huston (1984); Kim and Vanderploeg (1986b), Many et al. (1985), Kane and Levinson (1985), Huston (1990), etc). The concept of the *space of allowable motions* allows for a simple and general way to explain, on a unified background, many different ideas and formulations that have been introduced in the last years.

For rheonomous systems the analytical expression for the constraint equations is given by eq. (15). Differentiating this equation with respect to time once and twice, we obtain

$$\boldsymbol{\Phi}_{\mathbf{q}}(\mathbf{q}, t)\,\dot{\mathbf{q}} = -\,\boldsymbol{\Phi}_t \equiv \mathbf{b} \tag{18}$$

$$\boldsymbol{\Phi}_{\mathbf{q}}(\mathbf{q}, t)\,\ddot{\mathbf{q}} = -\,\dot{\boldsymbol{\Phi}}_t - \dot{\boldsymbol{\Phi}}_{\mathbf{q}}\,\dot{\mathbf{q}} \equiv \mathbf{c} \tag{19}$$

where the dot indicates total derivative and the sub index t partial derivative with respect to time. Eqs. (18) and (19) define vectors $\mathbf{b}$ and $\mathbf{c}$, which will be used in Section 3.

If all the degrees of freedom are controlled kinematically, that is, if the motion of all the input elements is known as function of time, eqs. (18) and (19) constitute two systems of m equations with m unknowns controlled by rank m matrices. From here on, however, it will be assumed that there are n dependent coordinates and $(n–m)$ free or kinematically undetermined degrees of freedom.

We will introduce now a large family of methods, in which the independent velocities $\dot{\mathbf{z}}$ can be defined as the projection of the dependent velocities $\dot{\mathbf{q}}$ on the rows of a constant (not time or position dependent) matrix $\mathbf{B}$

$$\dot{\mathbf{z}} = \mathbf{B}\;\dot{\mathbf{q}} \tag{20}$$

Eq. (19) can be augmented by eq. (20) to yield

$$\begin{bmatrix} \boldsymbol{\Phi}_{\mathbf{q}} \\ \mathbf{B} \end{bmatrix} \dot{\mathbf{q}} = \begin{Bmatrix} \mathbf{b} \\ \dot{\mathbf{z}} \end{Bmatrix} \tag{21}$$

Let us assume at this point that matrix $\mathbf{B}$ also fulfills the condition of having $f=n–m$ rows that are independent from one another, and also independent of the m rows of $\boldsymbol{\Phi}_{\mathbf{q}}$. With these assumptions, the matrix in eq. (21) can be inverted, and finding the vector $\dot{\mathbf{q}}$ involves the solution of the following equation

$$\dot{\mathbf{q}} = \begin{bmatrix} \mathbf{\Phi_q} \\ \mathbf{B} \end{bmatrix}^{-1} \begin{Bmatrix} \mathbf{b} \\ \dot{\mathbf{z}} \end{Bmatrix} \equiv \mathbf{S}\,\mathbf{b} + \mathbf{R}\;\dot{\mathbf{z}} \tag{22}$$

where **S** is a matrix constituted by the *m* first columns of the inverse matrix in eq. (22), and **R** is the matrix constituted by the $f=n-m$ last columns of the said inverse matrix. It is easy to show that the columns of **R** pertain to and generate the nullspace of matrix $\mathbf{\Phi_q}$. Regarding the linear system (18), which is undetermined as long as a value is not given to the input velocities, eq. (22) indicates that the general solution of the system is obtained as the sum of a *particular solution* of the complete equation (term **Sb**), plus the *general solution* of the homogeneous equation (term $\mathbf{R}\dot{\mathbf{z}}$).

The result of equation (22) may be compared with the terminology commonly used in Kane's method (Kane and Levinson (1985)). The columns of matrix **R** are the *partial velocities* with respect to the generalized coordinates **z**, and the term **Sb** constitutes the *partial velocities* with respect to time.

The acceleration equation can be obtained in a similar manner, augmenting eq. (19) with the derivative with respect to time of eq. (20),

$$\begin{bmatrix} \mathbf{\Phi_q} \\ \mathbf{B} \end{bmatrix} \{\ddot{\mathbf{q}}\} = \begin{Bmatrix} \mathbf{c} \\ \ddot{\mathbf{z}} \end{Bmatrix} \tag{23}$$

and the inversion of this matrix gives

$$\ddot{\mathbf{q}} = \begin{bmatrix} \mathbf{\Phi_q} \\ \mathbf{B} \end{bmatrix}^{-1} \begin{Bmatrix} \mathbf{c} \\ \ddot{\mathbf{z}} \end{Bmatrix} = \mathbf{S}\,\mathbf{c} + \mathbf{R}\;\ddot{\mathbf{z}} \tag{24}$$

This expression, analogously to eq. (22), indicates that matrix **R** can be calculated by triangularizing the leading matrix of systems (21) or (23), and performing *f* forward and backward substitutions, with the *f* last columns of unit matrix **I** as the RHS terms.

Some of the dynamic formulations that will be seen in Section 3, require the calculation of the term (**Sc**) in eq. (24). From this expression, it is concluded that this term is the dependent acceleration vector $\ddot{\mathbf{q}}$ when $\ddot{\mathbf{z}}$ is zero. Since the leading matrix of system (23) has been previously triangularized (when finding matrix **R**), the calculation of the term being considered requires very little additional effort.

Many methods currently used to determine a basis of the subspace of allowable motions, that is to say matrix **R**, can be considered inside a large group –the *projection* methods– which will be described next.

It is clear that eqs. (20)–(24) completely define the transformation between dependent and independent variables. This only leaves matrix **B** to be determined. Once this matrix is calculated, it can remain constant during a large range of motion of the system. The condition that matrix **B** must comply with is that its *(n–m)* rows must be independent from one another, and independent from the *m* rows of matrix $\mathbf{\Phi_q}$. At this point, we can identify and describe in this context, three methods that have been proposed in the literature to construct the matrix **B**.

1. *Method based on the Singular Value decomposition.* The SV decomposes a ($m \times n$) rectangular matrix, such as $\mathbf{\Phi_q}$, in the form

$$\mathbf{\Phi_q} = \mathbf{U}^{\mathrm{T}}\,\mathbf{D}\,\mathbf{V} \tag{25}$$

where matrix **U** is orthogonal of size ($m \times m$). Matrix **D** is composed of a diagonal matrix of size ($m \times m$) that contains the singular values, and a zero matrix given by $f=n-m$ last columns. Matrix **V** is orthogonal of size ($n \times n$), and can be decomposed into two sub-matrices $\mathbf{V}_d$ and $\mathbf{V}_i$ of sizes ($m \times n$) and ($f \times n$), respectively, according to the partition in **D**. The most important property of the SV decomposition that pertains to the problem at hand is that the rows of the matrix $\mathbf{V}_i$ constitute an *orthogonal basis* of the nullspace of matrix $\mathbf{\Phi_q}$. In other words, it is verified that

$$\mathbf{\Phi_q} \mathbf{V}_i^T = \mathbf{0} \tag{26}$$

In view of this expression, Singh and Likins (1985) proposed to construct matrix **R** directly from the SV decomposition. The problem is that the SVD is essentially an iterative process, which consumes a great deal of time, and it is absolutely impractical to carry out it at each position **q** of the system. Mani et al. (1985) have proposed using the SV decomposition to calculate the matrix **B**, whereby this operation only needs to be performed once or at most a few times throughout the entire range of the motion of the system. Bear in mind that many matrices **R**, corresponding to different positions **q** of the system, can be calculated from only one matrix **B**, that continues to be valid as long as its rows are independent from those of $\mathbf{\Phi_q}(\mathbf{q})$. Eq. (26) indicates that matrix $\mathbf{V}_i$ complies with the conditions required for matrix **B**, at least as long as no large changes are produced in the positions and in matrix $\mathbf{\Phi_q}$, so that the linear independence condition between the rows of the said matrix and those of matrix **B** is lost.

2. *Method based on the QR decomposition.* This method is similar to the previous one, but it uses the QR instead of the SV decomposition, because the QR decomposition is a direct and cheaper process (Kim and Vanderploeg (1986b)). The QR method decomposes the matrix $\mathbf{\Phi_q}$ in the form

$$\mathbf{\Phi_q^T} = \tilde{\mathbf{Q}} \tilde{\mathbf{R}} \tag{27}$$

where $\tilde{\mathbf{Q}}$ is an orthogonal ($n \times n$) matrix, and $\tilde{\mathbf{R}}$ is a rectangular ($n \times m$) matrix, formed by an upper triangular matrix ($m \times m$) and a zero matrix of order ($f \times m$). Note that a tilde has been used to distinguish the result of the QR decomposition from the matrix **Q** that symbolizes the external forces in dynamic analysis (Section 3), and the matrix **R** (basis of the Jacobian nullspace). The application of this decomposition to the problem at hand is straightforward when considering that the f last columns of $\tilde{\mathbf{Q}}$ constitute an orthogonal basis of the nullspace of the matrix $\mathbf{\Phi_q}$ that can be taken as matrix **B**, in the same way as in the SV decomposition.

3. *Method based on Gaussian triangularization.* This method, described by Serna et al. (1982), is based on the Gauss triangularization of matrix $\mathbf{\Phi_q}$ with total pivoting. This implies the decomposition of the Jacobian in sub-matrices, in the form

$$\mathbf{\Phi_q} \equiv \left[\mathbf{\Phi_q^d} \ \mathbf{\Phi_q^i}\right] \tag{28}$$

where matrix $\mathbf{\Phi_q^d}$ is a square matrix ($m \times m$) that contains the columns in which the pivots have appeared. Matrix $\mathbf{\Phi_q^i}$ contains the columns in which the pivots have not appeared, and has the size ($m \times f$). In the theory of linear equation, the variables associated with columns $\mathbf{\Phi_q^i}$ are called *independent variables*, and those associated with columns Φ_q^d are called *dependent variables*. Once matrix Φ_q is triangularized as in eq. (28), matrix $\mathbf{B}$ is a boolean matrix constructed as follows

$$\mathbf{B} \equiv [\ \mathbf{0} \mid \mathbf{I}\ ] \tag{29}$$

where $\mathbf{I}$ is a $(f x f)$ unit matrix. The matrix from whose inverse matrix $\mathbf{R}$ is calculated,

$$\begin{bmatrix} \Phi_q \\ \mathbf{B} \end{bmatrix} = \begin{bmatrix} \Phi_q^d & \Phi_q^i \\ \mathbf{0} & \mathbf{I} \end{bmatrix} \tag{30}$$

Since matrix Φ_q^d is triangularizable, it is guaranteed that the rows of matrix $\mathbf{B}$ are independent from those of Φ_q. Note that the triangularization of (30) is simpler than with the SV or QR decomposition, because in the part corresponding to matrix $\mathbf{B}$ no additional work is necessary. Eqs. (20) and (29) indicate that the independent velocities $\dot{\mathbf{z}}$ are chosen as a *sub-set* of the dependent velocities $\dot{\mathbf{q}}$.

2.3. TOPOLOGICAL SOLUTION METHOD

An obvious way to improve the efficiency of the finite displacement method previously explained is to take into account the topology of the system to be analyzed. It is known that open chain MBS driven by relative coordinates allow a simple and efficient recursive solution. Closed chains can also benefit from this open chain formulation, with some modifications. This is considered with more detail by Jiménez et al. (1993). Here we will give a very short, qualitative description of the topological methods.

2.3.1. *Open Chain Systems.* We will be consider an open chain system such as the one shown in figure 3. For such a system, the relative or joint coordinates also constitute a possible set of independent coordinates. If this is the case, the finite displacement problem can be solved directly, avoiding the solution of any system of linear or nonlinear equations.

In a system with a tree configuration, if a relative coordinate is incremented this motion affects only to the bodies that are upwards in the correrspondinq branch of the tree. It is very easy to compute the new Cartesian position of these bodies. If many relative coordinates are incremented it suffices to go over the tree recursively affecting to each body of the finite displacements of the joints that are backwards in the tree.

It is also interesting to consider that in an open chain system it is very easy to compute the matrix $\mathbf{R}$ that relates Cartesian with relative velocities. Instead of solving a system of linear equations as in (22), all columns of $\mathbf{R}$ can be computed by introducing a unit relative velocity in the corresponding joint (zero velocity in the other joint coordinates) and computing the Cartesian velocities in the upward bodies in the tree. This is a very simple and cheap recursive procedure.

2.3.2. *Closed Chain Systems: N-R.* Closed loop MBS, as the one shown in figure 4, can be converted in open chain systems by cutting the appropriate joints (see Kim and Vanderploeg (1986a)). The open chain system corresponding to the system in figure 4 is the one shown in figure 3. In the open chain system all the relative coordinates are independent. In the closed loop system only a subset of the relative coordinates can be taken as independent. In each loop that is open, as many relative coordinates as constraint equations the joint that has been open has shall be considered as dependent, so as to be able to impose the closure conditions.

In a closed loop system the finite displacement problem can be solved in the following way. First the system is converted in an open chain system by cutting some joints. Then the finite increments in the independent relative coordinates are applied in a direct process that violates the constraint conditions in the cut joints. Finally the closure constraint equations are enforced, keeping constant the independent relative coordinates and allowing the variation of the relative dependent ones. This represent the iterative solution of an small system of nonlinear equations. For the details of this procedure see Jiménez et al. (1993). This reference provides also quantitative data on the very important reductions in CPU time that can be obtained with respect to the global methods.

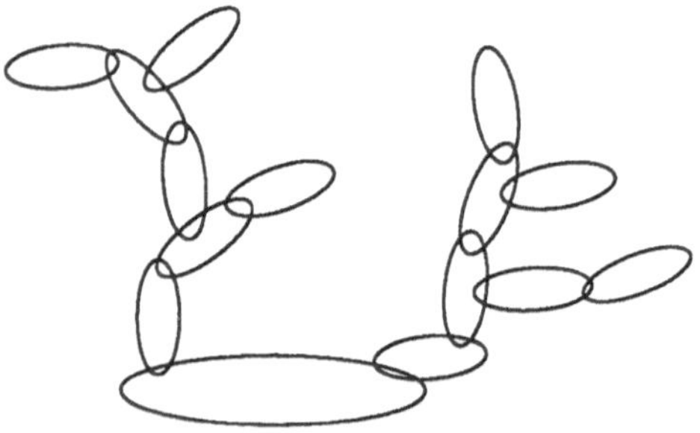

Figure 3. Open chain multibody system.

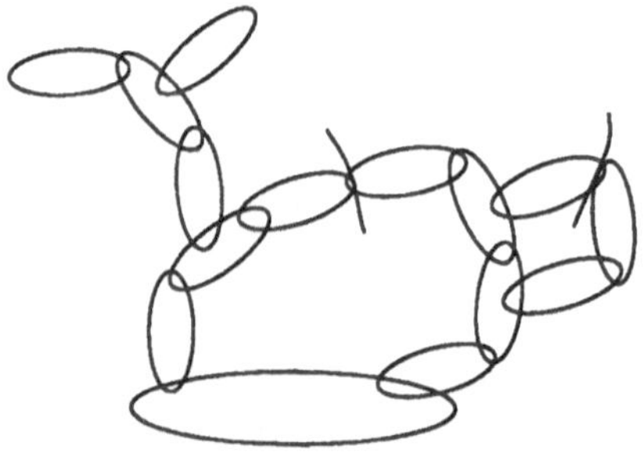

Figure 4. Closed chain multibody system.

2.3.3. *Topological Solution by Sparse Matrix Techniques*. It has been set previously that a way to increase the efficiency of global methods is improve the sparse matrix solvers by taking into account the topology of the system. We will introduce this subject with a very simple example. Figure 5 shows a four-bar planar system with a particular consideration for joint B. Both relative and Cartesian coordinates have been defined. Figure 6 shows the sparsity pattern of the Jacobian matrix after a suitable reordering of equations and variables. Angle φ_1 is the independent relative coordinate. Index "b" affects both the relative dependent coordinates and the constraint equations corresponding to the joint in B.

The Newton–Raphson iteration equations can be set in the partitioned form

$$\begin{bmatrix} \mathbf{\Phi}_{\mathbf{q}}^{aa} & \mathbf{\Phi}_{\mathbf{q}}^{ab} \\ \mathbf{\Phi}_{\mathbf{q}}^{ba} & \mathbf{\Phi}_{\mathbf{q}}^{bb} \end{bmatrix}_i \begin{Bmatrix} \Delta\mathbf{q}^a \\ \Delta\mathbf{q}^b \end{Bmatrix}_i = - \begin{Bmatrix} \mathbf{f}^a \\ \mathbf{f}^b \end{Bmatrix}_i \tag{31}$$

whose solution can be computed as

$$\begin{aligned} \Delta\mathbf{q}^a &= \mathbf{\Phi}_{\mathbf{q}}^{-aa}\left(-\mathbf{f}^a - \mathbf{\Phi}_{\mathbf{q}}^{ab}\,\Delta\mathbf{q}^b\right) \\ \left(\mathbf{\Phi}_{\mathbf{q}}^{bb} - \mathbf{\Phi}_{\mathbf{q}}^{ba}\,\mathbf{\Phi}_{\mathbf{q}}^{-aa}\,\mathbf{\Phi}_{\mathbf{q}}^{ab}\right)\Delta\mathbf{q}^b &= \mathbf{\Phi}_{\mathbf{q}}^{ba}\,\mathbf{\Phi}_{\mathbf{q}}^{-aa}\,\mathbf{f}^a \end{aligned} \tag{32}$$

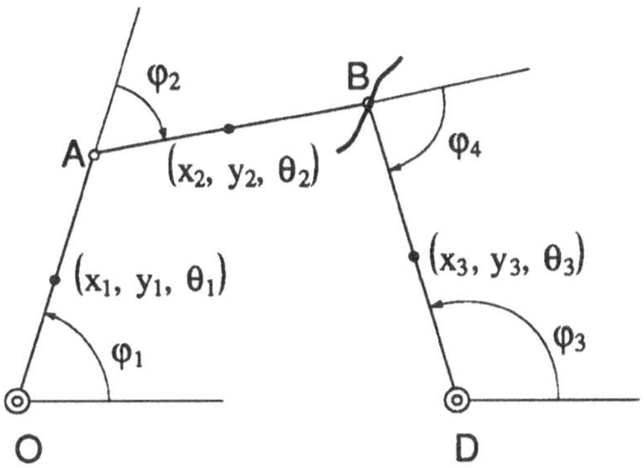

Figure 5. Four bar mechanism with a cutted joint.

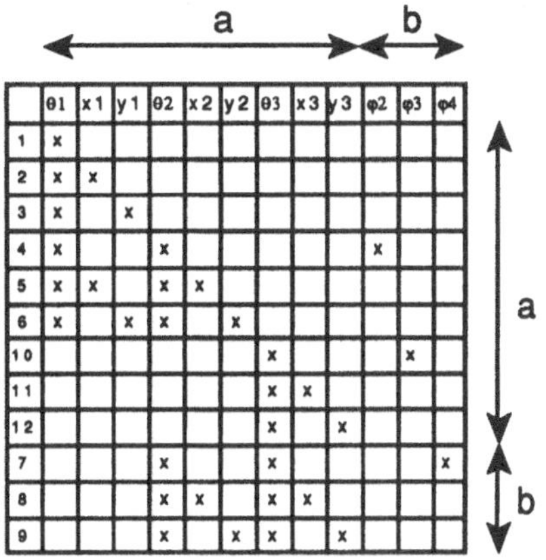

Figure 6. Sparsity pattern for the mechanism in figure 5.

It can be seen, in figure 6 and in eq. (32), that the larger system of linear equations to be solved is based on a lower triangular matrix $\mathbf{\Phi}_{\mathbf{q}}^{\mathbf{aa}}$; this matrix corresponds to the open chain system. The closure condition of the loop introduces a small system of three equations. It can be useful to remember that the Jacobian of an open chain system can always be arranged in a triangular form, that of course do not need to be factorized. It is also worth to remember that forward and backward substitutions with an sparse triangular matrix are computationally equivalent to forward and backward recursive processes.

3. Dynamic Analysis of Multi-Rigid-Body Systems

3.1. GLOBAL FORMULATIONS

This Section deals with the *direct dynamic problem*. The position of the multibody system is characterized by its dependent coordinates. However, at the time of formulating the equations of motion, it is possible to do it with dependent or independent coordinates. There is not a consensus among the experts as to which method is the best for all cases.

We discuss in this Section several methods of formulating and solving the direct dynamic problem with both dependent (Section 3.1.1) and independent coordinates (Section 3.1.2).

3.1.1. *Formulations in Dependent Coordinates.* The formulation the equations of motion with dependent coordinates can be obtained by either the Lagrange's equations or the method of virtual power. Hereinafter, the vector $\mathbf{q}$ will represent a set of n unknown dependent coordinates, m will be the total number of independent constraint equations (geometric and kinematic) and therefore $f=n-m$ will be the number of dynamic degrees of freedom. The constraint conditions are written in the following general form

$$\mathbf{\Phi}(\mathbf{q}, t) = \mathbf{0} \tag{33}$$

By using the Lagrange equations or the virtual power principle, it is possible to arrive to the following system of dynamic equilibrium equations

$$\mathbf{M}\ddot{\mathbf{q}} + \mathbf{\Phi}_{\mathbf{q}}^{\mathrm{T}}\boldsymbol{\lambda} = \mathbf{Q} \tag{34}$$

where vector $\mathbf{Q}$ contains the external and the velocity dependent inertia terms. The term $(\mathbf{\Phi}_{\mathbf{q}}^{\mathrm{T}}\boldsymbol{\lambda})$ corresponds to the constraint forces, that is, the forces necessary to enforce the constraint equations.

3.1.1.1. *Method of the Lagrange's Multipliers.* Eq. (34) has n equations and $(n+m)$ unknowns: the n elements of vector $\ddot{\mathbf{q}}$ and the m elements of vector $\boldsymbol{\lambda}$. In order to have a sufficient number of equations, it is necessary to supply m more equations. The obvious choice is to use the algebraic constraint equations (33) which along with (34) constitute a set of differential algebraic equations or DAEs of index three. In order to avoid DAEs one can use the acceleration kinematic equations, which are obtained by differentiating the constraint eqs. (33) twice with respect to time

$$\mathbf{\Phi}_{\mathbf{q}}\ddot{\mathbf{q}} = -\dot{\mathbf{\Phi}}_t - \dot{\mathbf{\Phi}}_{\mathbf{q}}\dot{\mathbf{q}} \equiv \mathbf{c} \tag{35}$$

this expression defines vector $\mathbf{c}$. By writing expressions (34) and (35) jointly, one obtains

$$\begin{bmatrix} \mathbf{M} & \mathbf{\Phi}_{\mathbf{q}}^{\mathrm{T}} \\ \mathbf{\Phi}_{\mathbf{q}} & \mathbf{0} \end{bmatrix} \begin{Bmatrix} \ddot{\mathbf{q}} \\ \boldsymbol{\lambda} \end{Bmatrix} = \begin{Bmatrix} \mathbf{Q} \\ \mathbf{c} \end{Bmatrix} \tag{36}$$

which is a system of $(n+m)$ equations with $(n+m)$ unknowns, whose matrix is symmetric, non positive definite, and sparse.

The main advantage of the dynamic formulation in dependent coordinates using Lagrange's multipliers, besides of the conceptual simplicity of the method, is that of permitting the calculation of forces associated with the constraints –which depend on the Lagrange's multipliers– with a minimum additional effort.

3.1.1.2. *Method Based on the Projection Matrix **R**.* A second possibility of formulating the motion differential equations with dependent coordinates is based on the matrix $\mathbf{R}$, introduced in Section 2.2. Remember that the $f=n-m$ columns of $\mathbf{R}$ represent a basis of the nullspace of $\mathbf{\Phi}_{\mathbf{q}}$; that is, a basis of the subspace of possible motions. If the dynamic equilibrium eq. (34) is pre-multiplied by $\mathbf{R}^{\mathrm{T}}$ and one takes into account that matrices $\mathbf{\Phi}_{\mathbf{q}}$ and $\mathbf{R}$ are orthogonal

$$\mathbf{R}^{\mathrm{T}}\mathbf{M}\ddot{\mathbf{q}} = \mathbf{R}^{\mathrm{T}}\mathbf{Q} \tag{37}$$

Eq. (37) contains $(n-m)$ equations with n unknowns. In order to have as many equations as unknowns, it is necessary to complete this system with the kinematic acceleration eqs. (35), resulting in

$$\begin{bmatrix} \mathbf{\Phi}_{\mathbf{q}} \\ \mathbf{R}^{\mathrm{T}}\mathbf{M} \end{bmatrix} \ddot{\mathbf{q}} = \begin{Bmatrix} \mathbf{c} \\ \mathbf{R}^{\mathrm{T}}\mathbf{Q} \end{Bmatrix} \tag{38}$$

which is a system of n equations with n unknowns, that can be solved for the dependent accelerations $\ddot{q}$. Note that the upper part of eq. (38), corresponding to matrix Φ_q, has been previously factorized in order to calculate the matrix **R**. Because of this, the system of eqs. (38) can be solved with very little additional effort, and this method is sometimes more efficient than the one based on eqs. (36) (Unda et al. (1987)). The dynamic formulation whose end result is eq. (38), was introduced by Kamman and Huston (1984), although they did not use a general matrix **R**, but a set of eigenvectors associated with the zero eigenvalues of the matrix $(\Phi_q^T\Phi_q)$.

Matrix **R** can be calculated by means of any of the methods referenced in Section 2.2, although in general the simplest is the projection method, based on the selection of the independent coordinates as a sub-set of the dependent ones.

Eq. (36) allows us to clearly distinguish the equations corresponding to the *kinematics* –the m first ones–, from the equations corresponding to the *dynamics* –the $(n-m)$ last ones–. Besides, system (38) does not explicitly contain any independent coordinates, rather they are implicitly considered via the matrix **R**. Each matrix **R** implies a choice of independent coordinates.

It is known that the time integration of the dependent accelerations that come from eqs. (36) or (37) leads to severe unstability problems. In order to avoid that it is necessary to integrate a mixed system of DAEs or to use special stabilization techniques as the one due to Baumgarte (1972).

3.1.1.3. *Penalty Formulations.* In this Section we will present an alternative formulation in dependent coordinates based on the penalty method, proposed by Bayo et al. (1988), This formulation eliminates the Lagrange's multipliers from the equations of motion, and leads to a set of n ordinary differential equations with $\ddot{q}$ as the only unknowns. In essence, this method directly incorporates the violation of constraints, penalized by a large factor, into the equations of motions. The larger the penalty factor the better the constraints will be achieved at the cost of introducing some numerical ill-conditioning. Theoretical studies of its convergence and stability have been carried out by Kurdila and Narcowich (1992). In this paper the penalty method will be introduced in a very simple way.

According to eq. (33) it can be considered that vectors Φ, $\dot{\Phi}$ and $\ddot{\Phi}$ represent the violations for the position, velocity and acceleration constraint equations. On the other hand, eq. (34) shows that the columns of Φ_q^T represent the direction of the constraint forces. So it is possible to formulate the penalty method from eq. (34) by introducing very big restoring forces, proportional to the constraint violation, on the direction of the constraint forces. Eq. (34) is transformed in

$$\mathbf{M}\,\ddot{\mathbf{q}} + \Phi_{\mathbf{q}}^{T}\,\alpha\left(\ddot{\Phi} + 2\,\Omega\,\mu\,\dot{\Phi} + \Omega^{2}\,\Phi\right) = \mathbf{Q} \tag{39}$$

where matrices α, Ω and μ are $(m\times m)$ diagonal matrices that contain the values of the penalty numbers, the natural frequencies and the damping ratios, of the 1 degree-of-freedom penalty system assigned to each constraint condition. If the same values are used for each constraint these matrices become identity matrices multiplied by the respective penalty numbers.

Note that the term $(\alpha\ddot{\Phi} + 2\alpha\Omega\mu\dot{\Phi} + \alpha\Omega^2\Phi)$ is an approximation to the true Lagrange's multipliers λ. The premultiplication by Φ_q^T projects the forces unto the space of the dependent coordinates. Substituting $\ddot{\Phi}$ in eq. (39) the following result is obtained,

$$(\mathbf{M}+\Phi_{\mathbf{q}}^{T}\,\alpha\,\Phi_{\mathbf{q}})\,\ddot{\mathbf{q}}=\mathbf{Q}-\Phi_{\mathbf{q}}^{T}\,\alpha\,(\dot{\Phi}_{\mathbf{q}}\,\dot{\mathbf{q}}+\dot{\Phi}_{t}+2\,\Omega\,\mu\,\dot{\Phi}+\Omega^{2}\,\Phi) \tag{40}$$

The condition of convergence for the penalty method is achieved by merely using large penalty factors. These in turn may produce numerical ill conditioning, which nevertheless may be avoided by the improved technique described below. It is well known that penalty methods bring forth the problem of choosing the right penalty number. It is very important that the analyst be supplied with a method that converges, regardless of the size of the penalty values, to the right solution within specified tolerances in the constraints. To this end, Bayo et al. (1988) extends the *augmented Lagrangian method* commonly used in optimization analysis to improve the numerical conditioning of the proposed penalty equations. Consider again the classical Lagrange's multipliers method as stated by eq. (34). Instead of following the standard approach, eq. (34) can be modified by adding the corresponding penalty terms, whose values will be zero if the constraints are satisfied. Therefore

$$\mathbf{M}\,\ddot{\mathbf{q}}+\Phi_{\mathbf{q}}^{T}\,\alpha\left(\ddot{\Phi}+2\,\Omega\,\mu\,\dot{\Phi}+\Omega^{2}\,\Phi\right)+\Phi_{\mathbf{q}}^{T}\,\lambda^{*}=\mathbf{Q} \tag{41}$$

This equation can also be viewed as a penalty method to which the Lagrange's multipliers are added. In the limit, the constraint conditions are satisfied, thus $\lambda=\lambda^{*}$ and eqs. (34) and (41) become equivalent, except for round off errors. In eq. (41) the Lagrange's multipliers λ^{*} play the role of *correcting terms*. By merely comparing eqs. (34) and (41) it can be inferred that

$$\lambda\equiv\lambda^{*}+\alpha\left(\ddot{\Phi}+2\,\Omega\,\mu\,\dot{\Phi}+\Omega^{2}\,\Phi\right) \tag{42}$$

The solution of (41) without the kinematic constraint eqs. (33) requires that the correct values of λ^{*} be known. Those values are not known in advance but it is possible to set up an iterative process that calculates them The iteration expression is easily established by taking advantage of eq. (42)

$$\lambda_{i+1}=\lambda_{i}+\alpha\left(\ddot{\Phi}+2\,\Omega\,\mu\,\dot{\Phi}+\Omega^{2}\,\Phi\right)_{i+1}\quad i=0,1,2,\ldots. \tag{43}$$

with $\lambda_{0}^{*}=0$ for the first iteration. Eq. (43) physically represents the introduction at iteration (i+1) of forces that tend to compensate the fact that the constraints are not exactly zero. It becomes now obvious how the penalty number does not need to be very large since the resulting error in the constraint equations will be eliminated by the Lagrange's terms during the iteration procedure.

The matrix formulation of (41), including the iterative process defined in (43), is given by the following expression

$$\begin{aligned}&\left(\mathbf{M}+\Phi_{\mathbf{q}}^{T}\,\alpha\,\Phi_{\mathbf{q}}\right)\ddot{\mathbf{q}}_{i+1}=\\&=\mathbf{M}\,\ddot{\mathbf{q}}_{i}-\Phi_{\mathbf{q}}^{T}\,\alpha\left(\dot{\Phi}_{\mathbf{q}}\,\dot{\mathbf{q}}+\dot{\Phi}_{t}+2\,\Omega\,\mu\,\dot{\Phi}+\Omega^{2}\,\Phi\right)\quad i=0,1,2,\ldots\end{aligned} \tag{44}$$

with $\mathbf{M}\,\ddot{\mathbf{q}}_{0}=\mathbf{Q}$ for the initial iteration. The subscript *i* represents the iteration number. The extra numerical effort to perform the iterations is not significant, since an iterative procedure is usually necessary to solve a system of nonlinear differential equations.

The penalty formulation has the advantage of having to solve a set of *n* equations, as compared to *(n+m)* needed by the Lagrange's multiplier method. In addition, constraint stabilization is implicitly considered within the algorithm, redundant constraint equations are considered automatically, the systems of linear equations has a positive-definite matrix and it is simpler to implement than the methods that use independent coordinates which are shown in the sequel. It has been shown (Bayo and Avello (1993)) that this formulation is also very robust with respect to singular positions.

3.1.2. *Formulations in Independent Coordinates.* Two important advantages of this type of coordinates are the reduction in the number of equations to be integrated, and the disappearance of the instability problem in the integration of the constraint equations using ODE solvers. However, this has a price in terms of computational effort (the position and velocity problems need be solved after the function evaluations) and difficulty in the implementation of some of the numerical integration methods, in particular the more stable implicit ones.

One point of great importance in these methods is the choice for the right set of independent coordinates; it is closely related to the methods to compute matrix $\mathbf{R}$ explained in Section 2.2 and will not be considered here. All that is important to point out is that, usually, no system of independent coordinates is adequate for the entire range of motion of the system. As a consequence, it is necessary to establish a double actuation procedure: on one hand a method must be developed that permits checking when a set of independent coordinates is becoming inadequate, and on the other hand, it is necessary to establish a method which will permit finding the most adequate new set of independent coordinates. Fortunately, there are mathematical properties of the Jacobian matrix $\mathbf{\Phi_q}$, that permit the solution of both problems satisfactorily.

One last important point should be remembered here: very often the numerical integration subroutines of ordinary differential equation are based on multistep methods; these methods are very efficient, but they require special techniques for starting the integration process. Due to the fact that each time it is necessary to change the independent coordinates, the numerical integration must be restarted again, it is recommended to carry out the minimum possible number of coordinate changes. On the other hand, when some determined coordinates start to be inadequate, the integration process becomes much slower. To summarize, it is necessary to arrive at a compromise solution, by making the minimum number of coordinate changes that guarantee quick and accurate numerical integration.

The numerical integration process with independent coordinates requires solving the position problem and performing the velocity analysis at each iteration. The latter does not constitute an important difficulty, however, the position problem does, because it requires an iterative solution that consumes an important amount of computational time. For this reason some authors as Paul (1975) have suggested the integration of the following extended set of variables

$$\dot{\mathbf{y}}_t^{\mathrm{T}} \equiv \left\{\ddot{\mathbf{z}}^{\mathrm{T}},\ \dot{\mathbf{q}}^{\mathrm{T}}\right\}_t \xrightarrow{\text{n.i.s.}} \mathbf{y}_{t+\Delta t}^{\mathrm{T}} \equiv \left\{\dot{\mathbf{z}}^{\mathrm{T}},\ \mathbf{q}^{\mathrm{T}}\right\}_{t+\Delta t} \tag{45}$$

where $\ddot{\mathbf{z}}$ are the independent accelerations. Because all the velocities have been integrated (and not only the independent ones), the new position of the multibody system is directly obtained as result of the numerical integration. In this numerical integration process, the constraint equation stabilization problem is not so critical as in Section 3.1.1., because the dependent variables that are integrated are the velocities, and not the accelerations. A lot of numerical experiments have shown that the numerical integration of (45), complemented with checking of constraint violations and the solution of the position problem when this violation is too large, provides an excellent compromise of speed and precision.

It is possible to compute dependent accelerations $\ddot{\mathbf{q}}$ by any of the methods explained in Section 3.1.1 and afterwards to integrate numerically only an appropriate subset of independent accelerations $\ddot{\mathbf{z}}$. This procedure has been called *coordinate partitioning method* (Wehage and Haug (1982)). We will describe here other family of methods based on the projection matrix **R**. In Section 2.2, the following transformation between dependent and independent accelerations has been introduced

$$\ddot{\mathbf{q}} = \mathbf{R}\,\ddot{\mathbf{z}} + (\mathbf{S}\,\mathbf{c}) \tag{46}$$

By introducing this equation in expression (37), it is obtained

$$\mathbf{R}^{\mathrm{T}}\,\mathbf{M}\,\mathbf{R}\,\ddot{\mathbf{z}} = \mathbf{R}^{\mathrm{T}}\,\mathbf{Q} - \mathbf{R}^{\mathrm{T}}\,\mathbf{M}\,(\mathbf{S}\,\mathbf{c}) \tag{47}$$

which constitutes the equations of motion in terms of independent coordinates. Eq. (48) represents a general matrix transformation from the vector spaces of dependent accelerations and forces to the vector space of independent accelerations and forces.

This formulation is valid for both scleronomous and rheonomous constraint equations. In addition, this layout is valid, irrespective of the method chosen to compute matrix **R**. Vector $\ddot{\mathbf{z}}$ doesn't need to be a subset of vector $\ddot{\mathbf{q}}$, but instead it can be a fully different set of variables.

3.1.3. *Comparative Remarks*. The penalty formulation defined by eqs. (41) and (44) has the advantage, over the formulations in independent coordinates, that the appearance or disappearance of constraints can be accommodated automatically without changing the coordinates, which in turn avoids the restarting procedure of the numerical integrator. The penalty formulation is also more suitable when the multibody system goes through a singular or bifurcation position, because in these cases the Jacobian changes its rank. As a consequence, and unless special provisions are made, the formulation in independent coordinates (and even the Lagrange's equations in dependent coordinates) tend to either crash the simulation or introduce sudden large errors, whereas with the penalty formulation the term $(\mathbf{M} + \Phi_{\mathbf{q}}^{\mathrm{T}}\,\alpha\,\Phi_{\mathbf{q}})$ in eq. (44) is free of singularities and makes it be very stable under these circumstances (for more details see Bayo and Avello (1993))

The penalty formulation (44) will tend to be more efficient numerically than the formulations in independent coordinates, simply because in eq. (44) the major computational burden is the formation, triangularization and one forward reduction and backsubstitution of $(\mathbf{M} + \Phi_{\mathbf{q}}^{\mathrm{T}}\,\alpha\,\Phi_{\mathbf{q}})$. Since the mass matrix does not modify the sparsity of the product $(\Phi_{\mathbf{q}}^{\mathrm{T}}\,\Phi_{\mathbf{q}})$, this operation is less costly than the formation, triangularization and f forward reductions and backsubstitutions of $(\Phi_{\mathbf{q}}^{\mathrm{T}}\,\Phi_{\mathbf{q}})$, required for the formation of the matrix **R** with the least squares formulation. Note, that these algorithms also include the formation and triangularization of $(\mathbf{R}^{\mathrm{T}}\mathbf{M}\mathbf{R})$ which represents an additional computational burden of these methods.

3.2. TOPOLOGICAL FORMULATIONS

The general purpose dynamic formulations described in Section 3.1 are simple, but they are not suitable for very fast dynamic simulation, that requires formulations that take into account the system's topology.

3.2.1. *Recursive Formulations.* Historically, most of the improvements in multibody dynamic formulations come from the robotics field. Walker and Orin (1982) shown that the solution of the inverse dynamics by recursive Newton-Euler method allows a very efficient formulation of the equations of motion. The *composite inertia* method seems to be the most efficient dynamic formulation for serial robots with N<10, which includes most practical cases. It is a $0(N^3)$ method.

Other authors (Featherstone (1987)) have developed fully recursive 0(N) algorithms for open-chain systems. Although they are not the most efficient in practice, the elegance and attractiveness of the Featherstone's formulation has exerted a strong influence on later developments that have generalized these ideas for non-serial (tree-configuration) systems and closed-loop systems (Bae and Haug (1987-88)). More recently, some interest has been placed in looking for improved efficiency using $0(N^3)$ variants of this method (Bae et al. (1988); Bae and Won (1990)).

Summarizing, the method of Featherstone proceeds with a triple recursion in the following way: 1) Knowing the relative position and velocity at the joints, the Cartesian position and velocity of all the links are computed recursively forward, from i=1 to i=N. 2) Equivalent or *articulated* inertias and forces are computed recursively backwards, from i=N to i=1. 3) Finally, the relative accelerations are computed recursively forward again, from i=1 to i=N.

These ideas have been extended to MBS with many branches on a tree-configuration, and afterwards to systems with closed loops. The consideration of branches in the kinematic chain is a simple task, and more difficulties are found for closed loop MBS.. These can be transformed into open chain systems by cutting a joint in each closed loop. (see Bae and Haug. (1987-88)). In a later work Bae et al. (1988) introduced a modification addressed to compute all the relative accelerations at once by solving a system of linear equations, thus becoming an $0(N^3)$ method.

3.2.2. *Velocity Transformations.* More recently, García de Jalón et al. (1989), Bae and Won (1990) and Bayo et al. (1991) have presented formulations well suited for real time analysis, that are based on *velocity transformations*, similar to the ones presented by Jerkovsky (1978) and Kim and Vanderploeg (1986a). We will study in this Section a general and simple method to formulate the dynamic equations of any open or closed chain MBS, and which can be parallelized even to the body (or element) level.

The dynamic formulation in independent coordinates described in Sections 3.1.2, based on the projection matrix **R**, is simple and general. It treat all systems in the same way, regardless of their topology and particular characteristics. A way to improve the efficiency of this formulation is to take advantage of the open chain configurations that the multibody systems may have or in which they may be transformed.

Let us consider an open chain multibody system that consists in one or several *branches*, which form a *tree structure*, as shown in figure 3. In open chain systems relative coordinates are also independent coordinates. It has been pointed out in Section 2.3.1. that for open chains matrix **R** can be computed directly, without forming and triangularizing the Jacobian matrix. In addition to this, the *sparsity* pattern of **R** becomes apparent and can be used in subsequent matrix operations. It is easy to see that the part of the matrix **R** that affects a particular link or element can be formed independently of the rest of the elements. This property leads to an *body–by–body* treatment of the equations of motion.

Finally, it is worth mentioning once again that, as the columns of matrix **R** thus calculated constitutes a basis for the nullspace of the Jacobian matrix, it can be written

$$\Phi_q \mathbf{R} = 0 \tag{48}$$

although the constraint equations are never explicitly calculated. Once the matrix **R** is known, we can use the method for the formulation of the equations of motion in independent coordinates explained in Section 3.1.2. In particular, we can use eq. (47). The matrix **R** may be obtained in an *body–by–body* basis, considering separately the rows that correspond to the dependent velocities of a particular body; this opens very good opportunities to carry out the computations in parallel. The matrix product ($\mathbf{R}^T\mathbf{M}\mathbf{R}$) and other terms that appear in eq. (47) can be computed on this *body–by–body* basis.

It was shown in figure 4 how a closed-loop system can be transformed to an open chain by simply eliminating certain constraint equations that enforce the closure of the loops. It is possible to divide the constraint equations into two groups: the first, denoted by the superscript 1, is formed by the constraints of the open chain system that result from the opening of the loops; the second, denoted by the superscript 2, will be formed by those constraints needed to close the loops previously opened. Consequently the Lagrange dynamic equations become

$$\begin{bmatrix} \mathbf{M} & \Phi_q^{1T} & \Phi_q^{2T} \\ \Phi_q^1 & 0 & 0 \\ \Phi_q^2 & 0 & 0 \end{bmatrix} \begin{Bmatrix} \ddot{\mathbf{q}} \\ \lambda^1 \\ \lambda^2 \end{Bmatrix} = \begin{Bmatrix} \mathbf{Q} \\ \mathbf{c}^1 \\ \mathbf{c}^2 \end{Bmatrix} \tag{49}$$

The key point in this formulation is the fact that the matrix $\mathbf{R}^1$, that defines a basis for the nullspace of Φ_q^1, *can be directly obtained* by the procedure explained previously. This obviously leads to large savings in computational costs, and even though the matrix Φ_q^1 is never formed explicitly, the following relationships will still be satisfied

$$\Phi_q^1 \mathbf{R}^1 = 0 \tag{50}$$

$$\dot{\mathbf{q}} = \mathbf{R}^1 \dot{\mathbf{z}}^1 \tag{51}$$

$$\ddot{\mathbf{q}} = \mathbf{R}^1 \ddot{\mathbf{z}}^1 + \mathbf{S}^1\mathbf{c}^1 \tag{52}$$

where the vector $\mathbf{z}^1$ is formed by the relative joint coordinates of the open chain system. Now, in the closed-loop system these coordinates $\mathbf{z}^1$ are not independent, because they are interrelated through the constraints Φ^2. The problem is that the constraints Φ^2 are not written in terms of $\mathbf{z}^1$ but in terms of **q**. However, this problem may be easily solved as follows. Substituting eq. (52) into eq. (49), premultiplying the first row by $(\mathbf{R}^1)^T$, and taking into account that the coefficient of λ^1 vanishes, we obtain

$$\begin{bmatrix} \mathbf{R}^{1T}\mathbf{M}\mathbf{R}^1 & \mathbf{R}^{1T}\Phi_q^{2T} \\ \Phi_q^2\mathbf{R}^1 & 0 \end{bmatrix} \begin{Bmatrix} \ddot{\mathbf{z}}^1 \\ \lambda^2 \end{Bmatrix} = \begin{Bmatrix} \mathbf{R}^{1T}\mathbf{Q} - \mathbf{R}^{1T}\mathbf{M}\mathbf{S}^1\mathbf{c}^1 \\ \mathbf{c}^2 - \Phi_q^2\mathbf{S}^1\mathbf{c}^1 \end{Bmatrix} \tag{53}$$

It is worth mentioning that the new projected mass matrix is much smaller than the original in eq. (49). A very important fact is again that the matrix transformation implied in eq. (53) may be performed in an *body–by–body* basis, and thus can be parallelized in an optimal manner. The equations of motion (54) may be solved by either one of the following methods, seen in Section 3.1: Lagrange's multipliers, penalty formulation and transformation to true independent coordinates. For the details see Jiménez et al. (1993).

4. Dynamics of Flexible Body Systems

So far, we have presented several approaches to the solution of the kinematics and dynamics of multi-rigid-body systems. There are some important cases, however, in which deformation plays an important role, as it happens, for instance, in light-weight spatial structures and manipulators or in high-speed machinery. The dynamics of those systems is influenced by the deformation and thus the formulations of the preceding Sections cannot be applied. The complexity of the equations of motion considering deformation grows considerably, and so does its size, since the variables defining the deformation must also be considered.

Due to strong space limitations, it is not possible to do here an overview on the methods presented in the literature for the analysis of flexible multibody systems. Instead, we will concentrate on the developments carried out by the authors in the last few years. In Section 4.1 we will describe a general method based on the *moving frame approach* with natural coordinates, that can be used when the elastic displacements are small and linear mode superposition can be applied. In Section 4.2 we will present a formulation for beam-like elements based on the large-displacement theory, that uses the same kind of Cartesian variables –points and unit vectors– described previously. Both methods can be used together, including also rigid bodies modeled as explained before. The use a common set of Cartesian coordinates is on the basis of such general approach.

4.1. THE CLASSICAL MOVING FRAME APPROACH

In this method two kinds of variables are considered. First, the *rigid body variables*, that express the large nonlinear overall motion of the moving frame attached to each body; second, the *deformation variables*, that express the state of deformation of each body with respect to its moving frame. The relative elastic displacements are assumed to be small, so that the linear theory of elasticity holds. It is possible to take as deformation variables the nodal displacements resulting from a finite element discretization of the flexible body, but this may lead to a large number of unknowns. One way of reducing the size of the problem consists in assuming that during the motion only a few deformation modes will be excited and in taking the amplitude of such modes as unknowns. This is the popular substructuring technique called *component mode synthesis*, described by Hurty (1965) and used for MBS by Shabana and Wehage (1983). For a general description of this technique see Shabana (1989). A major advantage of the moving frame approach is that it makes use of the classical linear finite element theory to introduce either the nodal variables or the assumed mode shapes. Some of the limitations of this method have been pointed out by Kane et al. (1987), who showed that the moving frame approach with linear elasticity fails to consider the rotational stiffening and other second order effects that appear at very fast speeds of operation.

Next we will describe the moving frame method using the *natural coordinates*. A slightly different approach can be found in Vukasovic et al. (1990).

4.1.1. *Kinematics of a Deformable Body*. Consider the flexible body shown in figure 7. The moving frame is rigidly attached to it at point O. The position vector of a generic point P can be expressed as

$$\mathbf{r} = \mathbf{r}_0 + \mathbf{A}\,\bar{\mathbf{r}} = \mathbf{r}_0 + \mathbf{A}\,(\bar{\mathbf{r}}_n + \bar{\mathbf{u}}) \tag{54}$$

where the local position vector $\bar{\mathbf{r}}$ is expressed as its value in the undeformed configuration $\bar{\mathbf{r}}_n$ plus the elastic displacement in the moving frame $\bar{\mathbf{u}}$. Matrix **A** is the rotation matrix. This elastic displacement can be expressed as a linear combination of *static* and *dynamic* modes, in the form

$$\bar{\mathbf{u}} = \sum_{i=1}^{N_s} \bar{\Phi}_i(\bar{\mathbf{r}})\,\eta_i + \sum_{j=1}^{N_d} \bar{\Psi}_j(\bar{\mathbf{r}})\,\xi_j \tag{55}$$

where $\bar{\Phi}_i$ and $\bar{\Psi}_i$ are the static and dynamic modes, and η_i and ξ_j the corresponding modal amplitudes (in number N_s and N_d, respectively). In this formulation the distinction between static and dynamic modes is very important because, as it will be seen in the sequel, they are managed in a completely different way.

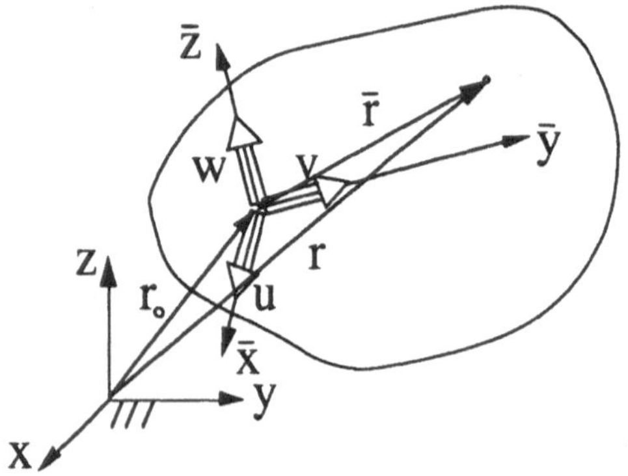

Figure 7. Deformable body with fixed and moving frames.

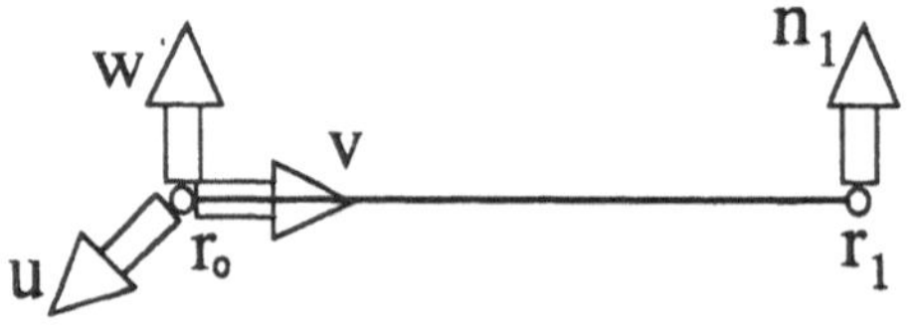

Figure 8. Flexible beam element in the undeformed configuration.

In this formulation for flexible MBS, points and unit vectors are defined on the joints and joint's axes exactly on the same way explained previously for rigid bodies, so as to be able to define the joints and joint's constraints in the same way (this is considered as essential so as to be able to mix rigid and flexible bodies in the analysis of a single MBS). The *static* modes are the deformation modes that result of introducing relative displacements between the points and vectors that belong to a deformable body. On the other hand, the *dynamic* modes describe internal deformation, i.e., deformation states that keep constant the relative position of points and unit vectors. We will present this formulation using a simple example: the beam element shown in figure 8.

In the element shown in figure 8 we will consider two points $\mathbf{r}_0$ and $\mathbf{r}_1$, and four unit vectors. Some of this unit vectors shall be chosen according with the axes orientation both joints. Point $\mathbf{r}_0$ is the origin of the moving frame and the vectors **u**, **v** and **w**, that are mutually orthogonal, define the orientation of the moving reference frame axes. The rotation matrix **A** can be expressed as

$$\mathbf{A} = [\mathbf{u} \mid \mathbf{v} \mid \mathbf{w}] \tag{56}$$

Let us consider the static modes of the body in figure 8. The modes $\overline{\Phi}_i$ (i = 1, 2, 3) are obtained by introducing unit displacements of point $\mathbf{r}_1$ on the directions $(\bar{x}, \bar{y}, \bar{z})$, respectively. These static modes are displayed in figure 9. Note that mode Φ_1 produce a deformation in the $(\bar{x}, \bar{y})$ plane; the deformation of mode Φ_2 is an axial deformation on the direction of axis $\bar{y}$; and the deformation of mode Φ_3 is contained in the plane $(\bar{y}, \bar{z})$. Note also the difference between modes Φ_1 and Φ_3 at point $\mathbf{r}_1$: in mode Φ_1 the beam is free to rotate around $\mathbf{n}_1$, while in mode Φ_3 rotation is forbidden because vector $\mathbf{n}_1$ shall maintain its direction.

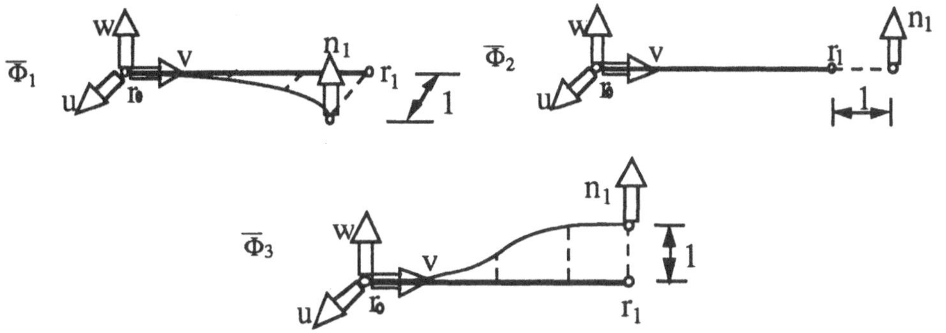

Figure 9. Static modes due to displacements of a point.

In figure 10 the two static modes that result from variation of the $(\bar{x}, \bar{y})$ components of $\mathbf{n}_1$ are shown. Mode Φ_4 is a torsion mode due to a variation in the component $\bar{x}$ of $\mathbf{n}_1$; mode Φ_5 is a bending mode in the plane $(\bar{y}, \bar{z})$ due to a variation in the component $\bar{y}$ of $\mathbf{n}_1$. Note that in this case a bending of the beam at point $\mathbf{r}_1$ with respect to vector $\mathbf{n}_1$ is not related with a variation in the natural coordinates, so it is considered as an internal deformation that shall be determined by the dynamic modes (see figure 11).

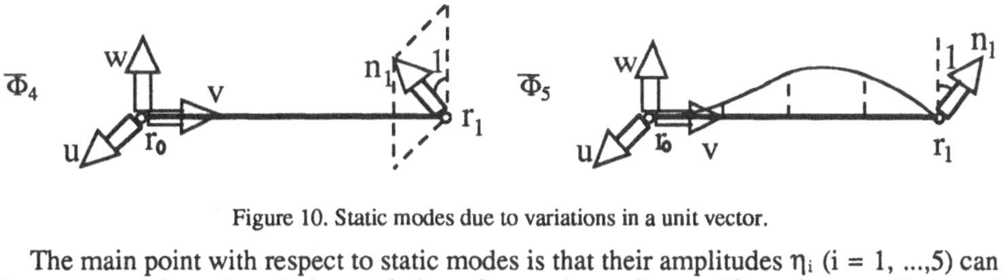

Figure 10. Static modes due to variations in a unit vector.

The main point with respect to static modes is that their amplitudes η_i (i = 1, ...,5) can be computed from the relative variation of natural coordinates (Cartesian coordinates of points and unit vectors), so they do not need to be considered as mechanism coordinates. We will see how this can be done for the flexible element we are considering.

Taking into account that static modes Φ_i (i = 1, 2, 3) have been computed by introducing unit displacements for point $\mathbf{r}_1$, the real amplitudes of these static modes can be easily computed in the moving frame from the expression,

$$\eta_{\mathbf{r}_1} = \bar{\mathbf{r}}_1 - \bar{\mathbf{r}}_{1n} \tag{57}$$

In order to compute these amplitudes as a function of the natural coordinates we can use the following coordinate transformation

$$\bar{\mathbf{r}}_1 - \bar{\mathbf{r}}_0 = \mathbf{A}^T (\mathbf{r}_1 - \mathbf{r}_0) \tag{58}$$

Taking into account that $\bar{\mathbf{r}}_0 = 0$, substituting eq. (58) into eq. (57) we obtain the expression of static modal amplitudes in terms of the natural coordinates,

$$\eta_{\mathbf{r}_1} = \mathbf{A}^T (\mathbf{r}_1 - \mathbf{r}_0) - \bar{\mathbf{r}}_{1n} \tag{59}$$

where $\bar{\mathbf{r}}_{1n}$ is a constant vector and matrix $\mathbf{A}$ is determined from eq. (56). In an analogous way it is possible to compute the modal amplitudes corresponding to the static modes that originate from the unit variation of the components of vector $\mathbf{u}_1$. It is obtained

$$\eta_{\mathbf{n}_1} = \mathbf{A}^T \mathbf{n}_1 - \bar{\mathbf{n}}_{1n} \tag{60}$$

where only two of the three components make sense in this particular case. It has been shown that static modes do not introduce additional global coordinates in the analysis.

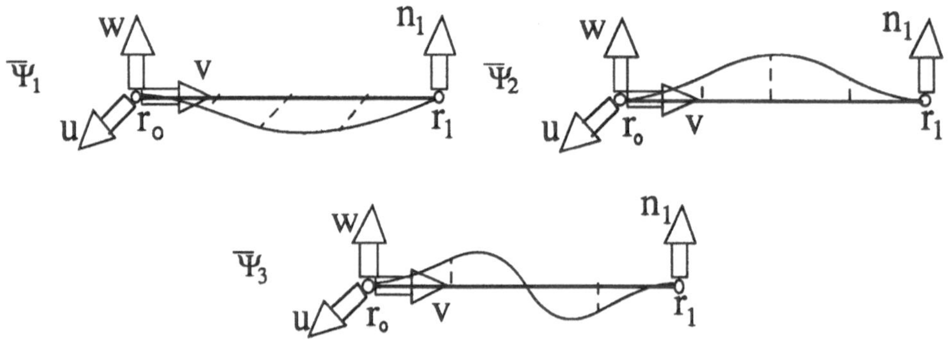

Figure 11. Some dynamic modes of the beam element.

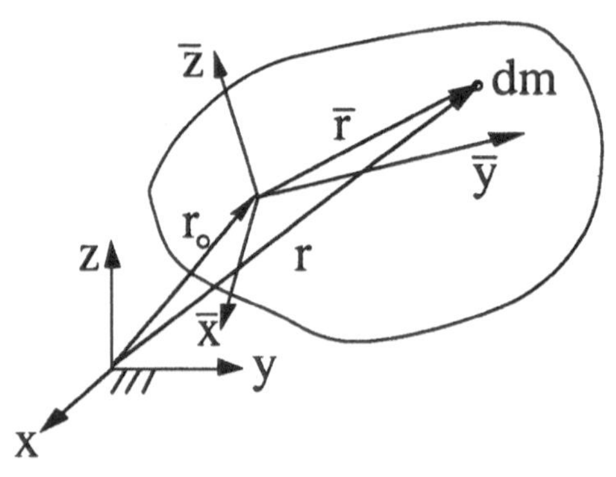

Figure 12. Differential mass element in a flexible body.

On the other hand, dynamic modes have been defined as internal deformation modes, i.e., modes that do not produce variation on the natural coordinates of the beam. Figure

11 shows three dynamic modes that have been computed as the eigenvectors (natural modes) of the beam with the boundaries clamped so as to do not allow variation in the Cartesian coordinates of the points and unit vectors. It can be seen that mode Ψ_1 is a bending mode in plane $(\bar{x}, \bar{y})$ with free rotation around unit vector $\mathbf{u}_1$; mode Ψ_2 is a bending mode in the plane $(\bar{y}, \bar{z})$ with both ends clamped and mode Ψ_3 is a second mode in the same plane with the same boundary conditions. Note that there are a finite number of static modes, but an infinity of dynamic modes, although only a few of them need to be included in the analysis (those that one expects that will be excited during the dynamic analysis). In some cases only some static modes –perhaps not all of them– will be enough to get the desired precision.

Of course, the amplitudes of dynamic modes ξ_j (i = 1, 2, ...) are independent of the Cartesian natural coordinates, and shall be included among the unknowns to be integrated during the dynamic simulation.

The extension of the concepts and ideas presented in this Section to more complex flexible bodies is straightforward. Static and dynamic modes can be computed with a finite element code by imposing the appropriate boundary conditions.

4.1.2. *Dynamic Equations and Constraint Equations.* Once the deformation modes have been defined, it is possible to formulate the motion differential equations. In this case we will use the virtual power principle. Consider the flexible body shown in figure 12. The virtual power of inertia forces of this body can be computed as

$$\widetilde{W} = \int_V \dot{\tilde{\mathbf{r}}}^T \ddot{\mathbf{r}}\, dm = \int_V \dot{\tilde{\mathbf{r}}}^T \ddot{\mathbf{r}}\, \rho\, dV \tag{61}$$

Taking time derivatives of eq. (54) it is obtained

$$\dot{\mathbf{r}} = \dot{\mathbf{r}}_0 + \dot{\mathbf{A}}\,\bar{\mathbf{r}} + \mathbf{A}\,\dot{\bar{\mathbf{r}}} \tag{62}$$

$$\ddot{\mathbf{r}} = \ddot{\mathbf{r}}_0 + \ddot{\mathbf{A}}\,\bar{\mathbf{r}} + 2\,\dot{\mathbf{A}}\,\dot{\bar{\mathbf{r}}} + \mathbf{A}\,\ddot{\bar{\mathbf{r}}} \tag{63}$$

and substituting in eq. (61),

$$\widetilde{W} = \int_V \left(\dot{\tilde{\mathbf{r}}}_0^T + \bar{\mathbf{r}}^T\dot{\tilde{\mathbf{A}}}^T + \dot{\tilde{\bar{\mathbf{r}}}}^T \mathbf{A}^T\right)\left(\ddot{\mathbf{r}}_0 + \ddot{\mathbf{A}}\bar{\mathbf{r}} + 2\dot{\mathbf{A}}\dot{\bar{\mathbf{r}}} + \mathbf{A}\ddot{\bar{\mathbf{r}}}\right)\rho\, dV \tag{64}$$

where $\dot{\bar{\mathbf{r}}}$ and $\ddot{\bar{\mathbf{r}}}$ can be obtained by differentiating eqs. (54) and (55), in which only η_i and $\xi_j$are functions of time. Although it is not possible to fully expand here this equation, it is interesting to consider one term and to remenber where are the problem variables (natural coordinates and dynamic modal coefficients); let us consider for instance the term,

$$\int_V \dot{\tilde{\mathbf{r}}}_0^T\, \dot{\mathbf{A}}\, \dot{\bar{\mathbf{r}}}\, \rho\, dV \tag{65}$$

In this integral term $\dot{\tilde{\mathbf{r}}}_0$ is a natural dependent velocity. Other dependent velocities appears in matrix $\dot{\mathbf{A}}$, that according to eq. (56) is the matrix $[\dot{\mathbf{u}}\,|\,\dot{\mathbf{v}}\,|\,\dot{\mathbf{w}}]$. From eq. (62), $\dot{\bar{\mathbf{r}}}$ contains the natural coordinates (**u**, **v**, **w**), the natural velocities $(\dot{\mathbf{r}}_0\,|\,\dot{\mathbf{u}}\,|\,\dot{\mathbf{v}}\,|\,\dot{\mathbf{w}})$ and the modal velocities $\dot{\eta}_i$ and $\dot{\xi}_j$. However, according to eqs. (59) and (60), the static modal velocities $\dot{\eta}_i$ can be expressed in terms of the natural velocities $(\dot{\mathbf{r}}_0\,|\,\dot{\mathbf{r}}_1\,|\,\dot{\mathbf{u}}\,|\,\dot{\mathbf{v}}\,|\,\dot{\mathbf{w}})$. The algebraic manipulations are straightforward, but too long to be reproduced here. The volume integration in eqs. (64) and (65) applies to the undeformed local coordinate $\bar{\mathbf{r}}_n$, to the modal shapes $\Phi_i(\bar{\mathbf{r}}_n)$ and $\Psi_j(\bar{\mathbf{r}}_n)$, and to the material density ρ.

Finally, we arrive to an expression in the form

$$\tilde{W}=\dot{\tilde{\mathbf{q}}}_e^T(\mathbf{M}_e\ddot{\mathbf{q}}_e+\mathbf{Qv}_e) \tag{66}$$

where $\dot{\tilde{\mathbf{q}}}_e$ is the vector that contains the dependent virtual velocities of the flexible body; $\mathbf{M}_e$ is the inertia matrix of the element.and $\mathbf{Qv}_e$the vector containing velocity dependent inertia forces.

Dynamic equations for the whole set of bodies can be obtained in an analogous way. Internal reaction forces do not produce virtual power. It is necessary to introduce the constraint equations that relate the natural coordinates. This can be carried out in the same way that for rigid bodies.

4.2. LARGE DEFORMATIONS

As mentioned previously, the classical moving frame approach is based on the assumption of small displacements. It assumes that the equilibrium condition is set in the undeformed configuration. Because of this, the method seen in the previous Section cannot handle larger deformations than those for which the linear finite element method and mode superposition yields accurate results.

When the second order effects become important and/or displacements become finite, the global or absolute method described in this section can be applied. We call it global or absolute because the entire motion of the body (finite rotation plus deformation) is all referred to a fixed frame. This produces a shifting of nonlinearity from the inertia terms in the moving frame approach, to the deformation terms in this approach. A formulation of this type was first presented by Simo and Vu-Quoc (1986 and 1988), for multibodies modeled as planar and three dimensional beams, respectively. See also Cardona (1989).

In this Section we will assume that the flexible bodies are long and slender and that they can be correctly modeled as beams. Timoshenko's beam theory will be used. With this basic assumption we will derive expressions for a simple nonlinear beam element that can be used to model flexible bodies in a multibody formalism. Perhaps the most attractive features of this formulation are its simplicity and the compatibility with the natural coordinates so far described in this paper, since the nodal variables of this beam element are also Cartesian coordinates of points and unit vectors.

4.2.1. *Kinematics of a Deformable Beam.* Figure 13 shows an initially straight prismatic beam of length L and constant cross section A. We introduce a fixed reference frame (X_1, X_2, X_3), with the X_1axis coincident with the centroidal line, and axes X_2 and X_3 coincident with the principal axes of inertia of the section. A cross section of the beam in this initial state can be described by the point (X_1, 0, 0) and by two mutually orthogonal vectors **M** and **N**, parallel to the X_2 and X_3 axes. We may think of **M** and **N** as *co-rotational vectors* that move rigidly attached to the cross section to which they belong.

After the beam has undergone finite displacements, we can define the position of its cross sections with position vector **r** and with the co-rotational vectors **m** and **n**. We will use upper case letters for the undeformed positions (*material coordinates*) and lower-case letters for deformed positions (*spatial coordinates*). The deformed positions can be expressed as a function of the undeformed ones. Since the undeformed beam is characterized by just the X_1 coordinate we can consider vectors **r**, **m** and **n**, as a function of X_1 and the time t, and the deformed coordinates $\mathbf{x}=(x_1, x_2, x_3)$ of a particle whose material coordinates are $\mathbf{X}=(X_1, X_2, X_3)$ can be written as

$$\mathbf{x}(\mathbf{X}, t) = \mathbf{r}(X_1, t) + X_2\, \mathbf{m}(X_1, t) + X_3\, \mathbf{n}(X_1, t) \tag{67}$$

where X_1 is not a function of time.

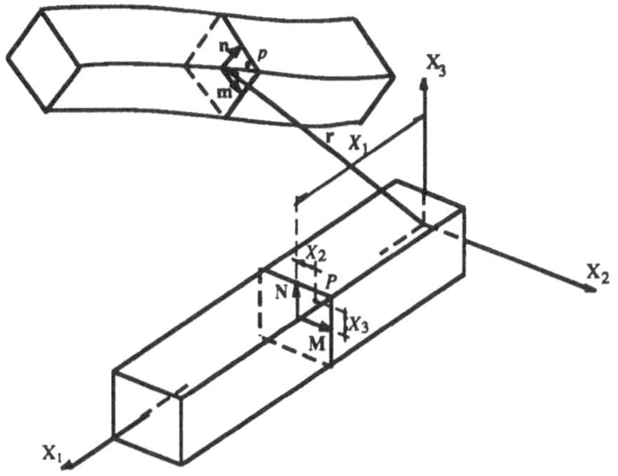

Figure 13. Deformed and undeformed prismatic 3-D beam.

Now *finite element* interpolation will be applied. Classic Timoshenko beam elements interpolate independently the displacements and rotations. The nodal variables are the three displacements u_i and three small rotations θ_i. In a similar way, we will assume an independent interpolation for the nodal variables, that will be different, in nature and number, from the classical ones. For this element the nodal variables are composed of the three coordinates of vector $\mathbf{r}^i$ and the six components of the orthogonal unit vectors $\mathbf{m}^i$ and $\mathbf{n}^i$, as we show in figure 14.

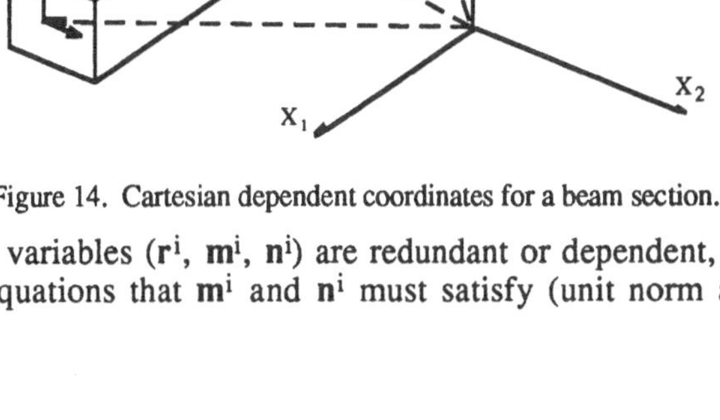

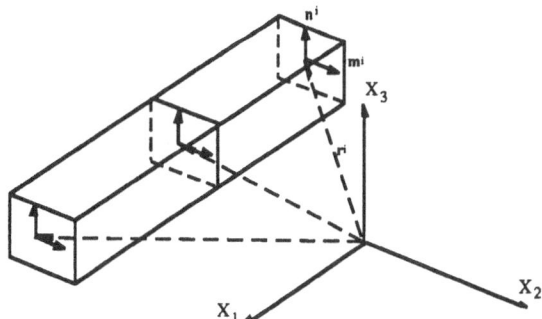

Figure 14. Cartesian dependent coordinates for a beam section.

The nine nodal variables ($\mathbf{r}^i$, $\mathbf{m}^i$, $\mathbf{n}^i$) are redundant or dependent, because there are three constraint equations that $\mathbf{m}^i$ and $\mathbf{n}^i$ must satisfy (unit norm and orthogonality conditions). Redundant variables have been extensively used in the analysis of multibody systems, but seldom in the finite element method. The use of redundant variables can reduce the complexity of the formulation. The cost that one has to pay is the introduction of constraint equations to enforce the satisfaction of the constraints at the nodes. For the sake of simplicity, only two-node elements will be considered here.

Let $(\mathbf{r}^i, \mathbf{m}^i, \mathbf{n}^i)$, i = 1, 2 be the values of $(\mathbf{r}, \mathbf{m}, \mathbf{n})$ in the nodes that belong to the beam element (e). The values of $(\mathbf{r}, \mathbf{m}, \mathbf{n})$ inside each finite element are obtained through the following interpolation scheme

$$\mathbf{r}^e = \sum_{i=1}^{2} N_i\, \mathbf{r}^i, \qquad \mathbf{m}^e = \sum_{i=1}^{2} N_i\, \mathbf{m}^i, \qquad \mathbf{n}^e = \sum_{i=1}^{2} N_i\, \mathbf{n}^i \tag{68}$$

where N_i are the standard finite element shape functions. Note that in eq. (68) also the unit vectors are interpolated. Since the shape functions are not required to preserve the norm, vectors $\mathbf{m}^e$ and $\mathbf{n}^e$ have no longer unit module and they are not orthogonal. This is a new source of discretization errors that is added to the standard error of the finite element method. A full discussion on how this error affects the accuracy of the solution goes beyond the scope of this paper, but it can be pointed out that the convergence of the finite element method is guaranteed, because this error decreases with element size, and in addition to this the numerical results obtained with this formulation are similar to the ones obtained with other nonlinear formulations (Avello et al. (1991)).

In order to obtain the inertia forces we first develop the expression for the kinetic energy, which can be obtained from the integral

$$T^e = \frac{1}{2}\int_{V^e} \dot{\mathbf{x}}^{eT}\, \dot{\mathbf{x}}^e\, dm \tag{69}$$

The velocity of a material point $\dot{\mathbf{x}}^e$ is obtained by differentiating expression (67) and by substituting the interpolation scheme given in (68), leading to

$$\dot{\mathbf{x}}^e = \sum_{i=1}^{2} N_i \left(\dot{\mathbf{r}}^i + X_2\, \dot{\mathbf{m}}^i + X_3\, \dot{\mathbf{n}}^i\right) \tag{70}$$

Substituting eq. (70) into (69) yields

$$\begin{aligned} T^e = \frac{1}{2}\int_{V^e} \sum_{i=1}^{2}\sum_{j=1}^{2} N_i N_j \Big(&\dot{\mathbf{r}}^{iT} \dot{\mathbf{r}}^j + X_2^2\, \dot{\mathbf{m}}^{iT} \dot{\mathbf{m}}^j + X_3^2\, \dot{\mathbf{n}}^{iT} \dot{\mathbf{n}}^j + \\ &+ 2\, X_2\, \dot{\mathbf{r}}^{iT} \dot{\mathbf{m}}^j + 2\, X_3\, \dot{\mathbf{r}}^{iT} \dot{\mathbf{n}}^j + X_2 X_3\, \dot{\mathbf{m}}^{iT} \dot{\mathbf{n}}^j\Big)\, dm \end{aligned} \tag{71}$$

where the only terms that depend on the integral variables are X_2 and X_3. Since X_2 and X_3 are principal axes of inertia, and X_1 coincides with the center of gravity of the cross section, the three last terms in the integral vanish After reordering eq. (71), the kinetic energy is obtained as

$$T^e = \frac{1}{2}\, \dot{\mathbf{q}}^{eT}\, \mathbf{M}^e\, \dot{\mathbf{q}}^e \tag{72}$$

where $\mathbf{q}^{eT} = \left\{\mathbf{r}^{1T}\, \mathbf{m}^{1T}\, \mathbf{n}^{1T}\, \mathbf{r}^{2T}\, \mathbf{m}^{2T}\, \mathbf{n}^{2T}\right\}$ is the vector that contains the nodal variables of element (e), and matrix $\mathbf{M}^e$ is a constant and symmetric matrix composed of sparse submatrices $\mathbf{M}_{ij}$ of size (9x9). In an homogeneous beam, $\mathbf{M}^e$ takes the form

$$\mathbf{M}^e = \begin{bmatrix} \mathbf{M}_{11} & \mathbf{M}_{12} \\ \mathbf{M}_{21} & \mathbf{M}_{22} \end{bmatrix}; \text{ with } \mathbf{M}_{ij} = \rho \begin{bmatrix} A\, c_{ij}\, \mathbf{I}_3 & \mathbf{0}_3 & \mathbf{0}_3 \\ \mathbf{0}_3 & I_2\, c_{ij}\, \mathbf{I}_3 & \mathbf{0}_3 \\ \mathbf{0}_3 & \mathbf{0}_3 & I_3\, c_{ij}\, \mathbf{I}_3 \end{bmatrix} \tag{73}$$

where ρ is the volumetric density, $\mathbf{I}_3$ the (3x3) unit matrix and c_{ij} the integral over the length of the element of the product of shape functions ($N_i\, N_j$).

This simple and constant expression for the mass matrix can be compared with the highly nonlinear matrix obtained in Section 4.1.1. However, the elastic potential energy is more complicated than with the moving frame method.

One of the basic assumptions often made in structural analysis is that displacements and displacement gradients are small. When this assumption holds, the *Cauchy strain tensor* can be used and gives accurate results. However it is known that the Cauchy strain tensor does not work for large displacements since it does not exhibit the proper invariance under rigid body rotations of the displacement field. The *Green strain tensor* has typically been used in nonlinear elasticity to characterize the deformation field of bodies undergoing large displacements. As the displacements and displacement gradients get smaller, the Green tensor tends to coincide with the Cauchy tensor.

Let us consider a continuous body and a fixed reference frame. We will use capital letters $\mathbf{X}=(X_1, X_2, X_3)$ to refer to the coordinates of a particle in the undeformed position and lower-case letters $\mathbf{x}=(x_1, x_2, x_3)$ for the deformed position. In the Lagrangian formulation $\mathbf{x}$ is taken as a function of $\mathbf{X}$ and time, in the form

$$\mathbf{x} = \mathbf{x}(\mathbf{X},\ t)) \tag{74}$$

The *deformation gradient* $\mathbf{F}$ is defined as the matrix that contains the partial derivatives of $\mathbf{x}$ with respect to $\mathbf{X}$. An infinitesimal vector in the deformed position $\mathbf{dx}$ can be expressed in terms of the deformation gradient and of its undeformed position $\mathbf{dX}$ as

$$dx = \frac{\partial x}{\partial X}\, dX = \mathbf{F}\, dX \tag{75}$$

The *Green deformation tensor* $\mathbf{C}$ is defined as the tensor that relates the square length $(ds)^2$ of vector $\mathbf{dx}$ with vector $\mathbf{dX}$. Thus

$$ds^2 = \mathbf{dX}^T\, \mathbf{C}\, \mathbf{dX} \tag{76}$$

The *Green strain tensor* $\mathbf{E}$ gives, by definition, the change in squared length between the deformed and the undeformed state of a vector $\mathbf{dX}$

$$ds^2 - dS^2 = 2\, \mathbf{dX}^T\, \mathbf{E}\, \mathbf{dX} \tag{77}$$

where $(dS)^2$ is the original length of vector $\mathbf{dX}$. From eqs. (75)-(77), it can be found

$$\mathbf{C} = \mathbf{F}^T\, \mathbf{F}; \qquad \mathbf{E} = \frac{\mathbf{C} - \mathbf{I}_3}{2} \tag{78}$$

The potential energy for a linearly elastic homogeneous material can be written in terms of the strain vector $E = \{E_{11}\, E_{22}\, E_{33}\, E_{12}\, E_{13} E_{23}\}^T$ as

$$V = \frac{1}{2} \int_{V^e} E^T\, \mathbf{D}\, E\; dV \tag{79}$$

where the integral is extended to the body in the undeformed configuration, and where **D** is a diagonal matrix defined in terms of *Lame's* constants λ and G.

From eq. (67) the deformation gradient **F** can be computed as

$$\mathbf{F} = \frac{\partial \mathbf{x}}{\partial \mathbf{X}} = [\, \mathbf{x}_{,1} \,|\, \mathbf{x}_{,2} \,|\, \mathbf{x}_{,3} \,] = [\, \mathbf{r}_{,1} + X_2\, \mathbf{m}_{,1} + X_3\, \mathbf{n}_{,1} \;\; |\, \mathbf{m} \,|\, \mathbf{n} \,] \tag{80}$$

where the vertical bars in equation indicate the separation between columns. We use the notation $(\text{-})_{,i}$ to represent $\partial(\text{-})/\partial X_i$. The Green strain tensor can be obtained by substituting eq. (80) into eqs. (78), as

$$\mathbf{E} = \frac{1}{2}\begin{bmatrix} \mathbf{x}_{,1}^T\, \mathbf{x}_{,1} - 1 & \mathbf{x}_{,1}^T\, \mathbf{m} & \mathbf{x}_{,1}^T\, \mathbf{n} \\ \mathbf{x}_{,1}^T\, \mathbf{m} & 0 & 0 \\ \mathbf{x}_{,1}^T\, \mathbf{n} & 0 & 0 \end{bmatrix}; \text{ with } \mathbf{x}_{,1} = \mathbf{r}_{,1} + X_2\, \mathbf{m}_{,1} + X_3\, \mathbf{n}_{,1} \tag{81}$$

Substituting eq. (81) into (80) and neglecting second order terms, after some algebraic manipulations the following expression for the strain vector **E** is obtained,

$$\begin{aligned} &E_{11} = \frac{1}{2}\left(\mathbf{r}_{,1}^T\, \mathbf{r}_{,1} - 1 + 2\, X_2\, \mathbf{r}_{,1}^T\, \mathbf{m}_{,1} + 2\, X_3\, \mathbf{r}_{,1}^T\, \mathbf{n}_{,1} \right); \quad E_{22} = E_{33} = 0 \\ &E_{12} = \frac{1}{2}\left(\mathbf{r}_{,1}^T\, \mathbf{m} + X_3\, \mathbf{n}_{,1}^T\, \mathbf{m} \right); \quad E_{13} = \frac{1}{2}\left(\mathbf{r}_{,1}^T\, \mathbf{n} + X_2\, \mathbf{m}_{,1}^T\, \mathbf{n} \right); \quad E_{23} = 0 \end{aligned} \tag{82}$$

which is in accordance with the strain distribution predicted by the strength of materials for a prismatic beam under axial, shearing, bending and torsion loads. For instance, the term $(\mathbf{r}_{,1}^T\, \mathbf{r}_{,1} - 1)/2$ in E_{11} represents a constant strain distribution corresponding to a pure axial load. Analogously, the term $(X_2\, \mathbf{r}_{,1}^T\, \mathbf{m}_{,1})$ in E_{11} represents a strain distribution that varies linearly with X_2 , with a zero value at the centroid and extreme values at the edges, as corresponds to a pure bending load.

Using eq. (82) the potential energy of a single element can be written as

$$\Pi^e = \frac{1}{2}\int_V \left[E\, A\, \Gamma_1^2 + E\, I_2\, \Gamma_2^2 + E\, I_3\, \Gamma_3^2 + G\, A_{s2}\, \Gamma_4^2 + G\, A_{s3}\, \Gamma_5^2 + G\, I_p\, \Gamma_6^2 \right] dX_1 \tag{83}$$

where Γ_1 represents the axial strain, Γ_2 and Γ_3 are the bending unit rotations per unit length, Γ_4 and Γ_5 are the shearing strains, and Γ_6 is the torsion rotation per unit length. Their expressions are

$$\Gamma_1 = \frac{\mathbf{r}_{,1}^T\, \mathbf{r}_{,1} - 1}{2}; \; \Gamma_2 = \mathbf{r}_{,1}^T\, \mathbf{n}_{,1}; \; \Gamma_3 = \mathbf{r}_{,1}^T\, \mathbf{m}_{,1}; \; \Gamma_4 = \mathbf{r}_{,1}^T\, \mathbf{m}; \; \Gamma_5 = \mathbf{r}_{,1}^T\, \mathbf{n}; \; \Gamma_6 = \mathbf{n}_{,1}^T\, \mathbf{m} \tag{84}$$

where A_{s2}, A_{s3} are the equivalent shear areas, and I_2, I_3 and I_p have the meaning

$$I_2 = \int_A X_3^2\, dA \qquad I_3 = \int_A X_2^2\, dA \qquad I_p = \int_A \left(X_2^2 + X_3^2 \right) dA \tag{85}$$

We can introduce the finite element interpolation given in eq. (68) into eqs. (83) and (84). After some algebraic manipulations and rearrangements, the following expressions for the strains Γ_i are obtained

$$\Gamma_i = \frac{1}{2}\, \mathbf{q}^{eT}\, \mathbf{G}^i\, \mathbf{q}^e - \beta_i, \qquad i = 1, \cdots, 6 \tag{86}$$

with $\beta_1 = 1$ and $\beta_i = 0$, $i = 2, \dots, 6$, and where $\mathbf{q}^e$ was defined previously. The matrices $\mathbf{G}^i$ are symmetric, sparse, and depend only on the shape functions and their derivatives with respect to X_1. Their expression can be found in Avello et al. (1991).The total potential energy for the beam is obtained by adding the potential energy of all the elements.

It can be pointed out that in this beam element the potential energy is obtained as a polynomial of order 4th in the position variables (Π^e depends on the square of Γ_i, and Γ_i depends on the square of $\mathbf{q}^e$), unlike the moving frame formulation of Section 4.1., in which the potential energy is a quadratic function of the position variables. Certainly, this complicates the implementation of the elastic forces, but recall that the mass matrix is constant and that it can be computed only once.

4.2.2. *Constraint Equations.* Since the position variables are not independent, it is necessary to introduce constraints at the finite element nodes and at the joints. The constraints at the nodes account for the unit norm and orthogonality conditions for the unit vectors. The constraints at the joints restrict the relative motion of adjacent bodies to the rotations or translations allowed by the kinematic joints.

The joint constraint equations at the joints can be written in terms of the nodal variables of the nodes next to the joint. This problem is fully analogous to the joint constraint equations for rigid bodies and will not be further developed here. Of course, the joint can link two flexible beams, but it can also link a beam and a rigid body or a beam and a flexible body computed with assumed deformation modes. The joint constraints would be developed in the same way, but different position variables shall be used for each kind of bodies.

4.2.3. *Dynamic Equations.* The equations of motion can be derived using any of the methods discussed previously. Here, we will use the Lagrange's multipliers method. The Lagrangian L can be written as

$$L = T - \Pi + \Phi^T \lambda \tag{87}$$

where Φ contains the constraints that arise from the unit norm and orthogonality condition for the nodal variables at the nodes, and the kinematic constraints imposed at the joints. The application of the Lagrange's equations leads to

$$\mathbf{M}\,\ddot{\mathbf{q}} + \Phi_{\mathbf{q}}^T \lambda = \mathbf{Q} - \mathbf{F} \tag{88}$$

where $\mathbf{M}$ is the mass matrix obtained by assembling the mass matrices $\mathbf{M}^e$ of each element, $\Phi_{\mathbf{q}}$ is the Jacobian of the constraint equations, $\mathbf{Q}$ is the vector of generalized external forces, and $\mathbf{F}$ the vector of elastic forces, that are obtained by differentiating eq. (83) with respect to $\mathbf{q}^e$.

5. Optimum Kinematic Synthesis of Linkages

The method presented in this Section is a contribution coming from Alvarez and Jiménez (1992). It is a good example of how natural coordinates can also be used to develop computer programs for the design of multibody systems.

Kinematic synthesis of mechanisms is mainly a geometric problem, about which much has been written in the past century and in the first half of the present one (Erdman and

Sandor (1978) and Suh and Radcliffe (1978)). During this time, many methods were developed (the majority of them were focused on the planar four bar mechanism), almost all of them graphic and containing a notable amount of ingeniousness and originality. The problems of *dimensional* synthesis are grouped together in three families: *function generation* synthesis, *path generation* synthesis and *rigid body guidance* synthesis.

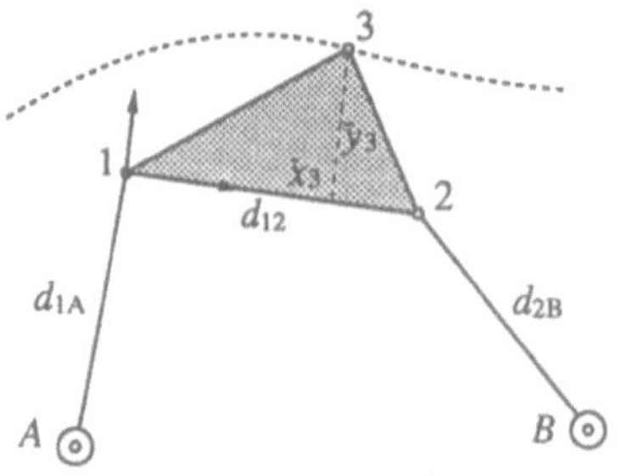

Figure 15. Path generated by a point of the coupling bar.

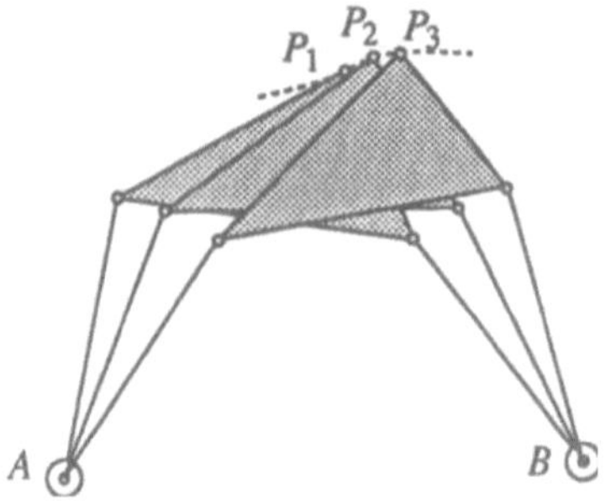

Figure 16. Design points for a path generation synthesis.

Graphic methods of kinematic synthesis are limited to simple mechanisms, they tend to be too specific, and at times difficult to use. In recent years, more general programs for *optimal synthesis* have been developed, and are applicable to many different types of planar and three-dimensional multibody systems, and include many different design conditions or specifications. Normally, these methods are based on numerical methods for optimization that seek the optimal solution with a minimum degree of error.

In this Section, we will describe a simple and general numerical method for the optimal kinematic synthesis of linkages. This method will be described with a path generation problem for a four-bar example, but it may be easily generalized for nearly any planar or three dimensional linkage and synthesis condition.

In order to carry out the optimum design of a multibody system for a defined set of design specifications, three steps shall be considered:

a) Choose the multibody system *topology*
b) Select the *design variables*
c) Define and minimize the *objective function*

We will consider two kinds of constraint equations: *geometric* constraints and *functional* constraints. The geometric constraints come from the multibody system topology –step (a)–, and are the constraints that we have considered in the previous Sections of this paper. The functional constraints come from the specific design requirements that the multibody systems must fulfill.

Let us consider the path generated by point 3 belonging to the coupler of the four bar mechanism in figure 15. In this case we will consider that points A and B can not be moved, thus the design variables are the elements of the following vector

$$\mathbf{b}^T = \{d_{1A}, d_{12}, d_{2B}, \bar{x}_3, \bar{y}_3\} \tag{89}$$

and the vector of dependent coordinates is

$$\mathbf{q}^T = \{x_1, y_1, x_2, y_2, x_3, y_3\} \tag{90}$$

In this example, the geometric constraints are the constraints that correspond to the four bar mechanism with three points in the coupler, that are

$$\phi_1 \equiv (x_1 - x_A)^2 + (y_1 - y_A)^2 - d_{1A}^2 = 0 \tag{91}$$

$$\phi_2 \equiv (x_1 - x_2)^2 + (y_1 - y_2)^2 - d_{12}^2 = 0 \tag{92}$$

$$\phi_3 \equiv (x_2 - x_B)^2 + (y_2 - y_B)^2 - d_{2B}^2 = 0 \tag{93}$$

$$\phi_4 \equiv x_3 - x_1 + \frac{(x_2 - x_1)}{d_{12}} \bar{x}_3 - \frac{(y_2 - y_1)}{d_{12}} \bar{y}_3 = 0 \tag{94}$$

$$\phi_5 \equiv y_3 - y_1 + \frac{(y_2 - y_1)}{d_{12}} \bar{x}_3 + \frac{(x_2 - x_1)}{d_{12}} \bar{y}_3 = 0 \tag{95}$$

In addition to the geometric constraints, the designer also specifies the functional constraints. For this example we will impose the conditions of the trajectory of point 3 passing as close as possible to a finite set of design points (P_1, P_2, P_3, ..., P_N), as shown in figure 16.

It is clear that each design point corresponds to a different value of the dependent coordinates vector $\mathbf{q}$. We will call these values ($\mathbf{q}^1$, $\mathbf{q}^2$, ..., $\mathbf{q}^N$). Now the functional constraints are imposed for each design point. For instance, for a generic point (i)

$$x_3^i - x_{P_i} = 0 \tag{96}$$

$$y_3^i - y_{P_i} = 0 \tag{97}$$

where i = 1, 2, ..., N.

In the general case, if $\mathbf{q}$ and $\mathbf{b}$ are the vectors of dependent coordinates and design variables, the geometric constraints equations can be expressed in vector form as

$$\mathbf{\Phi}(\mathbf{q}, \mathbf{b}) = 0 \tag{98}$$

It may be seen that, using natural coordinates, the constraints equations are very simple and the design variables $\mathbf{b}$ appear explicitly in $\mathbf{\Phi}$. The constraint eqs. (98) differ from the ones considered in previous Sections in the fact that the parameters in $\mathbf{b}$ are not constant as before, but true variables, because we are trying to finding their optimum values.

The whole set of constraints for the design point (i) –geometric and functional– can be written as

$$\mathbf{\Phi}^i(\mathbf{q}^i, \mathbf{b}) = 0 \qquad i = 1, 2, ..., N \tag{99}$$

Let us now introduce the *objective function*. We would like that point 3 of our four bar example goes exactly through the design points P_i. If it is not possible, we would like to get a four bar mechanism whose dimensions guarantee that the error in getting these design points is minimum in some sense. In other words, since exact solutions for the design problem may not exist, we will look for the optimal solution in the least square sense. Let us define an objective function of the form

$$\Psi(\mathbf{q}^1, \mathbf{q}^2, \ldots, \mathbf{q}^N, \mathbf{b}) = \frac{1}{2} \sum_{i=1}^{N} \Phi^{iT}(\mathbf{q}^i, \mathbf{b})\ \Phi^i(\mathbf{q}^i, \mathbf{b}) \tag{100}$$

or in a more compact form

$$\Psi(\bar{\mathbf{q}}, \mathbf{b}) = \frac{1}{2} \bar{\Phi}(\bar{\mathbf{q}}, \mathbf{b})^T \bar{\Phi}(\bar{\mathbf{q}}, \mathbf{b}) \tag{101}$$

where $\bar{\mathbf{q}}$ is the vector $\bar{\mathbf{q}}^T = \{\mathbf{q}^{1T}, \mathbf{q}^{2T}, \ldots, \mathbf{q}^{NT}\}$ and $\bar{\Phi}$ is a vector that contains all the geometrical and functional constraints. The optimum design problem consists in minimizing the objective function Ψ with respect to vectors $\bar{\mathbf{q}}$ and $\mathbf{b}$, that is

$$\min_{\bar{\mathbf{q}}, \mathbf{b}} \Psi(\bar{\mathbf{q}}, \mathbf{b}) = \frac{1}{2} \bar{\Phi}(\bar{\mathbf{q}}, \mathbf{b})^T \bar{\Phi}(\bar{\mathbf{q}}, \mathbf{b}) \tag{102}$$

Differentiating with respect to $\bar{\mathbf{q}}$ and $\mathbf{b}$, and equating to zero, the following system of non linear equations is obtained

$$\mathbf{J}(\bar{\mathbf{q}}, \mathbf{b})\ \bar{\Phi}(\bar{\mathbf{q}}, \mathbf{b}) = 0 \tag{103}$$

where

$$\mathbf{J}(\bar{\mathbf{q}}, \mathbf{b}) = \begin{bmatrix} \dfrac{\partial \bar{\Phi}(\bar{\mathbf{q}}, \mathbf{b})}{\partial \bar{\mathbf{q}}} \\ \dfrac{\partial \bar{\Phi}(\bar{\mathbf{q}}, \mathbf{b})}{\partial \mathbf{b}} \end{bmatrix} \tag{104}$$

We may now solve the nonlinear eqs. (103) by a quasi-Newton method. Expanding $\bar{\Phi}(\bar{\mathbf{q}}, \mathbf{b})$ in Taylor's series

$$\bar{\Phi}(\bar{\mathbf{q}} + \Delta\,\bar{\mathbf{q}}, \mathbf{b} + \Delta\,\mathbf{b}) = \bar{\Phi}(\bar{\mathbf{q}}, \mathbf{b}) + \mathbf{J}^T \begin{Bmatrix} \Delta\bar{\mathbf{q}} \\ \Delta\mathbf{b} \end{Bmatrix} + \cdots \tag{105}$$

and substituting in eq. (103), we obtain

$$\mathbf{J}(\bar{\mathbf{q}}, \mathbf{b})\ \bar{\Phi}(\bar{\mathbf{q}}, \mathbf{b}) + \mathbf{J}(\bar{\mathbf{q}}, \mathbf{b})\ \mathbf{J}^T(\bar{\mathbf{q}}, \mathbf{b}) \begin{Bmatrix} \Delta\bar{\mathbf{q}} \\ \Delta\mathbf{b} \end{Bmatrix} = 0 \tag{106}$$

from which the following iterative expression can be obtained

$$\begin{Bmatrix} \bar{\mathbf{q}} \\ \mathbf{b} \end{Bmatrix}_{k+1} = \begin{Bmatrix} \bar{\mathbf{q}} \\ \mathbf{b} \end{Bmatrix}_k - \left[\mathbf{J}(\bar{\mathbf{q}}, \mathbf{b})\ \mathbf{J}^T(\bar{\mathbf{q}}, \mathbf{b})\right]_k^{-1} \mathbf{J}(\bar{\mathbf{q}}, \mathbf{b})_k\ \bar{\Phi}(\bar{\mathbf{q}}, \mathbf{b})_k \tag{107}$$

This method is sufficiently simple and general to be applied to nearly any system topology (planar and three dimensional, open and closed chains, with any number and kind of joints and bodies), and can accommodate any kind of functional constraints, even a mixed set.

We will find now the complete set of constraint equations for the four-bar mechanism, considering five design points. Particularizing equations (91)–(95) and (96)–(97) for the generic design point P_i

$$(x_1^i - x_A)^2 + (y_1^i - y_A)^2 - d_{1A}^2 = 0 \tag{108}$$

$$(x_1^i - x_2^i)^2 + (y_1^i - y_2^i)^2 - d_{12}^2 = 0 \tag{109}$$

$$(x_2^i - x_B)^2 + (y_2^i - y_B)^2 - d_{2B}^2 = 0 \tag{110}$$

$$-(1 + \bar{x}_3/d_{12})\, x_1^i + (\bar{y}_3/d_{12})\, y_1^i + (\bar{x}_3/d_{12})\, x_2^i - (\bar{y}_3/d_{12})\, y_2^i + x_3^i = 0 \tag{111}$$

$$-(\bar{y}_3/d_{12})\, x_1^i - (1 + \bar{x}_3/d_{12})\, y_1^i + (\bar{y}_3/d_{12})\, x_2^i + (\bar{x}_3/d_{12})\, y_2^i + y_3^i = 0 \tag{112}$$

$$x_3^i - x_{P_i} = 0 \tag{113}$$

$$y_3^i - y_{P_i} = 0 \tag{114}$$

for i = 1, 2, ..., 5. There are then 35 constraint equations. The number of unknowns is also 35: five values of the 6-element dependent coordinates vector $\mathbf{q}^i$ plus the five elements of the design variables vector **b**. Then, with five design points it is possible to get a mechanism that exactly satisfies the functional constraints. If we have more than five design points, we only can get an optimal solution in the least square sense.

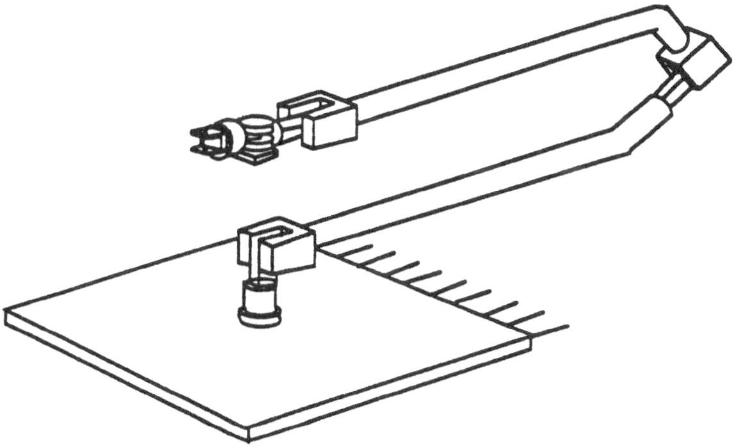

Figure 17. Complex multibody system with flexible bodies.

6. Numerical Example

Figure 17 shows a complex system consisting in a 6R spatial manipulator with two flexible links mounted on a clamped flexible plate. The manipulator's end-effector is grasping a lumped mass of 100 Kg. This example combines the three formulations with natural coordinates previously described. The plate has been modeled with 3 static and 6 dynamic modes obtained from a finite element discretization of 16 elements. Each one of the two slender bodies have been modeled with 4 nonlinear beam elements.

This manipulator undergoes a motion given by the following law:

$$\varphi(t) = \varphi_0 + \frac{\Delta\varphi}{2\pi}\left(2\pi\frac{t}{T} - \sin\left(2\pi\frac{t}{T}\right)\right) \quad 0 \le t \le T$$
$$\varphi(t) = 0 \quad T \le t \le 20 \text{ sec} \tag{115}$$

where T = 15 s and the angle increments for each joint are: $\Delta\varphi_1$= 1.650 rad, $\Delta\varphi_2$= 2.102 rad, $\Delta\varphi_3$= –1.200 rad, $\Delta\varphi_4$= 0.698 rad, $\Delta\varphi_5$= 0.0 rad and $\Delta\varphi_6$= 0.0. rad.

To evaluate the deviation of the manipulator's tip, the same motion law has been imposed to a rigid model of the manipulator and plate. The difference between the rigid

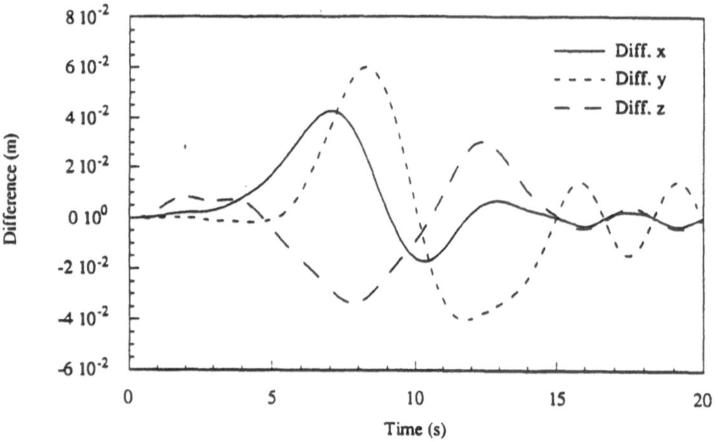

Figure 18. Difference between rigid and flexible tip's trajectories.

trajectory and the flexible one is illustrated in Figure 18. It can be seen that after the input motion stops, at t = 15 s, a free oscillation of constant amplitude remains.

7. References

Alvarez, G. and Jiménez, J.M. (1993) 'A Simple and General Method for Kinematic Synthesis of Spatial Mechanisms', work in preparation.

Amirouche, F.M.L. (1992) Computational Methods for Multibody Dynamics, Prentice-Hall.

Argyris, J. (1982) 'An Excursion into Large Rotations', Computer Methods in Applied Mechanics and Engineering, Vol. 32, pp. 85-155.

Avello, A. García de Jalón, J. and Bayo, E. (1991) 'Dynamics of Flexible Multibody Systems Using Cartesian Coordinates and Large Displacement Theory', International Journal for Numerical Methods in Engineering, Vol. 32, pp. 1543-1563.

Bae, D.-S. and Haug, E.J. (1987) 'A Recursive Formulation for Constrained Mechanical System Dynamics. Part I: Open Loop Systems', Mechanics of Structures and Machines, Vol. 15, pp. 359-382.

Bae, D.-S. and Haug, E.J. (1987-88) 'A Recursive Formulation for Constrained Mechanical System Dynamics. Part II: Closed Loop Systems', Mechanics of Structures and Machines, Vol. 15, pp. 481-506.

Bae, D.-S., Hwang, R.S. and Haug, E.J. (1988) 'A Recursive Formulation for Real-Time Dynamic Formulation', 1988 Advances in Design Automation, ed by S.S. Rao, ASME, pp. 499-508.

Bae, D.-S. and Won, Y.S. (1990) 'A Hamiltonian Equation of Motion for Real Time Vehicle Simulation', 1990 Advances in Design Automation, ed. by B. Ravani, ASME, pp. 151-157.

Baumgarte, J. (1972) 'Stabilization of Constraints and Integrals of Motion in Dynamical Systems', Computer Methods in Applied Mechanics and Engineering, Vol. 1, pp. 1-16.

Bayo, E. and Avello, A. (1993) 'Singularity Free Augmented Lagrangian Algorithms for Constraint Multibody Dynamics', to appear in the Journal of Nonlinear Dynamics.

Bayo, E., García de Jalón, J. and Serna, M.A. (1988) 'A Modified Lagrangian Formulation for the Dynamic Analysis of Constrained Mechanical Systems', Computer Methods in Applied Mechanics and Engineering, Vol. 71, pp. 183-195.

Bayo, E., García de Jalón, J., Avello, A. and Cuadrado, J. (1991) 'An Efficient Computational Method for Real Time Multibody Dynamic Simulation in Fully Cartesian Coordinates', Computer Methods in Applied Mechanics and Engineering, Vol. 92, pp. 377-395.

Cardona, A. (1989) An Integrated Approach to Mechanism Analysis, Ph. D. Thesis, Université de Liége, Belgium.

Duff, I.S., Erisman, A.M. and Reid, J.K. (1986) Direct Methods for Sparse Matrices, Clarendon Press, Oxford.

Erdman, A.G. and Sandor, G.N. (1984) Mechanism Design: Analysis and Synthesis, Volumes 1 and 2, Prentice-Hall.

Featherstone, R. (1987) Robot Dynamics Algorithms, Kluwer Academic Publishers..

García de Jalón, J., Avello, A. Jiménez, J.M, Martín, F. and Cuadrado, J. (1990) 'Real Time Simulation of Complex 3-D Multibody Systems with Realistic Graphics', Real-Time Integration Methods for Mechanical System Simulation, NATO ASI Series, pp. 265-292, Springer-Verlag.

García de Jalón, J. and Bayo, E. (1993) Kinematic and Dynamic Simulation of Multibody Systems -The Real-Time Challenge-, Springer-Verlag, New York.

García de Jalón, J., Serna, M.A. and Avilés, R. (1981) 'A Computer Method for Kinematic Analysis of Lower-Pair Mechanisms. Part I: Velocities and Accelerations and Part II: Position Problems', Mechanism and Machine Theory, Vol. 16, pp.543-566.

Hartenberg, R.S. and Denavit, I. (1964) Kinematic Synthesis of Linkages, McGraw-Hill, New York.

Haug, E.J. (1989) Computer-Aided Kinematics and Dynamics of Mechanical Systems, Volume I: Basic Methods, Allyn and Bacon.

Hurty, W.C. (1965) 'Dynamic Analysis of Structural Systems Using Component Modes', AIAA Journal, Vol. 3, pp. 678-685.

Huston, R.L. (1990) Multibody Dynamics, Butterworth-Heinemann.

Jerkovsky, W. (1978) 'The Structure of Multibody Dynamic Equations, Journal of Guidance and Control, Vol. 1, pp. 173-182.

Jiménez, J.M., Avello, An., Avello A. and García de Jalón, J. (1993) 'An Efficient Method Based on Velocity Transformations for Real Time Kinematic and Dynamic Simulation of Multibody Systems', NATO ASI Computer Aided Analysis of Rogid and Flexible Mechanical Systems, Troia (Portugal).

Kamman, J.W. and Huston, R.L. (1984) 'Dynamics of Constrained Multibody Systems', ASME Journal of Applied Mechanics, Vol. 51, pp. 899-903.

Kane, T.R. and Levinson, D.A. (1985) Dynamics: Theory and Applications, McGraw-Hill.

Kane, T.R., Ryan, R.R. and Banerjee, A.K. (1987) 'Dynamics of a Cantilever Beam Attached to a Moving Base', AIAAJournal of Guidance, Control and Dynamics, Vol. 10, pp. 139-151.

Kim, S.S. and Vanderploeg, M.J. (1986a) 'A General and Efficient Method for Dynamic Analysis of Mechanical Systems Using Velocity Transformations', ASME Journal of Mechanisms, Transmissions and Automation in Design, Vol. 108, pp. 176-182.

Kim, S.S. and Vanderploeg, M.J. (1986b) 'QR Decomposition for State Space Representation of Constrained Mechanical Dynamic Systems', ASME Journal on Mechanisms, Transmissions and Automation in Design, Vol. 108, pp. 183-188.

Kurdila, A. J. and Narcowich F.J. (1993) 'Sufficient Conditions for Penalty Formulation Methods in Analytical Dynamics', to appear inComputational Mechanics,

Mani, N.K., Haug, E.J. and Atkinson, K.E. (1985) 'Application of Singular Value Decomposition for Analysis of Mechanical System Dynamics', ASME Journal on Mechanisms, Transmissions and Automation in Design, Vol. 107, pp. 82-87.

Nikravesh, P.E. (1988) Computer-Aided Analysis of Mechanical Systems, Prentice-Hall.

Paul, B. (1975) 'Analytical Dynamics of Mechanisms - A Computer Oriented Overview', Mechanism and Machine Theory, Vol. 10, pp. 481-507.

Roberson, R.E. and Schwertassek, R. (1988) Dynamics of Multibody Systems, Springer-Verlag.

Schiehlen, W.O. (1990) Multibody System Handbook, Springer-Verlag.

Serna, M.A., Avilés, R. and García de Jalón, J. (1982) 'Dynamic Analysis of Planar Mechanisms with Lower-Pairs in Basic Coordinates', Mechanism and Machine Theory, Vol. 17, pp. 397-403.

Shabana, A.A. (1989) Dynamics of Multibody Systems, Wiley.

Shabana, A.A. and Wehage, R.A. (1983) 'A Coordinate Reduction Technique for Transient Analysis of Spatial Substructures with Large Angular Rotations', Journal of Structural Mechanics, Vol. 11, pp. 401-431.

Simo, J.C. and Vu-Quoc L. (1986) 'On the Dynamics of Flexible Beams Under Large Overall Motions - The Planar Case: Part I', Journal of Applied Mechanics, Vol. 53, pp. 849-854.

Simo, J.C. and Vu-Quoc L. (1988) 'On the Dynamics of Space Rods Undergoing Large Overall Motions', Computer Methods in Applied Mechanics and Engineering, Vol. 66, pp. 125-161.

Singh, R.P. and Likins, P.W. (1985) 'Singular Value Decomposition for Constrained Dynamic Systems', ASME Journal of Applied Mechanics, Vol. 52, pp. 943-948.

Suh, C.H. and Radcliffe, C.W. (1978) Kinematics and Mechanism Design, Wiley.

Unda, J., García de Jalón, J., Losantos, F. and Enparantza, R. (1987) 'A Comparative Study on Some Different Formulations of the Dynamic Equations of Constrained Mechanical Systems', ASME Journal of Mechanisms, Transmissions and Automation in Design, Vol. 109, pp. 466-474.

Vukasovic, N., Celigüeta, J.T., García de Jalón, J. and Bayo, E. (1990) 'Flexible Multibody Dynamics Based on a Fully Cartesian System of Support Coordinates', Flexible Mechanisms, Dynamics and Robot Trajectories, pp. 37-42, ed. by S. Derby, M. McCarthy and A. Pisano, ASME Press.

Walker, M.W. and Orin, D.E. (1982) 'Efficient Dynamic Computer Simulation of Robotic Mechanisms', ASME Journal of Dynamic Systems, Measurements and Control, Vol. 104, pp. 205-211.

Wehage, R.A. and Haug, E.J. (1982) 'Generalized Coordinate Partitioning for Dimension Reduction in Analysis of Constrained Dynamic Systems', ASME Journal of Mechanical Design, Vol. 104, pp. 247-255.

COMPUTER IMPLEMENTATION OF FLEXIBLE MULTIBODY EQUATIONS

A. A. SHABANA
Department of Mechanical Engineering
University of Illinois at Chicago
P. O. Box 4348
Chicago, Illinois 60680

ABSTRACT. A nonlinear finite element formulation for flexible multibody dynamics is summarized. In this formulation, it is required that the element shape functions can describe large rigid body translations. Of particular interest in the dynamics and control of flexible multibody systems is the concept of equivalent systems of forces. The formulation of the generalized forces and the nonlinear dynamic equations of substructures in flexible multibody dynamics is discussed and the validity of using the linear theory of elastodynamics in mechanical system applications such as tracked vehicles is reexamined.

1. Introduction

The fact that most of the element shape functions can be used to describe large translational displacements is crucial in the development of the nonlinear dynamic formulation of substructures in multibody dynamics. By using this fact and a set of coordinate systems that define the configuration of the finite element, the *nonlinear generalized Newton-Euler equations* of the substructures that undergo large rigid body displacements can be developed using the *principle of virtual work in dynamics* or *Lagrange's equation.* These equations can be expressed in terms of a unique set of *invariants of motion* that depend on the assumed displacement field and can be evaluated in advance in a preprocessor computer program.

In developing the equations of motion of the substructures in multibody dynamics, special attention must be paid to the definition of forces and moments. The concept of the equivalence of two systems of forces in rigid body dynamics is not applicable to deformable body dynamics. A force that acts at a point on a deformable body is equivalent to a system, defined at another point, that consists of the same force, a moment that depends on the relative displacement between the two points, and a set of generalized elastic forces that depends on the finite rotation of the body. This is a subject of particular interest in *control applications*, since in many cases the motion of the system is specified and the interest is focused on defining the *joint control forces* that produces the desired motion. Nonetheless, a close examination of

M. F. O. S. Pereira and J. A. C. Ambrósio (eds.),
Computer-Aided Analysis of Rigid and Flexible Mechanical Systems, 325–349.

the structure of the mass matrix and the forces in deformable body dynamics and the proper identification of the invariants leads to a systematic procedure for the automatic generation of the inertia and stiffness characteristics of deformable bodies in multibody systems.

Once the structure of the nonlinear dynamic equations that govern the unconstrained motion of deformable bodies is defined, two approaches can be used to formulate the multibody equations of motion. These are the *augmented* and the *recursive formulations*. In the augmented formulation, the multibody equations of motion are formulated in terms of a set of variables that include both the dependent and independent coordinates. In this type of formulation, constraints between the variables are formulated using a set of linear and/or nonlinear algebraic constraint equations that depend on the system coordinates and possibly on time. This leads to a mixed system of *algebraic* and *differential equations* that must be solved simultaneously using matrix and computer methods. In the recursive formulations, the equations of motion are formulated in terms of the joint variables or the system degrees of freedom. This leads to a smaller system of strongly coupled equations. In this case, one obtains only a set of differential equations that can be integrated numerically in order to define the state of the system.

Another topic of particular significance in the analysis of substructures in multibody dynamics is the coupling between the displacements. The coupling between the finite rotations and the deformation displacements has a significant effect on the dynamics of deformable bodies. Significant changes in the *wave phenomenon* occur as the result of the finite rotation. For example elastic waves in a perfectly elastic nonrotating rods propagate with the same phase velocity. Consequently, the group velocity is constant and is independent of the wave number or the dimension of the rod. *Dispersion*, however, occurs as the results of the finite rotation and its coupling with the deformation displacements. The *phase velocities* of harmonic waves are no longer equal and consequently the *group velocity* becomes dependent on the wave number.

2. Finite Rotations

In the transient finite element dynamic analysis, a *convected coordinate system* is attached to each finite element and hence it shares its rigid body motion. A sequence of fixed coordinate systems are introduced and at any instant of time it is assumed that the axes of the convected system coincide with the axes of one of the fixed coordinate systems. By assuming that there is a relatively sufficient number of fixed frames, the displacement of the element between two coordinate frames is described using the shape function and the nodal coordinates of the element. The current deformed state is used as the new reference state prior to the next incremental step in the transient dynamic solution. The updated Lagrangian formulation leads to a simple dynamic equations in which the element mass matrix defined in the convected coordinate system is constant. Furthermore, the use of the *lumped mass technique* leads to constant element mass matrix in the global coordinate system. Since several of the commonly used shape functions of *beams*, *plates* and *shells* can not be used to describe finite rotation, several of existing finite element formulations lead to a subtle linearization of the resulting dynamic equations. The limitations on the use of the commonly employed shape functions in the large displacement analysis of deformable bodies can be demonstrated. To this end, we use the shape function of the six degree of freedom, two node planar beam element. Each node is assumed to have three coordinates; two describe the translation and one describes the slope at this nodal point. The vector of nodal coordinates of the element j on the deformable body i can be written as

$$\mathbf{e}^{ij} = \begin{bmatrix} e_1^{ij} & e_2^{ij} & e_3^{ij} & e_4^{ij} & e_5^{ij} & e_6^{ij} \end{bmatrix}^{\mathrm{T}} \tag{1}$$

where e_1^{ij}, e_2^{ij}, e_4^{ij} and e_5^{ij} are the translational nodal coordinates, while e_3^{ij} and e_6^{ij} are the slopes at the two nodal points. An element shape function associated with this set of nodal coordinates is

$$\bar{\mathbf{S}}^{ij} = \begin{bmatrix} 1-\xi & 0 & 0 & \xi & 0 & 0 \\ 0 & 1-3\xi^2+2\xi^3 & l(\xi-2\xi^2+\xi^3) & 0 & 3\xi^2-2\xi^3 & l(\xi^3-\xi^2) \end{bmatrix}^{ij} \tag{2}$$

where $\xi = x/l$, and x is the spatial coordinate and l is the length of the element. A general rigid body translation can be described by the vector

$$\mathbf{R} = \begin{bmatrix} R_x & R_y \end{bmatrix}^{\mathrm{T}} \tag{3}$$

where R_x and R_y are the displacement of an arbitrary point on the element. As the result of this rigid body translation, the vector of nodal coordinates becomes

$$\begin{aligned} \mathbf{e}_t^{ij} &= \begin{bmatrix} e_1^{ij}+R_x & e_2^{ij}+R_y & e_3^{ij} & e_4^{ij}+R_x & e_5^{ij}+R_y & e_6^{ij} \end{bmatrix}^{\mathrm{T}} \\ &= \mathbf{e}^{ij} + \mathbf{R}_e \end{aligned} \tag{4}$$

where $\mathbf{e}^{ij}$ as defined by Eq. 1 is the vector of nodal coordinates before the rigid body translation and $\mathbf{R}_e$ is the vector

$$\mathbf{R}_e = \begin{bmatrix} R_x & R_y & 0 & R_x & R_y & 0 \end{bmatrix}^{\mathrm{T}}$$

By using simple matrix multiplication, it can be shown that

$$\bar{\mathbf{S}}^{ij}\,\mathbf{R}_e = \mathbf{R}$$

That is

$$\bar{\mathbf{S}}^{ij}\,\mathbf{e}_t^{ij} = \bar{\mathbf{S}}^{ij}\,\mathbf{e}^{ij} + \bar{\mathbf{S}}^{ij}\,\mathbf{R}_e = \bar{\mathbf{S}}\,\mathbf{e}^{ij} + \mathbf{R}$$

This implies that the element nodal coordinates can be used to describe an arbitrarily large rigid body translation. This is a basic assumption which is utilized in our formulation and its significance becomes apparent when the dynamic equations are formulated in terms of a minimum number of independent invariants of motion.

2.1. FINITE ROTATION

If the finite element undergoes a pure rotation defined by the angle θ, the position vector of an arbitrary point at a distance x from its end as the result of this rotation is

$$\mathbf{u} = \begin{bmatrix} u \\ v \end{bmatrix} = \begin{bmatrix} \cos\theta & -\sin\theta \\ \sin\theta & \cos\theta \end{bmatrix} \begin{bmatrix} x \\ 0 \end{bmatrix} = \begin{bmatrix} x\cos\theta \\ x\sin\theta \end{bmatrix} \tag{5}$$

Using this equation and the definition of the slope, one has

$$e_3^{ij} = e_6^{ij} = \frac{\partial v}{\partial x} = \sin\theta \tag{6}$$

In this case, the vector of nodal coordinates becomes

$$\mathbf{e}_r^{ij} = \begin{bmatrix} 0 & 0 & \sin\theta & l\cos\theta & l\sin\theta & \sin\theta \end{bmatrix}^{\mathrm{T}}$$

where l is the length of the element. It can be shown, by direct matrix multiplication, that

$$\bar{\mathbf{S}}^{ij}\,\mathbf{e}_r^{ij} = \begin{bmatrix} x\cos\theta \\ x\sin\theta \end{bmatrix}$$

That is, the shape function of the beam element, as defined by Eq. 2, can be used to describe finite rotations provided that the slopes at the nodal points can be defined using *trigonometric functions* as in Eq. 6. In the finite element formulation such definition can not be made since trigonometric functions lack any physical meaning. The use of trigonometric functions to define the nodal coordinates introduces technical difficulties in assembling the finite elements and in transforming the nodal coordinates from one coordinate system to another. On the other hand, if the rotation is assumed to be small, one has

$$e_3^{ij} = e_6^{ij} \approx \theta$$

Infinitesimal rotations can be treated as vectors. Therefore, the rule of transforming vectors from one coordinate system to another can be applied to the transformation of the nodal coordinates. Furthermore, the slopes as defined by the preceding equation have physical meaning and consequently no technical problems arise when the elements are assembled.

2.2. COORDINATE SYSTEMS

Using a similar procedure as the one described in this section, it can be shown that most of the commonly used shape functions can describe an arbitrary large rigid body translations. As demonstrated by the beam example presented in this section, some of the shape functions can not be used to describe an arbitrary finite rotation of the element. Even though in the cases where the element nodal coordinates can be used to describe finite rigid body rotations as in the case of *triangular*, *rectangular*, *solid* and *tetrahedral elements*, the use of the nodal coordinates is not convenient in describing the relative finite rotations between the components of the multibody system.

Using the fact that the element shape function can be used to describe an arbitrary large rigid body translation, the location of an arbitrary point on the element, as shown in Fig. 1, can be defined in an *intermediate element coordinate system* $\bar{\mathbf{X}}^{ij}\ \bar{\mathbf{Y}}^{ij}\ \bar{\mathbf{Z}}^{ij}$ whose axes are parallel to the axes of the *element coordinate system* $\mathbf{X}^{ij}\ \mathbf{Y}^{ij}\ \mathbf{Z}^{ij}$ as

$$\bar{\mathbf{u}}_i^{ij} = \bar{\mathbf{S}}^{ij}\left(\mathbf{e}_0^{ij} + \mathbf{e}_f^{ij}\right) \tag{7}$$

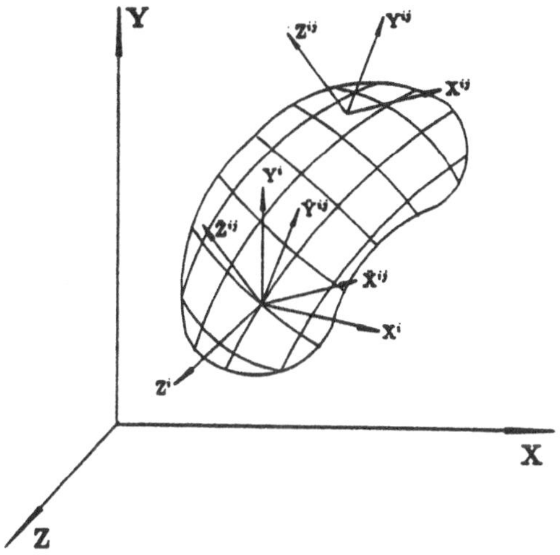

Figure 1. Coordinate systems

where $\bar{\mathbf{u}}_i^{ij}$ is the position vector of the arbitrary point on the element defined in the intermediate element coordinate system, $\mathbf{e}_0^{ij}$ is the vector of nodal coordinates in the undeformed state and $\mathbf{e}_f^{ij}$ is the vector of nodal deformations. The origin of the intermediate element coordinate system $\bar{\mathbf{X}}^{ij}\ \bar{\mathbf{Y}}^{ij}\ \bar{\mathbf{Z}}^{ij}$ is assumed to be rigidly connected to the origin of the *body coordinate system* $\mathbf{X}^i\ \mathbf{Y}^i\ \mathbf{Z}^i$. In this case, the global position vector of an arbitrary point on the element j on the deformable body i can be written as

$$\mathbf{r}^{ij} = \mathbf{R}^i + \mathbf{A}^i\,\bar{\mathbf{u}}^{ij} \tag{8}$$

where $\mathbf{R}^i$ is the global position vector of the origin of the body coordinate system, $\mathbf{A}^i$ is the transformation matrix from the body to the global coordinate system and $\bar{\mathbf{u}}^{ij}$ is the local position vector of the arbitrary point defined as

$$\bar{\mathbf{u}}^{ij} = \mathbf{S}^{ij}\,\mathbf{B}^{ij}\,\mathbf{B}^i\,\mathbf{q}_f^i \tag{9}$$

in which $\mathbf{B}^{ij}$ is the *Boolean matrix* that describes the element connectivity, $\mathbf{B}^i$ is the matrix of the reference conditions that eliminate the rigid body motion of the substructure with respect to its coordinate system, $\mathbf{q}_f^i$ is the vector of elastic coordinates of the deformable body i and $\mathbf{S}^{ij}$ is the shape matrix of the element j defined in the body coordinate system. Note that this description of motion does not imply any linearization of the kinematic relationships provided that the element rotations in the case of beams and plates with respect to the body coordinate system are assumed to be small. If the shape function of the element can be used to describe arbitrary finite rotations such as in the case of two dimensional triangular and rectangular elements and in the case of three dimensional solid and tetrahedral elements, the kinematic equations presented in this section can be used without any assumption of linearization in the large deformation analysis of flexible multibody systems.

3. Inertia Forces

Several techniques can be used to derive the dynamic equations of the deformable body i that undergoes large rigid body displacements. In the case of *unconstrained deformable body*, the application of the *principle of virtual work in dynamics* leads to

$$\mathbf{Q}_i^i = \mathbf{Q}_e^i \tag{10}$$

where $\mathbf{Q}_i^i$ is the vector of the *generalized inertia forces* and $\mathbf{Q}_e^i$ is the vector of *applied external* and *elastic forces.* In *Lagrange's equation* the generalized inertia forces are expressed in terms of the kinetic energy, while in *Gibbs-Appel equation* the generalized inertia forces are expressed in terms of the acceleration function. Both can be derived using the basic definition of the virtual work of the inertia forces defined as

$$\begin{aligned} \delta W_i &= \int_{V^i} \rho^i \, \ddot{\mathbf{r}}^{i\mathrm{T}} \, \delta \mathbf{r}^i \, dV^i = \int_{V^i} \rho^i \ddot{\mathbf{r}}^{i\mathrm{T}} \frac{\partial \mathbf{r}^i}{\partial \mathbf{q}^i} \, \delta \mathbf{q}^i \, dV^i \\ &= \mathbf{Q}_i^{i\mathrm{T}} \, \delta \mathbf{q}^i \end{aligned} \tag{11}$$

where ρ^i and V^i are, respectively, the mass density and volume of the deformable body i, $\ddot{\mathbf{r}}^i$ is the acceleration vector of an arbitrary point on the deformable body and $\mathbf{q}^i$ is the vector generalized coordinates of the body which can be defined using the *absolute reference* and the *elastic relative coordinates* as

$$\mathbf{q}^i = \left[\mathbf{R}^{i\mathrm{T}} \;\; \boldsymbol{\theta}^{i\mathrm{T}} \;\; \mathbf{q}_f^{i\mathrm{T}} \right]^{\mathrm{T}} \tag{12}$$

in which $\boldsymbol{\theta}^i$ is the set of rotational coordinates used to describe the orientation of the deformable body and $\mathbf{R}^i$ and $\mathbf{q}_f^i$ are as previously defined. It follows from Eq. 11 that the generalized inertia forces are

$$\mathbf{Q}_i^{i\mathrm{T}} = \int_{V^i} \rho^i \, \ddot{\mathbf{r}}^{i\mathrm{T}} \frac{\partial \mathbf{r}^i}{\partial \mathbf{q}^i} \, dV^i \tag{13}$$

which is the same as

$$\mathbf{Q}_i^i = \int_{V^i} \rho^i \, \ddot{\mathbf{r}}^{i\mathrm{T}} \frac{\partial \dot{\mathbf{r}}^i}{\partial \dot{\mathbf{q}}^i} \, dV^i \tag{14}$$

since

$$\frac{\partial \mathbf{r}^i}{\partial \mathbf{q}^i} = \frac{\partial \dot{\mathbf{r}}^i}{\partial \dot{\mathbf{q}}^i} \tag{15}$$

Using Eq. 10, the kinematic relationships presented in the preceding section, and the relationship between the angular acceleration $\boldsymbol{\alpha}^i$ of the coordinate system of the deformable body i and the time derivatives of the orientational coordinates, one obtains the *generalized Newton-Euler equations* for the deformable body as

$$\begin{bmatrix} \mathbf{m}_{RR}^i & \mathbf{m}_{R\theta}^i & \mathbf{m}_{Rf}^i \\ & \mathbf{m}_{\theta\theta}^i & \mathbf{m}_{\theta f}^i \\ \text{symmetric} & & \mathbf{m}_{ff}^i \end{bmatrix} \begin{bmatrix} \ddot{\mathbf{R}}^i \\ \boldsymbol{\alpha}^i \\ \ddot{\mathbf{q}}_f^i \end{bmatrix} = \begin{bmatrix} \mathbf{Q}_R^i \\ \mathbf{Q}_\alpha^i \\ \mathbf{Q}_f^i - \mathbf{K}_{ff}^i \, \mathbf{q}_f^i \end{bmatrix} + \begin{bmatrix} \mathbf{F}_R^i \\ \mathbf{F}_\alpha^i \\ \mathbf{F}_f^i \end{bmatrix} \tag{16}$$

where $\mathbf{m}_{RR}^i$, $\mathbf{m}_{R\theta}^i$, $\mathbf{m}_{Rf}^i$, $\mathbf{m}_{\theta\theta}^i$, $\mathbf{m}_{\theta f}^i$ and $\mathbf{m}_{ff}^i$ are the components of the mass matrix, $\mathbf{K}_{ff}^i$ is the stiffness matrix, $\mathbf{Q}^i = \left[\mathbf{Q}_R^{i\mathrm{T}} \; \mathbf{Q}_\alpha^{i\mathrm{T}} \; \mathbf{Q}_f^{i\mathrm{T}} \right]^{\mathrm{T}}$ is the vector of externally applied forces, and $\mathbf{F}^i = \left[\mathbf{F}_R^{i\mathrm{T}} \; \mathbf{F}_\alpha^{i\mathrm{T}} \; \mathbf{F}_f^{i\mathrm{T}} \right]^{\mathrm{T}}$ is a quadratic velocity vector that absorbs the *Coriolis* and the *centrifugal force components.*

4. Invariants of Motion

As the result of the finite rotation, the mass matrix of Eq. 16 is a nonlinear function of the coordinates while the Coriolis and centrifugal forces are nonlinear functions of the coordinates and velocities. It can be shown, however, that the nonlinear mass matrix and the nonlinear Coriolis and centrifugal forces can be expressed in terms of a set of invariants that depend on the assumed displacement field. These invariants can be developed for each finite element j on the deformable body i. The invariants of the deformable body i can then be obtained by assembling the invariants of its finite elements using a standard finite element assembly procedure. If the shape function of the finite element can be used to describe large rigid body translations in three orthogonal directions, it can be shown that the invariants of the element j on the deformable body i are

$$\mathbf{I}_1^{ij} = \int_{V^{ij}} \rho^{ij} \, \mathbf{S}^{ij} \, dV^{ij} \tag{17}$$

$$\mathbf{I}_{kl}^{ij} = \int_{V^{ij}} \rho^{ij} \, \mathbf{S}_k^{ij\mathrm{T}} \, \mathbf{S}_l^{ij} \, dV^{ij}, \qquad k, l = 1, 2, 3 \tag{18}$$

where ρ^{ij} and V^{ij} are, respectively, the mass density and volume of the element j and

$\mathbf{S}_k^{ij}$ is the kth row of the element shape function. The invariants of the body i can simply be obtained as

$$\mathbf{I}_1^i = \sum_{j=1}^{n_e} \mathbf{I}_1^{ij} \tag{19}$$

$$\mathbf{I}_{kl}^i = \sum_{j=1}^{n_e} \mathbf{I}_{kl}^{ij} \tag{20}$$

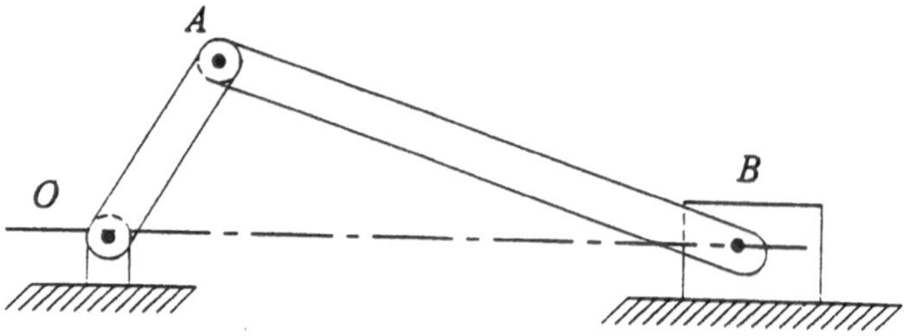

Figure 2. Slider crank mechanism

where n_e is the total number of the finite elements used to discretize the deformable body i.

The invariants of Eqs. 17 and 18 are given in their *consistent mass* form. These invariants can also be expressed in a *lumped mass* form. In this later case, the structure of the mass matrix does not change and it remains nonlinear function of the coordinates.

5. Equivalent Systems of Forces

In rigid body dynamics, a force that acts at a point on the body is *equivalent* or *equipollent* to a system of forces, defined at another point, that consists of the same force and a moment. Consequently, the force is defined by its magnitude, direction and its point of application. On the other hand, a moment in rigid body dynamics is a *free vector* which is independent of a point of application and is defined only by its magnitude and direction. In deformable body dynamics, however, a force that acts at a point, is equivalent to the same force, a moment that depends on the deformation of the body, and a set of generalized elastic forces that depend on the finite rotation and the assumed displacement field. Furthermore, a moment in flexible body dynamics is no longer a free vector, it is defined by its magnitude, its direction and its point of application.

In many *control applications*, the desired motion of a system is specified and the interest is focused on determining the *joint control forces* that produce this desired

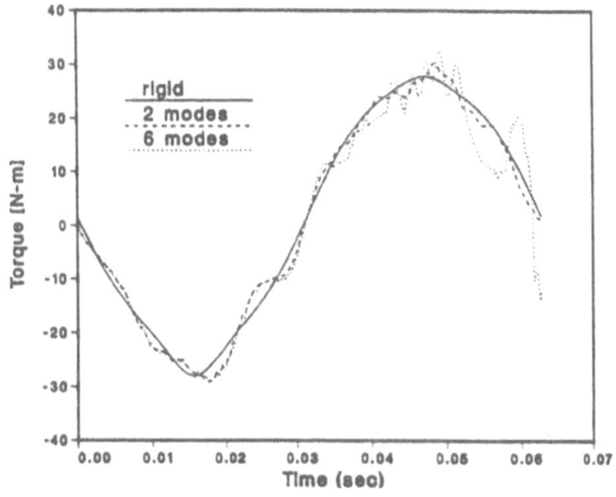

Figure 3. Solution of the inverse dynamics problem

motion. This *inverse dynamics problem* must be carefully handled in view of the definition of forces and moments in flexible body dynamics. Fig. 2 shows a model of a one degree of freedom slider crank mechanism. In this model, the connecting rod AB is assumed flexible. The motion of the slider block at B is assumed to be specified as a function of time and given by

$$x^4 \;=\; 0.35 \;-\; 0.8\, l^2 \,\sin \omega t \tag{21}$$

where $\omega \;=\; 150\, rad/s$ is the frequency of the motion of the slider block, $l^2 \;=\; 0.35m$ is the length of the connecting rod. Figure 3 shows the crankshaft torque that is required to produce the desired motion of the slider block. The results presented in Fig. 3 are obtained by solving the inverse dynamics problem in the case of rigid and flexible body dynamics. In the case of rigid body dynamics one needs to solve a system of algebraic equations. In the inverse dynamics of flexible multibody systems, one obtains a set of differential equations which must be, in general, integrated numerically because of the elastic degrees of freedom. In this case one obtains, in addition to the crankshaft torque, a set of generalized forces associated with the generalized elastic coordinates of the connecting rod. These generalized elastic coordinates depend on the boundary conditions of the flexible connecting rod. If the connecting rod is modeled as a simply supported beam and the motion of the slider block is prescribed, the generalized elastic forces are not equal to zero provided that at least one axial mode of vibration is included in the finite dimensional model.

6. Modal Coordinates

The generalized Newton-Euler equations as defined by Eq. 16 are formulated in terms of a coupled set of reference and elastic nodal variables. This is a *finite element formulation* which was obtained using the physical nodal coordinates of the finite element used to discretize the deformable body i. This matrix equation can be solved using matrix and computer methods for the reference motion and the elastic nodal coordinate of the elements. This formulation, therefore, should not be viewed as a *component mode* type of formulation. In a component mode formulation, the deformable body can be treated as one element whose deformation is described using a set of assumed modes. The differences between the finite element formulation and the assumed mode technique and the difference between the obtained invariants of motion in both cases must be clear. Component modes, however, can be used in a finite element formulation in order to reduce the number of elastic coordinates and eliminate insignificant high frequency modes. To this end, a set of assumed modes that can be determined by solving an eigenvalue problem or can be determined using *experimental modal analysis* techniques may be used. Let $\mathbf{B}_m^i$ be the modal matrix that contains a set of assumed modes that are determined experimentally or by solving the *eigenvalue problem.* A change from the space of the physical nodal coordinates to the space of modal coordinates can be achieved by using the modal transformation $\mathbf{B}_m^i$. In this case, one must realize that there is no change in the structure of Eq. 16; one only has to express the invariants of Eq. 19 in their modal form. These invariants can be transformed to their modal form according to

$$\left[\mathbf{I}_1^i\right]_m = \mathbf{I}_1^i \, \mathbf{B}_m^i \tag{22}$$

$$\left[\mathbf{I}_{kl}^i\right]_m = \mathbf{B}_m^{i^T} \, \mathbf{I}_{kl}^i \, \mathbf{B}_m^i \tag{23}$$

That is, the formulation remains the same and any change in the basis of the elastic nodal coordinates can be achieved by transforming the invariants of motion. Different sets of modes obtained using different sets of boundary conditions may lead to the same solution provided that linear combination of the modes produce similar shapes. Figure 4 shows the transverse deformation of the midpoint of the connecting rod of the slider crank mechanism shown in Fig. 2. The results presented in this figure are obtained using two different sets of boundary conditions. In the first case, mode shapes of the connecting rod are obtained using simply supported boundary conditions while in the second case the mode shapes are obtained using a body fixed coordinate system whose origin is rigidly attached to the center of the connecting rod.

The fact that the nonlinear dynamic equations of the deformable bodies that undergo large displacements can be expressed in terms of a set of invariants of motion suggests a two-stage computational strategy. In the first stage, the invariants of motion as well as the conventional stiffness matrix are evaluated in a *preprocessor computer program.* This program systematically constructs the invariants and stiffness matrices of the finite elements of each deformable body in the multibody system. These element matrices are then assembled in order to obtain the matrices of the deformable bodies in the system. If the modal coordinates are to be used to reduce

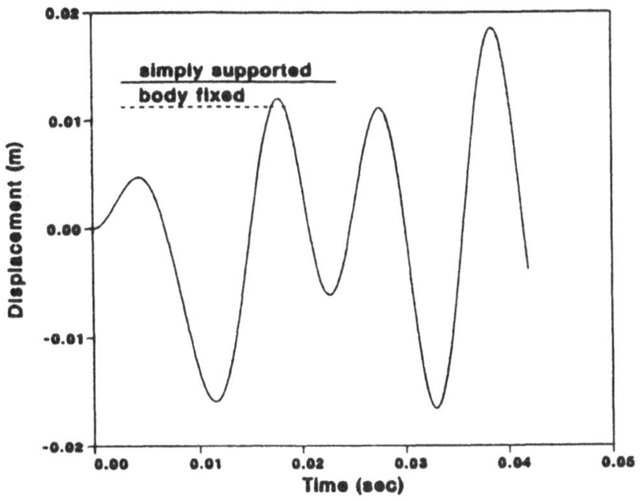

Figure 4. Transverse deformation of the midpoint of the connecting rod using different boundary conditions

the number of coordinates of some deformable bodies in the system, the invariants as well as the stiffness matrices of these bodies can be expressed in their modal form in the preprocessor computer program. The output of the preprocessor is a set of data that remain constant throughout the motion of the bodies. These data are used as part of the input data to the *main processor* used for the dynamic simulation. The computational algorithm of the main processor can be based on either the *augmented* or the *recursive formulation*. The same preprocessor can be used in both cases since the invariants of motion are characteristics of the deformable body and they do not depend on the approach used for formulating the dynamic equations of the multibody system.

7. Augmented Formulation

In the augmented formulation, the dynamic equations of the flexible multibody system is formulated in terms of a set of redundant coordinates. The relationships between these coordinates are formulated using a set of nonlinear algebraic constraint equations that describe mechanical joints and the specified motion trajectories in the multibody system. These kinematic constraint equations can be introduced to the dynamic formulation using the vector of kinematic constraint equations which can be written compactly as

$$\mathbf{C}(\mathbf{q}, t) = \mathbf{0} \tag{24}$$

where $\mathbf{C}$ is the vector of the kinematic constraint equations that can be linear or non-linear function of the system generalized coordinates $\mathbf{q}$ and time t. In the *augmented formulation*, the equations of motion can be written compactly as

$$\mathbf{M}\ddot{\mathbf{q}} + \mathbf{C}_{\mathbf{q}}^{\mathrm{T}}\boldsymbol{\lambda} = \mathbf{Q}_e + \mathbf{F} \tag{25}$$

where $\mathbf{M}$ is the system mass matrix, $\mathbf{C}_{\mathbf{q}}$ is the Jacobian matrix of the kinematic constraints, $\boldsymbol{\lambda}$ is the vector of Lagrange multipliers, $\mathbf{Q}_e$ is the vector of externally applied and elastic forces, and $\mathbf{F}$ is the vector of Coriolis and centrifugal forces.

7.1. COMPUTER FORMULATION OF THE JOINT CONSTRAINTS

Figure 5 shows examples of some of the mechanical joints that are often encountered in several industrial and technological applications. The *spherical joint* shown in Fig. 5a has three degrees of freedom which allow three independent relative rotations between the two bodies connected by this joint. The *cylindrical joint* shown in Fig. 5b has two degrees of freedom since it allows relative translation along, and relative rotation about the joint axis. The *revolute* and *prismatic* joints shown, respectively, in Figs. 5c and 5d have only one degree of freedom. The mathematical formulation of these joints can be expressed in the form of Eq. 24. For example in the case of the *spherical joint* we require that two points on body i and body j, which are connected by this joint, remain in contact throughout the motion of the two bodies. In terms of the absolute coordinates, this condition can be expressed in the form of Eq. 24 as

$$\mathbf{R}^i + \mathbf{A}^i\,\bar{\mathbf{u}}_P^i - \mathbf{R}^j - \mathbf{A}^j\,\bar{\mathbf{u}}_P^j = \mathbf{0} \tag{26}$$

where superscripts i and j refer, respectively, to bodies i and j and $\bar{\mathbf{u}}_P^i$ and $\bar{\mathbf{u}}_P^j$ are the local position vectors of the joint definition points on body i and body j, respectively. The vectors $\bar{\mathbf{u}}_P^i$ and $\bar{\mathbf{u}}_P^j$, in flexible body dynamics, are implicit functions of time since they depend on the deformation of the bodies.

In order to be able to formulate the kinematic constraints that describe the cylindrical, revolute and prismatic joints in flexible body dynamics a set of *intermediate body fixed joint coordinate systems* must be introduced. Figure 6 shows body i and body j that are connected by a *cylindrical joint* that allows relative translation and rotation between the two bodies. Let $\mathbf{X}^i\,\mathbf{Y}^i\,\mathbf{Z}^i$ and $\mathbf{X}^j\,\mathbf{Y}^j\,\mathbf{Z}^j$ be the coordinate systems of body i and body j, respectively. For the convenience of describing the large relative displacements between the two bodies, the intermediate body fixed coordinate systems $\mathbf{X}_P^i\,\mathbf{Y}_P^i\,\mathbf{Z}_P^i$ and $\mathbf{X}_P^j\,\mathbf{Y}_P^j\,\mathbf{Z}_P^j$ are introduced. These coordinate systems are assumed to have zero mass and inertia and their origins are assumed, in the finite element formulation, to be rigidly, attached to nodal points on the two bodies. The relative motion between the two bodies is assumed to be along the joint axis. We make the assumption that the joint axis can be described by a rigid line. Let $\mathbf{h}^i$ be a vector drawn on body i along the joint axis. Similarly, let $\mathbf{h}^j$ be a vector drawn on body j along the joint axis as shown in Fig. 6. As shown in the figure, the vector $\mathbf{s}^{ij}$ has a variable magnitude since it connects points P^i and P^j on bodies i and j, respectively. The kinematic constraint equations for the cylindrical joint can be written as

$$\left.\begin{aligned} \mathbf{h}^i \times \mathbf{h}^j &= \mathbf{0} \\ \mathbf{h}^i \times \mathbf{s}^{ij} &= \mathbf{0} \end{aligned}\right\} \tag{27}$$

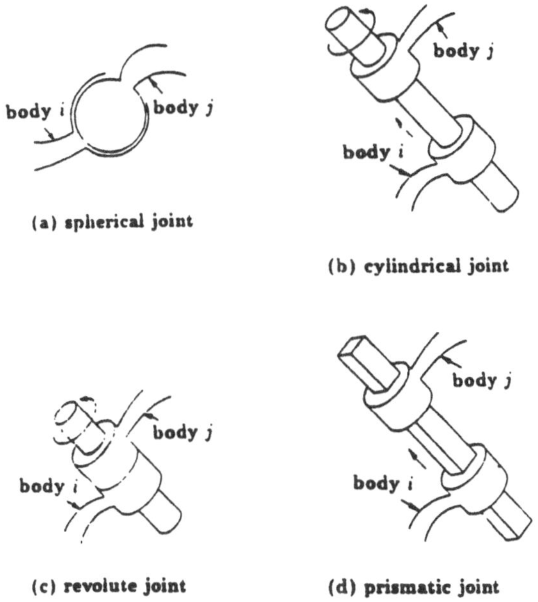

Figure 5. Joint constraints

where

$$\begin{aligned} \mathbf{s}^{ij} &= \mathbf{R}^i + \mathbf{A}^i \bar{\mathbf{u}}_P^i - \mathbf{R}^j - \mathbf{A}^j \bar{\mathbf{u}}_P^j \\ \mathbf{h}^i &= \mathbf{A}^i \mathbf{A}_P^i \bar{\mathbf{h}}^i \\ \mathbf{h}^j &= \mathbf{A}^j \mathbf{A}_P^j \bar{\mathbf{h}}^j \end{aligned}$$

in which $\bar{\mathbf{h}}^i$ and $\bar{\mathbf{h}}^j$ are constant vectors defined in the intermediate body fixed joint coordinate systems $\mathbf{X}_P^i\ \mathbf{Y}_P^i\ \mathbf{Z}_P^i$ and $\mathbf{X}_P^j\ \mathbf{Y}_P^j\ \mathbf{Z}_P^j$, respectively, $\mathbf{A}_P^i$ and $\mathbf{A}_P^j$ are the transformation matrices from the intermediate coordinate systems to the body coordinate systems. If the deformation of bodies i and j are assumed to be small, $\mathbf{A}_P^i$ and $\mathbf{A}_P^j$ are infinitesimal rotation matrices that can be expressed in terms of the slopes at the nodal points. The constraint equations of Eqs. 29 and 30 have only four independent algebraic equations which are nonlinear in the reference and elastic coordinates of the two bodies.

The *revolute joint* can be considered as a special case of the cylindrical joint. In this case, the length of the vector $\mathbf{s}^{ij}$ is constant. Therefore, the kinematic constraint equations of the revolute joint are

$$\left.\begin{aligned} \mathbf{h}^i \times \mathbf{h}^j &= \mathbf{0} \\ \mathbf{h}^i \times \mathbf{s}^{ij} &= \mathbf{0} \\ \mathbf{s}^{ij^{\mathrm{T}}} \mathbf{s}^{ij} &= c \end{aligned}\right\} \tag{28}$$

where c is a constant.

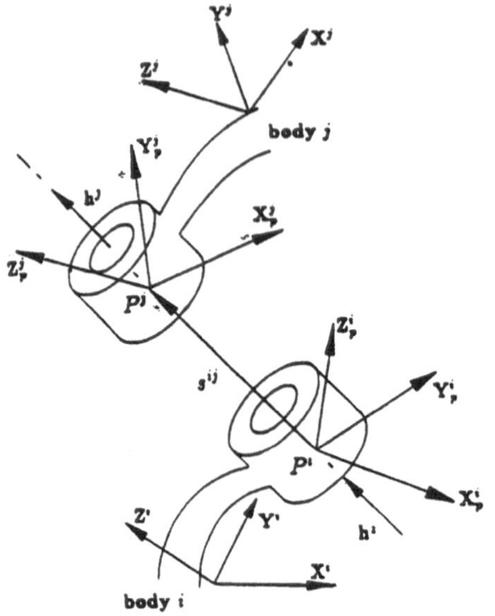

Figure 6. Intermediate body fixed joint coordinate systems

Similarly, the constraint equations of the *prismatic joint* are

$$\left.\begin{aligned} \mathbf{h}^i \times \mathbf{h}^j &= \mathbf{0} \\ \mathbf{h}^i \times \mathbf{s}^{ij} &= \mathbf{0} \\ \mathbf{n}^{i\mathrm{T}}\, \mathbf{n}^j &= 0 \end{aligned}\right\} \tag{29}$$

where $\mathbf{n}^i$ and $\mathbf{n}^j$ are two vectors drawn perpendicular to the joint axis on body i and body j, respectively. The vectors $\mathbf{n}^i$ and $\mathbf{n}^j$ can be defined as

$$\begin{aligned} \mathbf{n}^i &= \mathbf{A}^i\, \mathbf{A}^i_P\, \bar{\mathbf{n}}^i \\ \mathbf{n}^j &= \mathbf{A}^j\, \mathbf{A}^j_P\, \bar{\mathbf{n}}^j \end{aligned}$$

in which $\bar{\mathbf{n}}^i$ and $\bar{\mathbf{n}}^j$ are constant vectors defined in the intermediate joint coordinate systems $\mathbf{X}^i_P\ \mathbf{Y}^i_P\ \mathbf{Z}^i_P$ and $\mathbf{X}^j_P\ \mathbf{Y}^j_P\ \mathbf{Z}^j_P$. That is, the vectors $\mathbf{n}^i$ and $\mathbf{n}^j$ must be iteratively updated during the dynamic simulation.

7.2. SOLUTION FOR THE ACCELERATIONS

Equations 24 and 25 represent a system of algebraic and differential equations which can be solved using computer and numerical methods. In order to be able to numerically integrate this system, Eqs. 24 and 25 must be solved for the vector of accelerations. To this end, Eq. 24 is differentiated twice with respect to time. This leads to

$$\mathbf{C_q}\,\ddot{\mathbf{q}} \;=\; \mathbf{Q}_c \tag{30}$$

where $\mathbf{Q}_c$ is a vector that absorbs terms that are quadratic in the velocities. If Eqs. 25 and 30 are combined one obtains a system of matrix equation that is linear in the vectors of accelerations and Lagrange multipliers. This matrix equation can be written as

$$\begin{bmatrix} \mathbf{M} & \mathbf{C_q^T} \\ \mathbf{C_q} & \mathbf{0} \end{bmatrix} \begin{bmatrix} \ddot{\mathbf{q}} \\ \boldsymbol{\lambda} \end{bmatrix} = \begin{bmatrix} \mathbf{Q}_e + \mathbf{F} \\ \mathbf{Q}_c \end{bmatrix} \tag{31}$$

This system of equations can be solved for the generalized reference and elastic accelerations as well as the vector of Lagrange multipliers. The obtained solution contains both the dependent and independent accelerations. Lagrange multipliers can be used to determine the generalized forces of the joint constraints and specified trajectories.

Another alternate approach, but numerically different, is to use the generalized coordinate partitioning. In this case, the vector of system generalized coordinates can be written as

$$\mathbf{q} \;=\; \begin{bmatrix} \mathbf{q}_i^T & \mathbf{q}_d^T \end{bmatrix}^T \tag{32}$$

where $\mathbf{q}_i$ is the vector of system independent coordinates, and $\mathbf{q}_d$ is the vector of dependent coordinates. According to this coordinate partitioning, Eq. 30 can be written as

$$\mathbf{C}_{\mathbf{q}_i}\,\ddot{\mathbf{q}}_i + \mathbf{C}_{\mathbf{q}_d}\,\ddot{\mathbf{q}}_d \;=\; \mathbf{Q}_c \tag{33}$$

where $\mathbf{C}_{\mathbf{q}_i}$ and $\mathbf{C}_{\mathbf{q}_d}$ are the sub-Jacobians associated with the independent and dependent coordinates, respectively. The matrix $\mathbf{C}_{\mathbf{q}_d}$ is a square matrix and if the kinematic constraint equations are assumed to be linearly independent, the dependent coordinates can be selected such that the matrix $\mathbf{C}_{\mathbf{q}_d}$ is nonsingular. Wehage used the **LU** factorization method to identify the independent coordinates. Other techniques such as the *singular value decomposition* and the **QR** method that involves *Householder iterations* were also proposed. Equation 33 can then be used to write the dependent coordinates in terms of the independent ones. In this case one has

$$\ddot{\mathbf{q}}_d \;=\; \mathbf{B}_{di}\ddot{\mathbf{q}}_i + \mathbf{C}_{\mathbf{q}_d}^{-1}\,\mathbf{Q}_c$$

in which

$$\mathbf{B}_{di} \;=\; -\,\mathbf{C}_{\mathbf{q}_d}^{-1}\,\mathbf{C}_{\mathbf{q}_i}$$

Therefore, the total vector of system accelerations can be written in terms of the independent accelerations as

$$\ddot{\mathbf{q}} \;=\; \begin{bmatrix} \ddot{\mathbf{q}}_i \\ \ddot{\mathbf{q}}_d \end{bmatrix} \;=\; \mathbf{C}_{di}\ddot{\mathbf{q}}_i + \bar{\mathbf{Q}}_c \tag{34}$$

where

$$\mathbf{C}_{di} = \begin{bmatrix} \mathbf{I} \\ \mathbf{B}_{di} \end{bmatrix}, \quad \bar{\mathbf{Q}}_c = \begin{bmatrix} 0 \\ \mathbf{C}_{\mathbf{q}_d}^{-1}\mathbf{Q}_c \end{bmatrix}$$

Substituting Eq. 34 into Eq. 25, premultiplying by the transpose of the matrix $\mathbf{C}_{di}$, and using the fact that $\mathbf{C}_{di}^{\mathrm{T}}\mathbf{C}_{\mathbf{q}}^{\mathrm{T}} = \mathbf{0}$, the vector of Lagrange multipliers can be eliminated form Eq. 25. This leads to the reduced system of equations

$$\mathbf{M}_{ii}\ddot{\mathbf{q}}_i = \mathbf{R}_i \tag{35}$$

where $\mathbf{M}_{ii}$ is the generalized mass matrix associated with the independent coordinates and $\mathbf{R}_i$ is the vector of generalized forces associated with those coordinates.

The use of the *embedding technique* that leads to Eq. 35 is not computationally as efficient as the use of Eq. 31 to solve for the accelerations. For this reason, Eq. 31 is used with *sparse matrix techniques* in several commercially available multibody computer programs.

8. Recursive and Projection Methods

In the preceding section, the use of the augmented formulation in the computer aided analysis of flexible multibody systems is discussed. In this type of formulation, the kinematic and dynamic equations are formulated in terms of a mixed set of dependent and independent coordinates. In this case, one may introduce Lagrange multipliers, or use the embedding technique to reduce the number of dynamic equations to a minimum set. In this section, other alternate approaches that can be used in the analysis of flexible multibody systems are discussed. In these approaches the system kinematic and dynamic equations are formulated in terms of the system joint degrees of freedom. If two bodies are connected by a joint, the coordinates of one body can be expressed in terms of the coordinates of the other body as well as the joint degrees of freedom. Using these displacement relationships, the velocity and acceleration equations can be obtained by direct differentiation. For example, if two bodies are connected by a cylindrical joint as shown in Fig. 7, the relationship between the reference and elastic accelerations of body i and the reference and elastic accelerations of body j and the joint accelerations can be written as

$$\ddot{\mathbf{q}}^i = \mathbf{G}^i\ddot{\mathbf{q}}^j + \mathbf{H}^i\ddot{\mathbf{P}}^i + \gamma^i \tag{36}$$

where $\mathbf{G}^i$ and $\mathbf{H}^i$ are velocity influence coefficient matrices that depend on the coordinates of the two bodies, γ^i is a vector that absorbs terms that are quadratic in the velocities, $\ddot{\mathbf{q}}^i$ and $\ddot{\mathbf{q}}^j$ are the vectors of reference and elastic accelerations of bodies i and j, respectively, and $\ddot{\mathbf{P}}^i$ is the vector of the joint and elastic accelerations of body i. In the case of the constrained motion, the *generalized Newton-Euler equations* of Eq. 16 can be written for the deformable body i as

$$\mathbf{M}^i\ddot{\mathbf{q}}^i = \mathbf{Q}_e^i + \mathbf{Q}_v^i + \mathbf{Q}_R^i \tag{37}$$

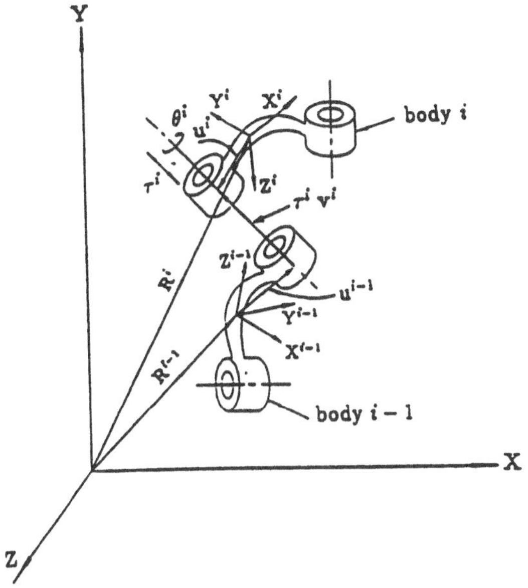

Figure 7. Relative motion

where $\mathbf{M}^i$ is the nonlinear symmetric mass matrix, $\mathbf{Q}^i_e$ is the vector of externally applied and elastic forces, $\mathbf{Q}^i_v$ is the vector of Coriolis and centrifugal force components, and $\mathbf{Q}^i_R$ is the vector of reaction forces and moments. Equation 36 can be used to eliminate the reference and elastic accelerations of body i from Eq. 37. This leads to a set of dynamic equations of body i expressed in terms of the reference and elastic accelerations of body j and the joint accelerations. Furthermore, in these equations, the joint reaction forces between the two bodies are automatically eliminated. This procedure can be continued from one body to another until the base body is reached, leading to a system of dynamic equations expressed in terms of the degrees of freedom of the multibody system.

Most existing recursive methods lead to small systems of strongly coupled equations. The coefficient matrix of the acceleration equation is dense and nonlinear as the result of the large relative displacement between the interconnected bodies. Decoupling the joint and elastic accelerations in these equations will require finding the inverse or the **LU** factorization of nonlinear matrices whose dimension depends on the number of elastic degrees of freedom. Consequently speaking of the order of an algorithm becomes meaningless since the number of elastic coordinates varies from one body to another. Recently, a recursive method that systematically decouple the joint and elastic accelerations was proposed. In this method, the generalized Newton-Euler equations, the relationship between the absolute, elastic and joint accelerations, and the reaction force equations are combined in order to form a system of loosely coupled equations which has a sparse matrix structure. By using matrix partitioning, the coupling between the joint and elastic accelerations can be eliminated. This leads to smaller system of equations expressed in terms of the joint accelerations and joint reaction forces. The dimension of the coefficient matrix in this system is independent of the number of elastic coordinates. This procedure can be demonstrated by utilizing Eq. 36 to write the absolute and elastic accelerations in terms of the joint and elastic accelerations as

$$\begin{bmatrix} \ddot{\mathbf{q}}_r \\ \ddot{\mathbf{q}}_f \end{bmatrix} = \begin{bmatrix} \bar{\mathbf{H}}_{PP} & \bar{\mathbf{H}}_{Pf} \\ \mathbf{0} & \mathbf{I} \end{bmatrix} \begin{bmatrix} \ddot{\mathbf{P}}_r \\ \ddot{\mathbf{q}}_f \end{bmatrix} + \begin{bmatrix} \boldsymbol{\gamma}_r \\ \mathbf{0} \end{bmatrix} \tag{38}$$

where subscripts r and f refer, respectively, to the absolute reference and elastic coordinates, $\mathbf{P}_r$ is the vector of system joint coordinates, $\bar{\mathbf{H}}_{PP}$ and $\bar{\mathbf{H}}_{Pf}$ are velocity influence coefficient matrices, and $\boldsymbol{\gamma}_r$ is a vector that absorbs terms which are quadratic in the velocities.

The equations of motion of the flexible multibody system expressed in terms of the absolute coordinates can be written as

$$\mathbf{M}\ddot{\mathbf{q}} = \mathbf{Q} + \mathbf{F} \tag{39}$$

where $\mathbf{M}$ is the system mass matrix, $\ddot{\mathbf{q}}$ is the vector of absolute accelerations, $\mathbf{Q}$ is the vector of forces that absorbs applied, centrifugal, Coriolis and elastic forces, and $\mathbf{F}$ is the vector of joint reaction forces. Equation 39 can also be written as

$$\begin{bmatrix} \mathbf{M}_{rr} & \mathbf{M}_{rf} \\ \mathbf{M}_{fr} & \mathbf{M}_{ff} \end{bmatrix} \begin{bmatrix} \ddot{\mathbf{q}}_r \\ \ddot{\mathbf{q}}_f \end{bmatrix} = \begin{bmatrix} \mathbf{Q}_r \\ \mathbf{Q}_f \end{bmatrix} + \begin{bmatrix} \mathbf{F}_r \\ \mathbf{F}_f \end{bmatrix} \tag{40}$$

where, as previously pointed out, subscripts r and f refer, respectively, to the rigid body and elastic coordinates.

The joint reaction forces must satisfy the identity

$$\begin{bmatrix} \bar{\mathbf{H}}^{\mathrm{T}}_{PP} & \mathbf{0} \\ \bar{\mathbf{H}}^{\mathrm{T}}_{Pf} & \mathbf{I} \end{bmatrix} \begin{bmatrix} \mathbf{F}_r \\ \mathbf{F}_f \end{bmatrix} = \begin{bmatrix} \mathbf{0} \\ \mathbf{0} \end{bmatrix} \tag{41}$$

and consequently

$$\bar{\mathbf{H}}^{\mathrm{T}}_{PP}\mathbf{F}_r = \mathbf{0} \tag{42}$$

$$\mathbf{F}_f = -\bar{\mathbf{H}}^{\mathrm{T}}_{Pf}\mathbf{F}_r \tag{43}$$

These equations show that the joint forces associated with the elastic coordinates do not introduce new independent variables. These forces can be determined by using the joint forces associated with the reference coordinates.

Substituting Eq. 43 into Eq. 40, one obtains

$$\mathbf{M}_{rr}\ddot{\mathbf{q}}_r + \mathbf{M}_{rf}\ddot{\mathbf{q}}_f = \mathbf{Q}_r + \mathbf{F}_r \tag{44}$$

$$\mathbf{M}_{fr}\ddot{\mathbf{q}}_r + \mathbf{M}_{ff}\ddot{\mathbf{q}}_f = \mathbf{Q}_f - \bar{\mathbf{H}}^{\mathrm{T}}_{Pf}\mathbf{F}_r \tag{45}$$

The first matrix equation in Eq. 38 can be written as

$$\ddot{\mathbf{q}}_r = \bar{\mathbf{H}}_{PP}^{i}\ddot{\mathbf{P}}_r + \bar{\mathbf{H}}_{Pf}\ddot{\mathbf{q}}_f + \gamma_r \tag{46}$$

Combining Eqs. 42, 44, 45, and 46, one obtains

$$\begin{bmatrix} \mathbf{M}_{rr} & \mathbf{M}_{rf} & \mathbf{I} & \mathbf{0} \\ \mathbf{M}_{fr} & \mathbf{M}_{ff} & -\bar{\mathbf{H}}_{Pf}^{\mathrm{T}} & \mathbf{0} \\ \mathbf{I} & -\bar{\mathbf{H}}_{Pf} & \mathbf{0} & -\bar{\mathbf{H}}_{PP} \\ \mathbf{0} & \mathbf{0} & -\bar{\mathbf{H}}_{PP}^{\mathrm{T}} & \mathbf{0} \end{bmatrix} \begin{bmatrix} \ddot{\mathbf{q}}_r \\ \ddot{\mathbf{q}}_f \\ -\mathbf{F}_r \\ \ddot{\mathbf{P}}_r \end{bmatrix} = \begin{bmatrix} \mathbf{Q}_r \\ \mathbf{Q}_f \\ \gamma_r \\ \mathbf{0} \end{bmatrix} \tag{47}$$

This system of equations has a dimension equal to $12n + n_f + n_r$ where n is the total number of bodies, n_f is the total number of elastic degrees of freedom, and n_r is the total number of joint coordinates. The coefficient matrix in Eq. 47 is symmetric and sparse. This system can be solved in order to obtain the absolute, joint and elastic accelerations as well as the joint reaction forces. Note that the joint and elastic accelerations are coupled in this equation. If the number of elastic degrees of freedom is large, the solution of Eq. 47 at every time step can be computationally expensive. The joint and elastic accelerations can, however, be decoupled leading to a smaller system of equations whose dimension is independent of the number of elastic degrees of freedom. To this end, one can utilize the fact that the matrix $\mathbf{M}_{ff}$ is a constant positive definite matrix. This is usually the case when a consistent mass formulation is used or when the modal coordinates are employed. Using $\mathbf{M}_{ff}$ as the pivot element in Eq. 47, one can use a simple Gauss-Jordan elimination procedure to obtain the following reduced system of equations

$$\begin{bmatrix} \left(\mathbf{M}_{rr} - \mathbf{M}_{rf}\mathbf{M}_{ff}^{-1}\mathbf{M}_{fr}\right) & \left(\mathbf{I} + \mathbf{M}_{rf}\mathbf{M}_{ff}^{-1}\bar{\mathbf{H}}_{Pf}^{\mathrm{T}}\right) & \mathbf{0} \\ \left(\mathbf{I} + \bar{\mathbf{H}}_{Pf}\mathbf{M}_{ff}^{-1}\mathbf{M}_{fr}\right) & -\bar{\mathbf{H}}_{Pf}\mathbf{M}_{ff}^{-1}\bar{\mathbf{H}}_{Pf}^{\mathrm{T}} & -\bar{\mathbf{H}}_{PP} \\ \mathbf{0} & -\bar{\mathbf{H}}_{PP}^{\mathrm{T}} & \mathbf{0} \end{bmatrix} \begin{bmatrix} \ddot{\mathbf{q}}_r \\ -\mathbf{F}_r \\ \ddot{\mathbf{P}}_r \end{bmatrix} = \begin{bmatrix} \mathbf{Q}_r - \mathbf{M}_{rf}\mathbf{M}_{ff}^{-1}\mathbf{Q}_f \\ \gamma_r + \bar{\mathbf{H}}_{Pf}\mathbf{M}_{ff}^{-1}\mathbf{Q}_f \\ \mathbf{0} \end{bmatrix} \tag{48}$$

The dimension of this system of equation is independent of the number of elastic coordinates of the system. Furthermore, the coefficient matrix remains symmetric. This system can be solved for the absolute reference and joint accelerations as well as the joint reaction forces. The elastic accelerations can then be obtained by solving Eq. 45. Since $\mathbf{M}_{ff}$ is a constant matrix, the solution for the elastic acceleration is trivial, especially in the case of using the modal coordinates because $\mathbf{M}_{ff}$ is a diagonal matrix in this case. It can be shown that the matrices $(\mathbf{M}_{rr} - \mathbf{M}_{rf}\mathbf{M}_{ff}^{-1}\mathbf{M}_{fr})$ and $\bar{\mathbf{H}}_{Pf}\mathbf{M}_{ff}^{-1}\bar{\mathbf{H}}_{Pf}^{\mathrm{T}}$ on the main diagonal of the coefficient matrix in Eq. 48 are block diagonal matrices. Consequently, a *recursive projection* procedure which has a computational advantage over existing order n algorithms, because it is independent of the number of elastic degrees of freedom of the system, can be applied.

9. Linear Theory of Elastodynamics

The dynamic equations of flexible multibody systems are highly nonlinear because of the finite rotation of the deformable body reference. A solution strategy that has been used in the past is to treat the multibody system first as a collection of rigid bodies. General-purpose multi-rigid-body computer programs can then be used to solve for the inertia and reaction forces. These inertia and reaction forces obtained from the rigid body analysis are then introduced to a linear elasticity problem to solve for the deflection of the bodies in the multibody systems. The total motion of a body is then obtained by superimposing the small elastic deformation on the gross rigid body motion. This approach is usually referred to as the *linear theory of elastodynamics.* In this approach, rigid body motion and elastic deformation are not solved for simultaneously. Furthermore, the effect of the elastic deformation on the rigid body motion is neglected. This assumption, however, may not be valid when high-speed, lightweight mechanical systems are considered. The effect of the coupling between the elastic deformation and the gross rigid body motion may be significant.

In order to understand the dynamic formulation based on the linear theory of elastodynamics we write the equations of motion of the deformable body in the following partitioned form:

$$\begin{bmatrix} \mathbf{M}^i_{rr} & \mathbf{M}^i_{rf} \\ \mathbf{M}^i_{fr} & \mathbf{M}^i_{ff} \end{bmatrix} \begin{bmatrix} \ddot{\mathbf{q}}^i_r \\ \ddot{\mathbf{q}}^i_f \end{bmatrix} + \begin{bmatrix} 0 & 0 \\ 0 & \mathbf{K}^i_{ff} \end{bmatrix} \begin{bmatrix} \mathbf{q}^i_r \\ \mathbf{q}^i_f \end{bmatrix} = \begin{bmatrix} \bar{\mathbf{Q}}^i_r \\ \bar{\mathbf{Q}}^i_f \end{bmatrix} \tag{49}$$

where $\mathbf{q}^i_r = \begin{bmatrix} \mathbf{R}^{i\mathrm{T}} & \boldsymbol{\theta}^{i\mathrm{T}} \end{bmatrix}^{\mathrm{T}}$ is the vector of reference coordinates of body i, subscripts r and f refer, respectively, to reference and elastic coordinates, and $\bar{\mathbf{Q}}^i$ is the vector of generalized forces, including the external, reaction, Coriolis and centrifugal forces. Equation 49 yields the following two matrix equations:

$$\mathbf{M}^i_{rr}\,\ddot{\mathbf{q}}^i_r + \mathbf{M}^i_{rf}\,\ddot{\mathbf{q}}^i_f = \bar{\mathbf{Q}}^i_r \tag{50}$$

$$\mathbf{M}^i_{fr}\,\ddot{\mathbf{q}}^i_r + \mathbf{M}^i_{ff}\,\ddot{\mathbf{q}}^i_f + \mathbf{K}^i_{ff}\,\mathbf{q}^i_f = \bar{\mathbf{Q}}^i_f \tag{51}$$

In the linear theory of elastodynamics, the term $\mathbf{M}^i_{rf}\ddot{\mathbf{q}}^i_f$ in Eq. 50 is neglected. Furthermore, the matrix $\mathbf{M}^i_{rr}$ and the vector $\bar{\mathbf{Q}}^i_r$ are assumed not to depend on the elastic deformation of the body. Using these assumptions, one can write Eqs. 50 and 51 as

$$\mathbf{M}^i_{rr}\,\ddot{\mathbf{q}}^i_r = \bar{\mathbf{Q}}^i_r \tag{52}$$

$$\mathbf{M}^i_{ff}\ddot{\mathbf{q}}^i_f + \mathbf{K}^i_{ff}\mathbf{q}^i_f = \bar{\mathbf{Q}}^i_f - \mathbf{M}^i_{fr}\ddot{\mathbf{q}}^i_r \tag{53}$$

Equation 52 can be solved for the reference coordinates, velocities, and accelerations using rigid multibody computer programs. The information obtained from solving Eq. 52 can then be substituted into Eq. 53 in order to obtain a linear structural problem. Equation 53 can then be solved for the vector $\mathbf{q}^i_f$ by using any of the existing linear structural dynamics programs.

Figure 8. Tracked vehicle

The linear theory of elastodynamics remains a viable approach in the dynamic analysis of many mechanical system applications. An example of these applications is *tracked vehicles* as the one shown in Fig. 8. A two dimensional planar model of this vehicle is shown in Fig. 9. The deformation mode of the chassis of the vehicle are of low frequency and consequently including the inertia coupling between these vibration modes of the chassis and the rigid body motion of the vehicle in the dynamic model does not lead to numerical problems. The track links on the other hand are very stiff and consequently the use of the vibration modes of the track links in the nonlinear dynamics leads to numerical difficulties in integration of the system equations of motion. In this case, the stresses in the track links can be efficiently predicted using the linear theory of elastodynamics.

10. Experimental Modal Analysis and Multibody Dynamics

One of the major problems that arise in flexible multibody dynamics is the selection of the mode shapes and the coordinate systems of the flexible components. Theoretically, there is an infinite number of arrangements for the deformable body coordinate systems. It is required only to eliminate the rigid body motion between the body and its coordinate system in order to be able to define a unique displacement field. In practical applications, however, only few arrangements are possible. The deformable body coordinate system must be carefully selected, otherwise numerical problems will be encountered. In many cases, floating frames of reference are used in order to define

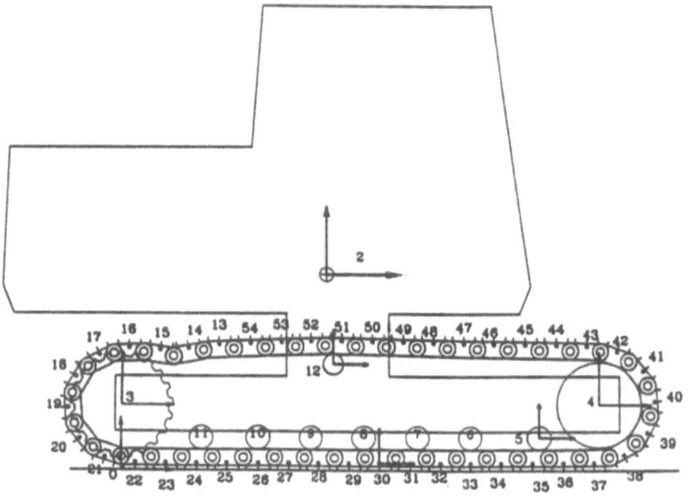

Figure 9. Two dimensional model

the proper shapes of deformation of the interconnected flexible bodies. Agrawal and Shabana [1] demonstrated that a poor selection of the deformable body coordinate system leads to erroneous results.

In complex mechanical systems such as vehicles, the determination of the proper set of deformation modes becomes a matter of judgment. Furthermore, a detailed finite element model for a vehicle structure may take several months or years. An alternate approach for introducing the deformation modes of the flexible components in multibody mechanical systems is to use experimentally identified modal parameters [7]. The advantage of using this approach is threefold. First, the mode shapes determined experimentally provide a better estimate of the vibration characteristics of the vehicle since the effect of the joint connections is taken into consideration. Second, determining the mode shapes experimentally takes much less efforts and time as compared to the efforts and time required for developing an idealized finite element model. Third, the analyst does not have to deal with the important and subtle problem of selecting the coordinate systems of the flexible components.

When the experimental modal test is performed on the assembled vehicle, one faces the problem of extracting the mode shapes of the flexible components. It is known in structural dynamics that it is difficult to extract the component modal parameters from the modal parameters of the entire system. The converse, however, is not true, that is if the modal parameters of the components are known, one can obtain the system coefficients. The problem of extracting the component modal characteristics from the modal characteristics of the assembled system represents one of the challenging problems in coupling modal analysis techniques with flexible multibody computer programs. Past and current research in experimental modal analysis does not address the problems in multibody dynamics. The coupling between experimental modal analysis techniques and multibody computational methods has great potential for providing a reliable and accurate method for the dynamic simulation of flexible mechanical systems.

Nakanishi and Shabana [5] demonstrated that in tracked vehicle applications, the component modes of the chassis can be easily extracted from the component modes of the assembled vehicle. The experimentally identified modal parameters are used in the dynamic simulation of the tracked vehicle shown in Figs. 8 and 9. In their analysis the joint articulations of the track were taken into consideration [5]

11. Summary and Conclusions

The computer implementation of the dynamic equations of flexible multibody systems [2, 3, 4, 6] is discussed in this investigation. Several issues related to the formulation of the nonlinear dynamic equations of flexible bodies are addressed. It was demonstrated that the shape functions of beam elements can not be used to describe large rigid body rotations. Consequently, the use of nodal coordinates to describe large displacements leads to a subtle linearization of the equations of motion. For that reason incremental methods have been used in the large displacement analysis of the finite elements.

In the multibody dynamic formulation presented in this paper, it is assumed that the element shape functions can be used to describe large translational displacements. Fortunately, this is the case in most existing finite elements. Four coordinate systems are used to describe the element configuration. The first coordinate system is the global coordinate system which is fixed in time. This coordinate system is used to describe the connectivities between the deformable bodies in the system. The second coordinate system is the body coordinate system which can be used to describe the connectivity between the finite elements of this body. The origin of this body coordinate system does not have to be rigidly attached to the body. Floating frames of reference [1, 4] can be used in flexible multibody dynamics provided that there is no rigid body motion between the body and its coordinate system. The third coordinate system is the element coordinate system whose origin is rigidly attached to the element. The fourth coordinate system is the *intermediate element coordinate system*. The origin of this coordinate system is rigidly attached to the origin of the body coordinate system. The axes of the intermediate element coordinate system are selected such that they are parallel to the axes of the element coordinate system in the undeformed state. Using these four coordinate systems, flexible bodies with complex geometrical shapes can be modelled using the finite element method. Furthermore, if the element shape function can be used to describe large rotation as in the case of triangular, rectangular, solid, and tetrahedral elements, the equations presented in this investigation can be used in the large deformation analysis of flexible multibody systems.

While the displacement field can be defined in a floating frame of reference, intermediate body fixed joint coordinate systems must be introduced at the joint definition points. These intermediate body fixed joint coordinate systems are used to formulate the joint constraints and to describe the relative displacements between deformable bodies. The formulation of the constraint equations of several joints were presented.

Using the principle of virtual work in dynamics, the equations of motion of the flexible body that undergoes large reference displacements were derived in their most general form. It was demonstrated that these equations are highly nonlinear as the result of the large rotation and also as the result of the nonlinear inertia coupling between the rigid body motion and the elastic deformations. The resulting nonlinear equations, however, can be expressed in terms of a set of invariants that depend on the assumed displacement field. These invariants which can be evaluated in a preprocessor computer program do not depend on the analytical approach used to formulate the flexible multibody equations of motions. Furthermore, the use of consistent or lumped masses, modal or physical coordinates, and finite element or experimentally identified modal parameters amounts to only change in the form of the invariants. Consequently, a main processor can be developed for the dynamic analysis of flexible multibody systems such that this processor does not depend on the type of element, the formulation of the inertia matrix (consistent or lumped masses), or the approach used to introduce the elastic degrees of freedom (finite element or experimentally identified modal parameters). In the main processor, the equations of motion can be formulated using the augmented formulation or recursive methods. In the augmented formulation, the nonlinear kinematic constraint equations of the joints and specified motion trajectories are adjoined to the system differential equations using the technique of Lagrange multipliers. The resulting mixed system of differential and algebraic equations is solved numerically using a direct numerical integration method coupled with a Newton-Raphson algorithm in order to check on the violation in the kinematic constraints. In the recursive methods, on the other hand, a minimum system of independent differential equations are obtained. These equations are expressed in terms of the joint and elastic coordinates. The use of the recursive methods leads to a dense generalized inertia matrix. Furthermore, the joint and elastic accelerations are strongly coupled. An alternate approach which can be used to systematically decouple the joint and elastic acceleration equations is the *amalgamated formulation*. In this formulation, the generalized Newton-Euler equations, the recursive kinematic equations, and the joint force equations are combined in order to form a large system of loosely coupled equations. Using matrix partitioning, the elastic accelerations can be eliminated leading to a system of equations whose dimension is independent of the number of elastic degrees of freedom [8, 9].

One important area which has been neglected for no obvious reason is the coupling between experimental modal analysis techniques and flexible multibody computer codes. While extensive research has been carried out in order to couple finite element techniques with multibody computer programs, surprisingly there is a very limited number of investigations on the use of experimental identification techniques in the multibody dynamics. Experimental modal analysis has several advantages as compared to the finite element method. Experimentally identified modal parameters provide a better estimate of the vibration characteristics of the system. The experimental identification of the system parameters requires much less time and efforts as compared to a detailed finite element model.

12. References

1. Agrawal, O.P., and Shabana, A.A. (1985) 'Dynamic Analysis of Multibody Systems Using Component Modes', Computers and Structures, 21, No. 6, 1301-1312.
2. Bhagat, B.M., and Willmert, K.D. (1973) 'Finite Element Vibration Analysis of Planar Mechanisms', Mechanism and Machine Theory, 8, 497-516.
3. Huston, R.L. (1981) 'Multibody Dynamics Including the Effects of Flexibility and Compliance', Computers and Structures, 14, No. 5-6, 443-451.
4. Koppens, W.P. (1989) 'The Dynamics of Systems of Deformable Bodies', Ph.D. Thesis, Technische Universiteit Eindhoven.
5. Nakanishi, T., and Shabana, A.A. 'On the Numerical Solution of Tracked Vehicle Dynamic Equations' Nonlinear Dynamics, to appear.
6. Shabana, A.A. (1989) Dynamics of Multibody Systems, John Wiley & Sons, New York.
7. Shabana, A.A. (1986) 'Dynamics of Inertia Variant Flexible Systems Using Experimentally Identified Parameters', ASME Journal of Mechanisms, Transmission, and Automation in Design, 108, 358-366.
8. Shabana, A.A., Hwang, Y.L., and Wehage, R.A. (1992) 'Projection Methods in Flexible Multibody Dynamics, Part I: Kinematics' International Journal for Numerical Methods in Engineering, 35, No. 10, 1927-1939.
9. Wehage, R.A., Shabana, A.A., and Hwang, Y.L. (1992) 'Projection Methods in Flexible Multibody Dynamics, Part II: Dynamics and Recursive Projection Methods' International Journal for Numerical Methods in Engineering, 35, No. 10, 1941-1966.

FLEXIBILITY EFFECTS IN MULTIBODY SYSTEMS

R.L. HUSTON and Y. WANG
Department of Mechanical, Industrial, and Nuclear Engineering
University of Cincinnati, Cincinnati, OH 45221-0072, USA

ABSTRACT. This paper summarizes procedures for studying flexible multibody systems using finite segment modelling. In these procedures flexible members of multibody systems are themselves modelled as multibody (or "lumped") systems. The flexibility is then modelled by springs and dampers between the bodies. Although the method has the disadvantage of being computationally intensive, the procedures presented are intended to ease the computational burden by efficient modelling and by efficient analytical formulations. It is believed that this approach combined with finite element and modal analysis methods can provide a comprehensive global and local analysis. Two examples are presented.

1. Introduction

Of all the features and phenomena associated with multibody systems the most difficult to model are the flexibility effects. Flexibility effects can significantly change the dimension of the governing dynamical equations and, hence, also the form of their solutions.

The modelling difficulties stem from several sources: First, for flexible bodies it is necessary to make both physical and geometrical approximations of the elastic, plastic, or viscoelastic effects. These approximations in turn raise issues regarding the consistency of the approximations and of their effects upon the accuracy and meaningfulness of subsequent analyses. Next, the flexibility effects greatly increase the number of variables required in the analysis, and thus the cost of the numerical analysis is greatly increased. Finally, the inclusion of flexibility effects involves a marriage of classical analysis (that is, with rigid bodies) and structural dynamics. This means that assumptions used in the classical analysis are violated - specifically those related to invariant geometry of the bodies. Thus, the implications of these violations need to be considered.

These difficulties have stimulated a vast variety of approaches in multi-flexible-body analyses. The references represent only a sampling of the many writings on the subject.

The methods of these analyses can generally be divided into two categories: 1) those which focus upon the flexible bodies while using the global multibody motion

M. F. O. S. Pereira and J. A. C. Ambrósio (eds.),
Computer-Aided Analysis of Rigid and Flexible Mechanical Systems, 351–376.

as a source of dynamic loading, and 2) those incorporating flexibility effects into the multibody dynamics analysis - the so called "lumped parameter" or "finite segment" approach. The method of analysis presented herein follows this second approach. We believe this approach has several advantages: First, it is intuitive and direct, resulting in a relatively simple formulation. Next, it is general and applicable with a broad class of multibody systems. Finally, the approach is "algorithmic" in that numerical procedures are readily developed from the governing dynamical equations. A disadvantage of the approach is that it can be computationally expensive. We believe, however, that this difficulty can be overcome by efficiencies in the formulation of the governing equations and by advances in computer technology.

In what follows we first review multibody system modelling and analysis procedures. We then consider the means of incorporating slender flexible bodies into the analysis. Finally, we present results of analyses of two simple systems.

2. Modelling

A multibody system is simply a collection of bodies with a given connection configuration. The bodies may be rigid or flexible. They may be physically connected (as with pins or spherical joints), or they may be separate (as with spring connections). Finally, the bodies may form a closed loop, or they may be open (as in a "tree").

The form and characteristics of a multibody system determines the complexity of a dynamical analysis of the system. Open and physically connected systems of rigid bodies are the easiest to study. An extension of such an analysis to accommodate separating bodies is relatively "straight forward". An extension to accommodate closed loops, however, is somewhat more difficult in that constraint equations then need to be developed. These equations are usually algebraic so that when they are combined with the dynamical equations a coupled system of differential and algebraic equations is obtained. Finally, extension to include flexibility effects are the most difficult in that assumptions about the flexibility need to be made. Such assumptions can dramatically affect the accuracy of the modelling and the subsequent numerical analyses.

Flexibility effects are generally important if the multibody system is physically large and massive, if it contains slender bodies, or if its members undergo high acceleration. Consider for example the three body system of Figure 1 consisting of two rigid bodies B_1 and B_3 connected by a slender flexible member B_2. Suppose the motion of B_1 is specified. Then the motion of B_3 depends not onlly upon the motion of B_1, but also upon the flexibility of B_2. This flexibility may be modelled by replacing B_2 by a chain of finite segments connected by springs as in Figure 2. In general these springs will represent the torsion, flexure and extension properties of the slender body. By such modelling a comprehensive representation of the behavior of the slender member can be obtained - including even large deformation effects.

The disadvantage (or "cost") of such modelling, however, is a dramatic increase in the number of degrees of freedom of the system. If the slender body is represented

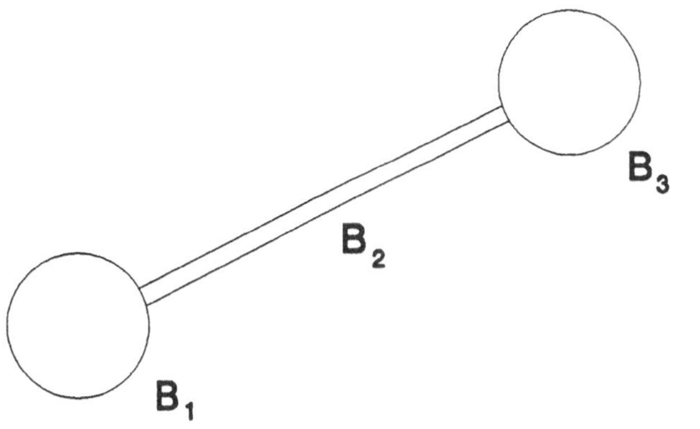

Figure 1. Two Bodies Connected by a Slender Flexible Member

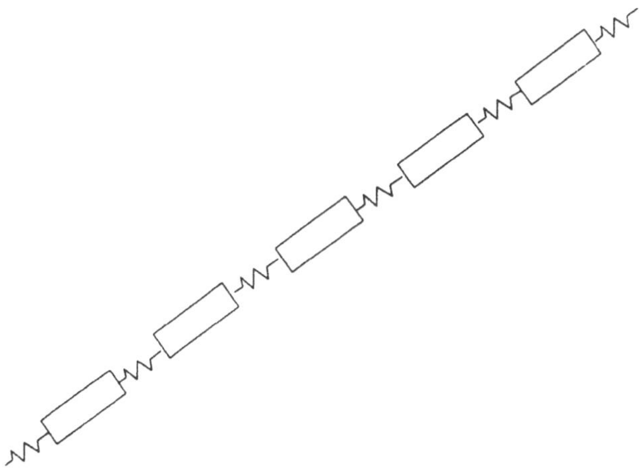

Figure 2. Finite Segment Model of a Slender Body

by say N segments then the number of degrees of freedom may be increased by 6N. The numerical effort to solve the ensuing governing equations is then greatly increased.

In the following part of the paper we present a method of analysis which is intended to minimize this numerical effort. The method is based upon established procedures of multibody dynamics and the procedure outlined in Reference [30].

3. Analysis

3.1 BODY ORIENTATIONS

Consider two typical adjoining bodies of the system as depicted in Figure 3. Let the bodies be called B_j and B_k and let $\mathbf{n}_{ji}$ and $\mathbf{n}_{ki}$ (i = 1,2,3) be sets of mutually perpendicular unit vectors fixed in B_j and B_k. Then the relative orientation of B_j and B_k may be measured by the relative inclinations of the unit vectors. Specifically, let SJK be the orthogonal matrix whose elements SJK_{im} are defined as:

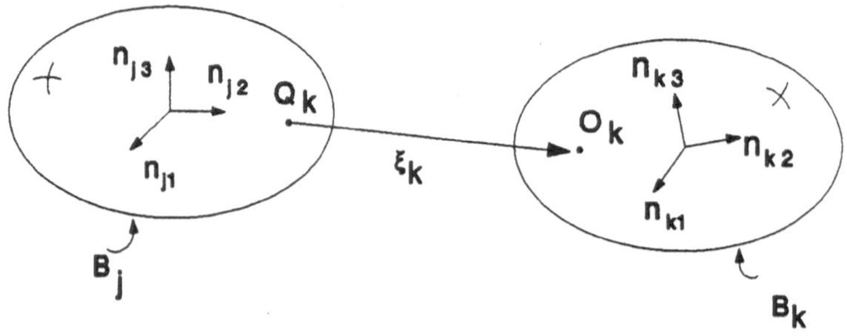

Figure 3. Two Typical Adjoining Flexible Bodies

$$SJK_{im} = \mathbf{n}_{ji} \cdot \mathbf{n}_{km} \tag{1}$$

The unit vectors are then related to each other through the equation

$$\mathbf{n}_{ji} = SJK_{im}\mathbf{n}_{km} \tag{2}$$

(Regarding notation, the J and K in SJK and the first subscripts on the unit vectors refer to the bodies B_j and B_k. Repeated indices such as the m in equation (1) represent a sum over the range of the index.)

3.2 EULER PARAMETERS

The elements SJK may be expressed in terms of relative orientation angles between the bodies in a variety of ways (see for example, Reference [51]). From a computational perspective, however, analysts have found it to be more convenient to express the elements of SJK in terms of Euler parameters [54, 58]. The advantage of using Euler parameters is that they are linearly related to the components of the relative angular velocity of the bodies. [Orientation angles are nonlinearly related to these components (through trigonometric functions). In some configurations of the bodies these nonlinear relations allow singularities to occur and these computational difficulties are experienced in the numerical integration of the governing differential equations.] A disadvantage of using Euler parameters is that four variables are required to define the relative orientation of the bodies whereas with orientation angles only three variables are needed. Therefore, with Euler parameters the number governing differential equations to be integrated is increased. However, this is generally preferable than contending with singularities in the equations.

To define the Euler parameters, let B_k have a general orientation relative to B_j. Then B_k may be brought into this orientation from a reference orientation by a single rotation about an appropriate axis. If $\boldsymbol{\lambda}_k$ is a unit vector parallel to this axis and if θ_k is the rotation angle, then the relative Euler parameters of B_k are:

$$\epsilon_{ki} = \lambda_{ki} \sin(\theta_k/2) \quad i = 1,2,3 \quad \text{and} \quad \epsilon_{k4} = \cos(\theta_k/2) \tag{3}$$

where the λ_{ki} are the $\mathbf{n}_{ji}$ components of $\boldsymbol{\lambda}_k$.

From a geometrical analysis outlined in Reference [58] the matrix SJK may be expressed in terms of the ϵ_{ki} as:

$$SJK = \begin{bmatrix} (\epsilon_{k1}^2 - \epsilon_{k2}^2 - \epsilon_{k3}^2 + \epsilon_{k4}^2) & 2(\epsilon_{k1}\epsilon_{k2} - \epsilon_{k3}\epsilon_{k4}) & 2(\epsilon_{k1}\epsilon_{k3} + \epsilon_{k2}\epsilon_{k4}) \\ 2(\epsilon_{k1}\epsilon_{k2} + \epsilon_{k3}\epsilon_{k4}) & (-\epsilon_{k1}^2 + \epsilon_{k2}^2 - \epsilon_{k3}^2 + \epsilon_{k4}^2) & 2(\epsilon_{k2}\epsilon_{k3} - \epsilon_{k1}\epsilon_{k4}) \\ 2(\epsilon_{k1}\epsilon_{k3} - \epsilon_{k2}\epsilon_{k4}) & 2(\epsilon_{k2}\epsilon_{k3} + \epsilon_{k1}\epsilon_{k4}) & (-\epsilon_{k1}^2 - \epsilon_{k2}^2 + \epsilon_{k3}^2 + \epsilon_{k4}^2) \end{bmatrix} \tag{4}$$

The ϵ_{ki} are not independent: From equations (3) they are seen to be related by the expression

$$\epsilon_{k1}^2 + \epsilon_{k2}^2 + \epsilon_{k3}^2 + \epsilon_{k4}^2 = 1 \tag{5}$$

Finally, using the procedures of Reference [58] the ϵ_{ki} are related to the relative angular velocity vector components as:

$$\dot{\epsilon}_{k1} = \frac{1}{2}(\epsilon_{k4}\hat{\omega}_{k1} + \epsilon_{k3}\hat{\omega}_{k2} - \epsilon_{k2}\hat{\omega}_{k3})$$

$$\dot{\epsilon}_{k2} = \frac{1}{2}(-\epsilon_{k3}\hat{\omega}_{k1} + \epsilon_{k4}\hat{\omega}_{k2} + \epsilon_{k1}\hat{\omega}_{k3})$$

$$\dot{\epsilon}_{k3} = \frac{1}{2}(\epsilon_{k2}\hat{\omega}_{k1} - \epsilon_{k1}\hat{\omega}_{k2} + \epsilon_{k4}\hat{\omega}_{k3}) \tag{6}$$

$$\dot{\epsilon}_{k4} = \frac{1}{2}(-\epsilon_{k1}\hat{\omega}_{k1} - \epsilon_{k2}\hat{\omega}_{k2} - \epsilon_{k3}\hat{\omega}_{k3})$$

and

$$\hat{\omega}_{k1} = 2(\epsilon_{k4}\dot{\epsilon}_{k1} - \epsilon_{k3}\dot{\epsilon}_{k2} + \epsilon_{k2}\dot{\epsilon}_{k3} - \epsilon_{k1}\dot{\epsilon}_{k4})$$

$$\hat{\omega}_{k2} = 2(\epsilon_{k3}\dot{\epsilon}_{k1} + \epsilon_{k4}\dot{\epsilon}_{k2} - \epsilon_{k1}\dot{\epsilon}_{k3} - \epsilon_{k2}\dot{\epsilon}_{k4}) \tag{7}$$

$$\hat{\omega}_{k3} = 2(-\epsilon_{k2}\dot{\epsilon}_{k1} + \epsilon_{k1}\dot{\epsilon}_{k2} + \epsilon_{k4}\dot{\epsilon}_{k3} - \epsilon_{k3}\dot{\epsilon}_{k4})$$

where the $\hat{\omega}_{k2}$ (i = 1,2,3) are the $\mathbf{n}_{ji}$ components of $\hat{\boldsymbol{\omega}}_k$, the angular velocity of B_k relative to B_j.

3.3 CONFIGURATION VARIABLES AND GENERALIZED SPEEDS

Consider again the two typical adjoining flexible bodies of Figure 3. Let Q_k and O_k be points of the respective bodies which are coincident with each other when the bodies are undeformed. Let ξ be expressed in terms of the unit vectors of B_j as

$$\xi_k = \xi_{ki}\mathbf{n}_{ji} \tag{8}$$

Let R_j and R_k be reference frames fixed in the undeformed states of B_j and B_k. Then the deformations of B_j and B_k may be measured locally in R_j and R_k. From a global perspective the movement of B_k relative to B_j may be measured by the translation and rotation of R_k relative to R_j. Thus for a system with N bodies 6N coordinates are needed to define its configuration and motion (6 for each body: 3 translation and 3 rotation). Let these coordinates be X_ℓ (ℓ = 1,...,6N). Next, let Y_ℓ (ℓ = 1,...,6N) be introduced as linear combinations of the derivatives of the X_ℓ as follows: Let the first 3N Y_ℓ be defined as

$$Y_{3(k-1)+i} = \hat{\omega}_{ki} \qquad i = 1,2,3;\; k = 1,\ldots,N \tag{9}$$

where now the $\hat{\omega}_{ki}$ measure the angular velocity of R_k relative to R_j. Let the remaining 3N Y_ℓ be defined as

$$Y_{3(N+k-1)+i} = \dot{\xi}_{ki} \qquad i=1,2,3;\ k=1,\ldots,N \tag{10}$$

where the ξ_{ki} are defined in equation (8).

The Y_ℓ are called "generalized speeds" [52]. They are not always integrable into elementary functions. That is, in general, there do not exist individual orientation functions $\hat{\theta}_{ki}$ whose derivatives are $\hat{\omega}_{ki}$. Instead, the $\hat{\omega}_{ki}$ are linear combinations of orientation angle derivatives, or of Euler parameter derivatives, as in equation (7).

3.4 GLOBAL KINEMATICS

The global kinematics of a multibody system may be developed using parameters outlined in Reference [30]. Specifically, the angular velocity $\boldsymbol{\omega}_k$ of a typical body B_k, in a Newtonian reference frame R, may be expressed in the form

$$\boldsymbol{\omega}_k = \hat{\boldsymbol{\omega}}_1 + \ldots + \hat{\boldsymbol{\omega}}_k \tag{11}$$

where the terms on the right are relative angular velocities through the adjoining bodies from B_1 to B_k. By using equations (9) and (11), $\boldsymbol{\omega}_k$ may be expressed in the form

$$\boldsymbol{\omega}_k = \omega_{k\ell m} Y_\ell \mathbf{n}_{om} \tag{12}$$

where the $\omega_{k\ell m}$ ($k = 1,\ldots,N$; $\ell = 1,\ldots,6N$; $m = 1,2,3$) form a block array of coefficients representing scalar components of the "partial angular velocity vectors" used with Kane's equations [49, 50, 52] and where the $\mathbf{n}_{om}$ ($m = 1,2,3$) are unit vectors fixed in R.

By differentiating, the angular acceleration of the bodies may be expressed as:

$$\boldsymbol{\alpha}_k = (\dot{\omega}_{k\ell m} Y_\ell + \omega_{k\ell m} \dot{Y}_\ell)\mathbf{n}_{om} \tag{13}$$

Explicit expressions for the $\omega_{k\ell m}$ and $\dot{\omega}_{k\ell m}$ arrays, together with efficient algorithms for computing them, are recorded in References [54, 58, 59, and 60].

Let G_k be the mass center of B_k and let $\mathbf{P}_k$ be a position vector locating G_k relative to a fixed point in R. Then $\mathbf{P}_k$ can be expressed in terms of vectors fixed in the bodies of the branches containing B_k and in terms of the displacements between the bodies. The derivatives of $\mathbf{P}_k$ produce the velocity and acceleration of G_k. These derivatives may be evaluated using vector products and procedures recorded in References [54, 58, 59, and 60]. The results may be expressed in the forms

$$\mathbf{v}_k = v_{k\ell m} Y_\ell \mathbf{n}_{om} \tag{14}$$

and

$$\mathbf{a}_k = (v_{k\ell m} \dot{Y}_\ell + \dot{v}_{k\ell m} Y_\ell)\mathbf{n}_{om} \tag{15}$$

where the $v_{k\ell m}$ ($k = 1,...,N$; $\ell = 1,...,6N$; $m = 1,2,3$) form a block array representing scalar components of the "partial velocity vectors" [49, 50, 52]. Explicit expressions for the $v_{k\ell m}$ and $\dot{v}_{k\ell m}$ arrays, together with algorithms for their computation, are also recorded in References [54, 58, 59, and 60].

3.5 KINETICS/EQUATIONS OF MOTION

The equations of motion are readily obtained using Kane's equations. Using the formulation of the foregoing kinematic analysis, Kane's equations are ideally suited for obtaining equations of motion for large lumped parameter systems. Kane's equations state that there is a balance (or "zero sum") of the generalized applied and inertia forces. These generalized forces are defined as a projection of forces and moments onto the partial velocity and partial angular velocity vectors.

Specifically, let the applied forces on a typical body B_k be equivalent to a force $\mathbf{F}_k$ passing through the mass center G_k together with a couple with torque $\mathbf{M}_k$. Thus the generalized applied (or "active") force on B_k, associated with the generalized speed Y_ℓ, becomes

$$F_\ell = v_{k\ell m} F_{km} + \omega_{k\ell m} m_{km} \tag{16}$$

where F_{km} and M_{km} are the $\mathbf{n}_{om}$ components of $\mathbf{F}_k$ and $\mathbf{M}_k$ and where, as before, there is a sum over the repeated indices.

Similarly, let the inertia forces on B_k be equivalent to a force $\mathbf{F}_k^*$ passing through G_k together with a couple with torque $\mathbf{M}_k^*$. Then, as in equation (16), the generalized inertia force on B_k, associated with the generalized speed Y_ℓ, is

$$F_\ell^* = v_{k\ell m} F_{km}^* + \omega_{k\ell m} M_{km}^* \tag{17}$$

where F_{km}^* and M_{km}^* are the $\mathbf{n}_{am}$ components of $\mathbf{F}_k^*$ and $\mathbf{M}_k^*$.

From the principles of classical mechanics $\mathbf{F}_k^*$ and $\mathbf{M}_k^*$ may be written in the forms:

$$\mathbf{F}_k^* = -m_k \mathbf{a}_k \quad \text{and} \quad \mathbf{M}_k^* = -\mathbf{I}_k \cdot \boldsymbol{\alpha}_k - \boldsymbol{\omega}_k \times (\mathbf{I}_k \cdot \boldsymbol{\omega}_k) \quad \text{(no sum)} \tag{18}$$

where m_k is the mass of B_k and $\mathbf{I}_k$ is the central inertia dyadic of B_k.

In equations (16) and (17) there is no sum on k. However, the generalized forces for the entire system are obtained by adding the contributions from the individual bodies. Hence, the generalized forces for the entire system are obtained by summing on k from 1 to N in equations (16) and (17).

The governing dynamical equations may be obtained using Kane's equations [50, 52] which state that

$$F_\ell + F_\ell^* = 0 \quad \ell = 1,\ldots,6N \tag{19}$$

By substituting from equations (12) through (18) into (19) the equations may be written in the form

$$a_{\ell p}\dot{Y}_p = f_\ell \quad (\ell,p = 1,\ldots,6N) \tag{20}$$

where the $a_{\ell p}$ and f_ℓ are

$$a_{\ell p} = m_k v_{kpm} v_{k\ell m} + I_{kmn}\omega_{kpm}\omega_{k\ell n} \tag{21}$$

and

$$f_\ell = F_\ell - (m_k v_{k\ell m}\dot{v}_{kqm} Y_q + I_{kmn}\omega_{k\ell m}\dot{\omega}_{kqn} Y_q$$
$$+ e_{rsm} I_{ksn}\omega_{k\ell m}\omega_{bqr}\omega_{kpn} Y_p Y_q \tag{22}$$

Equations (20), (6), (9), and (10) form a set of 13N first order differential equations for the 6N Y_p, the 3N ξ_{ki}, and the 4N ϵ_{ki}. Since the coefficients $a_{\ell p}$ and f_ℓ of equations (20) depend upon the four block arrays $\omega_{k\ell m}$, $\dot{\omega}_{k\ell m}$, $v_{k\ell m}$, and $\dot{v}_{k\ell m}$ and since efficient algorithms have been written for the computation of these arrays, algorithms can be written for the numerical development and solutions of equation (20). (One set of such algorithms forms the basis for the program DYNOCOMBS [60].)

The flexibility effects, which are the focus of this paper, enter the governing equations through the F_ℓ of equations (22). We take a closer look at these terms in the following part of the paper.

4. Flexibility Effects/Slender Members

4.1 GENERAL PROCEDURES

With our finite segment modelling the flexibility effects are modelled by springs and dampers between the bodies. With the generalized forces defined with generalized speeds which are relative angular velocity components and relative displacement derivatives, the spring and damper force and moment components occur singly in the governing equations. That is, with the generalized speeds defined as in equations (9) and (10), the governing equations are uncoupled in the spring and damper force and moment components.

To demonstrate this we follow the procedure outlined in Reference [30]. Specifically, consider again the two typical adjoining flexible bodies depicted in Figure 3. If ξ_k measures the displacement of O_k relative to Q_k, then the velocity of O_k in an inertial frame R may be expressed as:

$$\mathbf{v}_{O_k} = \mathbf{v}_{Q_k} + \boldsymbol{\omega}_j \times \boldsymbol{\xi}_j + \dot{\xi}_{ki}\mathbf{n}_{ji} \quad \text{(no sum on j)} \tag{23}$$

where $\mathbf{v}_{Q_k}$ is the velocity of Q_k in R. If ξ_j is small or if Q_k is that point of B_j (or B_j extended) which coincides with O_k, then equation (23) reduces to

$$\mathbf{v}_{O_k} = \mathbf{v}_{Q_k} + \dot{\xi}_{ki}\mathbf{n}_{ji} \tag{24}$$

Similarly, the angular velocities of the bodies are related by the expression

$$\boldsymbol{\omega}_k = \boldsymbol{\omega}_j + \hat{\boldsymbol{\omega}}_k = \boldsymbol{\omega}_j + \hat{\omega}_{ki}\mathbf{n}_{ji} \tag{25}$$

Let the force system exerted by the springs and dampers between the bodies be equivalent to a single force $\mathbf{f}_k$ passing through Q_k together with a couple with torque $\mathbf{m}_k$. If this is the force system exerted on B_j by B_k, then by the law of action-reaction [62] the force system exerted on B_k by B_j is equivalent to a single force - $\mathbf{f}_k$ passing through Q_k together with a couple with torque $-\mathbf{m}_k$.

Consider the contribution to F_ℓ from these force systems: From equation (16), this contribution F_ℓ^k may be expressed as [28, 29, 30]

$$\begin{aligned} F_\ell^k &= (\partial \mathbf{v}_{Q_k}/\partial Y_\ell)\cdot \mathbf{f}_k + (\partial \boldsymbol{\omega}_j/\partial Y_\ell)\cdot \mathbf{m}_k \\ &+ (\partial \mathbf{v}_{O_k}/\partial Y_\ell)\cdot(-\mathbf{f}_k) + (\partial \boldsymbol{\omega}_k/\partial Y_\ell)\cdot(-\mathbf{m}_k) \end{aligned} \tag{26}$$

Consider the following cases:

Case 1: Y_ℓ is not equal to either $\dot{\xi}_{ki}$ or $\hat{\omega}_{ki}$. Then from equations (24) and (25), the partial velocities and partial angular velocities of equation (26) are equal. That is,

$$\partial \mathbf{v}_{Q_k}/\partial Y_\ell = \partial \mathbf{v}_{O_k}/\partial Y_\ell \quad \text{and} \quad \partial \boldsymbol{\omega}_j/\partial Y_\ell = \partial \boldsymbol{\omega}_k/\partial Y_\ell \tag{27}$$

Hence, from equation (26) in this case $\hat{F}_\ell$ is zero.

Case 2: Y_ℓ is equal to one of the $\dot{\xi}_{ki}$. Then from equations (24) and (25), the partial velocities and partial angular velocities are

$$\partial \mathbf{v}_{Q_k}/\partial Y_\ell = \partial \boldsymbol{\omega}_j/\partial Y_\ell = \partial \boldsymbol{\omega}_k/\partial Y_\ell = 0 \tag{28}$$

and

$$\partial \mathbf{v}_{O_k}/\partial Y_\ell = \partial \mathbf{v}_{O_k}/\partial \dot{\xi}_{ki} = \mathbf{n}_{ji} \tag{29}$$

Hence, from equation (26), the contribution to F_ℓ is

$$\hat{F}_\ell = -f_{ji} \tag{30}$$

where f_{ji} is the $\mathbf{n}_{ji}$ component of $\mathbf{f}_k$.

Case 3: Y_ℓ is equal to one of the $\hat{\omega}_{ki}$. Then from equation (24) and (25), the partial velocities and partial angular velocities are

$$\partial \mathbf{v}_{Q_k}/\partial Y_\ell = \partial \mathbf{v}_{O_k}/\partial Y_\ell = \partial \boldsymbol{\omega}_j/\partial Y_\ell = 0 \tag{31}$$

and

$$\partial \boldsymbol{\omega}_k/\partial Y_\ell = \partial \boldsymbol{\omega}_k/\partial \hat{\omega}_{ki} = \mathbf{n}_{ji} \tag{32}$$

Hence, from equation (26), the contribution to F_ℓ is

$$\hat{F}_\ell = -\mathbf{m}_{ji} \tag{33}$$

where m_{ji} is the $\mathbf{n}_{ji}$ component of $\mathbf{m}_k$.

The contributions of equations (30) and (33) are to be inserted in the generalized forces of equation (16). It is seen that there is a one-to-one correspondence between the joint force and moment components and the individual F_ℓ. Thus, these components occur singly in the governing equations.

4.2 LOCAL KINEMATICS

Long slender members manifest flexibility effects more dramatically than nonslender bodies. Therefore, to illustrate the foregoing procedures and to provide an analysis for the most significant of the flexible bodies we will focus our discussion upon slender members.

Consider the flexible beam and its model as depicted in Figure 4. Consider two typical adjoining segments as in Figure 5. Let G_j and G_k be the mass centers of segments B_j and B_k and let R_j and R_k be reference frames fixed in B_j and B_k with origins at G_j and G_k. Let $\hat{O}_j$ and $\hat{O}_k$ be points at the connections between the springs and dampers of the adjoining bodies as shown (see also Reference [30]). Let $\hat{R}_j$ and $\hat{R}_k$ be reference frames with origins at $\hat{O}_j$ and $\hat{O}_k$ which are parallel to R_j and R_k when the beam is undeformed.

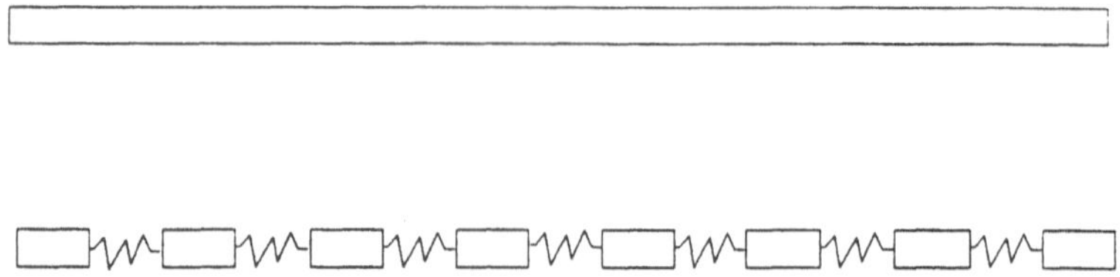

Figure 4. Flexible Beam and Model

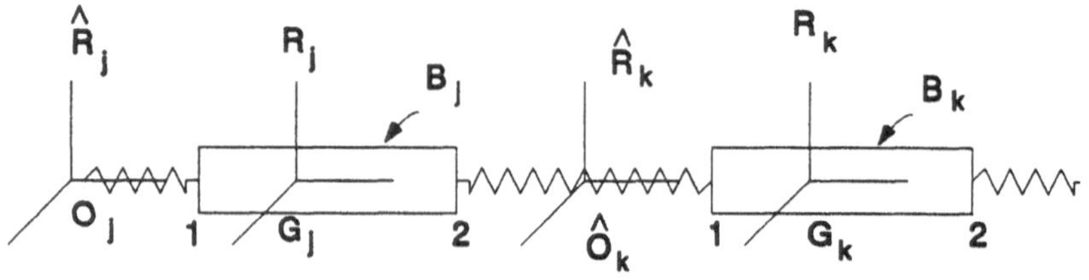

Figure 5. Typical Adjoining Beam Segments and Reference Frames

With this modelling and nomenclature there are 12 displacements and rotations associated with each segment - six at each end. When the beam is undeformed, all the reference frames are aligned. Then as the beam deforms, the reference frames translate and rotate relative to each other. For B_j the displacement of $\hat{O}_j$ relative to G_j may be expressed in terms of three parameters and the rotation of $\hat{R}_j$ relative to B_j (or R) may be described in terms of three parameters. Similarly, the displacement of $\hat{O}_k$ relative to G_j and the rotation of $\hat{R}_k$ relative to B_j (or R_j) may be described in terms of six parameters. Table 1. presents a summary listing of the parameters and notation (see References [30] and [58]). Regarding the notation, the first subscript of u_{j1x} refers to the segment, the second refers to the segment end and the third refers to the direction. The over hat of $\hat{u}_{kx}$ signifies a relative displacement.

Variable	Notation
Displacement of $\hat{O}_j$ relative to G_j	$u_{j1k}, u_{j1y}, u_{j1z}$
Rotation of $\hat{R}_j$ relative to R_j	$\theta_{j1k}, \theta_{j1y}, \theta_{j1z}$
Displacement of $\hat{O}_k$ relative to G_j	$u_{j2x}, u_{j2y}, u_{j2z}$
Rotation of $\hat{R}_k$ relative to R_j	$\theta_{j2x}, \theta_{j2y}, \theta_{j2z}$
Displacement of R_k (or B_k) relative to R_j (or B_j)	$\hat{u}_{kx}, \hat{u}_{ky}, \hat{u}_{kz}$
Rotation of R_k (or B_k) relative to R_j (or B_j)	$\hat{\theta}_{kx}, \hat{\theta}_{ky}, \hat{\theta}_{kz}$

Table 1. Displacement and Rotation Parameters and Notation

With this notation, the displacement and rotation of B_k relative to B_j may be expressed in terms of the end displacements and rotations by the expressions:

$$\hat{u}_{kx} = u_{j2x} - u_{k1x}, \quad \hat{u}_{ky} = u_{j2y} - u_{k1y}, \quad \hat{u}_{kz} = u_{j2z} - u_{k1z} \tag{34}$$

$$\hat{\theta}_{kx} = \theta_{j2x} - \theta_{k1x}, \quad \hat{\theta}_{ky} = \theta_{j2y} - \theta_{k1y}, \quad \hat{\theta}_{kz} = \theta_{j2z} - \theta_{k1z} \tag{35}$$

Observe that although there are 12 local displacement and rotation components associated with each segment (6 at each end), there are only 6 relative displacement and rotation components.

4.3 STIFFNESS COEFFICIENTS

For elastic segments the principles of structural analysis may be used to determine the relation between the force and moment components and the displacements and rotations. To illustrate this procedure, consider the tapered segments depicted in Figure 6. Let the segments be subjected to tension forces as with typical segment B_i. Let the extensibility be modelled by extension springs at each end with moduli k_{i1}^e and k_{i2}^e. Let C_i be the segment mass center and let ρ_{i1} and ρ_{i2} be the distances from C_i to the segment ends. Let A_{i1}, A_{i2}, and A_{ie} be the segment cross section areas at the ends 1 and 2 and at the mass center.

Let the tapered segment be subjected to extensive force F producing displacements u_{i1} and u_{i2} at the ends. Then relative to a reference frame fixed at the mass center u_{i1} and u_{i2} are

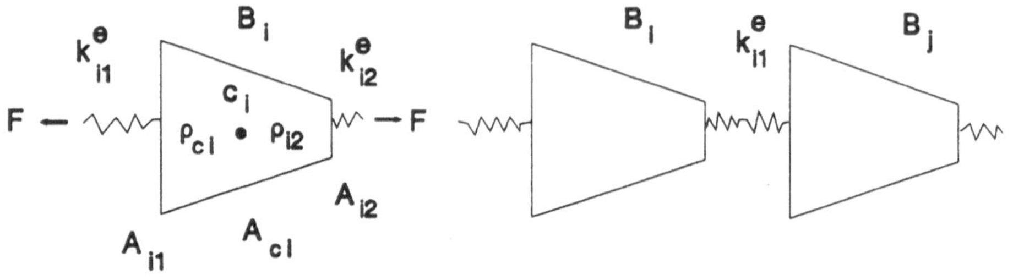

Figure 6. Tapered Flexible Segments

$$u_{i1} = -F\rho_{i1}\ln(A_{i1}/A_{ci})/[E_i(A_{i1} - A_{ci})] \quad \text{and} \quad u_{i2} = F\rho_{i2}\ln(A_{i2}/A_{ci})/[E_i(A_{i2} - A_{ci})] \tag{36}$$

where E_i is the elastic modulus.

The geometrical parameters A_{i1}, A_{i2}, A_{ci}, ρ_{i1}, and ρ_{i2} are not independent. Specifically, if A_{i1}, A_{i2}, and the segment length ℓ_i are known, then A_{ci}, ρ_{i1}, and ρ_{i2} are

$$\begin{aligned} A_{ci} &= (2/3)(A_{i1}^2 + A_{i1}A_{i2} + A_{i2}^2)/(A_{i1} + A_{i2}) \\ \rho_{i1} &= (\ell_i/3)(A_{i2} + 2A_{i1})/(A_{i1} + A_{i2}) \\ \rho_{i2} &= (\ell_i/3)(2A_{i2} + A_{i1})/(A_{i1} + A_{i2}) \end{aligned} \tag{37}$$

The spring moduli, the displacements, and the loading force are related by the expressions:

$$F = -k_{i1}^e u_{i1} \quad \text{and} \quad F = k_{i2}^e u_{i2} \tag{38}$$

By comparing equations (36) and (38) we obtain the moduli as:

$$k_{i1}^e = E_i(A_{i1} - A_{ci})/[\rho_{i1}\ln(A_{i1}/A_{ci})] \quad \text{and} \quad k_{i2}^e = E_i(A_{i2} - A_{ci})/[\rho_{i2}\ln(A_{i2}/A_{ci})] \tag{39}$$

If the segment is straight, the analogous moduli may be obtained from equation (39) by letting

$$A_{i1} = A_{i2} = A_{ci} = A_i \quad \text{and} \quad \rho_{i1} = \rho_{i2} = \ell_i/2 \tag{40}$$

The results are

$$k_{i1}^e = k_{i2}^e = 2E_iA_i/\ell_i \tag{41}$$

Finally, the equivalent spring modulus k_{ij} for the spring segments in series is:

$$k_{ij}^e = \frac{k_{j1}^e k_{i2}^e}{(k_{i2}^e + k_{j1}^e)} = \frac{E_iE_j(A_{j1} - A_{cj})(A_{i2} - A_{ci})}{\left[E_i(A_{i2} - A_{ci})\rho_{i1}\ell n\left(\frac{A_{j1}}{A_{cj}}\right) + E_j(A_{j1} - A_{cj})\rho_{i2}\ell n\left(\frac{A_{i2}}{A_{ci}}\right)\right]} \tag{42}$$

For straight segments k_{ij}^e then becomes:

$$k_{ij}^e = 2E_iE_jA_iA_j/(E_iA_i\ell_j + E_jA_j\ell_i) \tag{43}$$

Using similar procedures the spring moduli for segments in flexure and torsion may be obtained. The Appendix contains a comprehensive listing of the results.

5. Examples

For an example illustrating the efficacy of the method, consider a uniform flexible beam attached to a rotating hub of radius r as depicted in Figure 7. Let the beam be divided into 20 equal length segments connected by flexural springs with stiffness developed following the procedures of the foregoing section.

Consider first the case with the hub at rest. Then the efficacy of the modelling can be checked by comparing the natural frequencies of the finite segment model with those computed from a classical continuum model. Table 2 presents a comparison with a measurement of the error for the first 10 modes. The units of the natural frequencies are $(EI/\rho\ell^4)^{1/2}$, where E is the elastic modulus, I is the second moment of area of the beam cross-section, ρ is the mass density per unit length, and ℓ is the beam length.

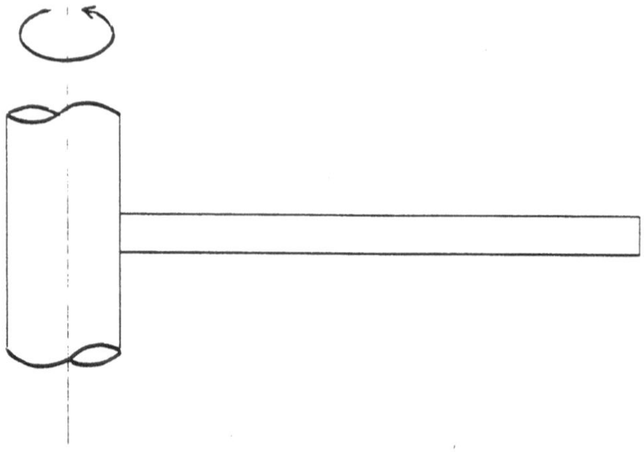

Figure 7. Beam Attached to a Rotating Hub

Mode	Continuum Model [66]	Finite Segment Model (percent error)	
1	3.5156	3.5120	(0.1%)
2	22.0337	21.9471	(0.39%)
3	61.7010	61.2960	(0.66%)
4	120.9027	119.7999	(0.91%)
5	199.8595	197.5090	(1.18%)
6	298.5555	294.2310	(1.45%)
7	416.9908	406.7571	(1.73%)
8	555.1652	534.8156	(2.04%)
9	713.0789	696.0192	(2.39%)
10	890.7318	865.7742	(2.80%)

Table 2. Comparison of Natural Frequencies from the Continuum and Finite Segment Models

Consider next the case of the hub starting to rotate from rest to an angular speed of 6 radians according to the expression

$$\dot{\theta} = \begin{cases} (2/5)[t - (7.5/\pi)\sin(\pi t/7.5) \text{ rad/sec} & 0 \le t < 15 \\ 6 \text{ rad/sec} & 15 \le t \le 30 \end{cases} \tag{44}$$

Let the hub radius r be zero; let the beam length ℓ be 10 m; let the elastic modulus E be 7×10^{10} N/m^2; let the second moment of area of the cross section I be 2×10^{-7}m^4; and let the mass density per unit length ρ be 1.2 kg/m. Let the beam be modelled by 10 segments.

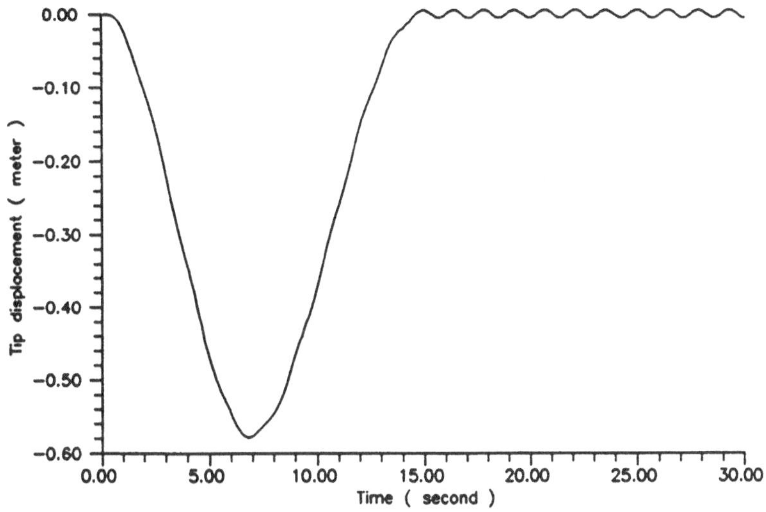

Figure 8. Rotating Beam Tip Displacement

Figure 8 shows the results for the tip displacement. The results are seen to compare favorably with those of Kane, et al. [34].

6. Discussion and Conclusions

These examples demonstrate the efficacy of the finite segment method for modelling flexible multibody systems. The examples show that accurate representations of flexibility - a phenomena of continuous, deformable bodies - can be simulated by discrete, rigid systems.

The success of flexibility modelling with finite segments then forms the basis for the simulation of large flexible multibody systems - particularly those with significant inertia loading.

The finite segment modelling method is not restricted to elastic systems. Indeed, the method may be used to model viscoelastic, plastic, and even nonlinearly elastic systems. The accuracy of the modelling increases as the number of segments is increased.

There are two principal disadvantages of the method. The first and most obvious is the computational burden. Although simulation accuracy is improved with an increasing number of segments, each additional segment can increase the number of equations to be solved by six second order or thirteen first order differential equations. This burden can be overcome somewhat by increased computer capability and by greater availability of super computer systems. The computational burden can also be reduced by more efficient modelling and by the development of efficient algorithms for numerical analysis. It is believed that the procedures presented herein form the basis for such efficient modelling and algorithm development.

The second disadvantage of the finite segment method is that it requires increased skill, insight, and intuition on the part of the analyst. These attributes are difficult to transmit from one analyst to another. Improved software [67] and greater experience of analysts are likely to diminish this disadvantage.

Finally, it is the writers' opinion that the full capabilities of the method are yet to be developed. For example, the combination of the method with the well established finite element method and with the emerging method of computer graphic modelling, is likely to lead to new analyses which are more comprehensive than heretofore deemed feasible.

7. Appendix: Stiffness Coefficients for Elastic Straight and Taper Segments

The following listings provide stiffness coefficients (moduli) for various combinations of elastic straight and tapered segments. The listed values could be of use in software development and in specific simulations.

7.1 STRAIGHT SEGMENTS

Extensional segment

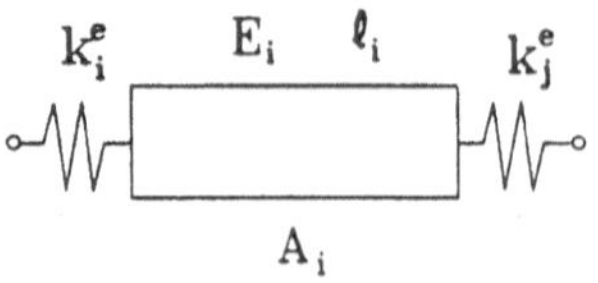

$k_i^e = k_j^e = 2E_iA_i/\ell_i$

ℓ_i: length of the segment

E_i: Young's modulus

A_i: cross sectional area

Torsional segment

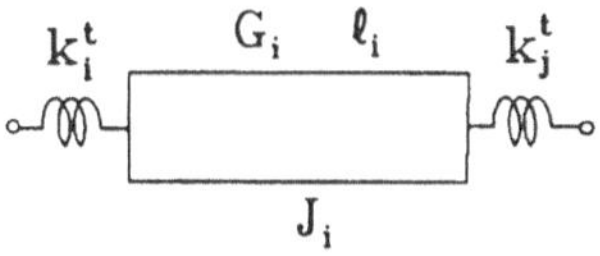

$$k_i^t = k_j^t = 2G_iJ_i/\ell_i$$

G_i: shear modulus

J_i: centroidal moment of inertia

Bending segment

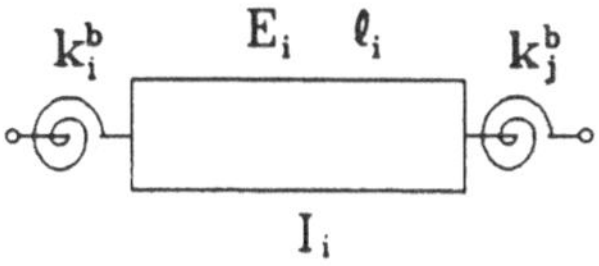

$$k_i^b = k_j^b = 2E_iI_i/\ell_i$$

I_i: moment of inertia of the cross section

7.2 COMBINED STRAIGHT SEGMENTS

Extension

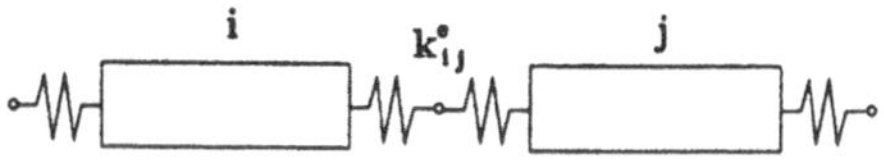

$$k_{ij}^e = 2E_iE_jA_iA_j/(E_iA_j\ell_j + E_jA_i\ell_i)$$

Torsion

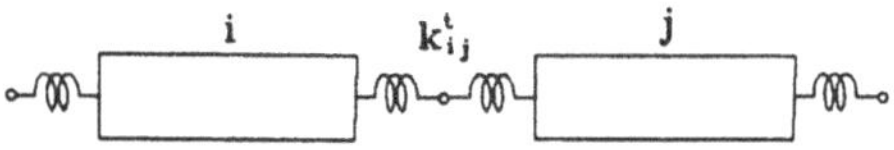

$$k_{ij}^t = 2G_iG_jJ_iJ_j/(G_iJ_j\ell_j + G_jJ_i\ell_i)$$

Bending

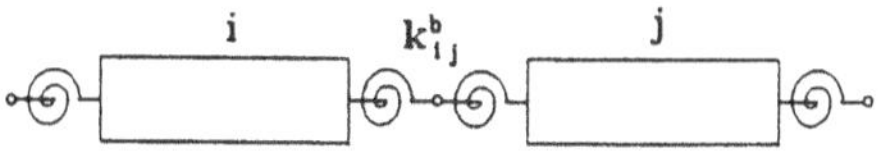

$$k_{ij}^b = 2E_iE_jI_iI_j/(E_iI_j\ell_j + E_jI_i\ell_i)$$

7.3 TAPERED SEGMENTS

Extensional segment

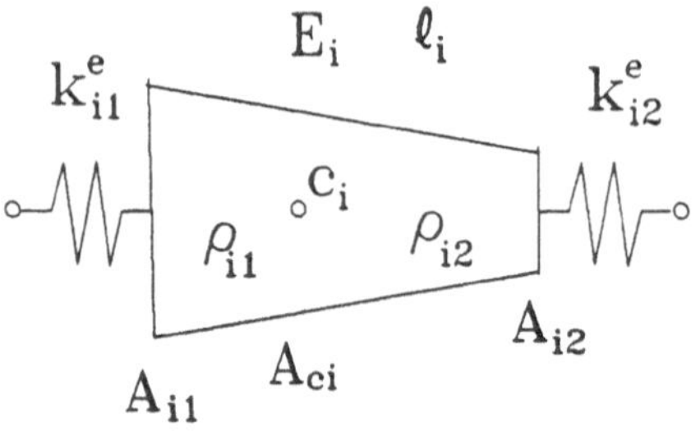

$k^e_{i1} = E_i(A_{i1}-A_{ci})/[\rho_{i1}\ln(A_{i1}/A_{ci})]$
$k^e_{i2} = E_i(A_{i2}-A_{ci})/[\rho_{i2}\ln(A_{i2}/A_{ci})]$

$A_{ci} = (2/3)(A^2_{i1} + A_{i1}A_{i2} + A^2_{i2})/(A_{i1} + A_{i2})$
$\rho_{i1} = (\ell_i/3)(A_{i2} + 2A_{i1})/(A_{i1} + A_{i2})$
$\rho_{i2} = (\ell_i/3)(2A_{i2} + A_{i1})/(A_{i1} + A_{i2})$

ℓ_i: length of the segment
E_i: Young's modulus of the segment
A_{i1}, A_{i2}: cross sectional areas at right and left ends respectively.
A_{ci}: cross sectional area at mass center c_i

Torsional segment

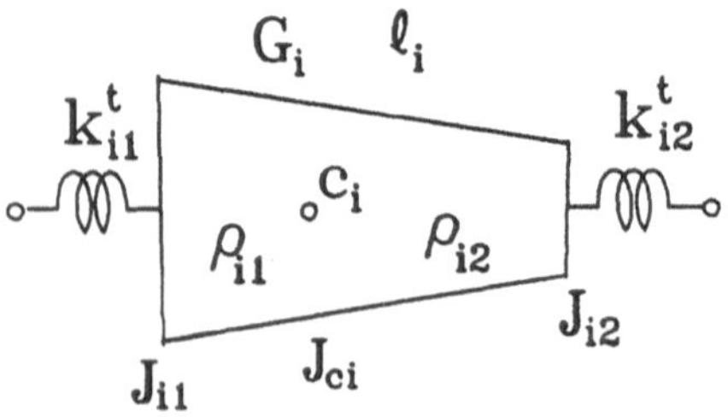

$k^t_{i1} = G_i(J_{i1}-J_{ci})/[\rho_{i1}\ln(J_{i1}/J_{ci})]$
$k^t_{i2} = G_i(J_{i2}-J_{ci})/[\rho_{i2}\ln(J_{i2}/J_{ci})]$

$J_{ci} = (2/3)(J^2_{i1} + J_{i1}J_{i2} + J^2_{i2})/(J_{i1} + J_{i2})$

G_i: shear modulus of the segment
J_{i1}, J_{i2}: centroidal moments of inertia at both ends
J_{ci}: centroidal moment of inertia at mass center c_i

Bending segment

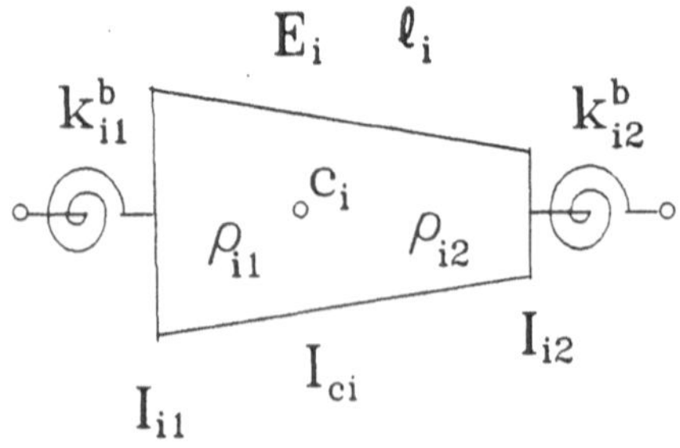

$k^b_{i1} = E_i(I_{i1}-I_{ci})/[\rho_{i1}\ln(I_{i1}/I_{ci})]$
$k^b_{i2} = E_i(I_{i2}-I_{ci})/[\rho_{i2}\ln(I_{i2}/I_{ci})]$

$I_{ci} = (2/3)(I^2_{i1} + I_{i1}I_{i2} + I^2_{i2})/(I_{i1} + I_{i2})$

I_i, I_j: moments of inertia at both ends
I_c: moment of inertia at mass center c

7.4 COMBINED TAPERED SEGMENTS

Extension

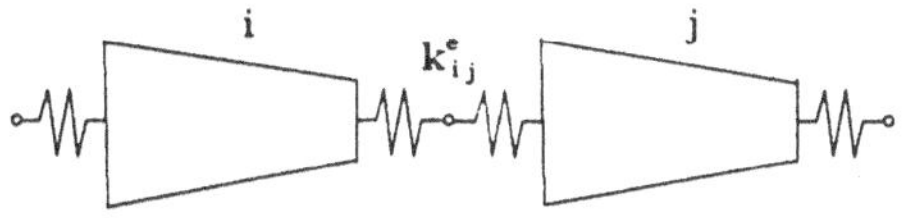

$$k^e_{ij} = E_i E_j (A_{j1}-A_{cj})(A_{i2}-A_{ci})/[E_i(A_{i2}-A_{ci})\rho_{j1}\ln(A_{j1}/A_{cj}) + E_j(A_{j1}-A_{cj})\rho_{i2}\ln(A_{i2}/A_{ci})]$$

Torsion

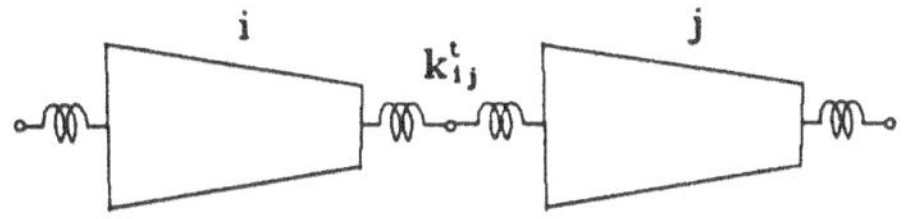

$$k^t_{ij} = G_i G_j (J_{j1}-J_{cj})(J_{i2}-J_{ci})/[G_i(J_{i2}-J_{ci})\rho_{j1}\ln(J_{j1}/J_{cj}) + G_j(J_{j1}-J_{cj})\rho_{i2}\ln(J_{i2}/J_{ci})]$$

Bending

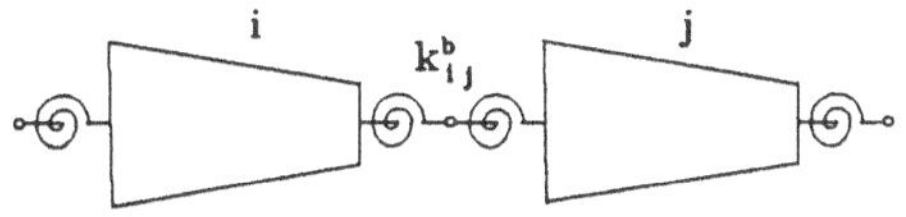

$$k^b_{ij} = E_i E_j (I_{j1}-I_{cj})(I_{i2}-I_{ci})/[E_i(I_{i2}-I_{ci})\rho_{j1}\ln(I_{j1}/I_{cj}) + E_j(I_{j1}-I_{cj})\rho_{i2}\ln(I_{i2}/I_{ci})]$$

7.5 COMBINED STRAIGHT AND TAPERED SEGMENTS

Extension

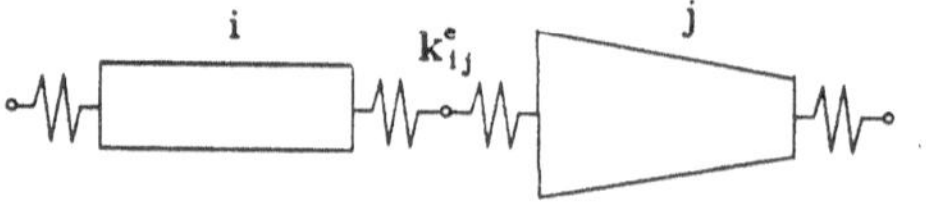

$$k^{e}_{ij}=2E_iE_jA_i(A_{j1}-A_{cj})/[l_iE_j(A_{j1}-A_{cj})+2E_iA_i\rho_{j1}\ln(A_{j1}/A_{cj})]$$

Torsion

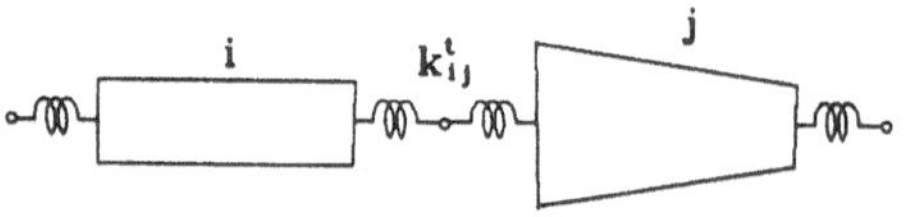

$$k^{t}_{ij}=2G_iG_jJ_i(J_{j1}-J_{cj})/[\ell_iG_j(J_{j1}-J_{cj})+2G_iJ_i\rho_{j1}\ln(J_{j1}/J_{cj})]$$

Bending

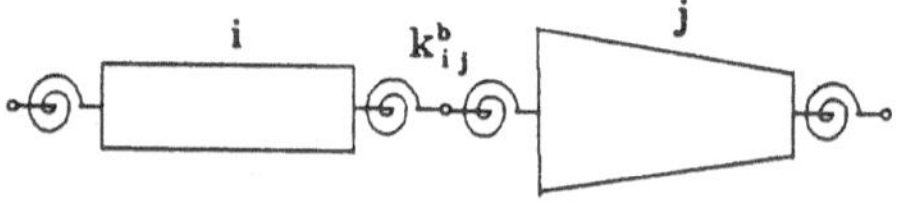

$$k^{b}_{ij}=2E_iE_jI_i(I_{j1}-I_{cj})/[\ell_iE_j(I_{j1}-I_{cj})+2E_iI_i\rho_{j1}\ln(I_{j1}/I_{cj})]$$

8. Acknowledgement

Research reported herein was partially supported by the National Science Foundation under grant MSS8912521 and is gratefully acknowledged.

9. References

1. O.P. Agrawal and A.A. Shabana, "Dynamic analysis of multibody systems using component modes", *Comp. Struct.*, **21**, 1303-1312 (1985).
2. M.L. Amirouche, "Flexibility effects - estimation of the stiffness matrix in the dynamics of large structures", *J. Vib. Acoust. Stress Reliab. Des.*, **109**, 283-288 (1987).
3. M.L. Amirouche, "Dynamic analysis of tree-like structure undergoing large motions - A finite segment approach", *Eng. Analysis*, **3**, 111-117 (1986).
4. F.M.L. Amirouche and R.L. Huston, "Dynamics of large constrained flexible structures", *J. Dyn. Syst. Meas. Control*, **110**, 78-83 (1988).
5. H. Ashley, "On passive damping mechanisms in large space structures", *J. Spacecraft Rockets*, **21**, 448-455 (1984).
6. P.M. Bainum, R. Krishna and V.K. Kuman, "The dynamics of large flexible earth painting structures with a hybrid control system", *J. Astronaut. Sci.*, **xxx**, 251-267 (1982).
7. E.M. Baker and A.A. Shabana, "Geometrically nonlinear analysis of multibody systems", *Comput. Struct.*, **23**, 739-751 (1986).
8. A.K. Banerjee and J.M. Dickens, "Dynamics of an arbitrary flexible body in large relations and translation", *J. Guid. Control Dyn.*, **13**, 221-227 (1990).
9. A.K. Banerjee and M.E. Lemak, "Multi-flexible body dynamics capturing motion-induced stiffness", *J. Appl. Mech.*, **58**, 766-775 (1991).
10. H. Baruh and S.S.K. Tadikonda, "Issues in the dynamics and control of flexible robot manipulators", *J. Guid. Control Dyn.*, **12**, 659-671 (1989).
11. C.E. Benedict and D. Tesar, "Dynamic response analysis of quasi-rigid mechanical systems using kinematic influence coefficients", *J. Mechanisms*, **6**, 383-403 (1971).
12. P. Boland, J.C. Samin and P.Y. Willems, "Stability analysis of interconnected deformable bodies in a topological tree", *AIAA J.*, **12**, 1025-1030 (1974).
13. H. Bremer, "Dynamics of flexible hybrid structures," *J. Guidance and Control*, **2**, 86-87 (1979).
14. R.H. Cannon, Jr. and E. Schmitz, "Initial experiments on the end-point control of a flexible one-link robot," *Int. J. Robotics Res.*, **3**, 62-75 (1984).
15. N.G. Chalhoub and A.G. Ulsoy, "Control of a flexible robot arm: Experimental and theoretical results", *ASME Paper 87-WA/DSC-3* (1987).
16. S. Dubowsky and T.N. Gardner, "Design and analysis of multilink flexible mechanisms with multiple clearance connections", *J. Eng. Industry ASME*, **99**, 88-96 (1977).
17. S. Dubowsky and M.F. Moening, "An experimental and analytical study of impact forces in elastic", *Mech. Mach. Theory*, **13**, 451-465 (1978).
18. A.G. Erdman, G.N. Sandor and R.G. Oakberg, "A general method for kineto-elastodynamic analysis and synthesis of mechanisms," *J. Eng. Industry ASME*, **94**, 1193-1205 (1972).

19. W.B. Gevarter, "Basic relations for control of flexible vehicles," *AIAA J.*, **8**, 666-678 (1970).
20. R.P.S. Han and Z.C. Zhao, "Dynamics of general flexible multibody systems", *Int. J. Numer. Methods Eng.*, **30**, 77-97 (1990).
21. J.Y.L. Ho, "Direct path method for flexible multibody spacecraft dynamics", *J. Spacecraft Rockets*, **14**, 102-110 (1977).
22. J.Y.L. Ho and R. Gluck, "Inductive methods for generating the dynamic equations of motion for multibody flexible systems Part 2", *Synthesis of Vibrating Systems*, ASME, New York, 1971.
23. J.Y.L. Ho and D.R. Herber, "Development of dynamics and control simulation of large flexible space systems", *J. Guid. Control Dyn.*, **8**, 374-383 (1985).
24. Y. Huang and C.S.G. Lee, "Generalization of Newton-Euler formulation of dynamic equations to nonrigid manipulators", *J. Dyn. Syst. Meas. Control*, **110**, 308-315 (1988).
25. P.C. Hughes, "Dynamics of a chain of flexible bodies", *J. Astronaut. Sci.*, **XXVII**, 359-380 (1979).
26. P.C. Hughes, "Dynamics of a flexible space vehicle with active attitude control", *Celestial Mech. J.*, **9**, 21-39 (1974).
27. R.L. Huston, "Flexibility effects in multibody systems", *Mech. Res. Commun.*, **7**, 261-268 (1980).
28. R.L. Huston, "Multibody dynamics including the effects of flexibility and compliance", *Comp. Struct.*, **14**, 443-451 (1981).
29. R.L. Huston, "Multibody dynamics: Analysis of flexibility effects", *Proc. Ninth U.S. National Congress of Applied Mechanics*, Cornell University, ASME, New York, 1982.
30. R.L. Huston, "Computer methods in flexible multibody dynamics", *Int. J. Numer. Methods Eng.*, **32**, 1657-1668 (1991).
31. I. Imam, G.N. Sandor and S.N. Kramer, "Deflection and stress analysis in high speed planar mechanisms with elastic links", *J. Eng. Industry ASME*, **95**, 541-548 (1973).
32. W. Jerkovsky, "Exact equations of motion for a deformable body", Aerospace Corporation, SA, *SO-TR-77-133* (1977.
33. J.J. Kalker and G.J. Olsder, "Robots, with flexible links: Dynamics, control and stability", *NTIS Report PB86-174885* (1985).
34. T.R. Kane, P.R. Ryan and A.K. Banerjee, "Dynamics of a cantilever beam attached to a moving base", *J. Guid. Control Dyn.*, **10**, 139-151 (1987).
35. D.A. Levinson, "Large motions of unrestrained structures", *Proc. Ninth U.S. National Congress of Applied Mechanics*, Cornell University, ASME, New York, 259-264 (1982).
36. P.W. Likins, "Dynamic analysis of a system of hinge-connected rigid bodies with nonrigid appendages", *Int. J. Solids Struct.*, **9**, 1473-1487 (1973).
37. P.W. Likins, "Quasi-coordinate equations for flexible spacecraft", *AIAA J.*, **13**, 524-526 (1975).
38. P.W. Likins and H.K. Bouvier, "Attitude control of nonrigid spacecraft",

Astonaut. Aeronaut., **9**, No. 5, 64-71 (1971).

39. K.W. Lipps and V.J. Modi, "Transient attitude dynamics of satellites with deploying flexible appendages", *Acta Astronaut.*, **5**, 797-815 (1978).
40. K.W. Lipps and V.J. Modi, "General dynamics of a large class of flexible satellite systems", *Acta Astronaut.*, **7**, 1349-1360 (1980).
41. V.J. Modi, "Attitude dynamics of satellites with flexible appendages - A brief review", *J. Spacecraft Rockets*, **11**, 743-751 (1974).
42. C.E. Padilla and A.H. Von Flotow, "Nonlinear-displacement relations and flexible multibody dynamics," *J. Guid. Control Dyn.*, **10**, 139-151 (1987).
43. M. Pascal, "Dynamical analysis of a flexible manipulator arm", *Acta Astronaut.*, **21**, 161-169 (1990).
44. S. Rajaram and J.L. Junkins, "Identification of vibrating flexible structures", *J. Guid. Control.*,**8**, 463-370 (1985).
45. J.P. Sadler and G.N. Sander, "A lumped parameter approach to vibration and stress analysis of elastic linkages", *J. Eng. Industry ASME*, **95**, 549-557 (1973).
46. A.A. Shabana, "On the use of the finite element method and classical approximation techniques in the non-linear dynamics of multibody system", *Int. J. Non-linear Mech.*, **25**, 153-162 (1990).
47. A.A. Shabana, "Substructure synthesis methods for dynamic analysis of multi-body systems", *Comp. Struct.*, **20**, 737-744 (1985).
48. R.C. Winfrey, "Elastic link mechanisms dynamics", *J. Eng. Industry ASME*, 268-272 (1971).
49. T.R. Kane, "Dynamics of nonholonomic systems", *J. Appl. Mech. ASME*, **28**, 574-578 (1961).
50. T.R. Kane, *Dynamics*, Holt, Rinehart and Winston, New York (1968).
51. T.R. Kane, P.W. Likins and D.A. Levinson, *Spacecraft Dynamics*, McGraw-Hill, New York (1983).
52. T.R. Kane and D.A. Levinson, *Dynamics: Theory and Application*, McGraw-Hill, New York (1985).
53. R.L. Huston and C.E. Passerello, "On multi-rigid body-systems dynamics", *Comp. Struct.*, **10**, 439-446 (1979).
54. R.L. Huston, C.E. Passerello and M.W. Harlow, "Dynamics of multirigid body systems", *J. Appl. Mech. ASME*, **45**, 889-894 (1978).
55. R.L. Huston and C.E. Passerello, "Multibody structural dynamics including translation between the bodies", *Comp. Struct.*, **18**, 999-1003 (1984).
56. R.L. Huston, "Multibody dynamics formulations via Kane's equations", in J.L. Junkins (ed.), *Mechanics and Control of Large Space Structures*, Vol. 129 of *Progress in Astronautics and Aeronautics*, AIAA, 71-86 (1990).
57. E.T. Whittaker, *Analytical Dynamics*, Cambridge (1957).
58. R.L. Huston, *Multibody Dynamics*, Butterworth-Heinmann, Stoneham, MA (1990).
59. R.L. Huston, "Computing angular velocity in multibody systems", *Eng. Computations*, **3**, 223-230 (1986).

60. J.W. Kamman, R.L. Huston and T.P. King, "UCIN-DYNOCOMBS - Software for the dynamic analysis of constrained multibody systems", in W. Schielen (ed.), *Multibody Systems Handbook*, Springer-Verlag, Berlin, 103-121 (1990).
61. T.R. Kane and C.F. Wang, "On the derivation of equations of motion", *J. Soc. Industrial Appl. Math.*, **13**, 487-492 (1965).
62. T.R. Kane, *Analytical Elements of Mechanics, 1*, Academic Press, New York (1959).
63. R.L. Huston and C.E. Passerello, *Finite Element Methods - An Introduction*, marcel Dekker, New York (1985).
64. Y. Wang and R.L. Huston, "Dynamic analysis of elastic beam-like mechanism systems", in A. Midha (ed.), *Trends and Developments in Mechanisms, Machines and Robotics - 1988*, DE-15-2, 457-460 (1988).
65. W. Weaver, Jr., S.P. Timoshenko and D.H. Young, *Vibration Problems in Engineering*, 5th edn, Wiley, New York, 427-428 (1990).
66. L.S. Jacobson and R.S. Ayre, *Engineering Vibration*, McGraw-Hill, 496 (1958).
67. W. Schiehlen (ed.), *Multibody Systems Handbook*, Springer-Verlag, Berlin (1990).

PART III
Kinematic Aspects of Multibody Systems

ON TWIST AND WRENCH GENERATORS AND ANNIHILATORS

Jorge Angeles
Department of Mechanical Engineering &
McGill Research Centre for Intelligent Machines
McGill University
817 Sherbrooke St. W.
Montreal, Quebec, CANADA
H3A 2K6
angeles@mcrcim.mcgill.ca

ABSTRACT. The concepts of *twist generator*, *wrench generator* and their counterparts, namely, *twist annihilator* and *wrench annihilator* are introduced in this paper. It is shown that twist annihilators allow the elimination of idle variables in the analysis of kinematic chains with multiple loops, thereby easing the formulation of the underlying kinematic relations. As examples of applications, the input-output velocity analysis of a four-bar spatial linkage and the Jacobians of a robotic mechanical system, pertaining either to a walking machine or a multi-fingered hand, are included. Furthermore, these concepts are extended in such a way that they find a straightforward application in the formulation of dynamics models of multi-body systems.

1 Introduction

The relations among the joint rates of *simple* kinematic chains, i.e., chains with links coupled to two other links at most, have been fully researched for some time. Pioneer work in this regard was reported by Freudenstein [1], who introduced the closure equations of a single-loop spatial kinematic chain as a linear combination of the screws associated with the axes of the kinematic pairs involved, the corresponding coefficients being the joint rates. Hence, the underlying differential relations can be written in a linear homogeneous form with the aid of a matrix that Freudenstein called the *functional matrix* of the chain. This matrix is formally identical to the Jacobian matrix of robotic manipulators with open kinematic-chain structure of the simple type.

Current developments in robotic technology have prompted the study of multi-loop, multi-degree-of-freedom kinematic chains. Such kinematic chains appear in robotic systems like parallel manipulators, walking machines and multi-fingered hands. The difference between kinematic chains with multiple loops and open kinematic chains with a simple structure, e.g., those occurring in serial manipulators, is the presence of a number of idle joints in the former. The feed-forward control of the robotic mechanical system at hand requires an explicit relation between the actuated joint rates and the Cartesian velocities

M. F. O. S. Pereira and J. A. C. Ambrósio (eds.),
Computer-Aided Analysis of Rigid and Flexible Mechanical Systems, 379–411.

of the system. This relation is known to be linear, the coefficient matrices involved being generically termed *Jacobians*. While in serial manipulators one single Jacobian appears, the presence of multiple loops brings about two *global* Jacobians, one multiplying the vector of joint rates, the other the twist of the controled link. Here, a distinction should be made between what we understand as a *global* and a *local* Jacobian. The former refers to one pertaining to the overall kinematic chain, while the latter to a particular subchain of the given chain. In simple kinematic chains this distinction is immaterial, but in multi-loop chains it is essential. While the derivation of the Jacobian of robots with serial kinematic-chain structure is a well-established subject, that of the Jacobians involved in multi-loop systems is still a research subject.

Here, we resort to the concept of *screw*, already discussed by Ball [2], to derive the desired relations. The approach to the analysis of kinematic chains usually encountered in the literature involves the calculation of *reciprocal screws*. Any screw multiplied by an *amplitude* with units of angular velocity yields a twist. If the screw is multiplied by an amplitude with units of force, a *wrench* is obtained. Moreover, if a given screw produces a wrench on a body moving with a twist produced by a second screw and the first wrench develops zero power onto the body under the aforementioned twist, the two screws are said to be *reciprocal*. A study on the duality between wrenches and twists in the context of reciprocal screws and its impact in the analysis of serial and parallel robotic manipulators was recently reported [3, 4]. Here, we show that, resorting to the concept of *twist annihilator*, not only one, but rather a set of reciprocal screws can be readily derived.

Applications of the concepts introduced here are anticipated in the area of hybrid or kinetostatic (feed-forward) control of manipulators. In this regard, our work can complement that reported by Lipkin and Duffy [5, 6].

For quick reference, we include below an outline of definitions and terminology pertaining to kinematic chains. The reader unfamiliar with this terminology will find in that section the necessary definitions of the various terms handled in the above discussion.

2 Background on Kinematic Chains

Under certain circumstances, a multi-body mechanical system can be accurately modeled as an assemblage of rigid bodies. When this is the case, the concept of *kinematic chain* becomes an essential tool in the kinematic analysis of the system. A kinematic chain is, then, an assemblage of rigid bodies, called *links*, that are coupled via *kinematic pairs*. We distinguish between *lower* kinematic pairs and *higher* kinematic pairs [7, 8]. Two links are coupled by a lower kinematic pair when the links undergo relative motion while maintaining *surface contact*. If such a motion occurs while the links maintain *point* or *line contact*, then we talk of a higher kinematic pair. The focus here will be on lower kinematic pairs. Upper kinematic pairs are equally important, for they occur in cam mechanisms, gear trains and in grasping with rolling contact, but deserve a special treatment. No unified treatment of upper kinematic pairs, like that pertaining to lower kinematic pairs, is available.

The six lower kinematic pairs are the *revolute pair* (R), the *prismatic pair* (P), the *screw pair* (H), the *cylindrical pair* (C), the *spherical pair* (S) and the *planar pair* (E). For completeness, we outline these pairs below.

The revolute pair allows a relative rotation about one fixed axis of one link with respect to the other, but no relative translation. The prismatic pair, in turn, allows a relative translation of one link with respect to the other, along a fixed direction, but no relative rotation. The screw pair allows both rotation about and translation along a fixed line, with the translation being proportional to the rotation; the proportionality constant is the *pitch* of the screw. Furthermore, the cylindrical pair allows a relative motion similar to that of the screw pair, but with both translation and rotation indpendent from each other. The spherical pair allows a pure rotation of one link with respect to the other about a fixed point, and no translation. Finally, the planar pair allows a relative motion that consists of two translations along distinct directions and a rotation in the plane defined by the two foregoing directions. Hence, the relative motion allowed by the first three pairs, revolute, prismatic and screw, has one degree of freedom (dof); that allowed by the cylindrical pair has two dof; and the one allowed by the spherical and planar pairs has three dof.

An essential attribute of a kinematic chain is its *topology*, which is fully described by the *graph representation* of the chain. In this representation, a link appears as a node and a pair as an edge. The *connectivity* of a link is given by an integer, which is the number of links to which the first link is coupled. In particular, we talk of a *simple* link if the link is connected to one other single link only; if the link is connected to two other links, then we have a *binary* link. *Ternary* links are coupled to three other links. In general, a *multiple* link is coupled to many other links. Links with various connectivities are sketched in Fig. 1. In Fig. 1a, links A and C are simple, while B is binary. Moreover, in Fig. 1b, link A is ternary, while links B, C and D are simple. Finally, link A of Fig. 1c is of connectivity five, or multiple, for brevity.

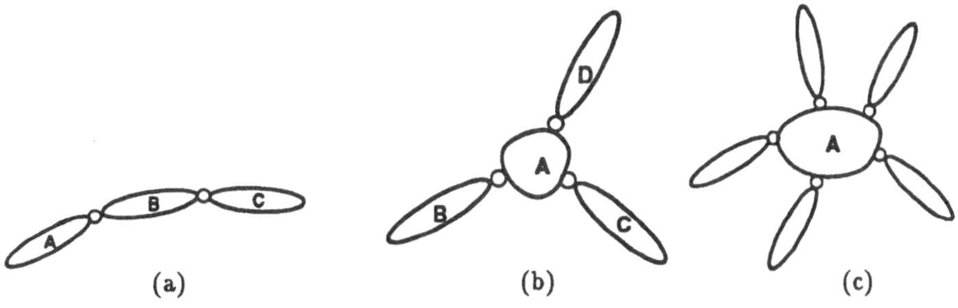

Figure 1: Link connectivities

Within the realm of the topology of kinematic chains, we distinguish between open and closed chains, as well as between simple and complex chains. A chain is open if it contains two extreme links that are simple, the intermediate ones being binary. The chain of Fig. 1a is an example of an open kinematic chain. If a chain contains only binary links, it forms a single loop and the chain is said to be closed and simple. A complex kinematic chain is composed of *subchains* that can be open or closed. If a complex chain has no closed subchains, then it has a *tree structure*. Examples of chains with tree structures are those of Figs. 1b & c. If the chain does contain various kinematic loops, it is called a *multi-loop* chain. In general, a complex kinematic chain has subchains with tree structure and closed subchains. A complex chain with a tree-structure subchain and two kinematic loops is illustrated in Fig. 2. In this figure, the hatched element represents a ternary link. Moreover, the subchain obtained when this link is removed has a tree structure. This chain thus contains one open subchain and two closed subchains.

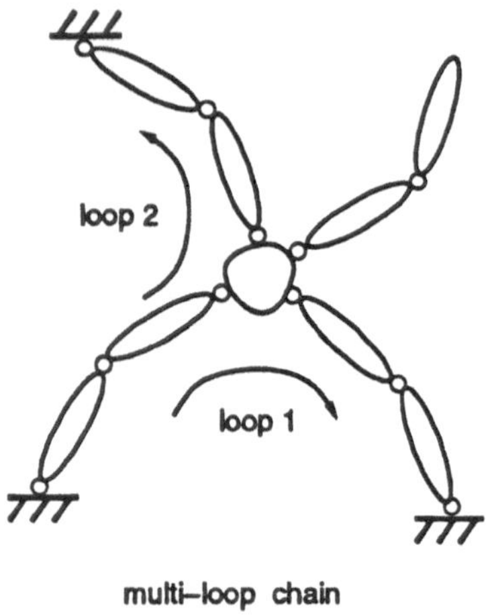

Figure 2: A complex chain

Furthermore, the kinematic chain of a manipulator with *parallel* topology and three degrees of freedom appears in the upper part of Fig. 3. In this figure, three joints, labeled M_k, for $k = 1, 2, 3$, are attached to ground, which is thus a ternary link. The actuators of the manipulators drive these joints. With the controled motion of these three joints, the motion of the central triangle is produced. This triangle, moreover, is connected to the ground via three *legs*, each consisting of an open, two-link subchain. Note that all joints are numbered successively from 1 to 9. In the lower part of this figure we show, from left to right, the graph representation of the same kinematic chain.

3 Background on Screws, Twists and Wrenches

We focus here on the screws associated with the foregoing lower kinematic pairs. To this end, we start by recalling the *Plücker coordinates* of a line $\mathcal{L}$, defined as an array of six real numbers, namely, the three components of a unit vector $\mathbf{e}$, parallel to $\mathcal{L}$, and the three components of its *moment* about a predefined point O that can be inside or outside the line. If P is a point of $\mathcal{L}$, and $\mathbf{p}$ is the vector directed from O to P, as shown in Fig. 4, then the moment $\mathbf{n}$ of the line is defined as

$$\mathbf{n} \equiv \mathbf{p} \times \mathbf{e} \tag{1}$$

Moreover, the *Plücker array* of the line is defined here as a six-dimensional array $\boldsymbol{\pi}_{\mathcal{L}}$, namely,

$$\boldsymbol{\pi}_{\mathcal{L}} \equiv \begin{bmatrix} \mathbf{e} \\ \mathbf{n} \end{bmatrix} \tag{2}$$

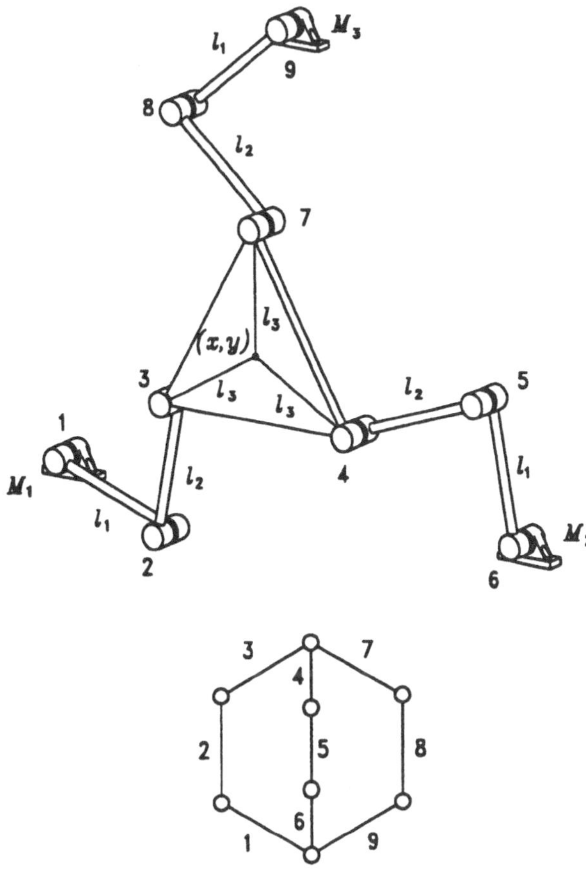

Figure 3: Kinematic chain of parallel topology and its graph

Note that the six entries of the Plücker array are not independent, for they must obey two conditions, namely,

$$\mathbf{e}\cdot\mathbf{e}=1, \quad \text{and} \quad \mathbf{e}\cdot\mathbf{n}=0 \tag{3}$$

Thus, the Plücker array of a line contains only four independent components, but these are enough to define the line. Now, if a pitch p is added as a fifth feature to the line or, correspondingly, to its Plücker array, we obtain a screw $\mathbf{s}$, namely,

$$\mathbf{s}\equiv\begin{bmatrix}\mathbf{e}\\ \mathbf{p}\times\mathbf{e}+p\mathbf{e}\end{bmatrix} \tag{4}$$

An amplitude is any scalar A multiplying the foregoing screw. It produces a twist or a wrench depending on its units. The twist or the wrench thus derived can be said to be in *canonical form*, for its representation involves explicitly the eight parameters defining

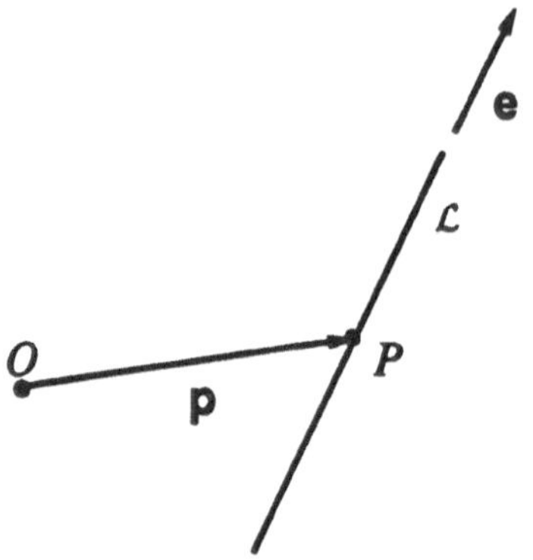

Figure 4: A line $\mathcal{L}$ passing through point P parallel to $\mathbf{e}$

it, namely, the amplitude, the pitch and the six Plücker coordinates of the associated line. Clearly, a twist or a wrench are defined completely by six independent real numbers. More generally, a twist can be regarded as a 6-dimensional array defining completely the velocity field of a rigid body and comprises the three components of the angular velocity and the three components of the velocity of any of the points of the body. The wrench can be regarded likewise, namely, as the 6-dimensional array defining completely the resultant of a system of forces and moments acting on a body.

Below we elaborate on the foregoing concepts. Upon multiplication of the screw appearing in eq.(4) by the amplitude A representing the magnitude of an angular velocity, we obtain a twist $\mathbf{t}$, namely,

$$\mathbf{t} \equiv \begin{bmatrix} A\mathbf{e} \\ \mathbf{p} \times (A\mathbf{e}) + p(A\mathbf{e}) \end{bmatrix}$$

where the product $A\mathbf{e}$ can be readily identified as the angular velocity $\boldsymbol{\omega}$ parallel to vector $\mathbf{e}$, of magnitude A. Moreover, the lower part of $\mathbf{t}$ can be readily identified with the velocity of a point of a rigid body. Indeed, if we regard the line $\mathcal{L}$ and point O as sets of points of a rigid body $\mathcal{B}$ moving with an angular velocity $\boldsymbol{\omega}$ and such that point P moves with a velocity $p\boldsymbol{\omega}$ parallel to the angular velocity, then $\mathbf{v}$ represents the velocity of point O, i.e.,

$$\mathbf{v} = -\boldsymbol{\omega} \times \mathbf{p} + p\boldsymbol{\omega}$$

We can thus express the twist $\mathbf{t}$ as

$$\mathbf{t} \equiv \begin{bmatrix} \boldsymbol{\omega} \\ \mathbf{v} \end{bmatrix} \tag{5}$$

Now, if we multiply the screw of eq.(4) by an amplitude A with units of force, what we will obtain is a wrench $\mathbf{w}$, i.e., a 6-dimensional array with its first three components having units of force and its last components units of moment. We would like to be able to obtain the power developed by the wrench on the body moving with the twist $\mathbf{t}$ by a simple inner product of the two arrays. However, because of the form the wrench $\mathbf{w}$ has taken, the inner product of these two arrays would be meaningless, for it would involve the sum of two scalar quantities with different units and, moreover, each of the two quantities without an immediate physical meaning. In fact, the first scalar would have units of force by frequency (angular velocity by force), while the second would have units of moment of moment multiplied by frequency (velocity by moment), thereby leading to a physically meaningless result. This inconsistency can be resolved if we redefine the wrench not simply as the product of a screw by an amplitude, but as a linear transformation of that screw, involving the 6×6 array $\mathbf{L}$ defined as

$$\mathbf{L} \equiv \begin{bmatrix} \mathbf{O} & \mathbf{1} \\ \mathbf{1} & \mathbf{O} \end{bmatrix} \tag{6}$$

where $\mathbf{O}$ and $\mathbf{1}$ denote, respectively, the 3×3 zero and identity matrices. Now we define the wrench as a linear transformation of the screw $\mathbf{s}$ defined in eq.(4). This transformation is obtained upon multiplying $\mathbf{s}$ by the product $A\mathbf{L}$, where the amplitude A has units of force, i.e.,

$$\mathbf{w} \equiv A\mathbf{L}\mathbf{s} \equiv \begin{bmatrix} \mathbf{p} \times (A\mathbf{e}) + p(A\mathbf{e}) \\ A\mathbf{e} \end{bmatrix}$$

Now, the first three components of the foregoing array can be readily identified with the moment of a force of magnitude A acting along a line of action given by the Plücker array of eq.(2), with respect to point O of Fig. 4, to which a moment parallel to that line and of magnitude pA is added. Moreover, the last three components of that array pertain apparently to a force parallel to the same line, of magnitude A. We denote here the said moment by $\mathbf{n}$ and the force by $\mathbf{f}$, i.e.,

$$\mathbf{f} \equiv A\mathbf{e}, \quad \mathbf{n} \equiv \mathbf{p} \times \mathbf{f} + p\mathbf{f}$$

The wrench $\mathbf{w}$ has then been defined as

$$\mathbf{w} \equiv \begin{bmatrix} \mathbf{n} \\ \mathbf{f} \end{bmatrix} \tag{7}$$

which can thus be interpreted as a representation of a system of forces and moments acting on a rigid body, with the force acting at point O of the body $\mathcal{B}$ defined above, and a moment $\mathbf{n}$. Under these circumstances, we say that $\mathbf{w}$ *acts* at point O of $\mathcal{B}$.

With the foregoing definitions it is now apparent that the wrench has been defined so that the *inner product* $\mathbf{t}^T\mathbf{w}$ will produce the power developed by $\mathbf{w}$ acting at O when $\mathcal{B}$ moves with a twist $\mathbf{t}$ defined at the same point.

4 The Twist- and Wrench-Transfer Formulas

The *twist-transfer formula*, which relates the twist of the same rigid body at two different points, is now derived. Here, we will need the *cross-product matrix* of a 3-dimensional vector $\mathbf{v}$, which is a 3×3 matrix $\mathbf{V}$. For any 3-dimensional vector $\mathbf{x}$, $\mathbf{V}$ is defined as

$$\mathbf{V} \equiv \frac{\partial(\mathbf{v} \times \mathbf{x})}{\partial \mathbf{x}} \tag{8}$$

which can be readily proven to be skew-symmetric. Indeed, from the above definition, the cross product $\mathbf{v} \times \mathbf{x}$ can be alternatively written as

$$\mathbf{v} \times \mathbf{x} = \mathbf{V}\mathbf{x} \tag{9}$$

Moreover, the product $\mathbf{Vx}$ vanishes whenever $\mathbf{x}$ is a multiple of $\mathbf{v}$, and hence, $\mathbf{V}$ is singular. Moreover, from the above relation,

$$\mathbf{x}^T\mathbf{V}\mathbf{x} = 0$$

for arbitrary $\mathbf{x}$. Now, the foregoing product can only vanish if $\mathbf{V}$ *i*) is a proper orthogonal matrix rotating vectors through 90° about a certain axis, or if *ii*) $\mathbf{V}$ is a multiple of the aforementioned orthogonal matrix, or if *iii*) $\mathbf{V}$ is skew-symmetric. However, $\mathbf{V}$ being singular, the first two possibilities are ruled out, and hence, $\mathbf{V}$ is skew-symmetric, q. e. d.

One further property of $\mathbf{V}$ is given below:

$$\mathbf{V}^2 = -\|\mathbf{v}\|^2\mathbf{1} + \mathbf{v}\mathbf{v}^T \tag{10}$$

where $\mathbf{1}$ denotes, again, the 3×3 identity matrix. The foregoing relation can be readily proven. Indeed, from the definition of $\mathbf{V}$, eq.(9), we have

$$\mathbf{V}^2\mathbf{x} \equiv \mathbf{v} \times (\mathbf{v} \times \mathbf{x})$$

and hence, upon expansion of the right-hand side,

$$\mathbf{V}^2\mathbf{x} \equiv \mathbf{v}(\mathbf{v} \cdot \mathbf{x}) - \|\mathbf{v}\|^2\mathbf{x}$$

which can be readily rewritten as

$$\mathbf{V}^2\mathbf{x} \equiv (\mathbf{v}\mathbf{v}^T - \|\mathbf{v}\|^2\mathbf{1})\mathbf{x}$$

the relation sought thus becoming apparent.

Now, let A and P be two arbitrary points of a rigid body. The twist of the body at these points is defined as

$$\mathbf{t}_A = \begin{bmatrix} \boldsymbol{\omega} \\ \mathbf{v}_A \end{bmatrix}, \quad \mathbf{t}_P = \begin{bmatrix} \boldsymbol{\omega} \\ \mathbf{v}_P \end{bmatrix} \tag{11}$$

where $\mathbf{v}_P$ can be rewritten as

$$\mathbf{v}_P = \mathbf{v}_A + (\mathbf{a} - \mathbf{p}) \times \boldsymbol{\omega} \tag{12}$$

with $\mathbf{a}$ and $\mathbf{p}$ defined as the position vectors of points A and P, respectively. Combining eq.(11) with eq.(12) yields

$$\mathbf{t}_P = \mathbf{T}\mathbf{t}_A \tag{13}$$

with the 6×6 matrix $\mathbf{T}$ defined as

$$\mathbf{T} \equiv \begin{bmatrix} \mathbf{1} & \mathbf{O} \\ \mathbf{A} - \mathbf{P} & \mathbf{1} \end{bmatrix} \tag{14}$$

in which the 3×3 matrices $\mathbf{A}$ and $\mathbf{P}$ are the cross-product matrices of vectors $\mathbf{a}$ and $\mathbf{p}$, respectively. Moreover, $\mathbf{1}$ and $\mathbf{O}$ denote, as before, the 3×3 identity and zero matrices.

Likewise, the *wrench-transfer formula* relates the wrench at two points on the same rigid body. We define the wrench at these points as

$$\mathbf{w}_A \equiv \begin{bmatrix} \mathbf{n}_A \\ \mathbf{f} \end{bmatrix}, \quad \mathbf{w}_P \equiv \begin{bmatrix} \mathbf{n}_P \\ \mathbf{f} \end{bmatrix} \tag{15}$$

where $\mathbf{n}_P$, the moment of the wrench about point P, is related to that about A, $\mathbf{n}_A$, by

$$\mathbf{n}_P = \mathbf{n}_A + (\mathbf{a} - \mathbf{p}) \times \mathbf{f} \tag{16}$$

and hence, $\mathbf{w}_P$ takes on the form

$$\mathbf{w}_P = \mathbf{U}\mathbf{w}_A \tag{17}$$

where U is the 6×6 matrix defined below:

$$\mathbf{U} \equiv \begin{bmatrix} \mathbf{1} & \mathbf{A} - \mathbf{P} \\ \mathbf{0} & \mathbf{1} \end{bmatrix} \tag{18}$$

and **A** and **P** were defined in eq.(13). Thus, $\mathbf{w}_P$ is a linear transformation of $\mathbf{w}_A$.

Multiplying the transpose of each side of eq.(13) by the corresponding side of eq.(17) yields

$$\mathbf{t}_P^T \mathbf{w}_P = \mathbf{t}_A^T \mathbf{T}^T \mathbf{U} \mathbf{w}_A \tag{19}$$

Upon expansion of the matrix product appearing in eq.(19), we obtain

$$\mathbf{T}^T\mathbf{U} = \begin{bmatrix} \mathbf{1} & -\mathbf{A} + \mathbf{P} \\ \mathbf{0} & \mathbf{1} \end{bmatrix} \begin{bmatrix} \mathbf{1} & \mathbf{A} - \mathbf{P} \\ \mathbf{0} & \mathbf{1} \end{bmatrix} = \mathbf{1}_{6\times 6} \tag{20}$$

with $\mathbf{1}_{6\times 6}$ denoting the 6×6 identity matrix. Hence, $\mathbf{t}_P^T \mathbf{w}_P = \mathbf{t}_A^T \mathbf{w}_A$, as expected, since the wrench develops the same amount of power, regardless of where the force is assumed to be applied. Also note that an interesting relation between **T** and **U** follows from eq.(20), namely,

$$\mathbf{U}^{-1} = \mathbf{T}^T \tag{21}$$

It is apparent that both det(**T**) and det(**U**) are equal to unity. Thus, the twist and the wrench at two different points of a rigid body are related by a linear transformation represented by a 6×6 matrix of the *unimodular group*, i.e., the group of 6×6 matrices of determinant equal to unity.

5 The Twist Generators of the Lower Kinematic Pairs

We define below the twist generators of the six lower kinematic pairs:

5.1 THE REVOLUTE

A revolute coupling of two rigid links, 1 and 2, appears in Fig. 5, which shows its attributes, namely, a line $\mathcal{L}$ passing through point O and parallel to the unit vector **e**. The screw of the revolute, denoted by the 6-dimensional array $\mathbf{s}_R$, is derived from the general expression for the screw given in eq.(4) with a pitch $p = 0$. Moreover, we let **p** denote the vector directed from point O of $\mathcal{L}$ to point P of body 2. The screw $\mathbf{s}_R$, then, takes on the form

$$\mathbf{s}_R = \begin{bmatrix} \mathbf{e} \\ \mathbf{e} \times \mathbf{p} \end{bmatrix} \tag{22}$$

Note that the foregoing array is identical to the Plücker array of $\mathcal{L}$ when written with a moment about point P. Also note that the order of **e** and **p** in the definition of the lower part of $\mathbf{s}_R$ has been reversed in eq.(22), but so have the roles of points O and P of Fig. 5 with respect to Fig. 4. In this case, since any motion of 2 with respect to 1 reduces to a rotation about $\mathcal{L}$ with point O of 2 coincident with point O of 1, the twist of 2 with respect to 1 can be simply expressed as $\dot{\theta}\mathbf{s}_R$, and hence, $\mathbf{s}_R$ is the twist generator of this pair.

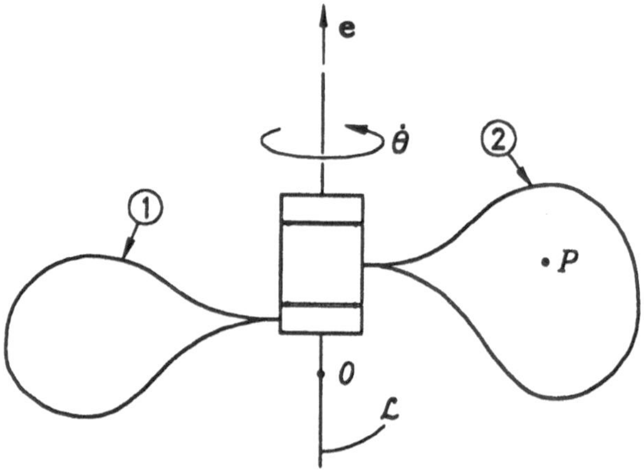

Figure 5: The revolute pair

5.2 THE PRISMATIC PAIR

A prismatic pair coupling bodies 1 and 2 is shown in Fig. 6, its sole attribute being the direction of the unit vector **e**. Here, no line can be defined, as in the case of the revolute, the associated screw, $\mathbf{s}_P$, being given as

$$\mathbf{s}_P = \begin{bmatrix} \mathbf{0} \\ \mathbf{e} \end{bmatrix} \tag{23}$$

where **0** is the 3-dimensional zero vector.

Thus, any motion of 2 relative to 1 reduces to a translation along the direction of **e**, the associated twist thus reducing to $\dot{b}\mathbf{s}_P$. The twist generator of the prismatic pair is then defined as $\mathbf{s}_P$.

5.3 THE SCREW PAIR

Shown in Fig. 7 is a screw pair coupling bodies 1 and 2, its attributes being a line $\mathcal{L}$ and a pitch p. Moreover, the line is defined by its direction parallel to the unit vector **e** and its moment about point P of body 2. The derivation of the associated screw, $\mathbf{s}_H$, from eq.(4), is straightforward, namely,

$$\mathbf{s}_H = \begin{bmatrix} \mathbf{e} \\ \mathbf{e} \times \mathbf{p} + p\mathbf{e} \end{bmatrix} \tag{24}$$

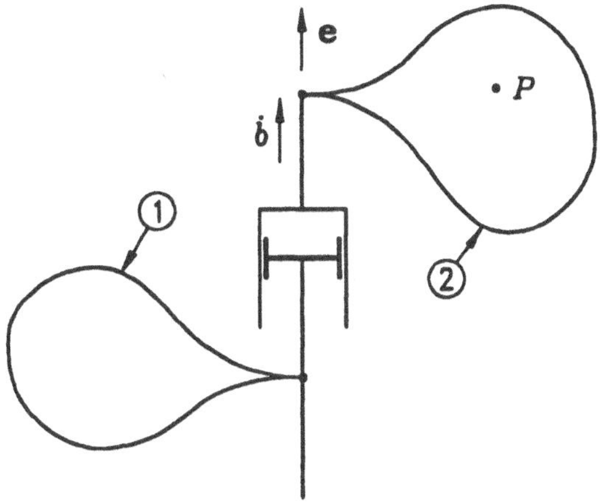

Figure 6: The prismatic pair

Thus, any motion of 2 with respect to 1 reduces to a rotation about and a translation along line $\mathcal{L}$, the associated twist thus becoming $\dot{\theta}\mathbf{s}_H$. The twist generator of the screw pair is thus $\mathbf{s}_H$.

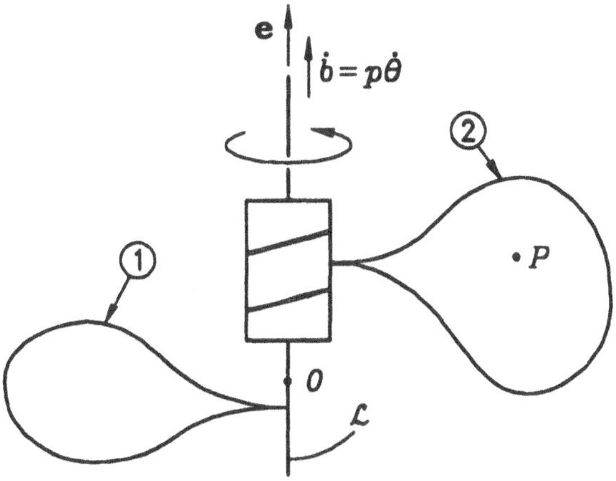

Figure 7: The screw pair

5.4 THE CYLINDRICAL PAIR

A cylindrical pair coupling bodies 1 and 2 appears in Fig. 8, which shows the attributes of the associated screw, namely, the line $\mathcal{L}$ and its two-dof capability, namely, a translation along and a rotation about $\mathcal{L}$, each independent from the other. Thus, the relative motion allowed by this pair has two degrees of freedom, the associated motion then being a combination of the motions allowed by a revolute and a prismatic pair, the latter with a direction parallel to that of the axis of the revolute. While in the first three cases the twist generator coincided with the screw associated with the pair at hand, and, hence, the twist generator reduced to a 6-dimensional array, in this case we need a 6×2 array, in light of the degree of freedom involved. In fact, the twist of any motion of 2 with respect to 1 can be expressed as a linear combination of the screws of the revolute and the prismatic pair, i.e.,

$$\mathbf{t} = \dot{\theta}\begin{bmatrix} \mathbf{e} \\ \mathbf{e} \times \mathbf{p} \end{bmatrix} + \dot{b}\begin{bmatrix} \mathbf{0} \\ \mathbf{e} \end{bmatrix}$$

which can be recast in the form

$$\mathbf{t} = \begin{bmatrix} \mathbf{e} & \mathbf{0} \\ \mathbf{e} \times \mathbf{p} & \mathbf{e} \end{bmatrix}\begin{bmatrix} \dot{\theta} \\ \dot{b} \end{bmatrix}$$

and hence, the twist generator sought is the 6×2 coefficient matrix $\mathbf{S}_C$ multiplying the two-dimensional vector of motion variables $[\,\dot{\theta},\, \dot{b}\,]^T$ in the above expression, i.e.,

$$\mathbf{S}_C = \begin{bmatrix} \mathbf{e} & \mathbf{0} \\ \mathbf{e} \times \mathbf{p} & \mathbf{e} \end{bmatrix} \tag{25}$$

where $\mathbf{0}$ is, again, the 3-dimensional zero vector.

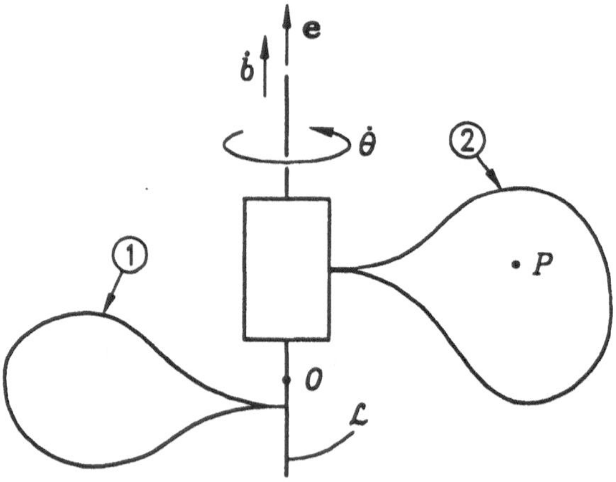

Figure 8: The cylindrical pair

5.5 THE SPHERICAL PAIR

Shown in Fig. 9 is a spherical pair coupling two bodies, 1 and 2, its sole attribute being point O common to the two bodies. Thus, any relative motion of 2 with respect to 1 maintains point O of 2 fixed to point O of 1, the motion thus being spherical. A spherical kinematic pair can be regarded as the combination of three revolutes in series, with axes intersecting at point O. Let these axes be parallel to the unit vectors $\mathbf{e}_k$, the associated motion variables being $\dot{\theta}_k$, for $k = 1,2,3$. Thus, any possible relative motion has a twist $\mathbf{t}$ given by

$$\mathbf{t} = \dot{\theta}_1 \begin{bmatrix} \mathbf{e}_1 \\ \mathbf{e}_1 \times \mathbf{p} \end{bmatrix} + \dot{\theta}_2 \begin{bmatrix} \mathbf{e}_2 \\ \mathbf{e}_2 \times \mathbf{p} \end{bmatrix} + \dot{\theta}_3 \begin{bmatrix} \mathbf{e}_3 \\ \mathbf{e}_3 \times \mathbf{p} \end{bmatrix}$$

which can be recast in the form

$$\mathbf{t} = \begin{bmatrix} \mathbf{e}_1 & \mathbf{e}_2 & \mathbf{e}_3 \\ \mathbf{e}_1 \times \mathbf{p} & \mathbf{e}_2 \times \mathbf{p} & \mathbf{e}_3 \times \mathbf{p} \end{bmatrix} \begin{bmatrix} \dot{\theta}_1 \\ \dot{\theta}_2 \\ \dot{\theta}_3 \end{bmatrix}$$

and hence, the twist generator sought is defined as the matrix coefficient $\mathbf{S}_S$ of the motion variables $\dot{\theta}_k$, for $k = 1,2,3$, i.e.,

$$\mathbf{S}_S = \begin{bmatrix} \mathbf{e}_1 & \mathbf{e}_2 & \mathbf{e}_3 \\ \mathbf{e}_1 \times \mathbf{p} & \mathbf{e}_2 \times \mathbf{p} & \mathbf{e}_3 \times \mathbf{p} \end{bmatrix} \tag{26}$$

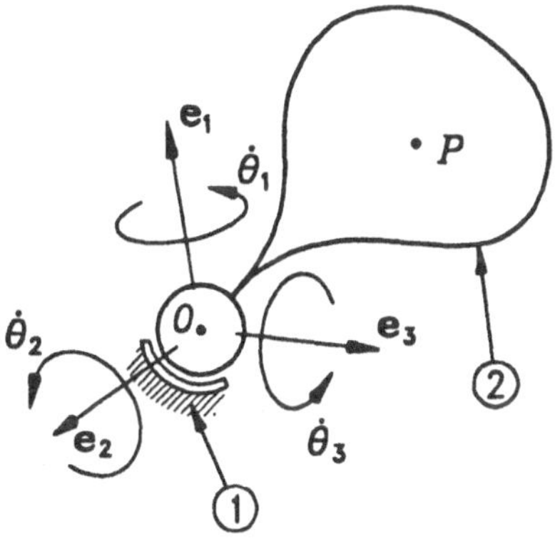

Figure 9: The spherical pair

5.6 THE PLANAR PAIR

Figure 10 below shows a planar kinematic pair, that can be regarded as the series combination of two prismatic pairs of non-parallel directions, given by unit vectors $\mathbf{e}_1$ and $\mathbf{e}_2$. Parallel to these two vectors and passing through a point O, we can define a plane Π, its normal being the unit vector $\mathbf{e}_3$. Thus, any motion of 2 relative to 1 has the twist

$$\mathbf{t} = \dot{b}_1 \begin{bmatrix} \mathbf{0} \\ \mathbf{e}_1 \end{bmatrix} + \dot{b}_2 \begin{bmatrix} \mathbf{0} \\ \mathbf{e}_2 \end{bmatrix} + \dot{\theta} \begin{bmatrix} \mathbf{e}_3 \\ \mathbf{e}_3 \times \mathbf{p} \end{bmatrix}$$

Alternatively, the above equation can be written in the form

$$\mathbf{t} = \begin{bmatrix} \mathbf{0} & \mathbf{0} & \mathbf{e}_3 \\ \mathbf{e}_1 & \mathbf{e}_2 & \mathbf{e}_3 \times \mathbf{p} \end{bmatrix} \begin{bmatrix} \dot{b}_1 \\ \dot{b}_2 \\ \dot{\theta} \end{bmatrix}$$

and hence, the twist generator sought is the 6×3 matrix coefficient $\mathbf{S}_E$ of the motion variables $\dot{b}_1$, $\dot{b}_2$ and $\dot{\theta}$, namely,

$$\mathbf{S}_E = \begin{bmatrix} \mathbf{0} & \mathbf{0} & \mathbf{e}_3 \\ \mathbf{e}_1 & \mathbf{e}_2 & \mathbf{e}_3 \times \mathbf{p} \end{bmatrix} \tag{27}$$

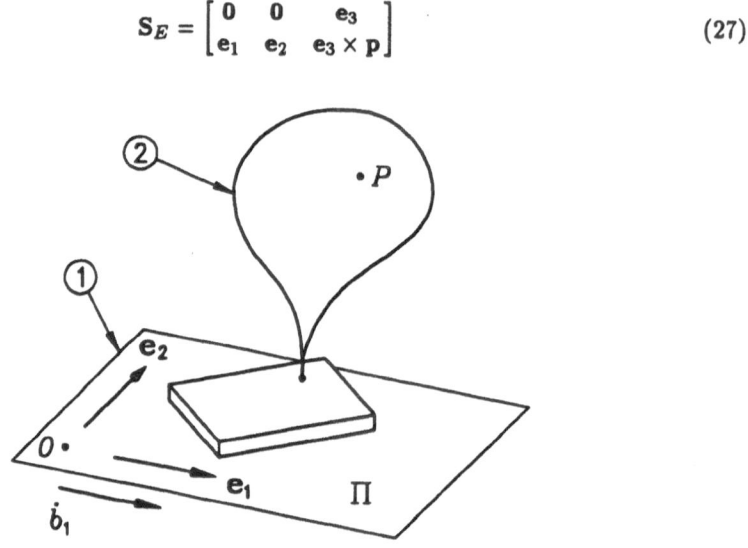

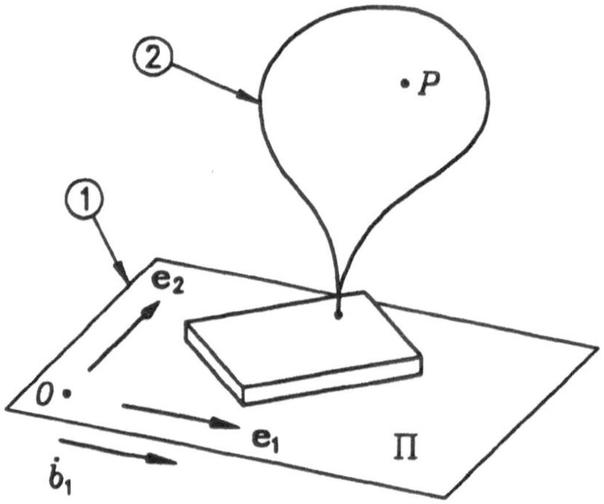

Figure 10: The planar pair

6 The Wrench Generators of the Lower Kinematic Pairs

The counterpart of a twist generator is a *wrench generator*. We define the wrench generator of a lower kinematic pair as a 6×6 matrix mapping an arbitrary wrench into a *feasible constraint wrench*, i.e., a wrench acting between two rigid bodies coupled by the aforementioned lower kinematic pair and that does not develop any power onto the system, its sole function being to keep the two bodies together.

Let the twist of the relative motion between two bodies coupled by a lower kinematic pair be denoted by $\mathbf{t}$. More precisely, we assume that two bodies, labeled 1 and 2, are coupled by a lower kinematic pair and denote the twist of the motion of body 2 with respect to body 1 by $\mathbf{t}$. Furthermore, the point of body 2 at which the twist is defined, is P, that is arbitrary.

Moreover, we denote the constraint wrench exerted by body 1 onto body 2 by $\mathbf{w}_P$, which is assumed to act at point P of body 2. That is, the force of $\mathbf{w}_P$ is assumed to be applied at point P. Furthermore, the wrench acting on body 1 due to contact with other bodies than 2 and to the environment is denoted by $\boldsymbol{\lambda}$ and consists of a moment $\boldsymbol{\mu}$ and a force $\boldsymbol{\nu}$ applied at point O of body 1 that is defined as in Figs. 5, 7–10. For the prismatic pair of Fig. 6, point O can be located anywhere, since this pair has only a direction, but no axis associated with it. Now, since bodies 1 and 2 are coupled via a lower kinematic pair, the wrench transmitted from 1 to 2 is not all of $\boldsymbol{\lambda}$, but a linear transformation of the latter, given by the 6×6 matrix $\mathbf{W}$ that we call the *wrench generator* associated with the pair at hand. We thus have

$$\mathbf{w}_P = \mathbf{W}\boldsymbol{\lambda} \tag{28}$$

Now, we call $\mathbf{S}$ the $6 \times r$ matrix that maps the r-dimensional vector of motion rates $\dot{\boldsymbol{\theta}}$ associated with the same lower kinematic pair into the twist $\mathbf{t}$, i.e.,

$$\mathbf{t} = \mathbf{S}\dot{\boldsymbol{\theta}} \tag{29}$$

Under the assumption that the kinematic pair is conservative, the foregoing wrench $\mathbf{w}_P$ develops no power onto body 2, and hence,

$$\mathbf{w}_P^T\mathbf{t} = 0$$

Upon substitution of the above expressions for the wrench and the twist into the latter expression, we have

$$\boldsymbol{\lambda}^T\mathbf{W}^T\mathbf{S}\dot{\boldsymbol{\theta}} = 0 \tag{30}$$

Now, the foregoing relation should hold for every value of $\dot{\boldsymbol{\theta}}$ and every value of $\boldsymbol{\lambda}$, and hence, we must have

$$\mathbf{W}^T\mathbf{S} = \mathbf{O}_{6r} \tag{31}$$

where $\mathbf{O}_{6r}$ is the $6 \times r$ zero matrix.

We derive below the wrench generators for all six lower kinematic pairs.

6.1 THE WRENCH GENERATOR OF THE REVOLUTE PAIR

Let $\mathbf{w}_O$ be the wrench applied by body 1 onto body 2 at point O of the revolute axis. Then, if $\boldsymbol{\lambda}$, as defined above, is the wrench exerted by other bodies and the environment on body 1 at the same point O, the only difference between $\boldsymbol{\lambda}$ and $\mathbf{w}_O$ is the moment, which, for the latter, is $\boldsymbol{\mu}$ minus its component along the revolute axis, and hence,

$$\mathbf{w}_O = \begin{bmatrix} (\mathbf{1} - \mathbf{e}\mathbf{e}^T)\boldsymbol{\mu} \\ \boldsymbol{\nu} \end{bmatrix} \equiv \begin{bmatrix} \mathbf{1} - \mathbf{e}\mathbf{e}^T & \mathbf{O} \\ \mathbf{O} & \mathbf{1} \end{bmatrix} \begin{bmatrix} \boldsymbol{\mu} \\ \boldsymbol{\nu} \end{bmatrix}$$

Now, in order to have the wrench at point P, all we do is recall the wrench-transfer formula of eq.(17), thereby obtaining

$$\mathbf{w}_P = \begin{bmatrix} \mathbf{1} & -\mathbf{P} \\ \mathbf{O} & \mathbf{1} \end{bmatrix} \begin{bmatrix} \mathbf{1} - \mathbf{e}\mathbf{e}^T & \mathbf{O} \\ \mathbf{O} & \mathbf{1} \end{bmatrix} \begin{bmatrix} \boldsymbol{\mu} \\ \boldsymbol{\nu} \end{bmatrix}$$

and hence, the wrench generator sought is the coefficient of the wrench $\boldsymbol{\lambda}$ in the last equation, i.e.,

$$\mathbf{W}_R = \begin{bmatrix} \mathbf{1} - \mathbf{e}\mathbf{e}^T & -\mathbf{P} \\ \mathbf{O} & \mathbf{1} \end{bmatrix} \tag{32}$$

6.2 THE WRENCH GENERATOR OF THE PRISMATIC PAIR

Here, $\mathbf{w}_O$ and $\boldsymbol{\lambda}$ are defined as before. The only difference between $\mathbf{w}_O$ and $\boldsymbol{\lambda}$ is now that, for the latter, the force is $\boldsymbol{\nu}$ minus its axial component along the direction of the pair, and hence,

$$\mathbf{w}_O = \begin{bmatrix} \boldsymbol{\mu} \\ (\mathbf{1} - \mathbf{e}\mathbf{e}^T)\boldsymbol{\nu} \end{bmatrix} \equiv \begin{bmatrix} \mathbf{1} & \mathbf{O} \\ \mathbf{O} & \mathbf{1} - \mathbf{e}\mathbf{e}^T \end{bmatrix} \begin{bmatrix} \boldsymbol{\mu} \\ \boldsymbol{\nu} \end{bmatrix}$$

Now, in order to have the wrench at point P, we apply the wrench-transfer formula of eq.(17), thereby obtaining

$$\mathbf{w}_P = \begin{bmatrix} \mathbf{1} & -\mathbf{P} \\ \mathbf{O} & \mathbf{1} \end{bmatrix} \begin{bmatrix} \mathbf{1} & \mathbf{O} \\ \mathbf{O} & \mathbf{1} - \mathbf{e}\mathbf{e}^T \end{bmatrix} \begin{bmatrix} \boldsymbol{\mu} \\ \boldsymbol{\nu} \end{bmatrix}$$

and hence, the wrench generator $\mathbf{W}_P$ of the prismatic pair is the above coefficient of the wrench $\boldsymbol{\lambda}$, i.e.,

$$\mathbf{W}_P = \begin{bmatrix} \mathbf{1} & -\mathbf{P}(\mathbf{1} - \mathbf{e}\mathbf{e}^T) \\ \mathbf{O} & \mathbf{1} - \mathbf{e}\mathbf{e}^T \end{bmatrix} \tag{33}$$

6.3 THE WRENCH GENERATOR OF THE SCREW PAIR

In order to find a suitable expression for $\mathbf{w}_O$, we proceed as follows: First, let $\mathbf{w}_O \equiv [\mathbf{n}^T,\ \mathbf{f}^T]^T$. The twist of body 2 with respect to body 1, at point O on the screw axis, is given by $\mathbf{t}_O = [\mathbf{e}^T,\ p\mathbf{e}^T]^T\dot{\theta}$. Now, the wrench does not develop any power onto body 2, and hence, for arbitrary $\dot{\theta}$,

$$\mathbf{w}_O^T\mathbf{t}_O \equiv (\mathbf{n}^T\mathbf{e} + p\mathbf{f}^T\mathbf{e})\dot{\theta} = 0$$

the wrench components $\mathbf{n}$ and $\mathbf{f}$ thus being subjected to the constraint

$$\mathbf{e}^T\mathbf{n} + p\mathbf{e}^T\mathbf{f} = 0$$

A suitable wrench that verifies the foregoing constraint is given below:

$$\mathbf{w}_O = \begin{bmatrix} \mathbf{1} - \mathbf{e}\mathbf{e}^T & -p\mathbf{e}\mathbf{e}^T \\ \mathbf{O} & \mathbf{1} \end{bmatrix} \begin{bmatrix} \boldsymbol{\mu} \\ \boldsymbol{\nu} \end{bmatrix}$$

and hence,

$$\mathbf{w}_P = \begin{bmatrix} \mathbf{1} & -\mathbf{P} \\ \mathbf{O} & \mathbf{1} \end{bmatrix} \begin{bmatrix} \mathbf{1} - \mathbf{e}\mathbf{e}^T & -p\mathbf{e}\mathbf{e}^T \\ \mathbf{O} & \mathbf{1} \end{bmatrix} \begin{bmatrix} \boldsymbol{\mu} \\ \boldsymbol{\nu} \end{bmatrix}$$

the desired wrench generator thus being derived as the product coefficient of $\boldsymbol{\lambda}$ in the foregoing equation, i.e.,

$$\mathbf{W}_H = \begin{bmatrix} \mathbf{1} - \mathbf{e}\mathbf{e}^T & -p\mathbf{e}\mathbf{e}^T - \mathbf{P} \\ \mathbf{O} & \mathbf{1} \end{bmatrix} \tag{34}$$

6.4 THE WRENCH GENERATOR OF THE CYLINDRICAL PAIR

Here, the difference between $\mathbf{w}_O$ and $\boldsymbol{\lambda}$ lies in both the moment and the force. That is, the moment and the force of $\mathbf{w}_O$ are those of $\boldsymbol{\lambda}$ minus the axial component of the moment and, correspondingly, of the force, i.e.,

$$\mathbf{w}_O = \begin{bmatrix} (\mathbf{1} - \mathbf{e}\mathbf{e}^T)\boldsymbol{\mu} \\ (\mathbf{1} - \mathbf{e}\mathbf{e}^T)\boldsymbol{\nu} \end{bmatrix} = \begin{bmatrix} \mathbf{1} - \mathbf{e}\mathbf{e}^T & \mathbf{O} \\ \mathbf{O} & \mathbf{1} - \mathbf{e}\mathbf{e}^T \end{bmatrix} \begin{bmatrix} \boldsymbol{\mu} \\ \boldsymbol{\nu} \end{bmatrix}$$

the wrench at point P, $\mathbf{w}_P$, now being obtained, as usual, by application of the wrench-transfer formula, i.e.,

$$\mathbf{w}_P = \begin{bmatrix} \mathbf{1} - \mathbf{e}\mathbf{e}^T & -\mathbf{P}(\mathbf{1} - \mathbf{e}\mathbf{e}^T) \\ \mathbf{O} & \mathbf{1} - \mathbf{e}\mathbf{e}^T \end{bmatrix} \begin{bmatrix} \boldsymbol{\mu} \\ \boldsymbol{\nu} \end{bmatrix}$$

Thus, the desired wrench generator is the matrix coefficient of the wrench $\boldsymbol{\lambda}$ in the foregoing expression, namely,

$$\mathbf{W}_C = \begin{bmatrix} \mathbf{1} - \mathbf{e}\mathbf{e}^T & -\mathbf{P}(\mathbf{1} - \mathbf{e}\mathbf{e}^T) \\ \mathbf{O} & \mathbf{1} - \mathbf{e}\mathbf{e}^T \end{bmatrix} \tag{35}$$

6.5 THE WRENCH GENERATOR OF THE SPHERICAL PAIR

The wrench $\mathbf{w}_O$ now contains only force and no moment. We can thus write

$$\mathbf{w}_O = \begin{bmatrix} \mathbf{O} & \mathbf{O} \\ \mathbf{O} & \mathbf{1} \end{bmatrix} \begin{bmatrix} \boldsymbol{\mu} \\ \boldsymbol{\nu} \end{bmatrix}$$

and hence, upon application of the wrench-transfer formula, we obtain

$$\mathbf{w}_O = \begin{bmatrix} \mathbf{O} & -\mathbf{P} \\ \mathbf{O} & \mathbf{1} \end{bmatrix} \begin{bmatrix} \boldsymbol{\mu} \\ \boldsymbol{\nu} \end{bmatrix}$$

from which the wrench generator can be readily identified, namely,

$$\mathbf{W}_S = \begin{bmatrix} \mathbf{O} & -\mathbf{P} \\ \mathbf{O} & \mathbf{1} \end{bmatrix} \tag{36}$$

6.6 THE WRENCH GENERATOR OF THE PLANAR PAIR

In this case the wrench $\mathbf{w}_O$ has a moment with zero component along the normal to the plane and a force that is normal to the plane. Thus, this wrench takes on the form

$$\mathbf{w}_O = \begin{bmatrix} \mathbf{1} - \mathbf{e}_3\mathbf{e}_3^T & \mathbf{O} \\ \mathbf{O} & \mathbf{e}_3\mathbf{e}_3^T \end{bmatrix} \begin{bmatrix} \boldsymbol{\mu} \\ \boldsymbol{\nu} \end{bmatrix}$$

and hence, the wrench at P becomes

$$\mathbf{w}_P = \begin{bmatrix} \mathbf{1} - \mathbf{e}_3\mathbf{e}_3^T & -\mathbf{P}\mathbf{e}_3\mathbf{e}_3^T \\ \mathbf{O} & \mathbf{e}_3\mathbf{e}_3^T \end{bmatrix} \begin{bmatrix} \boldsymbol{\mu} \\ \boldsymbol{\nu} \end{bmatrix}$$

from which the wrench generator for this pair is readily identified, i.e.,

$$\mathbf{W}_E = \begin{bmatrix} \mathbf{1} - \mathbf{e}_3\mathbf{e}_3^T & -\mathbf{P}\mathbf{e}_3\mathbf{e}_3^T \\ \mathbf{O} & \mathbf{e}_3\mathbf{e}_3^T \end{bmatrix} \tag{37}$$

7 The Twist Annihilators of the Lower Kinematic Pairs

A twist annihilator is defined here as a 6×6 singular matrix mapping any of the twist generators introduced in Section 4 into the 6-dimensional zero array. That is, the columns of the aforementioned twist generator lie in the nullspace of the corresponding twist annihilator. It should be apparent, then, that, for a given twist generator there are infinitely many twist annihilators. We will use the relation of eq.(31) to define the twist annihilators of the six lower kinematic pairs. From that equation it is apparent that we can choose the twist annihilator of any lower kinematic pair as the transpose of the corresponding wrench generator.

We list below the twist annihilators of all lower kinematic pairs.

7.1 THE TWIST ANNIHILATOR OF THE REVOLUTE

This annihilator is denoted by $\mathbf{A}_R$ and is given by

$$\mathbf{A}_R = \begin{bmatrix} \mathbf{1} - \mathbf{e}\mathbf{e}^T & \mathbf{O} \\ \mathbf{P} & \mathbf{1} \end{bmatrix} \tag{38}$$

Now it is a simple matter to show that

$$\mathbf{A}_R \mathbf{s}_R = \mathbf{0}$$

where $\mathbf{0}$ is the 6-dimensional zero vector.

7.2 THE TWIST ANNIHILATOR OF THE PRISMATIC PAIR

The twist annihilator of the prismatic pair is denoted by $\mathbf{A}_P$ and is given by

$$\mathbf{A}_P = \begin{bmatrix} \mathbf{1} & \mathbf{O} \\ (\mathbf{1} - \mathbf{e}\mathbf{e}^T)\mathbf{P} & \mathbf{1} - \mathbf{e}\mathbf{e}^T \end{bmatrix} \tag{39}$$

and hence, it is a simple matter to show that

$$\mathbf{A}_P \mathbf{s}_P = \mathbf{0}$$

7.3 THE TWIST ANNIHILATOR OF THE SCREW PAIR

This twist annihilator is given by

$$\mathbf{A}_H = \begin{bmatrix} \mathbf{1} - \mathbf{e}\mathbf{e}^T & \mathbf{O} \\ \mathbf{P} - p\mathbf{e}\mathbf{e}^T & \mathbf{1} \end{bmatrix} \tag{40}$$

and is related to the corresponding twist generator by

$$\mathbf{A}_H \mathbf{s}_H = \mathbf{0}$$

7.4 THE TWIST ANNIHILATOR OF THE CYLINDRICAL PAIR

The twist annihilator of the cylindrical pair is given by

$$\mathbf{A}_C = \begin{bmatrix} \mathbf{1} - \mathbf{e}\mathbf{e}^T & \mathbf{O} \\ (\mathbf{1} - \mathbf{e}\mathbf{e}^T)\mathbf{P} & \mathbf{1} - \mathbf{e}\mathbf{e}^T \end{bmatrix} \tag{41}$$

and thus,

$$\mathbf{A}_C\mathbf{S}_C = \mathbf{O}_{62}$$

where $\mathbf{O}_{62}$ is the 6×2 zero matrix.

7.5 THE TWIST ANNIHILATOR OF THE SPHERICAL PAIR

The twist annihilator of the spherical pair is given by

$$\mathbf{A}_S = \begin{bmatrix} \mathbf{O} & \mathbf{O} \\ \mathbf{P} & \mathbf{1} \end{bmatrix} \tag{42}$$

This matrix, then, maps $\mathbf{S}_S$ into the 6×3 zero matrix, i.e.,

$$\mathbf{A}_S\mathbf{S}_S = \mathbf{O}_{63}$$

7.6 THE TWIST ANNIHILATOR OF THE PLANAR PAIR

For the planar pair we have

$$\mathbf{A}_E = \begin{bmatrix} \mathbf{1} - \mathbf{e}_3\mathbf{e}_3^T & \mathbf{O} \\ \mathbf{e}_3\mathbf{e}_3^T\mathbf{P} & \mathbf{e}_3\mathbf{e}_3^T \end{bmatrix} \tag{43}$$

and so, $\mathbf{A}_E$ maps $\mathbf{S}_E$ into the 6×3 zero matrix, i.e.,

$$\mathbf{A}_E\mathbf{S}_E = \mathbf{O}_{63}$$

8 The Wrench Annihilators of the Lower Kinematic Pairs

The wrench annihilators of interest can be derived as in the previous section. Indeed, upon transposing eq.(31), we obtain

$$\mathbf{S}^T\mathbf{W} = \mathbf{O}_{r6} \tag{44}$$

from which it is apparent that the transpose of the twist generator of a given lower kinematic pair can indeed play the role of the corresponding wrench annihilator. As in the case of twist annihilators, wrench annihilators are not unique, but we will define them here uniquely as the transpose matrices of the corresponding twist generators. Henceforth, wrench annihilators are denoted by a subscripted $\mathbf{b}^T$ or $\mathbf{B}$, depending on whether we are talking of a row-vector or a matrix array, the subscript indicating the corresponding lower kinematic pair. For completeness, we list below the six wrench annihilators.

8.1 THE REVOLUTE

The wrench annihilator of the revolute pair, $\mathbf{b}_R^T$, is given by

$$\mathbf{b}_R^T = [\mathbf{e}^T \quad (\mathbf{e} \times \mathbf{p})^T] \tag{45}$$

which is apparently a 6-dimensional row-vector array.

8.2 THE PRISMATIC PAIR

The wrench annihilator of this pair, denoted by $\mathbf{b}_P^T$, is given by

$$\mathbf{b}_P^T = [\mathbf{0}^T \quad \mathbf{e}^T] \tag{46}$$

and is a 6-dimensional row-vector array as well.

8.3 THE SCREW PAIR

Likewise, the wrench annihilator of the screw pair, $\mathbf{b}_H^T$, is given by

$$\mathbf{b}_H^T = [\mathbf{e}^T \quad (\mathbf{e} \times \mathbf{p} + p\mathbf{e})^T] \tag{47}$$

and is, again, a 6-dimensional row-vector array.

8.4 THE CYLINDRICAL PAIR

Now, the wrench annihilator of the cylindrical pair is, from the above definition, the 2×6 array $\mathbf{B}_C$ given below:

$$\mathbf{B}_C = \begin{bmatrix} \mathbf{e}^T & (\mathbf{e} \times \mathbf{p})^T \\ \mathbf{0}^T & \mathbf{e}^T \end{bmatrix} \tag{48}$$

8.5 THE SPHERICAL PAIR

As expected, the wrench annihilator of a spherical pair is a 3×6 matrix array, denoted by $\mathbf{B}_S$, namely,

$$\mathbf{B}_S = \begin{bmatrix} \mathbf{e}_1^T & (\mathbf{e}_1 \times \mathbf{p})^T \\ \mathbf{e}_2^T & (\mathbf{e}_2 \times \mathbf{p})^T \\ \mathbf{e}_3^T & (\mathbf{e}_3 \times \mathbf{p})^T \end{bmatrix} \tag{49}$$

8.6 THE PLANAR PAIR

Finally, the wrench annihilator of the planar pair, $\mathbf{B}_E$, is given by a 3×6 matrix array as well, namely,

$$\mathbf{B}_E = \begin{bmatrix} \mathbf{0}^T & \mathbf{e}_1^T \\ \mathbf{0}^T & \mathbf{e}_2^T \\ \mathbf{e}_3^T & (\mathbf{e}_3 \times \mathbf{p})^T \end{bmatrix} \tag{50}$$

9 Applications

9.1 INPUT-OUTPUT ANALYSIS OF A SPATIAL FOUR-BAR LINKAGE

Here we use an *RSSR* linkage to illustrate our concepts because this is one of the simplest spatial mechanisms. This mechanism, shown in Fig. 11a, has the same input-output relation as the corresponding *RSUR* linkage shown in Fig. 11b, with *U* standing for *universal joint.* That is, if we denote by angles ψ and ϕ the input and output variables of the two linkages, they both have the same functional relation $f(\psi, \phi) = 0$. However, the *RSUR* linkage does not have the idle degree of freedom of the *RSSR* linage. The aforementioned idle degree of freedom pertains to a motion of the coupler link about the line of centers of its two spherical pairs. Moreover, in the *RSUR* linkage, one spherical pair is replaced by the concatenation of a universal joint and a revolute, the overall linkage thus having a total of seven revolutes. Here, the input variable is angle ψ, the output being ϕ. Now, the output link can be regarded as the end-effector (EE) of a six-axes serial manipulator that rotates about a fixed axis $\mathcal{A}$, while keeping fixed its point P, defined as the center of the revolute of the output axis. What we want is an expression between $\dot{\psi}$ and $\dot{\phi}$. This expression is derived below by relating first the joint rates of the serial manipulator with the twist of its EE, $\mathbf{t}$, namely,

$$\mathbf{J}\dot{\boldsymbol{\theta}} = \mathbf{t} \tag{51}$$

where the Jacobian matrix $\mathbf{J}$, the twist $\mathbf{t}$ and the 6-dimensional vector of joint rates $\dot{\boldsymbol{\theta}}$ take on the forms

$$\mathbf{J} \equiv \begin{bmatrix} \mathbf{e}_1 & \mathbf{e}_2 & \mathbf{e}_3 & \mathbf{e}_4 & \mathbf{e}_5 & \mathbf{e}_6 \\ \mathbf{e}_1 \times \mathbf{r}_1 & \mathbf{e}_2 \times \mathbf{r}_2 & \mathbf{e}_3 \times \mathbf{r}_2 & \mathbf{e}_4 \times \mathbf{r}_2 & \mathbf{e}_5 \times \mathbf{r}_5 & \mathbf{e}_6 \times \mathbf{r}_5 \end{bmatrix} \tag{52}$$

$$\mathbf{t} \equiv \begin{bmatrix} \mathbf{e}_7 \\ \mathbf{0} \end{bmatrix} \dot{\phi}, \qquad \dot{\boldsymbol{\theta}} \equiv \begin{bmatrix} \dot{\psi} \\ \dot{\theta}_2 \\ \vdots \\ \dot{\theta}_6 \end{bmatrix} \tag{53}$$

Thus, in order to derive the relation sought, we have to eliminate the five joint rates $\dot{\theta}_2, \ldots, \dot{\theta}_6$ from eq.(51), which is done in two steps. In the first step, we eliminate the joint rates $\dot{\theta}_2$, $\dot{\theta}_3$ and $\dot{\theta}_4$ associated with the spherical joint coupled to the input link. In the second step, we eliminate the joint rates associated with the remaining universal joint.

The first step is straightforward. All we need is multiply both sides of eq.(51) by the twist annihilator $\mathbf{A}_S$ of the spherical joint of interest. From our previous discussion, this annihilator is given by

$$\mathbf{A}_S = \begin{bmatrix} \mathbf{O} & \mathbf{O} \\ \mathbf{R}_2 & \mathbf{1} \end{bmatrix} \tag{54}$$

with $\mathbf{R}_2$ defined as the cross-product matrix of $\mathbf{r}_2$.

Upon multiplication of both sides of eq.(51) from the left by $\mathbf{A}_S$, we have

$$\mathbf{A}_S\mathbf{J}\dot{\boldsymbol{\theta}} = \mathbf{A}_S\mathbf{t}$$

where

$$\mathbf{A}_S\mathbf{J} = \begin{bmatrix} \mathbf{0} & \mathbf{0} & \mathbf{0} & \mathbf{0} & \mathbf{0} & \mathbf{0} \\ \mathbf{e}_1 \times (\mathbf{r}_1 - \mathbf{r}_2) & \mathbf{0} & \mathbf{0} & \mathbf{0} & \mathbf{e}_5 \times (\mathbf{r}_5 - \mathbf{r}_2) & \mathbf{e}_6 \times (\mathbf{r}_5 - \mathbf{r}_2) \end{bmatrix} \tag{55a}$$

$$\mathbf{A}_S\mathbf{t} = \begin{bmatrix} \mathbf{0} \\ \mathbf{r}_2 \times \mathbf{e}_7 \end{bmatrix} \dot{\phi} \tag{55b}$$

and hence, we end up with the simpler relation

$$\mathbf{e}_1 \times (\mathbf{r}_1 - \mathbf{r}_2)\dot{\psi} + \mathbf{e}_5 \times (\mathbf{r}_5 - \mathbf{r}_2)\dot{\theta}_5 + \mathbf{e}_6 \times (\mathbf{r}_5 - \mathbf{r}_2)\dot{\theta}_6 = \mathbf{r}_2 \times \mathbf{e}_7\dot{\phi} \tag{56}$$

Now it is apparent that the two terms containing the undesired joint rates, $\dot{\theta}_5$ and $\dot{\theta}_6$, will disappear from the above equation if its two sides are dot-multiplied by the cross product $\mathbf{k}$ of the two vector coefficients of these joint rates. The desired result, namely, $\dot{\phi}$ expressed as a function of $\dot{\psi}$, takes on the form

$$\dot{\phi} = \frac{N}{D}\dot{\psi} \tag{57a}$$

(57b)

with $\mathbf{k}$, N and D defined as

$$\mathbf{k} \equiv [\mathbf{e}_5 \times (\mathbf{r}_5 - \mathbf{r}_2)] \times [\mathbf{e}_6 \times (\mathbf{r}_5 - \mathbf{r}_2)] \tag{57c}$$

$$N \equiv \mathbf{k} \cdot \mathbf{r}_2 \times \mathbf{e}_7, \quad D \equiv \mathbf{k} \cdot \mathbf{e}_1 \times (\mathbf{r}_1 - \mathbf{r}_2) \tag{57d}$$

thereby completing all derivations.

9.2 JACOBIAN MATRICES OF A WALKING MACHINE

Now we study the multi-loop kinematic chain of a 3-dof robotic mechanical system, as depicted in Fig. 12a. This figure represents either a multi-legged walking machine with three feet in contact with the ground or a three-fingered hand grasping an object, whereby the ground of the walking machine becomes the object of the hand. From Euler's formula for graphs [9], it is apparent that the foregoing kinematic chain has two independent loops and three degrees of freedom. Now, if this system represents the kinematic chain of a walking machine, the spherical pairs are used to model the contact between feet and ground when no sliding is assumed. Note that, when one of the legs is in the swing phase, the leg becomes a two-dof serial manipulator, and hence, it requires two actuators to control it. Thus, one can assume that each of the revolute pairs is an actuated revolute, and the whole machine has six motors but only three dof, i.e., we have a redundantly-actuated machine. We want to derive a relation between the joint rates of the six revolutes and the twist of the body, namely, the triangular plate shown in the aforementioned figure.

Shown in Fig. 12b is the kinematic chain of the Jth leg, i.e., an open chain composed of a spherical pair and two revolutes, which can be regarded as a serial manipulator meant to position point C of the EE (the triangular plate) and to orient the latter. In this figure we assume that the spherical pair is the serial combination of three revolutes with concurrent axes. The leg is thus modeled as a 5-revolute serial manipulator. Let $\mathbf{e}_{Jk}$ denote the unit vector parallel to the axis of the kth revolute and $\dot{\theta}_{Jk}$ the associated joint variable, for $k = 1, \ldots, 5$ and $J = I, II, III$. Here, we number with a roman numeral the leg and with an arabic one the joint of a particular leg. Thus, the relation between the joint rates of the Jth leg and the twist of the EE takes on the usual form

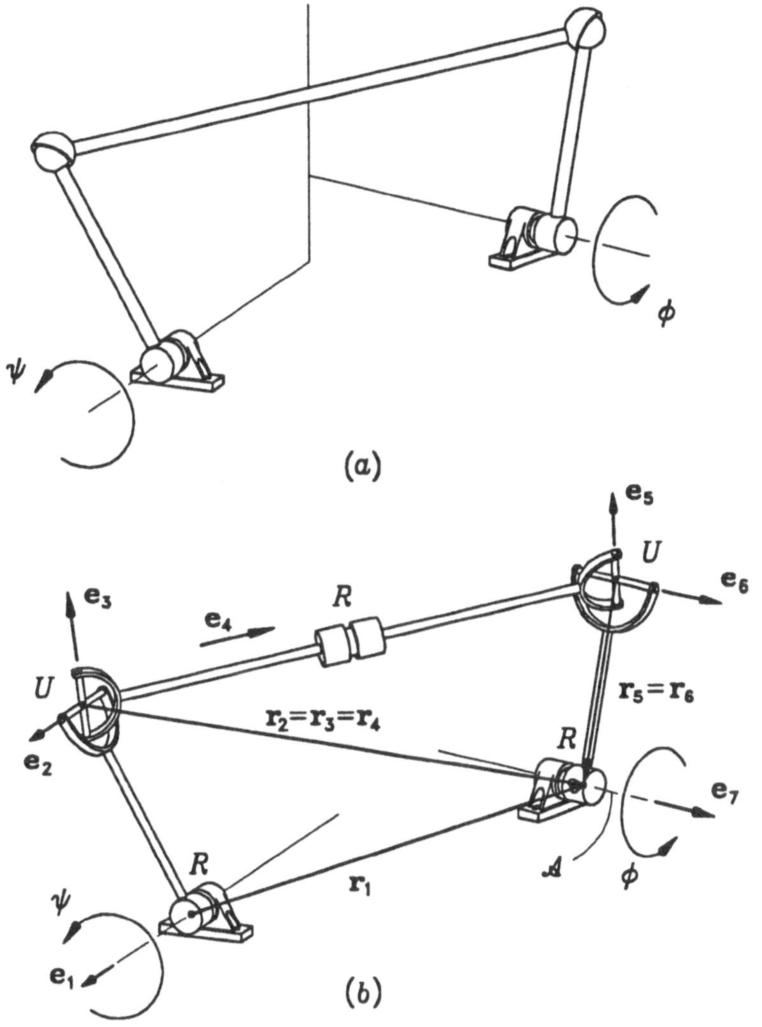

Figure 11: *RSSR* Linkage and its *RSUR* input-output equivalent

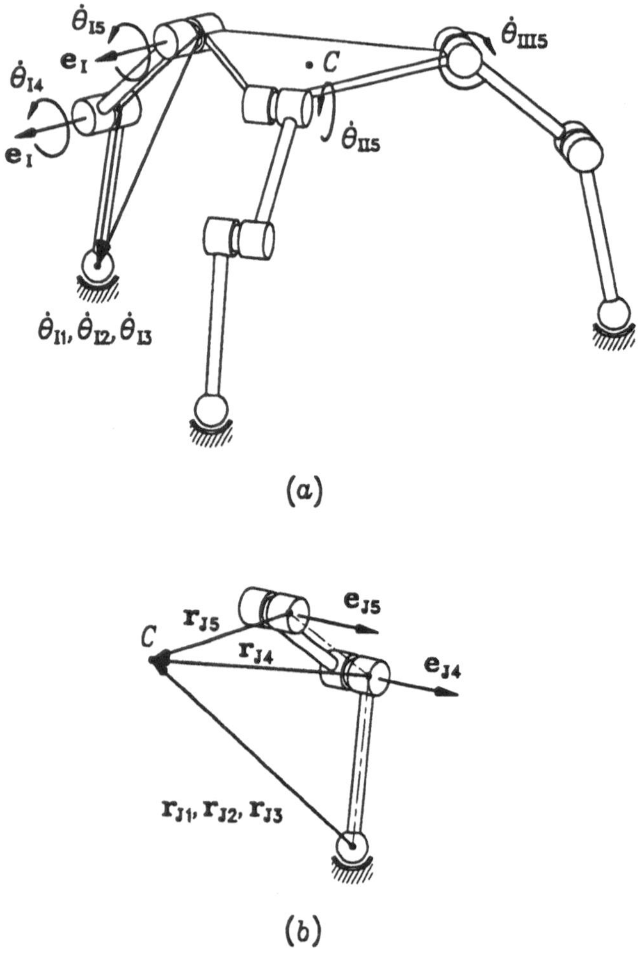

Figure 12: Multi-Loop Mechanical System

$$\mathbf{J}_J\dot{\boldsymbol{\theta}}_J = \mathbf{t} \tag{58}$$

where the leg Jacobian $\mathbf{J}_J$ is a 6×5 matrix, while $\dot{\boldsymbol{\theta}}_J$ and $\mathbf{t}$ are 5-dimensional and 6-dimensional vectors, respectively, defined below:

$$\mathbf{J}_J \equiv \begin{bmatrix} \mathbf{e}_{J1} & \mathbf{e}_{J2} & \mathbf{e}_{J3} & \mathbf{e}_{J4} & \mathbf{e}_{J5} \\ \mathbf{e}_{J1} \times \mathbf{r}_{J1} & \mathbf{e}_{J2} \times \mathbf{r}_{J1} & \mathbf{e}_{J3} \times \mathbf{r}_{J1} & \mathbf{e}_{J4} \times \mathbf{r}_{J4} & \mathbf{e}_{J5} \times \mathbf{r}_{J5} \end{bmatrix} \tag{59a}$$

$$\dot{\boldsymbol{\theta}}_J \equiv \begin{bmatrix} \dot{\theta}_{J1} \\ \dot{\theta}_{J2} \\ \vdots \\ \dot{\theta}_{J5} \end{bmatrix}, \qquad \mathbf{t} \equiv \begin{bmatrix} \boldsymbol{\omega} \\ \dot{\mathbf{c}} \end{bmatrix} \tag{59b}$$

The relation sought is now derived by annihilating the idle joint rates from relation (58). This is done by multiplying both sides of the said equation by the annihilator of the twist of the spherical joint, $\mathbf{A}_{SJ}$, defined as

$$\mathbf{A}_{SJ} \equiv \begin{bmatrix} \mathbf{O} & \mathbf{O} \\ \mathbf{R}_{J1} & \mathbf{1} \end{bmatrix} \tag{60}$$

We thus obtain

$$[\mathbf{0} \quad \mathbf{0} \quad \mathbf{0} \quad \mathbf{e}_{J4} \times (\mathbf{r}_{J4} - \mathbf{r}_{J1}) \quad \mathbf{e}_{J5} \times (\mathbf{r}_{J5} - \mathbf{r}_{J1})]\,\dot{\boldsymbol{\theta}}_J = \mathbf{r}_{J1} \times \boldsymbol{\omega} + \dot{\mathbf{c}}$$

If the system under study is redundantly actuated, i.e., if the two revolutes are powered, then the above relation is all we need, for that relation contains no idle joint rates. That is, if we let $\dot{\boldsymbol{\phi}}_J$ denote the 2-dimensional vector of actuated joint rates of the Jth leg and $\mathbf{K}_J$ the associated Jacobian of the same leg, we have

$$\mathbf{K}_J\dot{\boldsymbol{\phi}}_J = \mathbf{A}_{SJ}\mathbf{t} \tag{61}$$

where

$$\mathbf{K}_J \equiv [\mathbf{e}_{J4} \times (\mathbf{r}_{J4} - \mathbf{r}_{J1}) \quad \mathbf{e}_{J5} \times (\mathbf{r}_{J5} - \mathbf{r}_{J1})] \tag{62a}$$

$$\dot{\boldsymbol{\phi}}_J \equiv \begin{bmatrix} \dot{\theta}_{J4} \\ \dot{\theta}_{J5} \end{bmatrix}, \qquad \mathbf{A}_{SJ}\mathbf{t} \equiv \mathbf{r}_{J1} \times \boldsymbol{\omega} + \dot{\mathbf{c}} \tag{62b}$$

On the other hand, if the machine is driven with one single actuator per leg, namely, the one coupled to the triangular platform, then we should eliminate $\dot{\theta}_{J4}$ from the above expression, which is readily done if we rewrite the reduced relation, eq.(61), in the form

$$\mathbf{e}_{J4} \times (\mathbf{r}_{J4} - \mathbf{r}_{J1})\dot{\theta}_{J4} + \mathbf{e}_{J5} \times (\mathbf{r}_{J5} - \mathbf{r}_{J1})\dot{\theta}_{J5} = \mathbf{r}_{J1} \times \boldsymbol{\omega} + \dot{\mathbf{c}} \tag{63}$$

Thus, in order to obtain an expression containing only $\dot{\theta}_{J5}$, all we need is multiply both sides of the above equation by a suitable *annihilator*. For example, if we choose to eliminate $\dot{\theta}_{J4}$ from that equation, we can do this by dot-multiplying the two sides of the same equation by $\mathbf{e}_{J4}$, thereby deriving

$$\mathbf{e}_{J4} \times \mathbf{e}_{J5} \cdot (\mathbf{r}_{J5} - \mathbf{r}_{J1})\dot{\theta}_{J5} = \mathbf{e}_{J4} \times \mathbf{p}_{J1} \cdot \boldsymbol{\omega} + \mathbf{e}_{J4} \cdot \dot{\mathbf{c}} \tag{64}$$

which can be rewritten alternatively as

$$c_J\dot{\theta}_{J5} = \mathbf{k}_J^T\mathbf{t} \tag{65}$$

where the scalar c_J and the 6-dimensional vector $\mathbf{k}_J$ are defined below

$$c_J \equiv \mathbf{e}_{J4} \times \mathbf{e}_{J5} \cdot (\mathbf{r}_{J5} - \mathbf{r}_{J1}) \tag{66}$$

$$\mathbf{k}_J \equiv [(\mathbf{e}_{J4} \times \mathbf{p}_{J1})^T \quad \mathbf{e}_{J4}^T]^T \tag{67}$$

Under these conditions, the vector of actuated joint rates, namely, $\dot{\boldsymbol{\theta}}_a$, is related to the twist of the platform as indicated below:

$$\mathbf{J}\dot{\boldsymbol{\theta}}_a = \mathbf{K}\mathbf{t} \tag{68a}$$

where $\mathbf{J}$ and $\mathbf{K}$ are the 3×3 and 3×6 matrices defined as

$$\mathbf{J} \equiv \text{diag}(c_I, c_{II}, c_{III}) \tag{68b}$$

$$\mathbf{K} \equiv \begin{bmatrix} \mathbf{k}_I^T \\ \mathbf{k}_{II}^T \\ \mathbf{k}_{III}^T \end{bmatrix} \tag{68c}$$

$\mathbf{J}$ and $\mathbf{K}$ thus being the two global Jacobian matrices of the machine. Notice that the above relation allows the determination of the actuator joint rates for a given twist of the platform. Alternatively, if the above mechanical system is a multi-fingered hand, the same relations allow the determination of the actuator joint rates for a given twist of the manipulated object.

10 Extensions and Applications to Dynamics

In this section we will highlight an extension of the concept of twist annihilator, as applicable to the dynamics of multi-body mechanical systems. For conciseness, we will consider only systems with simple, open kinematic chains, like those appearing commonly in industrial manipulators. We will thus assume that the system is composed of r rigid links, coupled by revolute pairs. Shown in Fig. 13 is a six-axes industrial manipulator with a kinematic chain of the simple, open type, for which $r = 6$.

Let $\mathbf{M}_i$ be the 6×6 matrix of *extended mass* of the ith link and $\mathbf{W}_i$ be the matrix of *extended angular velocity* of the same link, defined as

$$\mathbf{M}_i \equiv \begin{bmatrix} \mathbf{I}_i & \mathbf{0} \\ \mathbf{0} & m_i\mathbf{1} \end{bmatrix}, \quad \mathbf{W}i \equiv \begin{bmatrix} \boldsymbol{\Omega}_i & \mathbf{0} \\ \mathbf{0} & \mathbf{0} \end{bmatrix}, \quad i = 1, \ldots, r \tag{69}$$

where $\boldsymbol{\Omega}_i$, the cross-product matrix of the angular-velocity vector $\boldsymbol{\omega}_i$, and $\mathbf{I}_i$ are the 3×3 angular-velocity and inertia matrices of the ith link, the latter being defined with respect to the mass center, C_i, of this link. Moreover, the mass of this link is denoted by m_i, whereas $\mathbf{c}_i$ and $\dot{\mathbf{c}}_i$ denote the position and the velocity vectors of C_i. Furthermore, let $\mathbf{t}_i$ denote the twist of the same link, the latter being defined in terms of the velocity of C_i. Furthermore, $\mathbf{w}_i^E$ and $\mathbf{w}_i^C$ are defined as the *external* and the *nonworking constraint* wrenches acting on

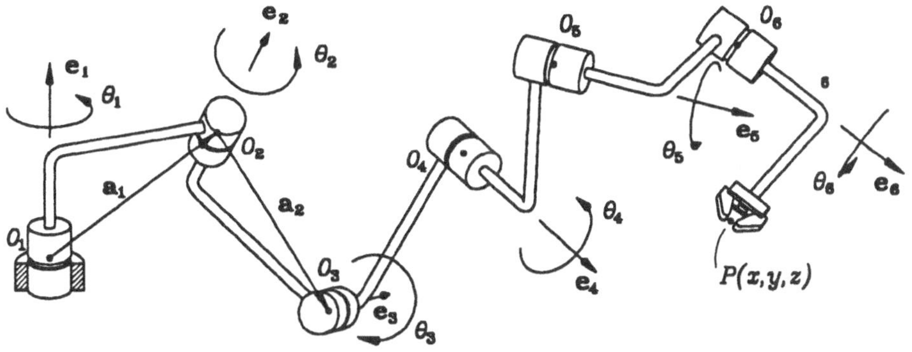

Figure 13: Industrial manipulator with simple, open kinematic chain

the ith link, in which forces are assumed to be applied at C_i. Thus, we have introduced the definitions below:

$$\mathbf{t}_i = \begin{bmatrix} \boldsymbol{\omega}_i \\ \dot{\mathbf{c}}_i \end{bmatrix}, \quad \mathbf{w}_i^E = \begin{bmatrix} \mathbf{n}_i^E \\ \mathbf{f}_i^E \end{bmatrix}, \quad \mathbf{w}_i^C = \begin{bmatrix} \mathbf{n}_i^C \\ \mathbf{f}_i^C \end{bmatrix}, \quad i = 1, \ldots, r \tag{70}$$

in which superscripted $\mathbf{n}_i$ and $\mathbf{f}_i$ stand, respectively, for the moment and the force acting on the ith body, the force being applied at the mass center C_i. Thus, whereas $\mathbf{w}_i^E$ accounts for forces and moments exerted by both the environment and the actuators, including driving forces as well as dissipative effects, $\mathbf{w}_i^C$ accounts for those forces and moments, generically termed *actions*, exerted by the neighboring links, which do not produce any mechanical work, its sole function being to keep the links together. Therefore, friction wrenches applied by the $(i-1)$st and the $(i+1)$st links onto the ith link are not included in $\mathbf{w}_i^C$; they are included rather in $\mathbf{w}_i^E$.

We recall now the Newton-Euler equations for a rigid body, namely,

$$\mathbf{I}_i\dot{\boldsymbol{\omega}}_i = -\boldsymbol{\omega}_i \times \mathbf{I}_i\boldsymbol{\omega}_i + \mathbf{n}_i^E + \mathbf{n}_i^C \tag{71a}$$

$$m_i\ddot{\mathbf{c}}_i = -m_i\boldsymbol{\omega}_i \times \mathbf{c}_i + \mathbf{f}_i^E + \mathbf{f}_i^C \tag{71b}$$

which can be written in *extended form* using the foregoing 6-dimensional twist and wrench vectors as well as the 6×6 matrices of extended mass and extended angular velocity. We thus obtain the Newton-Euler equations of the ith body in the form

$$\mathbf{M}_i\dot{\mathbf{t}}_i = -\mathbf{W}_i\mathbf{M}_i\mathbf{t}_i + \mathbf{w}^E i + \mathbf{w}_i^C, \quad i = 1, \ldots, r \tag{71c}$$

Further definitions are now introduced. These are the $6r$-dimensional vectors of *generalized twist*, $\mathbf{t}$, *generalized constraint wrench*, $\mathbf{w}^C$, *generalized active wrench*, $\mathbf{w}^A$ and *generalized dissipative wrench*, $\mathbf{w}^D$. Additionally, the $6r \times 6r$ matrices of *generalized mass*, $\mathbf{M}$, and *generalized angular velocity*, $\mathbf{W}$, are also introduced below:

$$\mathbf{t}=\begin{bmatrix}\mathbf{t}_1\\ \vdots\\ \mathbf{t}_r\end{bmatrix},\qquad \mathbf{w}^C=\begin{bmatrix}\mathbf{w}_1^C\\ \vdots\\ \mathbf{w}_r^C\end{bmatrix},\qquad \mathbf{w}^A=\begin{bmatrix}\mathbf{w}_1^A\\ \vdots\\ \mathbf{w}_r^A\end{bmatrix},\qquad \mathbf{w}^D=\begin{bmatrix}\mathbf{w}_1^D\\ \vdots\\ \mathbf{w}_r^D\end{bmatrix} \tag{72a}$$

$$\mathbf{M}=\mathrm{diag}\,(\mathbf{M}_1,\ldots,\mathbf{M}_r),\qquad \mathbf{W}=\mathrm{diag}\,(\mathbf{W}_1,\ldots,\mathbf{W}_r) \tag{72b}$$

We can now derive the model of the system. To this end, we first write the system of uncoupled Newton-Euler equations of the manipulator. In this case, we have $r+1$ links numbered from 0 to r, which are coupled by r revolutes. Moreover, the base link need not be an inertial frame but, for the sake of brevity, we will assume here that it is so.

Upon assembling all r systems of 6-dimensional uncoupled equations, eq.(71c), we obtain

$$\mathbf{M}\dot{\mathbf{t}}=-\mathbf{W}\mathbf{M}\mathbf{t}+\mathbf{w}^A-\mathbf{w}^D+\mathbf{w}^C \tag{73}$$

in which $\mathbf{w}^E$ has been decomposed into its active and dissipative components $\mathbf{w}^A$ and $\mathbf{w}^D$, respectively. The above equations are the *uncoupled* Newton-Euler equations of the overall system. The following step of this derivation consists of representing the coupling between every two consecutive links as a linear homogeneous system of algebraic equations on the link twists. If $\mathbf{K}$ denotes the $6r\times 6r$ matrix associated with the said system, then one has

$$\mathbf{K}\mathbf{t}=\mathbf{0} \tag{74}$$

where $\mathbf{0}$ is the $6r$-dimensional zero vector. The derivation of the foregoing equations, as pertaining to the systems of interest, will be described later. What is important to note at the moment is that the *kinematic constraint equations* or *constraint equations* for brevity, eqs.(74), are constituted by a system of $6r$ scalar equations. Moreover, when the system is in motion, $\mathbf{t}$ is different from zero and, hence, matrix $\mathbf{K}$ is singular. In fact, as shown in detail presently, the rank of $\mathbf{K}$ is exactly $5r$, which means that its nullspace is of dimension r, this being the degree of freedom of the system. Furthermore, since the constraint wrench $\mathbf{w}^C$ produces no work on the system, the power developed by this wrench on $\mathbf{t}$, for any possible motion of the system, is zero, i.e.,

$$\mathbf{t}^T\mathbf{w}^C=0 \tag{75}$$

On the other hand, if the two sides of eq.(74) are transposed and then multiplied by a $6r$-dimensional vector $\boldsymbol{\lambda}$, one has

$$\mathbf{t}^T\mathbf{K}^T\boldsymbol{\lambda}=0 \tag{76}$$

Upon comparing eqs.(75) and (76) it is apparent that $\mathbf{w}^C$ is of the form

$$\mathbf{w}^C=\mathbf{K}^T\boldsymbol{\lambda} \tag{77}$$

i.e., $\mathbf{K}^T$ is the wrench generator of the overall system. More formally, the inner product of $\mathbf{w}^C$ and $\mathbf{t}$, as stated by eq.(75), vanishes and, since $\mathbf{t}$ lies in the nullspace of $\mathbf{K}$, as stated by eq.(74), then $\mathbf{w}^C$ lies in the range of $\mathbf{K}^T$. The following step will be to represent $\mathbf{t}$ as a linear transformation of the joint rates, i.e., as

$$\mathbf{t}=\mathbf{T}\dot{\boldsymbol{\theta}} \tag{78}$$

which will be called the ***twist-rate relations***. Note that, in light of the form of eq.(78), the $6r \times r$ matrix $\mathbf{T}$ plays the role of the twist generator of the overall system. The derivation of expressions for matrices $\mathbf{K}$ and $\mathbf{T}$ will be described in detail below. Now, upon substitution of eq.(78) into eq.(74), we obtain

$$\mathbf{KT}\dot{\boldsymbol{\theta}} = \mathbf{0} \tag{79a}$$

Furthermore, since the degree of freedom of the system is r, the r joint rates $\{\dot{\theta}_i\}_1^r$ can be assigned arbitrarily. However, while doing this, eq.(79a) has to hold. Thus, the only possibility for this to happen is that the product $\mathbf{KT}$ vanish, i.e.,

$$\mathbf{KT} = \mathbf{O}_{6r \times r} \tag{79b}$$

where $\mathbf{O}_{6r\times r}$ denotes the $6r \times r$ zero matrix. Equation (79b) thus states that $\mathbf{T}$ is *an orthogonal complement* of $\mathbf{K}$, as defined by Huston and Passerello [10]. Because of the particular form of choosing this complement—see eq.(78)—, $\mathbf{T}$ was termed *the natural orthogonal complement* of $\mathbf{K}$ in [11].

Note that, from eq.(79b), we have

$$\mathbf{T}^T\mathbf{K}^T = \mathbf{O}_{r\times 6r}$$

with $\mathbf{O}_{r\times 6r}$ denoting the $r \times 6r$ zero matrix. Hence, if we premultiply both sides of eq.(77) by $\mathbf{T}^T$, we obtain

$$\mathbf{T}^T\mathbf{w}^C = \mathbf{0}_r$$

$\mathbf{0}_r$ denoting the r-dimensional zero vector. Thus, $\mathbf{T}^T$ is the wrench annihilator of the overall system.

In the final step of this formulation, $\dot{\mathbf{t}}$, of eq.(73) is obtained from eq.(78), namely,

$$\dot{\mathbf{t}} = \mathbf{T}\ddot{\boldsymbol{\theta}} + \dot{\mathbf{T}}\dot{\boldsymbol{\theta}} \tag{80}$$

Furthermore, the uncoupled dynamical equations, eqs.(73), are multiplied from the left by $\mathbf{T}^T$, thereby eliminating $\mathbf{w}^C$ from those equations and reducing these to a system of only r independent dynamical equations, free of nonworking constraint wrenches. These are the Euler-Lagrange equations of the system, namely,

$$\mathbf{I}\ddot{\boldsymbol{\theta}} = -\mathbf{T}^T(\mathbf{M}\dot{\mathbf{T}} + \mathbf{WMT})\dot{\boldsymbol{\theta}} + \mathbf{T}^T(\mathbf{w}^A - \mathbf{w}^D) \tag{81}$$

where $\mathbf{I}$ is the positive definite $r \times r$ *generalized inertia matrix* of the overall system, and is defined as

$$\mathbf{I} \equiv \mathbf{T}^T\mathbf{MT} \tag{82}$$

Now we let $\boldsymbol{\tau}$ and $\boldsymbol{\delta}$ denote the r-dimensional vectors of active and dissipative generalized force. Moreover, we let $\mathbf{C}(\boldsymbol{\theta}, \dot{\boldsymbol{\theta}})\dot{\boldsymbol{\theta}}$ be the r-dimensional vector of *quadratic* terms of inertia force. The foregoing vectors are defined as

$$\boldsymbol{\tau} \equiv \mathbf{T}^T\mathbf{w}^A, \quad \boldsymbol{\delta} \equiv \mathbf{T}^T\mathbf{w}^D, \quad \mathbf{C}(\boldsymbol{\theta}, \dot{\boldsymbol{\theta}})\dot{\boldsymbol{\theta}} \equiv (\mathbf{T}^T\mathbf{M}\dot{\mathbf{T}} + \mathbf{T}^T\mathbf{WMT})\dot{\boldsymbol{\theta}} \tag{83}$$

Clearly, the difference $\boldsymbol{\tau} - \boldsymbol{\delta}$ produces $\boldsymbol{\phi}$, the generalized force of the system. Thus, the Euler-Lagrange equations of the system take on the form

$$\mathbf{I}\ddot{\boldsymbol{\theta}} = -\mathbf{C}\dot{\boldsymbol{\theta}} + \boldsymbol{\tau} - \boldsymbol{\delta} \tag{84}$$

It is pointed out that the first term of the right-hand side of eq.(84) is *quadratic* in $\dot{\boldsymbol{\theta}}$ because the parenthesis multiplying $\dot{\boldsymbol{\theta}}$ in the definition of this term in eq.(83) is linear in $\dot{\boldsymbol{\theta}}$. In fact, the first term inside the parentheses is linear in a factor $\dot{\mathbf{T}}$ that is in turn linear in $\dot{\boldsymbol{\theta}}$. Moreover, the second term inside the parentheses is linear in $\mathbf{W}$, which is linear in $\dot{\boldsymbol{\theta}}$ as well. However, $\mathbf{C}$ is *nonlinear* in $\boldsymbol{\theta}$. Because of the quadratic nature of the aforementioned term of eq.(84), it is popularly known as the vector of *Coriolis and centrifugal force*, whereas the left-hand side of that equation is given the name of vector of *inertia force*. Properly speaking, both the left-hand side and the first term of the right-hand side of eq.(84) arise from inertia forces.

10.1 DERIVATION OF CONSTRAINT EQUATIONS AND TWIST-RATE RELATIONS

In order to derive the kinematic constraint equations, we first note that the relative angular velocity of the ith link with respect to the $(i-1)$st link, $\boldsymbol{\omega}_i - \boldsymbol{\omega}_{i-1}$, is $\dot{\theta}_i \mathbf{e}_i$. Thus, if matrix $\mathbf{E}_i$ is defined as the cross-product matrix of vector $\mathbf{e}_i$, then the angular velocities of two successive links obey a simple relation, namely,

$$\mathbf{E}_i(\boldsymbol{\omega}_i - \boldsymbol{\omega}_{i-1}) = \mathbf{0} \tag{85}$$

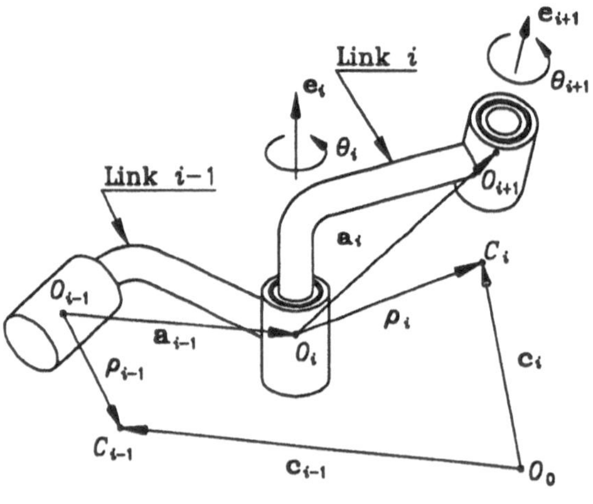

Figure 14: Revolute coupling of two successive links

Shown in Fig. 14 are the $(i-1)$st and ith links of the system. From this figure, we have the relations

$$\mathbf{c}_i = \mathbf{c}_{i-1} - \boldsymbol{\rho}_{i-1} + \mathbf{a}_{i-1} + \boldsymbol{\rho}_i \tag{86}$$

where $\boldsymbol{\rho}_i$ denotes the vector directed from point O_i on the axis of the ith joint to C_i, the ceentroid of this link. Upon differentiation of the two sides of the foregoing equation with respect to time, we obtain

$$\dot{\mathbf{c}}_i - \dot{\mathbf{c}}_{i-1} + \mathbf{S}_i\boldsymbol{\omega}_i + \mathbf{R}_{i-1}\boldsymbol{\omega}_{i-1} = \mathbf{0} \tag{87}$$

where $\mathbf{R}_i$ and $\mathbf{S}_i$ are defined as the cross-product matrices of vectors $\mathbf{a}_i + \boldsymbol{\rho}_i$ and $\boldsymbol{\rho}_i$, respectively. In particular, for $i = 1$, eqs.(85) and (87) reduce to

$$\mathbf{E}_1\boldsymbol{\omega}_1 = \mathbf{0} \tag{88a}$$

$$\dot{\mathbf{c}}_1 + \mathbf{S}_1\boldsymbol{\omega}_1 = \mathbf{0} \tag{88b}$$

Now, eqs.(85) and (87), as well as their counterparts for $i = 1$, eqs.(88a) and (88b), are further expressed in terms of the link twists, thereby producing the constraints below:

$$\mathbf{K}_{11}\mathbf{t}_1 = \mathbf{0} \tag{89a}$$

$$\mathbf{K}_{i,i-1}\mathbf{t}_{i-1} + \mathbf{K}_{ii}\mathbf{t}_i = \mathbf{0}, \quad i = 1, \ldots, r \tag{89b}$$

with $\mathbf{K}_{11}$ and $\mathbf{K}_{ij}$, for $i = 2, \ldots, r$ and $j = i-1, i$, defined as

$$\mathbf{K}_{11} \equiv \begin{bmatrix} \mathbf{E}_1 & \mathbf{0} \\ \mathbf{S}_1 & \mathbf{1} \end{bmatrix} \tag{90a}$$

$$\mathbf{K}_{i,i-1} \equiv \begin{bmatrix} -\mathbf{E}_i & \mathbf{0} \\ \mathbf{R}_{i-1} & -\mathbf{1} \end{bmatrix} \tag{90b}$$

$$\mathbf{K}_{ii} \equiv \begin{bmatrix} \mathbf{E}_i & \mathbf{0} \\ \mathbf{S}_i & \mathbf{1} \end{bmatrix} \tag{90c}$$

From eqs.(89a & b) and (90a–c), it is apparent that matrix $\mathbf{K}$ appearing in eq.(74) takes on the form

$$\mathbf{K} = \begin{bmatrix} \mathbf{K}_{11} & \mathbf{0} & \mathbf{0} & \cdots & \mathbf{0} & \mathbf{0} \\ -\mathbf{K}_{21} & \mathbf{K}_{22} & \mathbf{0} & \cdots & \mathbf{0} & \mathbf{0} \\ \vdots & \vdots & \ddots & \ddots & \vdots & \vdots \\ \mathbf{0} & \mathbf{0} & \mathbf{0} & \cdots & \mathbf{K}_{r-1,r-1} & \mathbf{0} \\ \mathbf{0} & \mathbf{0} & \mathbf{0} & \cdots & -\mathbf{K}_{r,r-1} & \mathbf{K}_{rr} \end{bmatrix} \tag{91}$$

Next, the link twists are expressed as linear combinations of the joint-rate vector $\dot{\boldsymbol{\theta}}$. To this end, we introduce further definitions below:

$$\mathbf{s}_{ij} \equiv \mathbf{a}_j + \mathbf{a}_{j+1} + \cdots + \mathbf{a}_{i-1} + \boldsymbol{\rho}_i, \quad i = 1, \ldots, r;\ j = 1, \ldots, i-1 \tag{92a}$$

$$\mathbf{s}_{ii} = \boldsymbol{\rho} i \tag{92b}$$

$$\mathbf{t}_{ij} \equiv \begin{bmatrix} \mathbf{e}_j \\ \mathbf{e}_j \times \mathbf{s}_{ij} \end{bmatrix}, \quad i = 1, \ldots, r;\ j = 1, \ldots, i \tag{92c}$$

Additionally, given the simple kinematic chain of the system, we can readily derive

$$\mathbf{t}_i = \dot{\theta}_1\mathbf{t}_{i1} + \cdots + \dot{\theta}_i\mathbf{t}_{ii}, \quad i = 1, \ldots, n \tag{93}$$

and hence, matrix $\mathbf{T}$ of eq.(78) is readily found to be

$$\mathbf{T} \equiv \begin{bmatrix} \mathbf{t}_{11} & \mathbf{0} & \cdots & \mathbf{0} \\ \mathbf{t}_{21} & \mathbf{t}_{22} & \cdots & \mathbf{0} \\ \vdots & \vdots & \ddots & \vdots \\ \mathbf{t}_{r1} & \mathbf{t}_{r2} & \cdots & \mathbf{t}_{rr} \end{bmatrix} \tag{94}$$

As a matter of verification, one can readily prove that matrix $\mathbf{T}$, as given by eq.(94), is an orthogonal complement of matrix $\mathbf{K}$, as given by eq.(91), in the sense of eq.(79b).

In order to set up eq.(84), then, all that is needed now is $\dot{\mathbf{T}}$, which is computed below. Upon differentiation of $\mathbf{t}_{ij}$ with respect to time, we obtain

$$\dot{\mathbf{t}}_{ij} = \begin{bmatrix} \boldsymbol{\omega}_j \times \mathbf{e}_j \\ (\boldsymbol{\omega}_j \times \mathbf{e}_j) \times \mathbf{s}_{ij} + \mathbf{e}_j \times \dot{\mathbf{s}}_{ij} \end{bmatrix} \tag{95}$$

where, from eq.(92a),

$$\dot{\mathbf{s}}_{ij} = \boldsymbol{\omega}_j \times \mathbf{a}_j + \ldots + \boldsymbol{\omega}_i \times (\mathbf{a}_i + \boldsymbol{\rho}_i) \tag{96}$$

11 Conclusions

We have introduced the concepts of twist generator and twist annihilator, their duals being the wrench generator and the wrench annihilator. For every lower kinematic pair we can define a $6 \times f$ matrix, where f is the degree of freedom of the kinematic pair, that produces a relative twist of the two coupled links when the f joint rates of the pair are specified. Likewise, the same kinematic pair transmits a relative constraint wrench between the coupled links, that does not develop any power onto the whole kinematic chain, its only role being to keep the two links together. Moreover, we have shown that the twist annihilator of a lower kinematic pair is an orthogonal complement of the corresponding twist generator. Also, the wrench generator of a lower kinematic pair is the transpose of the corresponding twist annihilator, or a multiple thereof. Furthermore, we extended these concepts and defined the twist generator and the wrench generator of an overall multi-body system with a simple, open kinematic chain. Concurrently, we defined the corresponding twist annihilator and wrench annihilator. We believe that these concepts can find extensive applications in the mechanics of grasping and in better understanding the problem of hybrid control, i.e., force and motion control of manipulators.

12 Acknowledgements

The research work reported here was made possible under NSERC (Natural Sciences and Engineering Research Council of Canada) Research Grants A4532, STR0100971 and EQP00-92729. The partial support of IRIS (Institute for Robotics and Intelligent Systems), under the C-3 Project, is highly acknowledged.

13 Bibliography

1. Freudenstein, F. (1962) "On the variety of motions generated by mechanisms," *J. of Engineering for Industry*, pp. 156–160.

2. Ball, R. S. (1875) "The theory of screws. A geometrical study of the kinematics, equilibrium, and small oscillations of a rigid body", *Trans. Royal Irish Acad.*, Vol. 25, pp. 157–217.

3. Samuel, A. E., McAree, P. R. and Hunt, K. H. (1991) "Unifying screw geometry and matrix transformations", *The International Journal of Robotics Research*, Vol. 10, No. 5, pp. 454–472.

4. Waldron, K. J. and Hunt, K. H. (1991) "Series-parallel dualities in actively coordinated mechanisms", *The International Journal of Robotics Research*, Vol. 10, No. 5, pp. 473–480.

5. Lipkin, H. and Duffy, J. (1985) "The elliptic polarity of screws", *ASME J. Mechanisms, Transmissions, and Automation in Design*, Vol. 107, pp. 377–387.

6. Lipkin, H. and Duffy, J. (1988) "Hybrid twist and wrench control for a robotic manipulator", *ASME J. Mechanisms, Transmissions, and Automation in Design*, Vol. 110, pp. 138–144.

7. Hartenberg, R.S., and Denavit, J. (1964) *Kinematic Synthesis of Linkages*, McGraw-Hill, New York.

8. Angeles, J. (1982) *Spatial Kinematic Chains*, Springer-Verlag, Berlin-Heidelberg-New York.

9. Harary, F. (1972) *Graph Theory*, Addison-Wesley Publishing Company Inc., Reading, MA.

10. Angeles, J. and Lee, S., 1989, "The modelling of holonomic mechanical systems using a natural orthogonal complement", *Transactions of the Canadian Society of Mechanical Engineering*, Vol. 13, No. 4, pp. 81–89.

11. Huston, R. L. and Passerello, C. E., 1974, "On constraint equations—a new approach", *Trans. ASME*, Vol. 96; *ASME J. Applied Mechanics*, Vol. 45, pp. 889–894.

APPLICATION OF COMPUTER AIDED KINEMATICS TO MODELLING OF CONTACTS IN ROBOTIC MANIPULATION

J. DE SCHUTTER, H. BRUYNINCKX and S. DUTRÉ
Katholieke Universiteit Leuven
Department of Mechanical Engineering, Division PMA
Celestijnenlaan 300B, B-3001 Heverlee, Belgium

ABSTRACT: This article presents a kinematic approach to the modelling and the motion specification of robotic manipulation tasks in which the manipulated object is constrained by contacts. The presented approach takes into account complex and time varying motion constraints, and is very appropriate to be integrated into CAD based task planning and control.

The description of the interaction between the manipulated object and other objects in its environment is based on the first and second order approximations of their geometry around the contact areas. From these geometric descriptions, the manipulated object's nominal motion freedom and its dual, the set of possible reaction forces, are then modelled using the similarity with the kinetostatics of kinematic chains.

The kinematic approach is illustrated with the important example of the classical peg-in-hole problem. The approach offers new tools to reliably model and specify the insertion motion of the peg, even in the case of very large misalignments between the axes of peg and hole.

1. Introduction

Robotic manipulation tasks very often involve contacts between the object attached to the manipulator (called the *manipulated object*, or "*MO*" for short) with other objects in its environment (called "*ENV*" for short). In many cases, the presence of contacts is indispensable for the execution of the manipulation, since 1) the contacts belong to the goal position of the *MO*, or 2) during the motion they reduce (part of) the inevitable uncertainties between the relative positions of *MO* and *ENV*. However, the contact interactions limit the motion freedom of the manipulated object, and generate contact forces. Hence, the manipulator needs some active or passive means to react safely to these forces, and at the same time to continue the desired manipulation action. Anyway, both the active and passive approaches (i.e., force control, respectively compliance at the end effector) can deal only with limited inaccuracies between the desired nominal motion of the *MO* on the one hand, and the real motion freedom of the *MO* as allowed by the *ENV* on the other hand. Therefore, a good nominal specification of the desired motion remains indispenable.

For the traditional robotic manipulation tasks, this specification relies on the human programmer's implicit mental model of the *MO*'s motion freedom. However, if the motion constraints become more complex, or if the task specification has to be generated by an

M. F. O. S. Pereira and J. A. C. Ambrósio (eds.),
Computer-Aided Analysis of Rigid and Flexible Mechanical Systems, 413–444.

automatic planning system, more explicit models are required, as well as a systematic, computer assisted approach. *Geometric models* are the appropriate building blocks with which to construct a computer aided task specification system for constrained robotic manipulation tasks: the description of the surfaces of *MO* and *ENV* around the contact areas determines the type of the contacts, as well as –up to non-ideal effects– the corresponding motion freedom of the manipulated object.

The intensely studied *peg-in-hole* task is a prime example of the difficulties which occur in the motion specification of a manipulation task, Fig. 1: most references, (following Whitney's seminal paper, (1982)) only discuss active or passive means to further insert the peg into the hole after an initial small insertion has already been established; the main problem is then to avoid extraneous forces, wedging and jamming. However, the question of how to reach this initial position is neglected. The kinematic approach presented in this paper offers a (partial) solution to this problem: once the bottom of the peg has made contact with the rim of the hole, a systematic method is presented to align the axes of peg and hole, even if the initial misalignment is very large. The same ideas are applicable to a wide range of motion constraints on the manipulated object.

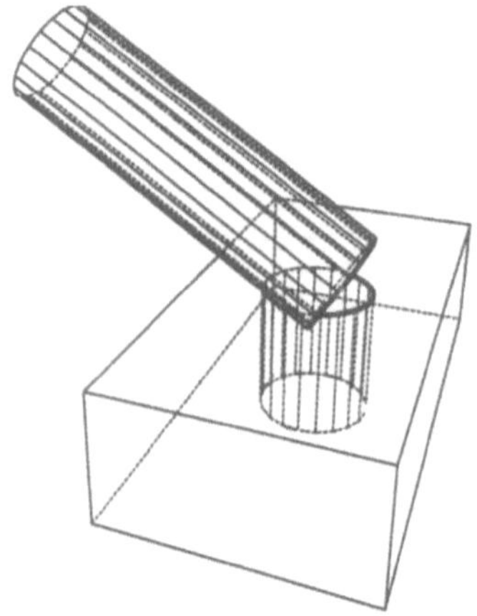

Figure 1: *Peg-in-hole.* If the axes of the peg and the hole are very badly aligned, it is not straightforward to specify the desired motion of the peg, especially since its instantaneous motion freedom is continuously changing during the task.

Force control (also called *compliant motion*) is undoubtedly the most intensely studied part of the constrained manipulation problem. This could give the impression that modelling and specification are trivial. Maybe this is the case for very simple force controlled tasks, as described by Mason (1981) and De Schutter and van Brussel (1988): peg in hole (with partly inserted peg), following a surface with a point contact, opening a door, turning a crank or a screw, ...

These simple, or *elementary*, compliant tasks can be specified with the *task frame* or *compliance frame* approach, De Schutter and Van Brussel (1988). The task frame with its *natural constraint directions* serves as a geometric model of the motion constraint. And indeed, for tasks with a simple contact geometry the relation between this contact geometry and the force controlled directions (i.e. the natural constraint directions) and the velocity controlled directions (i.e. the *artificial constraint* directions) in the task frame is quite straightforward and intuitively clear.

However, this is not the case for tasks with *complex* motion constraints. A motion constraint is complex if it is the *combination* of several simple, or *elementary*, constraints, and if this combination is *time varying*. See Fig. 2 for an example. Time variance means that the relative locations of some of the elementary constraints change during the execution of the task, see the peg-in-hole example.

This paper presents a *model based* compromise between off line modelling cost, on line modelling accuracy, and flexibility in the task specification: the compliant motion task relies on simple off line models of (a combination of) elementary motion constraints; these models are easily adaptable on line and they allow a user friendly task specification.

Other authors have already presented model based approaches to cope with motion constraints. Montana (1988) describes the kinematics of grasping objects with robotic fingers equipped with tactile sensors. Cai and Roth (1986, 1988) give a very thorough and complete description of the nominal kinematics for the relative motion of objects under point and line contacts. These approaches are of a much higher complexity than the one presented here. Moreover, they are more difficult to use on line, and not well suited for complex motion constraints.

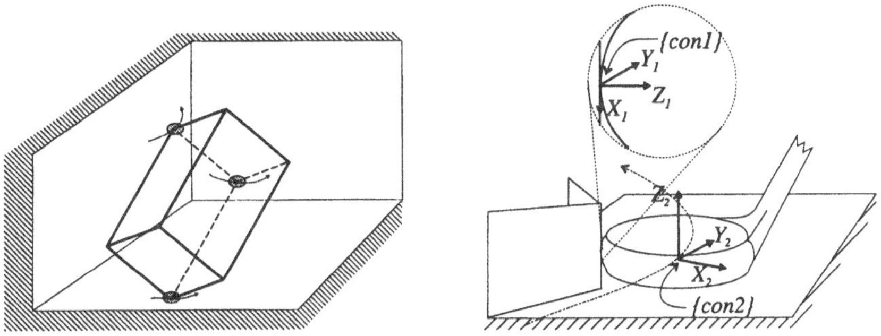

Figure 2: *Complex, time-varying motion constraints.* Left: moving a block subjected to two or three simultaneous contacts poses motion specification problems if the goal situation is a non-equilibrium position. Right: a *vertex-face* contact at the left side of the *MO*, plus a *face-face* contact at its bottom.

Motion constraint models exist on different levels of abstraction, and for a wide variety of applications. On the highest level the motion constraint is described as a list of elementary contacts, which act simultaneously on the manipulated object. This is the *topological* or *symbolical* model of the constraint. It contains the type of each elementary constraint, and a pointer to the geometric entities (face, edge, vertex, surface, ...) involved. This level of motion constraint models is generally used in off line, CAD based task planners for compliant motions, or *fine motions*, as they are often called in this context. The vocabulary of elementary contacts used at this level is rather uniform: *vertex-surface*, *edge-edge*, etc. Buckley (1989) and Laugier (1989) deal with the problem of automatically generating sequences of compliant actions to reach a user specified goal. Xiao and Volz (1989), Xiao (1992) and Desai and Volz (1989) also started to examine how to verify the current motion constraint, and *replan* the task whenever a deviation from the nominal plan is detected at execution time. The common limitation to all mentioned planners is that they offer no interface to an on line task controller: their plans are expressed as sets of elementary motion constraints, together with a purely *geometric* specification of the motion. Moreover, this motion specification invariably uses only a very limited subset of the total available motion freedom: pure translations or pure rotations.

This paper covers the modelling and motion specification aspects of constrained manipulation, within the framework of kinematics of complex chains, Angeles (1988). To this end, each elementary contact is replaced by an equivalent kinematic chain, called a *virtual (contact) manipulator*. The topology of the chain is given by the type of the contact; the local surface geometry around the contacts determines the numerical description of the chain: link lengths, current "joint" values and limits, etc. Simplicity of the models is emphasized, because:

1. The user interface of the computer assisted motion specification system must remain simple.
2. Exact modelling is practically impossible, or too costly.
3. It is assumed that the manipulator has some active or passive robustness against small uncertainties in the generated motion specification.

The kinematic approach permits the use of well-established tools, terminology and algorithms from kinematics: revolute and prismatic joints, twists and wrenches, Jacobian matrices, etc.

Local geometric models exist at different levels of detail:

1. *First order*, or polyhedral: the position of the *contact point* and the direction of the *contact normal*.
2. *Second order*: i.e., the first order description, plus *curvature* information at the contact point.
3. *Higher order*. This level is not discussed in this text, because it is incompatible with the requirement for simplicity.

Section 2. introduces the terminology to describe motion freedom and constraint reaction forces. This is followed in Section 3. by the kinematic models for elementary and complex motion constraints, and the corresponding mathematical representations. Finally, Section 4. applies this theoretical framework to the peg-in-hole example.

2. Motion Constraints: Concepts

This Section elaborates the concept of elementary motion constraint, which appeared already in the Introduction. Then follow the definitions of the twist and wrench systems of a motion constraint. They are the basic mathematical representations to link the geometric descriptions of the contacts on the *MO* to motion specification for the manipulating robot. The last subsection explains *reciprocity*, i.e., the physical duality relationship which always exists between the wrench and twist systems of any motion constraint.

2.1. ELEMENTARY MOTION CONSTRAINTS

The point contact (or surface-surface contact) between two rigid bodies is the simplest physical model of a motion constraint. Strictly speaking, it is the only really elementary motion constraint, in the sense that all other motion constraints consist of a (possibly continuous and infinite) set of point contacts. However, every field of application has its own particular set of "elementary" constraints, i.e. those that belong to the standard vocabulary of the field, and that are assumed to be the most basic and simple constraints needed to describe all systems in the field. Some examples:

Assembly The polyhedral contacts (*vertex-face*, *edge-face*, *edge-edge*, *face-face*), or the mixed polyhedral–non-polyhedral *vertex-surface* contact. A *face* is a planar part of an object; a *surface* is curved. An *edge* is the intersection of two faces, and a *vertex* is the intersection of two or more edges.

Robotic hands The constraints consist of *soft* and *hard finger* contacts, with sufficient friction to prohibit *slipping*.

Linkages The theory of machines and mechanisms makes use of all types of *joints*. The revolute and prismatic joints are the elementary ones.

Elementary motion constraints are the building blocks for all possible motion constraints in a particular field. Such a combination of motion constraints is called a *composite constraint*. This paper reformulates the elementary and composite motion constraints appearing in such fields of assembly or robotic hands as linkages of prismatic and revolute joints.

2.2. TWIST AND WRENCH SYSTEMS

To specify a constrained motion task, it is important to know how the *MO* can move, without breaking the desired contacts, and without generating excessive reaction forces. For any motion constraint on the manipulated object, either elementary or composite, mathematical models of these sets of possible motions and forces are given by the following two vector spaces:

Twist system This is the vector space of all instantaneous velocities, or infinitesimal displacements, that the *MO* can have with respect to the *ENV*, without breaking the contact, Lipkin and Duffy (1988). A *twist* t is the ordered combination of an angular velocity vector $\vec{\omega}$ and a linear velocity vector $\vec{v}$: $t = (\vec{\omega}, \vec{v})$. The notation for a twist system is $\boldsymbol{T}$.

Wrench system The vector space dual to $\boldsymbol{T}$ is the wrench system $\boldsymbol{W}$ of all ideal reaction forces that can be generated in the contact. A wrench w is the ordered combination of a force vector $\vec{f}$ and a moment vector $\vec{m}$: $w = (\vec{f}, \vec{m})$.

Twist and wrenches are kinetostatic applications of the geometric concept of a *screw*, i.e., the ordered combination of a *line vector* and a *free vector*, Lipkin and Duffy (1988).

The fact that $\boldsymbol{T}$ and $\boldsymbol{W}$ are vector spaces is mathematically appealing, because the systems are totally known if a *basis* for them is known. A possible drawback is the local validity of the models: in general, $\boldsymbol{T}$ and $\boldsymbol{W}$ change during the motion of the *MO*. It is also important to realize that:

- $\boldsymbol{T}$ and $\boldsymbol{W}$ are *first order* models, since they only describe the instantaneous motion freedom.
- $\boldsymbol{T}$ contains *velocities* or *infinitesimal displacements*, so that it gives no information about the motion freedom of the *MO* under *finite* displacements.
- $\boldsymbol{W}$ contains only the *ideal* and *static* reaction forces, i.e. no friction, elasticity or dynamic effects are modelled.

For elementary motion constraints it is straightforward to describe the instantaneous motion freedom of the *MO*. For composite constraints the modelling is more complicated. However, the following procedure leads to the desired result:

1. Construct the $\boldsymbol{T}$'s and $\boldsymbol{W}$'s of the composing elementary constraints.

2a. If a set of n_c elementary constraints on the *MO* act *in parallel*, then the twist system of the composite constraint is the vector space *intersection* of the twist systems of the composing constraints. The wrench system is the vector space *sum* of the wrench systems of the composing constraints. Formally one writes:

$$\boldsymbol{T}^{par} = \boldsymbol{T}^1 \cap \boldsymbol{T}^2 \cap \ldots \cap \boldsymbol{T}^{n_c}, \quad \boldsymbol{W}^{par} = \boldsymbol{W}^1 + \boldsymbol{W}^2 + \ldots + \boldsymbol{W}^{n_c}. \tag{1}$$

2b. If the elementary constraints act *in series*, the abovementioned relations are interchanged:

$$\boldsymbol{T}^{ser} = \boldsymbol{T}^1 + \boldsymbol{T}^2 + \ldots + \boldsymbol{T}^{n_c}, \quad \boldsymbol{W}^{ser} = \boldsymbol{W}^1 \cap \boldsymbol{W}^2 \cap \ldots \cap \boldsymbol{W}^{n_c}. \tag{2}$$

Numerical implementations of this procedure are supported by reliable and stable algorithms, based on Singular Value Decomposition of the matrices representing bases for $\boldsymbol{T}$ and $\boldsymbol{W}$, Golub and Van Loan (1989).

2.3. RECIPROCITY

The twist system $\boldsymbol{T}$ and the wrench system $\boldsymbol{W}$ are dual vector spaces. In a strict mathematical sense this means that to each wrench w there corresponds a *linear form* $R_{\mathsf{w}}(\mathsf{t})$ over the space of twists t. (A linear form is a linear mapping from a vector space to the line of real numbers.) Similarly, a linear form $R_{\mathsf{t}}(\mathsf{w})$ over the space of reaction forces is defined. The physical interpretation of this linear form is as follows: it represents the virtual power that the motion t generates against the wrench w:

$$R_{\mathsf{t}}(\mathsf{w}) = R_{\mathsf{w}}(\mathsf{t}) = \vec{\mathsf{m}} \cdot \vec{\omega} + \vec{\mathsf{f}} \cdot \vec{\mathsf{v}}. \tag{3}$$

If, with a small abuse of notation, t and w represent also the *coordinates* of a twist and a wrench with respect to a reference frame, the numerical equivalent of Eq. (3) is:

$$R_{\mathsf{t}}(\mathsf{w}) = R_{\mathsf{w}}(\mathsf{t}) = \mathsf{w}^T \tilde{\boldsymbol{\Delta}}\, \mathsf{t}, \tag{4}$$

with

$$\tilde{\boldsymbol{\Delta}} = \begin{bmatrix} O_{3\times 3} & I_{3\times 3} \\ I_{3\times 3} & O_{3\times 3} \end{bmatrix}. \tag{5}$$

For compliant manipulation tasks, the motion t of the *MO* does not break the contact with the *ENV*, and the virtual power generated against the ideal reaction wrench w vanishes. One says that t and w are *reciprocal*:

$$\mathrm{w}^T \tilde{\boldsymbol{\Delta}}\, \mathrm{t} = \vec{\mathrm{m}} \cdot \vec{\omega} + \vec{\mathrm{f}} \cdot \vec{\mathrm{v}} = 0. \qquad (6)$$

This reciprocity condition remains valid under serial and/or parallel composition of motion constraints.

3. Kinematic Models

As mentioned before, the surface geometry of the contacting rigid bodies determines the resulting motion constraint, because of the *mutual impenetrability* of the bodies. Hence, this Section starts with a simple geometric model of these surfaces, as given by their first and second order differential geometric properties. Second, it derives the relationship between this geometric description of each of the individual objects, and the features of their interaction, i.e., the contact. Third, it translates the concepts of motion freedom and constraint into equivalent kinematic models, and finally into their mathematical representations. Kinematic modelling of motion constraints has the following advantages:

- Standard terminology and algorithms of mechanisms and manipulators are available.
- There exists a close correspondence between, on the one hand, the first and second order geometry of a motion constraint, and, on the other hand, the kinematic model.
- Specification of the desired motion of the *MO* becomes intuitive and yet unambiguous: it boils down to deciding which are the *driving joints* in the kinematic model, and what are their desired speeds.

The kinematic model for a joint constraint is trivial: it is the joint itself. For a contact constraint only the *reciprocal* motion freedom is modelled, i.e. these motions allowed by the contact and which do not break the contact. This is, by definition, not a limiting assumption for the manipulation tasks discussed in this text. Moreover, it eliminates the distinction between (the models of) contact and joint constraints.

3.1. DESCRIPTION OF LOCAL GEOMETRY: GEOMETRIC FRAMES

In general, the surfaces of *MO* and *ENV* contact each other in one or more discrete contact points. For each of these points, the surface of both objects is of one of the following three types: a smooth surface, a smooth curve or a vertex. In differential geometry, the local geometry of a surface or a curve in the neighbourhood of a point is characterized by an orthogonal reference frame:

1. **Smooth surface**. The tangent plane (i.e., two independent tangent vectors) in any point on the surface is uniquely defined by the object's geometry. An orthogonal reference frame, the so-called *principal frame*, is defined as follows, O'Neill (1966): its origin lies in the contact point, one axis lies along the normal at the point, and the two other axes lie along the directions of minimum and maximum curvature.

These are called the *principal directions of curvature*, and are always orthogonal and uniquely defined, unless the object is *umbilic*, i.e. the curvature at the contact point is equal in all directions. This is the case for spheres and planes.

2. **Smooth curve**. The tangent to the object has a unique meaning in only one single direction. Yet, for each point on a curve, an orthogonal reference frame, the *Frenet frame*, is defined: one axis along the tangent, one along the normal (i.e., pointing to the centre of the osculating circle), and the third one, the *binormal* direction, orthogonal to the other two. The Frenet frame directions are uniquely defined, unless the curve is a straight line.

3. **Vertex**. A unique definition of tangent vector or tangent plane is impossible.

So, in a contact point, each of *MO* and *ENV* has its own orthogonal reference frame, which is, in general, fully determined by the *first* and *second order* geometric parameters of the surface: the tangents, normals and directions of curvature. The frames are called the *geometric frames* at the contact point, and denoted by $\{geo\}$ for the *MO* and $\{geo'\}$ for the *ENV*. However, the geometric parameters do *not completely* determine the geometric frames in the case of:

- Umbilic surfaces: the tangent plane is *uniquely defined*, but not the principal directions.

- Vertices and straight lines: the tangent plane is *not uniquely defined*.

3.2. DESCRIPTION OF CONTACT: CONTACT FRAMES

Geometric frames merely model the local geometry of the contacting object; they are independent of how *MO* and *ENV* are contacting each other. Hence, geometric frames do not unambiguously describe the contact geometry, especially in the case of curves and vertices. To this end *contact frames* are introduced. A contact frame is defined for both *MO* and *ENV*. They are named $\{con\}$ and $\{con'\}$, respectively. Their origin lies at the contact point, and one axis lies along the contact normal. Furthermore:

1. **Smooth surface**: the other two axes lie along the principal directions of curvature. Hence, for a smooth surface, the contact frame coincides with the geometric frame.

2. **Smooth curve**: one axis lies along the tangent. Hence, for a smooth curve, the contact frame and the geometric frame coincide, except for a rotation about the tangent.

3. **Vertex**: the contact frame and the geometric frame only have their origin in common.

The geometric frame $\{geo\}$ and the contact frame $\{con\}$ coincide for a smooth surface, irrespective of the geometry of the *ENV*, but not for a smooth curve or a vertex. This is because curves and vertices do not possess two independent tangent vectors in the point of contact, as explained in 3.1.. Hence, the definition of the contact frame for a curve or vertex of the *MO* requires the knowledge of one or two of the tangent vectors of the *ENV*, and vice versa. For the same reason, the *edge-vertex*, *vertex-vertex* or *edge-parallel edge* contacts are *unstable*: no two independent tangent vectors exist, so that contact normal and tangent plane are also not defined.

3.3. KINEMATIC MODEL OF POINT CONTACT: VIRTUAL MANIPULATORS

In principle, the equations describing the geometry of the surfaces of both *MO* and *ENV* are sufficient to model the instantaneous motion freedom of the *MO*. However, the use of these geometric models for the specification of the desired motion is not straightforward. Therefore, the geometric model is replaced by a kinematic model: this means that the *MO* and the *ENV* are linked by a *virtual manipulator*, which gives the same reciprocal motion freedom as allowed by the contact. The motion of the *MO* is then expressed as the "joint motions" of this virtual contact manipulator. The kinematic structure of the virtual contact manipulator is, as much as possible, determined by the local geometric properties of the contacting objects, as described in subsection 3.1., i.e., the tangent vectors and the principal directions (and centres) of curvature. The virtual contact manipulator consists of three connected sub-manipulators:

1. *SLIP*. This first sub-manipulator has two degrees of freedom, which correspond to the motion of the contact point on the surface of the *MO*. It links the *MO* to the $\{con\}$ frame as follows. First, choose a reference frame $\{base\}$ on the *MO*, which serves as the base of the virtual manipulator. $\{base\}$ is chosen arbitrarily. Furthermore:

   Smooth surface. Connect the base frame to a revolute joint at the centre of curvature corresponding to the largest radius of curvature, and with its joint axis parallel to the direction of maximum curvature; connect to this joint a second revolute joint, at the centre of curvature corresponding to the smallest radius of curvature, and with its joint axis parallel to the direction of minimum curvature. Connect the contact frame $\{con\}$ to this second joint, see Figs. 3 and 4.

   Smooth curve. Connect the base frame to a revolute joint at the centre of the osculating circle and with its axis parallel to the direction of the binormal. Attach it to a second revolute joint along the tangent at the contact point. Connect the contact frame $\{con\}$ to this second joint.

   Vertex. Connect the base frame to two revolute joints, both in the vertex, and with their axes, in principle, in arbitrary directions. So, one could prefer to make these axes parallel to some of the other axes in the total virtual contact manipulator. Connect the contact frame $\{con\}$ to the second joint.

2. *ROT*. A one degree of freedom manipulator, connecting *SLIP* and *SLID* by a fifth revolute joint along the common contact normal axes of $\{con\}$ and $\{con'\}$.

3. *SLID*. This third sub-manipulator, with base frame $\{base'\}$ is similar to *SLIP*. It has also two degrees of freedom, corresponding to the motion of the contact point on the surface of the *ENV*. It links the *ENV* to the $\{con'\}$ frame in a similar way as *SLIP* links the *MO* to $\{con\}$.

The twist system $\boldsymbol{T}$ of the *MO* due to these virtual manipulators is the union of the following subspaces:

1. $\boldsymbol{T}^{slip}$ **(slipping)**: motion of the *MO* due to the motion of the joints in *SLIP*. This does not move the contact point with respect to the *ENV*, but it does change the contact point on the surface of the *MO*. See Fig. 3.

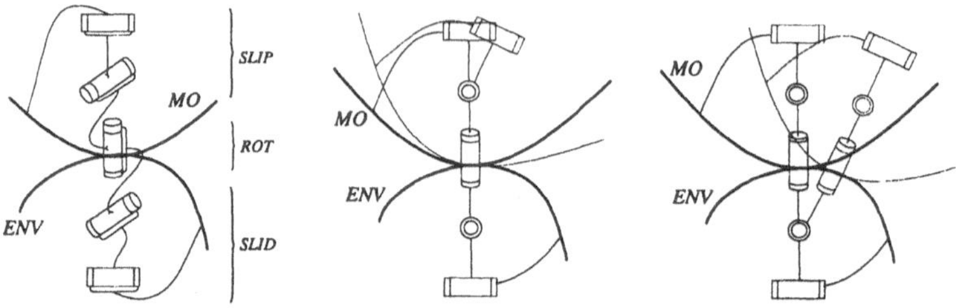

Figure 3: *Virtual manipulators*. Virtual manipulators for the case where both *MO* and *ENV* are smooth surfaces (left). Slipping (middle) moves the *MO* without changing the contact point on the *ENV*. Sliding (right) moves the *MO* without changing the contact point on the *MO*. For reasons of clarity, the figure shows only 2D motion.

2. T^{rot} (**rotation**): motion about the common contact normal axes of the contact frames, i.e. motion of the *ROT* manipulator. The contact point does not move in space.

3. T^{slid} (**sliding**): motion of the *MO* due to the motion of the joints in *SLID*. This does not move the contact point with respect to the *MO*, but it does change the contact point on the surface of the *ENV*. See Fig. 3.

The special combination of slipping and sliding such that the contact points on the surfaces of *MO* and *ENV* move with the same instantaneous velocity, is called **rolling**.

With these subsystems of motion freedom (each corresponding to parts of the virtual manipulator), the specification of the desired instantaneous velocity of the *MO* is performed in a model based and user friendly manner. One should keep in mind, however, that, in general, for a compliant motion task only *instantaneous velocities* or *infinitesimal displacements* are specifiable, but not *finite displacements*!

During on line execution of a compliant motion task, the *MO*'s measured motion t is decomposed into components of the abovementioned subspaces:

$$\mathsf{t} = \mathsf{t}^{slip} + \mathsf{t}^{rot} + \mathsf{t}^{slid}. \tag{7}$$

This decomposition gives the following *model update* information: both contact frames are rotated w.r.t. each other over the rotation component t^{rot}, the contact frame has moved over the surface of *MO* by the slipping component t^{slip}, and over the surface of the *ENV* by the sliding component t^{slid}. See 3.9. for more details.

3.4. MATHEMATICAL REPRESENTATION: JACOBIANS

Slipping, sliding and rotation completely define the five degrees of freedom of the manipulated object with respect to its environment, as allowed by the second order model of the contacting surfaces. This subsection presents a *numerical* description of this motion freedom, based on the reference frame definitions of the previous subsections. The only things that are still missing are: 1) a systematic naming convention for the reference frame axes (X, Y and Z), and 2) the matrices representing numerical bases for the twist and wrench systems of the motion constraints (i.e., the so-called *Jacobians*).

3.4.1. Geometric Frames First, the axes of the geometric frames $\{geo\}$ and $\{geo'\}$ are chosen in a systematic way, see Fig. 4:

1. **Smooth surface.** The Z axis lies along the surface normal; the X and Y axes along the directions of minimum, respectively maximum curvature. Freedom of choice for X and Y if the surface is umbilic in the contact point.

2. **Smooth curve.** The X, Y and Z axes lie along the tangent, binormal and normal directions, in this order. Freedom of choice for Y and Z if the curve is a straight line.

3. **Vertex.** Full freedom of choice. A practical choice is to make the geometric frame $\{geo\}$ coincide with the contact frame $\{con\}$ defined in the following subsection.

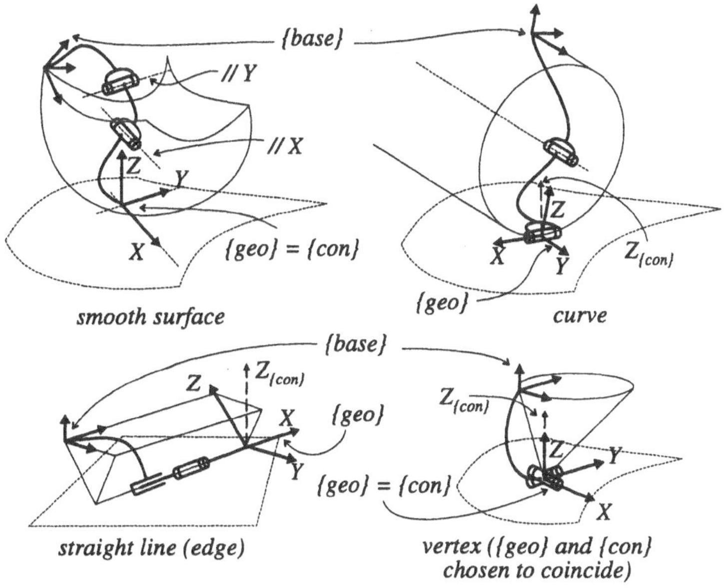

Figure 4: *Geometric and contact reference frames.* For each of the contact classes, the location and the relative motion freedom of the $\{geo\}$ and $\{con\}$ reference frames, with respect to the $\{base\}$ reference frame, are indicated.

3.4.2. Contact Frames Second, the contact frames are defined: the origins of $\{con\}$ and $\{con'\}$ lie at the contact point; their Z axes lie along the contact normal, and point "into" the *MO* and the *ENV*, respectively. This means that the Z axes of $\{con\}$ and $\{con'\}$ have opposite directions. Furthermore:

1. **Smooth surface**: X and Y coincide with the X and Y axes of the geometric frames.
2. **Smooth curve**: X coincides with the X axis of the geometric frame, i.e., the tangent to the curve. The Y axis is derived from the knowledge of X and Z.
3. **Vertex**: The contact frame for a vertex *MO* is fully determined by the contact frame of the *ENV*, and vice versa. The easiest choice is to let $\{geo\}, \{geo'\}, \{con\}$ and $\{con'\}$ all coincide (taking into account that the Z axes of the primed and unprimed frames are in the opposite directions).

3.4.3. Transformations Third, the transformations between all the frames are defined in terms of a simple, minimal and unambiguous set of geometrical parameters:

1. $\{con\} \rightarrow \{geo\}$: this transformation is the identity transformation, except for:

   Smooth curve: the Z axis of $\{con\}$ is rotated about the common X axis (tangent direction) over an angle μ_x.

   Vertex: in case the geometric frame and the contact frame are not *chosen* to coincide, the Z axis of $\{con\}$ is first rotated about the X axis over an angle μ_x, then about the Y axis over an angle μ_y. Remark that the order of the rotations matters, because μ_x and μ_y are in general of *finite* magnitude.

2. $\{con'\} \rightarrow \{geo'\}$: similar to the previous transformation, with only a notational difference: the possible rotation angles are called η_x and η_y.
3. $\{con'\} \rightarrow \{con\}$: these frames only differ by 1) a rotation over 180 degrees (about X or Y) since the Z axes have opposite direcions, and 2) a rotation about their common Z axis. This rotation angle is called ρ.

3.4.4. Jacobians A basis of the twist system $\boldsymbol{T}$ is given by the 6×5 *Jacobian matrix* $\boldsymbol{J}^{2nd}$, composed of slipping, rotation and sliding:

$$\boldsymbol{J}^{2nd} = \left[\boldsymbol{J}^{slip}\ \boldsymbol{J}^{rot}\ \boldsymbol{J}^{slid}\right]. \tag{8}$$

The meaning of the term "Jacobian" is exactly the same as in robot kinematics: each column of the Jacobian represents the velocity of the "end effector" of the virtual manipulator corresponding to the velocity in one of the manipulator's joints. The different sub-Jacobians are expressed in terms of the geometric parameters of the contact, and in the frame which results in the simplest representation. See Fig. 4 for the definition of the reference frames.

1. The *SLIP* Jacobian is expressed with respect to $\{geo\}$.

Smooth surface. Its first column represents the rotation about the first revolute joint of the *SLIP* virtual manipulator. This joint lies at the centre of curvature corresponding to the largest radius of curvature. Its axis is parallel to the Y axis. The distance between this joint and the contact point is noted r^{mo}_{max}. Similarly, the second column represents the rotation about the second revolute joint of the *SLIP* manipulator. Its axis is parallel to the X axis. r^{mo}_{min} is the distance between this joint and the contact point; it is the smallest radius of curvature at the contact point. Hence:

$$_{geo}J^{slip} = \begin{bmatrix} 0 & b \\ a & 0 \\ 0 & 0 \\ -r^{mo}_{max}a & 0 \\ 0 & r^{mo}_{min}b \\ 0 & 0 \end{bmatrix}, \quad a = \frac{1}{\sqrt{1+(r^{mo}_{max})^2}}, \; b = \frac{1}{\sqrt{1+(r^{mo}_{min})^2}}. \tag{9}$$

The particular form of the columns of this Jacobian allows limit transitions for the radii of curvature going to zero or to infinity.

Smooth curve. The direction of maximum curvature degenerates, and only the curvature along the curve remains. This curvature is represented by the radius r^{mo} of the osculating circle, and corresponds to a rotation about a revolute joint placed in the centre of this circle, and with its axis parallel to Y. The second joint is a rotation about the tangent X. Hence, the *SLIP* Jacobian becomes:

$$_{geo}J^{slip} = \begin{bmatrix} 0 & 1 \\ a & 0 \\ 0 & 0 \\ -r^{mo}a & 0 \\ 0 & 0 \\ 0 & 0 \end{bmatrix}, \quad a = \frac{1}{\sqrt{1+(r^{mo})^2}}. \tag{10}$$

This is consistent with the Jacobian for a smooth surface since Eq. (10) is the limit case of Eq. (9) for $r^{mo}_{min} \to 0$.

Vertex. The directions of both maximum and minimum curvature are degenerated. Since the two joints of the *SLIP* manipulator correspond to rotations about the contact point, the Jacobian becomes:

$$_{geo}J^{slip} = \begin{bmatrix} 0 & 1 \\ 1 & 0 \\ 0 & 0 \\ 0 & 0 \\ 0 & 0 \\ 0 & 0 \end{bmatrix}. \tag{11}$$

Once again, this is the limit of the previous Jacobians for $r^{mo}_{min}, r^{mo}_{max} \to 0$.

2. The *ROT* Jacobian is expressed with respect to $\{con\}$ and $\{con'\}$, and is very simple:

$$
{}_{con}\boldsymbol{J}^{rot} = {}_{con'}\boldsymbol{J}^{rot} = \begin{bmatrix} 0 \\ 0 \\ 1 \\ 0 \\ 0 \\ 0 \end{bmatrix}. \tag{12}
$$

3. The *SLID* Jacobian is expressed with respect to $\{geo'\}$: ${}_{geo'}\boldsymbol{J}^{slid}$ is completely similar to ${}_{geo}\boldsymbol{J}^{slip}$, i.e., replace all "*mo*"s by "*env*"s. For symmetry reasons, however, the order of the columns is interchanged.

The three uppermost rows of the Jacobians represent angular velocities (with physical units 1/*time*), the three rows at the bottom are translational velocities (with physical units *length/time*).
The coefficients a and b are chosen in such a way as to allow an easy transition to the limiting cases with zero or infinite curvature. These coefficients are not dimensionless: their units are 1/*time*. Hence, the 1s in the nominator and denominator also have the appropriate physical dimensions: the nominator has units of $1/time$, the 1 in the denominator has units of *length squared*. This should be kept in mind whenever a change of physical units is performed.

3.4.5. Transformation of Jacobians Finally, a Jacobian expressed with respect to a reference frame $\{a\}$ is transformed to another reference frame $\{b\}$ using a *screw transformation* matrix ${}^{a}_{b}\boldsymbol{S}$:

$$
{}_{b}\boldsymbol{J} = {}^{a}_{b}\boldsymbol{S}\, {}_{a}\boldsymbol{J}, \quad {}^{a}_{b}\boldsymbol{S} = \begin{bmatrix} {}^{a}_{b}\boldsymbol{R} & \boldsymbol{O}_{3\times 3} \\ [\boldsymbol{r}_{ba}\times]\, {}^{a}_{b}\boldsymbol{R} & {}^{a}_{b}\boldsymbol{R} \end{bmatrix}. \tag{13}
$$

$\boldsymbol{O}_{3\times 3}$ is the 3 by 3 zero matrix. ${}^{a}_{b}\boldsymbol{R}$ is the rotation matrix between both reference frames. $[\boldsymbol{r}_{ba}\times]$ is the matrix representing the vector product with $\boldsymbol{r}_{ba}$ linking the origins of the reference frames $\{b\}$ and $\{a\}$:

$$
[\boldsymbol{r}_{ba}\times] = \begin{bmatrix} 0 & -r_z & r_y \\ r_z & 0 & -r_x \\ -r_y & r_x & 0 \end{bmatrix}, \tag{14}
$$

with $r_i = (\boldsymbol{r}_{ba})_i, i = x, y, z$ the components of $\boldsymbol{r}_{ba}$ along the X, Y and Z unit vectors.

For example, when expressing all sub-Jacobians with respect to the $\{geo\}$ frame on the *MO*, the following transformations apply:

$$
{}_{geo}\boldsymbol{J}^{2nd} = \left[{}_{geo}\boldsymbol{J}^{slip} \;\vdots\; {}^{con}_{geo}\boldsymbol{S}(\mu_x, \mu_y)\, {}_{con}\boldsymbol{J}^{rot} \;\vdots\; {}^{con}_{geo}\boldsymbol{S}(\mu_x, \mu_y)\, {}^{con'}_{con}\boldsymbol{S}(\rho)\, {}^{geo'}_{con'}\boldsymbol{S}(\eta_x, \eta_y)\, {}_{geo'}\boldsymbol{J}^{slid} \right] \tag{15}
$$

Table 1 shows which of $\mu_x, \mu_y, \eta_x, \eta_y$ and ρ are zero in a specific case. For example, in case of a contact between two surfaces the geometric and contact frames coincide: $\{geo\} = \{con\}$ and $\{geo'\} = \{con'\}$. Hence Eq. (15) simplifies to:

$$_{geo}\boldsymbol{J}^{2nd} = _{con}\boldsymbol{J}^{2nd} = \left[_{geo}\boldsymbol{J}^{slip} : _{con}\boldsymbol{J}^{rot} : {}^{con'}_{con}\boldsymbol{S}(\rho)\ _{geo'}\boldsymbol{J}^{slid} \right]. \tag{16}$$

In case of a contact between a vertex and a surface, μ_x, μ_y and ρ may be chosen zero, by making the geometric and contact frames coincide. Eq. (15) then further reduces to:

$$_{geo}\boldsymbol{J}^{2nd} = _{con}\boldsymbol{J}^{2nd} = \left[_{geo}\boldsymbol{J}^{slip} : _{con}\boldsymbol{J}^{rot} : {}^{con'}_{con}\boldsymbol{S}(\rho = 0)\ _{geo'}\boldsymbol{J}^{slid} \right]. \tag{17}$$

In case of a contact between two curves Eq. (15) reduces to:

$$_{geo}\boldsymbol{J}^{2nd} = \left[_{geo}\boldsymbol{J}^{slip} : {}^{con}_{geo}\boldsymbol{S}(\mu_x)\ _{con}\boldsymbol{J}^{rot} : {}^{con}_{geo}\boldsymbol{S}(\mu_x)\ {}^{con'}_{con}\boldsymbol{S}(\rho)\ {}^{geo'}_{con'}\boldsymbol{S}(\eta_x)\ _{geo'}\boldsymbol{J}^{slid} \right] \tag{18}$$

MO / *ENV*	surface	curve	vertex
surface	$\mu_x = \mu_y = 0$ $\eta_x = \eta_y = 0$	$\mu_x = \mu_y = 0$ $\eta_y = 0$	$\mu_x = \mu_y = 0$ (Choose $\eta_x = \eta_y = \rho = 0$)
curve	$\mu_y = 0$ $\eta_x = \eta_y = 0$	$\mu_y = 0$ $\eta_y = 0$	not stable
vertex	$\eta_x = \eta_y = 0$ (Choose $\mu_x = \mu_y = \rho = 0$)	not stable	not stable

Table 1: *Frame transformation parameters.* The geometric and contact frames are linked by a set of transformation parameters. This set depends on the type of the contact. For the contact involving a vertex, the geometric and contact frames may be chosen coincident ($\rho = 0$, and $\mu_x = \mu_y = 0$ or $\eta_x = \eta_y = 0$): the contact frame of the vertex is chosen with its X and Y axes parallel to the corresponding axes of the contact frame of the other surface with which it is in contact.

With respect to the basis $\{a\}$, a twist $_a\mathrm{t}$ between the *MO* and the *ENV* is written as:

$$_a\mathrm{t} = _a\boldsymbol{J}^{2nd}\, \mathcal{T}, \tag{19}$$

where $\mathcal{T}$ is a column vector representing the (physically dimensionless!) magnitudes of the joint velocities (or infinitesimal displacements) in the virtual manipulator, due to the "end effector" twist t of the *MO*. The twist coordinate vector $\mathcal{T}$ remains unchanged under a change of reference frame as in Eq. (13), hence it carries no reference frame subscript.

3.5. polyhedral objects: edges and faces

The previous paragraphs describe a contact between two smooth curves and/or surfaces. However, contacts involving edges (i.e., straight lines) or faces are very common in industrial assembly. An edge has zero curvature along the curve; a face has zero curvature in all directions. Formally, the contact models for this type of motion constraint correspond to the limit cases of the smooth objects: the radii of curvature go to infinity, i.e. the revolute joint moves to infinity, so it becomes a prismatic joint. For example, Fig. 5 shows a *vertex-face* contact. The Jacobian expressed in the contact frame is derived from Eq. (17). Furthermore, ${}_{geo}\boldsymbol{J}^{slip}$ is given by Eq. (11); ${}_{con}\boldsymbol{J}^{rot}$ is given by Eq. (12), and ${}_{geo'}\boldsymbol{J}^{slid}$ is derived from Eq. (9) by interchanging the columns and taking the limits for $r^{env}_{max}, r^{env}_{min} \rightarrow \infty$. This results in:

$$ {}_{con}\boldsymbol{J}^{2nd}_{vertex-face} = \begin{bmatrix} 0 & 1 & 0 & 0 & 0 \\ 1 & 0 & 0 & 0 & 0 \\ 0 & 0 & 1 & 0 & 0 \\ 0 & 0 & 0 & 0 & 1 \\ 0 & 0 & 0 & 1 & 0 \\ 0 & 0 & 0 & 0 & 0 \end{bmatrix}. \tag{20} $$

Sliding is now the translational motion of the *MO* over the face of the *ENV*, *slipping* and *rotation* correspond to rotation around the contact point, about all three axes of the contact frame.

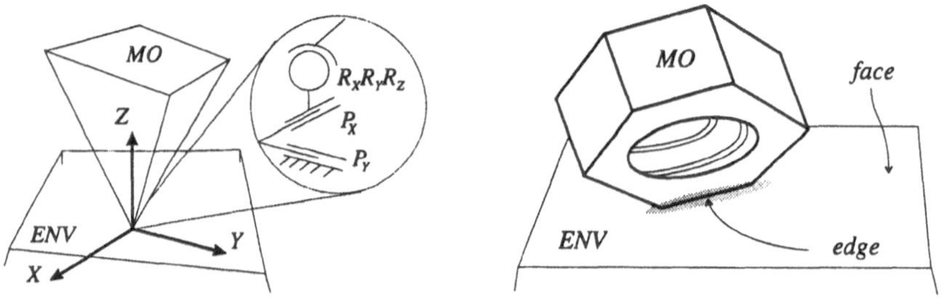

Figure 5: *Polyhedral contacts.* The *MO* consists of a *vertex* (left) or an *edge* (right); the *ENV* is a *face*. The drawing also gives an expanded view of the virtual manipulator for the *vertex-face* contact: the serial connection of a spherical joint (i.e. three intersecting revolutes, "*R*") and two prismatic joints ("*P*") with intersecting axes.

By taking similar limits, the other special cases of *edge-face* and *face-face* contacts result. For an *edge-face* contact, Fig. 5:

$$ {}_{con}\boldsymbol{J}^{2nd}_{edge-face} = \begin{bmatrix} 0 & 1 & 0 & 0 & 0 \\ 0 & 0 & 0 & 0 & 0 \\ 0 & 0 & 1 & 0 & 0 \\ 1 & 0 & 0 & 0 & 1 \\ 0 & 0 & 0 & 1 & 0 \\ 0 & 0 & 0 & 0 & 0 \end{bmatrix}. \tag{21} $$

The *MO* has four degrees of freedom in this motion constraint, hence the dimension of the twist system is four. This is reflected in the Jacobian since the first and fifth columns are dependent. There is also an ambiguity in selecting the origin of the contact frames: any point along the contact line will do. Similarly, for a *face-face* contact:

$$
{}_{con}\boldsymbol{J}^{2nd}_{face-face} = \begin{bmatrix} 0 & 0 & 0 & 0 & 0 \\ 0 & 0 & 0 & 0 & 0 \\ 0 & 0 & 1 & 0 & 0 \\ 1 & 0 & 0 & 0 & 1 \\ 0 & 1 & 0 & 1 & 0 \\ 0 & 0 & 0 & 0 & 0 \end{bmatrix}. \tag{22}
$$

The *MO* has three degrees of freedom in this motion constraint, hence the dimension of the twist system is three. This is reflected in the Jacobian since the first and fifth columns are dependent, as well as the second and fourth columns. Similarly, there is an ambiguity in selecting the origin of the contact frames: any point in the contact plane will do.

The procedure presented in Sect. 3.4. also allows the representation of mixed polyhedral/r polyhedral contacts. For example, the contact between an edge and a cylinder.

3.6. FIRST ORDER APPROXIMATION OF A POINT CONTACT

Even if the contact surfaces are not polyhedral, it is common practice to work with a *first order*, or *polyhedral*, approximation. For example, the contact in Fig. 3 could be modelled by the polyhedral model of Fig. 5 (left), if the curvature of the *MO* is significantly larger than that of the *ENV*.

First and second order models give the same instantaneous velocities for the *MO*, i.e. their Jacobian matrices span the same twist space $\boldsymbol{T}$. However, compared to the first order model, the second order model retains more information to interpret the executed motion of the *MO*, i.e., the subdivision into slipping, sliding and rotation. Moreover, it predicts the evolution of the contact frames during the motion more accurately since it models changes in the direction of the contact normal, which is impossible in a first order model. Hence, the usefulness of first order models relies more heavily on the ability of the compliant motion controller to identify on line the deviations between the model and the real world. These deviations are also larger than for second order models.

3.7. WRENCH SYSTEM

Besides a model of the allowed reciprocal motion, the robot controller has to know what reaction forces it can expect, i.e. it has to know the wrench system $\boldsymbol{W}$. For any motion constraint for which a kinematic model of the motion freedom $\boldsymbol{T}$ exists, a basis for $\boldsymbol{W}$, i.e. a wrench Jacobian $\boldsymbol{G}$, can be found numerically as follows: $\boldsymbol{G}$ is a basis of the vector space reciprocal to the twist system $\boldsymbol{T}$. This calculation is similar to the calculation of the orthogonal complement, but with the matrix $\tilde{\boldsymbol{\Delta}}$ of Eq. (5) as the “orthogonality” matrix instead of the unit matrix. Golub and Van Loan (1989) contains robust algorithms.

For a general point contact, with a twist Jacobian as in Eq. (8), this results in the following simple expression with respect to the contact frames:

$$G = \begin{bmatrix} 0 \\ 0 \\ 1 \\ 0 \\ 0 \\ 0 \end{bmatrix}. \tag{23}$$

For the *edge-face* and *face-face* contacts the wrench system has dimension two or three:

$$G_{edge-face} = \begin{bmatrix} 0 & 0 \\ 0 & 0 \\ 1 & 0 \\ 0 & 0 \\ 0 & 1 \\ 0 & 0 \end{bmatrix}, \quad G_{face-face} = \begin{bmatrix} 0 & 0 & 0 \\ 0 & 0 & 0 \\ 1 & 0 & 0 \\ 0 & 1 & 0 \\ 0 & 0 & 1 \\ 0 & 0 & 0 \end{bmatrix}. \tag{24}$$

With respect to the basis G, every (ideal) contact wrench w is expressed as

$$\mathrm{w} = G\,\Gamma, \tag{25}$$

where Γ is a column vector representing the (physically dimensionless) coordinates of the wrench w.

3.8. KINEMATIC MODEL FOR NEAR-CONTACT

Every compliant motion task starts with an *approach* move to bring the *MO* in contact with the *ENV*. The last part of this approach (i.e. the so-called *near-contact* phase) must take place under force control, and hence is also a compliant motion. The virtual manipulator linking the *MO* to the *ENV* in the general point contact model of Fig. 3 is extended with the *APP* virtual manipulator, consisting of a prismatic joint along the common normal. The contact frames on *MO* and *ENV* keep the same definition as in the case of real contact, but their origins are now chosen at the intersections of the common normal with the surfaces of *MO* and *ENV*. Accordingly, the Jacobian J^{2nd} is extended with one column:

$$J^{2nd,nc} = \left[J^{2nd}\ J^{app}\right]. \tag{26}$$

With respect to the $\{con\}$ frame, J^{app} is as follows:

$${}_{con}J^{app} = \begin{bmatrix} 0 \\ 0 \\ 0 \\ 0 \\ 0 \\ 1 \end{bmatrix} \tag{27}$$

Any twist t of the *MO* with respect to the *ENV* is now decomposed into *slipping*, *rotation*, *sliding*, and *approach* along the common normal:

$$\mathrm{t} = \mathrm{t}^{slip} + \mathrm{t}^{rot} + \mathrm{t}^{slid} + \mathrm{t}^{app}. \tag{28}$$

3.9. EVOLUTION OF THE CONTACT FRAME LOCATION: KINEMATIC DESCRIPTION

A good model not only describes the instantaneous reciprocal motion freedom of the *MO*, but also indicates how this motion freedom evolves due to the motion itself. Slipping and sliding move the point of contact over both the *MO* and the *ENV*, so the location of the contact normal and the tangent vectors change. Hence, the locations of the $\{geo\}, \{con\}, \{con'\}$ and $\{geo'\}$ reference frames change with respect to: 1) the base frames $\{base\}$ and $\{base'\}$, and 2) some reference frames $\{ref\}$ and $\{ref'\}$, which are chosen (arbitrarily, but fixed once and for all) on the *MO* and the *ENV*, respectively. This Section shows how to derive, from the motion of the virtual manipulators, the evolution of the contact and geometric frames with respect to the fixed reference frames $\{ref\}$ and $\{ref'\}$. To this end, the concept of the *evolution transformations EVOL* and *EVOL'* is introduced, see Fig. 6. These transformations complement the previously presented kinematic model of the instantaneous motion freedom as follows:

1. *EVOL* contains the current transformation between $\{ref\}$ and the $\{base\}$ of the virtual *SLIP* manipulator.
2. *EVOL'* contains the current transformation between $\{ref\}$ and the $\{base'\}$ of the virtual *SLID* manipulator.

Updating *EVOL* and *EVOL'* brings the virtual manipulators *SLIP* and *SLID*, which are used for the instantaneous motion specification, back to their "zero positions": whenever the specified desired "joint velocities" of *SLIP* and *SLID* have moved the joints of these virtual manipulators over a small angle (and hence the *MO* over some infinitesimal distance with respect to the *ENV*), this motion is incorporated into the *EVOL* and *EVOL'* transformations. This means that the base frames of *SLIP* and *SLID* are moved, in the same way as the contact frames.

The following list of the topology of the kinematic chains linking $\{ref\}$ to the contact frames, (see also Fig. 4), is used to determine which components of the executed motion are needed to update the evolution transformations. I is the identity transformation. (At the side of the *ENV*, the list is completed symmetrically.)

Smooth surface : $\{ref\} \overset{EVOL}{\longrightarrow} \{base\} \overbrace{\overset{R}{\to}\overset{R}{\to}}^{SLIP} \{geo\} \overset{I}{\to} \{con\} \overbrace{\overset{R}{\to}}^{ROT} \{con'\} \ldots (29)$

Smooth curve : $\overbrace{\{ref\} \overset{EVOL}{\longrightarrow} \{base\} \overset{R}{\to} \{geo\} \overset{R}{\to}}^{SLIP}\{con\} \overbrace{\overset{R}{\to}}^{ROT} \{con'\} \ldots (30)$

Straight line : $\{ref\} \overset{EVOL}{\longrightarrow} \overbrace{\{base\} \overset{P}{\to} \{geo\} \overset{R}{\to}}^{SLIP}\{con\} \overbrace{\overset{R}{\to}}^{ROT} \{con'\} \ldots (31)$

Vertex[1] : $\{ref\} \overset{EVOL}{\longrightarrow} \{base\} \overbrace{\overset{R}{\to}\overset{R}{\to}}^{SLIP} \{geo\} \overset{I}{\to} \{con\} \overbrace{\overset{R}{\to}}^{ROT} \{con'\} \ldots (32)$

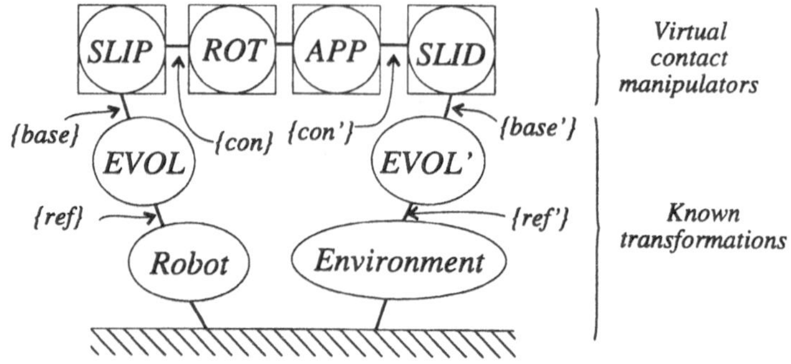

Figure 6: *Evolution transformations.* The boxed circles represent virtual manipulators, used to describe instantaneous motion freedom; the ellipses represent known transformations between reference frames. *EVOL* and *EVOL'* describe the actual locations of the base frames of *SLIP* and *SLID*, with respect to the known reference frames $\{ref\}$ and $\{ref'\}$.

3.9.1. Evolution of the Contact Frames In order to update the location of the contact frames, the base frames are chosen coincident with the contact frames in the "zero position" of the virtual manipulators. (Remember that the choice of $\{base\}$ and $\{base'\}$ is arbitrarily!) Eqs. (29)-(32) then show that, in order for $\{base\}$ to track the motion of $\{con\}$, the evolution transformation *EVOL* should be updated by the motion of the *SLIP* manipulator. Similary in order for $\{base'\}$ to track $\{con'\}$, the evolution transformation *EVOL'* should be updated by the motion of the *SLID* manipulator.

3.9.2. Evolution of the Geometric Frames Updating the location of the geometric frames proceeds along the same lines. The base frames are now chosen coincident with the geometric frames in the "zero position" of the virtual manipulators. Again, Eqs. (29)-(32) determine how the evolution transformations should be updated in order for the base frames to track the motion of the geometric frames : for smooth surfaces or vertices, there is no difference with the evolution of the contact frames, while for curves and edges only the first joint in the virtual manipulators (*SLIP* and *SLID*) is needed.

3.9.3. Validity Range of Kinematic Model Clearly, if the relative location of the contact frames $\{con\}$ and $\{con'\}$ with respect to the reference frames on the *MO* and the *ENV* changes over an extended range, the second order approximation of the surfaces may not remain valid. This is because, for arbitrarily curved surfaces, the principal directions and the corresponding curvatures vary with the location of the contact point on the surface. Hence, besides continuously updating the evolution transformations *EVOL* and *EVOL'*, also the kinematic construction of the virtual manipulator, (i.e., its link lengths and the orientation of its joints in space), has to be adapted in a second, less frequently applied update step.

3.10. EVOLUTION OF THE CONTACT FRAME LOCATION: MATHEMATICAL REPRESENTATION

3.10.1. Evolution of the Contact Frames Numerically, the evolution of the contact frames $\{con\}$ and $\{con'\}$ is expressed by the twists t_e and t_e':

$$\mathrm{t}_e = \boldsymbol{E}\,\epsilon\,,\ \mathrm{t}_e' = \boldsymbol{E}'\,\epsilon', \tag{33}$$

where $\boldsymbol{E}$ and $\boldsymbol{E}'$ are the *evolution Jacobians*. From the previous sections, it is clear that these evolution Jacobians correspond to the *SLIP* and *SLID* Jacobians, respectively. So, these are submatrices of $\boldsymbol{J}^{2nd,nc}$, with proper sign. ϵ and ϵ' are the corresponding subvectors of $\mathcal{T}$. $\mathcal{T}$ is determined by Eq. (19), with $\boldsymbol{J}^{2nd}$ replaced by $\boldsymbol{J}^{2nd,nc}$ in case of near-contact.

Alternatively, one can be interested in knowing the evolution of $\{con\}$ with respect to $\{ref'\}$, or $\{con'\}$ with respect to $\{ref\}$. Again, Eqs. (29)-(32) determine which columns from $\boldsymbol{J}^{2nd,nc}$ one needs. Table 2 lists the evolution Jacobians $\boldsymbol{E}$ and $\boldsymbol{E}'$ for these cases.

t_e and t_e' are used to update the transformations *EVOL* and *EVOL'*. Mathematically, at a given time instant *t*, *EVOL* and *EVOL'* are characterized by the screw transformation matrices ${}^{con}_{ref}\boldsymbol{S}(t)$ and ${}^{con'}_{ref'}\boldsymbol{S}(t)$, between, on the one hand, the contact frames $\{con\}$ and $\{con'\}$ and, on the other hand, their respective reference frames $\{ref\}$ and $\{ref'\}$. After some infinitesimal time interval Δt, the contact frames have moved with respect to their respective reference frames over infinitesimal displacements $\mathrm{t}_{e,\Delta}$ and $\mathrm{t}_{e,\Delta}'$:

$$\mathrm{t}_{e,\Delta} = \mathrm{t}_e\,\Delta t,\ \ \mathrm{t}_{e,\Delta}' = \mathrm{t}_e'\,\Delta t. \tag{34}$$

Hence, the evolution transformations have to be updated as:

$${}^{con}_{ref}\boldsymbol{S}(t+\Delta t) = {}^{con}_{ref}\boldsymbol{S}(t)\ \boldsymbol{S}(\mathrm{t}_{e,\Delta}),\ \ {}^{con'}_{ref'}\boldsymbol{S}(t+\Delta t) = {}^{con'}_{ref'}\boldsymbol{S}(t)\ \boldsymbol{S}(\mathrm{t}_{e,\Delta}'), \tag{35}$$

where $\boldsymbol{S}(\mathrm{t}_{e,\Delta})$ and $\boldsymbol{S}(\mathrm{t}_{e,\Delta}')$ are screw transformation matrices corresponding to the infinitesimal displacements $\mathrm{t}_{e,\Delta}$ and $\mathrm{t}_{e,\Delta}'$ (expressed in $\{con\}$ and $\{con'\}$, respectively): if t_Δ has Cartesian components $(\delta_x\ \delta_y\ \delta_z\ d_x\ d_y\ d_z)^T$, then $\boldsymbol{S}(\mathrm{t}_\Delta)$ is easily derived from Eqs. (13) and (14) as:

$$\boldsymbol{S}(\mathrm{t}_\Delta) = \begin{bmatrix} 1 & -\delta_z & \delta_y & 0 & 0 & 0 \\ \delta_z & 1 & -\delta_x & 0 & 0 & 0 \\ -\delta_y & \delta_x & 1 & 0 & 0 & 0 \\ 0 & -d_z & d_y & 1 & -\delta_z & \delta_y \\ d_z & 0 & -d_x & \delta_z & 1 & -\delta_x \\ -d_y & d_x & 0 & -\delta_y & \delta_x & 1 \end{bmatrix}. \tag{36}$$

In a second update step, the evolution Jacobians $\boldsymbol{E}$ and $\boldsymbol{E}'$ in Eq. (33) have to be adapted continuously, because, as explained in the previous subsection, the curvatures in the contact point may change, and hence also the corresponding columns of the Jacobians $\boldsymbol{J}, \boldsymbol{E}$ and $\boldsymbol{E}'$.

	$\boldsymbol{E}$ (evolution of $\{con\}$)	$\boldsymbol{E}'$ (evolution of $\{con'\}$)
w.r.t. $\{ref\}$	$-\boldsymbol{J}^{slip}$	$-\left[\boldsymbol{J}^{slip}\ \boldsymbol{J}^{rot}\ \boldsymbol{J}^{app}\right]$
w.r.t. $\{ref'\}$	$\left[\boldsymbol{J}^{slid}\ \boldsymbol{J}^{rot}\ \boldsymbol{J}^{app}\right]$	$\boldsymbol{J}^{slid}$

Table 2: *Evolution Jacobians*. The modelled evolution of the contact frames.

3.10.2. Evolution of the Geometric Frames Again, for smooth surfaces or vertices there is no difference with the evolution of the contact frames, while for curves and edges only one column, corresponding the first joint in the *SLIP* or *SLID* manipulator, is needed in the evolution Jacobians of Eq. (33).

3.11. KINEMATIC MODEL FOR MULTIPLE CONTACTS

In case several contacts exist between the *MO* and the *ENV*, as in Fig. 2, a virtual manipulator is used to describe every individual contact. This results in a complex kinematic chain connecting the *MO* and the *ENV*. The corresponding twist and wrench systems for the *MO* are calculated with the composition rules given in Section 2., i.e. the resulting twist system is the *intersection* of the individual twist systems, and the resulting wrench system is the *sum* of the individual wrench systems.

Figure 2 shows an example in which the *MO* is constrained by two elementary constraints, a *vertex-face* and a *face-face* contact. The twist Jacobian for the *vertex-face* contact, expressed in its own contact frame, is given by Eq. (20), and, similarly, the twist Jacobian for the plane contact, expressed in its own contact frame, is given by three independent columns of Eq. (22). The total twist Jacobian, expressed for example in $\{con1\}$, is given by:

$$_{con1}\boldsymbol{J} = \begin{bmatrix} 1 & 0 \\ 0 & 0 \\ 0 & 0 \\ 0 & 0 \\ 0 & 1 \\ 0 & 0 \end{bmatrix}. \qquad (37)$$

3.12. KINEMATIC MODEL FOR COOPERATING ROBOTS

The extension of the kinematic model to multiple cooperating robots is straightforward. For example, for two robots with contact between their manipulated objects, the role of the *ENV* is played by the manipulated object of the other robot, and the twist t is now really a *relative* twist between the two manipulated objects, i.e., different from the *absolute* twists

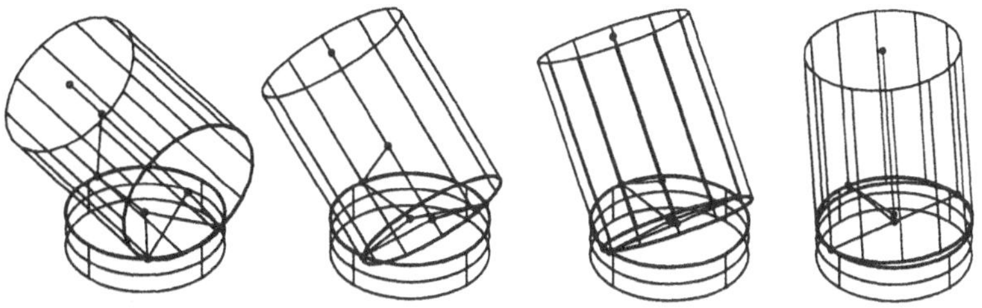

Figure 7: *Peg-in-hole*: insertion snapshots, from badly to almost perfectly aligned configurations.

of both manipulated objects.

4. Example: Alignment of Cylindrical Peg and Hole

This Section applies the presented kinematic approach to the peg-in-hole example. It describes the desired alignment motion between the axes of peg and hole starting from the initial position depicted in Fig. 1. This position is mostly not considered in literature, since the alignment between the peg and the hole is too bad to start the final insertion phase. Nevertheless, if the position of the hole is not exactly known, and hence the manipulator has to "look for it" by (stochastically or deterministically) moving the peg around in the neighbourhood of the current contact point, this is probably the most interesting relative position between the peg and the surface containing the hole: indeed, the passive compliance in the system helps the peg to "fall" into the hole once its bottom has crossed the rim of the hole. In other words, the relative position between peg and hole, shown in Fig. 1, corresponds to a very *stable* position, that can be used to start a further alignment motion before proceeding to the final insertion.

The next subsections show that an explicit kinematic model of the contact situation between peg and hole is very appropriate to derive the nominal alignment motion for the peg, given the time varying nature of the peg's motion freedom. Hence, it offers a user friendly interface for the motion specification.

4.1. MOTION CONSTRAINTS

The configuration of Fig. 1 contains three *point contacts* : one on the peg's surface, and two on the peg's bottom rim. Hereafter, these contacts are named *Surf, Rim1* and *Rim2*, respectively. Each of these point contacts removes one degree of freedom. Borrowing the terminology from Sect. 3.3., which was actually defined for a single elementary contact, the three remaining degrees of freedom could be termed as:

Slip. A pure rotation of the peg about its own axis does not move the position of the contact points in space, but it changes the contact areas on the peg's surface and bottom rim.

Slide. A pure rotation of the peg about the axis of the hole moves the contact points in space, but leaves them invariant with respect to the peg.

Insert. The proper combination of 1) a rotation about an axis through *Surf*, tangent to the hole's rim, and 2) a translation along the line through this contact point, parallel to the axis of the hole, aligns the axis of the peg better with the axis of the hole. At the same time, the *Surf* contact moves closer towards the peg's bottom, while the *RIM* contacts move towards the *Surf* contact point.

This description of the peg's motion freedom is at the same time intuitive, as well as compatible with the kinematic model of the constraint, as discussed in the next subsection.

4.2. NOMINAL KINEMATIC MODEL

Each of the point contacts *Surf*, *Rim1* and *Rim2* has five degrees of freedom. The discussion of Section 3. suggests the following set of virtual manipulators, consisting of only prismatic and revolute joints, and linking the peg to the hole with the same motion freedom as the contacts:

SURF. This virtual manipulator describes the reciprocal motion freedom of the peg w.r.t. the hole as allowed by the *Surf* contact, see Fig. 8. It consists of the following sub-manipulators:

- *SURF-SLIP*: one cylindrical joint, i.e., a revolute and a prismatic joint with their axes coinciding with the peg's axis.
- *SURF-ROT*: a revolute joint in the *Surf* contact point, with its axis along the contact normal.
- *SURF-SLID*: two revolute joints, one at the *Surf* contact point (with its axis along the tangent to the hole's rim), and one along the hole's axis.

RIM1. This virtual manipulator describes the reciprocal motion freedom given to the peg by the first *RIM* contact point, see Fig. 9. It consists of the following sub-manipulators:

- *RIM1-SLIP*: one revolute joint on the peg's axis, and a second revolute joint, tangent to the peg's bottom rim, and with its centre in the *Rim1* contact point.
- *RIM1-ROT*: a revolute joint in the *Rim1* contact point and with its axis along the contact normal.
- *RIM1-SLID*: two revolute joints, one at the *Rim1* contact point (with its axis along the tangent to the hole's rim), and one along the hole's axis.

RIM2. Completely similar to *RIM1*, but now at the *Rim2* contact point.

The three virtual manipulators form three parallel paths in the graph describing the topology of the motion constraint, see Fig. 10. Hence, the instantaneous motion freedom of the peg w.r.t. the hole is given by the *intersection* of the twist spaces corresponding to each of the virtual manipulators. A topological mobility analysis according to the Chebyshev–Grübler–Kutzbach formula, Angeles (1988), results in:

1. Number of degrees of freedom constrained in each joint: $d = 5$.

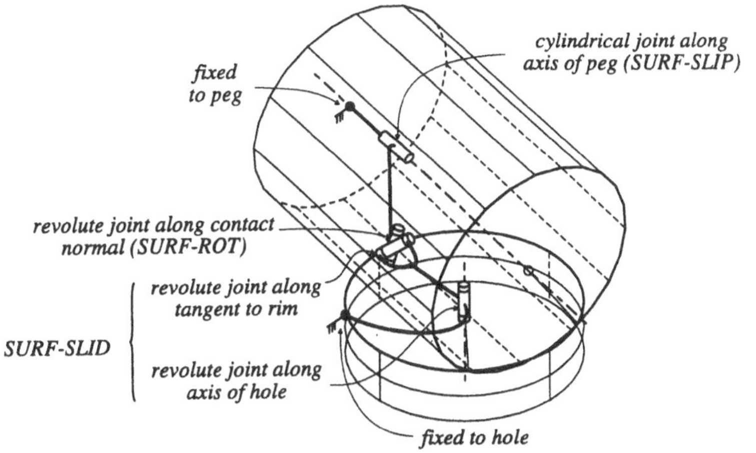

Figure 8: *Peg-in-hole: virtual* SURF *manipulator.* The depicted kinematic chain describes the reciprocal motion freedom of the peg w.r.t. the hole as allowed by the *Surf* point contact between the surface of the peg and the rim of the hole. The cylindrical joint forms the *SLIP* part of the *SURF* manipulator. At the contact point, there are two revolute joints: one has its axis along the contact normal (*SURF-ROT*), the other belongs to *SURF-SLID* (together with the revolute joint along the centerline of the hole) and has its axis along the rim's tangent.

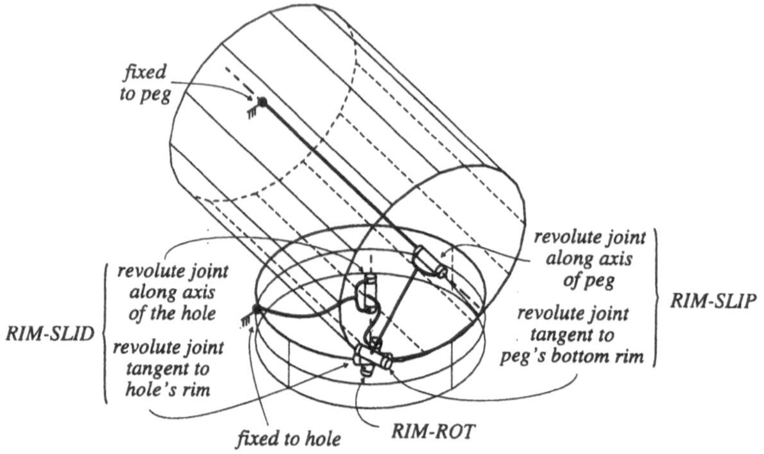

Figure 9: *Peg-in-hole: virtual* RIM *manipulator.* The depicted kinematic chain describes the reciprocal motion freedom of the peg w.r.t. the hole as allowed by one of the *RIM* point contacts between the bottom of the peg and the rim of the hole. The revolute joint along the peg's axis, together with the revolute joint at the contact point, with its axis along the tangent to the peg's bottom rim, form the *RIM-SLIP* manipulator. One of the two other revolute joints at the contact point (i.e., the one with its axis along the contact normal) is the *ROT* part of the *RIM* manipulator. The last two revolute joints form *RIM-SLID*.

2. Number of one-degree-of-freedom joints: $j = 15$.

3. Number of bodies (three chains of four bodies, plus *MO* and *ENV*): $n = 3 \times 4 + 2 = 14$.

4. Number of degrees of freedom $= 6 \times (n - 1) - j \times d = 3$.

This analysis breaks down at two *singular configurations*: the first one coincides with the desired end position (i.e., axes of peg and hole are parallel); the second one is the limit case of all possible initial configurations (i.e., axes of peg and hole are perpendicular). The three previously introduced degrees of freedom (**Slip**, **Slide** and **Insert**) allowed by the three point contacts correspond to three independent *drivers* for the complex kinematic chain formed by *SURF*, *RIM1* and *RIM2*: this means that if a motion is specified as a combination of **Slip**, **Slide** and **Insert**, the motion of all joints in the chain is uniquely defined.

In principle, the topology and the link lengths and link angles of the virtual manipulators remain unchanged during the complete alignment motion, since the geometry of both objects is perfectly described by a second order model. The definition of *RIM1* and *RIM2* uses three revolute joints at the contact points which have their axes aligned with some geometric features (tangents, normals) of the peg and the hole. However, these joint angles are not important to describe the motion freedom of the peg in an unambiguous and user friendly way. Hence, they can be replaced by one spherical joint for the purpose of motion specification. However, for modelling the evolution of the contact locations, the more detailed model with three revolute joints has to be retained (see next subsection).

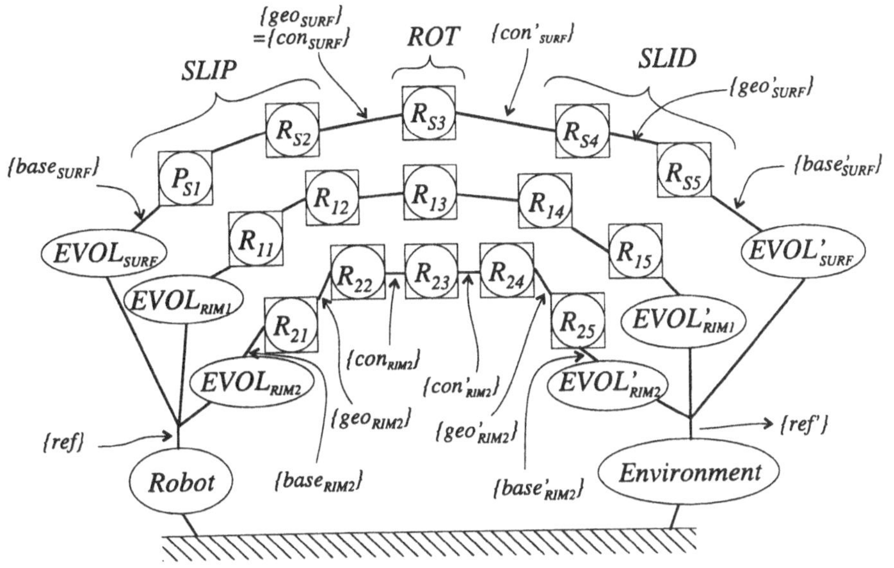

Figure 10: *Peg-in-hole*: topology of the virtual manipulators *SURF, RIM1* and *RIM2*, see also Figs. 6, 8 and 9. The prismatic joints (*P*) and the revolute joints (*R*) are numbered as follows : the first index represents the point contact : S for *SURF*, 1 for *RIM1* and 2 for *RIM 2*; the second index corresponds to the joint number within each virtual contact manipulator *1-5*.

4.3. MATHEMATICAL DESCRIPTION

A mathematical description of the instantaneous motion constraints acting on the *MO*, is needed to feed the robot controller with numerical data to execute the manipulation task. First, a numerical model of the instantaneous motion freedom has to be built (i.e., the twist and wrench Jacobians $\boldsymbol{J}$ and $\boldsymbol{G}$). Second, one has to choose from the available motion freedom the desired motion to be executed by the robot. And third, during the motion the robot controller checks whether the motion constraint models are still correct, and if not, it updates them (this is called *identification* of the *model uncertainties*).

This paper is only concerned about a world with limited uncertainties: this means that the inaccuracies of the models are too large to allow a direct alignment of the peg and the hole under pure position control, yet they are small enough so that the available passive compliance in the system is able to take up the possible small mismatches between model and reality. This is not an unrealistic assumption, since the insertion tolerance between the diameters of peg and hole is usually some order of magnitude smaller than the positioning accuracy of the robot. (Moreover, the authors have developed tools to identify the uncertainties during the motion, based on the nominal models (the same as the ones used in this text) on the one hand, and the measured motions and forces on the other hand, Bruyninckx, De Schutter and Dutré (1993).)

So, the following initial situation is assumed: the position of the three contact points *Surf, Rim1* and *Rim2* is approximately known. The derivation of the mathematical model proceeds as follows. From the known geometry of peg and the hole (radius = 10 mm; length = 25 mm) and their nominal relative position (angle between both axes, in this example $\theta = 0.5677$ rad), the position and orientation of the geometric frames $\{geo_{RIM1}\}$ and $\{geo_{RIM2}\}$ and the contact frame $\{con_{SURF}\}$ (coinciding with $\{geo_{SURF}\}$ are deduced with respect to a reference frame attached to the robot's end effector. This frame is located at the top of the pen with the axes parallel to the axes of the geometric frame $\{geo_{SURF}\}$ [2] :

$$
{}^{con_{SURF}}_{ref}\boldsymbol{S} = \begin{bmatrix} 1 & 0 & 0 & 0 & 0 & 0 \\ 0 & 1 & 0 & 0 & 0 & 0 \\ 0 & 0 & 1 & 0 & 0 & 0 \\ 0 & 10 & -19.1657 & 1 & 0 & 0 \\ -10 & 0 & 0 & 0 & 1 & 0 \\ 19.1657 & 0 & 0 & 0 & 0 & 1 \end{bmatrix}.
$$

$$
{}^{geo_{RIM1}}_{ref}\boldsymbol{S} = \begin{bmatrix} 0.0851 & 0 & 0.9964 & 0 & 0 & 0 \\ 0 & 0 & 0 & 0 & 0 & 0 \\ -0.9964 & 0 & 0.0851 & 0 & 0 & 0 \\ 24.9093 & 0.8509 & -2.1273 & 0.0851 & 0 & 0.9964 \\ -10 & 0 & 0 & 0 & 1 & 0 \\ 2.1273 & -9.9637 & 24.9093 & -0.9964 & 0 & 0.0851 \end{bmatrix}.
$$

[2]This position analysis is performed by modelling the virtual manipulators using the Applied Motion^{TM} software. The values of η_x ,μ_x and ρ are derived from this analysis.

$$
{}^{geo_{RIM2}}_{ref}\boldsymbol{S} = \begin{bmatrix} 0.0851 & 0 & -0.9964 & 0 & 0 & 0 \\ 0 & 1 & 0 & 0 & 0 & 0 \\ 0.9964 & 0 & 0.0851 & 0 & 0 & 0 \\ -24.9093 & 0.8509 & -2.1273 & 0.0851 & 0 & -0.9964 \\ -10 & 0 & 0 & 0 & 1 & 0 \\ 2.1273 & 9.9637 & -24.9093 & 0.9964 & 0 & 0.0851 \end{bmatrix}.
$$

Following the procedure described in section 3.4 the twist Jacobians of each contact are expressed with respect to each contact- or geometric frame. From Eq. (16) :

$$
{}_{con_{SURF}}\boldsymbol{J}^{2nd} = [\boldsymbol{J}_{S1} \ldots \boldsymbol{J}_{S5}]
$$

$$
= \left[{}_{geo_{SURF}}\boldsymbol{J}^{slip} : {}_{con_{SURF}}\boldsymbol{J}^{rot} : {}^{con'_{SURF}}_{con_{SURF}}\boldsymbol{S}(\rho = 0)\ {}^{geo'_{SURF}}_{con'_{SURF}}\boldsymbol{S}(\eta_x = 0.5677 - \pi)\ {}_{geo'_{SURF}}\boldsymbol{J}^{slid} \right]
$$

$$
= \begin{bmatrix} 0 & 0 & 0 & 0 & -1 \\ 0 & 1 & 0 & -0.8432 & 0 \\ 0 & 0 & 1 & -0.5377 & 0 \\ 0 & -10 & 0 & 10 & 0 \\ 1 & 0 & 0 & 0 & 0 \\ 0 & 0 & 0 & 0 & 0 \end{bmatrix}.
$$

From Eq. (18) :

$$
{}_{geo_{RIM1}}\boldsymbol{J}^{2nd} = [\boldsymbol{J}_{11} \ldots \boldsymbol{J}_{15}]
$$

$$
= \begin{matrix} \left[{}_{geo_{RIM1}}\boldsymbol{J}^{slip} : {}^{con_{RIM1}}_{geo_{RIM1}}\boldsymbol{S}(\mu_x = 0.2838)\ {}_{con_{RIM1}}\boldsymbol{J}^{rot} : \right. \\ \left. {}^{con_{RIM1}}_{geo_{RIM1}}\boldsymbol{S}(\mu_x = 0.2838)\ {}^{con'_{RIM1}}_{con_{RIM1}}\boldsymbol{S}(\rho = 0.5920)\ {}^{geo'_{RIM1}}_{con'_{RIM1}}\boldsymbol{S}(\eta_x = 0.2838)\ {}_{geo'_{RIM1}}\boldsymbol{J}^{slid} \right] \end{matrix}
$$

$$
= \begin{bmatrix} 0 & 1 & 0 & -0.8298 & 0.5357 \\ 1 & 0 & -0.2800 & -0.5357 & -0.8432 \\ 0 & 0 & -0.9600 & 0.1563 & -0.0458 \\ -10 & 0 & 0 & 0 & 8.2980 \\ 0 & 0 & 0 & 0 & 5.3573 \\ 0 & 0 & 0 & 0 & -1.5628 \end{bmatrix}.
$$

and :

$$
{}_{geo_{RIM2}}\boldsymbol{J}^{2nd} = [\boldsymbol{J}_{21} \ldots \boldsymbol{J}_{25}]
$$

$$
= \begin{matrix} \left[{}_{geo_{RIM2}}\boldsymbol{J}^{slip} : {}^{con_{RIM2}}_{geo_{RIM2}}\boldsymbol{S}(\mu_x = 0.2838)\ {}_{con_{RIM2}}\boldsymbol{J}^{rot} : \right. \\ \left. {}^{con_{RIM2}}_{geo_{RIM2}}\boldsymbol{S}(\mu_x = 0.2838)\ {}^{con'_{RIM2}}_{con_{RIM2}}\boldsymbol{S}(\rho = -0.5920)\ {}^{geo'_{RIM2}}_{con'_{RIM2}}\boldsymbol{S}(\eta_x = 0.2838)\ {}_{geo'_{RIM2}}\boldsymbol{J}^{slid} \right] \end{matrix}
$$

$$
= \begin{bmatrix} 0 & 1 & 0 & -0.8298 & -0.5357 \\ 1 & 0 & -0.2800 & 0.5357 & -0.8432 \\ 0 & 0 & -0.9600 & -0.1563 & -0.0458 \\ -10 & 0 & 0 & 0 & 8.2980 \\ 0 & 0 & 0 & 0 & -5.3573 \\ 0 & 0 & 0 & 0 & 1.5628 \end{bmatrix}.
$$

The columns of the Jacobians are numbered according to the corresponding joints in the topological diagram of Fig. 10. After the transformation to the end effector reference frame, the intersection of these Jacobians is calculated numerically (Golub and Van Loan, 1989). However, since at any time during the motion, the **Slip** and **Slide** components of this intersection are directly deduced from the nominal geometric model, only the **Insert** component has to be calculated numerically. So, one ends up with a 6×3 matrix, in which each column describes one of the three degrees of freedom:

$$J = \begin{bmatrix} J^{\mathbf{Slip}} & J^{\mathbf{Slid}} & J^{\mathbf{Insert}} \end{bmatrix} = \begin{bmatrix} 0 & 0 & 0.0521 \\ 1 & 0.8432 & 0 \\ 0 & 0.5377 & 0 \\ 0 & -11.8734 & 0 \\ 0 & 0 & 0.0443 \\ 0 & 0 & 0.9977 \end{bmatrix}. \tag{38}$$

The specification of the desired motion is then very simple: multiply each column with a scalar ($\tau^{Slip}, \tau^{Slid}, \tau^{Insert}$) indicating the desired magnitude of the corresponding motion component, and add these three twists together:

$$\mathbf{t}^{des} = J\,\mathcal{T}^{des} = \begin{bmatrix} J^{\mathbf{Slip}} & J^{\mathbf{Slid}} & J^{\mathbf{Insert}} \end{bmatrix} \begin{bmatrix} \tau^{Slip} \\ \tau^{Slid} \\ \tau^{Insert} \end{bmatrix}. \tag{39}$$

The *EVOL* transformations describing the current relative positions of peg and hole are easily updated on the basis of the *measured* joint motions in the three virtual manipulators *SURF*, *RIM1* and *RIM2*, see Eqs.(34) and (35). This is done using the evolution Jacobians :

$$E_{con_{SURF}} = [\ J_{S1} \quad J_{S2}\ ],\ E_{geo_{RIM1}} = [\ J_{11}\ ],\ E_{geo_{RIM2}} = [\ J_{21}\ ].$$

Notice that in this case the update procedure remains valid even for finite relative displacements between peg and hole (which maintain the three point contacts), because the virtual contact manipulators remain identical. Or: the surfaces of both objects are exactly described by second order models.

5. Conclusion

This paper shows how to model contacts between rigid objects using virtual manipulators, or kinematic chains, which have the same relative degrees of freedom as the contacting objects in the respective contact points. The kinematic composition of these virtual manipulators is derived from the first and second order geometry of the contacting surfaces in the neighbourhood of the respective contacts. Kinematic analogues based on the first order geometry (i.e. position of the contact point and orientation of the tangent plane) suffice to describe the instantaneous relative degrees of freedom between the contacting objects, but fail to model the relative motion accurately when the contact point moves over the contacting surfaces. On the other hand, kinematic analogues based on the second order geometry (i.e. first order geometry plus curvature information) remain accurate for a larger relative displacement between the objects.

It is shown how these kinematic analogues allow the use of well established tools in kinematics to specify the desired motion, and to analyse the resulting evolution of the contact locations. This approach is illustrated by means of the well known peg-in-hole assembly problem in case of a large initial misalignment.

In force controlled manipulation tasks, also called compliant motion, the compatibility of the specified motion with the constraints imposed by the environment onto the manipulated object, determines the quality of the task execution, both in case of passive and active force control. So, this paper rovides important new tools to improve the specification, and hence the execution, of force controlled robotic manipulation tasks.

ACKNOWLEDGEMENTS

The kinematic modelling, as well as most kinematic computations on the structures used in the peg-in-hole example of Section 4., have been generated with the *Applied Motion*^{TM} software from Rasna Corporation.

This work was sponsored by the *FIRST* and *SECOND* projects of the European Community (ESPRIT Basic Research Actions 3274 and 6769), and the Belgian Programme on Interuniversity Attraction poles initiated by the Belgian State – Prime Minister's Office – Science Policy Programming (IUAP-50). The scientific responsibility is assumed by its authors.

REFERENCES

1. Angeles, J. (1988) *Rational Kinematics*, Springer–Verlag.

2. Bruyninckx, H., De Schutter, J. and Dutré, S. (1993) 'The "reciprocity" and "consistency" based approaches to uncertainty identification for compliant motions', *IEEE Int. Conf. Rob. and Automation*, Atlanta.

3. Buckley, S. J. (1989) 'Planning Compliant Motion Strategies', *International Journal of Robotics Research*, Vol. 8, No. 5, pp. 28–44.

4. Cai, C. S. and Roth, B. (1986) 'On the planar motion of rigid bodies with point contact', *Mechanism and Machine Theory*, Vol. 21, No. 6, pp. 453–466.

5. Cai, C. S. and Roth, B. (1988) 'On the spatial motion of a rigid body with line contact', *IEEE Int. Conf. Rob. and Automation*, pp.1036–1041.

6. Desai, R. S. and Volz, R. A. (1989) 'Identification and verification of termination conditions in fine motion in presence of sensor errors and geometric uncertainties', *IEEE Int. Conf. Rob. and Automation*, pp.800–807.

7. De Schutter, J. and Van Brussel, H. (1988) 'Compliant Robot Motion', *Int. J. Rob. Res.*,Vol. 7, No. 4, pp. 3–33.

8. Golub, G.H. and Van Loan, C.F. (1989) *Matrix Computations*, The Johns Hopkins University Press.

9. Laugier, Ch. (1989) 'Planning fine motion strategies by reasoning in the contact space', *IEEE Int. Conf. Rob. Automation*, pp.653-661.

10. Lipkin, H. and Duffy, J. (1988) 'Hybrid Twist and Wrench Control for a Robotic Manipulator', *Trans. ASME J. Mech., Trans. and Aut. Design*, Vol. 110, pp.138–144.

11. Mason, M. T. (1981) 'Compliance and Force Control for Computer Controlled Manipulators', *IEEE Trans. on Systems, Man, and Cybernetics*, Vol. SMC–11, No. 6, pp. 418–432.

12. Montana, D.J. (1988) 'The kinematics of contact and grasp', *Int. J. Robotics Res.*, Vol. 7, No. 3, pp. 17–32.

13. O'Neill, B. (1966) *Elementary differential geometry*, Academic Press.

14. Whitney, D.E. (1982) 'Quasi-Static Assembly of Compliantly Supported Rigid Parts', *J. Dyn. Systems, Meas., and Control*, Vol. 104, pp. 65–77.

15. Xiao, J. and Volz, R.A. (1988) 'Design and Motion Constraints of Part-Mating Planning in the Presence of Uncertainties', *Proc. IEEE Conf. Rob. and Automation*, pp. 1260–1268.

16. Xiao, J. (1992) 'Replanning with compliant rotation in the presence of uncertainties', *Int. Symp. Intel. Control*, pp.102-108.

PART IV
Computational Issues and Numerical Methods

REDUCTION OF MULTIBODY SIMULATION TIME BY APPROPRIATE FORMULATION OF DYNAMICAL SYSTEM EQUATIONS

R. SCHWERTASSEK
DLR, Institute for Robotics and System Dynamics
German Aerospace Research Establishment
D-82230 Wessling, Germany

ABSTRACT: One of the most cited concerns of users applying multibody simulation codes to system design is computational speed. In this contribution two aspects of how to reduce simulation time are discussed: the application of parallel processing and the appropriate modelling of flexible bodies. To exploit the benefits of parallel computer architectures an interdisciplinary approach has been pursued, combining knowledge of the three disciplines dynamics, numerical mathematics and computer science. An analysis of the options available for the formulation and for the numerical solution of the system equations yields a surprising result. A method, initially proposed to solve the inverse problem of dynamics, is the best choice to generate the system equations required for solving the simulation problem when relying on implicit integration routines. The new $O(N)$ residual formalism has a high potential to benefit from parallel computer architectures. Its implementation on a Transputer network results in a significant reduction of execution time for a realistic multibody simulation problem. Flexible structures may be analyzed using finite element or multibody codes. Representing the motion of a flexible body as the superposition of large reference motions and small deformations requires to model so called geometric stiffening effects. This aspect of the derivation of flexible multibody equations is discussed here. The methodology is used to demonstrate the computational efficiency of the approach and to propose a standardization of flexible body models in multibody system codes.

1. Parallel Multibody Simulation

1.1. FORMULATION OF THE PROBLEM

In summer 1987 most of the multibody dynamics community met at the JPL, Pasadena, to discuss the needs and the open problems in multibody system simulation, especially for space applications. P. W. Likins stated in his survey [22]: "*Computational questions focused initially on the selection of subroutines for numerical integration, matrix inversion, or eigensystem analysis, and lately have shifted to preprocessors and postprocessors for user convenience. More fundamental issues are raised by the potential of symbolic manipulation and parallel processing, both of which present the possibility of revolutionizing the field.*" Concepts for symbolic implementation have been pursued at various places, e.g. [19, 38]. This paper presents results of our efforts to exploit the potential of parallel computer architectures for multibody simulation. It has its roots in an analysis of the status of knowledge at the time, the above statement was made.

Basic methods for multibody system simulation are provided by the disciplines of dynamics (the multibody formalisms), numerical mathematics (the solution techni-

M. F. O. S. Pereira and J. A. C. Ambrósio (eds.),
Computer-Aided Analysis of Rigid and Flexible Mechanical Systems, 447–482.

ques) and computer science (the design of simulation codes) – see boxes in Fig. 1. In the mid–eighties the formalisms for the generation of multibody system equations and the numerical methods for solving ordinary differential equations had been fully developed, but the interaction between the related areas of research was poor in most of the groups working on the development of multibody codes. The so–called $O(N)$–formalisms had been found at various places independently [37], yielding the state space representation of the system dynamics in explicit form with a number of operations, which grows linearly with the number N of system bodies. Solving the equations generated in such a way with numerical integration routines at hand was considered to be the most efficient approach to multibody system simulation. Other codes were based on the description of the system dynamics by a set of differential equations in terms of redundant variables accompanied by a set of algebraic constraint equations, i.e. the codes generated and solved the Lagrangian equations of type one. To improve the performance of such codes, the numerical solution of differential–algebraic equations was studied by an increasing number of mathematicians, generally without considering any special properties of the multibody system equations. One of the attempts to initialize communication between dynamicists and mathematicians was made by E. J. Haug when organizing the workshop on "Real-Time Integration Methods for Mechanical System Simulation" in 1989 [12]. The prime motivation for the meeting was the need to realize multibody real–time simulation in such applications as general purpose car simulators. Additional aspects for reducing multibody simulation time in vehicle dynamics, industrial applications have been described in [16], e.g. for usage in hardware–in–the–loop–tests of ABS and ASC (anti–slip–control). Design procedures for control systems in vehicles and spacecraft also called for a significant reduction of simulation–time. During the specialists meeting at the JPL in 1987 the simulation requirements in control system design were described as follows [15]: ***"Problem solutions must be run in large numbers to arrive at design decisions, and large systems must be studied. Computational speed therefore becomes the most important single consideration in code design."*** This necessity was even more emphasized at a conference in the summer of 1989 [27].

Most of the implementations of multibody codes were on serial computers. To reduce computational costs the usage of parallel computation was discussed, but again without any interdisciplinary considerations. In [2] the most advanced option of those days, the $O(N)$–formalism, was implemented on a parallel computer yielding little reduction of computer–time, even for large multibody systems.

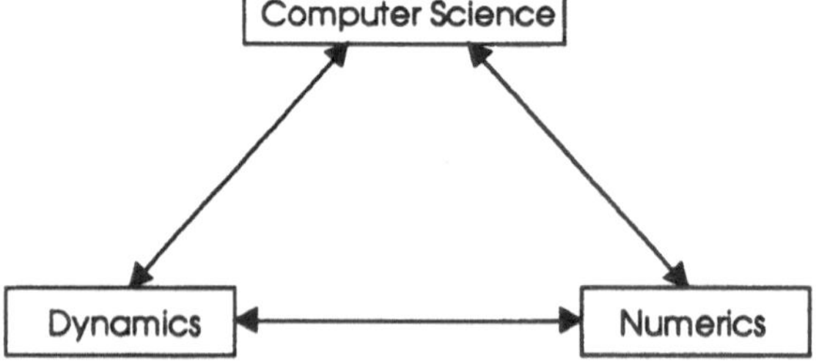

Figure 1: Areas of research required in multibody system simulation.

In this contribution the idea is pursued of how to combine the methods available in computer science, dynamics and numerical mathematics in an optimal way to obtain the most efficient solution of the simulation problem – in other words, it is proposed to exploit the potential of an interdisciplinary approach to the problem as visualized by the arrows in Fig. 1. The goal is a reduction of computer–time beyond the limits described in the references mentioned above. This goal is achieved by an appropriate tuning of multibody formalisms and numerical solution techniques resulting in a formulation of the simulation problem, which has a high potential for parallel computation. Its implementation on a network of Transputers yields reductions of simulation–time considerably higher than those found in previous approaches.

1.2. OPTIONS FOR SOLVING THE PROBLEM

In view of Fig. 1 the methodologies available from the branches of science contributing to multibody simulation are discussed now. Two forms of system equations may be generated by methods available from the first one of the three disciplines, *dynamics*. The two forms are the state space representation

$$\dot{y}_I = y_{II} \quad , \tag{1a}$$
$$M_R(y_I,t)\,\dot{y}_{II} = h_R(y_I,y_{II},t) \quad , \tag{1b}$$

and the descriptor form (Lagrangian equations of type one)

$$\dot{x}_I - x_{II} = 0 \quad , \tag{2a}$$
$$M(x_I,t)\,\dot{x}_{II} - G^T(x_I,t)\,\lambda - h(x_I,x_{II},t) = 0 \quad , \tag{2b}$$
$$g(x_I,t) = 0 \quad . \tag{2c}$$

Equations (1) are formulated in terms of the (independent) position- and velocity-state-variables y_I and y_{II}, whereas the coordinates x_I and the velocities x_{II} in (2) are redundant. Therefore, generalized constraint forces λ appear in (2b) and the differential equations (2a, 2b) are accompanied by a set of algebraic constraint equations (2c). Collecting coordinates y_I and velocities y_{II} in the state vector y, the two sets of equations (1) can be compacted into the explicit system of ordinary differential equations (ODE)

$$\dot{y} = f(y,t) \quad , \tag{3}$$

after the reduced mass matrix M_R has been inverted. Similarly, defining x to contain x_I, x_{II} and λ, the set of equations (2) can be abbreviated as

$$F(x,\dot{x},t) = 0 \quad , \tag{4}$$

yielding an implicit system of differential–algebraic equations (DAE).

A survey of methodologies in multibody dynamics shows, that the generation of the state space representation is straightforward only in the case of tree–configured systems, when using relative variables to represent the system motion. An application of the corresponding method to generate the equations of motion for systems with closed kinematic chains yields a set of partially reduced system equations in terms of redundant variables, i.e. a system representation of the general form (2). This is a *first option* when dealing with general multibody systems including loops. A *second choice* is to use absolute variables. Then the descriptor form (2) of the equations of motion can be obtained with very low effort. A *third alternative* is provided by the recursive $O(N)$–formulations. In case of tree–configured systems they yield the explicit form (3) in terms of relative variables with a number of operations, which grows linearly with the number N of system bodies. In case of systems with loops a so-called semi-explicit form [39] in terms of redundant relative variables can be obtained.

Numerical methods for solving the explicit equations are well developed. Such "explicit integration routines" for ODE require the evaluation of the right hand side of (3)

$$f_m = f(y_m, t_m) \tag{5}$$

given the state y_m at time t_m. These computations must be provided based on the multibody formalism. Unfortunately, explicit integrators break down in case of stiff systems. But these appear quite often in multibody simulation, e.g. when dealing with contact or control problems and with flexible bodies. This is why the usage of explicit integrators and of the corresponding form of the system equations as provided by the $O(N)$-formulations is excluded here.

For the numerical solution of the implicit equations (2) or (4) two approaches have been proposed. One possibility, the "coordinate partitioning method" [11], corresponds to a numerical reduction to the state space form (1). An improvement of this method has been proposed in [20]. The second option, to be pursued here, is to use implicit multistep methods as implemented in the code DASSL [30]. Applications of the code to solve mechanical system equations resulted in stability problems. They triggered the development of the derivative ODASSL [8] of DASSL, which avoids such instabilities. By contrast with explicit integration routines for ODE the implicit multistep methods for solving the DAE (4) require the computation of the residual

$$\Delta_m = F(x_m, \dot{x}_m, t_m) \tag{6}$$

given approximations for the variables x_m and the derivatives $\dot{x}_m$ together with time t_m. The residual Δ_m is nonzero as long as the approximations do not satisfy (4). Integration routines like ODASSL use values of $\Delta_m \neq 0$ to compute the solution of (4), corresponding to $\Delta_m = 0$.

A simple consequence of the *interaction between numerical mathematics and dynamics* in multibody simulation has been mentioned already: explicit integrators fail for stiff systems, which suggests to avoid the explicit form of the system equations in such cases. A more important aspect is related to the basic difference between the information required from a multibody formalism by implicit and explicit integration routines. In case of an explicit integration of ODE, the formalism must provide f_m. Because of this fact, all of the multibody formalisms presented previously headed towards an efficient generation of the right hand side f_m of the system equations. By contrast, implicit integrators for DAE need the residual Δ_m. In view of (2) the elements of Δ_m can be interpreted as (generalized) forces, strictly speaking as those forces, which must be added to the forces $G^T\lambda + h$ to satisfy the equations of motion for the given values $x_m, \dot{x}_m$ and t_m. The computation of forces given the system motion is known as the inverse problem of dynamics. It has been studied carefully for applications in robotics and efficient formalisms to compute the unknown forces – here the residuals – have been developed in this context [25].

In view of this interpretation there are three options to compute the residual Δ_m of the partially reduced system equations in terms of relative variables:

1. Computation of the system matrices in (2) using any suitable formalism. This yields a computational effort growing quadratically with the number N of system bodies.

2. Application of the recursive $O(N)$-formalism to compute accelerations $\dot{\xi}_m$ corresponding to the given values x_m and t_m. This yields the residual $\Delta_m = \dot{x}_m - \dot{\xi}_m$ with a computational effort depending linearly on N.

3. Usage of a formalism for direct computation of Δ_m as provided by the methods for solving the inverse problem of dynamics. Such a formulation – referred to here as a "residual algorithm" – is also of $O(N)$.

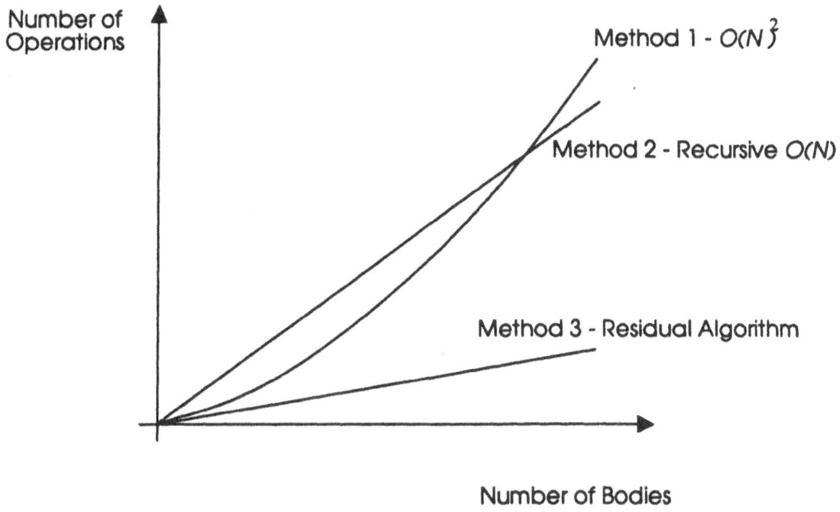

Figure 2: Computational costs for generating the residual.

The computational effort required in the three cases has been analyzed as a function of the number N of the system bodies. The result is summarized in Fig. 2. The diagram shows the well known advantage of $O(N)$-formalisms (method 2) as compared to classical formulations (method 1). A new and more important result is an additional reduction of the computational effort – see Fig. 2, method 3 – achieved when using instead of the $O(N)$-formalism the residual algorithm, which has been developed in [6]. The different slopes of the two curves for the methods 2 and 3 correspond to a factor 4.5 by which the residual algorithm is faster than the recursive $O(N)$-formulation (method 2).

To summarize, a combined consideration of dynamical and numerical aspects leads to the following conclusions: A generation of the explicit system equations (3) is excluded together with an application of the corresponding numerical solution techniques. Instead, the descriptor form (4) is used as a basis for simulation. Thus only two of the three options available in dynamics survive: implicit representation of the system motion in terms of absolute and relative variables. In both cases the residual Δ_m must be generated for numerical integration. For the system equations in terms of relative variables the residual is generated most efficiently with the residual algorithm, as demonstrated by Fig. 2.

Considering the interactions with the last corner of the triangle from Fig. 1, *computer science*, both, additional computational aspects resulting in additional options as well as a reduction of the options already available are obtained. Heading for an implementation on a parallel computer architecture the potential for parallelization must be explored. The computations required for multibody system simulation can be parallelized on three levels:

1. One may consider to parallelize basic mathematical operations, primarily matrix operations for solving problems of linear algebra, required to generate and solve the implicit system equations. This is what is called a parallelization on a "fine grain level".

2. On a "medium grain level" the residual Δ_m may be generated by computing those contributions to Δ_m in parallel, which result from the various elements of the multibody system (bodies, joints, etc.) [6].

3. Parallelism on a "coarse grain level" is obtained following another strategy. Repeatedly calling the residual algorithm one obtains the Jacobian of F, which is needed by the implicit integration scheme. The calls are independent of each other and can be organized in parallel.

The basic difference between a parallelization on the medium and the coarse grain level is as follows: on the medium grain level the residual is calculated in parallel combined with a serial computation of the Jacobian. Vice versa, on the coarse grain level a serial implementation of the residual algorithm is used and the Jacobian is computed in parallel.

In addition to these alternatives we also must consider the still remaining option of computing Δ_m based on equations (2) in terms of absolute variables. The generation of Δ_m is so simple in this latter case that its parallelization results in no significant benefits. The remaining part of the simulation problem, the solution of the system equations in terms of absolute variables based on the computation of Δ_m (which is the time–consuming part in this approach) involves linear algebra computations only. Numerical experience teaches that a parallelization of linear algebra operations does not pay off for orders of matrices typical in multibody system simulation [7]. This excludes two options, the usage of absolute variables at all and a fine grain parallelization of a simulation based on the partially reduced system equations in terms of relative variables. The two other candidates of medium and coarse grain parallelization, as mentioned above, remain competitive.

In summary, interdisciplinary considerations yield the result that implicit integration of the partially reduced system equations in terms of relative variables is the candidate for a parallelization of multibody system simulation. A numerical solution of the system equations requires the computation of the residual Δ_m. The implementation of the equations to compute the residual may be pursued following the two strategies of medium and coarse grain parallelization. The implicit formulation of the problem results in one of the advantages of the approach, its capability to deal with stiff system equations.

1.3. PARALLEL IMPLEMENTATION

The two alternatives of medium and coarse grain parallelization have been identified as candidates for the implementation of the multibody system equations as required for implicit numerical integration. Within the process of parallelization identical work packages are formulated, which can be treated simultaneously for both of the alternatives [6]. Having clarified such details depending on the structure of the equations of motion, two major points, a software– and a hardware–problem, require further consideration: The concept for parallelization, i.e. the software needed to distribute the work packages on a given topology of computing nodes[1] must be defined, and this topology of computing nodes, resulting in the best suited hardware architecture for solving the simulation problem, must be selected. Both of these problems will be discussed in more detail now.

The most important criterion for the selection of the concept for parallelization is the development of a "user-friendly" simulation code. Shifting any of those problems to the user, which can be handled by the computer, must be avoided. In particular,

- a user should not be occupied with the question of how to distribute work packages,
- the topology of the multiprocessor network should be independent of the multibody system topology,
- the problem of load balancing should be solved automatically, i.e. a user should not be asked to take care of how to distribute the computational load within the network in a uniform way,
- adding computing nodes to the network should result in a reduction of simulation time without additional programming effort.

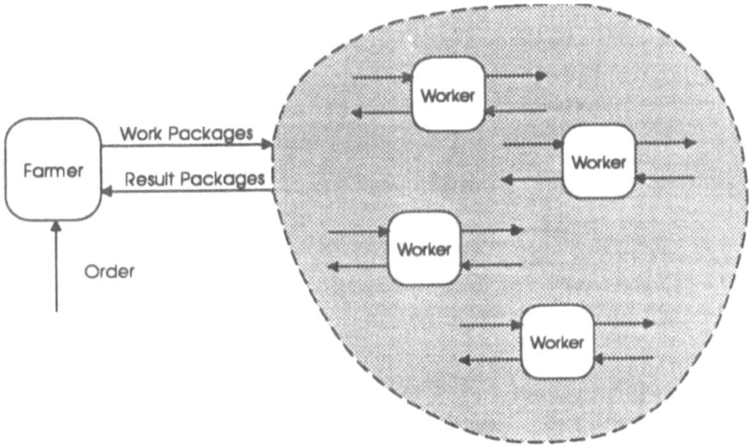

Figure 3: Organization of the distribution of work load in a computational network by means of the farming concept.

The farming concept, proposed in [14], meets all of the above mentioned requirements. As visualized in Fig. 3, the computing nodes are subdivided into the two groups of one single "farmer" and an arbitrary number of "workers". Whenever the farmer receives an order, he separates it into an appropriate number of identical work packages (e.g. the computation of a complete Jacobian into the computations of its individual columns). Then the farmer sends these work packages into the farm, where

[1]A computing node as used in this context is a more general term for a processor or a collection of processors.

the workers solve three problems: distribution of work packages, computation, i.e. execution of work packages and delivery of results back to the farmer. The farmer keeps control of the number of outgoing work packages and incoming result packages. This avoids an overload of the farm and ensures that the number of work packages is decoupled from the number of workers available. The distribution of work packages by the workers themselves guarantees both, a uniform distribution of the work load and the possibility to deal with an arbitrary processor topology. In particular, in case of multibody simulation the latter property avoids adapting the topology of the processor network to the topology of the multibody system. The parallelized code presents itself to the user in the same way as any conventional serial code running on a single processor.

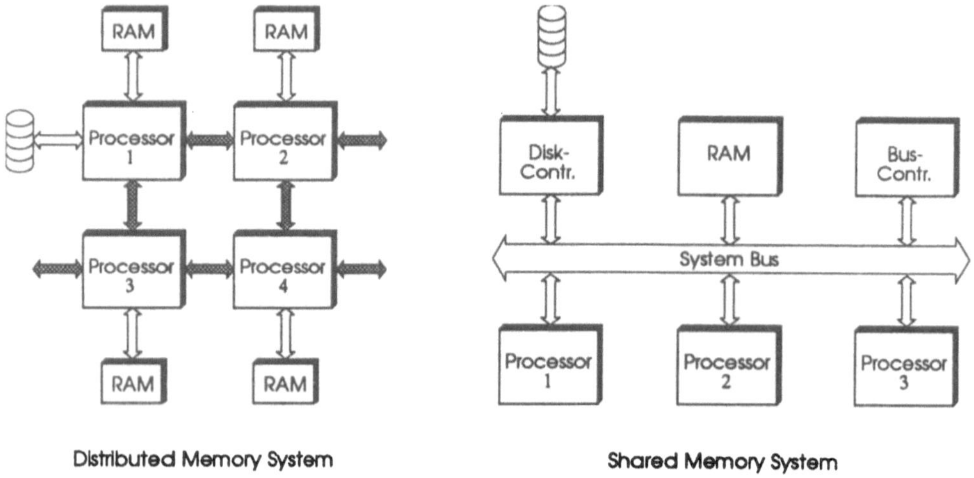

Figure 4: Different structures of multiprocessor systems.

Some hints dealing with the second point mentioned above, the choice of the hardware architecture for multibody simulation, are summarized now [6]. As in case of the first problem, the criteria for selecting a parallel computer architecture are collected first: The architecture should be well suited for applying the farming concept, it should not result in a conceptual bottleneck for communication and it should be easy to expand. Two groups of parallel computer architectures have been proposed, the shared memory systems and the distributed memory systems as represented in Fig. 4. In the first case the processors share one common memory and functions for network- (disc- and bus-) control. The interconnection of the processors, of the memory and of the control units by a single bus yields a limited communication bandwidth and results in a communication bottleneck when increasing the number of processors. By contrast, in a distributed memory system each single processor has its own memory and its own communication supports. The processors are interconnected to each other directly. As a result the communication bandwidth grows with the number of processors and the bottleneck resulting from communication by means of a single bus disappears.

A comparison of the computational performance of the two systems is shown in Fig. 5. As long as the number of processors is low, shared memory systems are slightly superior as compared to distributed memory systems, but when the number of processors has grown beyond a certain limit, the advantages of distributed memory

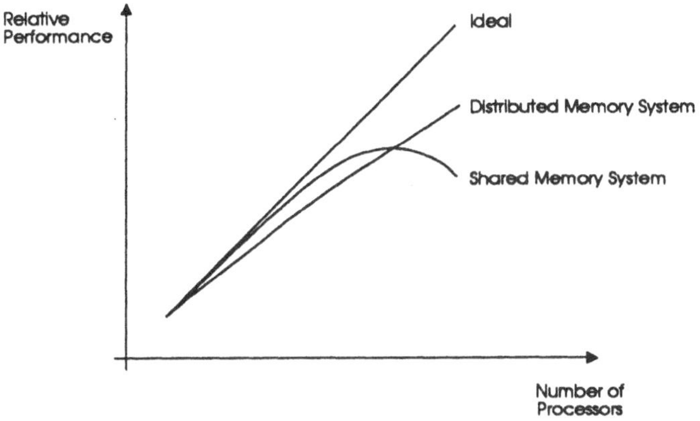

Figure 5: Relative performance of multiprocessor systems.

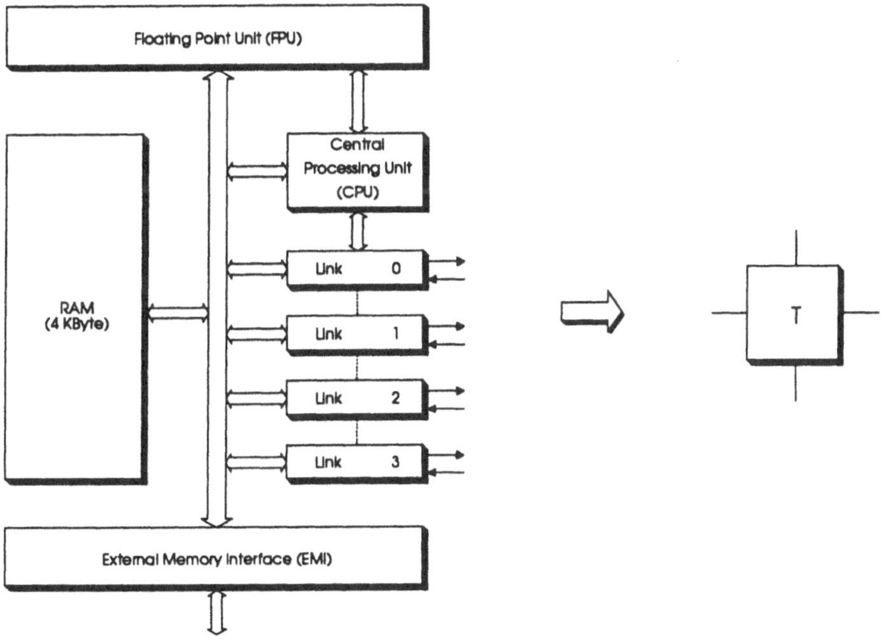

Figure 6: Transputer architecture and its representation by a symbol.

systems become more and more pronounced. The reason for this apparent disadvantage of shared memory systems lies in their conceptual bottleneck for communication by means of the system bus. By contrast, distributed memory systems avoid such a bottleneck, they can be expanded easily and they are well suited for applying the farming concept: one processor can be assigned to be the farmer and the others become the workers, all of them being interconnected directly with each other. Thus, in view of our criteria, distributed memory systems are well suited for multibody simulation.

The Transputer shown in Fig. 6 can be used as a building block to generate large distributed memory systems. The Transputer (the word is a combination of transmitter and computer) has been developed by INMOS [9]. It was the first microprocessor combining processors (CPU, FPU), memory (EMI, RAM), and communication support (Links) on one single chip.

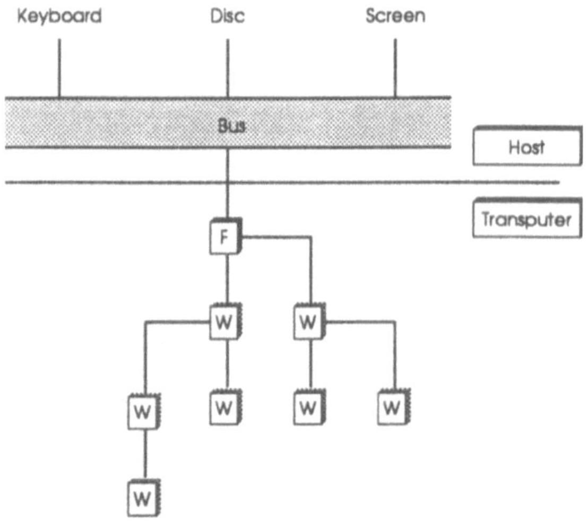

Figure 7: Host computer together with a Transputer network.

Because of the links Transputers can be interconnected easily in an arbitrary way to generate a network – see Fig. 7. When the network is linked to a host computer (personal computer or workstation) to provide such facilities as keyboard, disc-memory and screen, one obtains a powerful distributed memory multiprocessor system. In view of the farming concept the Transputer interconnected directly to the host computer can be identified with the farmer and the others become the workers. This is the parallel computer architecture to be used here for implementing the two alternatives of medium and coarse grain parallelization of multibody simulation as outlined in Sect. 1.2.

1.4. ARCHITECTURE AND CAPABILITIES OF THE SIMULATION PACKAGE

Three computer codes have been developed to test the computational efficiency of the methods described heretofore. A serial code SERSIM has been based on the residual algorithm described in Sect. 1.2. It serves as a standard to measure the speed-up[2] gained by parallelization and it has been used as well to test all of the serial parts in the parallel codes. Two of them are available, called PARSIM_1 and PARSIM_2. In PARSIM_1 medium grain parallelization is used to compute the residual, whereas PARSIM_2 follows the strategy of coarse grain parallelization as discussed at the end of Sect. 1.2. The entire simulation package including the three codes SERSIM,

[2]The speed–up is defined as the ratio of the computer–time for evaluation of an expression required by a number of processors greater than one as compared to the time needed by a single processor.

PARSIM_1 and PARSIM_2 for generating the residuals as required for numerical integration makes the following options available:

- generation of the equations of motion,
- determination of static equilibrium configurations,
- computation of a set of consistent initial conditions,
- kinematic analysis,
- simulation, including a simple on-line animation.

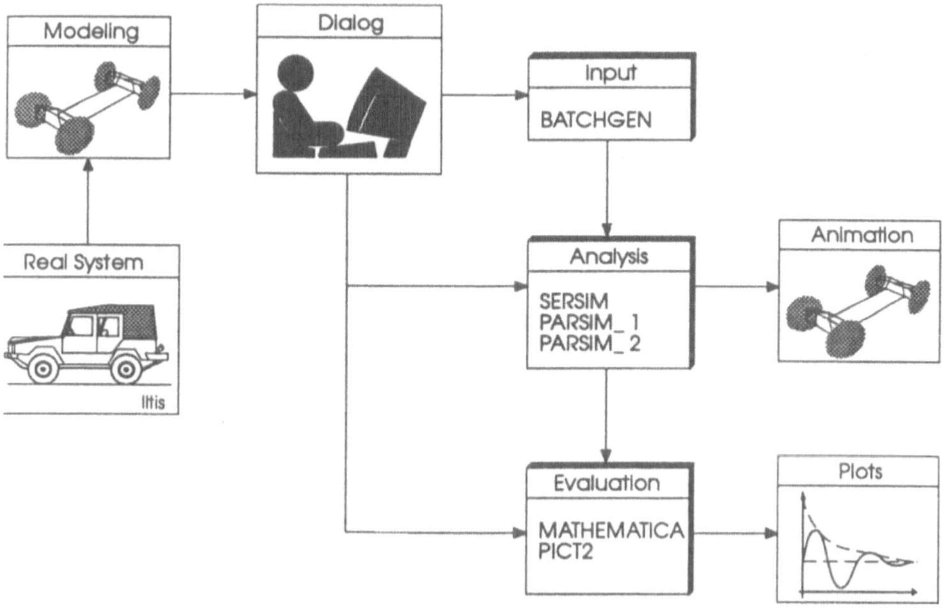

Figure 8: Structure of the simulation package.

A global representation of the structure of the package is given in Fig. 8. A main program BATCHGEN is used for dialog-oriented input of the system data (e.g. for the Iltis-vehicle shown in the diagram) and additional parameters for computational control. The program generates the system data in forms required for further computation. Together with initial values and control parameters the system data are used to start the analysis part, which produces the results for further evaluation together with an on-line animation. The analysis part yields the static equilibrium configuration, consistent initial conditions and the system motion for a given interval of time, using any of the three codes SERSIM, PARSIM_1 and PARSIM_2. With the third part of the package time histories can be plotted using the PICT2-code (available at DLR) or any standard features provided by MATHEMATICA.

1.5. RESULTS

Two typical applications of multibody system simulation – the analysis of the motions of an off–road vehicle and of a multiple body pendulum – are discussed now with regard to

- verification of the code,
- reduction of computer–time due to the new $O(N)$–residual algorithm and its parallel implementation,
- relative performance of medium and coarse grain parallelization.

The first example is the Iltis off–road vehicle depicted in the "Real System"–box of Fig. 8. Its multibody model, represented in Fig. 9, has been used as a benchmark problem for multibody simulation codes [18]. The model shows the car body with four identical McPherson suspensions and includes complex force laws for the leaf springs and the tires.

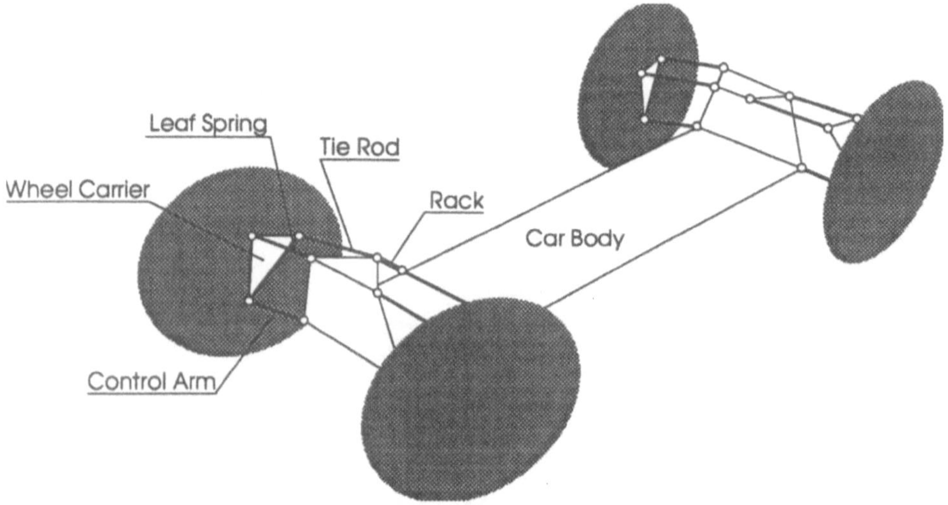

Figure 9: Multibody model of an off–road vehicle.

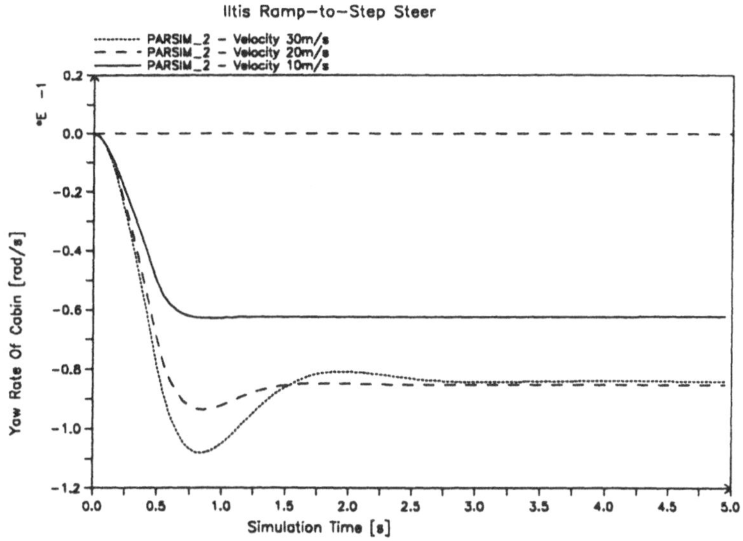

Figure 10: Yaw rate of the car body of the Iltis off-road vehicle.

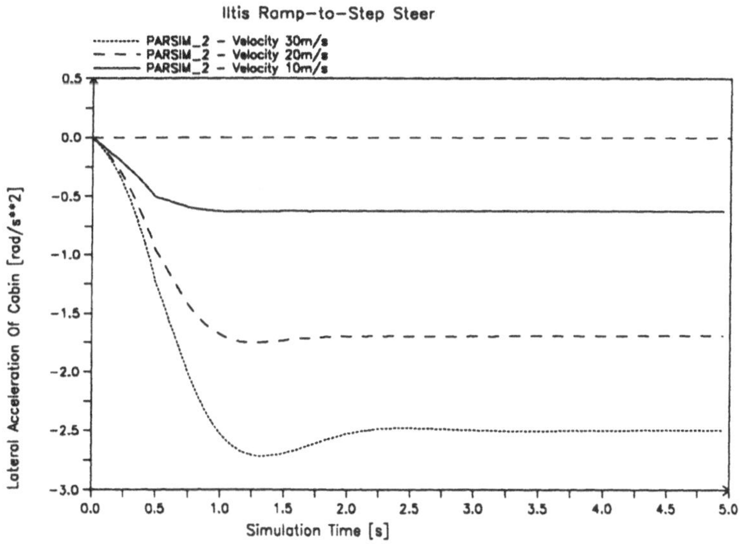

Figure 11: Lateral acceleration of the car body of the Iltis off-road vehicle.

code	integrator	processors	CPU-time
SIMPACK	ODASSL (impl.)	1	4.50
MEDYNA	DEABM (expl.)	1	0.91
SERSIM	ODASSL (impl.)	1	1.00
PARSIM_2	ODASSL (impl.)	2	0.71
PARSIM_2	ODASSL (impl.)	4	0.58
PARSIM_2	ODASSL (impl.)	8	0.46

impl. = implicit method; expl. = explicit method

Table 1: Computer-times required to simulate motions of the off-road vehicle.

Motions of the vehicle are excited by a "Ramp-to-Step Steer" maneuver due to a displacement of the rack given by

$$p(t) = \begin{cases} 0.4\,mm/s * t & : \quad 0.0\,s \le t \le 0.5\,s \\ 2.0\,mm & : \quad 0.5\,s < t \le 5.0\,s \end{cases}$$

It commands a steering motion of the front wheels of the vehicle resulting in a change of the direction of travel from straight track to a (circular) curve. A detailed description of the model may be found in [36] – here it is sufficient to know that this multibody system model has 10 bodies, 18 joints and 8 kinematically closed loops. In our approach the system motion is represented by a set of 52 differential-algebraic equations. Motions due to the rack displacement have been simulated for the three initial velocities of 10, 20 and 30 m/s of the vehicle. Typical examples of simulation results are given in Figs. 10 and 11. The diagram in Fig. 10 shows the time history of the yaw rate of the car body and the one in Fig. 11 its lateral acceleration. The plots demonstrate that the vehicle travels along a circle after some transients due to the maneuver have been damped out: Both, the lateral acceleration and the yaw rate become constant. The stationary values of the yaw rates of the vehicle traveling with 20 and 30 m/s are nearly identical – see Fig. 10. This is explained by the fact that the vehicle runs along circles having different radii in the two cases. The results are in good correspondence with those obtained using the SIMPACK- and the MEDYNA-codes described in [32] and [43].

The above results verify the new code, but its relative performance with respect to existing serial codes remains an open question. Two options, medium and coarse grain parallelization, are available within the simulation package. To select the one, which is best suited for simulating the problem under consideration, the diagram shown in Fig. 12 has been created. It shows the speed-up obtained for evaluating both, the residual Δ_m and the Jacobian of F when increasing the number of processors. The diagram shows clearly that a parallelization of the computation of the Jacobian results in a far better speed-up than the parallelization of the generation of the residual. For the simulation of the Iltis vehicle motions this suggests to use the coarse grain strategy for parallelization, i.e. the PARSIM_2 option of the simulation package.

Applying SERSIM and PARSIM_2 we now ask the following questions:

1. How does the parallelized code compare with the performance of the MEDYNA-code, which has been developed with special attention to vehicle dynamics applications? Such a comparison is of particular interest, because MEDYNA uses a system representation, in which the kinematics are linearized, resulting in considerable computational savings.

2. How does the new code, based on the new $O(N)$ residual algorithm compare to the SIMPACK-code, which uses the "classical" recursive $O(N)$-formulation for the generation of the system equations?

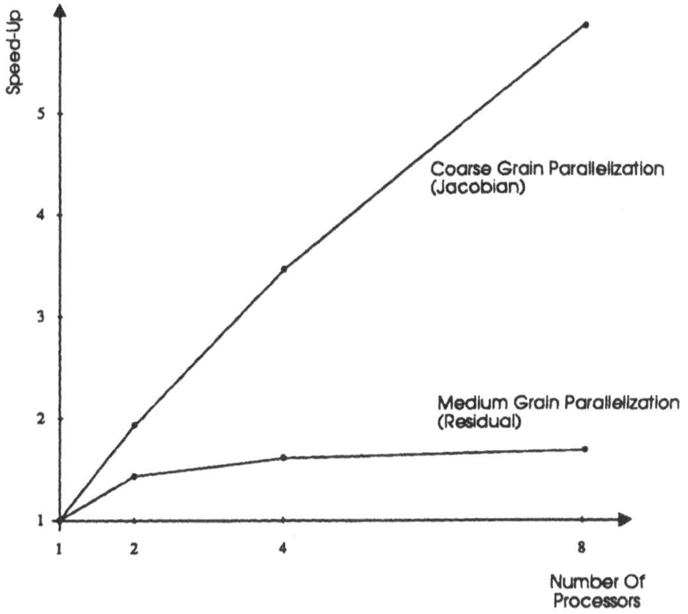

Figure 12: Speed–up in case of medium and fine grain parallelization.

3. What are the benefits of using several processors instead of one single processor only? This question can be answered by comparing the performance of PARSIM_2 and SERSIM.

Answers to the three questions are summarized in Table 1. It gives the CPU–times for running the same problem with the codes mentioned in the first column and with the numbers of processors shown in column 3. The CPU–times given in column 4 are normalized such that the time required by the SERSIM–code is 1. In all of the cases the same (implicit) integration routine ODASSL has been used, excluding MEDYNA. This code generates the explicit form of the state space representation and relies on the explicit integration routine DEABM [10]. The facts demonstrated by the table are as follows:

- The SERSIM–code, generating the nonlinear system equations, is nearly as fast as the MEDYNA–code, in which the kinematics have been linearized.
- The new $O(N)$ residual algorithm is much faster than the "classical" recursive $O(N)$– formulation used in the SIMPACK–code. The efficiency of the two approaches had been compared in Sect. 1.2. The factor 4.5 given in the table justifies the expectations suggested by Fig. 2.
- The simulation time is reduced considerably when using more than one processor. The speed–up is not as high as one might expect from Fig. 12. This is understood easily, when realizing that the Jacobian is re–evaluated in general only for 10 % of the integration steps.

- The combination of the new $O(N)$ residual algorithm and coarse grain parallelization yields a reduction of computer-time of an order of magnitude when compared with a serial implementation of the "classical" recursive $O(N)$-formulation (CPU-time 4.5 against 0.46).

Finally, it may be worth mentioning the real – not the relative – computer times. One single Transputer yields approximately the same computer-times as an apollo-Workstation DN 5500. The normalized time 1 in the table corresponds to a CPU-time of 18.3 s on the apollo, required to simulate the 5 s of the vehicle motion shown in the Figs. 10 and 11.

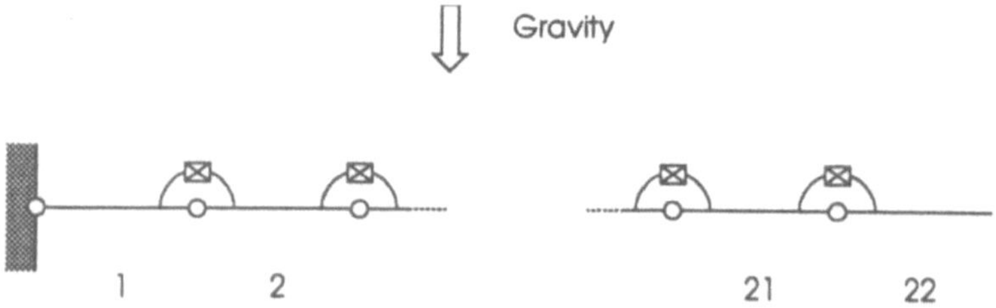

Figure 13: Multiple body pendulum.

The second example, a 22-body pendulum as shown in Fig. 13, demonstrates that the optimal strategy for parallelization depends on the problem. Multiple pendulum systems have been used as models for the simulation of the motions of cables [13]. In particular, the laws for the forces between the system bodies become quite complicated in such applications. The diagrams in Fig. 14 show the relative computer-times required for the simulation of the motions of the system using various codes and an increasing number of processors. Again the computer-time has been normalized to obtain the value 1 in the case of a simulation by the SERSIM-code.

The pendulum model has a tree configuration, in fact it is a simple chain. In such cases the SIMPACK-code generates the explicit form (3) of the system equations. Using the explicit integration routine DEABM one obtains a normalized simulation time of 4.46 as shown in Fig. 14, compared to 1 required by SERSIM when using the implicit integrator DASSL. The latter time is reduced as shown by the diagrams when using more than one processor and the two parallel codes PARSIM_1 and PARSIM_2. The example demonstrates that it is more advantageous to apply PARSIM_1 in this case, i.e. to follow the strategy of medium grain parallelization. With this concept and with 8 processors one obtains a reduction of simulation time by a factor of 23 with respect to the point 4.46 marked in the diagram for the "classical" $O(N)$-formulation implemented on a serial computer.

2. Modelling of Flexible Bodies in Multibody Systems

2.1. INTRODUCTION

The simulation of multibody systems with flexible bodies appeared as a problem in the late sixties already [24, 29, 23]. Although a quarter of a century with research on

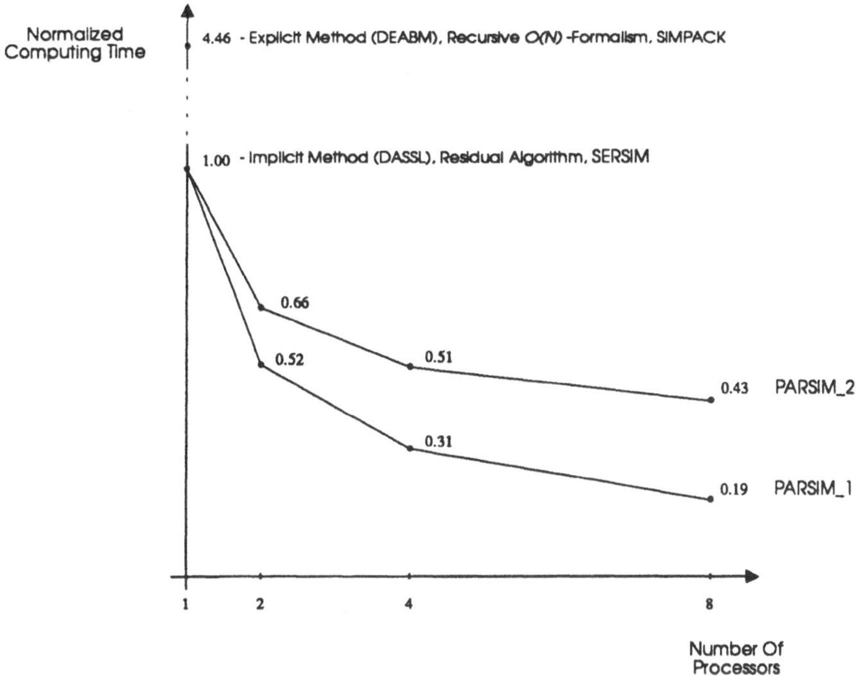

Figure 14: Computer–times required for the simulation of the multiple body pendulum.

the subject elapsed meanwhile the problem is still with us as shown by the proceedings of a recent series of conferences on "Aerospace Computational Control" [3, 26, 41] and by a survey given in [5]. Two types of approaches are used to analyze flexible structures: finite element and multibody system approaches. A simulation based on finite element models is straightforward, the corresponding codes are well developed, and include linear and nonlinear theory of elasticity. Unfortunately, the FEM codes have an important disadvantage: a dynamic analysis is very time consuming. By contrast, a dynamic simulation based on multibody system models is much more faster, at least in certain situations [44, 45].

The simulation of systems with flexible bodies is well understood when using a description in terms of absolute variables [40], but the computational burden seems to be high. In many applications one is confronted with system models, in which the deformations of the flexible bodies are small but superimposed on a large reference or rigid body motion. One can exploit this fact to reduce the computational burden by linearizing the equations of motion for small deflections. Flexible multibody codes using such an approach have been used frequently, but as shown in [17], not all mathematical models are correct. Because of forgetting so-called "geometric stiffening terms" some codes did yield wrong results as late as in the mid eighties. This is even more surprising, when recalling that geometric stiffening is a well known effect in dynamics and that its importance for multibody system simulation has been described clearly already in [21, 42].

Reduction of computer time and modelling of geometric stiffening is just one of the problems addressed by users of multibody codes [27]. Another concern deals with the fact that multibody codes are often used in conjunction with other software but that an integrated environment facilitating the exchange of data between the various software packages is not available. In particular, finite element codes are required to define the data describing flexible members in multibody systems. To facilitate the exchange of data between finite element and multibody codes a set of standard input data describing flexible bodies in multibody systems has been proposed [46]. Both of these aspects the reduction of simulation time by linearizing the equations of motion for small deformations which requires to include geometric stiffening and the standardization of flexible body data for multibody codes will be discussed in this second part of the paper.

The methods for solving the two problems have been used in the SIMPACK-code. As mentioned previously, the code is based on an $O(N)$-formalism as described in [32] and [37]. In the latter reference it has been shown of how to develop the explicit form of the dynamical equations of motion using the equation

$$J^i b^i = Q^i + C^i \Lambda^i - \sum_{b:b \neq i} S^{ib} D^{\kappa(b)} \Lambda^b, \quad i = 1,2,\ldots N \tag{7}$$

as a starting point. It represents the set of six equations of motion for a typical body i of a system of N interacting rigid bodies formulated in terms of the accelerations

$$b^i = \begin{bmatrix} \dot{\omega}^i \\ a^i \end{bmatrix}, \quad \dot{\omega}^i = \left[\dot{\omega}^i_\alpha\right], \quad a^i = \left[a^i_\alpha\right], \quad \alpha = 1,2,3 \tag{8}$$

as detailed in [37], p. 83. The 3×1-matrices $\dot{\omega}^i$ and a^i are the resolutions of the vectors $\underline{\dot{\omega}}^i$ and $\underline{a}^i$ in a frame fixed in body i. The vectors represent the angular acceleration of the body and the acceleration of its inboard attachment point, respectively. The 6×6-matrix J^i contains the generalized inertias corresponding to b^i, the 6×1-matrix Q^i summarizes the external forces and torques upon body i together with centrifugal terms. The internal forces and torques resulting from the joints and force elements between the system bodies are collected in the Λ-matrices. The 6×1-matrix Λ^i contains the interactions (forces and torques) across the inboard joint i on body i (it is convenient to give the same label i to both the body and its inboard joint) and the 6×1-matrices Λ^b represent the interactions across any joint b in the system. Numbers S^{ib} are the elements of the incidence matrix of the multibody system graph. Recalling its definition [31] one observes that the expressions under the summation sign represent the contributions resulting from the outboard joints attached to body i. Matrices C^i and D^κ are of dimensions 6×6. Matrix C^i transforms vectors resolved in a basis fixed in body i into the basis of the inboard body $\kappa(i)$ and matrix D^κ contains elements of displacement vectors connecting the inboard and all of the outboard attachment points on a body κ as required to form the contributions of forces to the torque with respect to the inboard attachment point. Details on the definitions of these matrices may be found in [37].

From (7) one concludes for a simple chain of bodies that three recursions are required to obtain the explicit form of the system equations in terms of relative variables, i.e. the relative accelerations, together with the unknown constraint forces resulting from the joints. Given the time t and a minimum set of position and velocity variables describing the motion of the bodies, the relative accelerations and constraint forces can be computed by the recursions with an effort growing linearly with the number N of system bodies, which suggests the name "$O(N)$-formalism" for the procedure. The first recursion, a forward recursion proceeding from the body $i = 1$ attached to the global reference frame to the terminal body $i = N$ of the chain generates the absolute motion of the system bodies from the given relative motion. By the second recursion running backwards from $i = N$ to $i = 1$ one generates generalized inertias $\hat{J}^i$ and forcing terms $\hat{Q}^i$ appearing in a set of modified equations (7) having the form

$$\hat{J}^i b^i = \hat{Q}^i + C^i \Lambda^i \tag{9}$$

By contrast with (7) the interactions across the outboard joints do no longer appear in (9) - they have been represented in terms of the accelerations b^i. Using matrices $\hat{J}^i$ and $\hat{Q}^i$ one obtains in a third recursion, again a forward recursion, the explicit form of the dynamical system equations together with the unknown constraint forces still incorporated in Λ^i. Details on the derivation of the recursions and on the computations to be performed may be found in [37, 32]. The procedure as described here applies for simple chains but it can be generalized easily to systems with tree configuration. In case of systems with loops one generates the explicit form of the partially reduced system equations and applies DAE-methods for their numerical solution.

2.2. GEOMETRIC STIFFENING

To generalize the $O(N)$-formalism for flexible bodies, the equations (7) must be reformulated considering the deformation of body i. Fig. 15 shows a typical body i of a multibody system in its reference configuration and in an actual, deformed configuration. An inertial frame is $\left\{O^I, \underline{e}^I\right\}$ and a frame associated with body i is $\{O^i, \underline{e}^i\}$. In the reference configuration the points of the body are identified by the material coordinates [3] R, i. e. the coordinates of vectors $\underline{R}$ in the body frame $\{O^i, \underline{e}^i\}$. The motion of the body is represented as a sum of a large reference motion and a deformation, which is assumed to be small. The reference motion is represented by the motion of the frame $\{O^i, \underline{e}^i\}$ as in case of a system of rigid bodies and the deformation is given by the displacement field $\underline{u} = \underline{u}(R,t)$. Representing the displacement vectors $\underline{u}$ in the body basis $\underline{e}^i(t)$ one obtains the coordinates

$$u = [u_\alpha], \quad u_\alpha = u_\alpha(R_1, R_2, R_3, t), \quad \alpha = 1,2,3. \tag{10}$$

The deformation of the body is described by the Green-Lagrange strain tensor having the elements

$$G_{\alpha\beta} = \frac{1}{2}\left(u_{\alpha,\beta} + u_{\beta,\alpha} + \sum_{\gamma=1}^{3} u_{\gamma,\alpha}u_{\gamma,\beta}\right) \text{ where } u_{\alpha,\beta} = \frac{\partial u_\alpha}{\partial R_\beta}. \tag{11}$$

To represent the stresses resulting from the deformation, the symmetric, second Piola stress tensor $\sigma_{\alpha\beta}$ is used. It is convenient to collect the six elements of both of these tensors in the 6×1-matrices

$$G = [G_{11}, G_{22}, G_{33}, 2G_{12}, 2G_{23}, 2G_{31}]^T \tag{12}$$

$$\sigma = [\sigma_{11}, \sigma_{22}, \sigma_{33}, \sigma_{12}, \sigma_{23}, \sigma_{31}]^T. \tag{13}$$

[3]For brevity, the index i used to identify a specific body in the multibody system is omitted in all of the quantities referring to the deformation of the body.

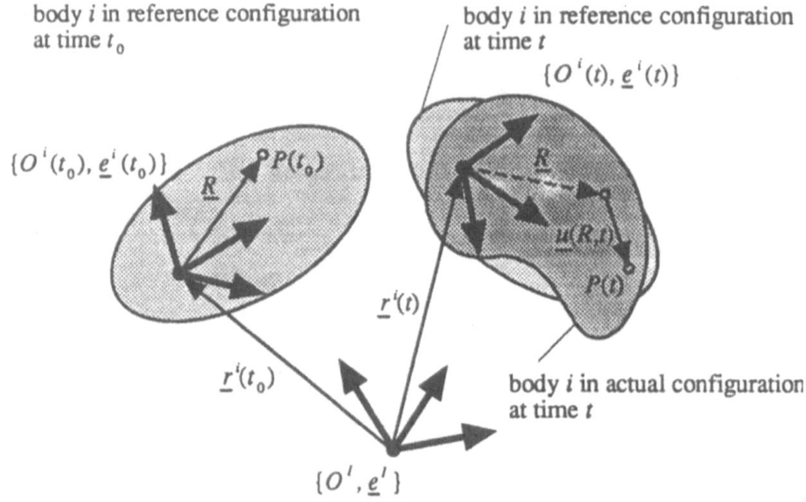

Figure 15: Notation to describe the motion of a flexible body

In these terms the stress-strain relations as given by a linear material law (Hooke) can be written in the form

$$\sigma = \sigma^r + \Delta\sigma, \qquad \Delta\sigma = H\,G. \tag{14}$$

Matrix H is of dimension 6×6 and σ^r characterizes the stress in the reference configuration at time t. The reference stresses are required to balance the forces acting upon the body in the reference configuration at time t , e. g. the inertial forces due to the large reference motion.

From (11) one concludes that the strain velocities can be written in the form

$$\dot{G} = L\dot{u} \quad \text{where} \quad L = L^0 + L^1(u) \tag{15}$$

is a differential operator defined in [45]. It may be written as a sum of an operator L^0, having elements of the form $\partial/\partial R_\alpha$, and an operator L^1, whose elements are sums of the products $u_{\alpha,\beta}\partial/\partial R\gamma$. The operator L^0 results from the linear terms in the strain-displacement relations whereas L^1 arises because of the nonlinear contributions in (11). After the displacement field has been represented by a Ritz-approximation

$$u\,(R,t) = \Phi\,(R)\,q\,(t)\,, \tag{16}$$

i. e. by a linear combination of known shape functions

$$\Phi\,(R) = [\Phi_{\alpha,j}]\,, \quad \alpha = 1,2,3, \quad j = 1,2,\ldots n_q, \tag{17}$$

and of unknown amplitudes

$$q(t) = [q_j(t)]\,, \quad j = 1,2,\ldots n_q, \tag{18}$$

the velocity of the body is represented by the velocity-variables ω^i and v^i describing the velocity of the body associated frame and by the derivatives $\dot{q}(t)$ of the amplitudes (18). Jourdain's principle is used to obtain the equations of motion of the flexible body corresponding to the equations (7) of a rigid body. The principle states that the virtual power δP^i of the forces upon body i is zero. Formulating δP^i in terms of the virtual velocities $\delta\omega^i, \delta v^i$ and $\delta\dot{q}$, one finds the equations of motion in terms of $a^i, \dot{\omega}^i$ and $\ddot{q}$ by realizing that the coefficients of the virtual velocities must disappear [45]. In particular, the coefficient of $\delta\dot{q}$ yields the subset of equations

$$\int_V [L\Phi]^T \sigma dV + F = 0. \tag{19}$$

Here V is the volume of the body, L is the differential operator introduced in (15) and σ is the stress-matrix satisfying (14). The integral in (19) represents the internal forces within body i due to its deformation. The external forces resulting from interconnections to other bodies are collected in matrix F together with all the other external and inertial forces. Now (14), (16) and (11) are used to represent σ in terms of σ^r and q. Assuming that the deformations, i. e. the amplitudes q_i are small and neglecting second order terms in q one obtains from (19) two sets of equations

$$\int_V \left[L^0\Phi\right]^T \sigma^r dV + F^r = 0, \tag{20}$$

$$\int_V \left[L^0\Phi\right]^T \Delta\sigma dV + \int_V \left[L^1\Phi\right]^T \sigma^r dV + \Delta F = 0. \tag{21}$$

Here, forces F^r are the reference-values of F, and ΔF are the deviations due to the deformation. The first equation (20) describes the conditions for the equilibrium of forces in the reference configuration: The internal forces represented by the stresses σ^r must balance the external actions summarized in F^r. Equations (21) are the linearized equations of motion. Because of the nonzero reference stresses σ^r and because of the nonlinearity of the strain displacement relations a term including the operator matrix L^1 appears. It is the geometric stiffness term, which had been shown in [17] to be of significant importance for the modelling of flexible bodies in multibody systems in cases in which the reference forces F^r are large.

The coefficients of the linearized equations of motion (21) can be precomputed, resulting in a description of flexible bodies in multibody systems, which yields computational savings. But the evaluation of the equations (21) requires some care. The reference forces F^r and thus the reference stresses σ^r depend on the reference motion, in general - for any reference configuration at time t one may have different reference stresses. On the other hand it has been assumed that the reference configuration, used to identify the points P of the body, remains unchanged - in Fig. 15 the reference configurations at the times t_0 and t are identical. This requires that the *large* reference stresses σ^r result in *small* deformations only. Otherwise the coefficients of the linearized equations of motion (21), which depend on σ^r, must be recomputed as given by equation (20), whenever the reference stresses have deformed the body that much that the assumptions used to derive the equations (21) are violated. Because of such calculations any computational savings resulting from the separation of the motion of the flexible body into a large reference motion and into small deformations disappear and the modelling becomes questionable. Examples for flexible bodies in which large reference stresses yield small or even zero deformations are beams and plates where specific deformations are locked or at least much smaller than other deformations. To clarify such details of geometric stiffening in multibody system simulation the example of a rotating beam will be discussed in more detail now.

2.3. MODELLING OF BEAMS IN MULTIBODY SYSTEMS

Fig. 16 shows the model of a beam to be used here: Rigid cross sections, having six degrees of freedom, are attached to a flexible axis. In the special case of an Euler-Bernoulli-beam the cross sections remain orthogonal to the axis resulting in a reduction of the number of degrees of freedom to four. Thus, beams as considered here, are models of flexible bodies in which the relative motion of all the points in a cross section is locked. Applying this model, the displacements $u_\alpha(R_1, R_1, R_3, t)$ of a representative point P of the beam can be represented by six deformation variables $y_d(R_1, t)$ depending on one material coordinate R_1 only. The variables y_d describe the location of the origin O^2 of a frame $\{O^2, \underline{e}^2\}$ fixed in the cross section and the orientation of the corresponding basis $\underline{e}^2$.

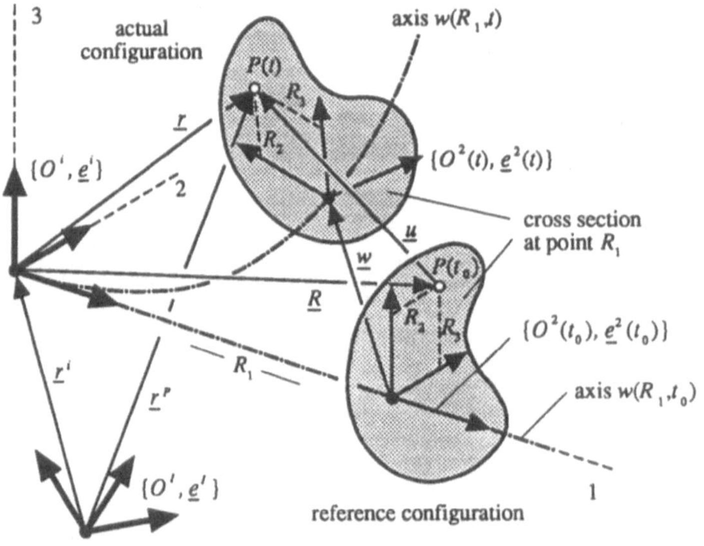

Figure 16: Notation to describe the motion of a flexible beam.

A convenient set of deformation variables $y_d(R_1, t)$ are the six coordinates $w_\alpha(R_1, t)$ of vector $\underline{w}$ in the basis $\underline{e}^i$ - see Fig. 16 - and three angles $\vartheta_\gamma(R_1, t)$ parameterizing the direction cosine matrix B, which describes the relative orientation of the triads $\underline{e}^i$ and $\underline{e}^2$ as

$$\underline{e}^2 = B\underline{e}^i, \quad \underline{e}^2 = [\underline{e}^2_\alpha], \quad \underline{e}^i = [e^i_\alpha], \quad B = [B_{\alpha\beta}], \quad \alpha, \beta = 1, 2, 3 \tag{22}$$

In these terms one obtains the representation of the displacement field in the form

$$u = \begin{bmatrix} w_1 & + & R_2 B_{21} & + & R_3 B_{31} \\ w_2 & + & R_2(B_{22}-1) & + & R_3 B_{32} \\ w_3 & + & R_2 B_{23} & + & R_3(B_{33}-1) \end{bmatrix} \tag{23}$$

Parameterizing the direction cosine matrix B by three angles, which correspond to three elementary rotations [31] about the 3-, 2- and 1-axis with angles ϑ_3, ϑ_2 and ϑ_1 and which carry $\underline{e}^i$ into $\underline{e}^2$ one obtains

$$B = \begin{bmatrix} +c_2c_3 & +c_2s_3 & -s_2 \\ -c_1s_3 + s_1s_2c_3 & +c_1c_3 + s_1s_2s_3 & +s_1s_2 \\ +s_1s_3 + c_1s_2c_3 & -s_1c_3 + c_1s_2s_3 & +c_1c_2 \end{bmatrix} \tag{24}$$

where $c_\alpha = \cos\vartheta_\alpha, s_\alpha = \sin\vartheta_\alpha$. Assuming small deformations and keeping 2nd order terms only yields [1]

$$\begin{aligned} u_1 &= w_1 + R_2(-\vartheta_3 + \vartheta_1\vartheta_2) + R_3(+\vartheta_2 + \vartheta_1\vartheta_3) \\ u_2 &= w_2 + R_2(-\frac{1}{2}\vartheta_1^2 - \frac{1}{2}\vartheta_3^2) + R_3(-\vartheta_1 + \vartheta_2\vartheta_3) \\ u_3 &= w_3 + R_2\vartheta_1 \qquad\qquad + R_3(-\frac{1}{2}\vartheta_1^2 - \frac{1}{2}\vartheta_2^2) \end{aligned} \tag{25}$$

In case of an Euler-Bernoulli beam the axis $\underline{e}_1^2$ (the first row of matrix B) is identical with the tangent vector of the beam's axis $w(R_1,t)$. Thus angles ϑ_2 and ϑ_3 can be represented by derivatives of the deformation-variables $w_\alpha(R_1,t)$. A second order approximation gives

$$\vartheta_2 = -w_3' - w_1'w_3' \qquad \vartheta_3 = w_2' - w_1'w_2'. \tag{26}$$

To represent strain, the Green-Lagrange strain tensor is used. For the specific case of a beam one finds

$$G_1 = \begin{bmatrix} G_{11} & G_{12} & G_{13} \\ G_{21} & 0 & 0 \\ G_{31} & 0 & 0 \end{bmatrix} \tag{27}$$

where in a second order approximation

$$\begin{aligned} G_{11} &= w_1' + R_3\vartheta_2' - R_2\vartheta_3' + \frac{1}{2}(w_1'^2 + w_2'^2 + w_3'^2) \\ &\quad + R_2(-w_1'\vartheta_3' + w_3'\vartheta_1' + \vartheta_1'\vartheta_2 + \vartheta_1\vartheta_2') + R_3(w_1'\vartheta_2' - w_2'\vartheta_1' + \vartheta_1'\vartheta_3 + \vartheta_1\vartheta_3') \\ &\quad + \frac{1}{2}R_2^2(\vartheta_1'^2 + \vartheta_3'^2) + \frac{1}{2}R_3^2(\vartheta_1'^2 + \vartheta_2'^2) - R_2R_3\vartheta_2'\vartheta_3', \qquad (28) \\ G_{12} &= G_{21} = \frac{1}{2}(w_2' - \vartheta_3 - w_1'\vartheta_3 + w_3'\vartheta_1 + \vartheta_1\vartheta_2 - R_3(\vartheta_1' - \vartheta_2\vartheta_3')), \\ G_{13} &= G_{31} = \frac{1}{2}(w_3' - \vartheta_2 + w_1'\vartheta_2 - w_2'\vartheta_1 + \vartheta_1\vartheta_3 + R_2(\vartheta_1' - \vartheta_2\vartheta_3')). \end{aligned}$$

In the expressions for $G_{12} = G_{21}$ and $G_{13} = G_{31}$ there appear two contributions. The first one is independent of the material coordinates R_2 and R_3 and it represents shear deformation. The second contribution contains the common term $\vartheta_1' - \vartheta_2\vartheta_3'$; it describes the effect of torsion. For a beam without torsion one obtains

$$\vartheta_1' = \vartheta_2\vartheta_3'. \tag{29}$$

Now new deformation variables $\varphi_\alpha(R_1, t)$ are introduced. They represent the orientation of $\underline{e}^2$ with respect to a triad fixed in the beams cross section, which is obtained in case of an Euler-Bernoulli-beam without torsion, i. e.

$$\begin{aligned} \vartheta_1' &= \vartheta_2\vartheta_3' + \varphi_1' \\ \vartheta_2 &= -w_3' + w_1'w_3' + \varphi_2 \\ \vartheta_3 &= w_2' - w_1'w_2' + \varphi_3. \end{aligned} \tag{30}$$

In these terms the off diagonal elements of the strain tensor reduce to

$$G_{12} = G_{21} = -\frac{1}{2}R_3\varphi_1', \qquad G_{13} = G_{31} = -\frac{1}{2}R_2\varphi_1'. \tag{31}$$

Using these kinematical relations, Hooke's law for the stress-strain relations and applying Hamilton's principle, one obtains the dynamical equations of motion and the boundary conditions as detailed in [1]. The derivation of the equations of motion shown below as equations (32) are based on the following assumptions

1. All the parameters appearing in the equations are of the order $O(1)$. All the deformation variables w_α and φ_α are of the order $O(\varepsilon)$ of a small parameter ε. All second order terms $O(\varepsilon^2)$ are neglected.

2. The frame $\{O^i, \underline{e}^i\}$ is an inertial frame - there is no superimposed reference motion.

3. The origin of the frame fixed in the cross section is the center of mass and the 2- and 3- axes of the basis are principal axes of inertia.

Based on these assumptions one finds the equations of motion:

$$\begin{aligned} &\rho_0 A\ddot{w}_1 - (\mathcal{E}Aw_1')' && - A_{w1} = 0 \\ &\rho_0 A\ddot{w}_2 - (\rho_0 J_{22}(\ddot{w}_2' + \ddot{\varphi}_3))' + (\mathcal{E}J_{22}(w_2'' + \varphi_3'))'' && - A_{w2} = 0 \\ &\rho_0 A\ddot{w}_3 - (\rho_0 J_{33}(\ddot{w}_3' - \ddot{\varphi}_2))' + (\mathcal{E}J_{33}(w_3'' + \varphi_2'))'' && - A_{w3} = 0 \\ &J_p(\rho_0\ddot{\varphi}_1 + (\mathcal{G}\varphi_1')') && - A_{\varphi 1} = 0 \\ &\rho_0 J_{33}(\ddot{\varphi}_2 - \ddot{w}_3') + \frac{1}{2}\mathcal{E}A\varphi_2 - (\mathcal{E}J_{33}(\varphi_2' - w_3''))' && - A_{\varphi 2} = 0 \\ &\rho_0 J_{22}(\ddot{\varphi}_3 + \ddot{w}_3') + \frac{1}{2}\mathcal{E}A\varphi_3 - (\mathcal{E}J_{22}(\varphi_3' + w_2''))' && - A_{\varphi 3} = 0 \end{aligned} \tag{32}$$

Here ρ_0 is the density, $\mathcal{E}$ is Young's modulus, $\mathcal{G}$ is the modulus of rigidity, A is the area of the cross section, J_{ii} are the principal moments of inertia of the cross section, J_p is the corresponding polar moment of inertia and the coefficients $A_{w\alpha}$

and $A_{\varphi\alpha}$ represent the external forces and torques upon the beam. It is easy to see that these equations reduce to the well known equations of motion for bending motions of an Euler-Bernoulli beam: Such motions satisfy $\varphi_2 \equiv \varphi_3 \equiv 0$ and the corresponding equations of motion are usually derived neglecting shear forces, which implies $w'''_{2,3} \equiv 0$. For this case one concludes from the fifth and sixth equation of the set (32)

$$\rho J_{22}\ddot{w}_2' = \rho_0 J_{33}\ddot{w}_3' = 0. \tag{33}$$

This is the well known result saying that neglecting the influence of shear on the bending motions also requires to neglect the rotational inertia of the cross sections in the second and third equation of the set (32) both of which describe bending.

We now use the same procedure to develop the equations for the planar motions of a beam which rotates about its 2-axis as shown in Fig. 17. The velocity $\dot{\alpha} = -\dot{\psi}$ and the acceleration $\ddot{\psi}$ of the reference motion are as given by Fig. 18. The derivation of the equations of motion is now based on slightly modified assumptions:

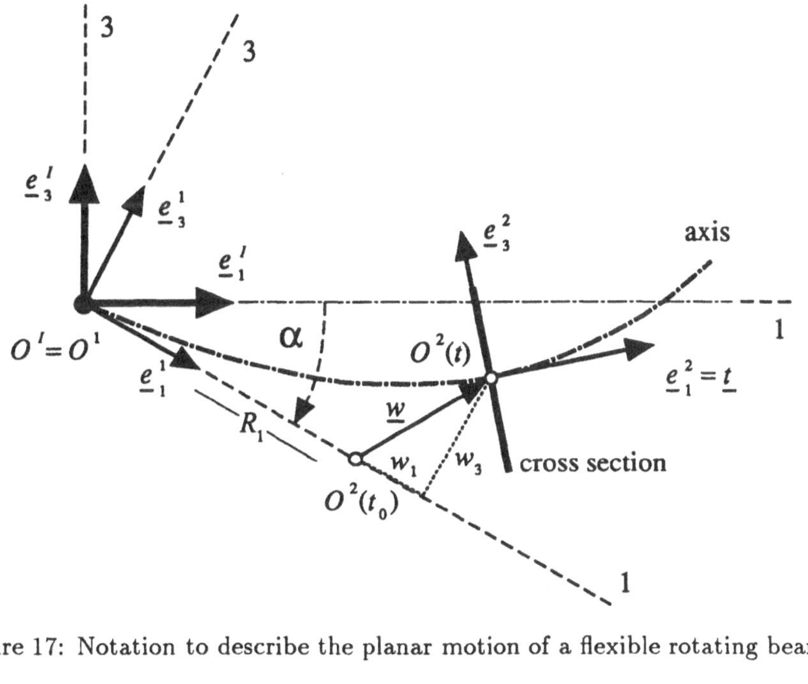

Figure 17: Notation to describe the planar motion of a flexible rotating beam.

1. The deformation variables are of the order $O(\varepsilon)$ and all the parameters in the equations of motion are of the order $O(1)$, excluding the longitudinal stiffness $\mathcal{E}A$, which is considered to be "very large", i. e. of the order $O(1/\varepsilon)$;

2. The reference motion and resulting centrifugal forces are of the order $O(1)$, but the acceleration $\ddot{\psi}$ is small, i. e. of the order $O(\varepsilon)$;

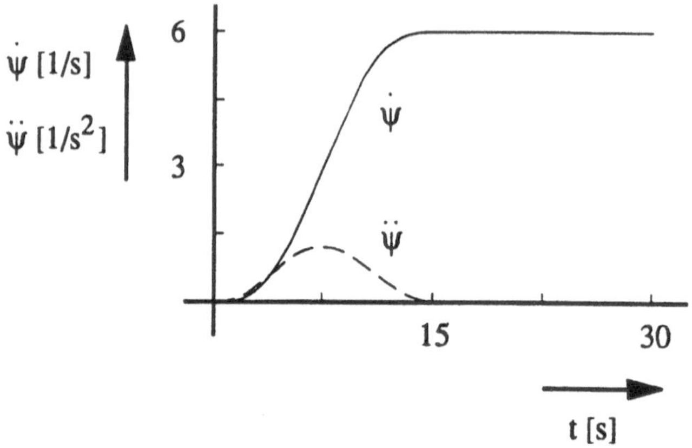

Figure 18: Velocity and acceleration of the reference motion of the rotating beam shown in Fig. 17

3. For the frame fixed in the cross section the assumptions mentioned above hold now as before.

Based on these conditions the procedure used previously yields the following linearized equations of motion

$$(\mathcal{E}Aw_1')' + \rho_0 A R_1 \dot{\psi}^2 - A_{w10} = 0 \tag{34}$$

$$\begin{aligned} \rho_0 A(\ddot{w}_3 + R_1\ddot{\psi} + 2\dot{\psi}\dot{w}_1 - \dot{\psi}^2 w_3) - (\rho_0 J_{33}(\ddot{w}_3' + \ddot{\psi} - w_3'\dot{\psi}^2))' \\ - (\mathcal{E}Aw_1'w_3')' + (\mathcal{E}J_{33}w_3'')'' - A_{w31} = 0 \end{aligned} \tag{35}$$

Assuming that the external actions are zero, i. e. $A_{w10} = A_{w31} = 0$, the first equation (34) describes of how to compute the elongation w_1 of the beam due to the centrifugal forces resulting from the reference motion. The term $(\mathcal{E}Aw_1')'$ represents the reference stresses balancing the centrifugal forces applied at the cross section. In the second equation (35) for the bending motion a term $(\mathcal{E}Aw_1'w_3')$ appears. It is interpreted as a nonlinear term sometimes, but this is not really true. Using equation (34) the $O(1)$-term $\mathcal{E}Aw_1'$ can be replaced by the centrifugal forces

$$\mathcal{E}Aw_1' = -\int_0^{R_1} \rho_0 A x \dot{\psi}^2 \, dx. \tag{36}$$

Introducing this representation of the reference stresses one obtains a linear equation (35) to describe the bending motion. The term including the reference stresses is the geometric stiffening term known from equation (21). Here, it results from the large centrifugal forces and from the nonlinearity representing the coupling between longitudinal and bending motions of a beam, which has its ultimate source in the nonlinearity of the stress-displacement relations (28). For catching this term a second order expansion of the kinematical relations was necessary.

The techniques described here have been implemented in the multibody code SIMPACK using a Ritz-approximation to solve the partial differential equations. For the example shown in Figs. 17 and 18 one obtains the result given in Fig. 19. It shows the elongation and the deflection of the tip of the beam due to the reference motion given in Fig. 18. The data of the beam are as follows:

length	L	$= 10$ m
cross section	A	$= 4 \cdot 10^{-4}$ m^2
moment of inertia	J_{33}	$= 2 \cdot 10^{-7}$ m^4
Young's modulus	$\mathcal{E}$	$= 7 \cdot 10^{10}$ N/m^2
density	ρ_0	$= 3 \cdot 10^3$ kg/m^3 .

The importance of considering the geometric stiffening terms in multibody system simulation has been demonstrated in [17]. Neglecting geometric stiffening yields a result in which the deflection of the beam grows to infinity. The computational savings resulting from the representation of the motion of the flexible body as a sum of a large reference motion and a superimposed small deformation as compared to a finite element analysis have been discussed in [45, 44]. For the example of a rotating beam the multibody code SIMPACK using the techniques described here is 70 times faster than FEM-approach using the code ANSYS.

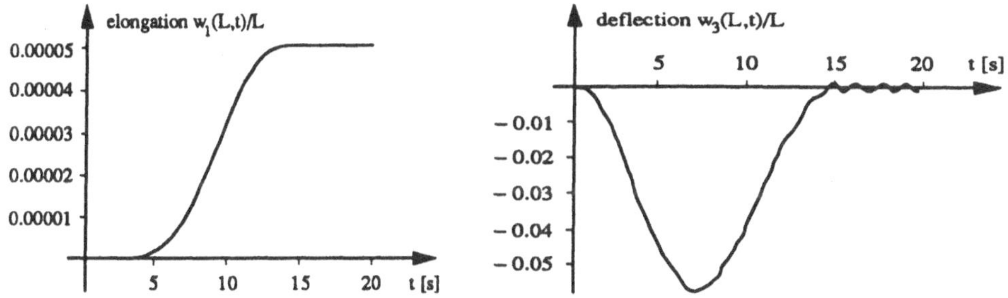

Figure 19: Displacement of the tip of the rotating beam shown in Fig. 17 .

In discussions of geometric stiffening it has been mentioned that catching the geometric stiffening terms is a question of the choice of variables to express strain energy and perform the linearization [33, p. 167]. Of course, this is not really true. A set of variables used frequently to represent the motions of Euler-Bernoulli-beams is provided by concepts of differential geometry of curves[4]. The axis of the beam is a curve in three dimensional space as shown in Fig. 20. It is given by the parameter representation

$$w_\alpha = w_\alpha(R_1),\ \alpha = 1,2,3 \tag{37}$$

and a point on the beam's axis is given by

$$\rho = \rho(R_1) = \begin{bmatrix} R_1 \\ 0 \\ 0 \end{bmatrix} + w(R_1) \tag{38}$$

when all the vectors are resolved in the frame $\{O^i, \underline{e}^i\}$. Instead of using (37) one can

[4]Such coordinates have been proposed in [33] to obtain the geometric stiffening terms.

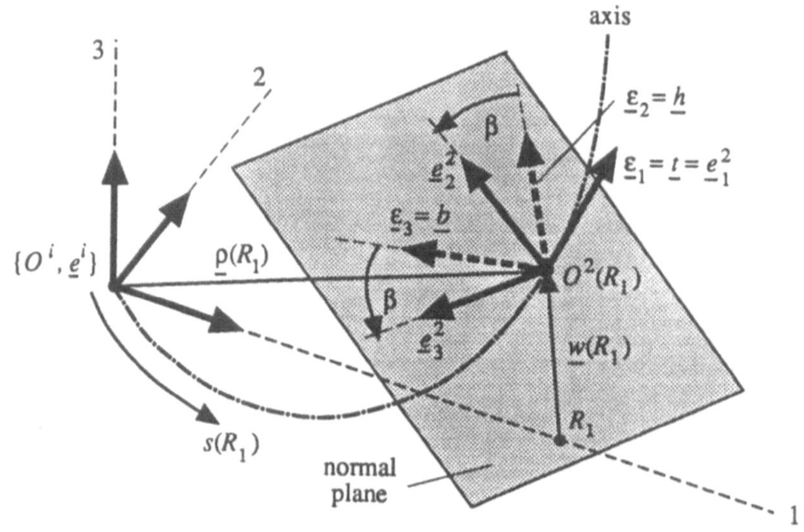

Figure 20: Notation to define a set of "natural" variables to represent the motion of an Euler-Bernoulli-beam.

represent the curve by the three functions

$$s = s(R_1) \qquad k_1 = k_1(R_1) \qquad k_3 = k_3(R_1) \tag{39}$$

where s is the arc length, and k_1 and k_3 are the two curvatures. All of these quantities can be computed from (37) using well known formulas given in any textbook on differential geometry of spatial curves. The representation of spatial curves is simplified considerably when introducing s instead of R_1 as a parameter, i. e. when using $k_1 = k_1(s)$ and $k_3 = k_3(s)$ to represent the curve.

A "natural frame" $\{O^2, \underline{\varepsilon}\}$ can be associated with the axis as shown in Fig. 20. Its orientation is given by the direction cosine matrix C as

$$\underline{\varepsilon} = C\underline{e}^i, \qquad C = [C_{\alpha\beta}], \qquad \alpha, \beta = 1, 2, 3 \tag{40}$$

When proceeding along the curve the orientation of $\underline{\varepsilon}$ is changed as described by Frenet's equations

$$\frac{d\underline{\varepsilon}}{ds} = -\tilde{k}\underline{\varepsilon}, \quad \underline{k} = \underline{\varepsilon}^T k, \quad k = \begin{bmatrix} k_1(s) \\ 0 \\ k_3(s) \end{bmatrix} \tag{41}$$

Here $k_i, i = 1, 3$ are the curvatures mentioned previously. The basis $\underline{\varepsilon}$ cannot be used as a triad fixed in the crossection of the beam: A straight beam may be twisted due to torsion, but both of the curvatures $k_i(s)$ are zero for a straight line. To represent torsion an angle β is introduced describing the orientation of a basis $\underline{e}^2$ with respect to $\underline{\varepsilon}$ where the basevectors $\underline{e}_2^2$ and $\underline{e}_3^2$ are aligned with the principal axes of inertia of the cross sections

$$\underline{e}^2 = A\underline{\varepsilon} \qquad A = \begin{bmatrix} 1 & 0 & 0 \\ 0 & \cos\beta & \sin\beta \\ 0 & -\sin\beta & \cos\beta \end{bmatrix}. \tag{42}$$

In these terms the matrix B from (22) becomes

$$B = AC. \tag{43}$$

Moreover by analogy with (41)

$$\frac{d\underline{e}^2}{ds} = \frac{dB}{ds}\underline{e}^i \tag{44}$$

$$\frac{dB}{ds} = \tilde{\kappa}B, \quad \underline{\kappa} = \underline{e}^{2^T}\kappa, \quad \kappa(s) = \begin{bmatrix} k_1(s) + d\beta/ds \\ k_3(s)\sin\beta(s) \\ k_3(s)\cos\beta(s) \end{bmatrix} \tag{45}$$

These relations give an interpretation of the curvatures κ_α often used in beam theory. The equations are in full correspondence with Poisson's equation, relating angular velocity coordinates and time derivatives of orientation matrices [31]. In terms of s and $\kappa_\alpha(s)$ one can introduce instead of $w_\alpha(R_1)$ and $\varphi_1(R_1)$ a new set of deformation variables representing the motion of an Euler-Bernoulli-beam. They are

$$\begin{aligned} v_1(R_1,t) &= s(R_1,t) - R_1 \\ &= \int_0^{R_1} \sqrt{(1+w_1'(\xi,t))^2 + (w_2'(\xi,t))^2 + (w_3'(\xi,t))^2}d\xi - R_1 \end{aligned} \tag{46a}$$

$$v_2(R_1,t) = +\int_0^{R_1}\int_0^{\xi} \kappa_3(\eta,t)d\eta d\xi \tag{46b}$$

$$v_3(R_1,t) = -\int_0^{R_1}\int_0^{\xi} \kappa_2(\eta,t)d\eta d\xi \tag{46c}$$

$$\tau(R_1,t) = \int_0^{R_1} \kappa_1(\xi,t)d\xi \tag{46d}$$

Variable v_1 represents elongations of the beam, v_2 and v_3 represent the two deflections and τ describes torsion. The variables have the particular advantage that a particular deformation can be locked just by setting the corresponding variable to zero. The new variables (46) can be represented in terms of w_α and φ_1 and vice versa. Assuming the deformation variables to be small one obtains in a second order approximation

$$w_1 = v_1 - \frac{1}{2}\int_0^{R_1}(v_2'^2 + v_3'^2)d\xi \tag{47a}$$

$$w_2 = v_2 + \int_0^{R_1} v_1'v_2'd\xi + \int_0^{R_1}\int_0^{\xi} v_1'v_2''d\eta d\xi - \int_0^{R_1}\int_0^{\xi} \tau v_3''d\eta d\xi \tag{47b}$$

$$w_3 = v_3 + \int_0^{R_1} v_1' v_3' d\xi + \int_0^{R_1}\int_0^{\xi} v_1' v_3'' d\eta d\xi + \int_0^{R_1}\int_0^{\xi} \tau v_2'' d\eta d\xi \tag{47c}$$

$$\varphi_1 = \tau_1 + \int_0^{R_1} v_1' \tau_1' d\xi \tag{47d}$$

These relations are found developing a second order approximation for B in terms of the variables (46) using equations (44), (45). Realizing that the first row of B corresponds to the tangent of the beam's axis in case of an Euler-Bernoulli-beam, one obtains the transformations (47) after the tangent vector has been represented by derivatives of the functions (37). In this context it is important to recognize that equation (45) is formulated in terms of $s = s(R_1)$, whereas the deformation variables w_α, φ_1 and v_α, τ depend on R_1. In particular, the relation

$$\frac{dB}{ds} = \frac{1}{s'}\frac{dB}{dR_1} = \frac{1}{s'}B' = (1 - v_1')B'$$

is used to manipulate (45) and it yields additional terms in (47), which may be forgotten easily. On the other side, representing the curvatures κ_α in terms of the variables w_α and ϑ_α one finds considering (30)

$$v_1 = w_1 + \frac{1}{2}\int_0^{R_1} (w_2'^2 + w_3'^2) d\xi \tag{48a}$$

$$v_2 = w_2 - \int_0^{R_1} w_1' w_2' d\xi - \int_0^{R_1}\int_0^{\xi} w_1' w_2'' d\eta d\xi + \int_0^{R_1}\int_0^{\xi} \vartheta_1 w_3'' d\eta d\xi \tag{48b}$$

$$v_3 = w_3 - \int_0^{R_1} w_1' w_3' d\xi - \int_0^{R_1}\int_0^{\xi} w_1' w_3'' d\eta d\xi - \int_0^{R_1}\int_0^{\xi} \vartheta_1 w_2'' d\eta d\xi \tag{48c}$$

$$\tau = \vartheta_1 - \int_0^{R_1} w_1' \vartheta_1' d\xi + \int_0^{R_1} w_3' w_2'' d\xi = \varphi_1 - \int_0^{R_1} w_1' \varphi_1' d\xi \tag{48d}$$

The transformations demonstrate that the deformation variables w_α and φ_1 are identical with the new variables v_α and τ in the linear approximation used to derive the equations of motion (34), (35). The appearance of geometric stiffening terms is not a question of the choice of variables - it depends on the magnitude of parameters in the system equations.

The two sets of coordinates have been compared in a Ph-D-thesis, which appeared recently [4]. In this publication one finds as well a comparison of the representation of the potential energy

$$U = \frac{1}{2}\int_0^{L} (\mathcal{E}Av_1'^2 + \mathcal{G}J_T\kappa_1^2 + \mathcal{E}J_{22}\kappa_2^2 + \mathcal{E}J_{33}\kappa_3^2)\, dR_1 \tag{49}$$

often used in beam theory as compared to a representation of U based on the description of deformation by the Green Lagrange strain tensor used previously. A remarkable result found in [4] is that both of the representations are identical in a fourth order approximation in the case of beams with circular cross sections.

2.4. STANDARDIZATION OF DATA TO DESCRIBE FLEXIBLE BODIES IN MULTIBODY SIMULATION CODES

One of the major concerns of users of multibody simulation codes, mentioned in [27] is the fact that an integrated environment for applying multibody codes together with other software is not available. In particular, a simple exchange of data with finite element software is needed. To facilitate the usage of data describing various finite element (or other) models of flexible bodies a set of standard input data is proposed in a recent paper [47, 48], using concepts as described above. The data are developed by generalizing equations (7) with b^i given by (8) for rigid body systems to equations for flexible body systems having the general form

$$J_f^i \begin{bmatrix} \dot{\omega}^i \\ a^i \\ \ddot{q} \end{bmatrix} = Q_f^i + C_f^i \Lambda^i - \sum_{b:b \neq i} S^{ib} D_f^{\kappa(b)} \Lambda^b \ , i = 1, 2, ... N \tag{50}$$

with variables q representing the deformations as suggested by (16). The system matrices appearing in (50) have additional submatrices due to the flexibility of the bodies. A standardization of the data required for the computation of their elements is proposed in [46]. It has been used to generalize a standard data set defined for rigid body systems in a joint effort supported by the German Research Council [35].

The set of Standard Input Data (SID) is used to describe the flexible truss structure (NASA Minimast) shown in Fig. 21 to the multibody code SIMPACK, [34]. A first analysis of a spin-up manoeuvre of the structure using an equivalent beam model, demonstrates - see Fig. 22 - that geometric stiffening is to be considered, when modelling the truss as a flexible body of a multibody system. The motion of the structure is given in this example by the functions which have been used previously and which are shown in Fig. 18.

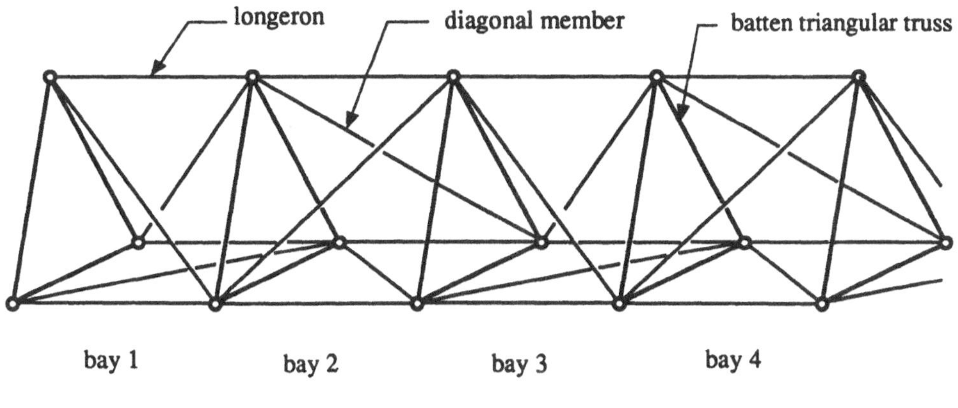

Figure 21: A flexible truss structure to be modelled as a single flexible body of a multibody system

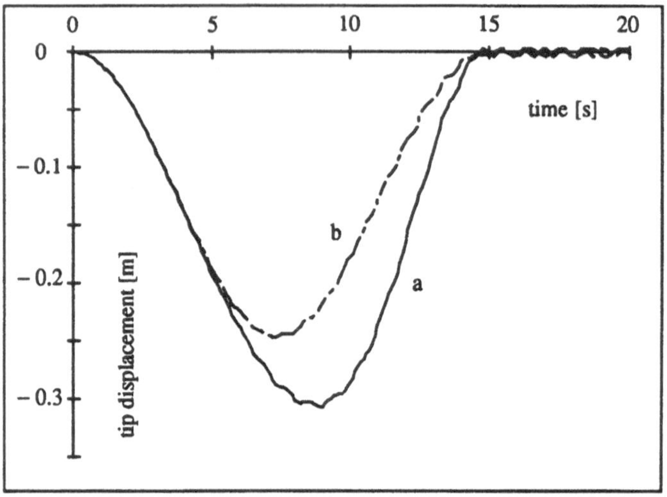

Figure 22: Tip displacement of the truss due to the planar spin-up manoeuvre from Fig. 18: a.) without and b.) with geometric stiffening

3. Conclusion

The formulation of multibody system equations affects the execution time of the corresponding codes considerably. To increase the computational speed of multibody codes two aspects of the formulation of multibody system equations have been discussed. First, it has been demonstrated that parallel computer architectures result in significant reductions of multibody simulation time when the methods available from multibody dynamics, numerical mathematics and computer science are tuned properly. A new residual formalism has been found, which has a high potential for parallelization when combined with implicit integration routines. Using the farming concept to implement the methods on a Transputer network results in a simulation program, which presents itself to the user in the same way as any serial code. In particular, the simulation program does not ask the user to distribute the workpackages in the network. For a realistic example a reduction of computer time by a factor of ten has been obtained.

A second field which can be exploited to reduce simulation time is adequate modelling of flexible bodies in multibody systems. In many applications the motion of a flexible body can be represented as a sum of a large reference motion and a small deformation. Linearizing the equations of motion in the deformation variables yields computational savings of multibody codes as compared to finite element codes dealing with more sophisticated models. A correct linearization requires to include so-called geometric stiffening terms, which have been discussed here. Based on this representation of flexible body motion in multibody systems, a standard set of input data for flexible bodies has been developed to facilitate the usage of flexible body data for a simulation with various multibody codes.

ACKNOWLEDGEMENT

Part of the research work reported here was made possible by the German Research Council (Deutsche Forschungsgemeinschaft). Its support is highly acknowledged. The author would like to thank A. Eichberger, C. Führer, and O. Wallrapp for their help in preparing this contribution.

4. References

1 S. Albrecht. Bewegungsgleichungen elastischer Balken und Platten für MKS-Simulationen. Technical Report, DLR IB 515-92-09, 1992.

2 D.S. Bae, J.G. Kuhl, and E.J. Haug. A recursive formulation for constrained mechanical system dynamics: Part iii, Parallel processor implementation. *MECH. STRUCT. &MACH.*, pp. 249–269, 1988.

3 D. E. Bernard, G. K. Man. *Proc. of the 3rd Annual Conference on Aerospace Computational Control.* JPL, Passadena, CA, Publ. 89-45, 1989.

4 M. Botz. *Zur Dynamik von Mehrkörpersystemen mit elastischen Balken*, Dissertation, TH Darmstadt, 1992.

5 K. Ebert, S. Graul, and R. Schwertassek. Simulation von Mehrkörpersystemen; Methode, Stand der Technik und Anforderungen für orbitale Systemdynamik. *DGLR-Jahrbuch*, pp. 226–235, 1988.

6 A. Eichberger. *Simulation von Mehrkörpersystemen auf parallelen Rechnerarchitekturen.* Dissertation, Universität - Gesamthochschule - Duisburg, Fachbereich Maschinenbau, Fortschrittberichte VDI, Reihe 8, Nr. 332, VDI-Verlag, Düsseldorf, 1993.

7 R. Fößmeier and U. Rüde. Operating System Support for Parallel Numerical Software Developement. Tum-info-10-87-i12-350/1-fmi, Mathematisches Institut und Institut für Informatik, Technische Univerität München, Arcisstr. 21, W-8000 München, 1987.

8 C. Führer. *Differential–algebraische Gleichungssysteme in mechanischen Mehrkörpersystemen.* Dissertation, Mathematisches Institut, Technische Universität München, 1988.

9 I. Graham and T. King. *The Transputer Handbook.* Prentice Hall International, 1990.

10 E. Hairer, S.P. Nørsett, and G. Wanner. *Solving Ordinary Differential Equations I, Nonstiff Problems.* Springer, 1987.

11 E. J. Haug. *Computer–Aided Kinematics and Dynamics of Mechanical Systems.* ALLYN and BACON, Boston, London, Sydney, Toronto, 1989.

12 E. J. Haug and R. C. Deyo, editors. *Real–Time Integration Methods for Mechanical System Simulation.* NATO ASI Series, Vol. F69, Springer-Verlag Berlin, 1990.

13 R.L. Huston. *Multibody Dynamics.* Butterworth–Heinemann, Boston, 1990.

14 The Transputer Application Notebook, Architecture and Software. INMOS Databook Series, INMOS Limited, 1000 Aztec West, Almondsbury Bristol BS12 4SQ, UK. INMOS document number: 72–TRN–205–00.

15 R. E. Jones. Multi–flex body dynamics for control design. In G. Man and R. Laskins, editors, *Proceedings of the Workshop on Multibody Simulation*, pp. 543–549. JPL D-5190, Pasadena, California, 1988.

16 C. Jung. *Reduktion nichtlinearer Systeme am Beispiel Fahrzeugsimulation.* Fortschrittberichte VDI, Reihe 8, Nr. 289, VDI-Verlag, Düsseldorf, 1992.

17 T.R. Kane, R.R. Ryan, and A.K. Banerjee. Dynamics of a beam attached to a moving base. In *AAS/AIAA Astrodynamics Specialist Conference, Vail, Colorado*, 1985.

18 W. Kortüm et al. Iltis benchmark proposal. Progress report to the 12th IAVSD symposium on a workshop and resulting activities. Technical report, DLR, Institute for Flight System Dynamics D-8031 Wessling, FRG, 1991.

19 K. Kreuzer and W. Schiehlen. *NEWEUL - Software for the Generation of Symbolical Equations of Motion*, pp. 181–202. Springer-Verlag, Berlin, 1990.

20 G. Leister. *Beschreibung und Simulation von Mehrkörpersystemen mit geschlossenen kinematischen Schleifen.* Fortschritt–Berichte VDI, Reihe 11, Nr. 167, VDI-Verlag, Düsseldorf, 1992.

21 P. W. Likins. Geometric stiffness characteristics of a rotating elastic appendage. *Int. J. Solids Structures*, 10, pp. 161-17, 1974.

22 P. W. Likins. Multibody dynamics – an historical perspective. In G. Man and R. Laskins, editors, *Proceedings of the Workshop on Multibody Simulation*, JPL D-5190, Pasadena, California, pp. 543–549, 1988.

23 P.W. Likins. Dynamics and control of flexible space vehicles. Technical Report, JPL-TR-32-1329, California Institute of Technology, 1970.

24 P. Lohmeier. Strukturanalyse und Stabilität rotierender elastischer Raumstationen, Dissertation, TU München, 1974.

25 J.Y.S. Luh, M.W. Walker, and R.P.C. Paul. On–line computational scheme for mechanical manipulators. *Journal of Dynamic Systems, Measurement and Control*, 102, pp. 69–76, 1980.

26 G. K. Man and R. A. Laskin. Proc. of the Workshop on Multibody Simulation, Technical Report, JPL D-5190, 1988.

27 G. K. Man and S. W. Sirlin. An assessment of multibody simulation tools for articulated spacecraft. In D. E. Bernhard and G. K. Man, editors, *Proc. 3rd Annual Conference on Aerospace Computational Control*, pp. 12–25. JPL, Pasadena, California, 1988.

28 L. Meirovitch and M.K. Kwak. A method for improving the convergence characteristics of substructure synthesis. In *Proc. Int. Symp. on Advanced Computer for Dynamics and Design*, Tsuchiura, Japan, 1989.

29 V.J. Modi. Attitude dynamics of satellites with flexible appendages - a brief review. *J. Spacecraft Rockets*, 11, pp. 743-751, 1974.

30 L. R. Petzold. A Description of DASSL - A Differential-Algebraic System Solver. In *IMACS Transactions Sci. Comp.*, p. 65, 1983.

31 R.E. Roberson and R. Schwertassek. *Dynamics of Multibody Systems*. Springer-Verlag, Berlin, 1988.

32 W. Rulka. SIMPACK - A Computer Program for Simulation of Large-Motion Multibody Systems. In W. Schiehlen, editor, *Multibody Systems Handbook*, pp. 265–284. Springer-Verlag, Berlin, 1990.

33 R. R. Ryan. Flexible multibody dynamics: Problems and solutions. In G. Man and R. Laskins, editors, *Proceedings of the Workshop on Multibody Simulation*, JPL D-5190, Passadena, Cal., pp. 103–190, 1988.

34 D. Sachau. Berücksichtigung elastischer Körper in Mehrkörpersimulationen. Technical Report, DLR-IB 515-93-19, Köln, 1993.

35 W. Schiehlen. Symbolic computations in multibody systems. In *NATO-Advanced Study Institute*, Lisboa, Portugal, 1993.

36 W. Schwartz. The IAVSD Road Vehicle Benchmark Bombardier Iltis. Technical Report ZB-64, Institut B für Mechanik, Prof. Dr.-Ing. W. Schiehlen, Universität Stuttgart, 1991.

37 R. Schwertassek and W. Rulka. Aspects of Efficient and Reliable Multibody System Simulation. In E.J. Haug and R.C. Deyo, editors, *Real-Time Integration Methods for Mechanical System Simulation*, pp. 115–126. NATO ASI Series, Vol. F69, Springer-Verlag, Berlin, 1990.

38 M. Sherman. The practical application of symbolic equation manipulation to multibody dynamics. In G. Man and R. Laskins, editors, *Proceedings of the Workshop on Multibody Simulation*, JPL D-5190,Passadena, Cal., pp. 952–982, 1988.

39 B. Simeon, C. Führer, and P. Rentrop. Differential–algebraic equations in vehicle systems dynamics. *Surveys on Mathematics for Industry*, pp. 1–37, 1991.

40 J.C. Simo and L. Vu-Quoc. On the dynamics of flexible beams under large overall motions - the plane case, part I. *J. Appl. Mech.*, 53, pp. 849–854, 1986.

41 L. Taylor. Proc. of the workshop on computational aspects in the control of flexible systems. Technical Report TM-101578-Pt-1, NASA, 1988.

42 A. Truckenbrodt. *Bewegungsverhalten und Regelung hybrider Mehrkörpersysteme mit Anwendung auf Industrieroboter*. Fortschritt–Berichte VDI, Reihe 8, Nr. 33, VDI-Verlag Düsseldorf, 1980.

43 O. Wallrapp and C. Führer. MEDYNA - An Interactive Analysis and Design Program for Geometrically Linear and Flexible Multibody Systems. In W. Schiehlen, editor, *Multibody Systems Handbook*, pp. 203–223, Springer–Verlag Berlin, 1990.

44 O. Wallrapp and R. Schwertassek. Geometric stiffening in multibody system simulation. In *Proc. 8th VPI&SU Symposium on Dynamics and Control of Large Structures*, L. Meirovitch, ed., Blacksburg, VA, pp. 405–416, 1991.

45 O. Wallrapp and R. Schwertassek. Representation of geometric stiffening in multibody system simulation. *Int. Journal for Numerical Methods in Engineering*, pp. 1833–1850, 1991.

46 O. Wallrapp. Standard input data of flexible bodies for multibody system codes. Technical Report, DLR-IB 515-93-04, Köln, 1993.

47 O. Wallrapp. Standardization of flexible body modelling in multibody codes, Part I: Definition of Standard Input Data, to appear in *Mechanics of Structures and Machines*, 1993.

48 O. Wallrapp. Standard input data of flexible members in multibody systems. In W. Schiehlen, editor, *Advanced Multibody System Dynamics*, pp. 445–450, Kluwer Academic Publ., Dordrecht, 1993.

COMPUTATIONAL CHALLENGES IN MECHANICAL SYSTEMS SIMULATION

LINDA R. PETZOLD [1]
Department of Computer Science
University of Minnesota
Minneapolis, Minnesota 55455

ABSTRACT: The numerical solution of the differential-algebraic equations of motion of mechanical systems offers many computational challenges. In this paper we describe progress which has been made in understanding the formulation of the equations of motion from the viewpoint of numerical stability, outline some of the difficulties which must be resolved for efficient and reliable numerical methods in real-time simulation of mechanical systems, and propose some solutions.

1. Introduction

In recent years much activity has been devoted to the development of numerical methods and underlying theory for the solution of differential-algebraic equation (DAE) systems. These types of systems occur frequently as initial value problems in the computer-aided design and modeling of mechanical systems subject to constraints, electrical networks, chemically reacting systems such as distillation, flow of incompressible fluids, and in many other applications. Differential-algebraic systems, which in general can take the form $F(t, y, y') = 0$, are different from standard-form ODE systems $y' = f(t, y)$ in that, while they include ODE systems as a special case, they also include problems which are quite different from ODEs.

In a sense, the more singular a DAE system is, the more difficult it is to solve numerically. The *index* of a DAE system is a measure of the degree of singularity of the system. Roughly speaking, ODE systems $y' = f(t, y)$ are index zero, differential equations coupled with algebraic constraints, $y' = f(y, z)$, $0 = g(y, z)$, where $\partial g / \partial z$ is nonsingular, are index one, and differential equations coupled with algebraic constraints where z cannot be solved for uniquely in g as a function of y are of index higher than one. The index can be defined also for systems which are not expressed in the semi-explicit form of differential equations coupled with algebraic constraints. Additional difficulties can occur for these systems because the singularity may be moving from one part of the system to another.

*This work was partially supported by the U.S. Army Research Office contract number DAAL03-89-C-0038 with the University of Minnesota Army High Performance Computing Research Center, and by ARO contract numbers DAAL03-92-G-0247 and DAAL03-92-G-0409 and DOE contract number DE-FG02-92ER25130, and by the Minnesota Supercomputer Institute.

M. F. O. S. Pereira and J. A. C. Ambrósio (eds.),
Computer-Aided Analysis of Rigid and Flexible Mechanical Systems, 483–499.

Much progress has been made on understanding the underlying structure and numerical solution of DAE systems. Fundamental concepts such as index and solvability have been extended to classes of DAEs describing a broad range of scientific and engineering problems. Convergence results have been given for numerical methods such as multistep and Runge-Kutta applied to several important classes of DAEs. Production-level computer codes such as DASSL [11] have been employed extensively for the solution of (index-one) engineering problems. Much of this work is described in the recent monographs [11, 29, 30].

There is much still that needs to be done for the effective solution of certain classes of DAEs. In this paper we will focus on the algorithms and analysis which are needed for the effective real-time simulation of mechanical systems. Real-time simulation of mechanical systems is needed in robotics, as well as in the design and simulation of vehicles, including automobiles, high-speed trains, tanks and construction equipment.

The modeling of multibody systems gives rise to Euler-Lagrange equations. Any effective numerical method for these systems must be very fast and extremely robust because the systems must often be solved repetitively by design engineers who do not have time to develop a working knowledge of complex computer software or numerical methods. For some important applications such as vehicle simulation and design, the systems must be solved in real-time.

Euler-Lagrange equations are usually posed initially in the form of a system of differential equations (Newton's laws of motion) coupled with nonlinear constraints which are enforced via a Lagrange multiplier. Direct discretization of this index-three system yields numerical methods which are often not very robust because of well-known[11] difficulties with error estimation and stepsize control, as well as severe ill-conditioning of linear systems at each time step and other problems. A wide variety of reformulations of the problem and associated numerical methods have been suggested in an attempt to find a system of equations describing the system which can be effectively solved numerically. However, each has some apparent disadvantage in terms of speed and/or robustness. Because the constraints are sometimes highly nonlinear and have a strong physical relevance, it is generally considered important that the constraints, and sometimes the time derivative of the constraints, be satisfied very accurately. In addition, there are other potential difficulties: the constraints can become rank-deficient or nearly rank-deficient, the solution may have components which are oscillating at a high frequency, and there is the possibility of frequent discontinuities which are especially troublesome because the solution of a high-index DAE can be less continuous than its input. Real-time simulation imposes severe requirements on the solution method. The solution must be computed extremely rapidly, necessitating the use of parallel computers and exploitation of the system structure. The challenge for multibody systems is to develop a problem formulation and associated class of numerical methods which preserves the stability of the system, ensures that the constraints are satisfied, adapts to possibly rapid or discontinuous changes in the solution and to nearly rank-deficient constraints, and accomplishes this task in an absolute minimum of computer time and extremely reliably. In Section 2 we will outline some recent results on the stable formulation of the equations of motion for numerical solution, and in Section 3 we will outline some of the computational challenges for efficient and reliable numerical methods, and propose some solutions.

2. Stable Formulations of the Equations of Motion

In this section we will be concerned with formulations and numerical methods for the Euler-Lagrange equations of constrained mechanical motion. These are systems of the form

$$M(t,p)\ddot{p} = f(t,p,v) - G(t,p)^T\lambda \tag{1a}$$
$$0 = g(t,p) \tag{1b}$$

where the positions and velocities satisfy p, $v \in \Re^{n_p}$, and $M(p)$ is a $n_p \times n_p$ regular (symmetric positive definite) mass matrix, f is a vector of applied forces, and λ represents the n_λ Lagrange multipliers or constraint forces coupled to the system by the $n_\lambda \times n_\lambda$ constraint matrix $G := \partial g/\partial p$. These types of systems arise frequently in the modeling of multibody systems[31], for example in vehicle simulation, computer-aided design of mechanical systems, and modeling of robotic manipulators.

The Euler-Lagrange system (1) poses difficulties for numerical methods in part because it is index-three. In particular, direct discretization of (1) yields numerical methods which are often not very robust because of the well-known[11] difficulties for higher-index systems with error estimation and stepsize control, as well as severe ill-conditioning of the linear systems at each time step, and a variety of other problems. In addition, for some problems the constraints can be poorly conditioned; in these cases methods applied to (1) and to some of its reformulations can behave numerically as if they were solving a problem for which the index is even higher.

To overcome the problems inherent in the direct numerical solution of the index-three form of the Euler-Lagrange equations, quite a number of reformulations of (1) have been suggested; some are in use in multibody codes [53]. Many of these formulations of the equations are based on differentiation of the constraints. The constraints

$$0 = g(p) \tag{2a}$$
$$0 = G(p)v \tag{2b}$$
$$0 = G(p)\dot{v} + v^T G_p(p)v =: G(p)\dot{v} + z(p,v) \tag{2c}$$

are called the *position, velocity, and acceleration-level constraints*, respectively. An index-two problem can be formed, for example, by replacing the position constraint in (1) with the velocity constraint. The resulting problem has, with appropriate initial conditions, the same solutions as (1) and is somewhat easier to solve numerically. However, the solutions can *drift* away from satisfying the position constraints because of numerical errors at each time step. This drift is often considered unacceptable in engineering problems because of the strong physical relevance of the position constraints (these are often holding components together), and because of their sometimes severe nonlinearity. An index-one problem can be formed by replacing the position constraint in (1) with the acceleration constraint. The resulting system is generally much easier to solve numerically, but now the solution can drift away from both the position and velocity constraints; the drift away from the position constraint can be quite significant.

Various formulations and solution procedures have been proposed to deal with or eliminate the problem of drift. Gear and others [26] have suggested that instead of replacing the position constraint with the velocity constraint, both constraints could be explicitly enforced by means of an additional Lagrange multiplier. This leads to a system of the form

$$\begin{aligned} \dot{p} &= v - G^T(t,p)\mu && \text{(3a)} \\ M(t,p)\dot{v} &= f(t,p,v) - G^T(t,p)\lambda && \text{(3b)} \\ 0 &= g(p) && \text{(3c)} \\ 0 &= G(p)v && \text{(3d)} \end{aligned}$$

The resulting problem is index-two. There is a similar formulation which enforces additionally the acceleration constraint by means of yet another Lagrange multiplier. These types of systems are generally called *stabilized formulations* of the Euler-Lagrange equations (as opposed to the unstabilized forms discussed earlier for which there may be drift). Although the stabilized formulations quite cleverly eliminate the drift problems, we have unfortunately found in recent numerical experiments that most ODE methods (including BDF and most implicit Runge-Kutta) applied to these equations may become very inefficient in certain situations, for example if the system is heterogeneous (includes components with widely disparate masses) or the constraints are poorly conditioned.

Equation (1) has $m = n_p - n_\lambda$ degrees of freedom. Using the constraints, we can reduce (1) locally to a system of m ODEs called a *state-space form*. The choice of coordinates is not unique; Haug and Wehage [59] use Cartesian coordinates. The resulting method is called *generalized coordinate partitioning*, and is the basis of the code DADS [53]. Potra and Rheinboldt[49, 50] suggest a different local parameterization. For the purposes of analysis, we will propose to define an *essential underlying ODE*, which is a certain class of state-space forms. By its construction, the original constraints are satisfied by a state-space form. The same set of coordinates may not work over an entire problem; thus the coordinates must be chosen adaptively.

Still another possible method for solving the Euler-Lagrange equations consists of appending the velocity and acceleration constraints to (1). The resulting system is called an *overdetermined DAE (ODAE)*, and has been investigated by Führer[21] and Leimkuhler[22], and others[45]. The ODAE is discretized by a numerical method such as BDF, and the resulting nonlinear system is solved by a Gauss-Newton iteration. In [22] it is shown that for a model problem where the constraints are linear with constant coefficients and under certain other conditions, the solution to the ODAE which is determined using a certain *ssf-iteration* to solve the nonlinear system is the same as that obtained by solving one of the stabilized forms, and that these solutions are equivalent to those obtained by numerically integrating the state-space form using the same discretization method. Unfortunately, these results do not appear to carry over to the more general case.

Various other methods have been proposed for solving the Euler-Lagrange equations, including the regularizations of Baumgarte[9], Lötstedt[33], Kalachev and O'Malley[32], and others. Sometimes these regularizations can be quite effective; variations of the method of Baumgarte are in use in many engineering codes. Unfortunately, it is not always easy to pick the regularization parameters which work.

As we have seen, a wide variety of formulations and associated numerical methods have been suggested for the solution of the Euler-Lagrange equations of constrained mechanical motion. In recent work [5], we have systematically evaluated the formulations and associated numerical methods from the standpoint of stability, to determine whether some formulations and methods are inherently better at preserving the conditioning of the original problem than others. The basic idea is to define a class of *essential underlying ODEs* (EUODE). The EUODE is defined for higher-index linear Hessenberg DAEs of the form

$$x^{(m)} = \sum_{j=1}^{m} A_j z_j + By + q \tag{4a}$$

$$0 = Cx + r \tag{4b}$$

where $z(x) = (x, x', \ldots, x^{(m_1)})^T$, A_j, B and C are smooth functions of t, $0 \leq t \leq 1$, $A_j(t) \in \Re^{n_x \times n_x}$, $j = 1, 2, \ldots, m$, $B(t) \in \Re^{n_x \times n_x}$, $n_y \leq n_x$ and CB nonsingular for each t (this assures that the DAE has index $m + 1$). All matrices involved are assumed to be uniformly bounded in norm by a constant of moderate size. The inhomogeneities are $q(t) \in \Re^{n_x}$ and $r(t) \in \Re^{n_y}$.

The EUODE is derived as follows. As in [4], there exists a smooth, bounded matrix function $R(t) \in \Re^{(n_x - n_y) \times n_x}$ whose linearly independent rows form a basis for the nullspace of B^T (R can be taken to be orthonormal). Thus, for each t, $0 \leq t \leq 1$,

$$RB = 0 \tag{5}$$

We assume that there exists a constant K_1 of moderate size such that

$$||(CB)^{-1}|| \leq K_1 \tag{6}$$

uniformly in t, and obtain (Lemma 2.1 in [4]) that there is a constant K_2 of moderate size such that

$$\left\| \begin{pmatrix} R \\ C \end{pmatrix} \right\| \leq K_2 \tag{7}$$

The constant K_2 depends, in addition to K_1, also on $||B||$, $||C||$ and $||R||$. Let K_3 be a moderate bound on R and its derivatives:

$$||R^{(j)}|| \leq K_3, \quad j = 0, 1, \ldots, m \tag{8}$$

Define new variables

$$u = Rx, \quad 0 \leq t \leq 1 \tag{9}$$

Then, using (4b), the inverse transformation is given by

$$x = \begin{pmatrix} R \\ C \end{pmatrix}^{-1} \begin{pmatrix} u \\ -r \end{pmatrix} = Su - Fr \tag{10}$$

where $S(t) \in \Re^{n_x \times (n_x - n_y)}$ satisfies

$$RS = I, \quad CS = 0 \tag{11}$$

and

$$F := B(CB)^{-1} \tag{12}$$

By our assumptions and (7) this mapping is well-conditioned. Both S and F are smooth and bounded. The first m derivatives of S and F are bounded by a constant involving K_2 and K_3. Taking m derivatives of (9) yields

$$u^{(m)} = (Rx)^{(m)} = \sum_{j=1}^{m}[RA_j + \binom{m}{j-1} R^{(m-j+1)}]z_j + Rq \tag{13}$$

Using $m-1$ derivatives of (10) we obtain the EUODE:

$$u^{(m)} = \sum_{j=1}^{m}[RA_j + \binom{m}{j-1} R^{(m-j+1)}][(Su)^{(j-1)} - (Fr)^{(j-1)}] + Rq \tag{14}$$

The EUODEs of a system are certain state-space forms which are uniquely defined up to a bounded, nonsingular change of variables. It is shown [5] that if the EUODE is stable, i.e. if its Green's function is bounded by a constant of moderate size, then a similar conclusion holds for the original DAE. Since the boundedness of the Green's function is invariant under bounded, nonsingular changes of variables, the question of stability for the EUODEs is well-defined. In [5], we used the EUODE to investigate the stability of some of the many equation formulations for Euler-Lagrange systems. We found that all of the formulations preserved the stability except unstabilized index reduction.

While several different equation formulations might equally preserve the conditioning of the Euler-Lagrange equations, the properties of numerical methods applied to these systems are often quite different. For example, it is well-known[11] that higher index systems are in a sense ill-posed, and can lead to difficulties for numerical methods with error control, ill-conditioning of linear systems at each time step, etc. For higher-index Hessenberg DAEs such as the Euler-Lagrange equations, there is a problem with *numerical* instability for many methods. Consider, for example, a linear homogeneous Hessenberg index-two system

$$x' = A(t)x + B(t)y \tag{15a}$$
$$0 = C(t)x \tag{15b}$$

This system has the EUODE

$$u' = RASu + R'Su \tag{16}$$

Now discretize with implicit Euler

$$x_{n+1} = x_n + hA_{n+1}x_{n+1} + hB_{n+1}y_{n+1} \tag{17a}$$
$$0 = C_{n+1}x_{n+1} \tag{17b}$$

Transforming back to the variables of the EUODE yields the discretization

$$u_{n+1} = u_n + hRASu_{n+1} + hR'Su_n \tag{18}$$

Comparing (18) with the EUODE (16) shows that the implicit Euler method corresponds to a discretization of the EUODE which handles the term $R'Su$ *explicitly*! Thus, although convergence results[11] predict that this method will converge globally to $O(h)$, there is a problem with respect to numerical stability which restricts the stepsize when $R'S$ is large. This problem is verified by experiment; there is very definitely a nonstiff behavior of methods ranging from BDF to most implicit Runge-Kutta for certain stable linear Hessenberg index-two systems. On the other hand, it is possible to argue, with a finer analysis, that under 'reasonable' conditions, this type of numerical instability should not occur for certain projections (for example, in (17) if $B = C^T$). The best cure for this numerical instability seems to be to reformulate the system in a form for which the instability cannot occur. This is done by reformulating the system in a form where the projection can be controlled, rather than dictated by the M matrix. In particular, we would like to formulate the system so that $B = C^T$. We call these formulations the methods of 'projected invariants'. The methods are constructed as follows:

1. Starting with the original Euler-Lagrange equation, use the acceleration constraint to eliminate λ and obtain an ODE in p, v which has as invariants the position and velocity constraints:

$$\dot{p} = v \tag{19a}$$
$$\dot{v} = (I - H)M^{-1}f - Fz(p, v) \tag{19b}$$

   where $F = M^{-1}G^T(GM^{-1}G^T)^{-1}$, and $H = FG$.

2. Project the solution onto the desired invariants using G^T or other stable projection. For example, project onto the position constraints:

$$\dot{p} = v - G^T\mu \tag{20a}$$
$$\dot{v} = (I - H)M^{-1}f - Fz(p, v) \tag{20b}$$
$$0 = g(p) \tag{20c}$$

3. Note that the above system has the same numerical solution as the following implicit formulation which can be implemented more efficiently:

$$\dot{p} = v - G^T\mu \tag{21a}$$
$$M\dot{v} = f(p, v) - G^T\lambda \tag{21b}$$
$$0 = G\dot{v} + z(p, v) \tag{21c}$$
$$0 = g(p) \tag{21d}$$

Depending on whether we do the projection onto the position constraints alone, or onto the position and velocity constraints, this leads us to two forms of projected invariants methods for constrained mechanical systems:

1. Project onto position constraint:

$$\dot{p} = v - G^T\mu \tag{22a}$$
$$M\dot{v} = f(p, v) - G^T\lambda \tag{22b}$$
$$0 = G\dot{v} + z(p, v) \quad \text{(acceleration)} \tag{22c}$$
$$0 = g(p) \quad \text{(position)} \tag{22d}$$

2. Project onto position and velocity constraints:

$$\begin{aligned}
\dot{p} &= v - G^T\mu - L^T\tau && \text{(23a)}\\
M\dot{v} &= f(p,v) - G^T\lambda - MG^T\tau && \text{(23b)}\\
0 &= G\dot{v} + z(p,v) + GG^T\tau \quad \text{(acceleration)} && \text{(23c)}\\
0 &= Gv \quad \text{(velocity)} && \text{(23d)}\\
0 &= g(p) \quad \text{(position)} && \text{(23e)}
\end{aligned}$$

where $L = G_p v$. These equations are studied in more detail in [45] and [6]. There is some controversy over whether it is really necessary to include the term $L^T\tau$, however numerical experiments in [6] seem to indicate that including this term is advantageous for numerical stability, in certain cases where the solution is oscillating at a high frequency.

There is also a nice geometrical interpretation for the method of projected invariants - it corresponds to the orthogonal projection onto the invariant constraints of the ODE.

Projected invariants is a subset of a class of methods called by numerical analysts *coordinate projection methods.* These methods are further analyzed in [1, 2] where some further modifications for stability and efficiency are proposed and tested.

3. Computational Challenges

3.1. Efficient solution techniques

Virtually all the proposed formulations for Euler-Lagrange equations have a similar structure with regard to the linear systems which must be solved at each time step. Even the solution of a state-space form, which at first glance might seem to have quite a different structure, can be expressed using Lagrange multipliers in a form with this structure [49, 50]. Thus it is important to be able to solve efficiently systems with this structure. There are several important cases:

3.1.1 *Nonstiff*

In the nonstiff case, half-explicit methods [30, 37] and/or iterations [22, 26, 48] can be devised which require much less work than in the stiff case. There is still a linear system, which arises because of the constraints, to be solved at each time step. However, the matrix, which has the form $\begin{pmatrix} M & G^T \\ G & 0 \end{pmatrix}$, has some nice properties: it is symmetric positive definite, and it does not depend on the stepsize or order of the discretization. Using the matrix as a preconditioner for iterative methods such as GMRES, the linear systems would not need to be solved very often. The mechanical systems have a special structure which can be further exploited; for example, the $O(n)$ methods [51, 37] can be used to solve the linear systems. However, this method leads to a recurrence which seems to be difficult to parallelize[8]. We are studying further possibilities for parallelizing the recurrence. Among

The problems are load balancing and exploiting parallelism when there are long chains. To some extent, it may be possible to use a block cyclic-reduction[40]-based algorithm to enhance the parallelization for long chains.

For many systems, for example if the stiffness arises because of a controller, only a small, readily identifiable part of the system may be stiff [58]. Here we expect that the GMRES iterative method [52], with the appropriate half-explicit methods as a preconditioner, should be effective.

3.1.2 *Stiff*

Stiff problems can arise for example in the modelling of flexible bodies subject to constraints. In the fully-stiff case, the linear systems to be solved at each time step still exhibit a special structure, but they are no longer symmetric and now depend on the stepsize. For example, the stabilized index-2 form of the equations of motion (3) leads to the linear system

$$\begin{pmatrix} \frac{1}{h\beta_0}I & -I & G^T & 0 \\ K & \frac{1}{h\beta_0}M + D & 0 & G^T \\ G & 0 & 0 & 0 \\ \Gamma & G & 0 & 0 \end{pmatrix} \begin{pmatrix} p \\ v \\ \mu \\ \lambda \end{pmatrix} = \begin{pmatrix} r_1 \\ r_2 \\ r_3 \\ r_4 \end{pmatrix}$$

The matrix above is rather large, of dimension $2n_p + 2n_\lambda$, and its LU decomposition is generally dense. If the number of constraints is of the same order of magnitude as the number of positions, methods which are analogous to the null-space method of numerical optimization [28] can be considered. At present, we do not have sufficient experience to determine whether this is preferable to other alternatives. In addition, for flexible structures, the considerable structure inherent in the linear system arising from the discretization should be exploited.

We have recently developed some new software for large-scale DAE systems. The new code DASPK [12], combines the time-stepping methods of DASSL with the preconditioned iterative method GMRES for solving the linear systems on each time step. There are also two new parallel versions, DASPKMP and DASPKF90, written for the Thinking Machines CM-5 in message-passing MIMD and data-parallel SIMD modes, respectively [39]. Preconditioners which can exploit the structure of multibody systems remain to be considered.

3.1.3 *Automatic stiffness detection*

Many problems in the simulation of multibody systems are nonstiff (or involve only a very few stiff components, as described above). However, stiff problems certainly do occur. A robust system for computer aided design should be able to treat both types of systems, hopefully with no intervention from the user. For example, it is possible to construct a method similar to those which have been proposed and implemented for ODEs [44, 54, 55], which monitors the convergence of the iteration and automatically switches to the appropriate method. To see how to do this, consider the index-two model system

$$y' = f(t, y) + G^T\lambda \tag{24}$$

$$0 \;=\; g(t,y) \tag{25}$$

where $\partial g/\partial y = G$. Suppose that the problem (25) is nonstiff. What does this mean, for a DAE of this form to be nonstiff? Differentiate the constraint to obtain $Gy' = 0$, and use the differentiated constraint to solve for λ in (25), to obtain

$$\lambda = -(GG^T)Gf$$

Now plug λ into the original equation (25) to obtain

$$y' = (I - H)f(t,y) \tag{26}$$

where $H = G^T(GG^T)^{-1}G$ is a projector. The system (26) is often called the ***underlying ODE*** of (25). We will say that (25) is stiff, if (26) is. A test of stiffness in (26) is: if

$$||h(I - H)f_y|| >> 1 \tag{27}$$

for a stepsize h that we would like to use, based on accuracy considerations, then the problem will be considered stiff.

Here is how the automatic method switching works. When the problem is nonstiff, we use an approximation to the iteration matrix (or, in combination with iterative methods, a preconditioner) which ignores the part of the matrix corresponding to f_y. For (25), this is

$$P_{nonstiff} = \begin{pmatrix} I & -G^T \\ G & 0 \end{pmatrix} \tag{28}$$

As mentioned in the section on nonstiff problems, depending on the structure of the DAE there are a number of ways to efficiently solve these kinds of linear systems. We start out by assuming the problem is nonstiff. When Newton, (or, in the case of iterative methods like GMRES, the iterative method) with the approximation to the iteration matrix given by (28), fails to converge for a current matrix approximation, it must be because the problem is stiff. To see this, note that the exact iteration matrix is given by

$$J = \begin{pmatrix} I - hf_y & -G^T \\ G & 0 \end{pmatrix}$$

Thus,

$$P^{-1}_{nonstiff}J = \begin{pmatrix} I - h(I - H)f_y & 0 \\ -hMf_y & I \end{pmatrix}$$

where $M = (GG^T)^{-1}G$. If the problem is nonstiff, then $(I - H)f_y$ is small, and the iteration should have no trouble converging. If the problem is determined to be stiff, the terms in the iteration matrix involving f_y will need to be approximated. Now, suppose that a problem has been determined to be stiff and that we are using an appropriate (but relatively expensive) iteration to solve it. How can we tell whether the problem later becomes nonstiff? One possibility is to monitor $||f_y||$.

3.2. Rank-deficient systems

It can sometimes happen that the constraint matrix G^T becomes rank-deficient or nearly rank-deficient[41, 18]. There are a number of possible situations. The matrix G^T

can be of constant, but reduced, rank. This is the case, for example, if you model a table with four legs. Then some of the Lagrange multipliers are not uniquely defined. However, the positions and velocities are well-defined [34]. For other problems, G^T may lose rank only locally, and then there are a number of possibilities. In some cases, the positions and velocities are well-defined, while some of the Lagrange multipliers are not unique. It can also happen that the solution fails to exist following the singularity. This latter situation is analogous to impasse points which have been studied in electrical engineering [15, 16].

There are a number of possibilities for dealing with singularities in the constraint matrix, in situations where the positions and velocities are well-defined. By considering the DAE in terms of a problem of minimizing the deviation of the constraint functions, subject to the differential equations, the well-known Baumgarte stabilization can be obtained [47], for well-conditioned constraints. This derivation also yields a strategy for selecting the Baumgarte parameter. When the constraints are poorly conditioned or not of full rank, a model trust-region approach [14] can be used for the optimization. This regularizes the DAE, introducing a term which is small except locally near the singularity. Our experience so far with the trust-region regularization has been favorable; there are some analytical results to justify this approach. Another regularization has been suggested by Park and Chiou [41]. Further possibilities, based on the augmented Lagrangian method of numerical optimization, are given by Ascher and Lin [3] and by Bayo and Avello [10].

3.3. Discontinuities

Frequent discontinuities are possible in a multibody system. Some of these discontinuities will be located very efficiently by a root-finder such as in DASSLRT[11]. However, others may arise from user-defined functions or other unanticipated situations, and need to be located automatically and handled efficiently. In the case of a collision, conservation properties of the solution should be preserved across the interface. The situation for DAEs presents difficulties in addition to the ODE case because the solution of a high-index system can be less continuous than the input, and singularities in the system can lead to numerical behavior which is quite similar to that caused by discontinuities. Impulsive solutions are possible.

3.4. Highly oscillatory systems

Often in multibody systems the solution may have components which are oscillating at a high frequency. This is a problem, for example, in vehicle suspension models. In a numerical method such as multistep or Runge-Kutta, which are based on approximating the solution locally, the stepsize must be chosen very small to resolve the oscillation in the solution, even if the amplitude of the oscillation is very small and does not significantly influence the long-term solution behavior. In collaboration with Jeng Yen, we have been investigating efficient numerical methods for these systems. On first glance, one might think that it would be possible to determine the local eigenstructure of the system, and then propagate the solution by methods based on matrix exponentiation. This would have a large cost per step but it could be made more efficient by using Krylov methods like GMRES to approximate the space of the high-frequency eigenvalues, rather than finding all of the eigenvalues of the system. However, in experiments with a stiff spring pendulum model problem, we found that unless one started almost exactly on the smooth (not the high-frequency) solution, the local eigenvalues do not lie on the imaginary axis, as you might expect for a high-frequency oscillation, but instead may have large positive and negative real parts, causing the method to go unstable. We are currently investigating the possibility of damping out the oscillation when its amplitude is small via BDF or other strongly damped methods. Lubich [38] has studied this problem for certain Runge-Kutta methods, and gives convergence results. The choice of formulation and method for highly oscillatory multibody systems is very important. For the large stepsizes one would like to take, there can be severe problems with Newton convergence for some methods and formulations. We have had some encouraging preliminary results with high-frequency bushing models where the number of steps has been brought down from about 50,000 for variable-stepsize Adams methods to less than 100 for low-order variable-stepsize BDF. These results will be reported in [46]. There are a number of difficulties in constructing a robust higher-order method, and it remains to be seen for problems in applications whether the cost of solving a nonstiff problem by an implicit method like BDF can be brought down sufficiently via iterative methods (although we suspect this will be true if the number of high-frequencies is substantially smaller than the size of the system and/or if the high frequencies appear in clusters). The approach of selectively damping high-frequency oscillations seems to be effective for short-time integrations like real-time vehicle simulations, however it is not appropriate for long-time simulation of highly oscillatory systems. Hence, different methods will need to be developed for these problems.

3.5. Parallel methods

Real-time simulation is required when it is necessary to simulate part of the system and use people or hardware in other parts of the system. An example is vehicle simulation. This introduces a number of additional requirements for numerical integration. For an introduction to some of the problems and challenges of real-time simulation, see [31]. In particular, it is essential that the solution be computed in an absolute minimum of computer time; if the computation is not complete in the allotted time, it will be useless.

Since an explicit method is generally much simpler than an implicit method to parallelize, it seems important to be able to identify the nonstiff parts of the system and treat them with explicit methods or functional iteration. For example, consider the Hessenberg index-2 system

$$x' = f(x,y) \qquad (29a)$$

$$0 = g(x) \qquad (29b)$$

where the matrix $g_x f_y$ is nonsingular. In the systems of interest, $f_y = -M^{-1}G^T y$, where $G = g_x$ and M is a symmetric positive definite mass matrix.

Following discretization with an explicit method, one obtains a nonlinear system which must be solved at each time step:

$$x = h\beta f(x,y) + b \qquad (30a)$$

$$0 = g(x). \qquad (30b)$$

For example, for implicit Euler discretization we would have $\beta = 1, b = x_n, x = x_{n+1}, y = y_{n+1}$. We will consider solving this with a half-explicit Newton method

$$x^{k+1} = b + h\beta f(x^k, y^{k+1}) \quad (31a)$$
$$0 = g(x^{k+1}). \quad (31b)$$

Alternatively, methods have been devised[30] which discretize explicitly in x and implicitly in y, but the nonlinear system for y looks pretty much the same. This leads to a nonlinear system in y to be solved at each time step, coupled to a functional iteration in x. Using Newton for y leads to:

$$y^{k+1} = y^k - (h\beta)^{-1}(g_x f_y)^{-1} g(b + h\beta f(x^k, y^k)) \quad (32a)$$
$$x^{k+1} = b + h\beta f(x^k, y^{k+1}). \quad (32b)$$

Now, noting that $g_x f_y$ is symmetric positive definite, we will update the inverse of this matrix using the parallel quasi-Newton methods (Broyden and BFGS) devised by Still [56, 57] for updating the matrix decomposition. Quasi-Newton methods have been tried previously in ODE solvers by Hindmarsh and others, but were not very successful because the matrix in ODEs, $(I - hJ)$, changes by as much as rank n whenever the stepsize h changes. However, in this DAE application the matrix does not depend on the stepsize and hence changes relatively slowly from step to step. Convergence of this iteration needs to be investigated, and its performance evaluated on the appropriate parallel computers. Several approaches to parallelization for multibody systems have been considered recently in the literature. Bae and Haug[7, 8] express the elimination as a recursion which is then distributed among processors. This method of using the recursion is the most efficient for serial computation but is difficult to parallelize; Bae and Haug give some results on an Alliant.

For fully-stiff systems, the iteration matrix is no longer symmetric, and it depends on the stepsize. Hence the quasi-Newton updating approach does not seem to be advantageous. The iteration matrix bears a strong resemblance to matrices encountered in constrained optimization, where null-space and range-space methods have proven to be useful in serial computation [28], depending on the relative dimensions of the number of positions n_p versus the number of constraints n_λ. For simulation of vehicles or rigid body mechanisms, where we usually expect n_λ to be of the same order of magnitude as n_p, a variant of the null-space method seems to be appropriate. Parallelization of this algorithm has been considered in [60] for medium-scale parallelism. Extension to massively parallel computers will require fast algorithms for Cholesky decomposition, QR factorization and backsolve [23]. Parallel solvers like DASPKMP or DASPKF90 [39] will be useful, but need to be combined with preconditioners which exploit the structure and parallelism in the system.

References

[1] U. Ascher, H. Chin, L. Petzold and S. Reich, *Stabilization of constrained mechanical systems with DAEs and invariant manifolds*, University of Minnesota, Dept. of Computer Science report 93-60, 1993.

[2] U. Ascher, H. Chin and S. Reich, *Stabilization of DAEs and invariant manifolds*, Tech. Rep. 92-17, Dept. of Computer Science, University of British Columbia, Vancouver, 1992.

[3] U. Ascher and P. Lin, *Sequential regularization methods for higher-index DAEs with constraint singularities: I. Linear index-2 case*, University of British Columbia, Dept. of Computer Science report 93-24, 1993.

[4] U. Ascher and L. Petzold, *Projected implicit Runge-Kutta methods for differential-algebraic equations*, SIAM J. Numer. Anal. 28 (1991), 1097-1120.

[5] U. Ascher and L. Petzold, *Stability of computational methods for constrained dynamics systems*, SIAM J. Sci. Stat. Comput. 14 (1993), 95-120.

[6] U. Ascher and L. Petzold, *Projected collocation for higher-order higher-index differential-algebraic equations*, to appear, Comp. Appl. Math.

[7] D. S. Bae and E. J. Haug, *A recursive formulation for constrained mechanical system dynamics. I and II*, Mech. Struct. & Mach. 15 (1987), pp. 359-382 and 481-506.

[8] D. S. Bae and E. J. Haug, *A recursive formulation for constrained mechanical system dynamics: Part III, Parallel processor implementation*, University of Iowa Report 1987.

[9] J. Baumgarte, *Stabilization of constraints and integrals of motion in dynamical systems*, Comp. Math. Appl. Mech. Eng. 1 (1976), 1-16.

[10] E. Bayo and A. Avello, *Singularity-free augmented Lagrangian algorithms for constrained multibody dynamics*, to appear in J. Nonlinear Dynamics.

[11] K. Brenan, S. Campbell and L. Petzold, *Numerical Solution of Initial-Value Problems in Differential-Algebraic Equations*, Elsevier Science Publishers, 1989.

[12] P. Brown, A. Hindmarsh and L. Petzold, *Using Krylov methods in the solution of large-scale differential-algebraic systems*, Lawrence Livermore National Laboratory Report, 1993.

[13] S. Campbell and B. Leimkuhler, *Differentiation of constraints in differential algebraic equations*, J. Mechanics of Structures and Machines, to appear.

[14] J. E. Dennis and R. B. Schnabel, *Numerical Methods for Unconstrained Optimization and Nonlinear Equations*, Prentice-Hall, 1983.

[15] L. Chua and A. Deng, *Impasse points. Part I: Numerical Aspects*, Int. J. of Circuit Theory and Applications 17 (1989), 213-235.

[16] L. Chua and A. Deng, *Impasse points. Part II: Analytical Aspects*, Int. J. of Circuit Theory and Applications 17 (1989), 271-282.

[17] E. Eich, C. Führer, B. Leimkuhler and S. Reich, *Stabilization and projection methods for multibody dynamics*, Research report, Inst Math, Helsinki Univ. of Technology, 1990.

[18] R. E. Ellis and S. L. Ricker, *Two numerical issues in simulating constrained dynamics*, Proc. 1992 IEEE Int. Conf. on Robotics and Automation, 312-318.

[19] W. H. Enright, K. R. Jackson, S. P. Norsett and P. G. Thomsen, *Effective solution of discontinuous IVPs using a Runge-Kutta formula pair with interpolants*, University of Toronto Numerical Analysis Report, 1986.

[20] W. H. Enright and M. S. Kamel, *Automatic partitioning of stiff systems and exploiting the resulting structure*, ACM Trans. on Math. Software 5 (1979), 374-385.

[21] C. Führer, *Differential-algebraische gleichungssysteme in mechanischen mehrkörpersystemen theorie, numerische ansätze und anwendungen*, Ph. D. Thesis, Technische Universität München, 1988.

[22] K. Führer and B. Leimkuhler, *Formulation and numerical solution of the equations of constrained mechanical motion*, Numerische Mathematik 59 (1991), 55-69.

[23] K. A. Gallivan, R. J. Plemmons and A. H. Sameh, *Parallel algorithms for dense linear algebra computations*, SIAM Review 32 (1990), 54-135.

[24] C. W. Gear, *Differential-algebraic equation index transformations*, SIAM J. Sci. Stat. Comput. 9 (1988), 39-47.

[25] C. W. Gear, *Maintaining solution invariants in the numerical solution of ODEs*, SIAM J. Sci. Stat. Comput. 7 (1986), 734-743.

[26] C. W. Gear, G. K. Gupta and B. Leimkuhler, *Automatic integration of Euler-Lagrange equations with constraints*, J. Comp. and Applied Math. 12 & 13 (1985), 77-90.

[27] C. W. Gear and O. Osterby, *Solving ordinary differential equations with discontinuities*, ACM Trans. on Math. Software 10 (1984), 23-44.

[28] P. E. Gill, W. Murray and M. H. Wright, *Practical Optimization*, Academic Press 1981.

[29] E. Griepentrog and R. März, *Differential-Algebraic Equations and Their Numerical Treatment*, Teubner-Texte zur Mathematik, Band 88, 1986.

[30] E. Hairer, C. Lubich and M. Roche, *The Numerical Solution of Differential-Algebraic Systems by Runge-Kutta Methods*, Springer Lecture Notes in Mathematics No. 1409, 1989.

[31] E. J. Haug and R. C. Deyo, Eds., Real-Time Integration Methods for Mechanical System Simulation, Springer-Verlag, 1989.

[32] L. Kalachev and R. O'Malley, *Regularization of linear differential algebraic equations*, preprint 1991.

[33] P. Lötstedt, *On a penalty function method for the simulation of mechanical systems subject to constraints*, Royal Institute of Technology TRITA-NA-7919, Stockholm, Sweden.

[34] P. Lötstedt, *Mechanical systems of rigid bodies subject to unilateral constraints*, SIAM J. Appl. Math. (1982), 281-296.

[35] Ch. Lubich, *h^2-Extrapolation methods for differential-algebraic systems of index 2*, IMPACT 1 (1989), 260-268.

[36] Ch. Lubich, *Extrapolation integrators for constrained multibody systems*, Technical Report, Universität Innsbruck, 1990.

[37] Ch. Lubich, U. Nowak, U. Pohle, Ch. Engstler, *MEXX - Numerical software for the integration of constrained mechanical multibody systems*, Konrad-Zuse-Zentrum fur Informationstechnik Berlin, Preprint 1992.

[38] Ch. Lubich, *Integration of stiff mechanical systems by Runge-Kutta methods*, Preprint, ETH, Zurich, 1992.

[39] R. S. Maier and L. R. Petzold, *User's guide to DASPKMP and DASPKF90*, University of Minnesota AHPCRC Technical Report, 1993.

[40] J. M. Ortega, *Introduction to Parallel and Vector Solution of Linear Systems*, Plenum, 1988.

[41] K. C. Park and J. Chiou, *Stabilization of computational procedures for constrained dynamical systems*, J. Guidance 11 (1988), 365-370.

[42] T. Park and E. J. Haug, *Ill-conditioned equations in kinematics and dynamics of machines*, Int. J. for Numerical Methods in Engineering 26 (1988), 217-230.

[43] L. R. Petzold, *An efficient numerical method for highly oscillatory ordinary differential equations*, SIAM J. Numer. Anal. 18 (1981), 455-479.

[44] L. R. Petzold, *Automatic selection of methods for solving stiff and nonstiff systems of ordinary differential equations*, SIAM J. Sci. Stat. Comput. 4 (1983), 136-148.

[45] L. R. Petzold and F. A. Potra, *ODAE methods for the numerical solution of Euler-Lagrange equations*, Applied Numerical Mathematics 10 (1992), 397-413.

[46] L. Petzold and J. Yen, in preparation.

[47] L. R. Petzold and Y. Ren, *Numerical solution of differential-algebraic equations with ill-conditioned constraints*, University of Minnesota, Dept. of Computer Science report, 1993.

[48] F. A. Potra, *Multistep method for solving constrained equations of motion*, University of Iowa, Dept. of Mathematics, 1991.

[49] F. A. Potra and W. C. Rheinboldt, *Differential-geometric techniques for solving differential algebraic equations*, in R. Deyo and E. Haug, eds., NATO Advanced Research Workshop in Real-Time Integration Methods for Mechanical System Simulation, Springer-Verlag, Heidelberg, 1990.

[50] F. A. Potra and W. C. Rheinboldt, *On the numerical solution of the Euler-Lagrange equations*, University of Iowa Center for Simulation and Design Optimization of Mechanical Systems, Technical Report R-81, 1990.

[51] D. E. Rosenthal, *An order n formulation for robotic systems*, Journal of the Astronautical Sciences 38 (1990), 511-529.

[52] Y. Saad and M. H. Schultz, *GMRES: A generalized minimum residual algorithm for solving nonsymmetric linear systems*, SIAM J. Sci. Stat. Comput. 7 (1986), 856-869.

[53] W. Schiehlen (Editor), *Multibody Systems Handbook*, Springer-Verlag, 1990.

[54] L. F. Shampine, *Type-insensitive ODE codes based on implicit $A(\alpha)$-stable formulas*, Math. Comp. 39 (1982), 109-123.

[55] L. F. Shampine, *Type-insensitive ODE codes based on extrapolation methods*, SIAM J. Sci. Stat. Comput. 4 (1983), 635-644.

[56] C. H. Still, *Parallel quasi-Newton methods for unconstrained optimization*, in Proceedings of the fifth conference on distributed memory concurrent computing, April 1990, 263-271.

[57] C. H. Still, *The parallel BFGS method for unconstrained minimization*, in Proceedings of the sixth conference on distributed memory concurrent computing, April 1991.

[58] R. Wehage, U.S. Army Tank Automotive Command Center, personal communication, 1991.

[59] R. A. Wehage and E. J. Haug, *Generalized coordinate partitioning for dimension reduction in analysis of constrained dynamic systems*, J. of Mechanical Design 104 (1982), 247-255.

[60] S. Wright, *A fast algorithm for equality-constrained quadratic programming on the Alliant FX/8*, Annals of Operations Research 14 (1988), 225-243.

NUMERICAL INTEGRATION OF SECOND ORDER DIFFERENTIAL–ALGEBRAIC SYSTEMS IN FLEXIBLE MECHANISM DYNAMICS

A. CARDONA[1] and M. GÉRADIN[2]
[1]*INTEC (CONICET/UNL)*
Güemes 3450, 3000 Santa Fé, ARGENTINA
[2]*LTAS, Univ. of Liège*
Rue E. Solvay 21, B-4000 Liège, BELGIUM

ABSTRACT. This paper studies second order accurate methods to numerically time-integrate the equations of motion for flexible mechanism dynamics. The aspects of stability, accuracy, conditioning of equations and time step control are discussed for the implicit scheme of Hilber, Hughes and Taylor (HHT).

1. Introduction

The equations of motion for a constrained mechanical system present the form of a mixed set of second order differential and algebraic equations called *a system of differential/algebraic equations* (DAE system). This kind of systems present particular characteristics which difficult their numerical treatment and difference them from systems of ordinary differential equations (ODE).

Probably one of the first approaches researchers have followed to solve a DAE system is to transform it to a system of second order ordinary differential equations (ODE), i.e. by differentiation of constraints or by using a penalty formulation. Actually, one popular technique in mechanisms analysis consists into differentiating constraints and introducing a stabilization term [1]. Then, the constraints are not satisfied exactly but oscillate with given stabilization period and damping constants about the exact verification point, tending to it in the long-time. The inconveniences of this approach are that the solution depends on some rather arbitrary constants to be selected by the user, and that the constraints are not verified exactly.

Gear, Petzold et al [2-6] developed a numerical theory of DAE systems. They showed under which conditions the use of integrators developed for treating ordinary differential equations may lead to acceptable solutions when applied to differential/algebraic systems. They advocated the use of techniques based on backward differentiation formulas to solve DAE systems, leading to schemes which preserve sparseness and are easy to implement.

Constrained dynamics equations can be seen as formed by two coupled subsystems: the first one describes the structural part, while the second subsystem describes the constraints acting on the structure. Thus, numerical algorithms for integrating these systems should be able to cope both with the structural part and with the constraints. In this sense, it seemed natural to us to look for an algorithm within the vast series of methods proposed to solve structural dynamics ODE's for over 30 years now (see [7] for a review on the subject), and introduce eventually to it the necessary modifications.

M. F. O. S. Pereira and J. A. C. Ambrósio (eds.),
Computer-Aided Analysis of Rigid and Flexible Mechanical Systems, 501–529.

Structural dynamics equations are a set of second order differential equations with the peculiarity of being "stiff", i.e. the system eigenfrequencies are distributed over a broad frequency range. Stiffness is produced either by the physical properties of the system or by the numerical technique followed to do the spatial discretization. For this reason, structural dynamics analysts seek algorithms which benefit from unconditional stability properties, which imply that the algorithm is stable regardless the relation between time step value and frequency of the oscillator. One-step methods are preferred to multistep ones, since they are self-starting. The cost of evaluating functions can be very important; thus, methods with a single function evaluation per time step are also preferred. Fully backward difference formulas may introduce an excess of artificial damping in the interesting part of the response, and tend to yield better results in first order than in second order problems.

Many well-known algorithms of numerical analysis are not used in structural dynamics because they do not possess one or the other of these properties (i.e. Runge-Kutta, Adams, Gear, ...). The most popular family of algorithms for the solution of problems in structural dynamics is probably the Newmark's one [8], which is based on the interpolation formulas:

$$\begin{aligned} q_{n+1} &= q_n + h\dot{q}_n + \frac{h^2}{2}[(1-2\beta)\ddot{q}_n + 2\beta\ddot{q}_{n+1}] \\ \dot{q}_{n+1} &= \dot{q}_n + h[(1-\gamma)\ddot{q}_n + \gamma\ddot{q}_{n+1}] \end{aligned}$$

where β, γ are the parameters that control the behavior of the method. Second order accuracy and unconditional stability, without energy dissipation in the whole frequency range, is reached by the trapezoidal rule ($\beta = 0.25, \gamma = 0.5$). For any other couple of parameter values, the algorithm accuracy falls to only first order.

The Newmark family has served as the basis for the development of many other algorithms. Researchers have tried to incorporate properties that ameliorate the performance of the algorithm. For instance, unconditional stability is not maintained for all nonlinear problems existing evidences of trouble with softening materials. One way to circumvent this inconvenience is by introducing some numerical dissipation at high frequencies in the algorithm, matching in some sense the real behavior of materials and structures [9-12]. Another recent proposal to warrant stability in the nonlinear regime are the so called "energy conserving" algorithms [13-14].

However, there exists wide concern in structural dynamics that algorithms of integration should provide at least a small amount of dissipation at high frequencies. A number of modifications of the classical Newmark time integrator have been proposed, introducing high frequency dissipation while retaining second order accuracy. Within these methods we can mention the α-method of Hilber [9-10], the α_B-method of Bossak [11] and the method by Hoff and Pahl [12]. This aspect has been found of upmost importance when solving differential algebraic systems [15-17]. The trapezoidal rule presents a weak instability which is excited for all values of the time step when applied to DAE's (the scheme becomes unconditionally unstable !) and numerical dissipation reestablishes stability to the scheme. It should be noted that this observation coincides in some sense with that of Gear and Petzold, since backward difference formulas completely filter-out the high frequencies.

In this paper, we discuss different aspects of the implementation of second order accurate algorithms for the integration of the equation of motion in constrained dynamics systems. We analyze first the system equations and determine a set of equivalent characteristic equations. These equations serve later to analyze stability and accuracy. The application of the implicit algorithm of Hilber, Hughes and Taylor (HHT) to the solution of constrained dynamics systems is studied and the aspects of stability, conditioning of equations and precision are analyzed with detail.

2. Constrained Dynamics Systems

2.1 DERIVATION OF THE EQUATIONS OF MOTION

The general form of the dynamic equilibrium equations for constrained dynamic systems is the following: n equations governing the dynamic behavior of the system, supplemented by m constraint equations introduced using the Lagrange multipliers technique.

$$\left\{ \begin{array}{l} \mathbf{M}\ddot{\mathbf{q}} - \mathbf{B}^T\boldsymbol{\lambda} + \mathbf{G}(\mathbf{q},\dot{\mathbf{q}},t) = \mathbf{0} \\ \boldsymbol{\Phi}(\mathbf{q},t) = \mathbf{0} \end{array} \right. \tag{1}$$

with $\mathbf{M}$ the structural mass matrix, $\mathbf{G}$ the nonlinear forces vector embodying both internal and external forces, $\boldsymbol{\Phi}$ the nonlinear holonomic constraints vector and $\mathbf{B} = -\frac{\partial \boldsymbol{\Phi}}{\partial \mathbf{q}}$ the matrix of constraint gradients (in a more general framework, we could have also included non holonomic constraints). This is a *semi-explicit system of second order differential-algebraic equations*, in the terminology usually employed in numerical analysis.

Our main objective in what follows is to stress on the specific difficulties encountered in the time integration of second order DAE systems characteristic of constrained dynamics systems. First, we will assess linear stability and convergence by analyzing the behavior of the integrator for the linearized homogeneous system of equations:

$$\begin{bmatrix} \mathbf{M} & \mathbf{0} \\ \mathbf{0} & \mathbf{0} \end{bmatrix} \begin{Bmatrix} \ddot{\mathbf{q}} \\ \ddot{\boldsymbol{\lambda}} \end{Bmatrix} + \begin{bmatrix} \mathbf{S} & -\mathbf{B}^T \\ -\mathbf{B} & \mathbf{0} \end{bmatrix} \begin{Bmatrix} \mathbf{q} \\ \boldsymbol{\lambda} \end{Bmatrix} = \begin{Bmatrix} \mathbf{0} \\ \mathbf{0} \end{Bmatrix} \tag{2}$$

with $\mathbf{S} = \frac{\partial \mathbf{G}}{\partial \mathbf{q}}$ the tangent stiffness matrix.

We will assume that results obtained from the linearized analysis can be extended to the nonlinear case without any further proof. Interested readers are referred to [6] for a more rigorous analysis. Some practical aspects of equations conditioning and error estimation for the full nonlinear case are discussed in sections 3.3 and 3.4.

We shall assume that the linearized system (2) is *solvable*. Solvability for a linear system like this will be characterized by the requirement that the matrix pencil

$$f(\lambda) = \lambda \begin{bmatrix} \mathbf{M} & \mathbf{0} \\ \mathbf{0} & \mathbf{0} \end{bmatrix} + \begin{bmatrix} \mathbf{S} & -\mathbf{B}^T \\ -\mathbf{B} & \mathbf{0} \end{bmatrix} \tag{3}$$

be *regular*, that is that the determinant of the matrix function $f(\lambda)$ is not identically zero.

We will also assume that the mass matrix $\mathbf{M}$ is positive definite, which is a reasonable assumption in structural dynamics problems. However, note that if this assumption is not verified, we may consider instead the following modified problem

$$\begin{bmatrix} \mathbf{M}^* & \mathbf{0} \\ \mathbf{0} & \mathbf{0} \end{bmatrix} \begin{Bmatrix} \ddot{\mathbf{q}} \\ \ddot{\boldsymbol{\lambda}} \end{Bmatrix} + \begin{bmatrix} \mathbf{S} & -\mathbf{B}^T \\ -\mathbf{B} & \mathbf{0} \end{bmatrix} \begin{Bmatrix} \mathbf{q} \\ \boldsymbol{\lambda} \end{Bmatrix} = \begin{Bmatrix} \mathbf{0} \\ \mathbf{0} \end{Bmatrix} \tag{4}$$

where $\mathbf{M}^* = \mathbf{M} + \mathbf{B}^T\mathbf{B}$ is positively definite for solvable systems. It is easy to verify that the solutions to both DAE systems (2) and (4) are strictly equivalent, since by differentiation of constraints we are able to verify that the computed accelerations should lie in the kernel of $\mathbf{B}^T\mathbf{B}$. It is also immediately verified that the corresponding matrix pencils are strictly equivalent.

2.2 EIGENVALUE PROBLEM FOR A CONSTRAINED SYSTEM

The homogeneous linear dynamics system leads to the following associated eigenvalue problem:

$$\omega_i^2 \begin{bmatrix} \mathbf{M} & \mathbf{0} \\ \mathbf{0} & \mathbf{0} \end{bmatrix} \begin{Bmatrix} \boldsymbol{\phi}_q \\ \boldsymbol{\phi}_\lambda \end{Bmatrix}_i = \begin{bmatrix} \mathbf{S} & -\mathbf{B}^T \\ -\mathbf{B} & \mathbf{0} \end{bmatrix} \begin{Bmatrix} \boldsymbol{\phi}_q \\ \boldsymbol{\phi}_\lambda \end{Bmatrix}_i \tag{5}$$

The first $(n-m)$ solutions of (5) is a set of finite frequencies ω_i^2 with eigenvectors $\langle \boldsymbol{\phi}_q^T \quad \boldsymbol{\phi}_\lambda^T \rangle_i^T$ such that $\boldsymbol{\phi}_{q\,i}$ verifies the equations of constraint and $\boldsymbol{\phi}_{\lambda\,i}$ gives the corresponding force of constraint. Next we show that the rest of the spectrum is composed by m couples of frequencies $+\infty$ and $-\infty$ associated to a unique eigenvector $\langle \mathbf{0}^T \quad \mathbf{e}_i^T \rangle^T$ with $\mathbf{e}_i$ being the unitary vector with a 1 at the i-th row. Therefore, the eigensystem (5) admits the following $n+m$ solutions:

$$\left\{ \left(\omega_1^2, \begin{Bmatrix} \boldsymbol{\phi}_q \\ \boldsymbol{\phi}_\lambda \end{Bmatrix}_1\right), \cdots \quad \left(\omega_{n-m}^2, \begin{Bmatrix} \boldsymbol{\phi}_q \\ \boldsymbol{\phi}_\lambda \end{Bmatrix}_{n-m}\right), \left(\infty, \begin{Bmatrix} \mathbf{0} \\ \mathbf{e}_1 \end{Bmatrix}\right), \cdots \right.$$
$$\left. \left(\infty, \begin{Bmatrix} \mathbf{0} \\ \mathbf{e}_m \end{Bmatrix}\right), \left(-\infty, \begin{Bmatrix} \mathbf{0} \\ \mathbf{e}_m \end{Bmatrix}\right) \right\} \tag{6}$$

Proof:

We introduce a small parameter ε into the eigensystem in order to eliminate the singularity :

$$\omega_i^2 \begin{bmatrix} \mathbf{M} & \mathbf{0} \\ \mathbf{0} & \varepsilon^2 \mathbf{1} \end{bmatrix} \begin{Bmatrix} \boldsymbol{\phi}_q \\ \boldsymbol{\phi}_\lambda \end{Bmatrix}_i = \begin{bmatrix} \mathbf{S} & -\mathbf{B}^T \\ -\mathbf{B} & \mathbf{0} \end{bmatrix} \begin{Bmatrix} \boldsymbol{\phi}_q \\ \boldsymbol{\phi}_\lambda \end{Bmatrix}_i \tag{7}$$

By making the transformations

$$\omega_i^2 = \frac{\omega_i'^2}{\varepsilon} \qquad \begin{Bmatrix} \boldsymbol{\phi}_q \\ \boldsymbol{\phi}_\lambda \end{Bmatrix}_i = \begin{Bmatrix} \varepsilon\boldsymbol{\phi}_q' \\ \boldsymbol{\phi}_\lambda \end{Bmatrix}_i \tag{8}$$

and by considering that $\varepsilon \ll 1$ the eigenproblem (7) is transformed into the eigensystem

$$\begin{bmatrix} \omega_i'^2\mathbf{M} & \mathbf{B}^T \\ \mathbf{B} & \omega_i'^2\mathbf{1} \end{bmatrix} \begin{Bmatrix} \boldsymbol{\phi}_q' \\ \boldsymbol{\phi}_\lambda \end{Bmatrix}_i = \begin{Bmatrix} \mathbf{0} \\ \mathbf{0} \end{Bmatrix} \tag{9}$$

Solvability of the DAE system assures that the matrix of constraint gradients $\mathbf{B}$ is well determined (its rows are linearly independent); then, the rank of $\mathbf{B}$ is m and there exist $(n-m)$ linearly independent eigenvectors with frequency $\omega_i'^2 = 0$ (which correspond to the already mentioned $(n-m)$ finite frequencies of (5)).

We will now compute the $2m$ eigenvectors of non-zero frequencies. Since the mass $\mathbf{M}$ (or the modified mass $\mathbf{M}^*$) is positive definite, we can express, from (9-a), that

$$\{\boldsymbol{\phi}_q'\}_i = -\frac{1}{\omega_i'^2}\, \mathbf{M}^{-1}\mathbf{B}^T\{\boldsymbol{\phi}_\lambda\}_i \tag{10}$$

After replacing the latter equation into (9-b), we get the $m-$dimensional fourth order auxiliary eigensystem:

$$(\omega_i'^4 \mathbf{1} \; - \; \mathbf{B}\mathbf{M}^{-1}\mathbf{B}^T)\{\boldsymbol{\phi}_\lambda\}_i \; = \; \mathbf{0} \tag{11}$$

The solution of (11) is given by the following $2m$ eigenpairs :

$$\{(\eta_1^2, \mathbf{v}_1) \; , \; (-\eta_1^2, \mathbf{v}_1) \; , \; (\eta_2^2, \mathbf{v}_2) \; , \; \ldots \qquad (-\eta_m^2, \mathbf{v}_m)\} \tag{12}$$

with $\mathbf{v}_i$ normalized to give:

$$\begin{aligned} \mathbf{v}_i^T \mathbf{v}_j \; &= \; \delta_{ij} \\ \mathbf{v}_i^T \mathbf{B}\mathbf{M}^{-1}\mathbf{B}^T \mathbf{v}_j \; &= \; \delta_{ij}\eta_i^4 \end{aligned} \tag{13}$$

By now using equation (10), we obtain the $2m$ solutions with non-zero frequency of the eigensystem (9):

$$\left\{ \left(\eta_1^2, \begin{Bmatrix} -\mathbf{w}_1 \\ \mathbf{v}_1 \end{Bmatrix}\right) , \left(-\eta_1^2, \begin{Bmatrix} \mathbf{w}_1 \\ \mathbf{v}_1 \end{Bmatrix}\right) , \left(\eta_2^2, \begin{Bmatrix} -\mathbf{w}_2 \\ \mathbf{v}_2 \end{Bmatrix}\right) , \ldots \quad \left(-\eta_m^2, \begin{Bmatrix} \mathbf{w}_m \\ \mathbf{v}_m \end{Bmatrix}\right) \right\} \tag{14}$$

where

$$\mathbf{w}_i \; = \; -\mathbf{M}^{-1}\mathbf{B}^T \mathbf{v}_i / \eta_i^2 \tag{15}$$

Using (8) we see that the full eigenspectrum of (7) is given by the set

$$\begin{aligned} \Bigg\{ \left(\omega_1^2, \begin{Bmatrix} \boldsymbol{\phi}_q \\ \boldsymbol{\phi}_\lambda \end{Bmatrix}_1\right) , \ldots \quad \left(\omega_{n-m}^2, \begin{Bmatrix} \boldsymbol{\phi}_q \\ \boldsymbol{\phi}_\lambda \end{Bmatrix}_{n-m}\right) , \left(\frac{\eta_1^2}{\varepsilon}, \begin{Bmatrix} \varepsilon \mathbf{w}_1 \\ \mathbf{v}_1 \end{Bmatrix}\right) , \ldots \\ \left(\frac{\eta_m^2}{\varepsilon}, \begin{Bmatrix} \varepsilon \mathbf{w}_m \\ \mathbf{v}_m \end{Bmatrix}\right) , \left(-\frac{\eta_m^2}{\varepsilon}, \begin{Bmatrix} -\varepsilon \mathbf{w}_m \\ \mathbf{v}_m \end{Bmatrix}\right) \Bigg\} \end{aligned} \tag{16}$$

Finally, by making $\varepsilon \to 0$, we see that the eigenspectrum of (5) is composed by $(n-m)$ finite eigenfrequencies ω_i^2 plus $2m$ eigenvalues that tend towards plus and minus infinity. The latter values have associated only m linearly independent eigenvectors spanning the subspace $\Re^m$ of Lagrange multipliers; then, by linear combinations between them we obtain the full set of solutions (6).

Remark:

- *Infinite elementary divisors of the matrix pencil $f(\lambda)$.* The "infinite frequencies" we have found for the constrained dynamic problem (5) are in fact infinite elementary divisors of the regular matrix pencil $f(\lambda)$ [18].

2.3 CHARACTERISTIC EQUATIONS FOR CONSTRAINED DYNAMICS SYSTEMS

Usually, the analysis of any integration method is made by following a two steps procedure: (i) reduction to a SDOF model problem; (ii) analysis of the equivalent SDOF equation. However, since the corresponding undamped eigensystem does not have $(n+m)$ linearly independent eigenvectors, we cannot transform as usual the equations of motion into $(n+m)$ independent equations. We demonstrate next that the linear DAE system (2) can be made equivalent to $(n-m)$ single DOF equations plus m systems of 2 equations with 2 unknowns of the form :

$$\left\{ \begin{array}{ll} \ddot{y}_i \; + \; \omega_i^2 y_i \; = \; 0 & i \; = \; 1, \ldots n-m \\ \begin{bmatrix} 0 & 0 \\ 1 & 0 \end{bmatrix} \begin{Bmatrix} \ddot{z}_1 \\ \ddot{z}_2 \end{Bmatrix} \; + \; \begin{Bmatrix} z_1 \\ z_2 \end{Bmatrix} \; = \; 0 & i \; = \; 1, \ldots m \end{array} \right. \tag{17}$$

Proof:

Let us consider the following set of $(n+m)$ vectors

$$\{\Psi\} = \left\{ \begin{Bmatrix} \phi_q \\ \phi_\lambda \end{Bmatrix}_1, \dots \begin{Bmatrix} \phi_q \\ \phi_\lambda \end{Bmatrix}_{n-m}, \begin{Bmatrix} \mathbf{w}_1 \\ \mathbf{0} \end{Bmatrix}, \begin{Bmatrix} \mathbf{w}_2 \\ \mathbf{0} \end{Bmatrix}, \dots \begin{Bmatrix} \mathbf{w}_m \\ \mathbf{0} \end{Bmatrix}, \begin{Bmatrix} \mathbf{0} \\ \frac{\mathbf{v}_1}{\eta_1^2} \end{Bmatrix}, \dots \begin{Bmatrix} \mathbf{0} \\ \frac{\mathbf{v}_m}{\eta_m^2} \end{Bmatrix} \right\} \tag{18}$$

These vectors are linear independent and form a basis. Then, we are able to expand the solution of (2) in the basis Ψ:

$$\begin{Bmatrix} \mathbf{q} \\ \boldsymbol{\lambda} \end{Bmatrix} = \Psi \mathbf{y} \qquad \begin{Bmatrix} \ddot{\mathbf{q}} \\ \ddot{\boldsymbol{\lambda}} \end{Bmatrix} = \Psi \ddot{\mathbf{y}} \tag{19}$$

We project the system matrices onto Ψ to get

$$\begin{aligned} \Psi^T \begin{bmatrix} \mathbf{M} & \mathbf{0} \\ \mathbf{0} & \mathbf{0} \end{bmatrix} \Psi &= \begin{bmatrix} \mathbf{1} & \\ & \begin{bmatrix} \mathbf{1} & \mathbf{0} \\ \mathbf{0} & \mathbf{0} \end{bmatrix}_{2m \times 2m} \end{bmatrix} \\ \Psi^T \begin{bmatrix} \mathbf{S} & -\mathbf{B}^T \\ -\mathbf{B} & \mathbf{0} \end{bmatrix} \Psi &= \begin{bmatrix} \boldsymbol{\Omega}^2 & \\ & \begin{bmatrix} \mathbf{A} & \mathbf{1} \\ \mathbf{1} & \mathbf{0} \end{bmatrix}_{2m \times 2m} \end{bmatrix} \end{aligned} \tag{20}$$

where $\mathbf{A}$ is an $m \times m$ symmetric matrix with coefficients

$$a_{ij} = \mathbf{w}_i^T \mathbf{S} \mathbf{w}_j = \frac{\mathbf{v}_i^T \mathbf{B} \mathbf{M}^{-1} \mathbf{S} \mathbf{M}^{-1} \mathbf{B}^T \mathbf{v}_j}{\eta_i^2 \eta_j^2} \tag{21}$$

with $\eta_i, i = 1, m$ the eigenvalues of the auxiliary problem (11), and where $\boldsymbol{\Omega}^2$ is a diagonal matrix formed by the squared elastic vibration eigenfrequencies. We see that in this basis the equations of motion are uncoupled into elastic deformation and constraint modes.

The constraints subsystem can be further simplified by similarity transformations (i.e. pre and post multiplying by $\mathbf{T}^T, \mathbf{T}$, with $\mathbf{T} = \begin{bmatrix} \mathbf{1} & \mathbf{0} \\ -\mathbf{A}/2 & \mathbf{1} \end{bmatrix}$). After doing so, and by appropriately scaling and reordering equations and variables, we can build new matrices Ψ_L^*, Ψ_R^* over which projection of the system matrices yields the result:

$$\begin{aligned} \Psi_L^{*T} \begin{bmatrix} \mathbf{M} & \mathbf{0} \\ \mathbf{0} & \mathbf{0} \end{bmatrix} \Psi_R^* &= \begin{bmatrix} \mathbf{1} & & & \\ & \begin{bmatrix} 0 & 0 \\ 1 & 0 \end{bmatrix}_1 & & \\ & & \ddots & \\ & & & \begin{bmatrix} 0 & 0 \\ 1 & 0 \end{bmatrix}_m \end{bmatrix} \\ \Psi_L^{*T} \begin{bmatrix} \mathbf{S} & -\mathbf{B}^T \\ -\mathbf{B} & \mathbf{0} \end{bmatrix} \Psi_R^* &= \begin{bmatrix} \boldsymbol{\Omega}^2 & & & \\ & \begin{bmatrix} 1 & 0 \\ 0 & 1 \end{bmatrix}_1 & & \\ & & \ddots & \\ & & & \begin{bmatrix} 1 & 0 \\ 0 & 1 \end{bmatrix}_m \end{bmatrix} \end{aligned} \tag{22}$$

Then, the behavior of the time integrators when dealing with constrained dynamic systems is characterized by their ability to treat a system of the form (17).

Remarks:

- A matrix $\mathbf{N}$ is said to have nilpotency index ν if $\mathbf{N}^\nu = \mathbf{0}$ and $\mathbf{N}^{\nu-1} \neq \mathbf{0}$. The preceding section has shown that the linear DAE system (2) is equivalent, through appropriate transformation matrices Ψ_L^* and Ψ_R^* to the quasi-diagonal system of differential equations

$$\begin{bmatrix} 1 & \\ & \mathbf{N} \end{bmatrix} \ddot{\mathbf{q}} + \begin{bmatrix} \Omega^2 & \\ & 1 \end{bmatrix} \mathbf{q} = \mathbf{0} \tag{23}$$

  where $\mathbf{N} = \mathrm{Diag}(\mathbf{N}_2, \mathbf{N}_2, \ldots)$ is an index-2 nilpotent matrix and $\mathbf{N}_2$ is the index 2 canonic nilpotent matrix $\begin{bmatrix} 0 & 0 \\ 1 & 0 \end{bmatrix}$. Usually, the degree of nilpotency of $\mathbf{N}$ is called the *index of the (second order) DAE system*. The system written in this form is called *canonical quasi-diagonal (Kronecker) form* [18].
- An alternative (and easier) way to compute the index of a DAE system is by successive differentiation of parts of the system (e.g. the constraints) up to transforming it into a system of ordinary differential equations. The minimal number of differentiations needed to get an ODE system is the index of the DAE [6].
- If the system equations describe also some other phenomena, like for instance system control laws, the nilpotency index can be higher. Thus, in a general case we can be faced to systems with index higher than two.

3. Implicit Time Integration of Constrained Dynamic Systems: the Hilber-Hughes-Taylor Algorithm

The integrator to be selected should be able to correctly integrate both the "structural" and "constraints" parts of the DAE system (1).

Several integration methods exist which have proven to give correct answers when dealing with structural dynamics equations, that is which solve accurately stiff second order ODE's systems. Explicit methods are recommended when high frequencies dominate because relatively short time steps are required. For the low-frequency response of multidegree of freedom systems, implicit methods with controlled numerical dissipation offer the advantage of suppressing the high-frequency modes of the numerical model which do not contribute significantly to the physical behavior. The numerical effort can thus be reduced without loss of accuracy by using large time steps.

The method of Hilber, Hughes and Taylor (HHT) is an implicit method widely used in structural dynamics. It consists on a slight modification of the Newmark algorithm, incorporating algorithmic dissipation at the high frequencies and retaining second order accuracy. The integration formulas can be summarized as follows, in the homogeneous single DOF case:

Find Δq such that:

$$\begin{aligned}
q_{n+1} &= q_n + h\dot{q}_n + \frac{h^2}{2}\ddot{q}_n + \Delta q \\
\dot{q}_{n+1} &= \dot{q}_n + h\ddot{q}_n + \frac{\gamma}{\beta h}\Delta q \\
\ddot{q}_{n+1} &= \ddot{q}_n + \frac{1}{\beta h^2}\Delta q
\end{aligned} \tag{24}$$

$$\ddot{q}_{n+1} + (1+\alpha)2\xi\omega\dot{q}_{n+1} - \alpha 2\xi\omega\dot{q}_n + (1+\alpha)\omega^2 q_{n+1} - \alpha\omega^2 q_n = 0$$

Parameters α, β, γ control the accuracy and numerical damping of the algorithm. In order to get second order accuracy, the following relations should be verified:

$$\beta = \frac{1}{4}(1-\alpha)^2 \qquad \gamma = \frac{1}{2}(1-2\alpha) \tag{25}$$

leaving only one parameter free which takes values of interest in the range $-0.3 \leq \alpha \leq 0$. Numerical dissipation is maximum for $\alpha = -0.3$, and for $\alpha = 0$ the canonic Newmark algorithm without dissipation ($\beta = \frac{1}{4}, \gamma = \frac{1}{2}$) is recovered.

3.1 STABILITY ANALYSIS

The Hilber-Hughes-Taylor algorithm is unconditionally stable with second order ODE's, for values of the parameter α lying in the range $[-0.3, 0]$ and β, γ computed according to equations (25). We will now analyze the stability of this algorithm with the characteristic DAE system (17-b), and see that not every value of α leads to stable computations.

We first regularize the system by introducing a small parameter ε that eliminates the singularity of the "mass" matrix:

$$\begin{bmatrix} 0 & \varepsilon^2 \\ 1 & 0 \end{bmatrix} \begin{Bmatrix} \ddot{z}_1 \\ \ddot{z}_2 \end{Bmatrix} + \begin{Bmatrix} z_1 \\ z_2 \end{Bmatrix} = \mathbf{0} \tag{26}$$

We make afterwards the change of variables:

$$\mathbf{y} = \mathbf{\Phi}_R \mathbf{z} \tag{27}$$

with

$$\mathbf{\Phi}_R = \begin{bmatrix} \varepsilon & -\varepsilon \\ 1 & 1 \end{bmatrix} \tag{28}$$

Let us also define the matrix

$$\mathbf{\Phi}_L = \frac{1}{2\varepsilon^2} \begin{bmatrix} 1 & \varepsilon \\ 1 & -\varepsilon \end{bmatrix} \tag{29}$$

such that the following properties hold

$$\begin{aligned}
\mathbf{\Phi}_L \begin{bmatrix} 0 & \varepsilon^2 \\ 1 & 0 \end{bmatrix} \mathbf{\Phi}_R &= \mathbf{1} \\
\mathbf{\Phi}_L \mathbf{\Phi}_R &= \begin{bmatrix} 1/\varepsilon & 0 \\ 0 & -1/\varepsilon \end{bmatrix}
\end{aligned} \tag{30}$$

Then, by pre and post multiplying the original equations by Φ_L, Φ_R, they are transformed to the uncoupled system of differential equations

$$\begin{cases} \ddot{y}_1 + \omega_1^2 y_1 = 0 \\ \ddot{y}_2 + \omega_2^2 y_2 = 0 \end{cases} \tag{31}$$

where $\omega_1^2 = 1/\varepsilon, \omega_2^2 = -1/\varepsilon$.

The discrete solution of (31) is given through the amplification matrix of the integrator, in the form:

$$\mathbf{Y}_{n+1} = \begin{Bmatrix} y_{1\,n+1} \\ h\dot{y}_{1\,n+1} \\ h^2\ddot{y}_{1\,n+1} \\ y_{2\,n+1} \\ h\dot{y}_{2\,n+1} \\ h^2\ddot{y}_{2\,n+1} \end{Bmatrix} = \begin{bmatrix} \mathbf{A}(\Omega_1) & 0 \\ 0 & \mathbf{A}(\Omega_2) \end{bmatrix} \mathbf{Y}_n \tag{32}$$

where Ω_i equals the product of the frequency of the oscillator i by the time step $h = t_{n+1} - t_n$ and where the amplification matrix $\mathbf{A}(\Omega_i)$ is a particular function of the numerical time integrator.

Projecting back to the original variables, we get the amplification matrix $\overline{\mathbf{A}}_\varepsilon$ of the regularized system:

$$\mathbf{Z}_{n+1} = \Phi_R \mathbf{Y}_{n+1} = \Phi_R \begin{bmatrix} \mathbf{A}(\Omega_1) & 0 \\ 0 & \mathbf{A}(\Omega_2) \end{bmatrix} \Phi_L \begin{bmatrix} 0 & \varepsilon^2 \\ 1 & 0 \end{bmatrix} \mathbf{Z}_n = \overline{\mathbf{A}}_\varepsilon \mathbf{Z}_n \tag{33}$$

The 6×6 amplification matrix $\overline{\mathbf{A}}$ of the characteristic DAE system is obtained by taking the limit of $\overline{\mathbf{A}}_\varepsilon$ when $\varepsilon \to 0$:

$$\overline{\mathbf{A}} = \lim_{\varepsilon \to 0} \frac{1}{2} \begin{bmatrix} \mathbf{A}(\Omega_1) + \mathbf{A}(\Omega_2) & (\mathbf{A}(\Omega_1) - \mathbf{A}(\Omega_2))\,\varepsilon \\ (\mathbf{A}(\Omega_1) - \mathbf{A}(\Omega_2))\,/\varepsilon & \mathbf{A}(\Omega_1) + \mathbf{A}(\Omega_2) \end{bmatrix} \tag{34}$$

In order to have convergence, the matrix

$$\mathbf{B} = \lim_{\varepsilon \to 0} \left(\frac{\mathbf{A}(\Omega_1) - \mathbf{A}(\Omega_2)}{2\varepsilon} \right) \tag{35}$$

should be bounded; this implies that $\mathbf{A}_\infty = \lim_{\Omega \to \infty} \mathbf{A}(\Omega) = \lim_{\Omega \to -\infty} \mathbf{A}(\Omega)$. Then, the amplification matrix results

$$\overline{\mathbf{A}} = \begin{bmatrix} \mathbf{A}_\infty & 0 \\ \mathbf{B} & \mathbf{A}_\infty \end{bmatrix} \tag{36}$$

The state vector at time t_n is obtained by successive applications of the amplification matrix to the initial state vector $\mathbf{Z}_0$:

$$\mathbf{Z}_n = \begin{Bmatrix} \mathbf{Z}_{1\,n} \\ \mathbf{Z}_{2\,n} \end{Bmatrix} = \begin{Bmatrix} \mathbf{A}_\infty^n \mathbf{Z}_{1\,0} \\ O(n\mathbf{A}_\infty^{n-1}\mathbf{B})\,\mathbf{Z}_{1\,0} + \mathbf{A}_\infty^n \mathbf{Z}_{2\,0} \end{Bmatrix} \tag{37}$$

Then, in order to get stability, the successive powers of $\mathbf{A}_\infty$ should be bounded.

Remarks:

- We see that matrix $\mathbf{B}$ affects the computation of values $\mathbf{Z}_2$, which correspond to the Lagrange multipliers amplitudes. This equation seems to indicate that in order to get stable results, powers of the amplification matrix $\mathbf{A}_\infty$ should necessarily go to zero to balance growing of the factor n in the term $(n\ \mathbf{A}_\infty^{n-1}\ \mathbf{B})$.
- Note that there exists a clear uncoupling between Lagrange multipliers and principal variables, reflected by the particular structure of the amplification matrix $\overline{\mathbf{A}}$. Previous computed values of the principal variables $\mathbf{Z}_1$ influence the value assumed by the Lagrange multipliers $\mathbf{Z}_2$ at the current step, but the previous values of $\mathbf{Z}_2$ do not influence the present value of $\mathbf{Z}_1$.

In the particular case of the HHT algorithm, the amplification matrix relating the state vectors computed by the algorithm at t_{n+1} and t_n can be written in the following form:

$$\left\{ \begin{array}{c} q_{n+1} \\ h\dot{q}_{n+1} \\ h^2\ddot{q}_{n+1} \end{array} \right\} = \mathbf{A} \left\{ \begin{array}{c} q_n \\ h\dot{q}_n \\ h^2\ddot{q}_n \end{array} \right\} \tag{38}$$

where

$$\mathbf{A} = \frac{1}{1+(1+\alpha)\beta\Omega^2} \begin{bmatrix} (1+\beta\alpha\Omega^2) & 1 & (\frac{1}{2}-\beta) \\ -\gamma\Omega^2 & (1+(\beta-\gamma)(1+\alpha)\Omega^2) & 1-\gamma+(\beta-\frac{\gamma}{2})(1+\alpha)\Omega^2 \\ -\Omega^2 & -(1+\alpha)\Omega^2 & (1+\alpha)(\beta-\frac{1}{2})\Omega^2 \end{bmatrix} \tag{39}$$

and where $\Omega = \omega h$.

The amplification matrix at infinity is directly deduced by computation in the limit when $\Omega \to \infty$, giving:

$$\mathbf{A}_\infty = \frac{1}{(1-\alpha)^2} \begin{bmatrix} \frac{\alpha(1-\alpha)^2}{1+\alpha} & 0 & 0 \\ \frac{4\alpha-2}{(1+\alpha)} & \alpha^2+2\alpha-1 & \alpha^2 \\ \frac{-4}{(1+\alpha)} & -4 & \alpha^2-2\alpha-1 \end{bmatrix} \tag{40}$$

where we have replaced the optimal values of parameters β, γ in terms of the dissipation parameter α.

An analysis of this matrix gives the three eigenvalues:

$$\lambda_1 = \frac{\alpha}{1+\alpha} \qquad \lambda_{2,3} = -\frac{1+\alpha}{1-\alpha} \tag{41}$$

For $\alpha = 0$ we get the algorithm of Newmark. Since in this case we have two eigenvalues of modulus equal to one with only one associated eigenvector, there will be some elements of $\mathbf{A}^n$ that grow as $O(n)$ when $n \to \infty$ and the algorithm will diverge (see its Jordan form below). For the other values of interest of α, for which the algorithm includes dissipation at $\Omega \to \infty$, stability is recovered since the eigenvalues $\lambda_i, i = 1,2,3$ are smaller than 1 in modulus ($\alpha < 0$).

Remarks:

- The *Jordan form of* $\mathbf{A}_\infty$ can be computed by projection of the amplification matrix $\mathbf{A}_\infty$ onto $\mathbf{W}_L, \mathbf{W}_R$, with

$$\mathbf{W}_L = \begin{bmatrix} \frac{1}{(1+3\alpha)^2} & 0 & 0 \\ -\frac{4}{(1-\alpha)^2(1+3\alpha)} & -\frac{2}{(1-\alpha)^2} & -\frac{\alpha}{(1-\alpha)^2} \\ -\frac{2(1+4\alpha)}{\alpha(1+3\alpha)^2} & -\frac{1}{\alpha} & 0 \end{bmatrix}$$
$$\mathbf{W}_R = \begin{bmatrix} (1+3\alpha)^2 & 0 & 0 \\ -2(1+4\alpha) & 0 & -\alpha \\ 4 & -\frac{(1-\alpha)^2}{\alpha} & 2 \end{bmatrix} \tag{42}$$

giving

$$\mathbf{J}_\infty = \mathbf{W}_L \mathbf{A}_\infty \mathbf{W}_R = \begin{bmatrix} \frac{\alpha}{1+\alpha} & 0 & 0 \\ 0 & -\frac{1+\alpha}{1-\alpha} & 0 \\ 0 & 1 & -\frac{1+\alpha}{1-\alpha} \end{bmatrix} \tag{43}$$

- The same procedure of stability analysis can be generalized to constrained canonic systems of any nilpotency index. To this end, define the index ν regularized matrix:

$$\mathbf{N}_\varepsilon^{(\nu)} = \begin{bmatrix} 0 & 0 & \dots & & \varepsilon^\nu \\ 1 & 0 & 0 & & \\ 0 & 1 & 0 & 0 & \\ & & \ddots & \ddots & \\ & & & 1 & 0 \end{bmatrix} \tag{44}$$

and verify that the change of basis matrices

$$\Phi_{R\,lm}^{(\nu)} = \varepsilon^{\nu-l}\left(\cos\frac{2\pi}{\nu} + i\sin\frac{2\pi}{\nu}\right)^{l(1-m)}$$
$$\Phi_{L\,lm}^{(\nu)} = \frac{\varepsilon^{m-\nu-1}}{\nu}\left(\cos\frac{2\pi}{\nu} + i\sin\frac{2\pi}{\nu}\right)^{(l-1)(m-1)} \tag{45}$$

transform the original system into the uncoupled equations

$$\ddot{y}_l + \omega_l^2 y_l = 0 \qquad l = 1, 2, \dots \nu \tag{46}$$

with

$$\omega_l^2 = \frac{1}{\varepsilon}\left(\cos\frac{2\pi}{\nu} + i\sin\frac{2\pi}{\nu}\right)^{(l-1)} \tag{47}$$

The amplification matrix of the HHT algorithm verifies, for ν arbitrary

$$\lim_{\varepsilon\to 0} \mathbf{A}\left(\frac{1}{\varepsilon}\left(\cos\frac{2\pi}{\nu} + i\sin\frac{2\pi}{\nu}\right)\right) = \mathbf{A}_\infty \tag{48}$$

Then, the global amplification matrix obtained after coming back to the original variables reads:

$$\overline{\mathbf{A}} = \begin{bmatrix} \mathbf{A}_\infty & & & & \\ \mathbf{B}_1 & \mathbf{A}_\infty & & & \\ \mathbf{B}_2 & \mathbf{B}_1 & \mathbf{A}_\infty & & \\ \vdots & & & \ddots & \\ \mathbf{B}_{\nu-1} & & & \mathbf{B}_1 & \mathbf{A}_\infty \end{bmatrix} \tag{49}$$

and the scheme is shown to be stable for arbitrary index ν whenever $\alpha \neq 0$.

3.2 CONVERGENCE OF THE ALGORITHM

The Hilber-Hughes-Taylor algorithm is globally second order accurate (locally third order accurate) for second order ODE's, with values of the parameter α lying in the range $[-0.3, 0]$ and β, γ computed according to equations (25). We will now analyze the convergence of the algorithm for the characteristic DAE system (17-b). First we regard the local truncation error and see that high index systems could be even locally divergent. However, we show afterwards that results are globally second order accurate (after several steps) when using constant step size and whenever the loads verify enough regularity assumptions.

Remark:

- It is easy to verify that the exact solution of the algebraic subsystem can be written in the form:

$$\mathbf{z}(t) = \sum_{i=0}^{\nu-1} (-1)^i \mathbf{N}^i \mathbf{f}^{(2i)} \tag{50}$$

  where ν is the nilpotency index of the DAE system. Note that the solution depends only on the current value of the loads and derivatives; note also that if the load is differentiable, but not continuously differentiable, $\mathbf{z}$ *can be discontinuous.*

3.2.1 Local Error. Let us now compute the local truncation error of our integration algorithm when applied to this system, i.e. the error of integration supposing that we start the step from exact values at time t_n.

The difference formulas of the HHT algorithm can be written:

$$\begin{aligned}
\mathbf{z}_{n+1} &= \mathbf{z}(t_n) + h\dot{\mathbf{z}}(t_n) + \frac{h^2}{2}\ddot{\mathbf{z}}(t_n) + \Delta\mathbf{z} \\
\dot{\mathbf{z}}_{n+1} &= \dot{\mathbf{z}}(t_n) + h\ddot{\mathbf{z}}(t_n) + \frac{\gamma}{\beta h}\Delta\mathbf{z} \\
\ddot{\mathbf{z}}_{n+1} &= \ddot{\mathbf{z}}(t_n) + \frac{1}{\beta h^2}\Delta\mathbf{z}
\end{aligned} \tag{51}$$

where $\mathbf{z}(t_n), \dot{\mathbf{z}}(t_n), \ddot{\mathbf{z}}(t_n)$ are the exact values at time t_n. Exact values at time t_{n+1} can be computed by adding error terms:

$$\begin{aligned}
\mathbf{z}(t_{n+1}) &= \mathbf{z}_{n+1} + \mathbf{e}_{z\,n+1} \\
\dot{\mathbf{z}}(t_{n+1}) &= \dot{\mathbf{z}}_{n+1} + \frac{\gamma}{\beta h}\mathbf{e}_{z\,n+1} + \boldsymbol{\tau}_{\dot{z}} \\
\ddot{\mathbf{z}}(t_{n+1}) &= \ddot{\mathbf{z}}_{n+1} + \frac{1}{\beta h^2}\mathbf{e}_{z\,n+1} + \boldsymbol{\tau}_{\ddot{z}}
\end{aligned} \tag{52}$$

Here, $\mathbf{e}_{z\,n+1}$ is the local truncation error (in displacements $\mathbf{z}$) at t_{n+1}, while $\boldsymbol{\tau}_{\dot{z}}$ and $\boldsymbol{\tau}_{\ddot{z}}$ are the discretization errors introduced by the approximation of difference formulas to compute velocities and accelerations.

The algorithm advances one step by solving the weighted equilibrium equation:

$$\mathbf{N}\left(\ddot{\mathbf{z}}(t_{n+1}) - \frac{1}{\beta h^2}\mathbf{e}_{z\,n+1} - \boldsymbol{\tau}_{\ddot{z}}\right) + (1+\alpha)\left(\mathbf{z}(t_{n+1}) - \mathbf{e}_{z\,n+1}\right) - \alpha\,\mathbf{z}(t_n) - (1+\alpha)\,\mathbf{f}_{n+1} + \alpha\,\mathbf{f}_n = \mathbf{0} \tag{53}$$

By taking into account that the equilibrium equation is exactly verified by actual values $\mathbf{z}(t_k)$, $\dot{\mathbf{z}}(t_k)$, $\ddot{\mathbf{z}}(t_k)$, i.e. :

$$\mathrm{N}\,\ddot{\mathbf{z}}(t_{n+1}) \;+\; \mathbf{z}(t_{n+1}) \;-\; \mathbf{f}_{n+1} \;=\; \mathbf{0} \tag{54}$$

we get the following equation for the local truncation error at t_{n+1}:

$$\left[\frac{1}{\beta h^2}\mathrm{N} \;+\; (1+\alpha)\mathbf{1}\right]\mathbf{e}_{z\;n+1} \;=\; -\alpha\mathrm{N}(\ddot{\mathbf{z}}(t_{n+1}) - \ddot{\mathbf{z}}(t_n)) \;-\; \mathrm{N}\tau_{\ddot{z}} \tag{55}$$

Next, we eliminate $\Delta\mathbf{z}$ between equations (51-a) and (51-c), and between equations (51-b) and (51-c) to get:

$$\begin{aligned}\dot{\mathbf{z}}(t_{n+1}) &= \dot{\mathbf{z}}(t_n) + h\ddot{\mathbf{z}}(t_n) + \frac{\gamma}{\beta h}\left(\mathbf{z}(t_{n+1}) - \mathbf{z}(t_n) - h\dot{\mathbf{z}}(t_n) - \frac{h^2}{2}\ddot{\mathbf{z}}(t_n)\right) + \tau_{\dot{z}} \\ \ddot{\mathbf{z}}(t_{n+1}) &= \ddot{\mathbf{z}}(t_n) + \frac{1}{\beta h^2}\left(\mathbf{z}(t_{n+1}) - \mathbf{z}(t_n) - h\dot{\mathbf{z}}(t_n) - \frac{h^2}{2}\ddot{\mathbf{z}}(t_n)\right) + \tau_{\ddot{z}}\end{aligned} \tag{56}$$

and using a Taylor expansion of the displacements $\mathbf{z}(t_{n+1})$, velocities $\dot{\mathbf{z}}(t_{n+1})$ and accelerations $\ddot{\mathbf{z}}(t_{n+1})$ around t_n, we get the expression of the discretization errors $\tau_{\dot{z}}, \tau_{\ddot{z}}$ in terms of the third derivative of displacements at t_n

$$\begin{aligned}\tau_{\dot{z}} &= \left(\frac{1}{2} - \frac{\gamma}{6\beta}\right) h^2\dddot{\mathbf{z}}(t_n) \;+\; O(h^3) \\ \tau_{\ddot{z}} &= \left(1 - \frac{1}{6\beta}\right) h\dddot{\mathbf{z}}(t_n) \;+\; O(h^2)\end{aligned} \tag{57}$$

Therefore, the local truncation error at t_{n+1} is given by the expression:

$$\left[\frac{1}{\beta h^2}\mathrm{N} \;+\; (1+\alpha)\mathbf{1}\right]\mathbf{e}_{z\;n+1} \;=\; -\left(1 + \alpha - \frac{1}{6\beta}\right) h\mathrm{N}\dddot{\mathbf{z}}(t_n) \tag{58}$$

After solving this system, we get

$$\begin{aligned}\mathbf{e}_{z\;n+1} &= -\left(1 - \frac{1}{6\beta(1+\alpha)}\right) h \sum_{i=1}^{\nu}\left(\frac{-1}{\beta h^2(1+\alpha)}\right)^{i-1} \mathrm{N}^i\dddot{\mathbf{z}}(t_n) = \\ &= -\left(1 - \frac{1}{6\beta(1+\alpha)}\right)\left\{\begin{matrix} 0 \\ h\dddot{z}_1(t_n) \\ \frac{-1}{\beta h(1+\alpha)}\dddot{z}_1(t_n) + h\dddot{z}_2(t_n) \\ \vdots \end{matrix}\right\}\end{aligned} \tag{59}$$

Thus, we see that the local truncation error for the i-th component of $\mathbf{z}$ for an index-ν DAE system is an $O(h^{5-2i})$ (except the first component which is verified exactly). Note that for systems of index higher than 2, computed values exhibit a divergent behavior.

3.2.2 Global Error. It is well known that for ordinary differential equations the order of the global error is one order smaller than the local error. For instance, the local error of the HHT algorithm being $O(h^3)$, then this algorithm is globally second order accurate for ODE's.

Although the local error estimates show a divergent behavior for algebraic systems with index equal or higher than 3, when applying the algorithm with constant step size the global error reaches second order accuracy after several steps (the same observation holds for other integrators used for DAE's, like BDF [6]).

We analyze next the global error of the Hilber-Hughes-Taylor algorithm for an index-2 algebraic system. Let us define the vector of local truncation errors $\mathbf{e}$ which consists on the displacements, velocities and accelerations errors:

$$\mathbf{e}^T = \langle e_{z\,1} \quad he_{\dot{z}\,1} \quad h^2 e_{\ddot{z}\,1} \quad e_{z\,2} \quad he_{\dot{z}\,2} \quad h^2 e_{\ddot{z}\,2} \rangle \tag{60}$$

The vector of local truncation errors $\mathbf{e}$ verifies the equation

$$\mathbf{Z}(t_{n+1}) = \overline{\mathbf{A}}\mathbf{Z}(t_n) + \mathbf{L}_n + \mathbf{e}_n \tag{61}$$

where $\mathbf{Z}(t) = \langle z_1(t) \quad h\dot{z}_1(t) \quad h^2\ddot{z}_1(t) \quad z_2(t) \quad h\dot{z}_2(t) \quad h^2\ddot{z}_2(t) \rangle$ is the state vector of exact values, $\overline{\mathbf{A}}$ is the amplification matrix of the algorithm (36) and $\mathbf{L}_n$ is a vector that depends on the applied loads $\langle f_1(t) \quad f_2(t) \rangle^T$. By using the exact solution (50) into the latter equation, and by retaining higher order terms than in the previous subsection we can show that

$$\mathbf{e}_n = \mathbf{v}_1 f_1^{(3)}(t_n) + \mathbf{v}_2 f_1^{(4)}(t_n) + \mathbf{v}_3 f_1^{(5)}(t_n) + \mathbf{v}_4 f_2^{(3)}(t_n) + \mathbf{o} \tag{62}$$

where

$$\mathbf{v}_1 = \left\{ \begin{array}{c} 0 \\ \frac{1}{2}\left(1 - \frac{\gamma}{3\beta}\right) h^3 \\ \left(1 - \frac{1}{6\beta}\right) h^3 \\ \left(\frac{1}{6\beta(1+\alpha)} - 1\right) h \\ \frac{\gamma}{\beta}\left(\frac{1}{6\beta(1+\alpha)} - 1\right) h \\ \frac{1}{\beta}\left(\frac{1}{6\beta(1+\alpha)} - 1\right) h \end{array} \right\} \qquad \mathbf{v}_2 = \left\{ \begin{array}{c} 0 \\ \frac{1}{6}\left(1 - \frac{\gamma}{4\beta}\right) h^4 \\ \frac{1}{2}\left(1 - \frac{1}{12\beta}\right) h^4 \\ \frac{1}{2}\left(\frac{1}{12\beta(1+\alpha)} - 1\right) h^2 \\ \frac{\gamma}{2\beta}\left(\frac{1}{12\beta(1+\alpha)} - 1\right) h^2 \\ \frac{1}{2\beta}\left(\frac{1}{12\beta(1+\alpha)} - 1\right) h^2 \end{array} \right\} \tag{63}$$

$$\mathbf{v}_3 = \left\{ \begin{array}{c} 0 \\ \frac{1}{24}\left(1 - \frac{\gamma}{5\beta}\right) h^5 \\ \frac{1}{6}\left(1 - \frac{1}{20\beta}\right) h^5 \\ \frac{1}{6}\left(\frac{1}{20\beta(1+\alpha)} - 1\right) h^3 \\ \frac{\gamma}{6\beta}\left(\frac{1}{20\beta(1+\alpha)} - 1\right) h^3 \\ \frac{1}{6\beta}\left(\frac{1}{20\beta(1+\alpha)} - 1\right) h^3 \end{array} \right\} \qquad \mathbf{v}_4 = \left\{ \begin{array}{c} 0 \\ 0 \\ 0 \\ 0 \\ \frac{1}{2}\left(1 - \frac{\gamma}{3\beta}\right) h^3 \\ \left(1 - \frac{1}{6\beta}\right) h^3 \end{array} \right\} \qquad \mathbf{o} = \left\{ \begin{array}{c} 0 \\ 0 \\ 0 \\ 0 \\ O(h^4) \\ O(h^4) \end{array} \right\}$$

By subtracting from (61) the integration equation in terms of the approximate values computed by the algorithm, we get the expression for the propagation of errors:

$$\mathbf{E}_{n+1} = \overline{\mathbf{A}}\mathbf{E}_n + \mathbf{e}_n = \overline{\mathbf{A}}^{n+1}\mathbf{E}_0 + \sum_{k=0}^{n} \overline{\mathbf{A}}^k \mathbf{e}_{n-k} \tag{64}$$

with $\mathbf{E} = \langle\, E_{z\,1} \quad hE_{\dot{z}\,1} \quad h^2E_{\ddot{z}\,1} \quad E_{z\,2} \quad hE_{\dot{z}\,2} \quad h^2E_{\ddot{z}\,2}\,\rangle$ the global errors vector.

If we assume mild enough conditions (i.e. smoothness of the applied forces and their derivatives), and since $\overline{\mathbf{A}}^n \to \mathbf{0}$ for $n \to \infty$, we may write that for $h \to 0$

$$\mathbf{E} \;\to\; \sum_{k=0}^{\infty} \overline{\mathbf{A}}^k \,(C_1\mathbf{v}_1 \;+\; C_2\mathbf{v}_2 \;+\; C_3\mathbf{v}_3 \;+\; C_4\mathbf{v}_4 \;+\; \mathrm{o}) \tag{65}$$

The amplification matrix of the index-2 algebraic system was given in equation (36). Note that for this matrix,

$$\sum_{k=0}^{\infty} \overline{\mathbf{A}}^k = \begin{bmatrix} \sum_{k=0}^{\infty} \mathbf{A}_{\infty}^k & \mathbf{0} \\ (\sum_{k=0}^{\infty} \mathbf{A}_{\infty}^k)\,\mathbf{B}\,(\sum_{k=0}^{\infty} \mathbf{A}_{\infty}^k) & \sum_{k=0}^{\infty} \mathbf{A}_{\infty}^k \end{bmatrix} \tag{66}$$

For $\alpha < 0$, the amplification matrix of the algorithm verifies

$$\sum_{k=0}^{\infty} \mathbf{A}_{\infty}^k \to \begin{bmatrix} 1+\alpha & 0 & 0 \\ -1 & \frac{1}{2} & \frac{\alpha^2}{4} \\ 0 & -1 & \frac{1}{2}-\alpha \end{bmatrix} \tag{67}$$

(this series can be easily evaluated using the Jordan form and projectors $\mathbf{W}_L, \mathbf{W}_R$ (42-43)). After replacement into (66) we get

$$\sum_{k=0}^{\infty} \overline{\mathbf{A}}^k \to \begin{bmatrix} 1+\alpha & 0 & 0 & 0 & 0 & 0 \\ -1 & \frac{1}{2} & \frac{\alpha^2}{4} & 0 & 0 & 0 \\ 0 & -1 & \frac{1}{2}-\alpha & 0 & 0 & 0 \\ 0 & \frac{1}{h^2} & \frac{1+2\alpha}{2h^2} & 1+\alpha & 0 & 0 \\ 0 & 0 & 0 & -1 & \frac{1}{2} & \frac{\alpha^2}{4} \\ 0 & 0 & 0 & 0 & -1 & \frac{1}{2}-\alpha \end{bmatrix} \tag{68}$$

Finally, replacement of (68) into equation (65) yields the expression of the global error of the algorithm

$$\mathbf{E} \to \begin{Bmatrix} 0 \\ C_1\frac{1+3\alpha^2}{12}h^3 \\ C_2\alpha h^3 \\ C_3\frac{h^2}{12} \\ C_4\frac{1+3\alpha^2}{12}h^3 \\ C_5\alpha h^3 \end{Bmatrix} + O(h^4) \tag{69}$$

where constants C_i differ from those in equation (65).

Remarks:

- Similar estimations can be obtained numerically for higher index systems.
- The global error estimates are correct provided the time step is kept constant and loads and derivatives are smooth enough. Changing the step size, or introducing a discontinuity in loads, generates a perturbation which affects the computations for some steps. After a while this perturbation is damped out and the results gain accuracy.

3.3 CONDITIONING OF EQUATIONS

The system equations in their original form can be very ill conditioned, thus causing divergence of the Newton iteration due to error cumulation. We analyze next the DAE equations, by regarding again the full system representation, and look for a way to improve the system condition.

The HHT algorithm can be written in the form:

Find $\Delta \mathbf{q}$ *such that:*

$$\begin{aligned} \mathbf{q}_{n+1} &= \mathbf{q}_n + h\dot{\mathbf{q}}_n + \frac{h^2}{2}\ddot{\mathbf{q}}_n + \Delta\mathbf{q} \\ \dot{\mathbf{q}}_{n+1} &= \dot{\mathbf{q}}_n + h\ddot{\mathbf{q}}_n + \frac{\gamma}{\beta h}\Delta\mathbf{q} \\ \ddot{\mathbf{q}}_{n+1} &= \ddot{\mathbf{q}}_n + \frac{1}{\beta h^2}\Delta\mathbf{q} \\ \mathbf{M}\ddot{\mathbf{q}}_{n+1} + (1+\alpha)\mathbf{G}(\mathbf{q}_{n+1}) &- \alpha\mathbf{G}(\mathbf{q}_n) - (1+\alpha)\mathbf{f}_{n+1} + \alpha\mathbf{f}_n = \mathbf{0} \end{aligned} \tag{70}$$

Equation (70-d) depends nonlinearly on the generalized displacements vector $\mathbf{q}_{n+1}$. Thus, a system of nonlinear algebraic equations has to be solved at each step to advance computations. This system of equations is solved iteratively using the Newton-Raphson method.

At each iteration of the Newton method, a system of linear algebraic equations is solved. Let us write this system for a typical iteration, i.e. when going from iteration k to iteration $k+1$:

$$\left(\frac{1}{\beta h^2}\begin{bmatrix}\mathbf{M} & \mathbf{0}\\ \mathbf{0} & \mathbf{0}\end{bmatrix} + (1+\alpha)\begin{bmatrix}\mathbf{K} & -\mathbf{B}^T\\ -\mathbf{B} & \mathbf{0}\end{bmatrix}\right)\begin{Bmatrix}\Delta\mathbf{q}^{k+1}\\ \Delta\boldsymbol{\lambda}^{k+1}\end{Bmatrix} = -\begin{Bmatrix}\boldsymbol{\eta}_q^k\\ \boldsymbol{\eta}_\lambda^k\end{Bmatrix} \tag{71}$$

where $\langle \boldsymbol{\eta}_q^{k\,T} \quad \boldsymbol{\eta}_\lambda^{k\,T}\rangle^T$ is the residual vector at iteration k, and where we have taken into account the presence of holonomic constraint equations.

The inverse of the coefficients matrix for $h \to 0$ can be computed in the form:

$$\left(\frac{1}{\beta h^2}\begin{bmatrix}\mathbf{M} & \mathbf{0}\\ \mathbf{0} & \mathbf{0}\end{bmatrix} + (1+\alpha)\begin{bmatrix}\mathbf{K} & -\mathbf{B}^T\\ -\mathbf{B} & \mathbf{0}\end{bmatrix}\right)^{-1} = \begin{bmatrix}\mathbf{S}_{qq} & \mathbf{S}_{q\lambda}\\ \mathbf{S}_{\lambda q} & \mathbf{S}_{\lambda\lambda}\end{bmatrix} \tag{72}$$

with

$$\begin{aligned} \mathbf{S}_{qq} &= \beta h^2\left(\mathbf{M}^{-1} - \mathbf{M}^{-1}\mathbf{B}^T(\mathbf{B}\mathbf{M}^{-1}\mathbf{B}^T)^{-1}\mathbf{B}\mathbf{M}^{-1}\right) + O(h^4) \\ \mathbf{S}_{q\lambda} &= -\frac{1}{(1+\alpha)}\mathbf{M}^{-1}\mathbf{B}^T(\mathbf{B}\mathbf{M}^{-1}\mathbf{B}^T)^{-1} + O(h^2) \\ \mathbf{S}_{\lambda q} &= \mathbf{S}_{q\lambda}^T \\ \mathbf{S}_{\lambda\lambda} &= -\frac{1}{(1+\alpha)^2\beta h^2}(\mathbf{B}\mathbf{M}^{-1}\mathbf{B}^T)^{-1} + O(1) \end{aligned} \tag{73}$$

We can see that this coefficients matrix is very ill-conditioned due to the presence of constraints (the condition number is an $O(\overline{m}^2/h^4)$, where $\overline{m}$ is the mean mass of the system). If we try to solve this problem without scaling, the Newton algorithm will not converge since round-off errors would become of the same order of the Newton correction itself.

A better conditioning of the coefficients matrix can be reached by solving the symmetrically scaled problem:

$$\begin{cases} \mathbf{M}\ddot{\mathbf{q}} \;-\; (fac\,\mathbf{B}^T)\left(\dfrac{1}{fac}\boldsymbol{\lambda}\right) \;+\; \mathbf{G}(\mathbf{q},\dot{\mathbf{q}},t) \;=\; \mathbf{0} \\ fac\;\boldsymbol{\Phi}(\mathbf{q},t) \;=\; \mathbf{0} \end{cases} \tag{74}$$

where the scaling factor *fac* is equal to $\overline{m}/h^2$. In this way, the condition of the coefficients matrix becomes independent of the time step and of the mean value of mass. The new system of equations to be solved writes:

$$\left(\frac{1}{\beta h^2}\begin{bmatrix}\mathbf{M} & \mathbf{0}\\ \mathbf{0} & \mathbf{0}\end{bmatrix} + (1+\alpha)\begin{bmatrix}\mathbf{K} & -\mathbf{B}^{*\,T}\\ -\mathbf{B}^{*} & \mathbf{0}\end{bmatrix}\right)\begin{Bmatrix}\Delta\mathbf{q}^{k+1}\\ \Delta\boldsymbol{\lambda}^{*\,k+1}\end{Bmatrix} \;=\; -\begin{Bmatrix}\boldsymbol{\eta}_q^k\\ \boldsymbol{\eta}_\lambda^{*\,k}\end{Bmatrix} \tag{75}$$

with

$$\mathbf{B}^* \;=\; \frac{\overline{m}}{h^2}\mathbf{B} \qquad \boldsymbol{\lambda}^* \;=\; \frac{h^2}{\overline{m}}\boldsymbol{\lambda} \qquad \boldsymbol{\eta}_\lambda^* \;=\; \frac{\overline{m}}{h^2}\boldsymbol{\eta}_\lambda \tag{76}$$

The drawback of this technique of equations balancing is that the scaling factor depends on the size of the time step, posing some practical inconveniences from the point of view of programming. We have also obtained good results using as scale factor a mean value of the stiffness of the system. Nevertheless, it should be noted that in very severe cases for which the time step is highly reduced the algorithm may fail and the user has to restart computations and increase this scale factor.

Remark:

- The Newton iteration is stopped when the norm of residual vector becomes smaller than a given threshold. The stopping criterion is based on comparing the residue to characteristic measures of force f_{char} (for $\boldsymbol{\eta}_q$) and of displacement ℓ_{char} (for $\boldsymbol{\eta}_\lambda$). Note that in order to be consistent, the threshold value for the constraints equations should also be affected by the scaling factor. Thus, the convergence criterion may look like:

$$\left(\frac{\|\boldsymbol{\eta}_q\|}{f_{char}}\right)^2 + \left(\frac{h^2}{\overline{m}}\right)^2\left(\frac{\|\boldsymbol{\eta}_\lambda^*\|}{\ell_{char}}\right)^2 \;=\; \left(\frac{\|\boldsymbol{\eta}_q\|}{f_{char}}\right)^2 + \left(\frac{\|\boldsymbol{\eta}_\lambda\|}{\ell_{char}}\right)^2 \;\le\; TOL_{eql}^2 \tag{77}$$

  If we do not scale the constraints threshold, the convergence requirements will be too stringent and the method would fail to find a solution.

3.4 CONVERGENCE ANALYSIS FOR THE FULL SYSTEM

An estimate of the local truncation error of the algorithm evaluated for the full system will be used to determine the new time step. The difference formulas of the HHT algorithm are as indicated in the preceding section. The weighted equilibrium equation is solved iteratively in the nonlinear case:

$$\begin{aligned}\mathbf{M}\left(\ddot{\mathbf{q}}(t_{n+1}) - \frac{1}{\beta h^2}\mathbf{e}_{n+1} - \boldsymbol{\tau}_{\ddot{q}}\right) \;+\; (1+\alpha)\mathbf{G}(\mathbf{q}(t_{n+1}) - \mathbf{e}_{n+1}) \;-\; \alpha\mathbf{G}(\mathbf{q}(t_n))\\ -\;(1+\alpha)\mathbf{f}_{n+1} \;+\; \alpha\mathbf{f}_n \;=\; \boldsymbol{\eta}\end{aligned} \tag{78}$$

In this equation $\boldsymbol{\eta}$ are the unbalanced forces remaining at the end of the iterative cycle used to solve the nonlinear problem (typically, a Newton method is used). The nonlinear forces vector evaluated at $(\mathbf{q}(t_{n+1}) - \mathbf{e}_{n+1})$ can be expanded, to a first order, as

$$\mathbf{G}(\mathbf{q}(t_{n+1}) - \mathbf{e}_{n+1}) = \mathbf{G}(\mathbf{q}(t_{n+1})) - \frac{\partial \mathbf{G}}{\partial \mathbf{q}} \mathbf{e}_{n+1} + O(\mathbf{e}^2) \tag{79}$$

After replacing this expression into (78), and by taking into account that the equilibrium equation is exactly verified by correct values $\mathbf{q}(t_k), \dot{\mathbf{q}}(t_k), \ddot{\mathbf{q}}(t_k)$, i.e. :

$$\mathbf{M}\ddot{\mathbf{q}}(t_{n+1}) + \mathbf{G}(\mathbf{q}(t_{n+1})) - \mathbf{f}_{n+1} = \mathbf{0} \tag{80}$$

and by replacing the expression of the discretization error $\boldsymbol{\tau}_{\ddot{q}}$, we get the following equation for the local truncation error at t_{n+1}:

$$\left[(1+\alpha)\frac{\partial \mathbf{G}}{\partial \mathbf{q}} + \frac{1}{\beta h^2}\mathbf{M}\right] \mathbf{e}_{n+1} = -\left(1+\alpha-\frac{1}{6\beta}\right) h\mathbf{M}\dddot{\mathbf{q}}(t_n) - \boldsymbol{\eta} \tag{81}$$

If we now compute explicitly the contribution of the constraints, we get the following system:

$$\begin{aligned}\left(\frac{1}{\beta h^2}\begin{bmatrix}\mathbf{M} & \mathbf{0}\\ \mathbf{0} & \mathbf{0}\end{bmatrix} + (1+\alpha)\begin{bmatrix}\mathbf{K} & -\frac{\overline{m}}{h^2}\mathbf{B}^T\\ -\frac{\overline{m}}{h^2}\mathbf{B} & \mathbf{0}\end{bmatrix}\right)\begin{Bmatrix}\mathbf{e}_{q\ n+1}\\ \mathbf{e}^*_{\lambda\ n+1}\end{Bmatrix} &= \\ -\begin{Bmatrix}\left(1+\alpha-\frac{1}{6\beta}\right) h\mathbf{M}\dddot{\mathbf{q}}(t_n)\\ \mathbf{0}\end{Bmatrix} &- \begin{Bmatrix}\boldsymbol{\eta}_q\\ \boldsymbol{\eta}^*_\lambda\end{Bmatrix}\end{aligned} \tag{82}$$

After computing the inverse of the coefficients matrix (equation (73)), the truncation errors can be expressed as follows:

$$\begin{aligned}\mathbf{e}_{q\ n+1} &= \left[\mathbf{1} - \mathbf{M}^{-1}\mathbf{B}^T(\mathbf{B}\mathbf{M}^{-1}\mathbf{B}^T)^{-1}\mathbf{B}\right]\left\{\left((1+\alpha)\beta - \frac{1}{6}\right) h^3\dddot{\mathbf{q}}(t_n) + \beta h^2 \mathbf{M}^{-1}\boldsymbol{\eta}_q\right\} + \\ &\quad + \frac{1}{1+\alpha}\mathbf{M}^{-1}\mathbf{B}^T(\mathbf{B}\mathbf{M}^{-1}\mathbf{B}^T)^{-1}\frac{h^2}{\overline{m}}\boldsymbol{\eta}^*_\lambda \\ &= O(h^3)\dddot{\mathbf{q}}(t_n) + O(h^2/\overline{m})\boldsymbol{\eta}_q + O(h^2/\overline{m})\boldsymbol{\eta}^*_\lambda \\ \mathbf{e}^*_{\lambda\ n+1} &= \frac{1}{1+\alpha}(\mathbf{B}\mathbf{M}^{-1}\mathbf{B}^T)^{-1}\mathbf{B}\left\{\left((1+\alpha) - \frac{1}{6\beta}\right)\frac{h^3}{\overline{m}}\dddot{\mathbf{q}}(t_n) + \mathbf{M}^{-1}\frac{h^2}{\overline{m}}\boldsymbol{\eta}_q\right\} + \\ &\quad + \frac{1}{(1+\alpha)^2\beta}(\mathbf{B}\mathbf{M}^{-1}\mathbf{B}^T)^{-1}\frac{h^2}{\overline{m}^2}\boldsymbol{\eta}^*_\lambda \\ &= O(h^3)\dddot{\mathbf{q}}(t_n) + O(h^2/\overline{m})\boldsymbol{\eta}_q + O(h^2/\overline{m})\boldsymbol{\eta}^*_\lambda\end{aligned} \tag{83}$$

The previous expressions shows that the truncation error depends uniquely on the derivatives of the displacements and on the error of the solution of the nonlinear problem and is independent of the Lagrange multipliers.

If we now consider that the tolerance for the constraints residue in the Newton iteration is scaled by $(h^2/\overline{m})$ (equation (77)), we get the following estimations for the norm of the truncation error in terms of the tolerance for the out-of-equilibrium forces TOL_{eql}:

$$\begin{aligned}\|\mathbf{e}_{q\ n+1}\| &\leq O(h^3)\,\|\dddot{\mathbf{q}}_n\| + O(1)\,\ell_{char}\,TOL_{eql} \\ \|\mathbf{e}^*_{\lambda\ n+1}\| &\leq O(h^3)\,\|\dddot{\mathbf{q}}_n\| + O(1)\,\ell_{char}\,TOL_{eql}\end{aligned} \tag{84}$$

The truncation error for the Lagrange multipliers is obtained by multiplying (84-b) by the scale factor $(\overline{m}/h^2)$:

$$\|\mathbf{e}_{\lambda\ n+1}\| \ \leq\ O(\overline{m}h)\ \|\dddot{\mathbf{q}}_n\|\ +\ O(\overline{m}/h^2)\ \ell_{char}\ TOL_{eql} \tag{85}$$

Thus, we see that the equations balancing only affects the Newton iteration and does not have any effect on the accuracy of results.

Remark:

- We have seen that the accuracy of displacements is not influenced by the truncation error of the Lagrange multipliers and that the main factor that could deteriorate the displacements convergence rate is a loss of equilibrium at the Newton iteration. Indeed, equation (83) points out the relation between the integration error and the loss of equilibrium, relation which we will take into account to establish appropriate values for the integration error and for the equilibrium tolerance.

3.5 TIME STEP CONTROL

The task of fixing the time step size at each instant of the algorithm by the user could be quite difficult in many situations, noting that :

- If the selected time step is too large, the error in the computed response will be large, masking important aspects of the response. A large time step would also increase the degree of nonlinearity of the algebraic system, with a consequent increment of the number of iterations per time step or even giving place to a divergence of the Newton-Raphson algorithm.
- If the selected time step is too small, the cost to obtain a solution would be increased with a waste of computer resources, even to the point of becoming practically unacceptable in many cases.

Several techniques have been proposed to control the time step in nonlinear dynamics:

- from a comparison of results between algorithms of different order [19-21];
- in terms of a dominant frequency of response [22]:

$$\omega_n^2\ =\ \frac{\Delta\mathbf{q}^T\mathbf{K}\Delta\mathbf{q}}{\Delta\mathbf{q}^T\mathbf{M}\Delta\mathbf{q}} \tag{86}$$

- after a measure of nonlinearity (*current stiffness parameter*, number of iterations, ...) [23];
- using the local truncation error [24-27].

The technique we follow to limit the time step is based on controlling an estimate of the local truncation error of the algorithm. In fact, the step will be controlled based on monitoring the norm of the local truncation error at the displacements terms only, and the error in the Lagrange multipliers will not be taken into account since their approximation order is worse than that of the displacements.

If we neglect the residue of the Newton iteration, an estimate of the local error is given by approximating the third derivative of displacements as the difference of accelerations:

$$\mathbf{e}_{q\ n+1}\ =\ \left[\mathbf{1}\ -\ \mathbf{M}^{-1}\mathbf{B}^T(\mathbf{B}\mathbf{M}^{-1}\mathbf{B}^T)^{-1}\mathbf{B}\right]\left((1+\alpha)\beta-\frac{1}{6}\right)h^2\Delta\ddot{\mathbf{q}} \tag{87}$$

with $\Delta\ddot{\mathbf{q}} = \ddot{\mathbf{q}}_{n+1} - \ddot{\mathbf{q}}_n$. This approximation will be valid provided the tolerance for the out-of-balance forces is small enough.

3.5.1 Analysis of the Local Error Estimation: the SDOF Oscillator: A standard practice for estimating a time step in structural dynamics, is based on comparing it to the period of oscillation of the structure. Usually, the time step is selected so that one period is integrated by using 10 integration steps. In this section, we establish a comparison with this criterion to determine an estimate of the local error that will give accurate results.

Let us consider a linear SDOF oscillator submitted to an initial displacement y_0:

$$\begin{aligned} \ddot{y} + \omega^2 y &= 0 \\ y(0) = y_0 \qquad \dot{y}(0) &= 0 \end{aligned} \tag{88}$$

Clearly, the exact solution to this problem is

$$y(t) = y_0 \cos(\omega t) \tag{89}$$

Let us suppose we are at time t, instant for which we know the exact solution $y(t)$, and that we integrate one step the dynamic equations using HHT with a time increment h. The change of displacements, velocities and accelerations from t to $(t+h)$ can be written in the form:

$$\begin{Bmatrix} \Delta y \\ h\,\Delta\dot{y} \\ h^2\,\Delta\ddot{y} \end{Bmatrix} = [\mathbf{A}(\Omega) - \mathbf{1}] \begin{Bmatrix} y_0 \cos(\omega t) \\ -\Omega y_0 \sin(\omega t) \\ -\Omega^2 y_0 \cos(\omega t) \end{Bmatrix} \tag{90}$$

After replacing the expression of the amplification matrix $\mathbf{A}(\Omega)$ into equation (90), we see that the local error measured by equation (87) results :

$$\frac{e}{|y_0|} = \frac{(1+\alpha)\left[(1+\alpha)\beta - \frac{1}{6}\right]\Omega^3}{1 + (1+\alpha)\beta\Omega^2} \left|\sin(\omega t) + \left(\frac{1}{2} - \beta\right)\Omega \cos(\omega t)\right| \tag{91}$$

The quotient $e/|y_0|$ can be seen as a non dimensional error, independent of the oscillator excitation. Note however that this measure still depends on time. To eliminate this dependence, we define the ***expected value of the non dimensional local error*** $\mathcal{E}$ in the form:

$$\mathcal{E}(\Omega) = \frac{\mathrm{E}[e]}{|y_0|} = \frac{\lim_{T\to\infty} \frac{1}{T}\left(\int_0^T e^2\,dt\right)^{1/2}}{|y_0|} \tag{92}$$

Using equation (91), we can see that

$$\mathcal{E}(\Omega) = \frac{(1+\alpha)\left[(1+\alpha)\beta - \frac{1}{6}\right]\Omega^3 \left(1 + \left(\frac{1}{2} - \beta\right)^2 \Omega^2\right)^{1/2}}{\sqrt{2}\,\left(1 + (1+\alpha)\beta\Omega^2\right)} \tag{93}$$

This function is plotted in figure 1. We can see that it is a monotonically increasing function of the non dimensional frequency Ω.

Remark:

- We can appreciate also in figure 1 the non dimensional cut-off frequency Ω_K. It indicates indirectly the maximum value of the time step h to integrate accurately the equations of motion of an oscillator of frequency ω. It is usually accepted that for a non dimensional cut-off frequency $\Omega_K = 0.6$ –value corresponding to a time step equal

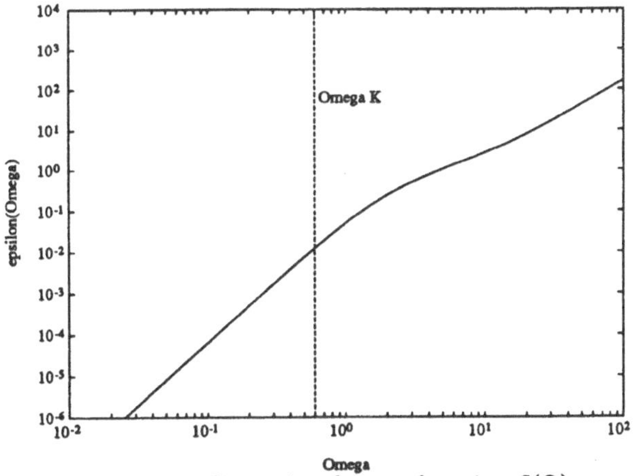

Figure 1 - Non dimensional error function $\mathcal{E}(\Omega)$

to one tenth of the oscillator period– the algorithm gives enough accurate results from an engineering point of view. The minimum spectral radius at this frequency equals

$$\rho(0.6)|_{\alpha=-0.3} = 0.9981$$

Let us define the constant $K_\Omega = \mathcal{E}(\Omega_K)$. From the definition of Ω_K, the quotient $\mathcal{E}(\Omega)/K_\Omega$ will be greater than or equal to 1 for values of frequency exceeding Ω_K. Therefore, if we accept that the expected and actual values of the local error are equal (in mean), we can write:

$$\frac{\left[(1+\alpha)\beta - \frac{1}{6}\right] h^2}{K_\Omega \; |y_0|} \, |\Delta \ddot{y}| \approx \frac{\mathrm{E}[e]}{K_\Omega \; |y_0|} = \frac{\mathcal{E}(\Omega)}{K_\Omega} \begin{cases} > 1 & \text{if } \Omega > \Omega_K \\ \leq 1 & \text{if } \Omega \leq \Omega_K \end{cases} \tag{94}$$

Then, by integrating the differential equation (88) with a time step that verifies:

$$\frac{\left[(1+\alpha)\beta - \frac{1}{6}\right] h^2}{K_\Omega \; |y_0|} \, |\Delta \ddot{y}| \leq 1 \tag{95}$$

the time step will be adjusted, in mean, to verify $\Omega \leq \Omega_K$. In other words, the time step h will take values for which the algorithm integrates correctly the equations of motion of the oscillator.

3.5.2 Analysis of the Local Error Estimation: the MDOF System: Let us compute the double product of the local truncation error $\mathbf{e}_q$ with the mass matrix:

$$\left(\mathbf{e}_q^T \mathbf{M} \mathbf{e}_q\right)^{1/2} = \left[(1+\alpha)\beta - \frac{1}{6}\right] h^2 \left(\Delta \ddot{\mathbf{q}}^T \mathbf{M} \Delta \ddot{\mathbf{q}} - \Delta \ddot{\mathbf{q}}^T \left[\mathbf{M} - \mathbf{B}^T (\mathbf{B}\mathbf{M}^{-1}\mathbf{B}^T)^{-1} \mathbf{B}\right] \Delta \ddot{\mathbf{q}}\right)^{1/2} \tag{96}$$

Since the second term on the right-hand-side is strictly negative, we get the inequality:

$$\left(\mathbf{e}_q^T \mathbf{M} \mathbf{e}_q\right)^{1/2} \leq \left[(1+\alpha)\beta - \frac{1}{6}\right] h^2 \left(\Delta \ddot{\mathbf{q}}^T \Delta \mathbf{G}_{iner}\right)^{1/2} \tag{97}$$

where $\Delta \mathbf{G}_{iner} = \mathbf{M}\Delta\ddot{\mathbf{q}}$. Let us now define a *relative error function* e_{rel} in the form

$$e_{rel} = \frac{\left[(1+\alpha)\beta - \frac{1}{6}\right] h^2}{K_\Omega\, \ell} \left(\Delta\ddot{\mathbf{q}}^T \Delta \mathbf{G}_{iner}\right)^{1/2} \tag{98}$$

where the reference length $\ell = \left(\mathbf{q}_0^T \mathbf{M} \mathbf{q}_0\right)^{1/2}$ is a mass-norm of the characteristic displacements $\mathbf{q}_0$.

The integration time step will be selected such that the relative error is smaller than a user-defined tolerance TOL_{int}. Thus, the relative mass-norm of the truncation error of displacements will be

$$\frac{\left(\mathbf{e}_q^T \mathbf{M} \mathbf{e}_q\right)^{1/2}}{K_\Omega \left(\mathbf{q}_0^T \mathbf{M} \mathbf{q}_0\right)^{1/2}} \leq e_{rel} \leq TOL_{int} \tag{99}$$

In this way, the time step will be adjusted to integrate (in mean) using 10 steps per period, for a single DOF system if the tolerance TOL_{int} is fixed equal to 1. In a MDOF system, the time step will be such that there will be 10 steps (in mean) per dominant period. See the remark below for a different interpretation of this criterion.

The tolerance TOL_{int} is independent of the problem under analysis. Computer experiments have shown that a value of TOL_{int} in the range $1\times10^{-2} - 1\times10^{-3}$ gives correct results for the engineer, with a good compromise between accuracy and economy of computation.

Remarks:

- It can be shown that if the time step is selected such that the relative error e_{rel} is below a given tolerance TOL_{int}, the sum of amplitudes of modal components exceeding the cut-off frequency Ω_K will be bounded by TOL_{int} :

$$e_{rel} \leq TOL_{int} \quad \Longrightarrow \quad \left(\frac{\sum_{\Omega_K}^{\Omega_m} \mathbf{y}_0^2}{\sum_0^{\Omega_m} \mathbf{y}_R^2}\right)^{1/2} \leq TOL_{int} \tag{100}$$

  Therefore, TOL_{int} limits indirectly the amount of energy dissipated by the algorithm.
- The estimations above will be valid provided the error coming from the out-of-equilibrium forces is small enough. The following relation between the integration tolerance and the equilibrium tolerance is based on demanding an equilibrium error one order of magnitude below the error of integration:

$$TOL_{eql} \leq \frac{TOL_{int} K_\Omega}{10} \tag{101}$$

- The analysis has been made considering the integration of linear systems with an almost constant time step h. The developed algorithms can be extended to the nonlinear case without any further difficulty.

3.5.3 Strategy for Changing the Time Step : The changing step strategy should be such that it keeps the integration step constant during long periods to avoid a deterioration of performance and to be in agreement with the already developed theory and with the stability and accuracy criteria of the HHT integrator. Therefore, the strategy should try to keep the time step unchanged unless strictly necessary.

By analyzing equation (93), we can estimate the effect of varying the time step on the computed relative error. By computing the quotient of errors for two different time steps h_1, h_2, we can note that

$$\frac{e_{rel}(h_1)}{e_{rel}(h_2)} \approx \left(\frac{h_1}{h_2}\right)^3 \tag{102}$$

The integration time step will be fixed according to the relation between the relative error and the user fixed tolerance. We follow a conservative criterion, similar to that presented in ref.[21] : at any time instant we will try to keep the relative error equal to half the tolerance. We distinguish between four cases :

i. If the error exceeds the tolerance, the computed step is rejected and recomputed using a time step equal to half the previous time increment.
ii. If the error is less than the tolerance, but greater than half its value, we accept the computed step but the next time step is decreased trying to make the relative error equal to $TOL/2$, using equation (102).
iii. Whenever the relative error lies between $TOL/2$ and $TOL/16$, we keep unchanged the time step. This criterion is based on considering that if the time step is doubled, the relative error will be greater than half the tolerance.
iv. If the relative error is less than $TOL/16$, we accept the step and double the time step.

4. Examples

4.1 CANONIC NILPOTENT SYSTEM

This first example treats the canonic nilpotent system of index-3:

$$\mathbf{N}_3\ddot{\mathbf{z}} + \mathbf{z} = \mathbf{f}(t)$$

with loads

$$\mathbf{f} = \begin{Bmatrix} \exp(t) \\ \cos(5t) \\ \exp(3t) \end{Bmatrix}$$

The exact solution is

$$\mathbf{z}(t) = \begin{Bmatrix} \exp(t) \\ \cos(5t) - \exp(t) \\ \exp(3t) + 25\cos(5t) + \exp(t) \end{Bmatrix}$$

Figures 2 and 3 show the computed displacements with a constant time step $h = 0.01$ and dissipation parameter $\alpha = -0.3$, compared to the exact solution. We note that z_1 and z_2 are in close agreement to the exact solution, while z_3 exhibits a perturbation at the initial steps which is damped out after $t \approx 0.2$.

Figures 4 and 5 show the evolution of velocities. Note that now, both $\dot{z}_2$ and $\dot{z}_3$ are perturbed at the beginning of computations. Note also that the amplitude of spurious oscillations of $\dot{z}_3$ is such that they mask the response at the subsequent steps.

Table 1 gives the global error at time $t = 3$ for different values of the time step. We can see verify that the error for the first component is always zero, and that the second and

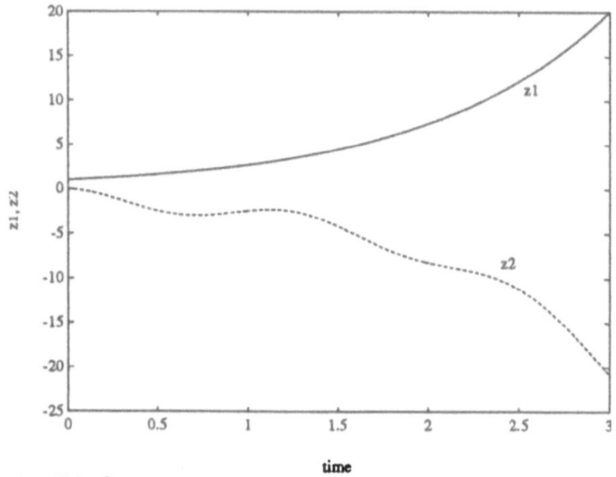

Figure 2 - Displacements z_1, z_2; nilpotent canonic index 3 system

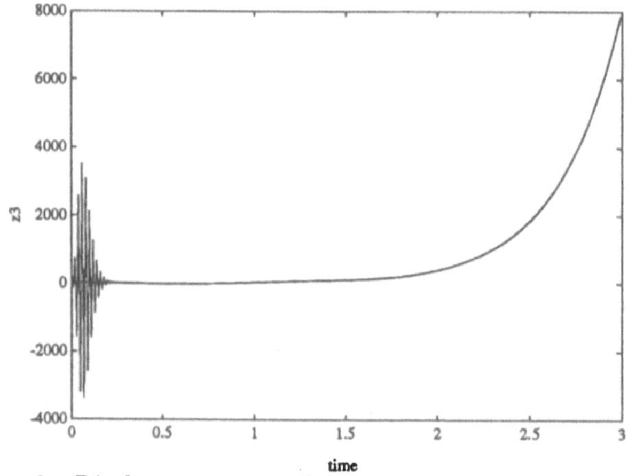

Figure 3 - Displacements z_3; nilpotent canonic index 3 system

third components converge with quadratic rate. However, we remark that the quadratic rate is lost for the third component for steps smaller than $h = 0.01$, most probably due to round-off errors cumulation.

4.2 SIMPLE PENDULUM

In this example we analyze the response of a simple rigid pendulum. The system has unit length, a unit mass at the extreme and oscillates freely from a horizontal initial position in

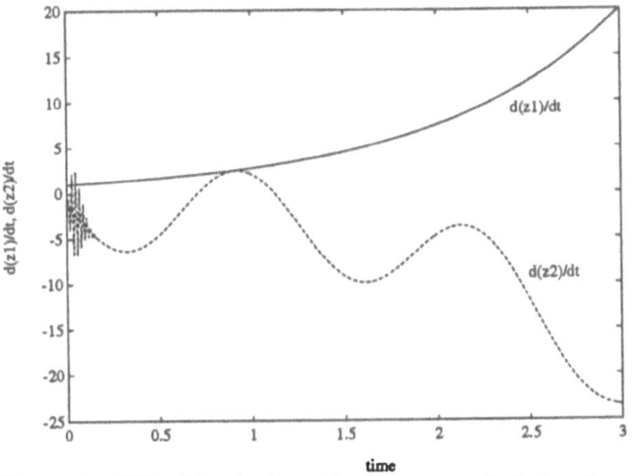

Figure 4 - Velocities $\dot{z}_1, \dot{z}_2$; nilpotent canonic index 3 system

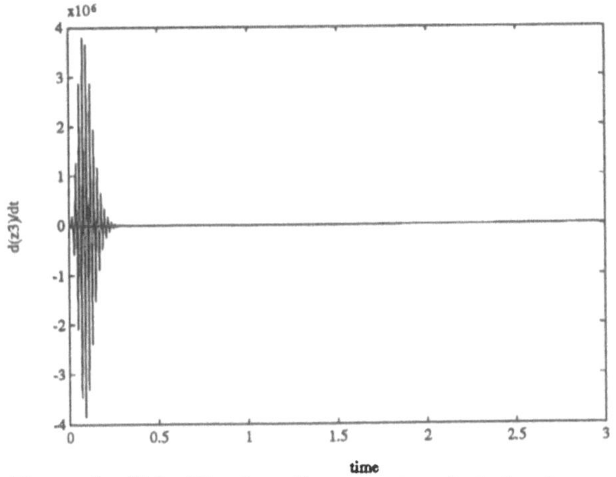

Figure 5 - Velocities $\dot{z}_3$; nilpotent canonic index 3 system

Table 1

h	E_1	E_2	E_3
0.1	3.553e-15	0.0445	0.631
0.05	3.553e-15	0.0118	0.238
0.01	0	0.0004947	0.0116
0.005	0	0.000124	0.0178

a $g = 10$ gravity field. The system of equations to be solved follows:

$$\begin{cases} \ddot{q}_1 - 2q_1\lambda = 0 \\ \ddot{q}_2 - 2q_2\lambda = -10 \\ q_1^2 + q_2^2 = 1 \end{cases}$$

The system was solved by using a variable step size, for various error tolerances ($TOL_{int} = 0.001, 0.01, 0.1, 1$). Table 2 shows the mean time step used in the different cases, together with the effectively computed mean integration error. We see that decreasing the error tolerance by a factor 10 implies almost halving the mean time step, in accordance to the predictions of equation (102) (actually, the exact relation is $10^{1/3} = 2.15$).

TOL_{int}	$\overline{e_{rel}}$	$\overline{h}$
1	0.3071	0.0738
0.1	0.0331	0.0349
0.01	0.0042	0.0177
0.001	0.000413	0.0082

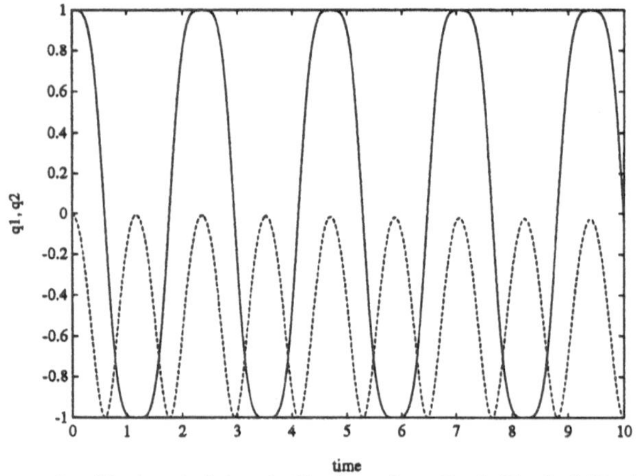

Figure 6 - Horizontal (cont. line) and vertical (dashed line) displacements at the pendulum extreme.

Figures 6 and 7 display the evolution of displacements and accelerations at the extreme of the pendulum. We can see that there appear some slight perturbations in the computed accelerations. In figure 8 we represent the evolution of the Lagrange multiplier, which evidence some perturbations at the same time instants than the accelerations do. If we analyze now the evolution of the time step (figure 9), we recognize that these perturbations are associated to changes in the step size, in complete accordance to the theoretical predictions.

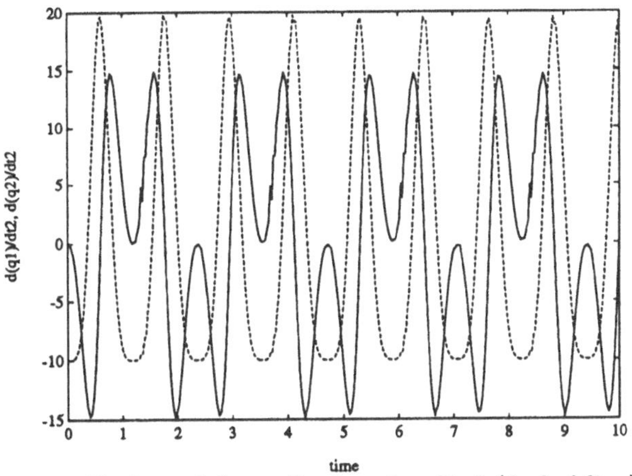

Figure 7 - Horizontal (cont. line) and vertical (dashed line) accelerations at the pendulum extreme.

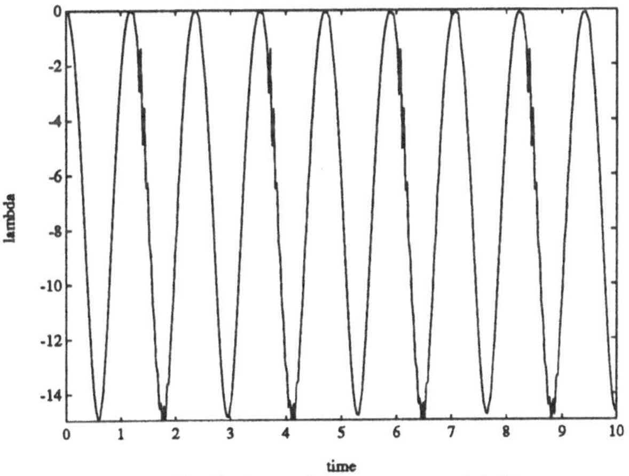

Figure 8 - Evolution of Lagrange multiplier.

5. Concluding Remarks

The paper discussed the application of second order implicit time integration schemes to the integration of the equations of motion in constrained dynamics simulation. The aspects of stability, conditioning of equations, accuracy and time step control have been analyzed in detail. Numerical examples describing the main issues of the developed algorithms have been shown. The algorithms evidenced global quadratic convergence rate in all variables, in complete accordance to the theoretical predictions.

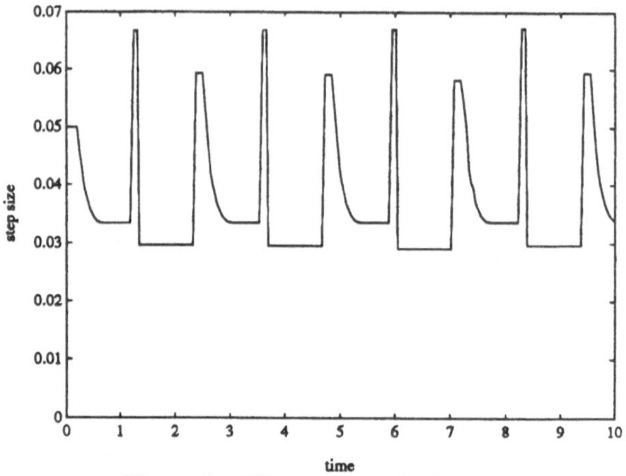

Figure 9 - Time step evolution.

References

1. Baumgarte J. (1972), 'Stabilization of constraints and integrals of motion in dynamic systems', Comp. Meth. Appl. Mech Engng., Vol.1, pp.1-16.

2. Gear C.W. and Petzold L.R. (1984), 'ODE methods for the solution of differential/algebraic systems', SIAM J. Numer. Anal. Vol.21, pp. 716-728.

3. Gear C.W. (1984), 'Differential-algebraic equations', in Haug E.J. (ed.), 'Computer aided analysis and optimization of mechanical system dynamics', NATO ASI Series Vol. F9, Springer-Verlag, pp. 323-334.

4. P. Lotstedt and L. Petzold (1986), 'Numerical solution of nonlinear differential equations with algebraic constraints I: convergence results for backward differentiation formulas', Math. of Computation, Vol.46, pp. 491-516.

5. P. Lotstedt and L. Petzold (1986), 'Numerical solution of nonlinear differential equations with algebraic constraints II: practical implications' SIAM J. Sci. Stat. Comput., Vol.7, pp. 720-733.

6. K.E. Brenan, S.L. Campbell and L.R. Petzold (1989), 'Numerical solution of initial-value problems in differential-algebraic equations', North-Holland.

7. T.J.R. Hughes (1987), 'Algorithms for hyperbolic and parabolic-hyperbolic problems' in "The Finite Element Method. Linear Static and Dynamic Finite Element Analysis", chap. 9, pp 490-569. Prentice - Hall, Englewood Cliffs.

8. Newmark N.M. (1959), 'A method of computation for structural dynamics', Journal of the Engineering Mechanics Division, ASCE, pp. 67-94.

9. H.M. Hilber, T.J.R. Hughes and R.L. Taylor (1977), 'Improved numerical dissipation for time integration algorithms in structural dynamics', Earthquake engineering and structural dynamics, vol. 5, 283-292.

10. H.M. Hilber and T.J.R. Hughes (1978), 'Collocation, dissipation and 'overshoot' for time integration schemes in structural dynamics', Earthquake engineering and structural dynamics, vol. 6, 99-117.

11. Wood W.L., Bossak M. and Zienkiewicz O.C. (1980), 'An alpha modification of Newmark's method', Int. J. Num. Meth. Engng., Vol.15, pp. 1562-1566.

12. C. Hoff and P.J.Pahl (1988), 'Development of an implicit method with numerical dissipation from a generalized single-step algorithm for structural dynamics', Comp. Meth. Appl. Mech. Engng., Vol 67, pp. 367-385.

13. J.C. Simo and K.K. Wong (1991), 'Unconditionally stable algorithms for rigid body dynamics that exactly preserve energy and momentum', Int. J. Numer. Meth. Engng., Vol.31, pp. 19-52.

14. J.C. Simo, N. Tarnow and K.K. Wong (1992), 'Exact energy-momentum conserving algorithms and symplectic schemes for nonlinear dynamics', Comp. Meth. Appl. Mech. Engng., Vol 100, pp. 63-116.

15. A. Cardona (1989), 'An Integrated approach to mechanism analysis', PhD. Thesis, Applied Sciences Faculty, University of Liège, Belgium.

16. A. Cardona, M. Géradin (1989) 'Time integration of the equations of motion in mechanism analysis', Computers and Structures, vol 33, 801-820.

17. J. García de Jalón and E. Bayo (1992), 'Computer assisted kinematic and dynamic analysis of multibody systems, Chapter 7, Centro de Estudios e Investigaciones Técnicas de Guipuzcoa, San Sebastián, España.

18. Gantmacher F.R. (1959), 'The theory of matrices', Vol.2, Chelsea Publishing Company.

19. R.M. Thomas, I. Gladwell (1988), 'Variable-Order Variable-Steps Algorithms for Second-Order Systems. Part 1: The Methods', Int. J. Num. Meth. Engng. vol 26, 39-53.

20. I. Gladwell, R.M. Thomas (1988), 'Variable-Order Variable-Steps Algorithms for Second-Order Systems. Part 2: The Codes', Int. J. Num. Meth. Engng. vol 26, 55-80.

21. L.F. Shampine, M.K. Gordon (1975), 'Computer solution of ordinary differential equations. The initial value problem', Freeman and Company.

22. S.H. Lee, S.S. Hsieh (1990), 'Expedient implicit integration with adaptive time stepping algorithm for nonlinear transient analysis', Comp. Meth. Appl. Mech. Engng., vol 81, 151-172.

23. C. Hoff, R. L. Taylor (1990), 'Step-by-step integration methods and time step control for systems with arbitrary stiffness', in Proc. of 'Second World Congress on Computational Mechanics', Stuttgart.

24. O.C. Zienkiewicz, W.L. Wood, N.W. Hine, R.L. Taylor (1984), 'A unified set of single step algorithms; part 1: General formulation and applications', International Journal for Numerical Methods in Engineering, vol 20, 1529-1552.

25. O.C. Zienkiewicz and Y.M. Xie (1991), 'A simple error estimator and adaptive time stepping procedure for dynamic analysis, Earthquake engineering and structural dynamics, vol. 20, pp. 871-887.

26. A. Cardona and A. Cassano (1992) ,'Integrador temporal de paso variable para análisis dinámico de estructuras y mecanismos', Revista Internacional de Métodos Numéricos Para Cálculo y Diseño en Ingeniería (Barcelona, Spain), Vol.8, pp. 439-461.

PART V
Man/Machine Interaction and Virtual Prototyping

DYNAMIC SIMULATION FOR VEHICLE VIRTUAL PROTOTYPING

J.S. FREEMAN, E.J. HAUG, J.G. KUHL, and F.F. TSAI
Center for Computer Aided Design, The University of Iowa,
Iowa City, Iowa 52242-1000, USA

ABSTRACT: Virtual prototyping is a key element in emerging simulation-based Integrated Product and Process Development (IPPD) frameworks for vehicles and other human-operated mechanical equipment. Virtual prototyping simulation can allow evaluation of design alternatives in the hands of a human operator at relatively early stages of the product development process. It thus provides a fundamentally new capability to investigate human factor-based design issues. A key requirement of operator-in-the-loop virtual prototyping is the ability to simulate the behavior of the mechanical system in real-time, at a level of fidelity sufficient to insure valid human-machine interaction and to support the associated engineering design and analysis functions. Modern recursive multibody dynamics formulations have been developed that can achieve real-time performance for detailed vehicle models, using advanced parallel processing computing methods. These dynamics formulations, augmented with appropriate models of vehicle subsystems (tires, steering system, powertrain, etc.) can achieve a level of fidelity and real-time performance sufficient to support engineering virtual prototyping applications. As such, they provide the foundation for advanced ground vehicle simulators that can provide a comprehensive operator-in-the-loop virtual prototyping capability to support the IPPD process.

1 Introduction

A revolutionary new capability is on the horizon for simulation based Integrated Product and Process Development (IPPD) of vehicles and equipment, based on three complementary levels of simulation;

(1) engineering performance and manufacturing process simulation,
(2) virtual prototyping simulation, and
(3) synthetic operational simulation, this being spearheaded by the military community in the form of distributed interactive battlefield simulation.

Interactions among these levels of simulation and data flow that permit a hierarchy of analyses and predictions of durability, reliability, maintainability, and life cycle costs are shown in Figure 1.

Developments in concurrent engineering and engineering performance and manufacturing process simulation are emerging [1] to provide the simulation-based design capabilities shown in dark background in Figure 1. Engineering performance and manufacturing process simulations are carried out, using a computer aided design (CAD) definition of mechanical system characteristics and assumed environments and duty cycles that may or may not have close resemblance to actual use of vehicles and equipment in the field. High fidelity simulations carried out in the engineering environment include fundamental physics of failure predictions of durability and simulation of personnel

M. F. O. S. Pereira and J. A. C. Ambrósio (eds.),
Computer-Aided Analysis of Rigid and Flexible Mechanical Systems, 533–554.

carrying out maintenance functions. The results are **engineering durability, reliability, and maintainability** predictions that are realistic, but based on assumed environments and use histories (duty cycles).

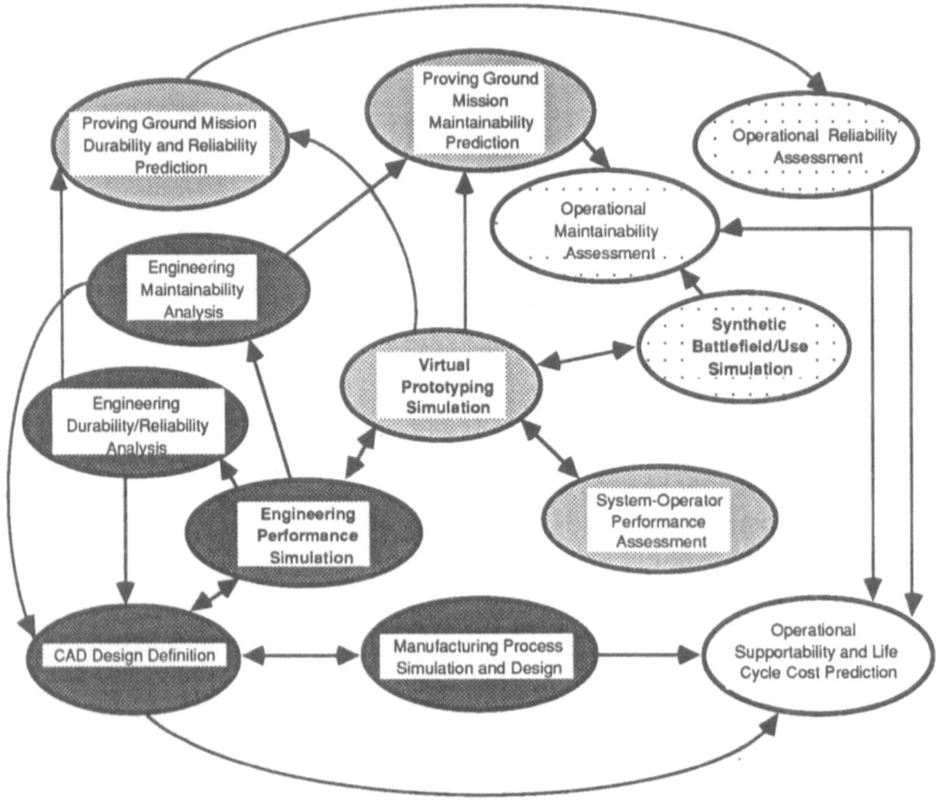

Figure 1. Integrated Product and Process Development (IPPD) Infrastructure for Engineering, Virtual Prototyping, and Synthetic Battlefield Simulation

Complementary efforts are demonstrating the feasibility of virtual prototyping simulation, shown in the gray background in Figure 1. Virtual prototyping simulations are carried out at a level of fidelity that is adequate to realistically account for human-system interaction in a simulated proving ground environment. This new capability accounts for driver/operator-system interactions that influence system performance. Data resulting from virtual prototyping simulation provide duty cycle information that is realistic regarding use of the vehicles and equipment by individual operators or crews. These performance data may then be used to define proving ground duty cycles that can be combined with durability, reliability, and maintainability data that are generated at the engineering level, to predict **proving ground mission durability, reliability, and maintainability**.

Finally, the synthetic battlefield simulation capability shown in dotted background in Figure 1 has been advanced by the military [2] and is now providing the capability to assess the impact of system characteristics on the outcome of battle. Synthetic battlefield simulation, or its civilian counterpart, is carried out at a level of fidelity that is adequate to realistically portray the capabilities and performance to be achieved by systems in an operational environment. This includes definition of how such systems would be used in tactics resulting from experience gained. These simulations yield operational duty cycle information that realistically portraying how systems are actually used by operators. The result will be an enhanced knowledge of tactical duty cycles that can be combined with proving ground predictions of mission durability, reliability, and maintainability to yield a fundamentally new assessment of vehicle and equipment **operational durability, reliability, and maintainability** in a tactical situation. All these data can then be combined with design, manufacturing, and field use information to realistically predict **operational supportability and life cycle costs.** Developments of this kind in the civilian sector are not nearly as advanced.

Creation of the infrastructure shown in Figure 1, with its unidirectional and bi-directional information flow channels, will allow rapid assessment of design alternatives, establishing trade-offs and providing a unique new capability for tuning the design of advanced vehicle systems that will function effectively in a broad range of applications. It is important to note that the three levels of simulation capability are advancing in parallel, but have not yet been fully integrated to provide the capability implied in Figure 1.

Technical distinctions between the three levels of simulation are highlighted in Figure 2, to emphasize differences in levels of fidelity involved, the knowledge of environments in which simulation is carried out, and data that are generated at each level of simulation. Results obtained in each level of simulation are complementary and can be combined to provide new information and knowledge that will impact vehicle and equipment design for integrity and basic elements of performance, as well as design for driver/operator-system effectiveness.

Engineering Performance and Manufacturing Process Simulation

- Design for product performance and manufacturing process effectiveness
- Analyze and design for durability, reliability, and maintainability
  - assumed system environment and use
  - highest fidelity, non real-time engineering simulation
  - durability, reliability, and maintainability models at highest level of fidelity

Virtual Prototyping Simulation

- Predict driver/operator-in-the-loop duty cycle and performance with proving ground fidelity
- Predict proving ground mission durability, reliability, and maintainability
  - realistic system use in proving ground environment
  - adequate fidelity for driver/operator-system interaction
  - build on engineering level of fidelity durability, reliability, and maintainability data

Synthetic Battlefield/Operational Simulation

- Predict battlefield duty cycle and performance
- Predict tactical durability, reliability, and maintainability
  - realistic system use in a tactical environment
  - adequate fidelity for tactical systems interaction
  - build on proving ground level of fidelity durability, reliability, and maintainability data

Figure 2. Engineering, Virtual Prototyping, and Synthetic Battlefield/ Operational Simulation

While each of the three levels of simulation is developing in parallel, a framework that is adequate to support the information flow and functional interactions shown in Figure 1 has not yet been created. Such a framework is needed to influence and integrate the three complementary, but somewhat independent, simulation environments being developed. Furthermore, operational supportability and life cycle cost prediction capabilities shown in white background in Figure 1 are needed. When implemented, they will exploit fundamental design information, manufacturing information, and operational performance data; e.g., usage rates of expendable materiel and time and cost to repair equipment, to obtain realistic life cycle cost information for system development decision making. Initial steps toward development of this capability are addressed here and in Refs. 1 and 3.

2 Virtual prototyping

Dynamic simulation of vehicle systems has seen a renaissance during the 1980s, due to a number of synergistic developments. The rapid increase in digital computer power has permitted an international community concerned with mechanical system dynamics to create new analytical and computational formulations that take advantage of emerging computer power and automate the process of forming and solving the differential-algebraic equations of vehicle system dynamics, using only engineering model data that can be naturally and effectively provided by the engineer. With this computational burden transferred from the engineer and analyst to the computer, the creative process of model development, design concept formulation and analysis, and testing of designs prior to fabrication of a prototype has revolutionized the process of vehicle system design for dynamic performance [4-9]. As an illustration of the explosive growth in the field of mechanical system dynamic simulation, six textbooks and advanced research monographs on the topic have been published since 1988 [10-15], whereas only two such books had been published prior to 1988 [16,17].

As impressive as has been the advancement in computer–based dynamic simulation of vehicle systems, computer times required for realistic simulation of dynamic performance have been extremely high. Even on the most powerful computers available in the late 1980s, the computer time required has typically been a factor of 10 to 100 greater than the clock-time that transpires during actual motion of the vehicle system. As a result, only off-line (non-real-time) vehicle dynamic simulation could be carried out in support of design applications, precluding applications in which the operator must interact with the vehicle to control its performance. Projections of increased computer performance and the emergence of revolutionary new dynamics formulations in the late 1980s suggested the potential for real-time simulation; i.e., computing the motion of a mechanical system in one unit of the computer time that corresponds to the same unit of time required for actual performance of the system. This led to a vision for operator-in-the-loop simulation in a broad range of applications in the late 1980s that is now coming to fruition. This new capability, permitting concurrent consideration of the human operator and the design of mechanical systems, in the conceptual design phase, might be thought of as prototyping and testing designs in a "virtual reality" operator-in-the-loop simulator, suggesting the term "virtual prototyping."

A precise definition of the term "virtual prototyping," especially as it applies to vehicle system design, is needed to avoid confusion with other concepts in design. As a foundation for such a definition, consider the following dictionary definitions:

Prototype: A first full-scale functional form of a new type or design of a construction (such as an airplane).

Virtual: Being such in essence or effect, although not in actual fact.

Reality: The quality or fact of being real.

While not yet in the dictionary, the term "virtual reality" has taken on the meaning of the "computer generated perception of reality on the part of an involved human." The term "virtual reality" motivated the emerging use of the term "virtual prototype," suggesting both computer and human involvement in "virtual prototyping."

A key concept that is implicit in each of the above definitions, but not explicitly stated, is that the functionality of the system or environment being addressed is clearly understood. The functionality of a prototype is central to the purpose for which it is fabricated and tested; e.g., assessment of dynamic performance, maintainability, manufacturability, and supportability. The expression "being such" in the definition of the word "virtual" implies some well understood form of functionality. The essence of the concept of "reality" is that some form of functionality should be, or appear to be, real. Thus, the central issue in defining and using the term "virtual prototyping" is making explicit the intended functionality of the prototype that is to be realized virtually. With this background, the following definitions have been proposed [1]:

Virtual Prototype: A computer based simulation of a prototype system or subsystem with a degree of functional realism that is comparable to that of a physical prototype.

Virtual Prototyping: The process of using a virtual prototype, in lieu of a physical prototype, for test and evaluation of specific characteristics of a candidate design.

These definitions are intended to **include** the following:

(1) The intended functionality of the prototype that is to be created virtually is clearly defined and at an engineering level of fidelity simulated; e.g., vehicle dynamic performance, vehicle maintainability, engine reliability, and vehicle component manufacturability.

(2) If human action is involved in the intended functionality of the prototype, then the human functions involved must be realistically simulated, or the human must be included in the simulation; i.e., real-time operator-in-the-loop simulation.

(3) If no human action is involved in the intended functionality of the prototype, then either off-line (non-real-time) computer simulation of the functions can be carried out; e.g., dynamic performance of an engine, stresses in its connecting rods, and fabrication of the connecting rods. Alternatively, a combination of computer and hardware-in-the-loop simulation can be carried out; e.g., vehicle dynamic performance prediction, laboratory durability testing for difficult to model failure modes, and manufacturing process analysis of critical components.

These definitions are intended to **exclude** the following:

(1) Partial simulation that does not include the full functionality intended for the prototype; e.g., geometric modeling with a CAD system that does not simulate dynamic performance, finite element stress analysis of a component that does not include system or subsystem performance simulation that defines loads on

the component, or manufacturing process planning that does not consider component performance or design constraints.

(2) Show-and-tell exercises that lack a prototype level of functional reality; e.g., goggles and gloves simulation with no underlying physical or mathematical simulation at an engineering or manufacturing level of reality.

To provide background on developments that have occurred recently in dynamic simulation, a brief summary of real-time recursive dynamics formulations that are well-suited to exploit the emerging capabilities of shared memory parallel processors is summarized, with references to further developments. Subsystem and component models for vehicle prototyping are detailed. Emerging graphics technologies and motion control for virtual prototyping applications are summarized, as the foundation for the Iowa Driving Simulator (IDS), an advanced ground vehicle driving simulator supporting a broad range of human factors research, including highway safety and vehicle and highway system design.

3 Real-time recursive multibody dynamics software for vehicle driving simulation

Real-time recursive dynamics (RTRD) software that has been developed for driver-in-the-loop real-time vehicle simulation is outlined in this section. The RTRD software is based on topology analysis of a rigid body mechanism, is formulated as a modified recursive dynamics formulation using relative coordinates, and is implemented with parallel computational algorithms. The mathematical details of the RTRD formulation appear in Refs. 18 to 25. The topology analysis utilizes graph theory to define the connectivity of a mechanism and to generate information necessary for the recursive dynamics formulation. This information includes identification of the base body, cut joints, decoupled loops, independent chains, junction nodes, and non-zero entries of generalized mass matrix. The topology analysis minimizes extreme chain length and the number of generalized coordinates in order to optimize computational efficiency on both serial and parallel processor computers.

The modified recursive dynamics formulation defines a body reference coordinate frame, x_j'-y_j'-z_j', at the inboard body joint position as shown in Figure 3. This formulation yields greater efficiency than traditional Cartesian formulations. These efficiency gains are yielded by taking advantage of invariant properties of generalized mass and generalized force in a velocity state multibody dynamics formulation.

The parallel computational algorithms used in the RTRD software exploit inherent parallelism in the modified recursive formulation and allow significant speed-up of the dynamics computation using parallel processing. The RTRD general purpose dynamic simulation software combines the topological analysis, recursive formulation, and parallel algorithms in a software design that is developed for real-time simulation efficiency on shared memory parallel processor computer systems.

The baseline RTRD software has the following capabilities:

(1) Rigid multibody mechanical systems can be modeled.

(2) Systems can have both open and closed kinematic chains.

(3) Bodies can be fixed to ground or float in space.

(4) The kinematic joint library contains revolute, translational, spherical, universal, spherical-translational, revolute-spherical, and translational-revolute joints. The joint library can be expanded to include other joint types, as required.(5)

Kinematic constraints include spherical cut joint constraints, revolute-spherical cut-joint constraints, and distance constraints. The constraint library can be expanded to include other cut joint types, if required.

(6) Force elements include translational spring-damper-actuators (TSDA), rotational spring-damper-actuators (RSDA) and more specialized coulomb friction, leaf spring, and anti-roll bar modules.

(7) Linear, piecewise-linear, and spline curve functions, as well as linear-linear surface plot interpolation functions, are provided.

(8) Constant step size integration algorithms are implemented for use in real-time simulation.

(9) Vehicle subsystem models -- tire, brake systems, steering system, powertrain, and aerodynamics -- are provided to characterize vehicle subsystem performance for real-time simulation.

(10) The RTRD code is written in Fortran-77 with Fortran-90 parallel processing extensions, it can be executed on both uniprocessor and parallel computers, using either Fortran-77 or Fortran-90 compilers.

(11) A translator program can be used to translate DADS input data files into RTRD input format [26].

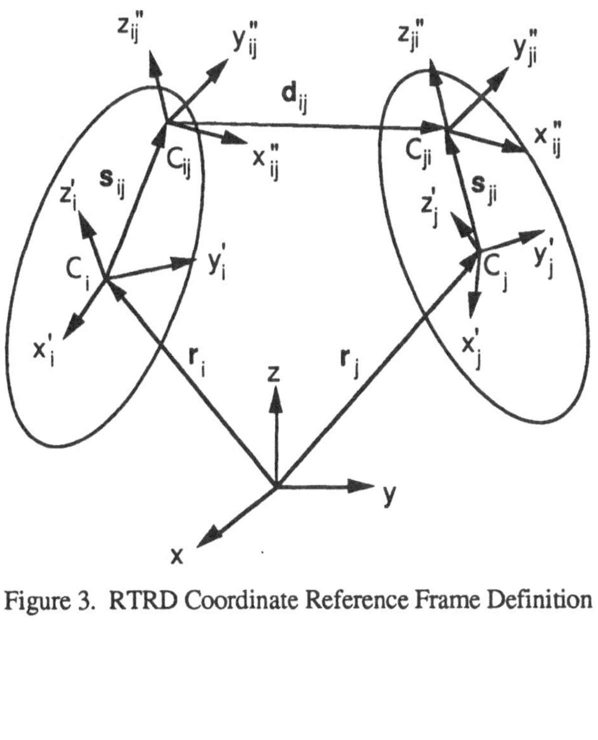

Figure 3. RTRD Coordinate Reference Frame Definition

The basic computational scheme of the RTRD software is diagrammed in Figure 4.

The steps on the left side of the diagram are referred to as the forward path of the algorithm, since they can be viewed as propagating relative position and velocity from the base body, down each kinematic chain. This is depicted in Figure 5. The steps on the right side of the diagram are referred to as the backward path, since once the forces are calculated, the equations of motion are recursively generated from the end bodies of the kinematic chains back to the base body. This is depicted in Figure 6.

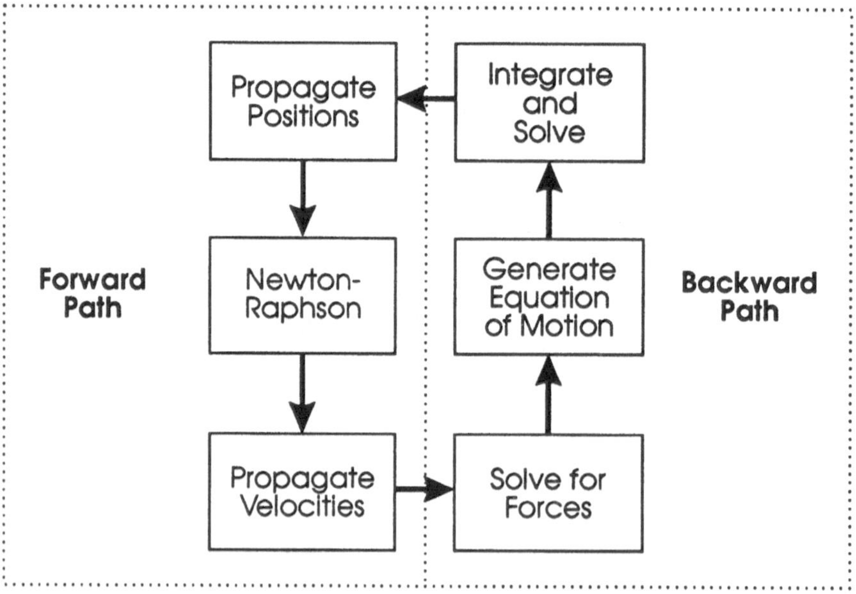

Figure 4. RTRD Basic Computational Scheme

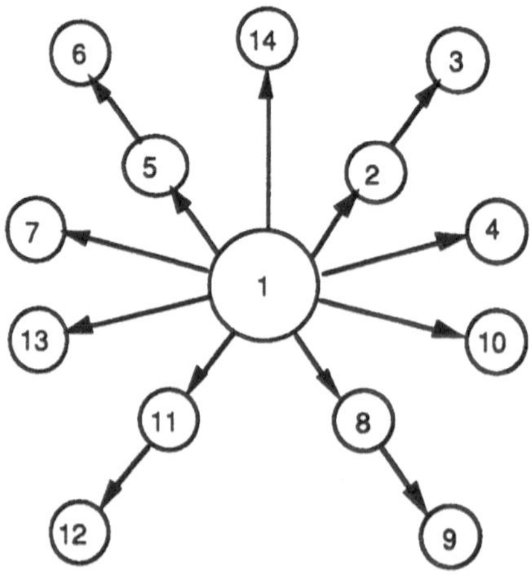

Figure 5. RTRD Forward Path Computational Sequence

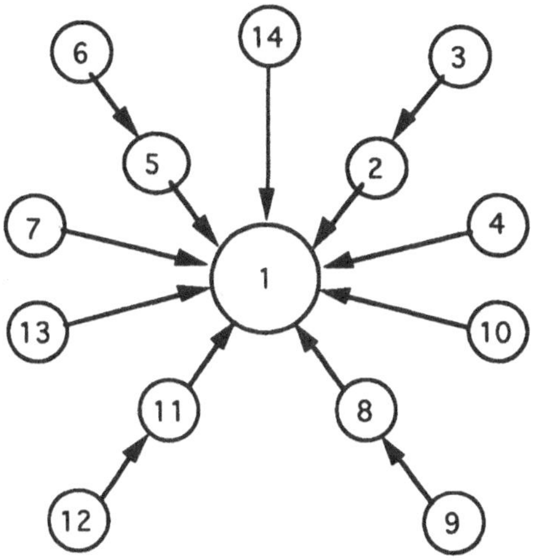

Figure 6. RTRD Backward Path Computational Sequence

4 Numerical integration

In real-time simulation, constant stepsize numerical integration algorithms are used to keep the sampling frequency constant, and maintain synchronization with other simulator subsystems. At the beginning of each time step, the generalized coordinates and velocities are known from the previous time step, and the generalized accelerations are computed based on the dynamic formulation. Generalized velocities and accelerations are then integrated to obtain the generalized coordinates and velocities to be used as initial conditions for the next time step. This procedure is repeated until the simulation is terminated.

A mechanical system, such as a vehicle, usually contains kinematic closed loops. In the RTRD recursive formulation, a joint is cut in each independent closed loop to obtain a spanning tree structure, and constraint equations associated with Lagrange multipliers are imposed for these cut joints to represent kinematically admissible motion. Therefore, in addition to second order dynamic equations of motion, nonlinear algebraic equations are introduced. As a result, a system of differential-algebraic equations (DAE) is formed. Since DAE are different and much more complex than ordinary differential equations (ODE), the stability of numerical integration methods is difficult to analyze. The RTRD software uses Newton-Raphson iteration to enforce kinematic constraints and implements second and third-order Adams-Bashforth integration algorithms to integrate the independent coordinates.

A mechanical system model can contain subsystems that are separately integratable. In vehicle dynamics, such subsystems are wheel spin and engine angular velocity. These subsystems are expressible using standard ODE's and integrable using standard algorithms comparable to those used for the DAE.

5 Subsystem modeling

The RTRD software can accurately compute the dynamic behavior of a vehicle chassis and suspension, applying detailed force input models of the vehicle mechanical components (struts, springs, dampers, tie rod, anti-roll bars, etc.). To create a complete vehicle model, the RTRD software must be interfaced with models of vehicle subsystems that directly act upon or are acted upon by the dynamics model. These subsystems include tire/wheel models, powertrain models, steering system models, brake system models, and aerodynamic load models. The subsystem modeling structure is shown in Figure 7.

Given the road/terrain data and the wind velocity vector, the operator maneuvers the vehicle by controlling the accelerator pedal, brake pedal, steering wheel, and gear shift setting. The RTRD software then generates and solves the equations of motion resulting from all these inputs and predicts the motion of the vehicle using numerical integration methods.

Extensions are required to the base RTRD software for modeling specialized systems found in vehicles. Extensions have been made for some specific component and subsystem models, including

- tire/wheel model
- powertrain model
- steering model
- brake model
- aerodynamics model

Each of these extensions incorporated within the RTRD software is detailed.

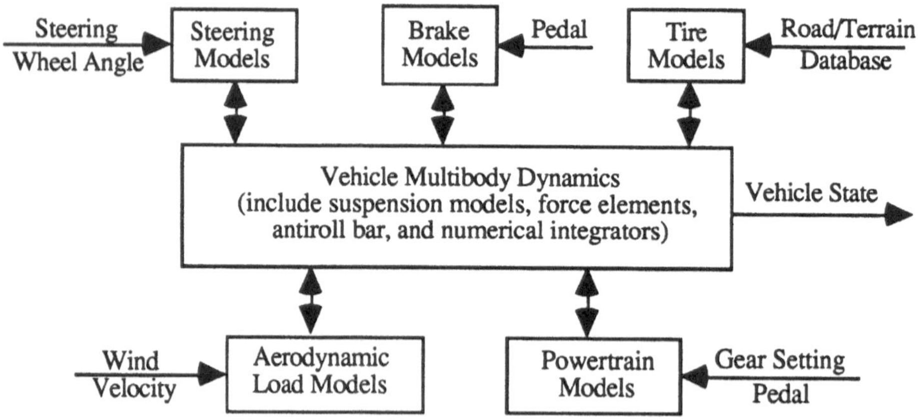

Figure 7. Vehicle Subsystem Modeling

TIRE/WHEEL MODEL

Wheels are a special extension of the RTRD multibody dynamics code that enable the calculation of tire forces without modeling the tire as a true body-type. The wheel element models the tire/wheel assembly as a combination of a spring, a slip velocity, and a point-contact tire tractive force model. In practice, the wheel element determines if the tire is in contact with the ground by calculating its position and orientation in the global coordinate reference frame and querying the terrain database. If the contact condition is true, the tire normal force is calculated from the compression of the spring.

The wheel slip velocity, rather than the wheel spin, is approximately integrated using the procedure of Bernard [27]. This simplification enables real-time simulation of tire dynamics. The wheel shown in Figure 8 is used for deriving the algorithm. In this model of the wheel, an accelerating torque is represented as being positive, while a braking torque is considered negative. Likewise, the force at the contact patch is considered positive for acceleration and negative for braking. However, the slip is taken as being negative for acceleration and positive for braking, the opposite of the force and torque conventions.

The equations of motion of the wheel, in its plane, can be expressed as

$$J\,\dot{\omega} = T - F_x r \tag{1}$$

where J is the mass moment of inertia of the wheel about its spin axis, ω is wheel angular velocity, T is the applied torque, F_x is the longitudinal tractive force, and r is the tire rolling radius. The longitudinal slip can be expressed as

$$s = 1 - \frac{\omega r}{v} \tag{2}$$

where v is the longitudinal velocity of the wheel spindle. This expression for slip can be differentiated with respect to time to obtain the following expression:

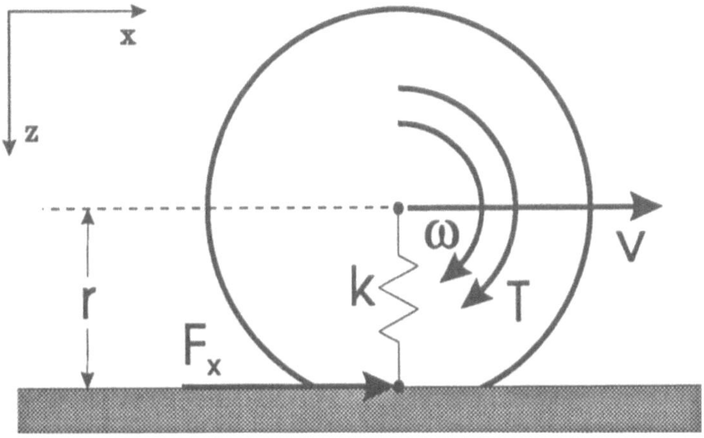

Figure 8. Wheel Model Coordinates

$$\dot{s} = \frac{r\dot{v}\omega - rv\dot{\omega}}{v^2} \tag{3}$$

which can be re-written as

$$\dot{\omega} = \frac{\dot{v}\omega}{v} - \frac{\dot{s}v}{r} \tag{4}$$

Substituting this expression into Equation (1) yields

$$J\left(\frac{v}{r}\right)\dot{s} = F_x r - T + J\left(\frac{\omega}{v}\right)\dot{v} \tag{5}$$

Following the procedure of Bernard [27], the coefficient of friction at any time step can be defined in terms of a Taylor series expansion with respect to wheel slip. That is,

$$\mu = \mu_0 + \frac{\partial\mu}{\partial s}(s - s_0) + O(s^2) \tag{6}$$

In this procedure, the Taylor series approximation is truncated with terms that are linear in slip. The longitudinal force on the tire can be expressed as

$$F_x = -\mu F_z = F_x^{old} - \frac{\partial\mu}{\partial s}(s - s_0)F_z \tag{7}$$

Making use of the slip equation for braking, Equation (2), and Equation (7), the equation for the wheel spin dynamics, Equation (5), can be re-written as

$$\dot{s} + Qs = F \tag{8}$$

where

$$Q \equiv \left(\frac{r^2 F_z}{J\,v}\right)\left(\frac{\partial\mu}{\partial s}\right) + \left(\frac{\dot{v}}{v}\right) \tag{9}$$

and

$$F \equiv \left(\frac{r}{J\,v}\right)\left(rF_x^{old} + \frac{\partial\mu}{\partial s} r\, s_0 F_z - T\right) + \left(\frac{\dot{v}}{v}\right) \tag{10}$$

Making the assumption that the values of Q and F are constant during the integration time step, the solution to the differential equation expressed by Equation (8) is given by

$$s = \left(s_0 - \frac{F}{Q}\right)e^{-Q(t-t_0)} + \frac{F}{Q} \tag{11}$$

Thus, by storing the values of slip, s_0, and longitudinal tractive force, F_x^{old}, from the previous time step, the longitudinal slip at the current time step can be calculated.

The tire model computes three forces (longitudinal, lateral, and vertical) and three torques (overturning moment, rolling resistance moment, and aligning torque) acting on the tire. Before the tire forces and torques are computed, the position, velocity, deflection, longitudinal slip, toe, camber, steer, and slip angles for each wheel are determined by the RTRD software. Based on these parameters, the tire model can compute the tire forces and torques. The choice of tire model depends on the available tire data. Several tire models from the published literature [28-32], in addition to a carpet plot model, are implemented in the RTRD software.

After tire forces and torques are computed by the selected tire model, they are transferred to the special wheel body within the multibody dynamics code. These forces and torques contribute to the generalized forces of the attached bodies and the rotational equations of motion of tires.

POWERTRAIN MODEL

Dynamic vehicle powertrain models have been formulated by a number of researchers [33-35] for particular applications, mainly the analysis of engine and transmission control. For real-time operator-in-the-loop simulation, the powertrain model does not need to be as complex as the control analysis model since it simulates the overall, or macroscopic, characteristics that the vehicle operator perceives. For virtual prototyping applications involving the powertrain, this model will have to be replaced, however it is adequate for other studies.

The powertrain model for a vehicle with an automatic or manual transmission must account for engine, torque converter or clutch, and differential component characteristics in addition to the transmission characteristics. Implemented within the RTRD software are three generic powertrain models. These models enable the simulation of vehicles using

- front wheel drive
- rear wheel drive, and
- four wheel drive.

The inputs to these models are the accelerator pedal position, gear shift selector position, and wheel angular velocity. The outputs are the torques on the wheels. The powertrain model formulation follows that presented by Freeman and Velinsky [36]. In this model, the angular velocities are transmitted backwards through each powertrain component from the wheels to the output shaft of the transmission. The computation of the manual and automatic transmission models differs, since the vehicle operator has control of the gear selection for the manual model, while the gear selection of the automatic model is controlled by the simulation software.

For the manual transmission model, the rotational velocity of the transmission output shaft is transformed using the current gear ratio to find the rotational velocity of the clutch component. A clutch torque capacity is calculated from the operator-controlled clutch plate pressure and other physical clutch parameters. The engine model is then evaluated to calculate the available engine torque. If the available engine torque is greater than the clutch torque capacity, then the clutch is slipping, otherwise it is engaged. This model has only one additional degree-of-freedom beyond the number of driven wheels.

For the automatic transmission model, a gear selection calculation is made, based on the rotational velocity of the transmission output shaft and the engine manifold vacuum pressure. The manifold vacuum pressure can be found by a table lookup process, based on the throttle angle and the engine speed. If the calculated gear of operation is the same as the current gear, then the torque converter turbine output rotational velocity is calculated from the transmission gear ratio. If the calculated gear of operation is different than the current gear, then a gear shift procedure is initiated. Once a gear shift is initiated, it must be completed before the model will test the shifting condition again. In the RTRD powertrain software, two shift quality functions are used to represent the relationships between effective speed ratio and effective torque ratio during a shift. During the shift, the effective speed ratio is used to find the torque converter turbine output velocity.

Independent of the shifting state of the automatic transmission model, the torque converter model depends only on the rotational velocities of its input and output connections. The input of the torque converter is connected to the engine, while the output is connected to the transmission input shaft. This model is a quasi-steady-state formulation, which assumes the torque converter to be fully-filled with fluid and operating at its steady-state torque capacity. The absorbed input torque, T_{TCi}, from the engine is calculated from the following equation

$$T_{TCi} = \frac{1}{K^2}\omega_E^2 \tag{12}$$

where K is the torque capacity factor, a characteristic of the converter size and geometry, and ω_E is the engine rotational velocity. This equation shows that the absorbed input torque at steady state is equal to the square of the inverse capacity factor multiplied by the square of the input speed. The output torque, passed from the torque converter to the transmission is found from the expression

$$T_{TCo} = \gamma T_{TCi} \tag{13}$$

where γ is the torque ratio. The torque ratio is found by a curve fit approximation, as a function of the speed ratio, which is the ratio of the of the output rotational velocity divided by the input rotational velocity.

STEERING SYSTEM MODEL

The steering column is not modeled in the RTRD software, but rather as part of the simulator's control loading subsystem. This subsystem models the steering system compliance, freeplay, coulomb friction, and power-assistance. The inputs to the RTRD software from the steering system is the steering rack displacement and velocity. The output from the RTRD software is the reaction torque acting on the column. This torque is calculated by recovering the forces from the tires acting on the steering rack through the tie rods. Since the tie rod is modeled as a cut joint distance constraint, this amounts to recovering the constraint forces, an easy task for cut joints.

BRAKE MODEL

The brake model computes braking torques applied to all wheels, as a function of the pressure in the brake master cylinder or the brake pedal displacement. In the simulator, a dummy set of brake calipers is fit to the vehicle cab to provide the operator with an appropriate pedal "feel". The computed brake torques are then fed to the tire model and contribute to the equations of wheel rotation. Braking torque as a function of brake line pressure and/or brake pedal displacement, either linear or non-linear, is defined for both front and rear wheels.

AERODYNAMICS MODEL

The aerodynamic load model generates aerodynamic forces acting on the aerodynamic center of the chassis. These forces are computed based on the relative velocity between the chassis and wind, aerodynamic sideslip angle, air density, and aerodynamic force and torque coefficients.

Required data for computing aerodynamic forces and torques, such as air density, vehicle front area, distance between the aerodynamic center and the center of gravity of the chassis, aerodynamic coefficients, and wheel base, are specified inside the model.

6 Vehicle modeling example

A 14-body model of a US Army's High Mobility Multipurpose Wheeled Vehicle (HMMWV) is used as the example vehicle model. The model, shown schematically in Figure 9, consists of a chassis, a front steering rack, and four double A-arm suspensions.Each double A-arm consists of a wheel assembly and an upper and a lower control arm. A listing of bodies and joint types corresponding to Figure 9 is as follows:

	Body:		Joint Type:
1	Chassis	R	Revolute joint
2	Right front upper control arm	T	Translational joint
3	Right front wheel spindle	S	Spherical joint
4	Right front lower control arm	D	Distance constraint
5	Left front upper control arm		
6	Left front wheel spindle		
7	Left front lower control arm		
8	Right rear upper control arm		
9	Right rear wheel spindle		
10	Right rear lower control arm		
11	Left rear upper control arm		
12	Left rear wheel spindle		
13	Left rear lower control arm		
14	Rack		

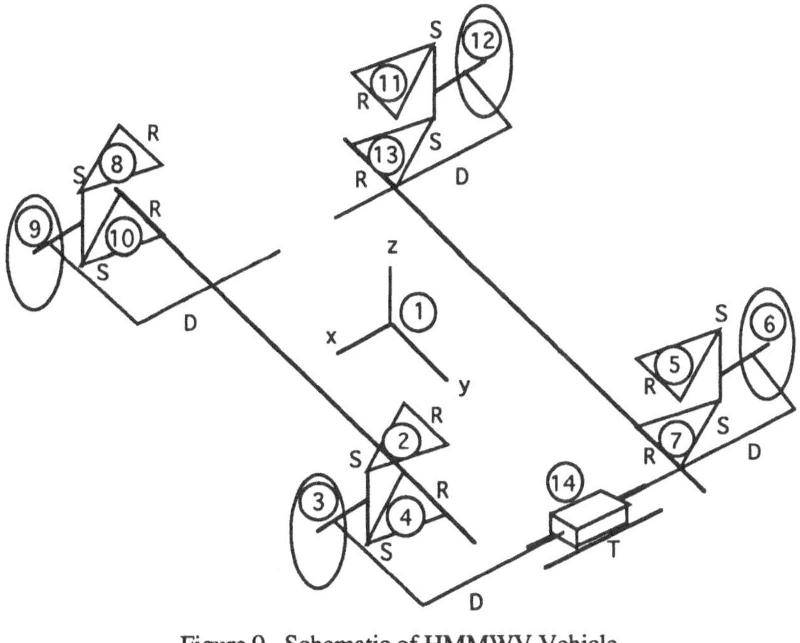

Figure 9. Schematic of HMMWV Vehicle

The chassis (body 1) is assigned as the base body by the RTRD topological analysis. The base body is unconstrained and has six degrees-of-freedom. While the steering system in the real vehicle consists of a steering gearbox and linkage assembly, for reasons of simplicity, it has been approximated in this model by a translational joint that models a rack-and-pinion steering assembly. Its translational motion determines the steering angles of the front left and right wheels via the tie rods, which are modeled as distance constraints. The input actuation has been compensated so the output characteristics of the model match the physical linkage system.

Defining rigid bodies as nodes and kinematic joints as edges, as in Ref. 18, a connected graph schematic of the HMMWV model can be represented as shown in Figure 10. The arrows in the graph indicate the joints that are cut in the topological analysis to obtain the open spanning tree. The spanning tree for this example was shown previously in Figures 5 and 6. The cut joints are chosen automatically by the RTRD topological analysis to optimize computational efficiency.

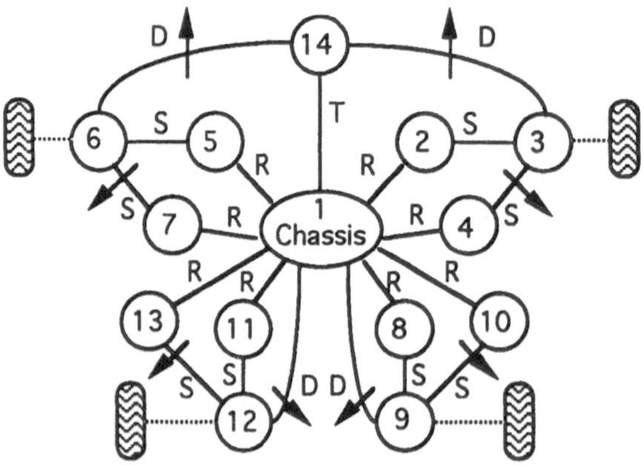

Figure 10: Topological Graph for HMMWV 14-Body Model

The powertrain model used with this example is a four-wheel-drive model with the unlocked transfer case, detailed in Ref. 36.

7 Interface with driving simulator subsystems

The RTRD and vehicle dynamics software is integrated with other subsystems of an advanced driving simulator for operator-in-the-loop simulation. A typical interface between dynamics and simulator subsystems is shown in Figure 11. The vehicle dynamics subsystem receives the driver's input via the control loading system, predicts the vehicle state, and generates necessary output required by other simulator subsystems such as visual, motion, audio, and scenario control.

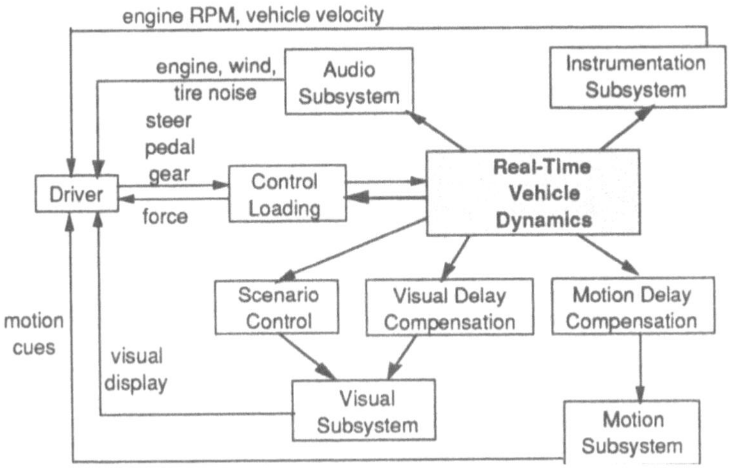

Figure 11: RTRD Interface with Simulator Subsystems

TERRAIN DATABASE

For real-time simulation, terrain (or road surface) data are provided by a global terrain database that is capable of supplying terrain information at a rate and resolution compatible with the requirements of the dynamics subsystem. At each time step, the vehicle dynamics provides the coordinates (x, y, and z) of wheel bottom positions for use in querying the terrain database. The database returns the terrain vertical height, z, the normal unit vector of the tangent plane, a nominal friction coefficient for the tire/terrain contact and a surface type index, all at the same x-y position. The z value is used to determine tire deflection, and thus the tire normal force, while the normal unit vector defines the z axis of the tire reference frame. Along with the y axis being defined as the tire rotational axis, the longitudinal x axis determines the tire reference frame.

INPUT/OUTPUT FOR ON-LINE SIMULATION

In addition to the vehicle modeling data described earlier, the dynamics software receives direct input and generates direct output at each time step for real-time simulation in a driving simulator. For efficiency, real-time input and output data are passed to and from the dynamics software via shared memory regions. These regions are implemented as shared global common blocks. Input data are placed into the input common area by another process or processes, prior to each time step of the dynamics software. At the completion of each time step, the dynamics software writes real-time output data to the output common block, where it can be read and used by other simulator processes.

The details of mapping shared regions vary among computer systems. As currently written, the RTRD software is configured for an Alliant FX/2800 multiprocessor. Modification of the shared region mapping for other multiprocessor systems is a straightforward process.

8 Real-Time Operator-in-the-loop Simulation

In a real-time operator-in-the-loop simulator application, the RTRD software, with associated vehicle subsystem models, computes updates to vehicle state at a fixed and constant rate that is compatible with the requirements of other simulator subsystems. Each incremental update of vehicle state requires one or more integration time-steps within the dynamics. At each time step, the RTRD software reads the real-time inputs from the shared region, forms and solves the equations of motion for the time step, integrates the result to obtain position and velocity for the next time step, and writes the updated vehicle state information to the shared output region.

The fixed integration time step specified for the RTRD software corresponds precisely to the synchronous rate of simulator cueing subsystems, so that real-time behavior is exhibited by the simulator. This requires that the RTRD software be capable of computing a time step within the real-time increment in vehicle state represented by the time step. To achieve sufficiently high computational performance, the RTRD software and associated subsystem models are hosted on a high performance multiprocessor computing system. Since operating system scheduler overheads, or other system interrupts will interface with deterministic real-time performance, the computing system must be able to assign dedicated control of processors to the RTRD software.

To achieve real-time performance on current computing platforms, the recursive formulation of multibody dynamics employs parallel processing. Since formulation and factoring of the equations of motion for each chain of bodies emanating from the base body can be computed simultaneously, parallel processing is accomplished in a relatively straightforward fashion. Although they contain little vectorizable computation, recursive dynamics formulations are well-suited for task-oriented parallel processing. Parallelism is expressed at the subroutine-level and the degree of parallelism for vehicle applications is generally equal to the number of independent chains or the number of bodies.

Currently, the RTRD software is implemented on a 26 processor Alliant FX/2800 parallel supercomputer and uses 8 processors for real-time simulation. The RTRD software can be straightforwardly ported to any multiprocessor computing platform that supports Fortran-77, with some Fortran-90 parallel extensions, and a Unix-based operating system that supports threads for low-overhead parallel processing.

9 The Iowa Driving Simulator

The Iowa Driving Simulator (IDS) is the culmination of the foregoing approach to virtual prototyping and IPPD. The IDS is shown in Figure 12. Visible through the open access wall of the projection dome is a HMMWV instrumented cab that is being used for validation studies.

The motion base of the IDS is a Link heavy payload system, employing a standard hexapod configuration. The system was originally developed for flight simulation applications and has been modified by the Center for Computer-Aided Design for ground vehicle simulations. Capabilities of the system are as follows

Figure 12: The Iowa Driving Simulator

Payload Capacity	18,000 lbs.
Linear Acceleration	1.1 g
Angular Acceleration	80 deg/sec^2
Position Limits	± 40 inches horizontal
	± 36 inches vertical
Angular Travel	± 30 deg.
Max. Frequency Response	3.3 Hz.

Since the low frequency, large amplitude motion that characterizes ground vehicle operation is impossible to reproduce on a range-limited motion base, an adaptive motion washout algorithm is used to calculate the control commands to the motion base actuators. The algorithm attempts to reproduce motion cues perceived by the vehicle operator, while minimizing the number and magnitude of false or missing motion cues. The vehicle linear accelerations and angular rates are high-pass filtered to maintain the high-frequency onset cues perceived by humans, while eliminating the low frequency cues, where the commanded motion can extend beyond the range-limits of the motion base. This results in a safety shutdown, as a result of the motion base reaching its safety limit switches. Tilt coordination, implemented in the motion base washout algorithm, can compensate for some missing longitudinal and lateral acceleration cues by utilizing roll and pitch motions of the simulator platform and the acceleration due to gravity. Tilt coordination is not a panacea for all cueing problems and can contribute to false cues, particularly for low speed, transient maneuvers. Care must be taken in selecting washout algorithm parameters and motion controller gains to ensure optimal performance.

10 Conclusions

Revolutionary new capabilities for simulation based IPPD of vehicles and equipment are in development. A key element in this process is virtual prototyping. Virtual prototyping simulation allows the evaluation of design alternatives in the hands of a human operator at early stages of the product development process where there is great latitude in optimizing design. As presented in this paper, the ability to simulate the characteristics of a ground vehicle at a level of fidelity sufficient to insure valid human-machine interaction responses has been created. In addition, the IDS implementation provides a level of fidelity and real-time performance that is sufficient to support engineering design applications. It provides the foundation for an advanced, operator-in-the-loop ground vehicle virtual prototyping capability in support of the IPPD process.

References

1. Haug, E.J. (ed.) (1993) Concurrent Engineering: Tools and Technologies for Mechanical System Design, Springer-Verlag, Heidelberg.
2. Beaver, R., O'Brien, S. and Riecken, M. (1992) 'Advanced Distributed Simulation Technology System Definition Document,' U.S. Army STRICOM Document #ADST/WDL/TR--92-03017-QTR2YR2.
3. 'Engineering in the Manufacturing Process,' Defense Science Board Task Force Report, Office of the Under Secretary of Defense for Acquisition, Washington D.C., March 1993.
4. Beck, R.R. (1993) 'Simulation Based Design of Off-Road Vehicles,' Concurrent Engineering: Tools and Technologies for Mechanical System Design (E. J. Haug ed.), Springer-Verlag, Heidelberg.
5. Ciarelli, K. (1989) 'Integrated CAE System for Military Vehicle Applications,' Proceedings of the First Annual Symposium on Mechanical System Design in a Concurrent Engineering Environment, Iowa City, Iowa, pp. 301-318, October 24-25.
6. Frisch, H.P. (1993) 'Man/Machine Interaction Dynamics and Performance Analysis,' Concurrent Engineering: Tools and Technologies for Mechanical System Design (E. J. Haug ed.), Springer-Verlag, Heidelberg.
7. Kuhl, J.G., Papelis, Y.E., Romano, R.A. (1993) 'An Open Software Architecture for Operator-in-the-Loop Simulator Design and Integration,' Concurrent Engineering: Tools and Technologies for Mechanical System Design (E. J. Haug ed.), Springer-Verlag, Heidelberg.
8. Schiehlen, W.O. (1993) 'Simulated Based Design of Automotive Systems,' Concurrent Engineering: Tools and Technologies for Mechanical System Design (E. J. Haug ed.), Springer-Verlag, Heidelberg.
9. Bestle, D. (1993) 'Optimization of Automotive Systems,' Concurrent Engineering Tools and Technologies for Mechanical System Design (E. J. Haug ed.), Springer-Verlag, Heidelberg.

10. Roberson, R.E., and Schwertassek, R. (1988) Dynamics of Multibody Systems, Springer-Verlag, Berlin.
11. Nikravesh, P.E. (1988) Computer-Aided Analysis of Mechanical Systems, Prentice-Hall, Englewood Cliffs, New Jersey.
12. Haug, E.J. (1989) Computer Aided Kinematics and Dynamics of Mechanical Systems, Vol. I: Basic Methods, Allyn and Bacon, Boston.
13. Shabana, A.A. (1989) Dynamics of Multibody Systems, John Wiley & Sons, New York.
14. Huston, R.L. (1990) Multibody Dynamics, Butterworth-Heinemann, Boston.
15. Amirouche, F.M.L. (1992) Computational Methods in Multibody Dynamics, Prentice-Hall, Englewood Cliffs, New Jersey.
16. Wittenburg, J. (1977) Dynamics of Systems of Rigid Bodies, Teubner, Stuttgart.
17. Haug, E.J.(ed.) (1984) Computer Aided Analysis and Optimization of Mechanical System Dynamics, Springer-Verlag, Berlin.
18. Tsai, F.F., and Haug, E.J. (1989) 'Automated Methods for High Speed Simulation of Multibody Dynamic Systems,' Technical Report R-47, Center for Computer Aided Design, The University of Iowa.
19. Tsai, F.F., and Haug, E.J.(1991) 'Real-Time Multibody System Dynamic Simulation, Part I - A Modified Recursive Formulation and Topological Analysis,' Mechanics of Structures and Machines, Vol. 19, No. 1, pp. 99-127.
20. Tsai, F.F., and Haug, E.J. (1991) 'Real-Time Multibody System Dynamic Simulation, Part II - A Parallel Algorithm and Numerical Results,' Mechanics of Structures and Machines, Vol. 19, No. 2, pp. 129-162.
21. Chung, Shu, and Haug, E.J. (1991) 'Real-Time Simulation of Multibody Systems on Shared-Memory Multiprocessors,' Technical Report R-113, Center for Computer Aided Design, The University of Iowa.
22. Bae, D.S., and Haug, E.J. (1987) 'A Recursive Formulation for Constrained Mechanical Systems, Part I - Open Loop,' Mechanics of Structures and Machines, Vol. 15, No. 3, pp. 359-382.
23. Bae, D.S., and Haug, E.J. (1987) 'A Recursive Formulation for Constrained Mechanical Systems, Part II - Closed Loop,' Mechanics of Structures and Machines, Vol. 15, No. 4,.
24. Bae, D.S., Hwang, R.S., and Haug, E.J. (1988) 'A Recursive Formulation for Real-Time Dynamic Simulation,' Proceedings 1988 ASME Design Automation Conference, pp. 499-508.
25. Hwang, R.S., Bae, D.S., Kuhl, J.G., and Haug, E.J. (1990) 'Parallel Processing for Real-Time Dynamic System Simulation,' J. Mech Des, Trans ASME 112(4), pp. 520-528.
26. Center for Computer Aided Design (1993) 'DADS Translator Example Manual,' The University of Iowa.
27. Bernard, J.E. (1973) 'Some Time-Saving Methods for the Digital Simulation of Highway Vehicles," Simulation, Vol. 21, No. 6, pp. 161-165.
28. Allen, Wade R., Rosenthal, Theodore J., Szostak, Henry T. (1987) 'Steady State and Transient Analysis of Ground Vehicle Handling,' SAE Paper No. 870495.
29. Bakker, E., Nyborg, L., and Pacejka, H.B. (1987) 'Tire Modelling for Use in Vehicle Dynamics Studies,' SAE Paper No. 870421.
30. Gim, G., and Nikravesh, P. (1991) 'An Analytical Study of Pneumatic Tire Dynamic Properties, Part 1,' International Journal of Vehicle Design, Vol. 11, No. 6.
31. Gim, G., and Nikravesh, P. (1992) 'An Analytical Study of Pneumatic Tire Dynamic Properties, Part 2,' International Journal of Vehicle Design, Vol. 12, No. 1.

32. Gim, G., and Nikravesh, P. (1992) 'An Analytical Study of Pneumatic Tire Dynamic Properties, Part 3,' International Journal of Vehicle Design, Vol. 12, No. 2.
33. Kotwicki, A.J. (1982) 'Dynamic Models for Torque Converter Equipped Vehicles,' SAE Paper No. 820393.
34. Karmel, A.M. (1986) 'A Methodology of Modeling the Dynamics of the Mechanical Automotive Drivetrains with Automatic Step-Transmission,' Proceedings of the 1986 American Control Conference, pp. 279-284
35. Cho, D. and Hedrick, J.K. (1988) 'Automotive Powertrain Modeling for Control,' ASME Paper No. 88-WA/DSC-35
36. Freeman, J.S. and Velinsky, S.A. (1991) 'Four-Wheel-Drive Powertrain Models for Real-time Simulation,' Proceedings of the Third Advanced Technologies Symposium, ASME Publication DE, Vol. 40, pp. 175-182.

MAN/MACHINE INTERACTION DYNAMICS AND PERFORMANCE ANALYSIS, BIOMECHANICS, AND NDISCOS MULTIBODY EQUATIONS

HAROLD P. FRISCH
Head, Robotics Applied Research
NASA/Goddard Space Flight Center
Greenbelt, MD 20771, USA

ABSTRACT. The Man/Machine Interaction Dynamics and Performance (MMIDAP) analysis project seeks to create an ability to study the consequences of machine design alternatives relative to the performance of both the machine and its operator. The MMIDAP problem highlights the conflicting needs and views of groups that focus on machine design and groups that focus on human performance, ergonomics, and cumulative injury potential. This chapter will overview and update ongoing MMIDAP capability development efforts being undertaken by a rather loose group of collaborating researchers. An attempt will also be made to highlight problems associated with using traditional multibody mechanical system analysis tools for musculoskeletal dynamics analysis at the level of fidelity needed by the biomechanics community. In particular problems associated with using traditional multibody system tools for viscoelastically restrained joints and multiple muscle systems will be discussed and enhanced solution approaches proposed. Additionally a definition of the capabilities and underlying theoretics of the multibody dynamics code NDISCOS (order N Dynamic Interaction Simulation of COntrols and Structure) will be provided. This code, originally referred to as DISCOS, has been significantly enhanced since its inital 1978 release. Capabilities and associated theoretics which have been previously reported in brief disconnected status papers are collected herein and concisely developed from end to end with appropriate source papers referenced.

1. Introduction

A confluence of diverse computer science, mechanical systems, and biosystem technologies is now forming. Advanced mechanical system dynamics analysis methodologies, biosystem measurement and modeling techniques, computer hardware configurations, concurrent engineering communications, database systems, anatomical, biomechanical, biodynamical, behavioral, and cognitive science research capabilities can, with reasonable effort and proper focus, be drawn together to create a Man/Machine Interaction Dynamics and Performance (MMIDAP) analysis capability. The envisioned capability is to build upon existing and readily extendable capabilities. Contained within this chapter is an abbreviated review of the MMIDAP project and an overview of work that has been initiated since the last summary paper on the subject [1]. This chapter also contains a complete development of the capabilities and theo-

M. F. O. S. Pereira and J. A. C. Ambrósio (eds.),
Computer-Aided Analysis of Rigid and Flexible Mechanical Systems, 555–608.

retics of the multibody program NDISCOS; which is evolving into the foundational dynamics modeling capability of the MMIDAP project.

The final report of the 1985 Integrated Ergonomic Modeling Workshop [2] contains a detailed review of pre-1988 software capability along with a list of recommendations for future research. It specifically remarks that "there is a paucity of dynamic interface models" and that "an integrated ergonomic model is needed, feasible, and useful." The report's review shows that some work exists under the general heading of optimization of sports motion; however, there is virtually nothing to support mechanical system designers that must evaluate machine operator interaction dynamics and performance with or without survival gear, in hostile environments, on-the-job, on earth, or in space.

The MMIDAP project supports the generic machine operator system design problem. It is directed toward machines that are controlled by a human operator's intelligent physical exertions. MMIDAP analysis tools will allow designers to introduce the physical and cognitive limitations of a specific operator or operator population class into the machine design process. The intent is to develop the MMIDAP analysis capability in as generic a manner as possible. This will enable its application within a broad range of aerospace, machine design, ergonomic, physical therapy and rehabilitation engineering problems. There is no desire to duplicate the statics and kinematics based software systems that now support human factors investigations such as those identified in [2] and [3]. Our intent is to complement these with new techniques that support "what if?" studies of problems that cannot ignore dynamics and human performance considerations.

2. Anthropometric and Biomechanical Databases

The National Library of Medicine (NLM) is currently undertaking a project that intends to build a digital image library of volumetric data representing a complete normal adult human male and female [4]. This "Visible Human Project" will include digital images derived from photographic images obtained from cryosectioning, computerized tomography, and magnetic resonance imaging, for example [5]. Several of the MMIDAP collaborating research groups have recognized the potential for the analysts to define data need to the anatomists while they determine if it is feasible with modern technology to provide the requested data as a by-product of the "Visible Human Project.

Kroemer [2] provides a review of currently supported anthropometric data bases and computer models used in the field of ergonomics. Winters [6] provides a source book for multiple muscle systems and movement organization along with a survey by Yamaguchi [7] listing human musculotendon actuator parameters from over 20 different published sources. Additionally Seireg in [8] provides the anthropometric and musculoskeletal data used to support musculo load sharing research carried on at The University of Wisconsin at Madison.

One major problem with existing biomechanical data is that it comes from so many different sources with almost as many different measurement reference frames. A quick scan of data provided by Yamaguchi in [7] reveals considerable numeric variation between reference sources for the same anatomical component. The data tables also reveal that there are considerable data gaps. The NLM's Visible Human Project is presenting the biomechanics community with a unique opportunity to fill these gaps and to obtain a consistent reference source of fundamental biomechanical data.

3. Human Performance Database

There is no lack of literature regarding the quantification of human function or performance. The literature as a whole can perhaps best be characterized by noting that it lacks a common conceptual framework upon which human performance quantification strategies can be based. As discussed by Kondraske in [9] this has made it difficult to organize previous work and compare methods. The approach that is advocated herein for resolving this problem introduces the concept of a *functional unit.* This entity is defined in such a manner that it must possess a measurable resource level to accomplish a highly focused task. Considering all functional units collectively leads to the realization of a finite set of basic elements of performance (BEPs). In mathematical terms, the BEPs define a set of basis vectors while associated measured resource level defines vector magnitude. To specify a BEP one must delineate both the functional unit and its dimensions of performance.

Human BEPs may be organized into three primary domains:

1. Central processing
2. Physical: Environmental interface
3. Physical: Life-sustaining

The collective set of all BEPs forms a *performance pool.* This performance pool may be defined for an individual or for a population group. It defines levels of resources available relative to all dimensions of performance associated with all functional units. To accomplish any task (physical or mental), humans draw upon appropriate BEPs from the performance pool in the required amount. Successful task performance is determined by the availability of required BEPs. If insufficient BEP resources are available from the performance pool, the task cannot be accomplished. If just enough exist, task performance will be stressful. If more than enough exist, task performance will be comfortable. Unfortunately one cannot assume that all BEPs are functionally independent of each other. There are dependencies that must be recognized and accounted for in the performance analysis process. As stated by Fitts in [10], we cannot study man's motor system at the behavioral level in isolation from its associated sensory mechanisms. We can only analyze the behavior of the entire receptor-neural-effector system. The implication here and the major challenge for MMIDAP is to develop and implement methods that automate the process of detecting and accounting for functional relationships between the sets of BEPs that must be simultaneously exercised during task performance.

The development of these functional relationships in a format compatible with being interfaced to the human performance database represents many cutting edge research projects at the Human Performance Institute (HPI) at The University of

Texas at Arlington. Kondraske in [9] and [11] provides a good overview of task decomposition via BEPs and the methods being used to database the BEP records of the 3000+ patients tested with systems developed at the Human Performance Institute (HPI).

4. System Performance Analysis

A theory is presented by Kondraske in [12] and [13] that develops a scientifically based conceptual framework for addressing many fields of concern relating to human performance. The theory involves the concepts of basic elements of performance and human resource economics.

Many questions regarding biomechanical behavior can and are being addressed without consideration of system performance; however, human performance questions associated with complex tasks cannot ignore biomechanics and biodynamics. Biomechanical models typically focus on the principles of materials and mechanical behavior, while a system performance model for a given subsystem recognizes dependency on components external to the biomechanical domain (vision, neuromotor control, etc.).

The basic difference between classic biomechanical analysis and performance analysis can be summarized as follows:

Biomechanical & Biodynamics Analysis - provide traditional static and dynamic analyses that depend upon the basic physical concepts of mass, inertia, geometry, stiffness, position, velocity, acceleration, musculotendon, and environmental loads.

Performance Analysis - uses biomechanical and biodynamics analysis information to quantify the qualitative parameters used to characterize a system's capacity to successfully accomplish a task. It focuses on providing analysts with an enabling capability to ask and quantify answers to the following 3 fundamental questions:

1. Can the task be accomplished? If not, why not.
2. How well can the task be accomplished?
3. What is the best way to accomplish the task?

These questions may be directed to either the machine operator, or the machine-operator system.

Performance analyses are simple in concept and yet powerful as total system design and evaluation tools. System performance can be modeled in terms of the *available performance resources* of the operator while quantitative task characterization can be expressed in terms of the *performance resource demands* required of the operator. A detail development of these concepts is found in [9], [12] and [13].

5. Prediction of Human Motion

One fundamental difference between repetitively testing human subjects and repetitively testing mathematical models is that the human's response is nonrepeatable, [14]. The modeling goal for the prediction of human performance can therefore only be that the predicted motion be physically reasonable. Predictions and reasonable variations around them should be viewed as defining an envelop of possible human response. With this realization in mind, simplified motion prediction algorithms can justifiably be introduced into the motion prediction model. Physical realizability can be checked by viewing animated response, monitoring joint rates, acceleration, jerk, loading, and comparing these with norms in the BEP database.

The program JACK [15] has several unique features that make it ideal for MMI-DAP application. Figure positioning by multiple constraints [16] is a capability that allows users to specify trajectories at several body fixed points (hand, feet, torso) and to then have motion trajectories for all other points predicted. Strength guided motion [17] is a capability that allows for human strength and comfort data to be used in the motion prediction process. The creators of JACK make note of the fact that others such as Wilhelms in [18] have used forward dynamics for human motion prediction. The JACK development team argues that utilization of a forward dynamics approach to human animation is difficult for the user to control because users must provide all joint torques. For a 3D system, this is a near impossible task. Kinematic and inverse kinematic approaches are easier to manipulate but suffer from the potential of unrealistic joint motion. JACK uses a blend of kinematic, dynamic, and biomechanical information when planning and executing a path. The task only needs to be described by a starting point, ending position, and external loads such as gravity and weights to be transported. An excellent review of the program JACK and computer graphics research as applied to the animation of human figures is provided by Badler in [15], [19], [20] and [21].

6. Man-in-the-Loop Simulators for Complex Mechanical Systems

Major advances in formulating the multibody equations of motion needed to simulate complex mechanical equipment, along with the availability of low cost parallel processor computers, have provided a unique opportunity to create low cost real-time simulators for complex mechanical equipment with man in the control loop. Simulators accept real-time man-in-the-loop commands, graphically create a simulated visual environment, and drive other laboratory devices to create a simulated vibro-thermal-acoustic-shock environment. Stress and load information associated with machine components can be obtained directly from the predictive capabilities of the multibody dynamics software system that drives the simulator. Qualitative measures of system feel, cognition, and hand-foot-eye-ear coordination information can be obtained from operator comments. Human performance data can be obtained by monitoring operator response in the simulated environment. Machine performance can be obtained directly from the simulator. The ability to simulate in real-time all operator experienced loads at the operator machine interface and all other loads, such as sound and machine vibration that operator's subconsciously use for cueing purposes, sets this new effort apart from such systems as aircraft flight trainers. These loads are an integral part of the machine- operator feedback control cueing process.

The first step toward developing such a simulator capability for complex mechanical systems was taken at The University of Iowa with its development of a simulator for the J.I. Case backhoe [22]. The second step was the creation of the Iowa Driving Simulator [23]. The next step will be to develop the Department of Transportation's National Advanced Driving Simulator [24], [25] at The University of Iowa in the mid 1990's.

7. Integrated Musculoskeletal and Machine Dynamics

The creation of mathematical models for the characterization of system dynamics is a fundamental part of engineering analysis. Both mechanical and biomechanical groups frequently make use of lumped parameter models. These models consist of hinge connected rigid and flexible bodies, i.e., multibody systems. An excellent overview of existing automated methods for developing simulation models for complex mechanical systems via multibody dynamics analysis software systems is provided by Schiehlen in [26] and [27]. In the late 1980's, several international groups discovered that equations of motion could be rederived in a manner that would greatly improve computational speed [28], [29], and [30]. New implementations of these and analogous methods with improved speed and modeling capability are now in use [31], [33], [34], and [35].

Multibody simulation models have been successfully used to model certain classes of musculoskeletal systems. However, modeling weaknesses exist and these must be recognized before one attempts to use multibody tools for general biomechanical and biodynamical application. The following deficiencies associated with vertebrate biodynamics application have been recognized and plans are now underway to enhance the program NDISCOS, [31] and [32] accordingly:

- Inverse dynamics for deterministic systems. This capability is necessary to predict resultant joint loads associated with the dynamic interaction between machine and operator.

- Intermittent loop closure and range of motion constraints. This capability is needed to model the interface between man and machine and to routinely include range of motion limits for anatomical joints.

- Biomechanical joints must now be approximated by conventional mechanical joints. To support more detailed analysis an enhanced joint modeling capability that includes the full complement of human joints defined by Norkin in [64] must be developed.

- Rolling/sliding contact of penetrating surfaces. This capability is needed to model the details of joint motion and loading and too adequately model the soft tissue interface contact between an operator's hand and the manipulandum used for machine control.

- Flexible body modeling. This capability is currently available within NDISCOS. It is required for stress distribution determination within the skeletal structure being stressed by physical exertion. This is an important capability needed to support the joint prothesis design problem and the effectiveness of exercise equipment for long duration manned space flight.
- Body clustering. This capability is needed to model joints such as the ankle and wrist. It is also needed to model the spine. In each case, relatively small bones are tied together by ligaments into a cluster that has limited range of motion. The desire is to develop a general cluster capability that will be applicable for generic mechanical system dynamics, biodynamic, and molecular dynamics research projects.

The dynamics analysis of mechanical systems is dominated by the need to solve the forward dynamics problem. That is, given a prescribed set of internal and external loads, predict system response. Attempts to perform forward dynamics analysis with neuro-musculo-skeletal systems are usually stopped by ones inability to mathematically characterize the human's cognitive processes that generate the neural activation signals that stimulate the body's musculo actuator system. Forwards dynamics studies however do provide the framework for the study of underlying principles controlling how individuals optimize the natural motion of their musculoskeletal system.

The author has been recently made aware of an active research program on the subject of forward dynamics in the Netherlands. Much of this work makes use of multibody dynamics technology to study the action of musculo-skeletal subsystems in action. Relevant PhD dissertations are: [36], [37], [38], [39], [40], [41], [42], [43] and [44].

8. Man/Machine Dynamic Interaction

Figure 1 provides a flow diagram of the proposed closed loop man/machine interaction dynamics and performance assessment process. The output of the program JACK is animated human system response. As for any engineering analysis study, the physical realizability of predicted response must always be checked. This is done by viewing animated response and resultant joint behavior. Performance parameters such as joint stiffness and comfort level within the JACK program allow users to tune predictions to bring them into the realm of physical realizability for the particular population group under study (old, young, normal, obese, handicapped, etc.) As a further check, JACK's predicted joint response information can be used as input to the program NDISCOS. This program offers a functionally complete capability for analyzing models of arbitrary complexity. NDISCOS can be used to create a detail dynamics model for the machine and the machine operator's musculoskeletal system. The associated equations of motion for the multibody model are exact, relative to the laws of Newtonian mechanics. The inverse dynamics capability of NDISCOS can be used to obtain a refined prediction for resultant joint loading. Differences between JACK and NDISCOS resultant load predictions stem from the simplifying assumptions within JACK's motion prediction algorithms and man/machine interaction dynamics effects.

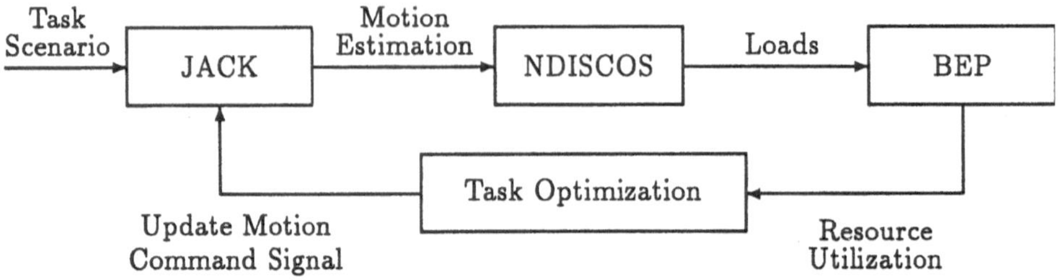

Figure 1: Closed-Loop Man/Machine Interaction Dynamics and Performance Analysis Assessment Process

The resultant joint load predictions made via NDISCOS's inverse dynamics capability can be used as input to a capability that addresses the nondeterministic muscle load sharing problem. If resultant joint loading and muscle load sharing predictions are acceptable, then motion and load prediction information is ready to be used as input to the human performance BEP database at the HPI. The output of this step provides another assessment of physical realizability. If results violate physical realizability, JACK performance parameters can be adjusted and the process repeated.

If muscle allocation studies are required, the skeletal system model and associated computational theoretics will require non-trivial enhancement to include a detailed three dimensional characterization of critical joint complexes. An understanding of detail muscle load sharing is needed to explain, in a quantifiable sense, exactly why certain design options or operational scenarios have the potential of causing machine induced discomfort, fatigue, pain, or trauma.

In application, the assessment process will use an iterative refinement process that can be used until the successive approximations strategy converges to acceptable results. The predictions either confirm that man/machine interaction is acceptable or that some human performance parameters have exceeded database norms. If human performance requirements are excessive, machine design changes or operational scenarios can be refined until acceptable performance measures are achieved for the machine operator's population group. It is also possible to incrementally change population group by selecting different sets of anthropometric and BEP data from the database. Normally once an acceptable set of JACK performance parameters are obtained, they should be rather insensitive to modest changes in machine design, anthropometric, or BEP information.

The critical issue associated with the determination of an optimal scenario is the selection of a physically meaningful optimization criteria. This problem is compounded with the need and desire to minimize time, fatigue, and machine complexity while maximizing throughput and efficiency. The next key issue centers on how to develop a systematic means for determining the sensitivity of performance relevant motion parameters of interest to design variables. Design variables are the system variables that the engineer alters during the design optimization process.

9. Multibody Methods for Biomechanics

The inclusion of musculoskeletal loading effects for detail biomechanical analysis of action and reaction loading effects at joints can become exceedingly complex.

The modeler must address the problem of modeling both joints and actuators. At the coarse fidelity level joints are modeled as conventional rotational type mechanical joints, and actuators are modeled as torque producing motors. At this level of modeling fidelity second and higher order biomechanical properties of joints and muscle actuation are ignored. The capabilities defined within this section are designed to enable the biomechanical modeling of the details of musculoskeletal system dynamics and associated joint loading effects. The objective is to provide a generic capability that will allow the biomechanics modeler to hypothesize theories to explain laboratory observations and to computationally investigate these relative to the first principles of kinematics and dynamics.

9.1. MUSCLE MODELING AND LOAD SHARING

Detailed neuro-musculo-skeletal modeling of the detail human system or any of its subsystems is an extremely complex problem that is beyond today's state-of-the-art capability. The First World Biomechanics Congress in August 1990 had over 80 oral presentations on the subjects of multiple muscle systems, biomechanics and movement organization. Formal reports on 46 of these presentations have been collected by Winters in [6]. From these reports and others presented at the Congress it is clear that muscle dynamics, neuro-musculo-skeletal organization and movement modeling is a subject that will occupy researchers for many more years.

There is great deal known about how nerve cells transmit signals, how these signals are put together, and how out of this integration higher functions emerge [45]. Nerve cells are connected through their synapses to form functional circuits; these are organized into the multineuronal circuits and assemblies that provide the basis for neural organization [46]. Muscles are controlled by nerves at neuromuscular junctions, and at these points activation signals are biochemically processed to initiate the muscle contraction process [47].

The details of muscle modeling have several more layers of complexity. For example, muscles are composed of muscle fibers that are differentiated by the biochemical properties that dictate their respective response speeds and resistance to fatigue. When the muscle is innerviated, select sets of muscle fibers called motor units contract while others remain in a rest or in an energetics recovery state. This motor unit apportionment issue further complicates the mathematical modeling of muscle contraction dynamics as can be seen from Hatze's complex mathematical formulation of the problem [48].

In spite of these outlined complexities in muscle dynamics modeling, progress is being made at a level compatible with real world application. Relatively simple mathematical approximations are appearing in the literature that are providing a basis for understanding how musculotendon systems produce force as a function of the associated reaction dynamics of the biochemical processes that produce muscle contraction, for example [49], [50], [51], [52], [53], [54], [55], and [56]. Also [57] contains a rather detail review of the complexity associated with using system identification techniques to obtain the data needed to support studies associated with joint dynamics modeling.

The incorporation of muscle dynamics into the framework of a multibody simulation capability is a rather straight forward process if the physics of muscle contraction can be assumed known. This has been demonstrated by Hatze in [58] and Morris in [59]. Never the less forward dynamics can still be used when known musculo innerviation is imposed, for example, by functional neuromuscular stimulation systems, as discussed by Chizeck in [60]. It can also be used for well defined structured motion such as reflex response actions. The availability of measures for the biochemical dynamics of calcium ion concentration within the system of defining equations for muscle contraction dynamics as defined in [48],[52], [53], and [54], provides an avenue to an understanding of the process of fatigue, discomfort, and pain. An extensive review of this connection is available from several papers published in a special issue on "Occupational Muscle Pain and Injury" by the European Journal of Applied Physiology, [61] and [62].

Complexities associated with modeling muscle contraction dynamics are matched by the problem of resolving muscle load sharing and kinematic redundancy. The presence of redundant muscle actuators at virtually every anatomical joint implies that rules must exist for defining how muscles share the work load. Kinematic redundancy within the upper and lower extremety systems also presents mathematical modeling problems. Redundancy in the physical system to be modeled leads to a mathematical problem with an infinite set of solutions. This problem is resolved by optimization techniques that find the unique solution that minimizes a user defined cost function. Zajac in [63] provides an indepth review of the complexity associated with modeling multijoint muscle systems. Seireg in [8] provides an extensive summary of cost functions relevant to ongoing research in muscle load sharing at The University of Wisconsin at Madison.

9.2. MUSCULOSKELETAL JOINTS

In general, biomechanical joints can be modeled as conventional mechanical joints only as a first order approximation. Nearly all joints have complex concave/convex surfaces and compliant interfacing tissue that acts to lubricate, cushion, and limit the range of relative motion between the interfacing surfaces, [64], [65], [66] and [67].

The intent of this section is to outline modeling procedures that can be used to go beyond first order joint modeling restrictions. The basic idea is to forget about trying to create a variety of complex mechanical joints with an associated set of kinematically constrained degrees of freedom. Rather, simply accept the fact that musculoskeletal joints have six kinematically **restrained** degrees of freedom; they normally do not have kinematically **constrained** degrees of freedom (dof). In NDISCOS biomechanical joints are to be modeled as restrained 6 dof joints. It is up to the modeler to decide if it is more appropriate to kinematically **constrain** or viscoelastically **restrain** motion relative to each of the 6 degrees of freedom defined at each joint. This is a decision that cannot be made a priori, it is a function of the study at hand. For example, assumptions made for the study of muscle allocation and resultant joint loading effects during natural motion optimization may not be valid for the study of joint motion trauma, such as fracture or dislocation.

The human body has three types of joints: **synarthrosis (fibrous)**, **amphiathro-**

sis (cartaginous), and **diathrosis (synovial)**. From a multibody modeling point of view these have the following characteristics:

- Contiguous bones joined together at **synarthrosis** joints are connected with fibrous tissue. The bone plates of the skull and the teeth in the jaw are connected at synarthrosis joints. These allow virtually no relative movement and are normally not modeled as joints for kinematic or dynamics analysis.

  The fibula and tibia of the lower leg and the ulna and radius of the lower arm are connected along their entire length by a fibrous interosseus membrane. This interface is classified as an anatomical joint but it is not viewed as a biomechanical joint within the context of a multibody modeling capability such as NDISCOS. It is best to model the coupling effects of such connective membrane as a set of lumped straight line passive viscoelastic couplers, each with a well defined point of insertion and origin, defined along the length of the membrane connection.

- Contiguous bones joined together at **amphiarthrosis** joints are connected by either fibrocartilage or hyaline cartilage. This cartilage joins one boney surface to another. For example, the 2 pubic bones in the pelvis and the first rib and the sternum. Normally there is very little relative motion allowed at these joints and relative motion effects here are normally ignored. If constraint loads at these joints are of interest then each bone can be modeled as a single body and the connection modeled as a joint with 6 kinematically constrained degrees of freedom. The Lagrange multipliers developed within NDISCOS to enforce these kinematic constraints provide the desired joint loading information. Another option is to model the joint as a combination of kinematic constraints and nonlinear viscoelastic restraints. Again, kinematic constraint loads and viscoelastic restraint loads are computed by straight forward computation. Cross axis viscoelastic coupling could also be modeled if desired but this is probably not needed and support data would be difficult to obtain.

- Contiguous bones joined together at **diathrosis** or **synovial** joints are the only joints that allow a wide range of relative motion. At these **synovial** joints contiguous boney surfaces are not in direct physical contact. This means that there are no kinematically constrained degrees of freedom. However for modeling purposes it is frequently appropriate to make the justifiable assumption that several relative joint degrees of freedom are kinematically constrained. Although the boney surfaces are not in physical contact there exists a variety of different forms of tissue connection that both cushion and limit range of motion. All of these connectivity situations are important. It is necessary to understand available modeling options so that the effects under investigation can be accurately captured by the mathematical model. It should be clear that the modeler must define the physics of the problem. NDISCOS simply provides the ability to study the consequences of a hypothesis relative to the first principles of viscoelasticity, kinematics, statics and dynamics.

9.3. SUBCLASSIFICATIONS OF SYNOVIAL JOINTS

Diathrosis or **synovial** joints are anatomically subclassified into three main categories on the basis of the motions that are available. These subclassifications are: **uniaxial, biaxial, and triaxial**. While these subclassifications are adequate for gross motion discussions they need to be further subclassified if the objective is detail kinematics and dynamics analysis via multibody methods.

- **Uniaxial diathrodial** joints are of two types:
  - **Hinge joints** - These are one degree of freedom joints that allow only flexion and extension about a well defined axis. The outer joints of all fingers and toes are uniaxial diathrodial hinge joints. If included in a simulation one restrained rotational degree of freedom and five kinematically constrained degrees of freedom would normally be appropriate. The investigation of joint dislocation and hand trauma for peak regular and irregular grasping problems would probably require some of the 5 kinematically constrained degrees of freedom at some of the joints to be modeled as non-linear viscoelastically restrained degrees of freedom.
  - **Pivot joints** - These are one degree of freedom joints constructed so that one component is shaped like a ring and the other shaped so that it can rotate within the ring. The **Atlas** is the first vertebrae of the cervical (neck) region of the spine. It supports the skull and rests on the **Axis**, the second cervical vertebrae. The joint between Atlas and Axis is classified as a pivot joint. While motion here is primarily rotation, motion about other axes is frequently important to model. In many situation this joint should be modeled as a restrained six degree of freedom joint.
- **Biaxial diathrodial** joints are free to move around two axes. These are subclassified as:
  - **Condyloid** joints are constructed so that a concave surface of one bone slides over the convex surface of the interfacing bone. Knuckle joints of the hand and foot are examples.
  - **Saddle** joints are constructed so that each interfacing surface is both concave and convex. The knuckle joint of the thumb is an example

  These are normally modeled for first order analysis as 2 restrained and 4 kinematically constrained degrees of freedom. This modeling assumption forces the user to accept the limitations associated with modeling relative rotational motion via an Euler angle rotation sequence. This implies that the intersection of the two restrained rotation axes is constant over the full range of joint motion. This modeling limitation may not be acceptable for some problems. In these situations the restrained six degree of freedom joint will be the appropriate modeling option.
- **Triaxial** or **Multiaxial diathrodial** joints are joints that allow three or more degrees of relative freedom. These are subclassified as:

- **Ball-and-socket** joints are formed by a ball-like convex surface fitted within a concave socket. The hip joint and the glenohumeral (shoulder) joint are examples. The ball-and-socket joints are normally modeled as three restrained and three kinematically constrained degrees of freedom. This modeling assumption forces the user to accept the limitations associated with modeling relative motion via an Euler angle rotation sequence between the two contiguous body fixed reference frames defined at the joint. This implies that the intersection of the three restrained degrees of freedom axes is constant over the full range of joint motion. This modeling limitation may not be acceptable for some problems. In these situations the restrained six degree of freedom joint will be the appropriate modeling option.
- **Plane** joints permit gliding between interfacing surfaces. The joints used to interface the eight **Carpus** or **wrist-bones** and the joints used to interface the seven **Tarsus** bones of the foot are examples. The specification of relative motion of reference frames fixed within the bones of the **Carpus** and **Tarsus** is non-trivial. The complexity of the problem is evident from the research papers presented in [67].

  The relative motion of bones interfacing in the **Carpus** region of the hand and in the **Tarsus** region of the foot are normally not even attempted. NDISCOS provides the capability to study this region if supporting data can be developed. The key to the ability to support this problem is the capability of NDISCOS to accept a definition of systems that include topological loops. The ability to define both topological loops and restrained six dof joints provide the basis for this generic modeling capability.

9.4. RESTRAINED SIX DEGREE OF FREEDOM JOINTS

Restrained 6 degree of freedom joints are designed to be compatible with the needs of biomechanical joint modeling. There is a class of problems that must take into account bone flexibility, however, we make the assumption herein that all interfacing bones at restrained 6 dof joints are rigid. This assumption could be relaxed but at this time it does not appear to be worth the effort.

In the terminology of the program NDISCOS relative body motion at joints is defined by computing the relative motion of 2 body fixed reference frames. These are referred to as the $p-$ and $q-$ frames. The user locates these so that joint motion can be computed in a manner compatible with motion specification needs. They are located and oriented relative to their respective body fixed reference frames. Joint motion variables are developed to define their position and orientation relative to each other. The restrained 6 dof joint capability will allow the user to specify a continuous surface fixed relative to the $p-$frame and a set of at least three points fixed relative to the $q-$frame. The continuous surface relative to the $p-$frame lies on the undeformed surface of the tissue that is fixed to the bone at the joint interface. The set of surface contact points fixed in the contiguous body relative to the q-frame forms a coarse but yet "adequate" representation of the adjacent surface. The surface is assumed to be compliant and its linear or nonlinear viscoelastic properties definable by the user.

The position of each point fixed in the $q-$ frame is computed relative to its position along a normal to the surface that is fixed in the $p-$frame. If the point is inside the surface there is surface contact and the interfacing tissue undergoes compression with appropriate viscoelastic loads applied equal and opposite at the contact point. If the point is outside of the surface, the surfaces are not in contact at the surface contact point and the associated viscoelastic loading there is taken to be zero. There is no restriction on how many surface contact points may be in contact. The user is responsible for making sure that enough surface contact points are defined, and the user defines "enough". In some modeling situations it may be important to include linear or non-linear extensional viscoelastic loading, this too can easily be incorporated.

The following list outlines associated theory. Let:

- (u, v) - Surface coordinates used to locate a point on the surface fixed relative to the $p-$frame.
- $\vec{C}(u, v)$ - Vector from the origin of the $p-$frame to the surface point identified by surface coordinates (u, v)
- ${}_p\vec{d}_q$ - Vector from the origin of the $p-$frame to the origin of the $q-$frame
- $\vec{C}_s$ - Vector from the origin of the $q-$frame to the surface contact point s
- $\vec{\Delta}_s(u, v)$ - Shortest vector from the $p-$frame fixed surface to the $q-$frame fixed surface contact point s, that is,

$$\vec{\Delta}_s(u, v) = Min[{}_p\vec{d}_q + \vec{C}_s - \vec{C}(u, v)] \tag{1}$$

  for $u, v \in S$ where

- S - region of the surface to be searched for minimum contact distance. This is introduced to allow physical insight to limit the search region.

Once $\vec{\Delta}_s(u, v)$ is computed it defines both the $+$ or $-$ penetration distance and the direction of surface interaction load application. This penetration distance is used with a viscoelastic load determination function to determine contact load. Normally a negative value would signal penetration and hence a compression of the interfacing tissue. A positive value would signal no contact and hence tension within connective tissue. The location vectors $\vec{C}_s$ and $\vec{C}(u, v)$ define load application points and the vector $\vec{\Delta}_s(u, v)$ defines the direction of application. The assumption here is that sliding friction is effectively zero. In biomechanics this is a reasonably good assumption since the synovial fluid that lubricates joints is better than teflon on teflon, except in the very aged.

9.5. VISCOELASTIC RESTRAINT LOADS

The resultant nonlinear viscoelastic restraint loads acting at joints are the vector sum of a number of different effects. These must be separated and modeled individually. From the perspective of modeling two generic situation classes exist:

- Hinge load is a linear or non-linear function of the relative displacement and relative rate between $p-$ and $q-$ reference frames fixed in the contiguous bodies at the hinge point associated with the biomechanical joint. Relative displacement and relative rate data are computed within NDISCOS. This modeling option would be the appropriate choice for modeling the joint loading contributions associated viscoelastic properties of **menisci** and **discs**. The program NDISCOS is only interested in obtaining a bottom line resultant (6 long) force/torque vector. This will be applied in an equal opposite manner to the contiguous bodies at the associated hinge point. The degree of non-linear equation complexity associated resultant load computation is of no interest to NDISCOS.

- Hinge load is a linear or non-linear function of points of insertion, origin, line of action and of the relative displacement and relative rate of the $p-$ and $q-$ reference frames fixed in the contiguous bodies at the hinge point. The modeling of this type of hinge load is most appropriately done via the specification of passive viscoelastic couplers.

The modeling of range of motion limits take special consideration. These may be caused by a boney obstruction within the joint or by connective tissue at its elastic limit along some line of action. The user must decide if the range of motion restraint is best modeled as a resultant viscoelastic restraint acting within the joint, or as passive viscoelastic couplers acting between points of insertion and origin.

9.6. STRAIGHT LINES OF ACTION

Straight lines of action act between the point o, the point of origin on one body and the point i, the point of insertion on the connected body. The vector between origin and insertion defines both line of action length and direction. System equilibrium conditions require that the sum of the load vectors applied at the origin and the insertion points equal zero. Two load vectors are developed, one at the origin and one at the insertion. They must act equal and opposite so that system equilibrium conditions are satisfied.

9.7. CURVED LINES OF ACTION

Musculotendon tissue and other connective tissue between bones wrap over each other, around and over boney protrusions. Loads are exerted not only at the points of origin and insertion but along the entire length in a direction normal to the line of action and on the structure at the points where connective tissue contacts the surface that it is wrapped around. Several layers of modeling complexity can be introduced to investigate this effect. If the descriptive mathematics can be developed the effects can be incorporated.

- The simplest form of a curved line of action is a straight line with a single sharp bend point. The bend point can be fixed on either body or on another body in the system. The vectors from the origin and from the insertion points to the bend point define the direction of load application at the points of origin and insertion. The reaction load at the bend point is of such magnitude and direction that the resultant of the three load vectors sum to zero. In a manner analogous to the straight line of action case the triad of external loads act on the system. Equilibrium conditions require that their vector sum equals zero.

- More complex curved lines of action can be defined and their loading effects upon the bodies that they wrap around are more difficult to define mathematically. From the standpoint of NDISCOS, that's still the user's job. The only thing that NDISCOS wants to see is an external vector load set that sums to zero.

9.8. PASSIVE VISCOELASTIC COUPLING

This includes all non-contractile tissue. Membrane connections resist extension but not compression, cartilage, discs and menisci resist compression and not extension. Depending upon the situation they may act along or about any one or all six degrees of relative joint freedom. In many situations viscoelastic loads acting relative to one degree of freedom are uncoupled from all other joint degrees of freedom. An exception is hyaline articular cartilage. This tissue cushions and distributes joint loads, it can be considered to be porous and fluid filled. It acts somewhat like a fluid filled sponge. This situation can also be modelled, however it gets a bit complex. The net result is that the resultant joint loading vector becomes a non-linear function of all relative displacement and relative rate coordinates. Again the only thing that NDISCOS wants to see is the resultant load vector of length 6. How it is developed is the user's problem.

9.9. MUSCULOTENDON COUPLING

Numerous theories exist for the prediction of muscle contractile dynamics. It is a user decision to decide what's best. NDISCOS provides the user with the ability to define a set of first order nonlinear differential equations. These are normally used in the spacecraft world to define controller dynamics. In the biomechanical world these are used to define the dynamics associated with the biochemical processes that control muscle contraction, for example, via a Zahalak, Zajac, or Hatze model. NDISCOS simultaneously integrates these equations with the equations of motion that define multibody system dynamics. The user makes use of the muscle state variables to define the muscle contraction loads that are to be applied to the system along either a straight or curved line of action. Again the user must define the physics of muscle contraction, muscle apportionment, and motor unit recruitment problems. NDISCOS just wants to have a resultant load and a line of action defined.

10. General Purpose Multibody Dynamics Simulation via NDISCOS

The multibody dynamics program DISCOS and its predecessor NBOD2, [68] and [69], were the first major general application multibody dynamics software system made available for public distribution. DISCOS was placed into the public domain by NASA in 1978 [70]. Since its original release there have been several major enhancements and its application domain has expanded significantly beyond its initial spacecraft focus.

Today, it is not only applied to spacecraft control structure interaction dynamics problems, but it is also being extended into application areas that did not even exist when the program was being created. In addition to supporting human performance and biodynamics applications as discussed herein, researchers at MOLDYN Inc.[1] are finding extensive application in the subject area of computational chemistry [71] and [72].

The purpose of this section is to provide the multibody dynamics community with a detail overview of existing capability and associated theoretics. The source material for the theoretics outlined in this section is contained in references [31], [32], [70], [73], [74], [75], and [76]. NDISCOS source code is available for distribution to groups seriously interested in pursuing the possibilities associated with expanding core capability via collaboration[2]. The text provided within this section is a capsulated summary of that provided in [76].

The program NDISCOS may be viewed as old DISCOS with a new computational dynamics heart and expanded capability. Its input, output and kinematics support infrastructure has been kept nearly unchanged. The following list summarizes enhanced modeling capability:

- System topology - Both tree and loop type systems may be defined without any body or hinge modeling restrictions.

- Body characterization - Traditionally, multibody systems are composed of a mix of rigid and flexible bodies. NDISCOS expands this list to include the wheel. Both rigid and flexible bodies may have embedded wheels. Spacecraft momentum wheel controller design and performance analysis investigations required this capability for computational efficiency. The NDISCOS collaboration group has also developed a particle body for use in the version of NDISCOS tailored for computational chemistry application. Additionally, efforts are currently underway to gain support for the development of a variety of fluid body capabilities. These will be used to support the attitude control system design and analysis of spacecraft that carry cryogenic superfluid for scientific instrument cooling and liquid propellant for station keeping and orbit adjustment.

- Hinges - In addition to having the ability to create a full compliment of classic mechanical hinge configurations, the ability to defined hinge gearing is also provided. As defined herein the ability to define highly irregular biomechanical joints with the degree of fidelity needed for biomechanical analysis is under development.

- Rheonomic Constraints - The ability to rheonomically constrain motion relative to any hinge or wheel degree of freedom is provided.

- Contact - Applications associated with robot manipulation have forced the development of an impulsive contact capability and an intermittent branch closure

[1]Attention Dr. Hon Chun, MOLDYN Inc, Subsidiary of Photon Research Associates, 1033 Massachusetts Ave. Cambridge, MA 02138, Phone (617) 354-3124, FAX (617) 491-4522.

[2]Ibid.

capability. These options provide for the simulation of robot dual arm cooperative work and pick and put operations. This capability is somewhat limited for non-zero coefficient of restitution problems with non- zero sliding friction since the ability to develop tangential contact loads is not yet developed. These are assumed to be zero herein. This is usually an adequate assumption for biomechanical problems but it does have limitations in mechanical system modeling application. The development of an enhanced contact capability that can be used to accurately model both normal and tangential contact loads for a variety of rolling and sliding friction conditions on rigid and deformable surfaces is viewed as a major research challenge.

- Constraint Stabilization - One of the problems that all precision multibody dynamics simulation software codes must address is constraint stabilization. Lagrange multipliers are almost always used to enforce kinematic constraints. If special care is not taken numerical drift occurs. The same generic problem occurs when one attempts to conserve system momentum during impact and intermittent branch closure. Kinematic constraint drift and conservation of momentum problems are addressed by a momentum impulse correction capability that stems from the initial version of DISCOS.

- Singular Loops - Simulation of topological loop system dynamics can become very difficult when the system passes through a singular loop configuration. Singularities must be anticipated; if not correctly done, numerical solutions will diverge in a physically unrealizable manner. Anticipation is possible by monitoring the associated matrix inversion condition number. When it gets high a switch to a singular value decomposition solution method is automatically made.

- High Speed Multiplication - The order N algorithm is dominated by the need to multiply large numbers of matrices that are normally of order less than or equal to 6. General purpose high speed vector processing algorithms are normally tuned for "large" matrix multiplies, much greater than 6. These may be marginally efficient for the low order matrix multiplies that dominate the software implementation of order N recursive algorithms. Special attention to this problem is made within the version of NDISCOS currently used for computational chemistry work.

- Computational Efficiency - The underlying theoretics of all multibody formalisms create a built in bias toward certain problem classes. NDISCOS is no exception. Its computational bias is toward systems that can be viewed as trees with a modest number of loops within its branches. For systems that are dominated by large numbers of complex intertwined loops other capabilities will most probably be found to be more efficient. For example, see [77].

NDISCOS begins with the reading of system sizing information:

- N_B, Total number of rigid and flexible bodies.
- N_H, Total number of hinges.
- $NCUT = N_H - N_B$, Total number of cut joints.
- N_S, Total number of "sensor points," i.e. points of interest.
- N_W, Total number of constant and variable speed momentum wheels.
- N_δ, Total number of user defined first order ordinary differential equations. These are simultaneously integrated with the multibody equations of motion.

A series of integer arrays are used to define topology, characterize bodies, wheels, and hinges, define hinge point kinematic constraint conditions, and locate reference frames. These are:

- Topology, A 2 by N_H integer array is used to define the directed graph needed to define system topology, see [78]. Columns are associated with hinge numbers while the pair of associated row entries define contiguous body pair. At each hinge point a pair of body fixed reference frames are defined. The p-frame is fixed in the body defined in row 1 and the q-frame is fixed in the body defined in row 2. Motion of the q-frame relative to the p-frame defines relative body motion at the hinge point.
- Cut Joints, If $N_H > N_B$ topological loops exist and the user must identify the $NCUT = N_H - N_B$ hinge points to be used as "cut joints," see [79]. A bit of numerical experimentation to determine a "best" set may be required but normally this is not a very solution sensitive selection.
- Body type, A 1 by N_B integer array is used to identify which bodies are flexible. Zeros identify rigid bodies. Positive integers identify flexible bodies and define the number of flexible body modes to be used for deformation computation.
- Sensor points, A 1 by N_S integer array is used to define the body on which each respective senor point is located. These are all points, other than hinge points, at which kinematic information is required by the user.
- Wheels, Integer codes are used to define associated sensor point, spin axis and wheel rate constraint classification.

NDISCOS, unlike most other multibody codes does not provide users with a library of classical mechanical joints. Users are required to define mechanical joints via an integer array that enhances information contained in the directed graph used to define topology. This array provides all information needed to define the type of kinematic constraint that is to be imposed relative to each of the 6 degrees of freedom defined at each hinge point.

At each hinge point the coordinates axes of the p-frame are used to measure q-frame origin translation. The user selects one of 12 possible Euler angle rotation sequences and then uses the the respective Euler rotation axes as a basis for measuring relative rotational motion.

Relative to each of the $6N_H$ hinge degrees of freedom NDISCOS provides the following kinematic constraint modeling options:

- Fixed, motion is kinematically constrained
- Free, motion is not kinematically constrained
- Intermittent, relative motion at hinge points may be subject to impulsive contact or the associated kinematic constraint may switch between free and fixed.
- Rheonomic, a relative acceleration profile will be user provided

Momentum wheels are embedded within bodies. These are located at sensor points and there axis of rotation is defined to be aligned with one of the sensor frame axes. Angular wheel position is normally wasteful computation and accordingly is not computed. The following wheel rate constraint modeling options are provided:

- Constant speed
- Variable speed
- Rheonomically constrained speed
- Gear constrained

The geared constraint capability can be used to model harmonic drives, rack-and-pinion drives, screw drives, and other such types of mechanisms. The capability can be used within topological loops and at hinges subject to intermittent contact.

10.2. MULTIBODY SYSTEM REFERENCE FRAMES

The details associated with the underlying theoretics of multibody software codes rests upon the methodologies used to define relative body kinematics. The NDISCOS formalism uses the following reference frames:

- Body fixed reference frame - All vectors and tensors associated with a body are defined relative to their respective body fixed reference frame.
- Hinge point reference frames - Each hinge point has an associated pair of reference frames, referred to as the p-frame and the q-frame. Body flexibility allows hinge frame motion relative to the body fixed frame. In the undeformed state hinge frames are fixed relative to their associated body fixed reference frame. Sliding contact via use of a hinge point reference frame that is constrained to a curved or a deformable surface is not easily implemented within the current version of NDISCOS.

- Sensor point reference frames - Points of interest are referred to as sensor points. Each has an associated reference frame that is located and oriented relative to the associated body fixed reference frame.

10.3. PHYSICAL PROPERTIES OF BODIES

Bodies are either rigid or flexible and wheels are rigid and symmetric about their spin axis. The following body characterization data is required:

- Rigid Bodies - Mass, moment of inertia tensor, center of mass location and the location and orientation of each hinge point and sensor point reference frame on the body, relative to the body fixed reference.
- Flexible Bodies - A lumped mass model consisting of grid point location vectors, a 6x6 lumped inertia matrix for each grid point, and a modal displacement matrix consisting of a 6x1 modal displacement vector at each grid point for all modes. Modal stiffness and modal damping matrices are also needed. All resultant mode dependent parameters are created on the first pass through. There are no mode normalization or orthogonality restrictions. A consistent mass modeling capability is also available however it is rarely used and poorly documented.
- Wheels - These may be embedded in both rigid and flexible bodies. Spin inertia and spin axis are defined relative to the associated sensor point reference frame.

10.4. USER SUPPLIED SUBROUTINES

Users define all non-gyroscopic loads acting on the system. Users have access to all system state information and from this they create all load and sensory information needed for the accurate simulation of the system at hand. Users with a modest degree of experience find this to be a rather trivial exercise. The following interface subroutines are available for user interface:

- CONTRL, All differential equations needed to characterize non-gyroscopic loading are defined here. Computed loads are then normally sent to the other user supplied subroutines via labeled common.
- EXTOR, All external loads acting on the system are entered here.
- SHAFTT All momentum wheel torques are entered here.
- KHINGE All hinge loads are entered here. These include loads associated with motors that control relative body motion, along with any linear or nonlinear spring and damping effects that act in equal and opposite pairs at hinge points.
- DLMBRT Dynamics measured relative to user specified accelerating reference.

- ADDT Used to define the acceleration constraint function for rheonomically constrained hinge degrees of freedom.
- ADDWT Used to define acceleration constraint function for rheonomically constrained momentum wheel degrees of freedom.

10.5. BODY INERTIA TENSOR

The order N recursive algorithm is structured to avoid the need to ever develop and invert a full order system inertia matrix. The algorithm develops an inertia matrix for each body. It never inverts an inertia matrix greater than the number of velocity variables associated with the body. The format of the body inertia matrix with momentum wheels is:

$$[m] = \begin{bmatrix} J & -S & d & f \\ S & M & a & 0 \\ d^T & a^T & e & g \\ f^T & 0 & g^T & J_W \end{bmatrix} \tag{2}$$

where

$[J]$ - Deformation dependent moment of inertia matrix, measured relative to the body fixed reference frame.

$$[J] = [J_0] + [J_1]_m \xi_m + [J_2]_{m,n} \xi_m \xi_n \tag{3}$$

where the repeated index summation convention is used and ξ_m is the m-th generalized flexible body displacement coordinate.

$[S]$ - Deformation dependent skew symmetric static mass moment matrix

$$[S] = [S_0] + [S_1]_m \xi_m \tag{4}$$

$[M]$ - Rigid body mass matrix

$[d]$ - deformation dependent matrix to define the inertia coupling between rigid body rotational motion and deformation relative to the undeformed state

$$[d] = [d_0] + [d_1]_m \xi_m \tag{5}$$

$[a]$ - deformation dependent matrix to define the inertia coupling between rigid body translational motion and deformation relative to the undeformed state

$$[a] = [a_0] + [a_1]_m \xi_m \tag{6}$$

f - Coupling between wheels and body angular acceleration.

g - Coupling between wheels and modal acceleration.

J_W - Diagonal matrix of wheel spin inertias.

e - Modal mass matrix. This may be nonorthogonal

These coefficient matrices can be used to define the contribution of flexible body deformation to the linear and angular momentum vectors associated with the deforming body. These contributions are:

$[d]\{\dot{\xi}\}$ - vector components of body angular momentum due to deformation about the body fixed reference point measured relative to the undeformed state.

$[a]\{\dot{\xi}\}$ - vector components of body linear momentum due to deformation, measured relative to the undeformed state.

10.6. FULL ORDER FINITE ELEMENT MODELS

The above mode dependent coefficient matrices must be determined from the output data of a finite element modeling (FEM) program such as NASTRAN. One of the first steps within a FEM program is to assemble the full order mass and stiffness matrices M_{gg} and K_{gg}. The FEM equations for dynamic equilibrium are:

$$M_{gg} \overset{oo}{U_g} + K_{gg} U_g = B_g L \tag{7}$$

The link between FEM output data and multibody input data requirements is the velocity transformation matrix which relates the absolute velocity of each grid point fixed reference frame to the absolute velocity of the body fixed triad and all generalized modal velocity coordinates. To arrive at the coefficient matrices 3 through 6, the velocity transformation matrix must be expressed in terms of a deformation independent part and a part which is linearly dependent upon modal deformation. The transformation matrix can be partitioned as:

$$\left[\begin{array}{cc} \phi_0 & \psi \end{array} \right] + \left[\begin{array}{cc} \phi_1 & 0 \end{array} \right] \tag{8}$$

where ϕ_0 contains grid point location vectors, ϕ_1 contains grid point deformation vectors, and ψ contains modal displacement vectors.

This form allows for the definition of a set of matrix triple products that immediately lead to the required data. It follows that:

$$\left\{ \begin{array}{c} (\phi_0 + \phi_1)^T \\ \psi^T \end{array} \right\} \left[M_{gg} \right] \left[\begin{array}{cc} (\phi_0 + \phi_1) & \psi \end{array} \right] \tag{9}$$

Detail development yields:

$$\left[\begin{array}{cc} J_0 & -S_0 \\ S_0 & M \end{array} \right] = \phi_0^T M_{gg} \phi_0 \tag{10}$$

$$[e] = \psi^T M_{gg} \psi \tag{11}$$

$$\begin{bmatrix} J_1(m) & -S_1(m) \\ S_1(m) & 0 \end{bmatrix} \xi(m) = \left[\phi_0^T M_{gg} \phi_1(m) + \phi_1^T(m) M_{gg} \phi_0\right] \xi(m) \tag{12}$$

$$\begin{bmatrix} J_2(m,n) & 0 \\ 0 & 0 \end{bmatrix} \xi(m)\xi(n) = \phi_1^T(m) M_{gg} \phi_1(n) \xi(m)\xi(n) \tag{13}$$

$$\begin{bmatrix} d_0 \\ a_0 \end{bmatrix} = \phi_0^T M_{gg} \psi \tag{14}$$

$$\begin{bmatrix} d_1(m) \\ a_1(m) \end{bmatrix} \xi(m) = \phi_1^T(m) M_{gg} \psi \xi(m) \tag{15}$$

The equations for the momentum wheel partitions are also derivable, see reference [76].

10.7. SYSTEM KINEMATICS

This section provides a definition of all matrices used for the transformation of displacement and velocity vectors between system reference frames. Detail development is contained in references [70] and [76].

The relative orientation of all body fixed reference frames, all hinge point p- and q-frames and all sensor point frames is done via a triplet of Euler angles $\bar{\beta}$. This triplet along with a Euler angle rotation sequence identifier is all that is needed to define transformation matrices via Euler angle or quaternion methods. Quaternion methods are used to create the small angle transformation matrices associated with body deformation at hinge and sensor points. The following operators are implemented:

$\mathcal{R}_i(\bar{\beta})$ - Uses the Euler angle triplet $\bar{\beta}$ and Euler angle rotation sequence identifier to define an associated 3x3 rotation transformation matrix. The associated theoretics is provided in [70] equations B-1 through B-9.

$\mathcal{V}_i(\bar{\beta}) = \pi$ - Velocity transformation matrix π as defined in [70] by equations B-10 through B-13. This transformation is used to transform angular velocity vectors from the skew frame associated with the Euler angle rotation sequence to the orthogonal body fixed frame.

$\mathcal{V}_i^{-1}(\bar{\beta}) = \pi^{-1}$ - Inverse velocity transformation matrix π^{-1} as defined in [70] by equations B-14 through B-17.

$\mathcal{Q}_i(\bar{\Delta})$ - Uses the small angle deformation triplet $(\Delta_1, \Delta_2, \Delta_3)$ in $\bar{\Delta}$ and quaternion techniques to define an associated 3x3 rotation transformation matrix.

10.7.1. *Hinge Point Reference Frames:*

For hinge point k, the following position vectors and transformations matrices are developed:

$[{}_nR_q]$ - Transformation body n fixed, to q-frame on deformed body n.

$\{{}_n\vec{d}_q\}$ - Position vector body n fixed origin, to q-frame on deformed body n.

$[_mR_p]$ - Transformation body m fixed, to p-frame on deformed body m.

$[_m\vec{d_p}]$ - Position vector body m fixed origin, to p-frame on deformed body m.

$[_pR_q]$ - Transformation q-frame on deformed body n, to p-frame on deformed body m.

$[_mR_n]$ - Transformation body n fixed coordinates, to body m fixed coordinates.

$\{_p\vec{d_q}\}$ - Position vector p-frame, to q-frame on contiguous deformed bodies m and n.

$\{_m\vec{d_n}\}_m$ - Position vector body m origin, to body n origin, in body m fixed coordinates.

10.7.2. Hinge Frame Velocity Transformation: Underlying theoretics requires the ability to transform between relative velocity and absolute velocity coordinates. The basic relation is given in [70] by equation II-81 as

$$\{\dot{\beta}\}_k = [b_p]_{k,m}\{U\}_m + [b_q]_{k,n}\{U\}_n \tag{16}$$

At hinge point k let the p-frame be on body m, and the q-frame be on body n, then

$\vec{\omega}_q$ - Absolute angular velocity of the q-frame.

$\vec{\omega}_p$ - Absolute angular velocity of the p-frame.

$\vec{U}_q$ - Absolute translational velocity of the q-frame.

$\vec{U}_p$ - Absolute translational velocity of the p-frame.

$\dot{\vec{\theta}}_{pq}$ - Angular velocity of the q-frame, relative to the p-frame.

$\overset{o}{\vec{\Delta}}_{pq}$ - Translational velocity of the q-frame, relative to the p-frame.

super(.) - derivative relative to absolute coordinates.

super(o) - derivative relative to local coordinates.

The velocity transformation matrix is the concatenation of q- frame angular and translational velocity equations expressed relative to the p-frame and the skewed Euler rotation axes. It follows that:

$$\vec{\omega}_q = \vec{\omega}_p + \dot{\vec{\theta}}_{pq} \tag{17}$$

$$\vec{U}_q \;=\; \vec{U}_p + \vec{\omega}_p \times \vec{\Delta}_{pq} + \overset{o}{\vec{\Delta}}_{pq} \tag{18}$$

The transformation from skew Euler rotation axes to q-frame coordinates is defined in [70] by equation II-19 as

$$\left\{\dot{\vec{\theta}}_{pq}\right\}^{q} = [\pi]\left\{\dot{\vec{\theta}}_{pq}\right\}^{s} \tag{19}$$

where the super q and super s symbols outside to enclosing braces define the coordinate frames (q-frame, skew-frame) relative to which the components of the enclosed vector are measured. The velocity transformation matrix $[\pi]$ is a function of the Euler angles and the associated Euler rotation sequence selected.

From [70] equation II-78 it follows that:

$$\{\vec{\omega}_q\}^q = [{}_qR_n]\{\vec{\omega}_n\}^n + [{}_qR_n][\sigma_q]\left\{\dot{\xi}_n\right\} \tag{20}$$

$$\{\vec{\omega}_p\}^p = [{}_pR_m]\{\vec{\omega}_m\}^m + [{}_pR_m][\sigma_p]\left\{\dot{\xi}_m\right\} \tag{21}$$

$$\{\vec{\omega}_p\}^q = [{}_qR_p]\{\vec{\omega}_{np}\}^p \tag{22}$$

where $[\sigma_p]$ and $[\sigma_p]$ respectively contain modal rotation amplitudes at the p- and q-frame origins. Direct substitution of equations 19 through 22 yields the $\dot{\beta}$ coordinates associated with the rotation variables of equation 16, that is

$$\begin{aligned}\left\{\dot{\vec{\theta}}_{pq}\right\}^{s} &= [b_q(1,1)]\{\vec{\omega}_n\}^n + [b_q(1,3)]\left\{\dot{\xi}_n\right\} + \\ &\quad [b_p(1,1)]\{\vec{\omega}_m\}^m + [b_p(1,3)]\left\{\dot{\xi}_m\right\}\end{aligned} \tag{23}$$

To arrive at the $\dot{\beta}$ coordinates associated with translation in equation 16, equation 18 must be evaluated relative to the p-frame. Again from [70] equation II-78 it follows that

$$\begin{aligned}\left\{\vec{U}_q\right\}^p &= [{}_pR_q][{}_qR_n]\left[S^n_{nq}\right]\{\vec{\omega}_n\}^n + [{}_pR_q][{}_qR_n]\left\{\vec{U}_n\right\}^n + \\ &\quad [{}_pR_q][{}_qR_n][h_q]\left\{\dot{\xi}_n\right\}\end{aligned} \tag{24}$$

$$\left\{\vec{U}_p\right\}^p = [{}_pR_m]\left[S^m_{mp}\right]\{\vec{\omega}_m\}^m + [{}_pR_m]\left\{\vec{U}_m\right\}^m + [{}_pR_m][h_p]\left\{\dot{\xi}_m\right\} \tag{25}$$

where $[h_p]$ and $[h_q]$ respectively contain modal displacement amplitudes at the p- and q-frame origins. The elements of $\left[S^m_{mp}\right]$ are defined in [70] by equation II-79 and the skew of the vector $\vec{\Delta}_{pq}$ expressed relative to the p-frame is symbolized as

$$\left[S^p_{pq}\right] = \begin{bmatrix} 0 & \Delta_{pq}(3) & -\Delta_{pq}(2) \\ -\Delta_{pq}(3) & 0 & \Delta_{pq}(1) \\ \Delta_{pq}(2) & -\Delta_{pq}(1) & 0 \end{bmatrix} \tag{26}$$

This yields:

$$\{\vec{\omega}_p\}^p = [{}_pR_m]\{\vec{\omega}_m\}^m + [{}_pR_m][\sigma_p]\left\{\dot{\xi}_m\right\} \tag{27}$$

$$\left\{\vec{\omega}_p \times \vec{\Delta}_{pq}\right\}^p = \left[S^p_{pq}\right]\left[_pR_m\right]\{\vec{\omega}_m\}^m + \left[S^p_{pq}\right]\left[_pR_m\right][\sigma_p]\left\{\dot{\xi}_m\right\} \tag{28}$$

Direct substitution of equations 24 through 28 into equation 18 yields the $\dot{\beta}$ coordinates associated with translation in equation 16, that is

$$\begin{aligned}\left\{\overset{o}{\vec{\Delta}}_{pq}\right\}^p &= [b_q(2,1)]\{\vec{\omega}_n\}^n + [b_q(2,2)]\left\{\vec{U}_n\right\}^n + [b_q(2,3)]\left\{\dot{\xi}_n\right\} + \\ &\quad [b_p(2,1)]\{\vec{\omega}_m\}^m + [b_p(2,2)]\left\{\vec{U}_m\right\}^m + [b_p(2,3)]\left\{\dot{\xi}_m\right\}\end{aligned} \tag{29}$$

10.7.3. Sensor Point Reference Frames: For sensor point s on body n the following position vectors and transformations matrices are developed:

$[_nR_s]$ - Transformation from s-frame on deformed body n, to body n fixed.

$[_n\vec{d}_s]$ - Position vector from body n fixed origin, to s-frame on deformed body n.

10.7.4. Derivatives of Kinematic Coefficients: The multibody equations of motion use Lagrange multipliers to define forces and torques of constraint. These must be evaluated and substituted into the equations of motion as load vectors. Their determination requires the evaluation of the first time derivative of the velocity transformation matrices. The theoretics associated with taking the time derivative of these matrices are provided in Appendix C of [70].

10.8. ORDER N RECURSIVE ALGORITHM, WITH GEAR CONSTRAINTS

On the first forward pass the 6 absolute rate coordinates associated with each body are developed as functions of the state variables used to define hinge rate, modal rate and the absolute body rate of the inboard body. The pass starts with body 1 and proceeds outward along all branches of the topological tree created by the introduction of "cut-joints." Topological loops will be reconnected via application of the constraint loads defined within the vector of Lagrange multipliers.

10.8.1. Recursive Velocity Equations: The key step in the development of the recursive form of the acceleration equations follows from the definition of the kinematic expression that relates velocity and acceleration components of bodies connected through a hinge. This is done by differentiation of the hinge kinematics equation. It is defined in [75] by equations (11, 12, and 31), in [74] by equation 2, in [32] by equation 2, and in [31] by equation 3. These equations are extended in NDISCOS to include the option to model geared hinges, each of which has an associated momentum wheel. This capability enables one to easily model complex gearing mechanisms, such as the harmonic drive.

It follows that the velocity transformation at hinge point Ii (hinge inboard of body i) with gear-constraint in condensed matrix notation is

$$[\Phi_{Ii}]\left\{\begin{array}{c}\{\dot{\beta}^f_{Ii}\}\\ \{\dot{\theta}^f_i\}\end{array}\right\} + [\bar{\Phi}_{Ii}]\left\{\begin{array}{c}\{\dot{\alpha}_{Ii}\}\\ \{\dot{\theta}^c_i\}\end{array}\right\} = [\mathbf{B}_{O,i-1}]\{U_{i-1}\} + [\mathbf{B}_{Ii}]\{U_i\} \tag{30}$$

The following definitions are used in the above equations, and those that follow:

$\{\dot{\beta}^f_{Ii}\}$ - Vector of unconstrained (free) relative velocities at the body i inboard hinge .

$[\phi_{Ii}]$, $[\phi_{Oi}]$ - selection matrices composed of 0's and 1's. These map the free coordinates at the directly inboard hinge and the directly outboard hinge of the i-th body into the appropriate hinge degrees of freedom.

$[\bar{\phi}_{Ii}]$, $[\bar{\phi}_{Oi}]$ - selection matrices composed of 0's and 1's. These map the fixed and rheonomic constraints at the directly inboard hinge and the directly outboard hinge of the i- th body into the appropriate hinge degrees of freedom. The following orthogonality relations apply:

$$[\phi_{Ii}]^T[\phi_{Ii}] = [1] \tag{31}$$
$$[\bar{\phi}_{Ii}]^T[\bar{\phi}_{Ii}] = [1] \tag{32}$$
$$[\bar{\phi}_{Ii}]^T[\phi_{Ii}] = [0] \tag{33}$$

$[V_i]$ - selection matrix composed of 0's and 1's. These map the free momentum wheels in body i to all momentum wheels embedded in body i.

$[\bar{V}_i]$ - selection matrix composed of 0's and 1's. These map the constrained momentum wheels embedded in body i to all body i momentum wheels. The following orthogonality relations apply:

$$[V_i]^T[V_i] = [1] \tag{34}$$
$$[\bar{V}_i]^T[\bar{V}_i] = [1] \tag{35}$$
$$[\bar{V}_i]^T[V_i] = [0] \tag{36}$$

$[\Phi_{Ii}], [\bar{\Phi}_{Ii}]$ - Union of hinge and wheel selection matrices, defined as

$$[\Phi_{Ii}] = \begin{bmatrix} [\phi_{Ii}] & 0 \\ 0 & [V_i] \end{bmatrix} \tag{37}$$

$$[\bar{\Phi}_{Ii}] = \begin{bmatrix} [\bar{\phi}_{Ii}] & 0 \\ 0 & [\bar{V}_i] \end{bmatrix} \tag{38}$$

$\{\dot{\theta}^f_i\}$ - Vector partition containing the wheel rate state variables for all free wheels in body i, wheel rates associated with geared wheels not subject to rheonomic constraints are included herein.

$\{\dot{\theta}^c_i\}$ - Vector partition containing the rates for all rheonomically constrained wheels in body i, wheel rates associated with geared wheels that are subject to rheonomic constraints are included herein. This is a user provided function.

$\{\dot{\alpha}_{Ii}\}$ - Vector of fixed or rheonomically constrained relative velocities at the inboard hinge of body i. This is a user provided function.

$[b_{Ii}]$ - Partition of the velocity transformation matrix associated with the hinge inboard of body i. For notational and equation development purposes it is convenient to partition this matrix as

$$[b_{Ii}] = \left[R_{Ii} \mid F_{Ii} \mid \mathbf{W}_{Ii} \right] \tag{39}$$

The elements of this partition are the $[b_p(...)]$ and $[b_q(...)]$ coefficient matrices defined within equations 23 and 29.

R_{Ii} - Partition associated with rigid body motion coordinates for body i and inboard hinge Ii.

F_{Ii} - Partition associated with flexible body motion coordinates for body i and inboard hinge Ii.

$[\mathbf{W}_{Ii}]$ - Maps the vector of all body i momentum wheel rates to the rate vector for the 6 hinge degrees of freedom inboard of body i.

$[\mathbf{B}_{Ii}], [\mathbf{B}_{O,i-1}]$ - Union of hinge and wheel velocity transformation matrices for the hinge that is both inboard of body i and outboard of body i-1. These are defined as

$$[\mathbf{B}_{Ii}] = \begin{bmatrix} [b_{Ii}] \\ [0|0|1] \end{bmatrix} \tag{40}$$

$$[\mathbf{B}_{O,i-1}] = \begin{bmatrix} [b_{O,i-1}] \\ 0 \end{bmatrix} \tag{41}$$

10.8.2. First Forward Pass, Absolute Velocity Vectors: The first forward pass of the order N algorithm recursively computes the first 6 elements of the body i velocity state vector $\{U_i\}$. These 6 elements $\dot{\gamma}_i$ are expressed as functions of $\{U_{i-1}\}$ the absolute velocities of the inboard body, $\dot{\beta}^f_{Ii}$ the relative hinge velocities, $\dot{\xi}_i$ the body modal amplitude velocities, and $\dot{\theta}^f_i$ the free momentum wheel rates. The recursion progresses from the base body forward to the end body of each chain of the topological tree formed by the introduction of cut-joints. Let

$$\{U_i\} = \begin{Bmatrix} \{\dot{\gamma}_i\} \\ \{\dot{\xi}_i\} \\ \{\dot{\theta}_i\} \end{Bmatrix} \tag{42}$$

and use all definitions provide in section to obtain

- For the base body, i=1.

$$\{\dot{\gamma}_1\} = [R_{I1}]^{-1} \left([\phi_{I1}]\{\dot{\beta}^f_{I1}\} + [\bar{\phi}_{I1}]\{\dot{\alpha}_{I1}\} - [F_{I1}]\{\dot{\xi}_1\} - [\mathbf{W}_{I1}]\{\dot{\theta}_1\} \right) \tag{43}$$

where the $[F_{I1}]\{\dot{\xi}_1\}$ contribution has been included in all theoretical developments, however, it is null and so recognized within the NDISCOS code.

- For all other bodies the recursion proceeds outboard from body 1. The subscript i-1 implies "body inboard of body i".

$$\{\dot{\gamma}_i\} = [R_{Ii}]^{-1}\left([\phi_{Ii}]\{\dot{\beta}^f_{Ii}\} + [\bar{\phi}_{Ii}]\{\dot{\alpha}_{Ii}\} - [F_{Ii}]\{\dot{\xi}_i\} - [\mathbf{W}_{Ii}]\{\dot{\theta}_i\} - [b_{0,i-1}]\{U_{i-1}\}\right) \tag{44}$$

10.8.3. Closed Loops Cut-Joint Kinematics: The velocity transformation is used to compute the relative hinge velocity at each cut-joint degree of freedom. Cut-joint kinematics is defined by

$$[\bar{\phi}_c]\{\dot{\alpha}_c\} + [\phi_c]\{\dot{\beta}^f_c\} = \sum_{k=1}^{2}[b^{(k)}_c]\{U^{(k)}_c\} \tag{45}$$

where c is the cut-joint hinge number. Superscript (k) indicates branch 1 or 2 of the cut-joint. Let the p-frame be fixed in body m on branch 1, and let the q-frame be fixed in body n on branch 2. In the notation of section this equation becomes;

$$[\bar{\phi}_c]\{\dot{\alpha}_c\} + [\phi_c]\{\dot{\beta}^f_c\} = [b_p]_{c,m}\{U\}_m + [b_q]_{c,n}\{U\}_n \tag{46}$$

The computed values for the variables on the left hand side of the equation are evaluated and stored. These are computed estimates of the relative velocity vector at cut-joint c. This data is subject to correction since the effects of cut-joint constraints have not yet been accounted for. This computed information will be compared to data defined by known constraints. The difference is constraint violation error. This must be compensated for in a manner consistent with the fundamental laws of motion.

10.8.4. Hinge Kinematics, Recursive Acceleration Equations: The kinematics defined by equation 30 can be differentiated and rearranged to obtain the recursive equations that define the absolute acceleration of the i-th body in terms of the absolute acceleration of the directly inboard body and the relative acceleration at the inboard hinge. It is

$$\begin{aligned}[\phi_{Ii}]\{\ddot{\beta}^f_{Ii}\} + [\bar{\phi}_{Ii}]\{\ddot{\alpha}_{Ii}\} &= [\dot{b}_{O,i-1}]\{U_{i-1}\} + [\dot{b}_{Ii}]\{U_i\} \\ &+ [b_{O,i-1}]\{\dot{U}_{i-1}\} + [b_{Ii}]\{\dot{U}_i\} \qquad (47)\\ [V_i]\{\ddot{\theta}^f_i\} + [\bar{V}_i]\{\ddot{\theta}^c_i\} &= \{\ddot{\theta}_i\} \qquad (48)\end{aligned}$$

where

$\{\ddot{\theta}_i\}$ - Vector of all body i wheel rate state variables.

A unique solution for $\{\dot{U}_i\}$ is obtained from recognition that the rigid body component $[R_{Ii}]$ within the velocity transformation matrix $[b_{Ii}]$ is non-singular.

The desired kinematic coordinate transformation for $\{\dot{U}_i\}$ is defined in [75] by equation 14, in [32] by equation 3, in [74] by equation 4 and in [31] by equation 4 and extended herein for geared hinges. It follows that:

$$[\phi_{Ii}]\{\ddot{\beta}^f_{Ii}\} + \{\zeta_i\} = [R_{Ii}]\{\ddot{\gamma}_i\} + [F_{Ii}]\{\ddot{\xi}_i\} + [\mathbf{W}_{Ii}][V_i]\{\ddot{\theta}^f_i\} + [b_{O,i-1}]\{\dot{U}_{i-1}\} \tag{49}$$

where the identities

$$[b_{Ii}]\{\dot{U}_i\} = [R_{Ii}]\{\ddot{\gamma}_i\} + [F_{Ii}]\{\ddot{\xi}_i\} + [\mathbf{W}_{Ii}]\{\ddot{\theta}_i\} \tag{50}$$

$$[\mathbf{W}_{Ii}]\{\ddot{\theta}_i\} = [\mathbf{W}_{Ii}][V_i]\{\ddot{\theta}_i^f\} + [\mathbf{W}_{Ii}][\bar{V}_i]\{\ddot{\theta}_i^c\} \tag{51}$$

$$\{\zeta_i\} = [\bar{\phi}_{Ii}]\{\ddot{\alpha}_{Ii}\} - [\dot{b}_{O,i-1}]\{U_{i-1}\} - [\dot{b}_{Ii}]\{U_i\} - [\mathbf{W}_{Ii}][\bar{V}_i]\{\ddot{\theta}_i^c\} \tag{52}$$

have been used in creating equation 49. It follows that

$$\{\ddot{\gamma}_i\} = [R_{Ii}]^{-1}\left\{[\phi_{Ii}]\{\ddot{\beta}_i^f\} + \{\zeta_i\} - [F_{Ii}]\{\ddot{\xi}_i\} - [\mathbf{W}_{Ii}][V_i]\{\ddot{\theta}_i^f\} - [b_{O,i-1}]\{\dot{U}_{i-1}\}\right\} \tag{53}$$

and from equation 48

$$\{\ddot{\theta}_i\} = [V_i]\{\ddot{\theta}_i^f\} + [\bar{V}_i]\{\ddot{\theta}_i^c\} \tag{54}$$

By combining these in matrix format the following acceleration equation is obtained in recursive format.

$$\{\dot{U}_i\} = [\Psi_i]\{\dot{\mathbf{P}}_i\} + \{\mathbf{A}_{ci}\} - [\Gamma_i]\{\dot{U}_{i-1}\} \tag{55}$$

where

$$\{\dot{U}_i\} = \left\{\begin{array}{c}\{\ddot{\gamma}_i\}\\ \{\ddot{\xi}_i\}\\ \{\ddot{\theta}_i\}\end{array}\right\} \tag{56}$$

$$[\Psi_i] = \begin{bmatrix} [R_{Ii}]^{-1}[\phi_{Ii}] & -[R_{Ii}]^{-1}[F_{Ii}] & -[R_{Ii}]^{-1}[\mathbf{W}_{Ii}][V_i] \\ 0 & 1 & 0 \\ 0 & 0 & [V_i] \end{bmatrix} \tag{57}$$

$$\{\dot{\mathbf{P}}_i\} = \left\{\begin{array}{c}\ddot{\beta}_{Ii}^f\\ \ddot{\xi}_i\\ \ddot{\theta}_i^f\end{array}\right\} \tag{58}$$

$$\{\mathbf{A}_{ci}\} = \left\{\begin{array}{c}[R_{Ii}]^{-1}\{\zeta_i\}\\ 0\\ [\bar{V}_i]\{\ddot{\theta}_i^c\}\end{array}\right\} \tag{59}$$

$$[\Gamma_i] = \left\{\begin{array}{c}[R_{Ii}]^{-1}[b_{O,i-1}]\\ 0\\ 0\end{array}\right\} \tag{60}$$

Equation 55 is derived strictly from the kinematics relationships between relative and absolute coordinates. These will be used to develop a recursive format for the Newtonian laws of motion that relate relative body motion to applied loads.

Within multibody equations of motion there is a vector of unknown constraint loads. The equations of motion cannot be integrated until these are solved for and eliminated. To support this computational step it is convenient to differentiate equation 30 again and condense the results into another notational format, that is

$$[\Phi_{Ii}]\left\{\begin{array}{c}\{\ddot{\beta}_{Ii}^f\}\\ \{\ddot{\theta}_i^f\}\end{array}\right\} + \{\mathbf{Z}_i\} = [\mathbf{B}_{O,i-1}]\{\dot{U}_{i-1}\} + [\mathbf{B}_{Ii}]\{\dot{U}_i\} \tag{61}$$

where the intermediate acceleration vector $\{\mathbf{Z}_i\}$ is defined as:

$$\{\mathbf{Z}_i\} = [\bar{\Phi}_{Ii}]\left\{ \begin{array}{c} \{\ddot{\alpha}_{Ii}\} \\ \{\ddot{\theta}_i^c\} \end{array} \right\} - [\dot{\mathbf{B}}_{O,i-1}]\{U_{i-1}\} - [\dot{\mathbf{B}}_{Ii}]\{U_i\} \tag{62}$$

10.8.5. Multibody Equations of Motion for End Body: The equations of motion for an end body e are presented in [75] by equation 9 without cut-joint constraints, in [32] by equation 1 with cut-joint constraints, in [31] by equation 2 and in [75] by equation 23 with provision for geared hinges and wheels. The union of all this theoretical source material plus recent work yields the following set of motion equations for the end bodies.

$$[M_e]\{\dot{U}_e\} = \{G_e\} + [\mathbf{B}_{Ie}]^T[\bar{\Phi}_{Ie}]\{\Lambda_{Ie}\} + \sum_{ec}[\tilde{\mathbf{b}}_{ec}]^T\{\lambda_c\} \tag{63}$$

where

$$[\tilde{\mathbf{b}}_{ec}]^T = [\mathbf{b}_{Oec}]^T[\bar{\phi}_c] \tag{64}$$

The definitions of $[\mathbf{B}_{Ie}]$ and $[\bar{\Phi}_{Ie}]$ are consistent with the notation used in equations 30, and 38. Furthermore:

$[M_e]$ - Body e inertia matrix, arranged as in equation 2.

$\{G_e\}$ - Vector of all gyroscopic and external loads acting on body e.

$\{\Lambda_{Ie}\}$ - Vector of Lagrange multipliers associated with fixed and rheonomic constraints at the directly inboard hinge of body e and all fixed or rheonomically constrained momentum wheels in body e.

$\{\lambda_c\}$ - Vector of Lagrange multipliers associated with the fixed, geared, and rheonomic constraints at cut-joint c.

$[\mathbf{b}_{Oec}]^T$ - Union of hinge and wheel transformations for all body e outboard cut-joints.

Substitute kinematics equation 55 into motion equation 63, and premultiply through by $[\Psi_e]^T$. Then condense notation, recognize that the projection of the hinge constraint loads acting at the inboard hinge of body e onto the set of relative coordinates is a null vector and obtain the body e motion equation as

$$\{\dot{\mathbf{P}}_e\} = [\bar{M}_e]^{-1}[\Psi_e]^T\left\{\{G_e\} - [M_e]\{\mathbf{A}_{ce}\} + [M_e][\Gamma_e]\{\dot{U}_{e-1}\} + \sum_{ec}[\tilde{\mathbf{b}}_{ec}]^T\{\lambda_c\}\right\} \tag{65}$$

10.8.6. Computation of Constraint Loads at Body e Inboard Hinge: During the system design process it is frequently necessary to predict constraint loads at hinge points. This information is contained with the vector $\{\Lambda_{Ie}\}$. To solve for it define $[J_e]$, the reduced order body e inertia matrix. This is the body e inertia matrix projected onto the set of constrained coordinate degrees of freedom at the inboard hinge.

$$[J_e] = \left([\bar{\Phi}_{Ie}]^T[\mathbf{B}_{Ie}][M_e]^{-1}[\mathbf{B}_{Ie}]^T[\bar{\Phi}_{Ie}]\right)^{-1} \tag{66}$$

It follows from equation 61 with body index i set equal to e and equation 63 that

$$\begin{aligned}\{\Lambda_{Ie}\} &= [J_e][\bar{\Phi}_{Ie}]^T\left[\{\mathbf{Z}_e\} - [\mathbf{B}_{O,e-1}]\{\dot{U}_{e-1}\}\right] \\ &- [J_e][\bar{\Phi}_{Ie}]^T[\mathbf{B}_{Ie}][M_e]^{-1}\left(\{G_e\} + \sum_{ec}[\tilde{\mathbf{b}}_{ec}]^T\{\lambda_c\}\right)\end{aligned} \tag{67}$$

Equation 67 is the same as that provided in [32] by equation 6, in [75] by equation 17, without wheels or cut-joints and in [31] by equation 6 with wheels and without cut-joints.

10.8.7. Multibody Equations of Motion for Intermediate Body: The motion equations for an intermediate body i are presented in [75] by equation 10 without cut-joint constraints, in [31] by equation 7 and in [75] by equation 28 with provision for a geared wheels. The union of all theoretical source material yields the following set of equations for intermediate body i. This equation is a straight forward extension of body e equation 63.

$$[M_i]\{\dot{U}_i\} = \{G_i\} + [\mathbf{B}_{Ii}]^T[\bar{\Phi}_{Ii}]\{\Lambda_{Ii}\} + \sum_{Oi}[\mathbf{B}_{Oi}]^T[\bar{\Phi}_{Oi}]\{\Lambda_{Oi}\} + \sum_{ic}[\mathbf{b}_{Oic}]^T[\bar{\phi}_c]\{\lambda_c\} \tag{68}$$

where from the definition of inboard "I" and outboard "O" it follows that

$$\{\Lambda_{Oi}\} = \{\Lambda_{I,i+1}\} \tag{69}$$

$$\{\Lambda_{Ii}\} = \{\Lambda_{O,i-1}\} \tag{70}$$

The solution for both end and intermediate body acceleration state requires that two sets of unknowns be determined. The first consists of the Lagrange multipliers used to define interaction constraint loads acting between bodies and the second set of unknowns is the acceleration state for body i.

The recursive algorithm presented here creates implicit sets of equations for the equations of motion transformed to recursive format, and Lagrange multipliers. The kinetics equations 63 and 68 are relative to the dimensional space that allows 6 rigid body degrees of freedom at each hinge. The recursion algorithm maps these into a lower dimensional space. This is the space in which motion is kinematically allowed. This lower dimensional space is consistent with the kinematic constraint conditions defined at all system hinges. The effect of this mapping is to eliminate all Lagrange multipliers $\{\Lambda_{Ii}\}$ and $\{\Lambda_{Oi}\}$. The mapping however does not eliminate the Lagrange multipliers associated with cut-joints. These must be solved for and substituted into the equations of motion that will be integrated.

10.8.8. Recursive Form of Multibody Equations of Motion: The right hand side of equation 65 contains information that is strictly dependent upon body e absolute velocity and inertia, the absolute acceleration of the inboard body, gyroscopic and external loads on body e, and loads associated with cut-joint constraints.

The objective now to express the right hand side of the recursive kinetics equation for intermediate body i in terms of information that is strictly dependent upon body i absolute velocity, the inertia of the system of bodies outboard of body i, the absolute acceleration of the body inboard to body i, and loads associated with all cut-joint constraints acting on body i. It follows that

$$[\tilde{M}_i]\{\dot{U}_i\} = \{\tilde{G}_i\} + [\mathbf{B}_{Ii}]^T[\bar{\Phi}_{Ii}]\{\Lambda_{Ii}\} + \sum_{ic}[\tilde{\mathbf{b}}_{ic}]^T\{\lambda_c\} \tag{71}$$

where

$\{\dot{\mathbf{P}}_i\}$ - Relative acceleration vector at the inboard hinge of body i.

$$\{\dot{\mathbf{P}}_i\} = [\bar{M}_i]^{-1}[\Psi_i]^T\left\{\{\tilde{G}_i\} - [\tilde{M}_i]\{\mathbf{A}_{ci}\} + [\tilde{M}_i][\Gamma_i]\{\dot{U}_{i-1}\} + \sum_{ic}[\tilde{\mathbf{b}}_{ic}]^T\{\lambda_c\}\right\} \tag{72}$$

$\{\Lambda_{Ii}\}$ - Lagrange multiplier vector of hinge and wheel constraint loads acting at the inboard hinge of body i.

$$\begin{aligned}\{\Lambda_{Ii}\} &= [J_i][\bar{\Phi}_{Ii}]^T\left(\{\mathbf{Z}_i\} - [\mathbf{B}_{O,i-1}]\{\dot{U}_{i-1}\}\right)\\ &- [J_i][\bar{\Phi}_{Ii}]^T\left[[\mathbf{B}_{Ii}][\tilde{M}_i]^{-1}\left(\{\tilde{G}_i\} + \sum_{ic}[\tilde{\mathbf{b}}_{ic}]^T\{\lambda_c\}\right)\right]\end{aligned} \tag{73}$$

$[J_i]$ - The reduced order inertia matrix of body i relative to the constrained degrees of freedom at body i's inboard hinge.

$$[J_i] = \left([\bar{\Phi}_{Ii}]^T[\mathbf{B}_{Ii}][\tilde{M}_i]^{-1}[\mathbf{B}_{Ii}]^T[\bar{\Phi}_{Ii}]\right)^{-1} \tag{74}$$

$[\tilde{M}_i]$ - The equivalent inertia matrix of body i and all outboard bodies relative to the inboard hinge of body i.

$$[\tilde{M}_i] = [M_i] + \sum_{Oi}[\mathbf{B}_{Oi}]^T[\bar{\Phi}_{Oi}][J_{i+1}][\bar{\Phi}_{Oi}]^T[\mathbf{B}_{Oi}] \tag{75}$$

$[\bar{M}_i]$ - The reduced order inertia matrix of body i and all outboard bodies relative to the unconstrained degrees of freedom at the inboard hinge of body i.

$$[\bar{M}_i] = [\Psi_i]^T[\tilde{M}_{i+1}][\Psi_i] \tag{76}$$

$\{\tilde{G}_i\}$ - The equivalent force vector of body i and all outboard bodies relative to the inboard hinge of body i.

$$\begin{aligned}\{\tilde{G}_i\} &= \{G_i\}\\ &+ \sum_{Oi}[\mathbf{B}_{Oi}]^T[\bar{\Phi}_{Oi}][J_{i+1}][\bar{\Phi}_{Oi}]^T\left(\{\mathbf{Z}_{i+1}\} - [\mathbf{B}_{I,i+1}][\tilde{M}_{i+1}]^{-1}\{\tilde{G}_{i+1}\}\right)\end{aligned} \tag{77}$$

$[\tilde{\mathbf{b}}_{ic}]$ - The equivalent constraint kinematics matrix for body i and all outboard bodies. relative to the inboard hinge of body i.

$$
\begin{aligned}
[\tilde{\mathbf{b}}_{ic}] &= [\mathbf{b}_{Oic}]^T[\bar{\phi}_c] \\
&- \sum_{Oi}[\mathbf{B}_{Oi}]^T[\bar{\Phi}_{Oi}][J_{i+1}][\bar{\Phi}_{Oi}]^T[\mathbf{B}_{I,i+1}][\tilde{M}_{i+1}]^{-1}[\tilde{\mathbf{b}}_{i+1,c}]^T
\end{aligned}
\tag{78}
$$

10.9. TOPOLOGICAL LOOPS ABD CUT-JOINT CONSIDERATIONS

There are two fundamental problems that must be addressed for problems that involve topological loops.

1. Constraint stabilization. There is always a tendency towards numerical drift. For topological loop problems this has the net effect of allowing bodies to drift away from each other. This happens when the cut-joint constraint loads computed via Lagrange multipliers produce "equal and opposite" constraint loads that do not exactly cancel. Several methods exist for resolving this well known problem, see reference [80] and references cited therein. At every cycle through the integration procedure NDISCOS uses body velocity rate information to compute how well velocity constraints are satisfied. Inertia times the difference between computed and defined relative velocity is viewed as a momentum pulse that came from an extraneous physical impulse to the system. If the momentum pulse is small then associated impulse Lagrange multipliers can be computed. From a knowledge of these impulse Lagrange multipliers system state velocity data can be corrected in a manner consistent with the laws of conservation of momentum.

2. Constraint Loads. The order N algorithm requires that the vector of hinge constraint loads be computed for cut-joints and geared hinges, these are the Lagrange multipliers.

There are advantageous correlations between the defining equations for impulse Lagrange multipliers and the Lagrange multipliers defined at cut-joints and geared hinges. These will be fully utilized in the computational process.

10.10. CUT-JOINT KINEMATICS

The first forward pass for the closed loop system order N algorithm uses the velocity transformation array to compute relative hinge velocities at each cut-joint degree of freedom. This operation was expressed by equation 45. Additionally it is possible to define geared hinge constraints at cut-joints. The associated kinematics equation is expressed by equation 48. This equation is not used in the determination of the impulse Lagrange multipliers, but it will be used in the development of the Lagrange multipliers used to enforce cut-joint and geared hinge constraints.

Let the p-frame be fixed in body m on branch 1, and let the q- frame be fixed in body n on branch 2. In the notation of section equation 45 becomes;

$$[\phi_c]\{\dot{\beta}_c^f\} + [\bar{\phi}_c]\{\dot{\alpha}_c\} = [b_p]_{c,m}\{U\}_m + [b_q]_{c,n}\{U\}_n \tag{79}$$

In notation more supportive of the development of the set of simultaneous algebraic equations used to define the Lagrange multipliers, express this kinematic equation for the relative velocity at cut-joint $c(j)$ as:

$$[\phi_{c(j)}]\{\dot{\beta}^f_{c(j)}\} + [\bar{\phi}_{c(j)}]\{\dot{\alpha}_{c(j)}\} = [b^{(1)}_{c(j)}]\{U^{(1)}_{c(j)}\} + [b^{(2)}_{c(j)}]\{U^{(2)}_{c(j)}\} \tag{80}$$

Differentiate and define the intermediate acceleration vector $\{\zeta_{c(j)}\}$ for cut-joint $c(j)$ as

$$\{\zeta_{c(j)}\} = [\bar{\phi}_{c(j)}]\{\ddot{\alpha}_{c(j)}\} - [\dot{b}^{(1)}_{c(j)}]\{U^{(1)}_{c(j)}\} - [\dot{b}^{(2)}_{c(j)}]\{U^{(2)}_{c(j)}\} \tag{81}$$

Direct substitution and rearrangement yields

$$[\bar{\phi}_{c(j)}]^T\{\zeta_{c(j)}\} = [\tilde{\mathbf{b}}^{(1)}_{c(j)}]\{\dot{U}^{(1)}_{c(j)}\} + [\tilde{\mathbf{b}}^{(2)}_{c(j)}]\{\dot{U}^{(2)}_{c(j)}\} \tag{82}$$

where

$$[\mathbf{b}^{(k)}_{c(j)}]^T = \begin{bmatrix} [b^{(k)}_{c(j)}] & 0 \end{bmatrix} \tag{83}$$

$$[\tilde{\mathbf{b}}^{(k)}_{c(j)}] = [\bar{\phi}_{c(j)}]^T[\mathbf{b}^{(k)}_{c(j)}] \tag{84}$$

10.10.1. Cut-joint Kinetics Equations: Equations that define unknown cut-joint multipliers can be obtained by successive use of the kinetics recursion equation 71 for each end body acceleration vector $\{\dot{U}^k_{c(j)}\}$.

$$\{\dot{U}^k_{c(j)}\} = \{\dot{U}_c\} = [\tilde{M}_c]^{-1}\left(\{\tilde{G}_c\} + [\mathbf{B}_{Ic}]^T[\bar{\Phi}_{Ic}]\{\Lambda_{Ic}\} + \sum_d [\tilde{\mathbf{b}}_{cd}]^T\{\lambda_d\}\right) \tag{85}$$

where the summation is over all cut-joints d, attached to cut joint body c. This summation is never null, since it is associated with bodies located at cut-joints. From equation 73 the Lagrange multiplier vector for the joint inboard of body c is

$$\begin{aligned}\{\Lambda_{Ic}\} &= [J_c][\bar{\Phi}_{Ic}]^T \Big[\{\mathbf{Z}_c\} - [\mathbf{B}_{O,c-1}]\{\dot{U}_{c-1}\} \\ &\quad - [\mathbf{B}_{Ic}][\tilde{M}_c]^{-1}\Big(\{\tilde{G}_c\} + \sum_{cd}[\tilde{\mathbf{b}}_{cd}]^T\{\lambda_d\}\Big)\Big]\end{aligned} \tag{86}$$

and from equation 62

$$\{\mathbf{Z}_c\} = [\bar{\Phi}_{Ic}]\left\{\begin{array}{c}\{\ddot{\alpha}_{Ic}\}\\ \{\ddot{\theta}^c_c\}\end{array}\right\} - [\dot{\mathbf{B}}_{O,c-1}]\{U_{c-1}\} - [\dot{\mathbf{B}}_{Ic}]\{U_c\} \tag{87}$$

Direct substitution of equation 86 into 85 with notational condensation yields:

$$\{\dot{U}_c\} = \{\mathbf{U}_c\} + [\mathbf{N}_c]\sum_d[\tilde{\mathbf{b}}_{cd}]^T\{\lambda_d\} - [M^*_c][\tilde{J}_c][\mathbf{B}_{O,c-1}]\{\dot{U}_{c-1}\} \tag{88}$$

Successive substitutions of equation 88 into itself yields

$$
\begin{aligned}
\{\dot{U}_c\} &= \{\mathbf{U}_c\} + [\mathbf{N}_c]\sum_d [\tilde{\mathbf{b}}_{cd}]^T\{\lambda_d\} \\
&- [M_c^*][\tilde{J}_c][\mathbf{B}_{O,c-1}]\left[\{\mathbf{U}_{c-1}\} + [\mathbf{N}_{c-1}]\sum_d [\tilde{\mathbf{b}}_{c-1,d}]^T\{\lambda_d\}\right. \\
&\vdots \\
&- [M_2^*][\tilde{J}_2][\mathbf{B}_{O,1}]\left[\{\mathbf{U}_1\} + [\mathbf{N}_1]\sum_d [\tilde{\mathbf{b}}_{1,d}]^T\{\lambda_d\}\right]\cdots\Big]
\end{aligned}
\tag{89}
$$

The desired recursion is developed by defining

$$
\{U_{i+1}^*\} = \{\mathbf{U}_{i+1}\} + [\mathbf{N}_{i+1}]\sum_d [\tilde{\mathbf{b}}_{i+1,d}]^T\{\lambda_d\} - [M_{i+1}^*][\tilde{J}_{i+1}][\mathbf{B}_{Oi}]\{U_i^*\} \tag{90}
$$

where

$$
\begin{aligned}
\{\dot{U}_i\} &= \{\mathbf{U}_i\} + [\mathbf{N}_i]\sum_d [\tilde{\mathbf{b}}_{id}]^T\{\lambda_d\} \\
&- [M_i^*][\tilde{J}_i][\mathbf{B}_{O,i-1}]\left[\{\mathbf{U}_{i-1}\} + [\mathbf{N}_{i-1}]\sum_d [\tilde{\mathbf{b}}_{i-1,d}]^T\{\lambda_d\}\right. \\
&\vdots \\
&- [M_2^*][\tilde{J}_2][\mathbf{B}_{O,1}]\left[\{\mathbf{U}_1\} + [\mathbf{N}_1]\sum_d [\tilde{\mathbf{b}}_{1,d}]^T\{\lambda_d\}\right]\cdots\Big]
\end{aligned}
\tag{91}
$$

and

$$\{\dot{U}_1\} = \{\mathbf{U}_1\} + [\mathbf{N}_1]\sum_d [\tilde{\mathbf{b}}_{1d}]^T\{\lambda_d\} \tag{92}$$

$$[M_i^*] = [\tilde{M}_i]^{-1}[\mathbf{B}_{Ii}]^T \tag{93}$$

$$
\begin{aligned}
[\mathbf{N}_i] &= [\tilde{M}_i]^{-1}\left([\mathbf{1}] - [\mathbf{B}_{Ii}]^T[\bar{\Phi}_{Ii}][J_i][\bar{\Phi}_{Ii}]^T[\mathbf{B}_{Ii}][\tilde{M}_i]^{-1}\right) \\
&= [\tilde{M}_i]^{-1} - [M_i^*][\bar{\Phi}_{Ii}][J_i][\bar{\Phi}_{Ii}]^T[\mathbf{B}_{Ii}][\tilde{M}_i]^{-1}
\end{aligned}
\tag{94}
$$

$$[\tilde{J}_i] = [\bar{\Phi}_{Ii}][J_i][\bar{\Phi}_{Ii}]^T \tag{95}$$

$$\{\mathbf{U}_i\} = [\mathbf{N}_i]\{\tilde{G}_i\} + [M_i^*][\tilde{J}_i]\{\mathbf{Z}_i\} \tag{96}$$

The final result is

$$\{\dot{U}_i\} = \{U_i^*\} \tag{97}$$

where each $\{U_i^*\}$ is recursively computed according to the above relations. The values of the index i spans the set of all body labels associated with all bodies on one or more paths from the origin to a cut joint.

10.10.2. Cut-joint Lagrange Multiplier Equations: From section , cut-joint kinematics was defined by equation 82. It can be expressed as

$$[\bar{\phi}_{c(j)}]^T\{\zeta_{c(j)}\} = \sum_{k=1}^{2}[\tilde{\mathbf{b}}_{c(j)}^{(k)}]\{\dot{U}_{c(j)}^{(k)}\} \tag{98}$$

This relation holds for every cut-joint. Recall that superscript k, subscript $c(j)$ implies the end body on branch k, k=1,2 at the j-th cut-joint identified as hinge point $c(j)$, j=1,2,...,NCUT.

From section , cut-joint kinetics was defined by equation 89. A simple notational change in equation 97 to more precisely identify the associated body yields

$$\{\dot{U}_{c(j)}^{(k)}\} = \{U_{c(j)}^{*(k)}\} \tag{99}$$

where $\{U_{c(j)}^{*(k)}\}$ is computed in accordance with the recursion equations 90 through 96. Substitution of equation 99 into 98 yields one cut-joint Lagrange multiplier matrix equation for each cut-joint. It follows that

$$[\bar{\phi}_{c(j)}]^T\{\zeta_{c(j)}\} = \sum_{k=1}^{2}[\tilde{\mathrm{b}}_{c(j)}^{(k)}]\{U_{c(j)}^{*(k)}\} \tag{100}$$

Equation 100 defines all NCUT cut-joint kinetics equations used to compute impulse Lagrange multipliers.

10.11. IMPULSE LAGRANGE MULTIPLIERS

Cut-joint velocity constraint conditions are maintained during the numerical integration process by the imposition of impulses that enforce satisfaction of constraint conditions and conservation of momenta laws. During the evaluation of the function to be integrated, each velocity constraint condition is compared to the velocity constraint condition as computed from a knowledge of system position and velocity state. The difference reflects itself into the simulation as a pulse in the computed value of system momenta. This must be corrected. NDISCOS theoretics chooses to view this pulse to system momenta as the result of an imaginary vector of impulses applied to the set of cut-joint constrained degrees of freedom. By the application of the associated set of impulse Lagrange multipliers in an equal and opposite manner NDISCOS insures that all cut-joint fixed and rheonomic velocity constraints and all system momenta conservation laws are satisfied.

From classic impulsive motion theory, let P be the force applied to a particle of mass m and let v be its velocity. These are related by

$$m\frac{dv}{dt} = P \tag{101}$$

where integration of both sides of the equation from time t_1 to t_2 yields

$$m(v_1 - v_2) = \int_{t_1}^{t_2} P dt \tag{102}$$

In the limit as $\Delta t = t_2 - t_1 \rightarrow 0$

- $m(v_1 - v_2)$ = pulse in the particle's linear momentum, and
- $\int_{t_1}^{t_2} P dt$ = associated impulse.

In NDISCOS theoretics we perform the same operation, it's notation is just more complex. The impulse Lagrange multipliers are obtained from the equation that results after both sides of kinetics equation 100 are integrated with respect to time and the limit taken as the range of integration is allowed to approach zero. That is

$$\lim_{\Delta t\to 0}\int_{\Delta t}[\bar{\phi}_{c(j)}]^T\{\zeta_{c(j)}\}dt = \lim_{\Delta t\to 0}\int_{\Delta t}\sum_{k=1}^{2}[\tilde{\mathbf{b}}^{(k)}_{c(j)}]\{U^{*(k)}_{c(j)}\}dt \tag{103}$$

From equation 81 let

$$\begin{aligned}\{v_{c(j)}\} &= \lim_{\Delta t\to 0}\int_{\Delta t}[\bar{\phi}_{c(j)}]^T\{\zeta_{c(j)}\}dt\\ &= \lim_{\Delta t\to 0}\int_{\Delta t}\left(\{\ddot{\alpha}_{c(j)}\} - [\bar{\phi}_{c(j)}]^T[\dot{b}^{(1)}_{c(j)}]\{U^{(1)}_{c(j)}\} - [\bar{\phi}_{c(j)}]^T[\dot{b}^{(2)}_{c(j)}]\{U^{(2)}_{c(j)}\}\right) dt\end{aligned} \tag{104}$$

and from equation 100 let

$$\sum_{i=1}^{NCUT}[A_{c(j),c(i)}]\{h_{c(i)}\} = \lim_{\Delta t\to 0}\int_{\Delta t}\sum_{k=1}^{2}[\tilde{\mathbf{b}}^{(k)}_{c(j)}]\{U^{*(k)}_{c(j)}\}dt \tag{105}$$

where

$$\lim_{\Delta t\to 0}\int_{\Delta t}\{\lambda_{c(i)}\}dt = \{h_{c(i)}\} \tag{106}$$

In the above equations $\{h_{c(i)}\}$ is the vector of impulse Lagrange multipliers acting on the set of fixed and rheonomically constrained degrees of freedom at the i-th cut-joint that is user defined to be hinge point $c(i)$.

Integration of the right hand side of equation 104 and the use of equation 84 leads to

$$\{v_{c(j)}\} = \{\dot{\alpha}_{c(j)}\} - [\tilde{\mathbf{b}}^{(1)}_{c(j)}]\{U^{(1)}_{c(j)}\} - [\tilde{\mathbf{b}}^{(2)}_{c(j)}]\{U^{(2)}_{c(j)}\} \tag{107}$$

In this equation

- $\{\dot{\alpha}_{c(j)}\}$ defines the vector of fixed and rheonomic velocity constraints acting at the j-th cut joint (hinge point $c(j)$),
- $[\tilde{\mathbf{b}}^{(1)}_{c(j)}]\{U^{(1)}_{c(j)}\} + [\tilde{\mathbf{b}}^{(2)}_{c(j)}]\{U^{(2)}_{c(j)}\}$ defines the vector of fixed and rheonomic velocity constraints computed from a knowledge of system state.
- $\{v_{c(j)}\}$ defines the cut-joint velocity pulse that contributes to the pulse in system momenta associated with computational error.

Integration of the right hand side of equation 105 is a bit less straight-forward. From equation 91 and 99 the vector $\{U^{*(k)}_{c(j)}\}$ can be expressed by the recursion:

$$\begin{aligned}\{U^*_i\} &= \{\mathbf{U}_i\} + [\mathbf{N}_i]\sum_d[\tilde{\mathbf{b}}_{id}]^T\{h_d\}\\ &- [M^*_i][\tilde{J}_i][\mathbf{B}_{O,i-1}]\Bigg[\{\mathbf{U}_{i-1}\} + [\mathbf{N}_{i-1}]\sum_d[\tilde{\mathbf{b}}_{i-1,d}]^T\{h_d\}\\ &\vdots\\ &- [M^*_2][\tilde{J}_2][\mathbf{B}_{O,1}]\left[\{\mathbf{U}_1\} + [\mathbf{N}_1]\sum_d[\tilde{\mathbf{b}}_{1,d}]^T\{h_d\}\right]\cdots\Bigg]\end{aligned} \tag{108}$$

where index i is equal to the body number associated with the end body on branch k, k=1,2 at the j-th cut-joint (hinge point $c(j)$), j=1,...,NCUT.

From equation 96 given by

$$\{\mathbf{U}_i\} = [\mathbf{N}_i]\{\tilde{G}_i\} + [M_i^*][\tilde{J}_i]\{\mathbf{Z}_i\} \tag{109}$$

it is seen that $\{\mathbf{U}_i\}$ is a function of the equivalent load vectors $\{\tilde{G}_i\}$ and inertial loads associated with the intermediate acceleration vector $\{\mathbf{Z}_i\}$. Since there are no impulsive loads defined within this function, it follows that

$$\lim_{\Delta t \to 0} \int_{\Delta t} \{\mathbf{U}_i\} dt = \{0\} \tag{110}$$

$$\lim_{\Delta t \to 0} \int_{\Delta t} \{\lambda_d\} dt = \{h_d\} \tag{111}$$

and therefore

$$\begin{aligned} \lim_{\Delta t \to 0} \int_{\Delta t} \{U_i^*\} dt &= [\mathbf{N}_i] \sum_d [\tilde{\mathbf{b}}_{id}]^T \{h_d\} \\ &- [M_i^*][\tilde{J}_i][\mathbf{B}_{O,i-1}] \left[[\mathbf{N}_{i-1}] \sum_d [\tilde{\mathbf{b}}_{i-1,d}]^T \{h_d\} \right. \\ &\vdots \\ &- [M_2^*][\tilde{J}_2][\mathbf{B}_{O,1}] \left[[\mathbf{N}_1] \sum_d [\tilde{\mathbf{b}}_{1,d}]^T \{h_d\} \right] \cdots \Bigg] \end{aligned} \tag{112}$$

It is now a matter of notation adjustment to set up the relationship

$$\lim_{\Delta t \to 0} \int_{\Delta t} \sum_{k=1}^{2} [\tilde{\mathbf{b}}_{c(j)}^{(k)}]\{U_{c(j)}^{*(k)}\} dt = \sum_{i=1}^{NCUT} [A_{c(j),c(i)}]\{h_{c(i)}\} \tag{113}$$

There are exactly NCUT sets of unknown Lagrange multiplier vectors defined by these equations. All other terms appearing in the equation are known. It follows that the matrix form of the impulse Lagrange multiplier equation is

$$\begin{bmatrix} [A_{c(1),c(1)}] & \cdots & [A_{c(1),c(NCUT)}] \\ \vdots & \ddots & \vdots \\ [A_{c(NCUT),c(1)}] & \cdots & [A_{c(NCUT),c(NCUT)}] \end{bmatrix} \begin{Bmatrix} \{h_{c(1)}\} \\ \vdots \\ \{h_{c(NCUT)}\} \end{Bmatrix} = \begin{Bmatrix} \{v_{c(1)}\} \\ \vdots \\ \{v_{c(NCUT)}\} \end{Bmatrix} \tag{114}$$

If there is only one closed loop, the cut-joint Lagrange multipliers can be solved for directly. However, if there are multiple closed-loops, then the resultant Lagrange multipliers equations are extensively coupled. The solution of the set of simultaneous linear algebraic equations defines the desired set of cut-joint impulse Lagrange multipliers. These equations are solve by standard LU decomposition and solution methods if the coefficient matrix is non-singular. If it is near singular, as measured by its condition number, singular value decomposition methods are used to obtain a least squares solution.

10.11.1. Impulse Corrections to Velocity State Variables: Body i absolute acceleration is defined by equation 55. The associated impulse equation is obtained by integrating both sides of it with respect to time and then taking the limit as the time interval approaches zero. The net result is

$$\{\Delta U_i\} = [\Psi_i]\{\Delta \mathbf{P}_i\} - [\Gamma_i]\{\Delta U_{i-1}\} \tag{115}$$

where

$$\lim_{\Delta t \to 0} \int_{\Delta t} \{\dot{U}_i\} dt = \{\Delta U_i\} \tag{116}$$

$$\lim_{\Delta t \to 0} \int_{\Delta t} \{\dot{\mathbf{P}}_i\} dt = \{\Delta \mathbf{P}_i\} \tag{117}$$

and from the fact that there are no impulse elements within $\{\mathbf{A}_{ci}\}$

$$\lim_{\Delta t \to 0} \int_{\Delta t} \{\dot{\mathbf{A}}_{ci}\} dt = 0 \tag{118}$$

Let

- $\{U_i^-\}$ = Uncorrected vector of body i velocity state variables.
- $\{U_i^+\}$ = Corrected vector of body i velocity state variables.
- $\{\Delta U_i\}$ = Impulse correction factor for vector of body i velocity state variables.

It follows from these definitions that

$$\{U_i^+\} = \{U_i^-\} + \{\Delta U_i\} \tag{119}$$

and similarly

$$\{\mathbf{P}_i^+\} = \{\mathbf{P}_i^-\} + \{\Delta \mathbf{P}_i\} \tag{120}$$

Body i relative motion is defined by equation 72 and its associated impulse equation is

$$\{\Delta \mathbf{P}_i\} = [\bar{M}_i]^{-1}[\Psi_i]^T \left\{ [\tilde{M}_i][\Gamma_i]\{\Delta U_{i-1}\} + \sum_{ic} [\tilde{\mathbf{b}}_{ic}]^T \{h_c\} \right\} \tag{121}$$

where $\{h_c\}$ is the impulse Lagrange multiplier vector defined by equation 106. Direct application of the above equations leads to the following recursion relations for the corrected body i velocity state variables.

$$\{U_i^+\} = \{U_i^-\} + [\Psi_i]\{\Delta \mathbf{P}_i\} - [\Gamma_i]\{\Delta U_{i-1}\} \tag{122}$$

10.12. COLLECT GYROSCOPIC AND USER SUPPLIED LOADS

The load vector is developed by sequentially stepping through all types of admissible loading effects. This is the $\{G\}$ vector defined in [70] by equation II-63. It correlates with the $\{G_e\}$ and $\{G_i\}$ vectors of equations 63 and 68 herein. Its general format is

$$\{G\} = \{G_{ex}\} - \begin{bmatrix} 0 \\ k \end{bmatrix} \{\xi\} - \begin{bmatrix} 0 \\ C \end{bmatrix} \{\dot{\xi}\} + \left[\tilde{\Omega}\right] [m]\{U\}$$
$$+\frac{1}{2}\left\{\{U\}^T [m_{,j}]\{U\}\right\} - [\dot{m}]\{U\} \tag{123}$$

Term 1 on the right hand side of equation 123 defines all external loads acting on body i. Terms 2 and 3 are define modal stiffness and modal damping respectively. The coefficient matrices k and C may be full and even nonsymmetric, NDISCOS does a blind matrix multiply to determine associated contributions. Terms 4, 5 , and 6 are inertial forces that involve velocities and displacements of the bodies. These include all Coriolis and centrifugal effects.

10.13. ORDER N RECURSIVE ALGORITHM, BACKWARD SWEEP

The backward sweep is used to recursively compute the equivalent inertia tensors and equivalent generalized force vectors which account for all outboard bodies. The recursion starts from the end body of each limb and proceeds inward to body 1.

10.13.1. Open Loop Backward Sweep: The implementation of this sweep is very difficult. To gain computational efficiency the large sparse arrays used in the theoretics must not be created by the straight forward matrix multiplications implied in the theoretics. Computational speed requirements demand use of matrix sparsity knowledge.

Several definitions are required:

$$[A_1] = [\tilde{M}_i]^{-1}[\mathbf{B}_{Ii}]^T[\bar{\Phi}_{Ii}] \tag{124}$$

$$[A_2] = \left([\bar{\Phi}_{Ii}]^T[\mathbf{B}_{Ii}][A_1]\right)^{-1} = [J_i] \tag{125}$$

From these the backward recursion relation for the equivalent inertia matrix of body i-1 and all outboard bodies is

$$[\tilde{M}_{i-1}] := [\tilde{M}_{i-1}] + [\mathbf{B}_{O,i-1}]^T[\bar{\Phi}_{Ii}][J_i][\bar{\Phi}_{Ii}]^T][\mathbf{B}_{O,i-1}] \tag{126}$$

and

$$[\bar{M}_i] = [\Psi_i]^T[\tilde{M}_i][\Psi_i] \tag{127}$$

Also, the backward recursion relation for the equivalent force vector acting on body i-1 and all outboard bodies is

$$\begin{aligned}\{\tilde{G}_{i-1}\} &:= \{\tilde{G}_{i-1}\} \\ &+ [\mathbf{B}_{O,i-1}]^T[\bar{\Phi}_{Ii}][J_i][\bar{\Phi}_{Ii}]^T\left(\{\mathbf{Z}_i\} - [\mathbf{B}_{Ii}][\tilde{M}_i]^{-1}\{\tilde{G}_i\}\right)\end{aligned} \tag{128}$$

where the symbol := is used to imply "accumulation".

10.14. OPEN LOOP SECOND FORWARD PASS

During the second forward pass for the order N algorithm the relative hinge accelerations and modal accelerations are computed. The recursion begins with the base body and terminates with the end body or bodies.

10.14.1. Body 1 Acceleration Variables & Hinge Constraint Loads: The recursion starts with the development of associated acceleration coefficients for body 1 as defined by equation 72, with the cut-joint Lagrange multiplier and inboard body acceleration terms deleted, that is

$$\{\dot{\mathbf{P}}_1\} = [\bar{M}_1]^{-1}[\Psi_1]^T \left(\{\tilde{G}_1\} - [\tilde{M}_1]\{\mathbf{A}_{c1}\}\right) \tag{129}$$

and from equations 58 and 59

$$\left\{ \begin{array}{c} \ddot{\beta}^f_{I1} \\ \ddot{\xi}_1 \\ \ddot{\theta}^f_1 \end{array} \right\} = [\bar{M}_1]^{-1}[\Psi_1]^T \left(\{\tilde{G}_1\} - [\tilde{M}_1] \left\{ \begin{array}{c} [R_{I1}]^{-1}\{\zeta_1\} \\ 0 \\ [\bar{V}_1]\{\ddot{\theta}^c_1\} \end{array} \right\} \right) \tag{130}$$

Using acceleration coefficients $\{\ddot{\beta}^f_{I1}\}$, $\{\ddot{\xi}_1\}$, $\{\ddot{\theta}_1\}^T$ and equations 55 through 58 all elements of $\{\ddot{\gamma}_1\}$ are computed from

$$\left\{ \begin{array}{c} \ddot{\gamma}_1 \\ \ddot{\xi}_1 \\ \ddot{\theta}^f_1 \end{array} \right\} = [\Psi_1] \left\{ \begin{array}{c} \ddot{\beta}^f_{I1} \\ \ddot{\xi}_1 \\ \ddot{\theta}_1 \end{array} \right\} + \left\{ \begin{array}{c} [R_{I1}]^{-1}\{\zeta_1\} \\ 0 \\ [\bar{V}_1]\{\ddot{\theta}^c_1\} \end{array} \right\} \tag{131}$$

At every output data step the Lagrange multiplier vector $\{\lambda_1\}$ is computed. It is used for output reporting. It is only needed for the development of the equations of motion if the intermittent surface contact capability is used. If the intermittent surface contact capability is used then the elements of this vector are used to determine the instant of contact release. Equation 73 without cut-joint and inboard body acceleration contributions defines the body 1 Lagrange multiplier vector, that is,

$$\{\Lambda_{I1}\} = [J_1][\bar{\Phi}_{I1}]^T(\{\mathbf{Z}_1\} - [\mathbf{B}_{I1}][\bar{M}_1]^{-1}\{\tilde{G}_1\}) \tag{132}$$

10.14.2. Body i Acceleration Variables & Hinge Constraint Loads: After the development of body 1 acceleration data, the system sequentially steps from body 1 to all limb end bodies. The associated acceleration coefficients for open loop body i are defined by equation 72 by deleting the term associated with closed loop cut-joint constraints; that is

$$\{\dot{\mathbf{P}}_i\} = [\bar{M}_i]^{-1}[\Psi_i]^T \left(\{\tilde{G}_i\} - [\tilde{M}_i]\{\mathbf{A}_{ci}\} + [\tilde{M}_i][\Gamma_i]\{\dot{U}_{i-1}\}\right) \tag{133}$$

and from equations 58, 59 and 60

$$\left\{ \begin{array}{c} \ddot{\beta}^f_{Ii} \\ \ddot{\xi}_i \\ \ddot{\theta}^f_i \end{array} \right\} = [\bar{M}_i]^{-1}[\Psi_i]^T \left(\{\tilde{G}_i\} - [\tilde{M}_i] \left\{ \begin{array}{c} [R_{Ii}]^{-1}(\{\zeta_i\} - [b_{O,i-1}]\{\dot{U}_{i-1}\}) \\ 0 \\ [\bar{V}_i]\{\ddot{\theta}^c_i\} \end{array} \right\} \right) \tag{134}$$

The acceleration coordinates for body i are obtained from equations 55 through 58 as

$$\left\{\begin{array}{c}\ddot{\gamma}_i\\ \ddot{\xi}_i\\ \ddot{\theta}_i\end{array}\right\} = [\Psi_i]\left\{\begin{array}{c}\ddot{\beta}^f_{Ii}\\ \ddot{\xi}_i\\ \ddot{\theta}^f_i\end{array}\right\} + \left\{\begin{array}{c}[R_{Ii}]^{-1}(\{\zeta_i\} - [b_{O,i-1}]\{\dot{U}_{i-1}\})\\ 0\\ [\bar{V}_i]\{\ddot{\theta}^c_i\}\end{array}\right\} \tag{135}$$

and the Lagrange multipliers associated with fixed kinematic hinge constraints are determined from equation 73 as

$$\{\Lambda_{Ii}\} = [J_i][\bar{\Phi}_{Ii}]^T\left(\{\mathbf{Z}_i\} - [\mathbf{B}_{Ii}][\tilde{M}_i]^{-1}\{\tilde{G}_i\} - [\mathbf{B}_{O,i-1}]\{\dot{U}_{i-1}\}\right) \tag{136}$$

10.14.3. Lagrange Multipliers: Cut-joint constraints are maintained by an associated set of constraint loads. These loads are defined by the to be determined set of Lagrange multipliers.

Lagrange multipliers are obtained from equation 100. The associated definition of $\{\zeta_{c(j)}\}$ is given by equation 81 as

$$[\bar{\phi}_{c(j)}]^T\{\zeta_{c(j)}\} = \{\ddot{\alpha}_{c(j)}\} - [\bar{\phi}_{c(j)}]^T[\dot{b}^{(1)}_{c(j)}]\{U^{(1)}_{c(j)}\} - [\bar{\phi}_{c(j)}]^T[\dot{b}^{(2)}_{c(j)}]\{U^{(2)}_{c(j)}\} \tag{137}$$

The associated definition of $\{U^{*(k)}_{c(j)}\}$ is given by equation 90 as

$$\{U^*_{i+1}\} = \{\mathbf{U}_{i+1}\} + [\mathbf{N}_{i+1}]\sum_d[\tilde{\mathbf{b}}_{i+1,d}]^T\{\lambda_d\} - [M^*_{i+1}][\tilde{J}_{i+1}][\mathbf{B}_{Oi}]\{U^*_i\} \tag{138}$$

From the format of this equation it is seen that it is possible to express the right hand side of equation 100 as

$$\sum_{k=1}^{2}[\tilde{\mathbf{b}}^{(k)}_{c(j)}]\{U^{*(k)}_{c(j)}\} = \sum_{i=1}^{NCUT}[A_{c(j),c(i)}]\{\lambda_{c(i)}\} - \{\mu^*_{c(j)}\} \tag{139}$$

In the above equations $\{\lambda_{c(i)}\}$ is the vector of Lagrange multipliers acting on the set of fixed and rheonomically constrained degrees of freedom at the i-th cut-joint which is located at hinge point $c(i)$. Furthermore $\{\mu^*_{c(j)}\}$ is the vector that contains all components of the vector recursion $\{U^{*(k)}_{c(j)}\}$ defined by equation 138 that are not Lagrange multiplier coefficients.

The evaluation of the right hand side of equation 138 is a bit less than straightforward. From equation 91 and 99 the vector $\{U^{*(k)}_{c(j)}\}$ can be expressed by the expanded recursion:

$$\begin{aligned}\{U^*_i\} &= \{\mathbf{U}_i\} + [\mathbf{N}_i]\sum_d[\tilde{\mathbf{b}}_{id}]^T\{\lambda_d\}\\ &- [M^*_i][\tilde{J}_i][\mathbf{B}_{O,i-1}]\Bigg[\{\mathbf{U}_{i-1}\} + [\mathbf{N}_{i-1}]\sum_d[\tilde{\mathbf{b}}_{i-1,d}]^T\{\lambda_d\}\\ &\quad\vdots\\ &- [M^*_2][\tilde{J}_2][\mathbf{B}_{O,1}]\left[\{\mathbf{U}_1\} + [\mathbf{N}_1]\sum_d[\tilde{\mathbf{b}}_{1,d}]^T\{\lambda_d\}\right]\cdots\Bigg]\end{aligned} \tag{140}$$

where index i is equal to the body number associated with the end body on branch k, k=1,2 at the j-th cut-joint (hinge point $c(j)$, j=1,...,NCUT).

It is now a matter of notation adjustment to set up the matrix from of the relationship defined by equation 139 so that the set of unknown Lagrange multipliers can be determined via conventional linear equation solution techniques.

There are exactly NCUT sets of unknown Lagrange multiplier vectors defined by these equations. All other terms appearing in the equation are known. It follows that the matrix form of the Lagrange multiplier equation is

$$\begin{bmatrix} [A_{c(1),c(1)}] & \cdots & [A_{c(1),c(NCUT)}] \\ \cdot & \cdot & \cdot \\ \cdot & \cdot & \cdot \\ \cdot & \cdot & \cdot \\ [A_{c(NCUT),c(1)}] & \cdots & [A_{c(NCUT),c(NCUT)}] \end{bmatrix} \begin{Bmatrix} \{\lambda_{c(1)}\} \\ \cdot \\ \cdot \\ \cdot \\ \{\lambda_{c(NCUT)}\} \end{Bmatrix} = \begin{Bmatrix} \{\mu_{c(1)}\} \\ \cdot \\ \cdot \\ \cdot \\ \{\mu_{c(NCUT)}\} \end{Bmatrix} \tag{141}$$

The vector $\{\mu_{c(j)}\}$ is obtained from the substitution of equation 100 into equation 139 to obtain

$$\{\mu_{c(j)}\} = \{\mu^{*}_{c(j)}\} + [\bar{\phi}_{c(j)}]^T\{\zeta_{c(j)}\} \tag{142}$$

It should be noted the the elements of the Lagrange multiplier coefficient matrix are identical to the elements of the impulse Lagrange multiplier coefficient matrix defined by equation 114. From this it is noted that the coefficient matrix evaluated and decomposed to solve for the impulse Lagrange multipliers is also usable for the determination of the Lagrange multipliers used to enforce loop closure and gear constraints. The solution of this set of simultaneous linear algebraic equations defines the desired set of cut-joint Lagrange multipliers.

10.14.4. Second Forward Pass for Closed Loop Systems: On the second forward pass of the closed loop order N algorithm the elements of equation 141 are recursively propagated from the base body towards each of the cut-joint end bodies. The equations for the cut-joint Lagrange multipliers are solved at the end of the recursion.

10.14.5. Closed Loop Second Pass, Obtain Accelerations: During this last pass of the order N algorithm all relative hinge and modal accelerations are computed.

10.14.6. Body 1 Acceleration Variables & Hinge Constraint Loads: The recursion starts with the development of associated acceleration coefficients for body 1 as defined by equation 72, that is,

$$\{\dot{\mathbf{P}}_1\} = [\tilde{M}_1]^{-1}[\Psi_1]^T\left\{\{\tilde{G}_1\} - [\tilde{M}_1]\{\mathbf{A}_{c1}\} + \sum_{1c}[\tilde{\mathbf{b}}_{1c}]^T\{\lambda_c\}\right\} \tag{143}$$

where it is noted that constraints associated with cut-joints are now included.

Using acceleration coefficients $\{\ddot{\beta}^f_{I1}\}$, $\{\ddot{\xi}_1\}$, $\{\ddot{\theta}_1\}^T$ and equations 55 through 60 all elements of $\{\ddot{\gamma}_1\}$ are computed from

$$\begin{Bmatrix} \ddot{\gamma}_1 \\ \ddot{\xi}_1 \\ \ddot{\theta}^f_1 \end{Bmatrix} = [\Psi_1]\begin{Bmatrix} \ddot{\beta}^f_{I1} \\ \ddot{\xi}_1 \\ \ddot{\theta}_1 \end{Bmatrix} + \begin{Bmatrix} [R_{I1}]^{-1}\{\zeta_1\} \\ 0 \\ [\bar{V}_1]\{\ddot{\theta}^c_1\} \end{Bmatrix} \tag{144}$$

At every output data step elements of the Lagrange multiplier vector $\{\Lambda_{I1}\}$ are computed. These are used for output reporting. They are needed for the development of the equations of motion only if the intermittent surface contact capability is used. If the intermittent surface contact capability is used then the elements of this vector are used to determine contact release. For body 1 these are developed from equation 73:

$$\begin{aligned}\{\Lambda_{I1}\} &= [J_1][\bar{\Phi}_{I1}]^T\{\mathbf{Z}_1\} \\ &- [J_1][\bar{\Phi}_{I1}]^T\left[[\mathbf{B}_{I1}][\tilde{M}_1]^{-1}\left(\{\tilde{G}_1\}+\sum_{1c}[\tilde{\mathbf{b}}_{1c}]^T\{\lambda_c\}\right)\right]\end{aligned} \tag{145}$$

10.14.7. Body i Acceleration Variables & Hinge Constraint Loads: After the development of body 1 data the system cycles outboard from body 1 to each end body. The associated relative acceleration coefficients for body i are defined by equation 72: that is,

$$\{\dot{\mathbf{P}}_i\} = [\tilde{M}_i]^{-1}[\Psi_i]^T\left\{\{\tilde{G}_i\} - [\tilde{M}_i]\{\mathbf{A}_{ci}\} + [\tilde{M}_i][\Gamma_i]\{\dot{U}_{i-1}\} + \sum_{ic}[\tilde{\mathbf{b}}_{ic}]^T\{\lambda_c\}\right\} \tag{146}$$

The absolute acceleration coordinates for body i are obtained from equation 55 as

$$\{\dot{U}_i\} = [\Psi_i]\{\dot{\mathbf{P}}_i\} + \{\mathbf{A}_{ci}\} - [\Gamma_i]\{\dot{U}_{i-1}\} \tag{147}$$

and the Lagrange multipliers associated with fixed and rheonomic hinge constraints are determined from equation 73 as

$$\begin{aligned}\{\Lambda_{Ii}\} &= [J_i][\bar{\Phi}_{Ii}]^T\left(\{\mathbf{Z}_i\} - [\mathbf{B}_{O,i-1}]\{\dot{U}_{i-1}\}\right) \\ &- [J_i][\bar{\Phi}_{Ii}]^T\left[[\mathbf{B}_{Ii}][\tilde{M}_i]^{-1}\left(\{\tilde{G}_i\}+\sum_{ic}[\tilde{\mathbf{b}}_{ic}]^T\{\lambda_c\}\right)\right]\end{aligned} \tag{148}$$

10.15. INTERMITTENT LOOP CLOSURE AND SURFACE CONTACT

This capability will be significantly altered when a full rolling/sliding intermittent contact capability is introduced for elastically deformable surfaces with contact friction. Never-the-less the developmental framework created herein has been created in a generic manner so that future expansion needs can be readily supported.

10.15.1. Introduction and Assumptions: Intermittent surface contact is used to simulate the condition when the surfaces of two bodies come into contact and separate when contact forces are no longer compressive. If a hinge may be closed because of surface contact, the topology is set up to simulate loop topology for all time. This means that when the surfaces are separated, there are no constrained degrees of freedom at the cut-joint. When surfaces are in contact and the coefficient of restitution is defined to be 0.0 associated contacting degrees of freedom are kinematically constrained. This logic framework allows the system to change from loop to tree topology, or vice-versa, in a straight-forward manner.

The intermittent contact capability has been extended to handle both topological loops and gear constraints. Intermittent contact is now allowed to occur across cut-joint hinges and gear-constrained hinges. In fact, the gear-constrained degrees-of-freedom themselves can be subject to intermittent contact.

Assumptions

1. For translational degrees-of-freedom, one of the contacting surfaces is assumed to be flat. The p-frame of the associated hinge should be attached to the flat surface. For both translational and rotational degrees-of- freedom, the hinge frame placement and orientation should be defined so that $\beta = 0$ corresponds to the contact condition.
2. It is assumed that $\beta \geq 0$ implies no contact. A negative value of β means that one body has penetrated the surface of the other and that there is contact. The orientation of the p- and q-frames, defined by the user in the input data file, **must** be set so that this assumption holds true.
3. As a consequence of assumption 2, the value of the associated Lagrange multiplier λ should be positive when surfaces are in contact. Internal logic is based upon the assumption that compressive loads during contact are positive and negative when surfaces separate.
4. It is assumed that the contacting surfaces are smooth and frictionless, so that there are no tangential force components at the instant of impact.
5. It is assumed that the use of the coefficient of restitution is valid for bodies that are not necessarily point masses or perfectly smooth spheres.
6. If the coefficient of restitution equals zero the associated surface contact degree of freedom becomes kinematically fixed during contact and returns to a free state at the instant of separation.

Approach Surface contact is determined by monitoring the appropriate free hinge displacement β to see when it changes sign. Once this has occurred, the algorithm finishes up the current integration step, and applies an impulse at the impacting hinge degree-of-freedom. The calculation of this impulse is based on the desired post-impact velocity that is consistent with the specified coefficient of restitution between the two surfaces. If the coefficient of restitution is zero, the code also modifies the hinge constraint at the contacting surface to maintain the contact condition for subsequent integration steps. These fixed constraints hold until it has been determined that the condition for separation has been met. Separation occurs when associated Lagrange multipliers undergo a sign change. The hinge constraint is then released by reintroducing appropriate β and $\dot{\beta}$'s into the system state vector for that degree of freedom, and resetting the hinge constraint definitions.

10.15.2. Check Surface Contact & Separation: A check for surface contact and newly separating surfaces is made at the end of each Runge Kutta integration step. If there are any new surface contacts, instantaneous velocity changes are introduced into the system state vector for all associated surface contact degrees of freedom. Associated theoretics is analogous to that discussed in the section on impulse Lagrange multipliers, . The algorithm used can be summarized as follows:

1. Compute the desired velocity change for each surface contact direction using velocity information that is available before impulse application. Let:

$$\Delta\dot{\beta}_{di} = -(1 + e_i)\dot{\beta}_i^- \tag{149}$$

   where

   - $\Delta\dot{\beta}_{di}$ - Desired instantaneous velocity change for the i-th surface contact degree of freedom
   - e_i - Coefficient of restitution for the i-th surface contact degree of freedom
   - $\dot{\beta}_i^-$ - Associated relative velocity before applying the impulse for the i-th surface contact degree of freedom

2. For a unit impulse acting along the i-th surface contact direction compute the instantaneous velocity changes of all state vector velocity variables. Let

   n_c - Number of newly established surface contacts

   $\{\Delta U^{(i)}\}$ - Vector of all instantaneous body absolute, body modal, and embedded wheel velocity changes associated with a unit impulse acting along the i-th surface contact degree of freedom.

   $\{\Delta\dot{\beta}^{f(i)}\}$ - Vector of all instantaneous relative velocity changes associated with a unit impulse acting along the i-th surface contact degree of freedom.

3. Compute the scale factors for each impulse so that superposition of impulses would result in the desired velocity change. Solve the following set of equations for scale factors, S_j, j=1,...,n_c

$$\Delta\dot{\beta}_{di} = \sum_{j=1}^{n_c}\{\Delta U_i^{(j)}\}S_j \tag{150}$$

4. Apply the appropriate velocity changes to the system level velocity vector to yield post-impact velocities:

$$\{U^+\} = \{U^-\} + \sum_{j=1}^{n_c} S_j\{\Delta U^{(j)}\} \tag{151}$$

5. Check if coefficient of restitution equals zero for each surface contact degree of freedom. If true then these degrees of freedom must be switched from free to fixed.

10.16. INVERSE DYNAMICS

It is oftentimes desirable to compute the force and torque loads at hinges that would result from specified motions of the system. In NDISCOS there is no need to develop special coding logic to provide this inverse dynamics capability. The existing rheonomic constraint capability is used to specify the motion of the joints of interest, and the resulting joint loads are extracted from the vector of Lagrange multipliers that are normally computed for all of the constraint loads acting on the system.

11. Summary

The Man/Machine Interaction Dynamics and Performance (MMIDAP) project seeks to create an ability to study the consequences of machine design alternatives relative to the performance of both the machine and its operator. The envisioned MMIDAP capability is to be used for mechanical system design, human performance assessment, extrapolation of man/machine interaction test data, biomedical engineering, and soft prototyping within a concurrent engineering system. This chapter has reviewed the existing methodologies and techniques needed to create such capability. It has attempted to outline ongoing efforts to integrate both human performance and musculoskeletal databases with the host of analysis capabilities necessary for the early design analysis of dynamic actions, reactions, and performance assessment of coupled machine-operator systems. The multibody system dynamics software program NDISCOS of GSFC and Photon Research Associates can be used for machine and fine grain detail musculoskeletal dynamics modeling; a detail exposition of its underlying theoretics has been provided herein. The program JACK from The University of Pennsylvania can be used for estimating and animating whole body human response to given loading situations and motion constraints. The basic elements of performance (BEP) task decomposition methodologies associated with The University of Texas at Arlington's Human Performance Institute's BEP database can be used for human performance assessment. Techniques for resolving the statically indeterminant muscular load sharing problems can be used for a detailed understanding of potential musculotendon or ligamentous fatigue, pain, discomfort, and trauma problems. The MMIDAP problem as defined herein highlights the conflicting needs and views of groups that focus on machine design and groups that focus on musculoskeletal biodynamics, on human performance and cumulative injury potential. An attempt has been made to show that there is a critical need to integrate design and simulation tools and to establish multidisciplinary lines of communication. Futhermore an outline is provided of planned integration efforts for human performance analyses, associated databases, and mechanical system design capabilities. This integration effort is expected to provide an ability to perform both stand alone studies and the early system trade studies needed to assess man/machine interaction dynamics and performance.

References

1 Frisch, H.P., "Man/Machine Interaction Dynamics and Performance," Concurrent Engineering Tools and Technologies for Mechanical System Design, Edt. E.J. Haug, NATO ASI Series F, Springer-Verlag, pp 901-928, 1993.

2 Kroemer, K.H.E., et al. (Editors), "Ergonomic Models of Anthropometry, Human Biomechanics and Operator Equipment Interfaces," Proceedings, Workshop on Integrated Ergonomic Modeling, 1988.

3 "Proceedings of the AFHRL Workshop on Human-Centered Design Technology for Maintainability," Air Force Human Resources Laboratory, Wright Patterson Air Force Base, Dayton Ohio, Sept 12-13, 1990.

4 "Electronic Imaging," Report of the Board of Regents, National Library of Medicine Long Range Plan, NIH Publication Number 90-2197, U.S. Department of Health and Human Services, April 1990.

5 Whitlock, D. and Spitzer, V., "Video 3D Atlas of the Human Knee in Cross Sections," From the Departments of Cellular and Structural Biology and Radiology at The University of Colorado School of Medicine, Denver CO., The C.V. Mosby Company, 1990.

6 Winters, J.M. and Woo, S.L. (Editors), "Multiple Muscle Systems Biomechanics and Movement Organization," Springer Verlag, 1990

7 Yamaguchi, G.T., Sawa, A.G.U., Moram, D.W., Fessler, M.J., and Winters, J.M., "A Survey of Human Musculotendon Actuator Parameters," Appendix of [6]

8 Seireg, A. and Arvikar, R., "Biomechanical Analysis of the Musculoskeletal Structure for Medicine and Sports," Hemisphere Publishing Corporation, 1989.

9 Kondraske, G.V. "Quantitative Measurement and Assessment of Performance," Chapter 6 of "Rehabilitation Engineering", Smith R.V. and Leslie J.H. (Editors), CRC Press, 1990.

10 Fitts, P.M., "The Information Capacity of the Human Motor System in Controlling the Amplitude of Movement," Journal of Experimental Psychology, Vol 47, No 6, pp 381-391, 1954

11 Kondraske, G.V., et al., "Measuring Human Performance: Concepts, Methods, and Applications," SOMA: Engineering for the Human Body (ASME), pp 6-13, January 1988.

12 Kondraske, G.V., "Human Performance: Science or Art?," In: Foster, K. edt., 13th Northeast Bioengineering Conference, Philadelphia PA Proceedings, pp 44-47, 1987.

13 Kondraske, G.V., "An Elemental Resource Model for the Human-Task Interface," International Journal of Technology Assessment in Health Care, special issue on "Technology and Disability", to be published.

14 Hatze, H., "A Method for Describing the Motion of Biological Systems," Journal of Biomechanics, Vol 9, pp 101-104, 1976

15 Badler, N.L., "Human Factors Simulation Research at the University of Pennsylvania," Proceedings of the AFHRL Workshop on Human-Centered Design Technology for Maintainability," Air Force Human Resources Laboratory, Wright Patterson Air Force Base, Dayton Ohio, Sept 12-13, 1990.

16 Badler, N.I., "Articulated Figure Positioning by Multiple Constraints," IEEE Journal on Computer Graphics and Application, pp 26-37, 1987

17 Lee, P., Wei, S., Zhao, J., and Badler, N.I., "Strength Guided Motion," Computer Graphics, Vol 24, pp 253-262, 1990

18 Wilhelms, J., "Using Dynamic Analysis for Realistic Animation of Articulated Bodies," IEEE Journal on Computer Graphics and Application, pp 12-27, 1987

19 Badler N.L., Lee, P., Phillips, C., and Otani, E.M., "The JACK Interactive Human Model," Proceedings of the First Annual Symposium on Mechanical System Design in a Concurrent Engineering Environment," The University of Iowa, Iowa City, Iowa, pp 179-198, 1989

20 Badler, N.L., Barsky, B.A., and Zeltzer, D., "Making Them Move, Mechanics, Control, Animation of Articulated Figures," Morgan Kaufmann Publishers, Inc. 1990

21 Badler, N.L., Phillips, C.B., and Webber, B.L., "Simulating Humans: Computer Graphics Animation and Control," Oxford University Press, 1993

22 Chang, J.L., Kim, S.S., and Haug, E.J., "Real- Time Operator in the Loop Simulation of Multibody Systems," Technical Report R-72, The University of Iowa, Center for Simulation and Design Optimization of Mechanical Systems, 1990.

23 Stoner, J.W., et al., "Introduction to the Iowa Driving Simulator and Simulation Research Program," University of Iowa, Center for Simulation and Design Optimization of Mechanical Systems, Technical Report R-86.

24 Gunby, P., "Simulator Designed for Advanced Traffic Research," JAMA (Journal of the American Medical Association), Contempo Issue, Vol 268, No 3, pp 303, July 15, 1992

25 Haug, E.J., et al., "Feasibility Study and Conceptual Design of a National Advanced Driving Simulator," US Department of Transportation, DOT HS-807-596, 1990.

26 Schiehlen, W. (Editor), "Multibody Systems Handbook," Springer Verlag, 1990.

27 Schiehlen, W. (Editor), "Advanced Multibody System Dynamics," Kluwer Academic Publishers Group, 1993

28 Bae, D.S. and Haug, E.J., "A Recursive Formulation for Constrained Mechanical Systems, Part 1 - Open Loop," Mechanics of Structures and Machines, Vol 15, No 4, 1987.

29 Bae, D.S. and Haug, E.J., "A Recursive Formulation for Constrained Mechanical Systems, Part 2 - Closed Loop," Mechanics of Structures and Machines, Vol 15, No 4, 1987.

30 Bae, D.S., Haug, E.J., and Kuhl, J.G., "A Recursive Formulation for Constrained Mechanical Systems, Part 3 - Parallel Processor," Mechanics of Structures and Machines, Vol 16, No 2, 1988.

31 Chun, H.M, Turner, J.D. and Frisch, H.P., "Order (N) DISCOS for Multibody Systems with Gear Reduction," Guidance and Control Conference, Portland Oregon, Aug 20-22, 1990, AIAA Paper-90- 3385

32 Chun, H.M, Turner, J.D. and Frisch, H.P., "A Recursive Order (N) Formulation for DISCOS with Topological Loops and Intermittent Surface Contact," AAS/AIAA Astrodynamics Specialists Conference, Durango, Colorado, Aug 19-22, 1991.

33 Hwang, R.S., and Haug, E.J., "A Recursive Multibody Dynamics Formulation for Parallel Computation," Technical Report R-13, The University of Iowa, Center for Simulation and Design Optimization of Mechanical Systems, 1988.

34 Kim, S.S., et al., "New General Purpose Dynamics Simulation Code (NGDC)," The University of Iowa, Center for Simulation and Design Optimization of Mechanical Systems, 1990.

35 Tsai, F.F. and Haug, E.J., "Automated Methods for High Speed Simulation of Multibody Dynamic Systems," Technical Report R-39, The University of Iowa, Center for Simulation and Design Optimization of Mechanical Systems, 1989.

36 Bobbert, M.F., "Vertical jumping: A study of muscle functioning and coordination," Free University Amsterdam, 1988.

37 van Soest, A.J., "Jumping from structure to control. A simulation study of explosive movements," Free University Amsterdam, 1992.

38 Pronk, G., "The shoulder girdle," Delft University of Technology, 1991.

39 van der Helm, F.C.T., The shoulder mechanism. A dynamic approach," Delft University of Technology, 1991.

40 Happee, R., "The control of shoulder muscles during goal directed movements," Delft University of Technology, 1992.

41 Spoor, C.W., "Mechanical models of selected parts of the human musculoskeletal system," University of Leiden, 1992.

42 Jaegers, S.M.H.J., "The morphology and functions of the muscles around the hip joint after a unilateral transfemoral amputation," University Groningen, 1989.

43 Koopman, B. "The three-dimensional analysis and prediction of human walking," University of Twente, Enschede, 1989.

44 Thunnissen, J. "Muscle force prediction during human gait," University of Twente, Enschede, 1993.

45 Kuffler, W.W., Nicholls, J.G., and Martin, A.R., "From Neuron to Brain: A Cellular Approach to the Function of the Nervous System," Second Edition, Sinauer Associates Inc., 1984

46 Shepherd, G.M., "Neurobiology," Oxford University Press, 1988

47 Bridgeman, B., "The Biology of Behavior and Mind," John Wiley & Sons, 1988

48 Hatze, H., "Myocybernetic Control Models of Skeletal Muscle, Characteristics and Applications," University of South Africa, Muckleneuk, Pretoria, 1981

49 Hatze, H., "The Charge-Transfer Model of Myofilamentary Interaction: Prediction of Force Enhancement and Related Myodynamic Phenomena," pp 46-56 of [6], 1990

50 Hoy, M.G., Zajac, F.E., and Gordon, M.E., "A Musculoskeletal Model of the Human Lower Extremety: The Effect of Muscle, Tendon, and Moment Arm on the Moment-Angle Relationship of Musculotendon Actuators at the Hip, Knee, and Ankle," Journal of Biomechanics, Vol 23, pp 157-169, 1990.

51 Ma S.P. and Zahalak G.I., "A Distribution- Moment Model of Energetics in Skeletal Muscle," Journal of Biomechanics, Vol 24, pp 21-35, 1991

52 Zahalak, G.I., "A Distribution-moment Approximation for Kinetic Theories of Muscular Contraction," Math. Biosci. Vol 55, pp 89-114, 1981.

53 Zahalak, G.I., "A Comparison of the Mechanical-Behavior of the Cat Soleus Muscle with a Distribution- Moment Model," Journal of Biomechanical Engineering, Vol 108, pp 131-140, 1986

54 Zahalak, G.I. and Ma, S.P., "Muscle Activation and Contraction: Constitutive Relations Based Directly on Cross- Bridge Kinetics," Journal of Biomechanical Engineering, Vol 112, pp 52-62, 1990

55 Zajac, F.E., Topp, E.L., and Stevenson P.J., "A dimensionless Musculotendon Model," IEEE/8th Annual Conference of the Engineering in Medicine and Biology Society, pp 601-604, 1986.

56 Zajac, F.E., Topp, E.L., and Stevenson P.J., "Musculotendon Actuator Models for Use in Computer Studies and Design of Neuromuscular Stimulation Systems," RESNA 9th Annual Conference on Rehabilitation Engineering, Minneapolis Minnesota, pp 442-444, 1986.

57 Kearney, R.E. and Hunter I.W., "System Identification of Human Joint Dynamics," Critical Reviews in Biomedical engineering, Vol 18, pp 55-87, 1990.

58 Hatze, H., "A Comprehensive Model for Human Motion Simulation and its Application to the Take-off Phase of the Long Jump," Journal of Biomechanics, Vol 14, pp 135-142, 1981.

59 Morris, G.R., Douglas, A.S. and Guoan, L. "A Comparison of Two Muscle Models for Simulating Human Saccadic Eye Motion," Final Report, 1990. Copy available from H.P. Frisch, Code 714.1 NASA/GSFC, Greenbelt, MD 20771.

60 Chizeck, H.J., et al., "Control of Functional Neuromuscular Stimulation Systems for Standing and Locomotion in Paraplegics," Proceedings of the IEEE, Vol 79, pp 1155-1165, Sept 1988.

61 Sejersted, O.M. and Westgaard, R.H., "Occupational Muscle Pain and Injury; Scientific Challenge," editorial, pp 271-274, special issue on Occupational Muscle Pain and Injury, European Journal of Applied Physiology, Vol 57, pp 271-372, 1988

62 Vollestad, N.K. and Sejersted, O.M. "Biochemical Correlates of Fatigue, A Brief Review," European Journal of Applied Physiology, Vol 57, pp 322-326, 1988

63 Zajac, F.E. and Gordon, M.E., "Determining Muscle's Force and Action in Multi-Articular Movement," Exercise and Sports Science Reviews, Vol 17, pp 187-230, 1989.

64 Norkin, C.C and Levangie, P.A, "Joint Structure & Function, A Comprehensive Analysis," F.A. Davis Company, 1983

65 Norkin, C.C and Frankel, V.H., "Basic Biomechanics of the Musculoskeletal System", Lea & Febiger, 1989

66 Berme, N., Engin, A.E. and Correia da Silva, K.M. "Biomechanics of Normal and Pathological Human Articulating Joints," NATO ASI Series, Martinus Nijhoff Publishers, 1985

67 An, K-N, Berger, R.A. and Cooney, W.P., "Biomechanics of the Wrist Joint," Springer-Verlag, 1991

68 Frisch, H.P., "A Vector Dyadic Development of the Equations of Motion for N-coupled Rigid Bodies and Point Masses," NASA Technical Report TN D-7767, 1974.

69 Frisch, H.P., "The NBOD2 User's and Programmer's Manual," NASA Technical Paper 1145, 1978.

70 Bodley, C.S., Devers, A.D., Park, A.C., and Frisch, H.P., "A Digital Computer Program for the Dynamic Interaction Simulation of Controls and Structure (DISCOS)," NASA Technical Paper 1219, Vols 1 and 2, May 1978.

71 Turner, J., Chun, H., Lupi, V., Weiner, P.., Gallion, S, and Singh, C., "Researchers Apply Variable Reduction Techniques to Molecular Dynamics Simulations, Part 1", Chemical Design Automation News, Vol 7, No. 12, Dec 1992.

72 Turner, J., Chun, H., Lupi, V., Weiner, P.., Gallion, S, and Singh, C., "Researchers Apply Variable Reduction Techniques to Molecular Dynamics Simulations, Part 2", Chemical Design Automation News, Vol 8, No. 1, Jan 1993.

73 Chun, H.M, Turner, J.D. and Frisch, H.P., "Experimental Validation of Order(N) DISCOS," AAS/AIAA Astrodynamics Specialist Conference, Aug 7-10, 1989.

74 Chun, H.M, Turner, J.D. and Frisch, H.P., "Recursive Multibody Formulations for Robotics Applications with Harmonic Drives," Presented to: International Conference on Dynamics of Flexible Structures in Space, Cranfield England, May 15-18, 1990.

75 Chun, H.M, and Turner, J.D. "DISCOS Upgrade for Recursive Dynamics," Final Report from Photon Research Associates to the University of Iowa, Contract No YO3525, Feb 1991 - CDR191.

76 Frisch, H.P. and Chun, H.M., "NDISCOS - User's & Programmer's Manual," Availability, Attention Dr. Hon Chun, MOLDYN Inc, Subsidiary of Photon Research Associates, 1033 Massachusetts Ave. Cambridge, MA 02138, Phone (617) 354-3124, FAX (617) 491- 4522.

77 Hiller, M., "Dynamics of Multibody Systems with Minimal Constraints," this book.

78 Wittenburg, J., "Dynamics of Systems of Rigid Bodies," B.G. Teubner, Stuttgart, 1977.

79 Haug, E.J., "Intermediate Dynamics," Prentice Hall, 1992.

80 Ostermeyer, G.P., "On Baumgarte Stabilization for Differential Algebraic Equations," NATO Series, Vol F69, Real-Time Integration Methods for Mechanical System Simulation, Edt Haug, E.J. and Deyo, R.C., Springer-Verlag, 1990.

Index

A

B

C

D

E

F

G

H

I

J

K

L

M

N

O

P

Q

R

S

T

U

V

W